Algebra

Authors

Susan A. Brown
R. James Breunlin
Mary Helen Wiltjer
Katherine M. Degner
Susan K. Eddins
Michael Todd Edwards
Neva A. Metcalf
Natalie Jakucyn
Zalman Usiskin

Director of Evaluation

Denisse R. Thompson

UChicagoSolutions

Authors

3rd EDITION AUTHORS

Susan A. Brown, *Mathematics Department Chair*
York High School, Elmhurst, IL

R. James Breunlin, *Mathematics Department Chair*
Schaumburg High School, Schaumburg, IL

Mary Helen Wiltjer, *Mathematics Teacher*
Oak Park and River Forest High School, Oak Park, IL

Katherine M. Degner, *Mathematics Teacher*
Williamsburg Comm. High School, Williamsburg, IA

Susan K. Eddins, *Mathematics Teacher (retired)*
IL Mathematics & Science Academy, Aurora, IL

Michael Todd Edwards, *Assistant Professor of Mathematics Education*
Miami University, Ohio, Oxford, OH

Neva A. Metcalf, *Mathematics Teacher*
Evanston Township High School, Evanston, IL

Natalie Jakucyn, *Mathematics Teacher*
Glenbrook South High School, Glenview, IL

Zalman Usiskin, *Professor of Education*
The University of Chicago

AUTHORS OF EARLIER EDITIONS

John W. McConnell, *Instructional Supervisor of Mathematics*
Glenbrook South High School, Glenview, IL

Sharon Senk, *Professor of Mathematics*
Michigan State University, East Lansing, MI

Ted Widerski, *Mathematics Teacher*
Waterloo High School, Waterloo, WI

Cathy Hynes Feldman, *Mathematics Teacher*
The University of Chicago Laboratory Schools

James Flanders, UCSMP

Margaret Hackworth, *Mathematics Supervisor*
Pinellas County Schools, Largo, FL

Daniel Hirschhorn, UCSMP

Lydia Polonsky, UCSMP

Leroy Sachs, *Mathematics Teacher (retired)*
Clayton High School, Clayton, MO

Ernest Woodward, *Professor of Mathematics*
Austin Peay State University, Clarksville, TN

http://ucsmp.uchicago.edu/secondary/overview

UChicagoSolutions

Copyright © 2016 by The University of Chicago

All rights reserved. Except as permitted under the United States Copyright Act, no part of this publication may be reproduced or distributed in any form or by any means, or stored in a database or retrieval system, without the prior written permission from the publisher, unless otherwise indicated.

Printed in the United States of America.

Send all inquiries to:
UChicagoSolutions
1427 E. 60th Street
Chicago, IL 60637

ISBN 978-1-943237-24-1
ISBN 1-943237-24-7

1 2 3 4 5 6 7 8 9 RRDW 21 20 19

Director of Evaluation
Denisse R. Thompson, *Professor of Mathematics Education*
University of South Florida, Tampa, FL

Evaluation Assistants
Gladys Mitchell, Zhuo Zheng

Editorial Staff
Catherine Ballway, Grant Owens, Asaf Hadari

Evaluation Consultant
Sharon L. Senk, *Professor of Mathematics*
Michigan State University, East Lansing, MI

Executive Managing Editor
Clare Froemel

Manuscript Production Coordinator
Benjamin R. Balskus

Since the first two editions of *Algebra* were published, millions of students and thousands of teachers have used the materials. Prior to the publication of this third edition, the materials were again revised, and the following teachers and schools participated in evaluations of the trial version during 2005–2006:

Shannon Johnson
Junction City Middle School
Junction City, KS

Julie Pellman
Hyman Brand Hebrew Academy
Overland Park, KS

Dan Kramer
Highlands High School-Ft. Thomas
Fort Thomas, KY

Craig Davelis, Megan Mehilos, Sue Nolte, Lynette TeVault
York High School
Elmhurst, IL

Jan Boudreau
Rosemont Middle School
La Crescenta, CA

Tammy Anderson
Ashland High School
Ashland, OR

Dennis Massoglia
Washington Middle School
Calumet, MI

Erica Cheung
Stone Scholastic Academy
Chicago, IL

The following schools participated in field studies in 1992–1993, 1987–1988, or 1986–1987 as part of the first edition or the second edition research.

Rancho San Joaquin Middle School
Lakeside Middle School
Irvine High School
Irvine, CA

D.W. Griffith Jr. High School
Los Angeles, CA

Mendocino High School
Mendocino, CA

Chaffey High School
Ontario, CA

Eagleview Middle School
Colorado Springs, CO

Lincoln Junior High School
Lesher Junior High School
Blevins Junior High School
Fort Collins, CO

Bacon Academy
Colchester, CT

Rogers Park Jr. High School
Danbury, CT

Clearwater High School
Clearwater, FL

Safety Harbor Middle School
Safety Harbor, FL

Aptakisic Junior High School
Buffalo Grove, IL

Austin Academy
Bogan High School
Disney Magnet School
Hyde Park Career Academy
Von Steuben Metropolitan Science Center
Washington High School
Chicago, IL

Morton East High School
Cicero, IL

O'Neill Middle School
Downers Grove, IL

Elk Grove High School
Elk Grove Village, IL

Glenbrook South High School
John H. Springman School
Glenview, IL

Mendota High School
Mendota, IL

Carl Sandburg Jr. High School
Winston Park Jr. High School
Palatine, IL

Grant Middle School
Springfield, IL

McClure Junior High School
Western Springs, IL

Hubble Middle School
Wheaton, IL

Central Junior High School
Lawrence, KS

Old Rochester High School
Mattapoisett, MA

Fruitport High School
Fruitport, MI

Sauk Rapids-Rice Schools
Sauk Rapids, MN

Parkway West Middle School
Chesterfield, MO

Taylor Middle School
Van Buren Middle School
Albuquerque, NM

Crest Hills Middle School
Shroder Paideia Middle School
Walnut Hills High School
Cincinnati, OH

Lake Oswego Sr. High School
Lake Oswego, OR

Springfield High School
Springfield, PA

R.C. Edwards Jr. High School
Central, SC

Easley Junior High School
Easley, SC

Liberty Middle School
Liberty, SC

Northeast High School
Clarksville, TN

Hanks High School
El Paso, TX

Robinson Middle School
Maple Dale Middle School
Fox Point, WI

Glen Hills Middle School
Glendale, WI

UCSMP The University of Chicago School Mathematics Project

The University of Chicago School Mathematics Project (UCSMP) is a long-term project designed to improve school mathematics in grades pre-K through 12. UCSMP began in 1983 with a 6-year grant from the Amoco Foundation. Additional funding has come from the National Science Foundation, the Ford Motor Company, the Carnegie Corporation of New York, the Stuart Foundation, the General Electric Foundation, GTE, Citicorp/Citibank, the Exxon Educational Foundation, the Illinois Board of Higher Education, the Chicago Public Schools, from royalties, and from publishers of UCSMP materials.

From 1983 to 1987, the director of UCSMP was Paul Sally, Professor of Mathematics. Since 1987, the director has been Zalman Usiskin, Professor of Education.

UCSMP *Algebra*

The text *Algebra* has been developed by the Secondary Component of the project, and constitutes the core of the third year in a seven-year middle and high school mathematics curriculum. The names of the seven texts around which these years are built are:

- *Pre-Transition Mathematics*
- *Transition Mathematics*
- *Algebra*
- *Geometry*
- *Advanced Algebra*
- *Functions, Statistics, and Trigonometry*
- *Precalculus and Discrete Mathematics*

Why A Third Edition?

Since the second edition, there has been a general increase in the performance of students coming into middle school due, we believe, to a combination of increased expectations and the availability of improved curricular materials for the elementary grades. These materials are more ambitious and take advantage of the knowledge students bring to the classroom.

In addition, increased expectations for the performance of all students in both middle schools and high schools and the increased levels of testing that have gone along with those expectations are requiring a broad-based, reality-oriented, and easy-to-comprehend approach to mathematics. UCSMP third edition is being written to better accommodate these factors.

The writing of the third edition of UCSMP was also motivated by the recent advances in technology both inside and outside the classroom, coupled with the widespread availability of computers with internet access at school and at home.

Another factor for the continued existence of UCSMP is the increase in the number of students taking a full course in algebra before the ninth grade. These students will have four years of mathematics beyond algebra before calculus and other college-level mathematics. UCSMP is the only secondary curriculum to make such a sequence available.

Thousands of schools have used the first and second editions and have noted success in student achievement and in teaching practices. Research from these schools shows that the UCSMP materials really work. Many of these schools have made suggestions for additional improvements in future editions of the UCSMP materials. We have attempted to utilize all of these ideas in the development of the third edition.

UCSMP *Algebra*–Third Edition

The content and questions of this book integrate geometry, probability, and statistics together with algebra. Pure and applied mathematics are also integrated throughout. The earlier editions of *Algebra* introduced many features that have been retained in this edition. There is **wider scope**, including significant amounts of geometry and statistics, and some combinatorics and probability. These topics are not isolated as separate units of study or enrichment. A **real-world orientation** has guided both the selection of content and its applications. Applications are essential because being able to do mathematics is of little use to an individual unless he or she can apply that content. We require **reading mathematics** because students must read to understand mathematics in later courses and learn to read technical matter in the world at large. The use of **new and powerful technology** is integrated throughout. *Graphing calculator* use is assumed while *spreadsheets* and *computer algebra systems* are used periodically throughout the materials to develop patterns and practice skills.

Four dimensions of understanding are emphasized: skill in carrying out various algorithms; developing and using mathematics properties and relationships; applying mathematics in realistic situations; and representing or picturing mathematical concepts. We call this the SPUR approach: **S**kills, **P**roperties, **U**ses, and **R**epresentations.

The **book organization** is designed to maximize the acquisition of both skills and concepts. Ideas introduced in a lesson, as well as ideas from prior chapters, are reinforced through Review questions in the succeeding lessons. This daily review feature allows students several nights to learn and practice important new concepts and skills and increase retention of old ones. Then, at the end of each chapter, a carefully focused Self-Test and a Chapter Review are used to solidify performance of skills and concepts from the chapter so that they may be applied later with confidence. The Self-Test and Chapter Review, which are keyed to objectives in all the dimensions of understanding, aid student self-assessment.

Those familiar with the earlier editions will note a rather significant reorganization of the content in the third edition, particularly restructuring of the beginning of the course so that some ideas are introduced one or two months earlier than before. We were encouraged to do this because a high percentage of *Algebra* students enter this course with a better mathematical background than could have been expected when we wrote the earlier editions. There are also a number of instructional features new to this edition, including the following: **Activities** are more extensive and have been incorporated within lessons to enable students to take a more active approach to learning and developing concepts. There are **Guided Examples** that provide partially completed solutions to encourage active learning. **Quiz Yourself** stopping-point questions ask students to periodically check their understanding. There are many more questions requiring **writing** because writing helps students clarify their own thinking. Also, writing is an important aspect of communicating mathematical ideas to others.

Comments about these materials are welcomed. Please address them to:

The University of Chicago School Mathematics Project
http://ucsmp.uchicago.edu
ucsmp@uchicago.edu
773-702-1130

Contents

Getting Started 1

Chapter 1 — 4
Using Algebra to Describe

1-1	Evaluating Expressions	6
1-2	Describing Patterns	13
1-3	Equivalent Expressions	20
1-4	Picturing Expressions	27
1-5	Using a Graphing Calculator	33
1-6	Absolute Value and Distance	42
1-7	Data and Spread	47
	Projects	55
	Summary and Vocabulary	57
	Self-Test	58
	Chapter Review	60

Chapter 2 — 64
Using Algebra to Explain

2-1	The Distributive Property and Removing Parentheses	66
2-2	The Distributive Property and Adding Like Terms	72
2-3	Explaining Number Puzzles	79
2-4	Opposites	85
2-5	Testing Equivalence	91
2-6	Equivalent Expressions with Technology	98
2-7	Explaining Addition and Subtraction Related Facts	105
2-8	Explaining Multiplication and Division Related Facts	112
	Projects	121
	Summary and Vocabulary	123
	Self-Test	124
	Chapter Review	125

Chapter 3 128
Linear Equations and Inequalities

3-1	Graphing Linear Patterns	130
3-2	Solving Equations with Tables and Graphs	135
3-3	Exponential Decay Equivalent Equations	139
3-4	Solving $ax + b = c$	144
3-5	Using the Distributive Property in Solving Equations	149
3-6	Inequalities and Multiplication	155
3-7	Solving $ax + b < c$	162
3-8	Solving Equations by Clearing Fractions	167
▶	Projects	174
▶	Summary and Vocabulary	176
▶	Self-Test	177
▶	Chapter Review	178

Chapter 4 180
More Linear Equations and Inequalities

4-1	Solving Percent Problems Using Equations	182
4-2	Horizontal and Vertical Lines	188
4-3	Using Tables and Graphs to Solve	196
4-4	Solving $ax + b = cx + d$	202
4-5	Solving $ax + b < cx + d$	210
4-6	Situations That Always or Never Happen	216
4-7	Equivalent Formulas	221
4-8	Compound Inequalities, *AND* or *OR*	227
4-9	Solving Absolute Value Equations and Inequalities	234
▶	Projects	240
▶	Summary and Vocabulary	242
▶	Self-Test	243
▶	Chapter Review	245

Chapter 5 — 250
Division and Proportions in Algebra

5-1	Multiplication of Algebra Fractions	252
5-2	Division of Algebraic Fractions	258
5-3	Rates	263
5-4	Multiplying and Dividing Rates	269
5-5	Ratios	274
5-6	Probability Distributions	280
5-7	Relative Frequency and Percentiles	289
5-8	Probability without Counting	296
5-9	Proportions	301
5-10	Similar Figures	308
●	Projects	315
●	Summary and Vocabulary	317
●	Self-Test	318
●	Chapter Review	320

Chapter 6 — 324
Slopes and Lines

6-1	Rate of Change	326
6-2	The Slope of a Line	333
6-3	Properties of Slope	341
6-4	Slope-Intercept Equations for Lines	348
6-5	Equations for Lines with a Given Point and Slope	356
6-6	Equation for Lines through Two Points	361
6-7	Fitting a Line to Data	368
6-8	Standard Form of the Equation of a Line	374
6-9	Graphing Linear Inequalities	381
●	Projects	387
●	Summary and Vocabulary	389
●	Self-Test	390
●	Chapter Review	392

Chapter 7   396
Using Algebra to Describe Patterns of Change

Chapter 8   456
Powers and Roots

7-1	Compound Interest	398
7-2	Exponential Growth	404
7-3	Exponential Decay	411
7-4	Modeling Exponential Growth and Decay	419
7-5	The Language of Functions	425
7-6	Function Notation	432
7-7	Comparing Linear Increase and Exponential Growth	439
	Projects	447
	Summary and Vocabulary	449
	Self-Test	450
	Chapter Review	452

8-1	The Multiplication Counting Principle	458
8-2	Products and Powers of Powers	464
8-3	Quotients of Powers	469
8-4	Negative Exponents	474
8-5	Powers of Products and Quotients	481
8-6	Square Roots and Cube Roots	488
8-7	Multiplying and Dividing Square Roots	497
8-8	Distance in a Plane	505
8-9	Remembering Properties of Powers and Roots	511
	Projects	517
	Summary and Vocabulary	519
	Self-Test	520
	Chapter Review	521

Chapter 9   524
Quadratic Equations and Functions

9-1	The Function with Equation $y = ax^2$	526
9-2	Solving $ax^2 = b$	532
9-3	Graphing $y = ax^2 + bx + c$	537
9-4	Quadratics and Projectiles	544
9-5	The Quadratic Formula	552
9-6	Analyzing Solutions to Quadratic Equations	558
9-7	More Applications of Quadratics: Why Quadratics Are Important	565
▶	Projects	571
▶	Summary and Vocabulary	573
▶	Self-Test	574
▶	Chapter Review	576

Chapter 10   580
Linear Systems

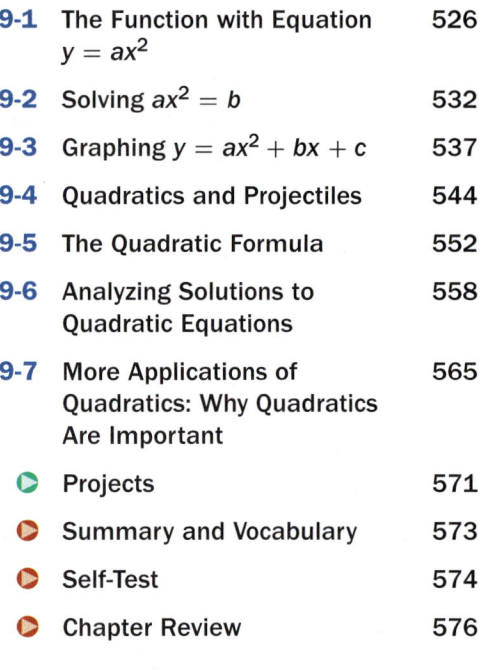

10-1	An Introduction to Systems	582
10-2	Solving Systems Using Substitution	589
10-3	More Uses of Substitution	594
10-4	Solving Systems by Addition	601
10-5	Solving Systems by Multiplication	608
10-6	Systems and Parallel Lines	616
10-7	Matrices and Matrix Multiplication	622
10-8	Using Matrices to Solve Systems	629
10-9	Systems of Inequalities	635
10-10	Nonlinear Systems	640
▶	Projects	645
▶	Summary and Vocabulary	647
▶	Self-Test	648
▶	Chapter Review	650

Chapter 11 — Polynomials — 654

11-1	Investments and Polynomials	656
11-2	Classifying Polynomials	663
11-3	Multiplying a Polynomial by a Monomial	669
11-4	Common Monomial Factoring	675
11-5	Multiplying Polynomials	680
11-6	Special Binomial Products	685
11-7	Permutations	691
11-8	The Chi-Square Statistic	697
▶	Projects	703
▶	Summary and Vocabulary	705
▶	Self-Test	706
▶	Chapter Review	708

Chapter 12 — More Work with Quadratics — 712

12-1	Graphing $y - k = a(x - h)^2$	714
12-2	Completing the Square	723
12-3	The Factored Form of a Quadratic Function	729
12-4	Factoring $x^2 + bx + c$	736
12-5	Factoring $ax^2 + bx + c$	742
12-6	Which Quadratic Expressions Are Factorable?	748
12-7	Graphs of Polynomial Functions of Higher Degree	754
12-8	Factoring and Rational Expressions	761
▶	Projects	768
▶	Summary and Vocabulary	770
▶	Self-Test	771
▶	Chapter Review	773

Chapter 13   776
Using Algebra to Prove

13-1	If-Then Statements	778
13-2	The Converse of an If-Then Statement	784
13-3	Solving Equations as Proofs	788
13-4	A History and Proof of the Quadratic Formula	795
13-5	Proofs of Divisibility Properties	802
13-6	From Number Puzzles to Properties of Integers	809
13-7	Rational Numbers and Irrational Numbers	816
13-8	Proofs of the Pythagorean Theorem	823
▶	Projects	829
▶	Summary and Vocabulary	831
▶	Self-Test	832
▶	Chapter Review	833

Selected Answers	S2
Glossary	S60
Index	S70
Photo Credits	S85
Symbols	S87

To the Student: Getting Started

Welcome to *Algebra*! We hope you enjoy this book; it was written for you.

Why is Algebra Important?

In almost every country of the world, students learn algebra. But you seldom see algebra when you go into a store, or when you read a newspaper, or when you watch a television show. So why is algebra so important? It is because algebra is the language of generalization. It is used to describe how quantities relate to each other and helps to solve countless numbers of problems.

For those reasons, with a knowledge of algebra:

- You become eligible for more jobs or training programs that will help you get a job.
- You are able to take courses that would help you determine your career path.
- You are able to understand many ideas discussed in business, science, psychology, politics, and in many other areas.
- You are more likely to make wise decisions about money and other personal matters.

Learning From This Book

We want you to be able to understand the mathematics around you, in newspapers and magazines, on television, in a job, and in school. To accomplish this goal, you should try to learn from the reading in each lesson as well as from your teacher and your classmates. The authors, who are all experienced teachers, offer the following advice:

1. You can watch basketball hundreds of times on television. Still, to learn how to play basketball, you must have a ball in your hand and practice dribbling, shooting, and passing. Mathematics is no different. You cannot learn much mathematics just by watching other people do it. You must participate in mathematics. Some teachers have a slogan that has served their classes well in the past:

    Mathematics is not a spectator sport.

2. You are expected to read each lesson. Here are some ways to improve your reading comprehension:

    - Read slowly, paying attention to each word and symbol.
    - Look up the meaning of any word you do not understand.
    - Work examples yourself as you follow the steps in the text.
    - Reread sections that are unclear to you.
    - Discuss difficult ideas with a fellow student or your teacher.

3. Writing can help you understand mathematics, too. So you will sometimes be asked to explain your solution to a problem or to justify an answer. Writing good explanations takes practice. You can look at the solutions to the examples in each lesson as a guide for your own writing.

4. If you cannot answer a question immediately, don't give up! Read the lesson again. Read the question again. Look for examples. If you can, get away from the question and come back to it a little later. Ask questions in class and talk to others when you do not understand something. School is designed so that you do not have to learn everything by yourself. Here is another slogan that many teachers have used:

None of us is as smart as all of us.

What Tools Do You Need for this Book?

You need to have some tools to do any mathematics. The most basic tools are paper, pencil, and erasers. For this book, you will also need the following equipment:

- a **ruler** with both centimeter and inch markings,
- a **protractor,**
- **graph paper,** and
- a **graphing calculator**

At times in this book, there are activities involving a *computer algebra system* (CAS). Your teacher may want you to have a calculator that has CAS capability, or such calculators may be available in your classroom or school.

Getting Off to a Good Start

One way to get off to a good start is to spend some time getting acquainted with your textbook. The questions that follow are part of an activity designed to help you become familiar with *Algebra*.

We hope you join the millions of students who have enjoyed this book. We wish you much success.

Questions

COVERING THE IDEAS

1. Why is algebra important?
2. List four things, besides paper and pencil, you will need for your work in *Algebra*.
3. Explain what is meant by the statement "Mathematics is not a spectator sport."
4. Of the five things listed that you can do to improve reading comprehension, list the three you think are most helpful to you.
5. Where can you look for a model to help you in writing an explanation to justify your answer to a question?

KNOWING YOUR TEXTBOOK

In 6 and 7, refer to the Table of Contents beginning on page vi.

6. Algebra involves finding the slope of a line. In which chapter would you find the definition of slope?

7. In which lesson would you first learn how to use a graphing calculator?

In 8–15, refer to other parts of the book.

8. Look at several lessons. Where can you find the answers to the Quiz Yourself questions? Where can you find the answers to the Guided Examples?

9. Suppose you have just finished doing Activity 2 in Lesson 2-6. How can you check your answers?

10. What are the four categories of Questions at the end of each lesson?

11. Suppose you have just finished the Questions in Lesson 1-7. On what page can you find answers to check your work? For which Questions are answers given?

12. Refer to a Self-Test at the end of a chapter. When you finish the test, what is it recommended that you do?

13. What kinds of questions are in the Chapter Review at the end of each chapter?

14. Where is the Glossary and what does it contain?

15. Locate the Index.
    a. According to the Index, where can you find information about the Carlsbad Caverns?
    b. In what state are the Caverns located?

Chapter 1
Using Algebra to Describe

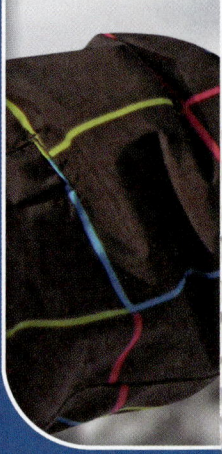

Contents

- 1-1 Evaluating Expressions
- 1-2 Describing Patterns
- 1-3 Equivalent Expressions
- 1-4 Picturing Expressions
- 1-5 Using a Graphing Calculator
- 1-6 Absolute Value and Distance
- 1-7 Data and Spread

A picture is worth a thousand words.
In mathematics, pictures and symbols are used to summarize mathematical concepts that normally require many words to describe, as shown below.

For many years you have multiplied fractions. You know that $\frac{5}{8} \cdot \frac{3}{4} = \frac{15}{32}$. Here is the general rule for multiplying fractions.

> To multiply two fractions, multiply their numerators to get the numerator of the product, and then multiply their denominators to get the denominator of the product.

Wow! This rule is quite a mouthful. It is 26 words long. Here is the same rule in the language of algebra.

$$\frac{a}{b} \cdot \frac{c}{d} = \frac{ac}{bd},\ b \neq 0,\ d \neq 0$$

The description in algebra is much shorter than the description in words, just like showing the photograph above is "shorter" than trying to describe it in words. The algebraic description also shows the arithmetic of fractions. *Algebraic descriptions can make it easier to understand relationships among numbers and quantities.*

In the algebraic description, *a*, *b*, *c*, and *d* are *variables*. Except for the fact that *b* and *d* cannot equal zero (you cannot divide by 0), these variables can stand for any numbers. They could be whole numbers, fractions, decimals, or percents. They could be positive or negative.

The algebraic description works for any situation in which you need to multiply fractions. Algebra is a powerful language that is used throughout the world. This is why almost all students are required to learn some algebra.

In this book you will see how algebra can be used to describe patterns, to explain why numbers act as they do, to solve problems, and to discover and prove relationships between quantities. This first chapter describes patterns with algebra and illustrates them with graphs.

Chapter 1

Lesson 1-1

Evaluating Expressions

Vocabulary

variable
algebraic expression
evaluating the expression

▶ **BIG IDEA** The order of operations is used to evaluate expressions with variables and expressions with numbers.

Algebraic Expressions

Adults are roughly twice as tall as they were when they were 3 years old, as shown in the photograph at the right. This suggests a simple rule of thumb for predicting the adult height of a 3-year-old child: multiply his or her current height by 2. In the language of algebra, if c represents the height of a 3-year-old child, then $2 \cdot c$, or $2c$ for short, is the child's predicted height as an adult. The letter c is a *variable* and $2c$ is an *algebraic expression*. A **variable** is a letter or other symbol that can be replaced by any number (or other object) from a set. When numbers and variables are combined using the operations of arithmetic, the result is called an **algebraic expression**. Some examples of algebraic expressions are $7y$, $5x - 9$, and $\frac{4n^3p + 16}{z}$. Finding the numerical value of an expression is called **evaluating the expression**. To evaluate an algebraic expression, substitute numbers for its variables. The order of operations for algebraic expressions is the same as in numerical expressions.

Mental Math

a. For how long did a gym class play softball if they played from 10:45 A.M. to 11:35 A.M.?

b. For how long did a cross country team run if they ran from 11:10 A.M. to 1:40 P.M.?

c. If a 1 hour, 45 minutes long gym class starts at 11:30 A.M., at what time does it end?

Order of Operations in Evaluating Expressions

1. Perform operations within parentheses or other grouping symbols.
2. Within grouping symbols, or if there are no grouping symbols:
    a. Evaluate all powers from left to right.
    b. Next multiply and divide from left to right.
    c. Then add and subtract from left to right.

The father is approximately 2 times the height of his daughter.

Activity

Suppose a and b are whole numbers with $a + b = 100$. What is the largest possible value of $a + ab - b$?

6 Using Algebra to Describe

Lesson 1-1

Example 1
a. If $a = 5$, find $3a^2$.
b. If $a = 5$, calculate $(3a)^2$.

Solution

a. Substitute 5 for a. There are no grouping symbols, so the power is the first operation to perform. Then multiply.
$$3 \cdot 5^2 = 3 \cdot 25 = 75$$

b. Substitute 5 for a. Perform the operation within the parentheses first. Then square the product.
$$(3 \cdot 5)^2 = 15^2 = 225$$

Subtraction and Division Expressions

Subtracting a number yields the same result as adding the opposite of that number. For example, substitute 12 for x in the following equation.

$$
\begin{aligned}
35 - 4x &= 35 - 4 \cdot 12 &&\text{Substitute 12 for } x.\\
&= 35 - 48 &&\text{Multiply first.}\\
&= 35 + -48 &&\text{Rewrite using addition.}\\
& &&-48 \text{ is the opposite of } 48.\\
&= -13 &&\text{Add.}
\end{aligned}
$$

For some people, rewriting $35 - 48$ as $35 + -48$ makes the computation easier.

When subtracting a negative number, rewriting the expression is helpful.

$$
\begin{aligned}
-30 - -6 &= -30 + 6 &&\text{6 is the opposite of } -6.\\
&= -24
\end{aligned}
$$

The relationship between addition and subtraction is true for all real numbers. It is known as the *algebraic definition of subtraction*.

Algebraic Definition of Subtraction
For all real numbers a and b, $a - b = a + -b$.

There is a similar relationship between multiplication and division. Dividing a number by b is the same as multiplying by the reciprocal of b, or $\frac{1}{b}$. For example, $32 \div 5 = 32 \cdot \frac{1}{5} = 6.4$.

Algebraic Definition of Division
For all real numbers a and b with $b \neq 0$, $a \div b = a \cdot \frac{1}{b}$.

Evaluating Expressions

Chapter 1

Remember that there are several symbols for division. The operation $a \div b$ can also be written as $\frac{a}{b}$ or a/b. In all cases, b (the denominator, or divisor) cannot be zero. For example:

$$5 \div 12 = \frac{5}{12} = 5/12 = 5 \cdot \frac{1}{12}$$

$$24 \div 6 = \frac{24}{6} = 24/6 = 24 \cdot \frac{1}{6}$$

$$\frac{1}{4} \div 3 = \frac{\frac{1}{4}}{3} = (1/4)/3 = \frac{1}{4} \cdot \frac{1}{3}$$

🛑 See Quiz Yourself at the right.

Quiz Yourself (QY) questions are designed to help you follow the reading. You should try to answer each Quiz Yourself question before reading on. The answer to the Quiz Yourself is found at the end of the lesson.

> **QUIZ YOURSELF**
>
> Complete the pattern as shown in the three examples above and at the left.
>
> $a \div b = \frac{a}{b}$
>
> $ = \frac{?}{}$
>
> $ = \frac{?}{}$

Evaluating with Technology

Scientific and graphing calculators use the algebraic order of operations. However, they can differ in the keys they use for squares, exponents, and square roots. Some possible key sequences are shown below. The order of entry is similar to the order you use to write the symbols by hand.

Key	Example
squaring key	For 5^2 enter 5 $\boxed{x^2}$ to get 25.
exponent key	For 5^3 enter 5 $\boxed{\wedge}$ 3 to get 125.
square root key	For $\sqrt{1{,}156}$ enter $\boxed{\sqrt{}}$ 1156 to get 34.

With most calculators, you must input fractions on one line. Therefore, you must include grouping symbols to show calculations within fractions. For example, to evaluate $\frac{24}{8-2}$, you must enter $24/(8-2)$ to get the correct answer 4. If you enter $24/8 - 2$, the calculator will follow the order of operations doing the division $24 \div 8$ first and then calculating $3 - 2$ to get 1 for the answer.

GUIDED

Example 2

A Guided Example is an example in which some, but not all, of the work is shown. You should try to complete the example before reading on. Answers to Guided Examples are in the Selected Answers section at the back of the book.

Use a calculator to evaluate $\left(\frac{6.8 - w}{n + w}\right)^3$ when $n = 21$ and $w = 0.5$.

8   Using Algebra to Describe

Solution

Step 1 Write the expression. $\left(\frac{?}{?}\right)^3$

Step 2 Substitute given values for the variables. $\left(\frac{6.8 - ?}{? + ?}\right)^3$

Step 3 Enter into a calculator. Here is a start.

$((6.8 - \underline{\ ?\ })/\underline{\ ?\ }$

Step 4 Round the result to the nearest thousandth.

In Example 2 above, notice that w appears in the expression two times. If the same variable appears more than once in an expression, the same number must be substituted for it every time the variable appears.

Evaluating Expressions in Formulas

Sometimes you will need to evaluate an expression as part of working with a formula. For example, the formula $V = \ell w h$ is used to find the volume V of a rectangular solid with length ℓ, width w, and height h as shown in Example 3 below.

Example 3

Find the volume of the box shown at the right.

Solution Substitute the given dimensions for ℓ, w, and h in the expression. Let $\ell = 2$ in., $w = 3$ in., and $h = 5$ in., and evaluate the expression $\ell \cdot w \cdot h$.

Volume $= \ell \cdot w \cdot h = 2 \cdot 3 \cdot 5 = 6 \cdot 5 = 30$

So, the volume is 30 in³.

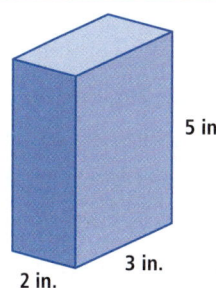

Three Important Properties

In Example 3, we followed the order of operations and multiplied $2 \cdot 3 \cdot 5$ from left to right. However, you could first multiply 3 and 5 and then multiply the product by 2 to get the same answer, 30. This illustrates that $(\ell \cdot w) \cdot h = \ell \cdot (w \cdot h)$. This important general property, true for all real numbers, is called the *Associative Property of Multiplication*. Notice that the product of numbers a and b can be written as ab, $a \cdot b$, $a(b)$, or (ab). The product is usually not written as $a \times b$ in algebra because x is such a common letter for a variable.

Associative Property of Multiplication

For any real numbers a, b, and c, $(ab)c = a(bc)$.

Evaluating Expressions

A similar property is true for addition. You can regroup numbers being added without affecting the sum. For example, the sum $(67 + 98) + 2$ is easier to calculate in your head if the numbers being calculated are regrouped.

$$(67 + 98) + 2 = 67 + (98 + 2)$$
$$= 67 + 100$$
$$= 167$$

This is an example of the *Associative Property of Addition*.

> **Associative Property of Addition**
>
> For any real numbers a, b, and c, $(a + b) + c = a + (b + c)$.

Another property of algebra, called the *Transitive Property of Equality*, has been used throughout this lesson in evaluating expressions. It allows us to look at calculations like $20 - 46 = 20 + -46$ and $20 + -46 = -26$, and deduce that $20 - 46 = -26$.

> **Transitive Property of Equality**
>
> For any real numbers a, b, and c, if $a = b$ and $b = c$, then $a = c$.

Questions

COVERING THE IDEAS

These questions cover the content of the lesson. If you cannot answer a Covering the Ideas question, you should go back to the reading for help in obtaining an answer.

In 1–6, evaluate the expression using the order of operations.

1. $36 - 3 \cdot 4$
2. $36 \div 3 \cdot 4 - (-2)$
3. $(4 + 5)^2$
4. $4^2 + 5^2$
5. $10 - 70 - 5^{(2+1)} + 4(20)$
6. $-16 + 3(4 - 5) \div \frac{9 - 3}{17 - 13}$

In 7 and 8, use the order of operations to evaluate the expression for the given values of x.

7. $2(4x - 5) + 2$
   a. $x = 5$
   b. $x = -1$
8. $2^x - (1 + x)$
   a. $x = 1$
   b. $x = 3$

9. Evaluate each expression when $t = 10$.
   a. $5t^2$
   b. $(5t)^2$

10. Evaluate $(-1 + r)^3$ when $r = \frac{1}{3}$. Write your answer as a fraction.

11. Evaluate $\frac{a + 2b}{5}$ when $a = 11.6$ and $b = 9.2$.

In 12–14, use the Associative Properties of Addition and Multiplication to classify each statement as true or false. Then check your answer by doing the arithmetic.

12. $5 \cdot (2 \cdot 7) = (5 \cdot 2) \cdot 7$
13. $-2 \cdot \left(\frac{1}{2} + 6\right) = \left(-2 + \frac{1}{2}\right) \cdot 6$
14. $13.5 + (-2 + 3 + -4) = (13.5 + -2 + 3) + -4$

APPLYING THE MATHEMATICS

These questions extend the content of the lesson. You should study the examples and explanations if you cannot answer the question. For some questions, you can check your answers with the ones in the back of this book.

15. Joshua found that $\frac{63}{225} = \frac{7}{25}$. Sabrina realized that $\frac{7}{25} = \frac{28}{100}$.
    a. What can be deduced from Joshua's and Sabrina's results using the Transitive Property of Equality?
    b. $\frac{28}{100} = 28\%$. Is it true that $\frac{63}{225} = 28\%$? Explain your answer.

In 16 and 17, which expression is *not* equal to the others?

16. $27 - 3$ $\quad -27 + 3$ $\quad 27 + -3$ $\quad -3 + 27$
17. $-9 \div 4$ $\quad -9 \cdot \frac{1}{4}$ $\quad \frac{1}{-9} \cdot 4$ $\quad \frac{-9}{4}$

In 1926, Robert H. Goddard launched the first liquid-fueled rocket and laid the foundation for a technology that would eventually take humans to the moon.
Source: NASA

18. When an object is shot from the ground into the air, the formula $h = -16t^2 + vt$ gives the height h in feet of the object t seconds later, where v is the velocity of the object in feet per second when it first leaves the ground. If a toy rocket is launched with a velocity of 80 feet per second, find its height 2 seconds later.

19. Use the rule $\frac{a}{b} \cdot \frac{c}{d} = \frac{ac}{bd}$ to multiply the fractions $\frac{-2}{11}$ and $\frac{-3}{5}$. Write your answer as a single fraction.

20. Elias and Marissa are the same height. Samuel and Elias are equally tall.
    a. What conclusion can be made based on the Transitive Property of Equality?
    b. Write another real-world situation that uses the same property.

Chapter 1

REVIEW

Every lesson contains review questions to practice ideas you have studied earlier.

In 21–23, compute in your head. (Previous Course)

21. $-4 \cdot \$1.25$

22. $1{,}000 \cdot 11.4$

23. $3\frac{1}{2} \cdot 20$

24. Put these numbers in order from least to greatest. (Previous Course)

$$12 \quad 5.3 \quad 2 \quad -\tfrac{2}{7} \quad -7 \quad 5.39 \quad -\tfrac{1}{6} \quad \pi$$

25. Nikki's Bike Shop gives customers a free helmet with the purchase of any new bike from the store. Last week, 19 new bikes were purchased from Nikki's Bike Shop. If the cost of each helmet was $32.50, what was the total cost of all the helmets that the store gave away last week? (Previous Course)

In 26–29, compute without a calculator. (Previous Course)

26. $5 + -9 - 22$

27. $4 + 11 - -7$

28. $\tfrac{3}{4} + \tfrac{2}{3} - \tfrac{1}{6}$

29. $-3.52 - 11.4 + 30$

Wearing a bicycle helmet while riding helps reduce injuries by up to 88%.

Source: Bicycle Helmet Safety Institute

EXPLORATION

These questions ask you to explore topics related to the lesson. Sometimes you will need to use references found in a library or on the Internet.

30. Suppose a, b, c, and d are different positive integers whose sum is 100, and $a - c = 5$. What is the greatest possible value of $ab - cd$?

QUIZ YOURSELF ANSWER

$a/b, \; a \cdot \tfrac{1}{b}$

Lesson 1-2

Lesson 1-2

Describing Patterns

Vocabulary

pattern
instance
define a variable
term
factor

▶ **BIG IDEA** Patterns in tables are often described by expressions with variables.

Using Tables to Look at Patterns

Ian wants to save money to buy a new bike. He already has $25, and he decides he can save $15 each week. If the bike costs $220, will Ian have saved enough to buy the bike after 12 weeks?

To answer this question it may help to look at a table.

Mental Math

a. During a 25%-off sale, what is the cost of a pair of shoes that normally costs $80?

b. How much do you save on that pair of shoes?

Number of Weeks (w)	Calculation	Pattern	Money Saved
0	25	25 + 15 · 0	$25
1	25 + 15	25 + 15 · 1	$40
2	25 + 15 + 15	25 + 15 · 2	$55
3	25 + 15 + 15 + 15	25 + 15 · 3	$70
4	25 + 15 + 15 + 15 + 15	25 + 15 · 4	$85
5	25 + 15 + 15 + 15 + 15 + 15	25 + 15 · 5	$100
6	25 + 15 + 15 + 15 + 15 + 15 + 15	25 + 15 · 6	$115

The table shows the amount Ian has saved after 0, 1, 2, 3, 4, 5, and 6 weeks. The key to the table is the *Pattern* column. In that column, the repeated adding of 15 is rewritten as multiplication. A **pattern** is a general idea for which there are many **instances.** The last row of the table shows one instance, stating that after 6 weeks, Ian has saved $25 + 15 \cdot 6$, or $115. The pattern lets you write an algebraic expression to describe the amount of money Ian has after *any* number of weeks. If the variable w is used to represent the number of weeks that have passed, an expression that describes the pattern is $25 + 15 \cdot w$, as shown in the table on page 14.

Describing Patterns 13

Number of Weeks	Calculation	Pattern	Money Saved
w	25 + 15 + 15 + ... + 15 (w addends)	25 + 15 · w	25 + 15w

The expression 25 + 15w can be used to find out how much money Ian will have after 12 weeks. Replace w with 12 and evaluate 25 + 15 · 12. In 12 weeks, Ian will have 25 + 15(12), or $205. Since the bike costs $220, he will not have enough money after 12 weeks.

🛑 See Quiz Yourself at the right.

Notice that an important step is to define the variable used in the algebraic expression. To **define a variable** means to describe the quantity the variable represents. Defining a variable is often signaled by the word *let*, as in "Let $x = \ldots$."

> ▶ QUIZ YOURSELF
>
> Assuming the pattern continues, will Ian have enough money after 14 weeks? How much money will he have saved up?

GUIDED

Example 1

At the start of the school year, a school's library had a total of 3,600 individual magazines that it had collected over time. Each month 22 new magazines are added to the collection.

a. Complete the table to show the number of magazines in the library each month.

b. Define a variable and write an algebraic expression for the number of magazines.

Solution

a.

Months Since Start of School Year	Calculation	Pattern	Magazines in Library
0	?	3,600 + 22(0)	3,600
1	?	?	?
2	3,600 + 22 + 22	?	3,644
3	?	3,600 + 22(?)	?

b. Let $m = $ ___?___.
   The number of magazines after *m* months is ___?___.

Patterns Having Two or More Variables

Some patterns have two or more variables and can also be represented by tables and expressions.

Example 2
A family consisting of 2 adults and 3 children was planning a vacation. They looked in a tour book to find the cost of some activities.

a. Complete the table to show the family's cost for each activity.

Activity	Adult Ticket	Child Ticket	Family's Cost
Movie	$15.00	$8.00	2 · 15 + 3 · 8
Ferryboat	$22.00	$10.00	?
Water Park	$10.00	$4.50	?

b. How much does each activity cost the family?
c. Let a = the adult ticket price for an activity and let c = the child ticket price. Write an algebraic expression for the family's total cost for this activity.

Solution
a. Multiply the cost of an adult ticket by the number of adults (2), and the cost of a child ticket by the number of children (3).

Activity	Adult Ticket	Child Ticket	Family's Cost
Movie	$15.00	$8.00	2 · 15 + 3 · 8
Ferryboat	$22.00	$10.00	2 · 22 + 3 · 10
Water Park	$10.00	$4.50	2 · 10 + 3 · 4.50

b. Movie: 2 · 15 + 3 · 8 = 30 + 24 = $54
Ferryboat: 2 · 22 + 3 · 10 = 44 + 30 = $74
Water Park: 2 · 10 + 3 · 4.50 = 20 + 13.50 = $33.50

c. The total cost is 2 · a + 3 · c, or $2a + 3c$.

Approximately 73 million people visited North American water parks during the summer 2004 season.
Source: World Waterpark Association

The Commutative Properties of Addition and Multiplication

Numbers or expressions that are added are called **terms**. Numbers or expressions that are multiplied are called **factors**. In the expression $2a + 3c$ of Example 2, there are two terms: $2a$ and $3c$. 2 and a are factors of $2a$, and 3 and c are factors of $3c$.

Chapter 1

The expression $2a + 3c$ stands for "twice the cost of an adult ticket added to three times the cost of a child ticket." The order of operations indicates to do both multiplications before adding. But you can switch the terms being added because the total cost is the same whether you buy adult tickets first or child tickets first. The expression $3c + 2a$ gives the same values as $2a + 3c$.

It is also the case that you can switch the order of the factors. $2 \cdot a$ gives the same values as $a \cdot 2$. These are examples of the *commutative properties*. The word *commutative* comes from the French word *commutatif*, which means "switchable."

> **Commutative Property of Addition**
>
> For all real numbers a and b, $a + b = b + a$.

> **Commutative Property of Multiplication**
>
> For all real numbers a and b, $a \cdot b = b \cdot a$.

Examples of the commutative properties can also be numerical. They are true for all real numbers. For example, when $a = 72$ and $b = 10$, $72 \cdot 10 = 10 \cdot 72$.

Questions

COVERING THE IDEAS

In 1–4, give two instances of each pattern.

1. $y - y = 0$
2. $x + x = 2x$
3. $x \cdot x = x^2$
4. $\frac{x}{y} = x \cdot \frac{1}{y}$

In 5 and 6, describe the given pattern using one variable.

5. $(3 + 9) - 2 = 1 + 9$
   $(3 + 4) - 2 = 1 + 4$
   $(3 + 90) - 2 = 1 + 90$

6. $15 + 2 \cdot 15 = 3 \cdot 15$
   $\frac{1}{3} + 2 \cdot \frac{1}{3} = 3 \cdot \frac{1}{3}$
   $47.1 + 2 \cdot 47.1 = 3 \cdot 47.1$

7. Pearl is starting a lawn mowing business. She plans to spend $1,200 on advertising during the first few weeks that she is in business. Each week she spends $45 to place an ad in a newspaper.
   a. Give the amount of money she will have left to spend on advertising after 1, 2, and 3 weeks of advertising.
   b. Write an algebraic expression for the amount she will have left to spend after w weeks of advertising.

Using Algebra to Describe

8. A group consisting of 1 adult and 5 children went to a special Kid's Day baseball game that offered special prices for children. Some prices are shown in the table below.

JACKSON FIELD
Home of the Barracudas

Item	Adult Price	Child Price	Group's Cost
Ticket	$17.00	$10.00	?
Hot Dog	$5.00	$3.50	?
Drink	$3.00	$2.00	?

   a. Fill in the group's cost for each item.
   b. Write an algebraic expression that would describe the group's cost for *any* item in terms of the adult price P and the child price p.

9. **Multiple Choice** Which expression below has three terms?

   A $4ab + 25$ B pqr C $2n + 6y + 15$ D $3x$

APPLYING THE MATHEMATICS

10. The following describes a pattern. A number is multiplied by 12, the product is divided by 2, and then 3 is subtracted from the result.
    a. Give three instances of the pattern.
    b. Write an algebraic expression to describe the pattern if the original number is n.

11. Istu is 5 years older than his sister Christine.
    a. Copy the table at the right and fill in Christine's age.
    b. Let i = Istu's age. Write an expression for Christine's age.
    c. Flor is 3 years older than Christine. Use your answer to Part b to write an expression for Flor's age.

Istu's Age	Christine's Age
9	?
16	?
25	?
89	?

12. Use the following information. When m is positive and n is greater than m, there exists a right triangle with side lengths given by the expressions $n^2 - m^2$, $2nm$, and $n^2 + m^2$, as shown in the diagram at the right.
    a. Let $n = 2$ and $m = 1$. Find the lengths of the three sides of the right triangle.
    b. Let $n = 3$ and $m = 2$. Find the lengths of the three sides of the right triangle.

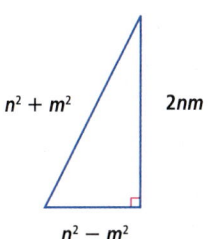

13. A pizza is cut into pieces by making each cut pass through the center.

| 1 cut | 2 cuts | 3 cuts |

   a. Describe the number of pieces made for 1, 2, 3, 4, and 5 cuts.
   b. Use two variables to describe the pattern.

14. a. Evaluate each expression.

   $\frac{2}{3} \cdot \frac{3}{2}$ $\frac{23}{11} \cdot \frac{11}{23}$ $\frac{-7}{5} \cdot \frac{5}{-7}$ $\frac{0.03}{6} \cdot \frac{6}{0.03}$

   b. Describe the pattern.

   c. **Do you think the pattern is true for all numbers? Why or why not?**

15. Suppose $x = 58$ and $y = 31$.
   a. Evaluate $xy - yx$.
   b. Suppose the values of x and y are changed. Will the answer to Part a change? Why or why not?

REVIEW

In 16–17, evaluate when $x = 2$, $y = 8$, and $z = 4$. (Lesson 1-1)

16. $x^3 + \frac{y}{z}$

17. $y + z + \frac{x+z}{x} - z(z-1)^4$

18. The planet Neptune is approximately a sphere with radius (r) of 25,000 kilometers. Use the formula $V = \frac{4}{3}\pi r^3$ to estimate the volume V of Neptune in scientific notation. (Lesson 1-1, Previous Course)

19. Evaluate each of the following without a calculator. (Previous Course)
   a. $3 \cdot -5$
   b. $-70 \cdot -6$

20. Round π to the nearest thousandth. (Previous Course)

21. Often, 20% of a restaurant bill is left for a tip. If a bill is $34.76, what would be a 20% tip rounded to the nearest dollar? (Previous Course)

22. **Skill Sequence** Find the sum. (Previous Course)
   a. $\frac{3}{5} + \frac{4}{5}$
   b. $\frac{3}{5} + \frac{4}{15}$
   c. $\frac{3}{5} + \frac{4}{17}$

Lesson 1-2

EXPLORATION

23. A monthly calendar contains many patterns.

July

Sun	Mon	Tue	Wed	Thu	Fri	Sat
				1	2	3
4	5	6	7			
11	12	13	14			
18	19	20	21			
25	26	27	28	29	30	31

a. Consider a 3 × 3 square such as the one drawn on the calendar. Copy the square and insert the nine dates. Then add the numbers along the diagonals. What is the relationship between the sums? Try this again with a different 3 × 3 square. Does it always seem to work?

b. In a 3 × 3 square portion of the calendar, if the middle date is expressed as N, then the date above would be $N - 7$ because it is 7 days earlier. Copy the chart at the right and fill in the other blanks.

?	$N - 7$	?
?	N	?
?	?	?

c. Show how your result from Part b can be used to explain your conclusion in Part a.

QUIZ YOURSELF ANSWER

yes; $235

Describing Patterns

Chapter 1

Lesson 1-3

Equivalent Expressions

Vocabulary

sequence

term

equivalent expressions

counterexample

▶ **BIG IDEA** If different expressions describe the same patterns, then the expressions are equivalent.

Finding Equivalent Expressions

The picture below shows a *sequence* of designs made from toothpicks. A **sequence** is a collection of numbers or objects in a specific order. The objects are called the **terms** of the sequence.

Mental Math

a. If international stamps cost 75¢ each, how much does a pack of 20 stamps cost?

b. If the price of the stamps increases by 2¢ per stamp, by how much will a pack of 20 stamps increase?

c. If a pack of 20 stamps cost $2.00 less two years ago, how much less did one stamp cost?

Activity 1

Refer to the toothpick sequence below.

| Term Number | 1 | 2 | 3 | 4 |

Term

1. Use toothpicks to find how many are used to make the pattern for terms 1 through 10?

2. In words, explain how to use the term number to find the number of toothpicks needed to make that pattern.

3. Use *n* to represent the term number. Write an algebraic expression for the number of toothpicks needed for the *n*th term.

4. Use your algebraic expression to complete the table below.

Term Number	Calculation	Number of Toothpicks
11	?	?
12	?	?
15	?	?
100	?	?

5. Compare your expression with others in your class. Are the expressions the same or different? Write down all the expressions that you think are correct.

This sequence of designs is made from bricks.

20   Using Algebra to Describe

Lesson 1-3

Activity 2

Now consider a different toothpick sequence. The picture at the right shows the number of toothpicks required to construct a sequence of rectangles. The first and second terms are shown. Each rectangle is one toothpick wider and one toothpick taller than the previous rectangle.

1. How many toothpicks are needed for the 1st term?
2. How many toothpicks are needed for the 2nd term?
3. Build the 3rd term in the sequence. How many toothpicks are needed?
4. In words, describe a method to find the number of toothpicks it takes to make the nth term.
5. Write an algebraic expression to find the number of toothpicks it takes to make the nth term.
6. Fill in the table below.

Term Number	Calculation	Number of Toothpicks
4	?	?
5	?	?
6	?	?
13	?	?
58	?	?

In the activities, you looked at sequences of toothpicks and the number of toothpicks needed to make them. For each sequence, you used the term number n to determine the number of toothpicks needed for the nth term. You and your classmates may have found that there can be more than one expression to describe the nth term.

Consider the sequence of dots at the right. Students were asked to describe the number of dots needed to make the nth term.

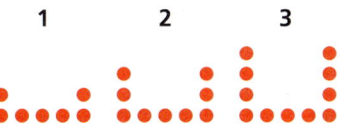

Alf explained that in each term he saw 2 columns of dots and 3 dots between the columns. He wrote the following.

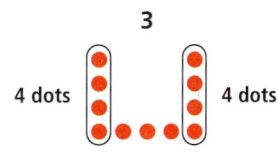

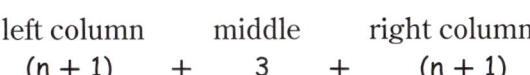
left column   middle   right column
   $(n + 1)$ +   3   + $(n + 1)$

Beth saw a row of 5 dots at the bottom and 2 columns above the row. She wrote the following.

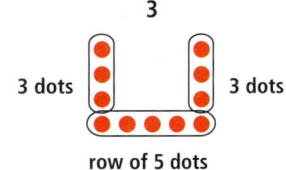

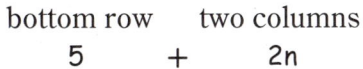
bottom row   two columns
   5   + $2n$

Equivalent Expressions

Chapter 1

Are both Alf and Beth correct? One way to tell is by substituting a value for n and testing whether the two expressions equal the same value. For instance, let $n = 11$.

Alf's expression: $(n + 1) + 3 + (n + 1) = (11 + 1) + 3 + (11 + 1)$
$= 12 + 3 + 12$
$= 27$

Beth's expression: $5 + 2n = 5 + 2(11)$
$= 5 + 22$
$= 27$

However, it is risky to test only one instance. Sometimes a pattern works for a few numbers and fails for all others. For example, suppose you are not sure whether $2w$ and w^2 have the same meaning.

STOP See Quiz Yourself 1 at the right.

The Quiz Yourself shows that $2w$ and w^2 are not equal for all values of w. When you evaluate an expression for several values of the variable, a table of values helps to organize your work.

> ▶ QUIZ YOURSELF 1
>
> a. Show that $2w$ and w^2 have the same value when $w = 2$.
> b. Show that $2w$ and w^2 have different values when $w = 5$.

GUIDED

Example 1

Compare Alf's and Beth's expressions for several instances by filling in the tables below.

Alf	
n	$(n + 1) + 3 + (n + 1)$
1	?
2	?
3	?
10	?
20	?
35	?

Beth	
n	$5 + 2n$
1	?
2	?
3	?
10	?
20	?
35	?

Solution Comparing the second columns of the two tables, you should have found that when the expressions $(n + 1) + 3 + (n + 1)$ and $5 + 2n$ are evaluated for these values of n, the same result is obtained. Therefore, $(n + 1) + 3 + (n + 1)$ and $5 + 2n$ appear to be *equivalent*.

Equivalent expressions are expressions that have the same value for *every* number that can be substituted for the variable(s). If two expressions produce different results when evaluated for the same number, then the expressions are not equivalent.

Lesson 1-3

Example 2

Suppose Azami and Haley both wrote expressions to describe the same pattern. Azami wrote $k - 8 + 5$ and Haley wrote $k - 13$. Are the expressions equivalent?

Solution Pick a number to substitute for k. Let $k = 10$.

Azami's expression: $k - 8 + 5 = 10 - 8 + 5 = 7$

Haley's expression: $k - 13 = 10 - 13 = -3$

When $k = 10$ the expressions have different values, so the expressions are not equivalent.

To show that $k - 8 + 5$ and $k - 13$ are not always equal, you only need to show one instance in which the expressions have different values. The situation in which $k = 10$ is a *counterexample*. A **counterexample** is an instance which shows that a general statement is not always true. It is not true that for all values of k, $k - 8 + 5 = k - 13$.

 See Quiz Yourself 2 at the right.

> **QUIZ YOURSELF 2**
>
> Which two of these expressions seem to be equivalent? How do you know?
>
> $(n + 3)^2$
>
> $n^2 + 9$
>
> $n^2 + 6n + 9$

Questions

COVERING THE IDEAS

1. Consider the sequence created with square tiles shown below.

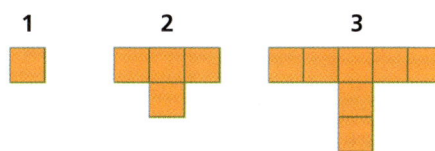

   a. Evaluate the expression $3n - 2$ for various values of n to show that it describes the number of squares in the nth term for the 1st, 2nd, and 3rd terms.
   b. Use the expression to find the number of tiles in the 100th term.

2. Consider the sequence of dots below.

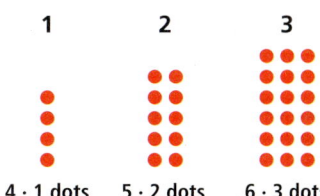

4 · 1 dots   5 · 2 dots   6 · 3 dots

   a. Draw the 4th and 5th terms. For each, write the multiplication expression for the number of dots.
   b. In words, describe how to find the number of dots in the nth term if you know the term number.
   c. Write an expression for the number of dots in the nth term.

Equivalent Expressions   23

3. Two different expressions were used to generate each table of values below. Use the tables to decide whether the expressions seem to be equivalent. Explain how you know.

x	3x − 17
10	?
9	?
8	?
7	?
6	?
5	?

x	x − 6 − (11 − 2x)
10	?
9	?
8	?
7	?
6	?
5	?

4. Consider the sequence of toothpicks shown at the right. A student describes the number of toothpicks required to make the nth pattern in the following way: I split the toothpicks up into three kinds: (1) top toothpicks; (2) bottom toothpicks; and (3) vertical toothpicks.

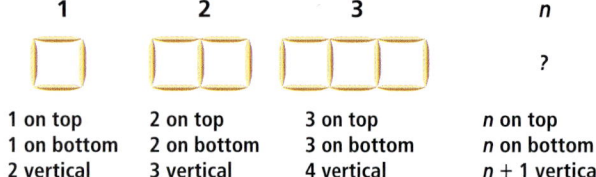

1	2	3	n
1 on top	2 on top	3 on top	n on top
1 on bottom	2 on bottom	3 on bottom	n on bottom
2 vertical	3 vertical	4 vertical	n + 1 vertical

In the nth figure, there were n toothpicks on top, n on the bottom, and $n + 1$ vertical ones. So, in the nth figure, there are $n + n + (n + 1)$ toothpicks.

a. Substitute for n to verify that the expression $n + n + (n + 1)$ describes the number of toothpicks needed for the first three terms.

b. Use the expression to find the number of toothpicks in the 100th term.

In 5–7, two expressions are given.

a. Using a table, evaluate each expression for four different values of the variable.

b. Based on your results, do the two expressions appear to be equivalent?

5. $25 + (x − 5)(x + 5)$ and x^2

6. $\frac{8n - 4}{4}$ and $8n − 1$

7. $x^2 − 4x − 3$ and $(x − 3)(x + 1)$

APPLYING THE MATHEMATICS

8. Tile patterns are often used in bathroom flooring. Consider this sequence created with green and yellow hexagonal tiles.

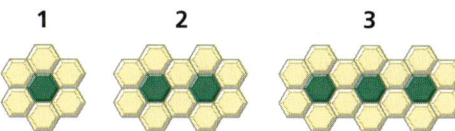

   a. Write an expression describing the number of green hexagonal tiles in the nth term of the sequence.
   b. Write an expression describing the number of yellow hexagonal tiles in the nth term of the sequence.
   c. Use your answers from Parts a and b to write an expression describing the total number of hexagonal tiles in the nth term of the sequence.
   d. How many hexagonal tiles are in the 100th term?

9. Give a counterexample to show that the equation $(a - 5) + b = a - (5 + b)$ is not true for all real numbers a and b.

10. An airplane manufacturer sells small planes that have seats arranged so there is a center aisle. In each row, there are 3 seats on the left side of the aisle and 2 seats on the right side of the aisle. If there are r rows of seats, find two different expressions for the total number of seats in a plane.

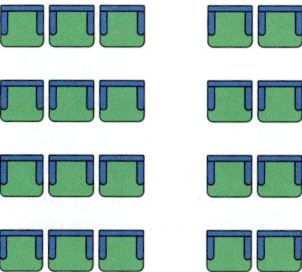

REVIEW

11. Describe the general pattern using one variable. **(Lesson 1-2)**

$$8^2 - 8 = 8(8 - 1)$$
$$30^2 - 30 = 30(30 - 1)$$
$$6.5^2 - 6.5 = 6.5(6.5 - 1)$$

12. Each morning, Crystal buys a cup of coffee for $2.25. She uses a table to record her total coffee expenditures. **(Lesson 1-2)**

Day	Calculation	Cost
1	$2.25	$2.25
2	$2.25 + $2.25	$4.50
3	$2.25 + $2.25 + $2.25	$6.75
4	?	?
5	?	?

   a. Complete the table.
   b. After one year how much will Crystal have spent on coffee?
   c. After d days, how much will Crystal have spent on coffee?

Newer airplanes have seats that are made with a lightweight carbon fiber-reinforced frame that helps reduce weight and fuel costs.
Source: www.boeing.com

Chapter 1

13. Give two instances of each pattern. (**Lesson 1-2**)
    a. $x^2 \cdot x = x^3$ b. $3g - g - g = g$ c. $n(3 + 8) = 3n + 8n$

14. a. Evaluate $\frac{1}{10}x^2$ when $x = 200$.
    b. Evaluate $\left(\frac{1}{10}x\right)^2$ when $x = 200$.
    c. Find a value of x so that the value of $\frac{1}{10}x^2$ is the same as the value of $\left(\frac{1}{10}x\right)^2$. (**Lesson 1-1**)

In 15–17, fill in the blank with =, <, or >. (**Lesson 1-1**)

15. $(-25)(-16)$ __?__ $-25 + -16$

16. $3^3 + 3^3$ __?__ 3^6

17. $(8 \cdot 6) \cdot 3 + 1$ __?__ $8 \cdot (6 \cdot 3) + 1$

18. Raul's Video Store has certain DVDs on sale for $9.95, but all others are priced at $14.95. Suppose a customer wishes to purchase 7 DVDs, 2 of which are on sale. What is the total cost? (**Lesson 1-1**)

19. In 2006, the city of Los Angeles charged a sales tax of 8.25%. If you bought a pair of jeans in Los Angeles that cost $35 before tax, what was the total cost after tax? (**Previous Course**)

A DVD has about 26 times the storage capacity of a CD.

Source: about.com

EXPLORATION

20. In this sequence of dots, two rows are added to each term to get the next term.

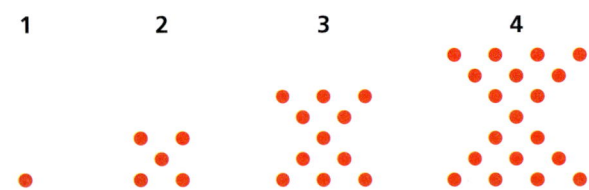

  a. Find the number of dots in the 5th and 6th terms.
  b. Find the number of dots in the 20th term.
  c. Try to find an expression for the number of dots in the nth term.

21. Two terms in a sequence of squares are shown.
  a. Draw the 3rd and 4th terms.
  b. Give the number of toothpicks in the 1st, 2nd, 3rd, 4th, and 5th terms.
  c. In words, describe a method to find the number of toothpicks it takes to make the nth term.
  d. Write an algebraic expression to find the number of toothpicks in the nth term.

QUIZ YOURSELF ANSWERS

1a. When $w = 2$, $2w = 2 \cdot 2 = 4$ and $w^2 = 2^2 = 2 \cdot 2 = 4$.

1b. When $w = 5$, $2w = 2 \cdot 5 = 10$ and $w^2 = 5^2 = 5 \cdot 5 = 25$.

2. $(n + 3)^2$ and $n^2 + 6n + 9$ are equivalent. $(n + 3)^2$ and $n^2 + 9$ are not equivalent since $(1 + 3)^2 = 4^2 = 16$, $1^2 + 9 = 1 + 9 = 10$. $n^2 + 9$ and $n^2 + 6n + 9$ are not equivalent since $1^2 + 9 = 10$, $1^2 + 6 \cdot 1 + 9 = 1 + 6 + 9 = 16$. Thus, we know that $(n + 3)^2$ and $n^2 + 6n + 9$ are equivalent.

Lesson 1-4

Picturing Expressions

Vocabulary

scatterplot

domain of the variable

▶ **BIG IDEA** Graphing the ordered pairs (value of variable, value of expression) helps you understand the relationship between these values.

Activity 1 in Lesson 1-3 discussed this toothpick sequence.

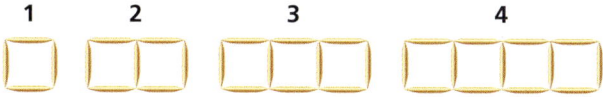

Suppose a student looks at each instance as one vertical toothpick on the left and 3 more toothpicks being added for each additional square.

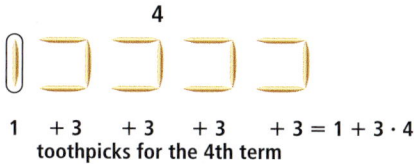

$1 \;\; + 3 \;\; + 3 \;\; + 3 \;\; + 3 = 1 + 3 \cdot 4$
toothpicks for the 4th term

Then the student might write the expression $1 + 3n$ to describe the sequence.

Mental Math

a. If cantaloupes cost $2.98 each, estimate how much 5 cantaloupes cost.

b. If you paid for the 5 cantaloupes with $15, how much change would you expect?

Scatterplots

The table shows values for the term number n and for $1 + 3n$, the number of toothpicks in the nth term. Each row can be written as an ordered pair. These pairs can then be graphed on the coordinate plane.

Term Number	Number of Toothpicks	(Term Number, Number of Toothpicks)
1	4	(1, 4)
2	7	(2, 7)
3	10	(3, 10)
4	13	(4, 13)
n	$1 + 3n$	$(n, 1 + 3n)$

Every point in the plane of the graph can be identified with *coordinates*. A graph like this, in which individual points are plotted, is called a **scatterplot,** as shown on the next page.

Picturing Expressions **27**

Chapter 1

If the points on the graph had been connected with a line, then the numbers between 1, 2, 3, and 4 would be allowed for n, and therefore the numbers for $1 + 3n$ between 4, 7, 10, and 13. But this is not possible in the toothpick sequence because n represents the term number. There is a third term and a fourth term, but for $n = 3.5$ there is no toothpick term. In this situation, n must be a positive integer. All the values that may be meaningfully substituted for a variable make up the **domain of the variable.** For the toothpick sequence on page 27, the domain is the set of positive integers $\{1, 2, 3, 4, …\}$.

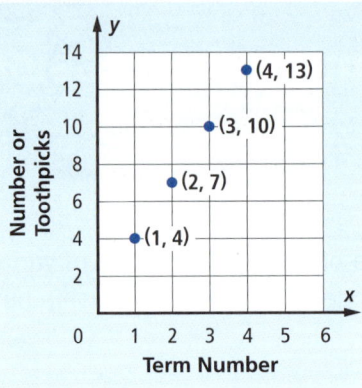

Connected Graphs

When writing expressions for real-world problems, you often must decide what domain makes sense for a situation.

Example 1

Suppose Rebecca drives to her grandmother's house, which is 500 miles away. Her average speed is 65 miles per hour. Fill in the table of values to calculate her remaining distance after each hour. Write an expression to represent her remaining distance, and then plot the points on a graph.

Hours of Driving	Remaining Distance (mi)
0	
1	
2	
3	
h	

Solution Let h represent the number of hours spent driving.

Hours of Driving	Calculation and Pattern	Remaining Distance (mi)
0	500	500
1	500 − 65	435
2	500 − 65 − 65 = 500 − 65(2)	370
3	500 − 65 − 65 − 65 = 500 − 65(3)	305
h	500 − 65 − ... − 65 = 500 − 65h (h terms)	500 − 65h

California drivers consume about 11% of the fuel used in the United States.

Source: U.S. Department of Transportation

So after h hours, Rebecca is $500 - 65h$ miles away from her grandmother's house. The graph is shown on the next page.

28 Using Algebra to Describe

Connecting the points is appropriate because time and distance are both measures and do not need to be integers. Rebecca could drive for a half hour or an hour and fifteen minutes. Substituting these numbers into the expression $500 - 65h$ will result in a distance that is not a whole number. The points (0.5, 467.5) and (1.25, 418.75) are plotted here. By connecting the points on the graph, you are showing that all nonnegative real numbers make sense in this situation. Therefore, the domain of h is the set of nonnegative real numbers.

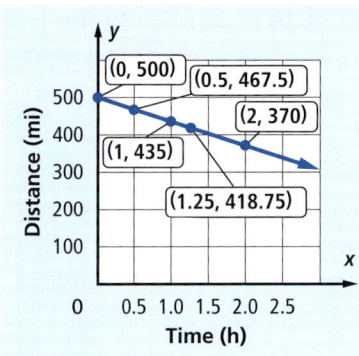

Common Domains of Variables

The following domains are frequently used in arithmetic and algebra.

Name of Set	Description	Examples of Elements
whole numbers	{0, 1, 2, 3, …}	Five; $\frac{16}{2}$; 2,007; 7 million
integers	{…, −2, −1, 0, 1, 2, …} whole numbers and their opposites	$\frac{-21}{3}$, −17.00, negative one thousand
real numbers	the set of all numbers that can be represented as terminating or nonterminating decimals	5, 0, π, −0.0042, $-3\frac{1}{3}$, $0.\overline{13}$, $\sqrt{2}$, one hundred thousand
nonnegative real numbers	{$x: x \geq 0$} the set consisting of 0 and all positive real numbers	2.56; 3,470; $\frac{1}{100}$; 0

In the above table, the sets of whole numbers and integers are described with a *roster,* or list, of elements. Set-builder notation is used to describe the set of nonnegative real numbers. In {$x: x \geq 0$}, the symbol ":" is read "such that." It is followed by an expression that describes the set. The set {$x: x \geq 0$} is read "the set of numbers x such that x is greater than or equal to 0." Some people write {$x | x \geq 0$}, using a single vertical bar to mean "such that."

Questions

COVERING THE IDEAS

1. Each term in the following sequence is a rectangular array of squares. The nth term has n rows and $n + 2$ columns and therefore $n(n + 2)$ squares.

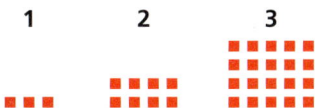

   a. Make a table of values for $n(n + 2)$ using $n = 1, 2, 3, 4,$ and 5.
   b. Make a scatterplot to graph the values from Part a.

2. The graph below shows the number of toothpicks used to make each term in a sequence.

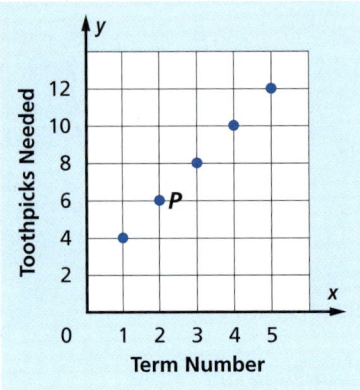

   a. Give the coordinates of P.
   b. Describe the situation that corresponds to P, giving the term number and number of toothpicks needed.
   c. Explain why the domain of n is the set of positive integers.

3. Make a table and scatterplot for $4n - 5$ using the following numbers for n: −3, −2, 0, 2, and 3.

4. Make a table and graph for values of the expression $n \cdot (-1)^n$ when $n = 1, 2, 3, 4,$ and 5.

In 5–7, values of a variable x and an expression are graphed. The coordinates of each point are integers.

5. Fill in the table based on the graph.

x	Value of Expression
−3	?
0	?
2	?
?	0
?	6
?	−4

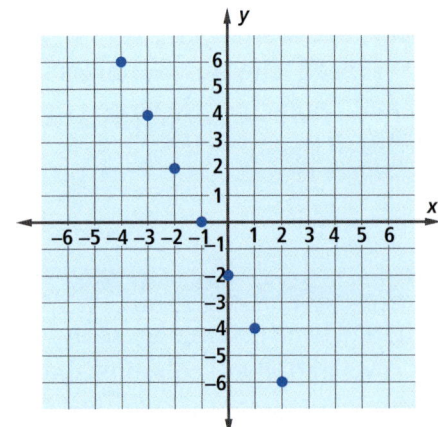

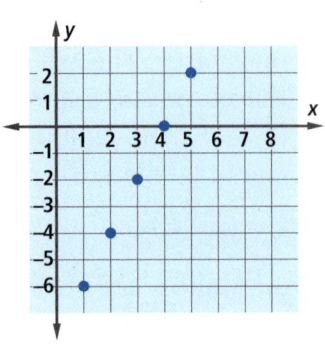

6. Use the graph at the right.
   a. What is the value of the expression when x is 2?
   b. What value(s) of x makes the value of the expression equal to 2?

30   Using Algebra to Describe

7. a. If $x = 5$, what is the value of the expression graphed below?
   b. What value(s) of x makes the value of the expression equal to 5?

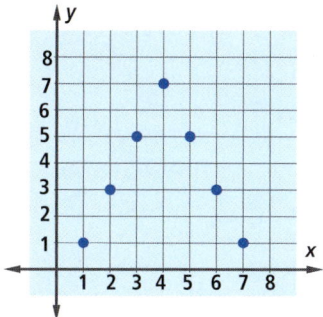

8. Express the set $\{x: x$ is an integer and $x \geq 20\}$ in roster form.

9. Express the set $\{-9, -10, -11, \ldots\}$ in set-builder notation.

APPLYING THE MATHEMATICS

In 10 and 11, consider these two situations:

a. A landscaper sells black dirt for $20 per cubic yard, plus a $40 delivery fee. Let $x =$ number of cubic yards ordered.

b. Arrangements for special dinners can be made at a restaurant. There is a $40 rental fee for a special room, plus the $20 per person cost for the food. Let $x =$ number of people at the dinner.

10. For which situation does it make sense to have $x = 2\frac{1}{2}$? In that situation, find the cost when $x = 2\frac{1}{2}$.

11. Match each graph below with the situation it represents.

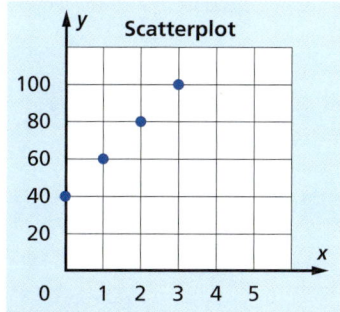
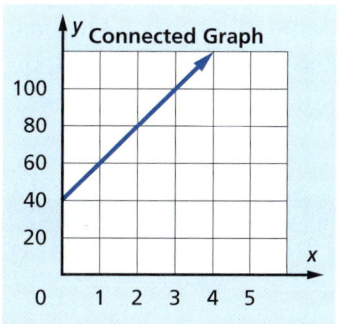

Topsoil is sold in 40-pound bags or in bulk measured in cubic yards.

In 12–14, choose the most reasonable domain for the variable.
a. set of whole numbers   b. set of real numbers
c. set of integers   d. set of positive real numbers

12. $n =$ the number of people at a restaurant on Friday night

13. $t =$ time it takes to do your homework

14. $E =$ elevation of a place in the United States

Chapter 1

REVIEW

15. Complete the table of values for each expression. Use the table to conclude whether the expressions seem to be equivalent. Explain your reasoning. (**Lesson 1-3**)

n	$2(n-3)$
−5	?
−3	?
1	?
2	?
5	?

n	$2n-3$
−5	?
−3	?
1	?
2	?
5	?

16. Find a counterexample to show that the equation $(n + 10) \cdot x = n + 10 \cdot x$ is not true for all real numbers n and x. (**Lesson 1-3**)

In 17 and 18, an equation describing a pattern is given.
a. Give three instances of the pattern. (**Lesson 1-2**)
b. Do you think the equation is true for all real numbers? Explain your reasoning. (**Lesson 1-3**)

17. $t - 2t + 3t = 2t$ 18. $9 - (2x - 4) = -2x + 13$

19. Using the formula $V = \ell wh$, find the volume of the figure at the right. (**Lesson 1-1**)

20. Evaluate the expression $a^2 + b^2 + 2ab$ for the following values of a and b. (**Lesson 1-1**)
    a. $a = 2, b = 3$ b. $a = -2, b = 3$ c. $a = -2, b = -3$

21. Suppose there are 273 students in a school. Of these, $\frac{3}{7}$ play on an athletic team, and of those that are on an athletic team, $\frac{2}{13}$ play on the soccer team. How many students play on the soccer team? (**Previous Course**)

EXPLORATION

22. Here is a pattern of dots.

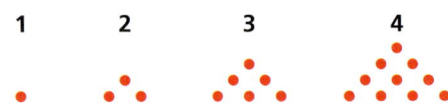

a. Find an expression for the nth term in the pattern.
b. Make a scatterplot.

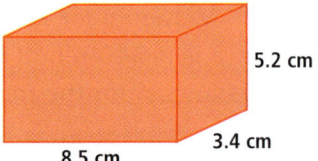

In 2004–2005, there were a total of 670,691 high school soccer players, including both boys and girls.

Source: The National Federation of State High School Associations

Lesson 1-5

Using a Graphing Calculator

Vocabulary

window
Xmin
Xmax
Ymin
Ymax
Xscl
Yscl
standard window

Mental Math

Evaluate
$3n + (5 - n) - 2n$ when:
a. $n = 3$.
b. $n = 0$.
c. $n = -2$.
d. $n = 1{,}000$.

▶ **BIG IDEA** By letting $y =$ the value of an expression with variable x, a graphing calculator or computer can automatically generate a graph of the relationship between x and y.

Graphing with Technology

An algebraic expression such as $x^2 - 15x$ shows calculations that are performed on a variable. In Lesson 1-4, you evaluated expressions by substituting variables with numbers taken from tables. In much the same way, graphing calculators find values and graph the points they describe. These machines were first introduced in 1985 and are popular today because they perform repeated calculations quickly and allow a person to focus on the mathematical relationships shown by graphs and tables.

Activity 1

To graph values of an expression on a graphing calculator, a second variable y is used. It represents the value of the expression that is calculated for each x-value.

Step 1 Go to the [Y=] menu on your graphing calculator.

Step 2 Type in $x^2 - 15x$.

Since the calculator can display several graphs in the same window, it uses names like Y₁, Y₂, and Y₃ in its equations.

When you draw a graph by hand, you must decide what portion of the coordinate grid to display. The same decision must be made when using a graphing calculator. The part that is shown is called the **window**. The window can be described by two inequalities.

Step 3 Go to the WINDOW menu on your calculator.

Step 4 Enter the same values as in the screen at the right. The information in this window can be expressed with the inequalities $-10 \leq x \leq 25$ and $-70 \leq y \leq 35$.

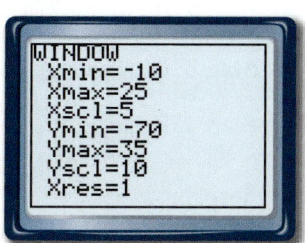

(continued on next page)

Using a Graphing Calculator **33**

Chapter 1

Step 5 Press the GRAPH command. Your graph should look like the one at the right.

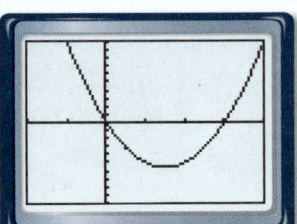

Most graphing calculators use the variables x and y for graphing. If a problem uses other letters for variables, you must replace those letters with x and y. On a calculator, the four numbers that describe the boundaries of the window are each given a name. For the x-coordinates, the minimum edge, or least value displayed, is called **Xmin** or x-min (the left edge of the screen) and the maximum edge, or greatest value, is **Xmax** or x-max (the right edge of the screen). Similarly, **Ymin** (or y-min) and **Ymax** (or y-max) are the least (bottom edge) and greatest (top edge) values for the y-coordinates.

The calculator does not put coordinates by the tick marks on the axes, so you need to look carefully at the window description to understand what numbers are being shown. In the graph above, the tick marks are spaced 5 units apart on the x-axis and 10 units apart on the y-axis. This scale is described by **Xscl** (x-scale) and **Yscl** (y-scale) on the window settings screen.

Activity 2

Step 1 Go to the [Y=] menu and enter $-7x^2 + 3x - 4$.

Step 2 Go to the WINDOW menu and enter the settings $-3 \leq x \leq 3$ with x-scale of 1 and $-25 \leq y \leq 0$ with a y-scale of 5.

Step 3 Copy your window screen and sketch the graph your calculator displays.

Activity 3

Step 1 Enter $y = \dfrac{5}{x^2 + 1}$ into the [Y=] menu. (Type in as $5 \div (x^2 + 1)$.)

Step 2 Go to the ZOOM menu. Press STANDARD.

Step 3 Copy your graph down on a sheet of paper.

Step 4 Describe the window by completing the inequalities below.

$\underline{} \leq x \leq \underline{}$ and $\underline{} \leq y \leq \underline{}$
x-scale = $\underline{}$ y-scale = $\underline{}$

Anytime you use the **standard window,** your calculator will display in the above window.

34 Using Algebra to Describe

Lesson 1-5

You may find that when you try to graph an equation, no graph appears. This may mean that your window settings are not appropriate for your equation. Deciding on a good window is an important skill, so learn to think about it each time you graph!

Activity 4

Step 1 Enter $y = 5x(x + 4)(x - 3)$ into the [Y=] menu. (*Caution:* You may need to enter multiplication symbols that are not shown in the expression.)

Step 2 Use your calculator to match the windows below and graphs of $y = 5x(x + 4)(x - 3)$ at the right.

Window 1
$-6 \leq x \leq 6$ with *x*-scale of 1
$-100 \leq y \leq 150$ with *y*-scale of 25

Window 2
$-2.8 \leq x \leq -2.1$ with *x*-scale of 0.1
$40 \leq y \leq 145$ with *y*-scale of 10

Window 3 (Standard window)
$-10 \leq x \leq 10$ with *x*-scale of 1
$-10 \leq y \leq 10$ with *y*-scale of 1

Window 4
$-30 \leq x \leq 30$ with *x*-scale of 5
$-500 \leq y \leq 500$ with *y*-scale of 100

Graph 1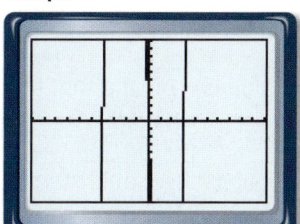

Graph 2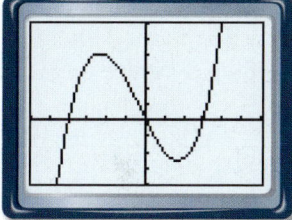

Graph 3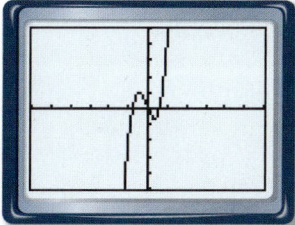

Graph 4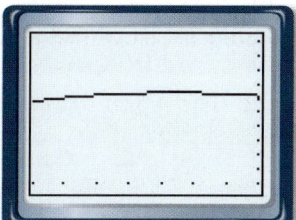

Activity 5

Step 1 The graph at the right shows $y = x^2 - 15x$ graphed on the window $-10 \leq x \leq 20$ and $-100 \leq y \leq 100$. Use the tick marks on the graph to find the coordinates of the points where the curve crosses the *x*-axis.

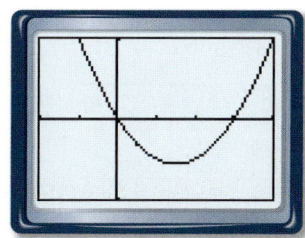

Step 2 Use the **TRACE** or **VALUE** commands to verify your answers to Step 1.

Step 3 Substitute those *x*-values into $x^2 - 15x$ and evaluate.

Step 4 Do your results match the *y*-coordinates?

Using a Graphing Calculator

Chapter 1

Tables from a Graphing Calculator

In Lesson 1-4, you drew graphs from tables you had created. The graphing calculator will also make a table of values for an expression entered in the Y= menu.

Activity 6

Step 1 Enter $y = x^2 - 15x$ into the Y= menu.

Step 2 Go to TABLE SETUP.

Step 3 Start table at −10 and have the table increment or △TABLE equal 1.

Step 4 Go to TABLE. Your screen should match the one at the right.

Step 5 Notice that the first x-value is −10 and that the x-values go up by 1 each step.

Step 6 In TABLE SETUP menu, change table start to 2 and △TABLE to 3. Write down the x- and y-values that appear in the table.

Step 7 Scroll up to where $x = -4$. What is the corresponding y-value?

Step 8 Scroll down to where $x = 35$. Give the corresponding y-value.

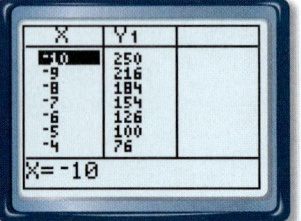

Comparing Expressions with Technology

You can also use your graphing calculator to see how the values of the two expressions are similar or different. Be careful in setting the window so you can see the important features of both graphs.

Activity 7

Compare the expressions $(x - 3)^2$ and $(x + 3)^2$ by graphing.

Step 1 Graph $y = (x - 3)^2$ and $y = (x + 3)^2$ in a standard window. Sketch your results.

Step 2 Give the coordinates of the intersection of the graph of $y = (x - 3)^2$ and the x-axis.

Step 3 Give the coordinates of the intersection of the graph of $y = (x + 3)^2$ and the x-axis.

Since the graphs are different, the two expressions $(x - 3)^2$ and $(x + 3)^2$ are not equivalent.

🛑 See Quiz Yourself 1 at the right.

> ▶ **QUIZ YOURSELF 1**
>
> Use your calculator to determine which equation below is represented by the given graph.
> $y = \sqrt{x}$
> $y = \sqrt{-x}$
> $y = -\sqrt{-x}$

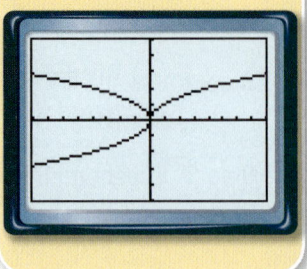

36 Using Algebra to Describe

Lesson 1-5

Creating Scatterplots with Technology

A graphing calculator can plot individual points. To do this, enter a table of values into lists. Many calculators use names L1, L2, and L3 for the lists (just as Y1, Y2, and Y3 are used for the graphing of equations).

Activity 8

Step 1 Clear any equations in the Y= menu and turn on the STAT PLOT function.

Step 2 Enter the table below in the STAT lists. Then graph the points.

x	y
2	6
5	−1
−7	−4
3	10

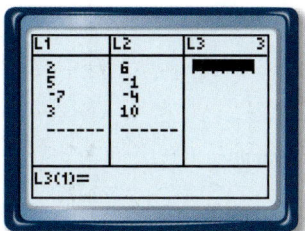

To decide on a window for a scatterplot, look at the values you are graphing. Find the least x-value in the table of values. Your x-minimum in the window should always be *less than* that value. Likewise, find the greatest x-value in the table and set the x-maximum in the window to a *greater* value. Follow the same procedure for the y-minimum and y-maximum in the window.

Step 3 Complete the following using the table above.

least x-value __?__   a lesser x-value for x-min __?__
greatest x-value __?__   a greater x-value for x-max __?__
least y-value __?__   a lesser y-value for y-min __?__
greatest y-value __?__   a greater y-value for y-max __?__

Step 4 Use your results to graph the points. One possible graph is shown below.

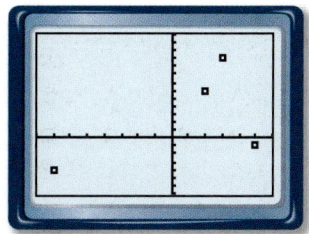

 See Quiz Yourself 2 at the right.

> **QUIZ YOURSELF 2**
>
> Plot the following table of values using lists and a proper window.
>
> | x | y |
> |---|---|
> | 61 | −681 |
> | 64 | −661 |
> | 58 | −661 |
> | 67 | −651 |
> | 55 | −651 |
> | 70 | −661 |
> | 52 | −661 |
> | 50.5 | −681 |
> | 71.5 | −681 |
> | 70 | −705 |
> | 52 | −705 |
> | 67 | −725 |
> | 55 | −725 |
> | 64 | −745 |
> | 58 | −745 |
> | 61 | −765 |

Using a Graphing Calculator

Chapter 1

Questions

COVERING THE IDEAS

1. Write the two inequalities that describe the window for the graph at the right.
   ___?___ ≤ x ≤ ___?___ and ___?___ ≤ y ≤ ___?___

2. a. Write the two inequalities that describe the window below.
   b. Find the distance between tick marks on the x-axis and on the y-axis.

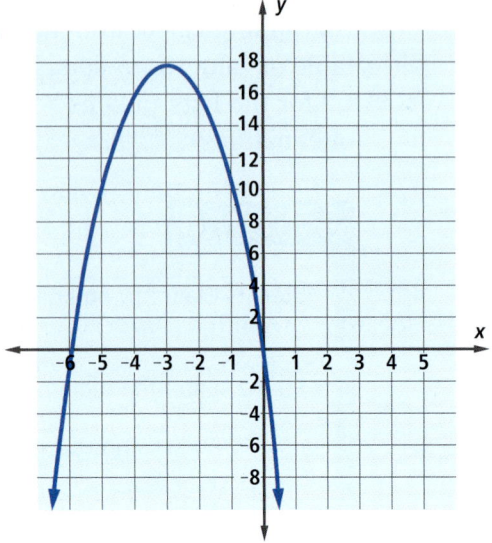

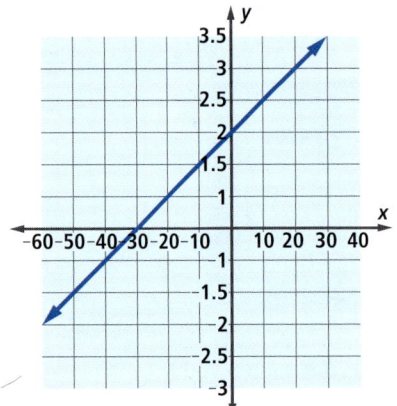

3. The window at the right is described by $-10 \leq x \leq 25$ and $-10 \leq y \leq 10$. Copy the window and label the tick marks with their coordinates.

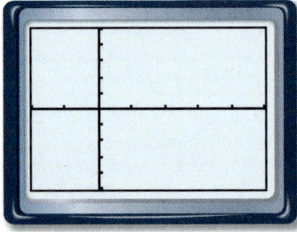

4. Use the graphing calculator screen below to make a table of values for the expression $2x + 1$. List the first five ordered pairs in the table.

raphing calculator to make a table of values and graph $^2 + 3$ using the window $-6 \leq x \leq 6$ and $0 \leq y \leq 40$. ur table settings so that x increases by 2 for each row 'e.

37

6. a. Use lists on a calculator to graph the ordered pairs in the table below.

x	−59	−41	67	13	103	58	31	85	−23	4	121	49	31	−5
y	371	375	383	387	375	379	391	379	379	379	371	387	379	383

  b. Use inequalities to show a window that contains all the points.
  c. Sketch the graph using your window from Part b.

7. a. Graph $y = x^3 - 6x^2 - 9x + 4$ on a graphing calculator in a standard window.

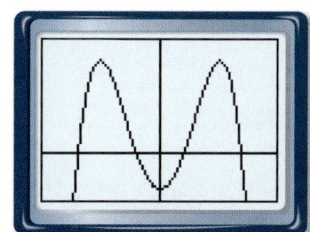

  b. Now change the values in the window so the graph looks like the image at the right. Give the values of your window's Xmin, Xmax, Ymin, and Ymax.

8. a. Graph the equation $y = -0.04x^4 + 2.12x^2 - 7.84$ on a graphing calculator's standard window. Sketch the graph on paper.

  b. Change the y-max value in the window so the graph looks like the letter M, as shown at the right.

  c. Change the x-values and the y-values in the window so only the right "bump" of the graph appears, as shown below.

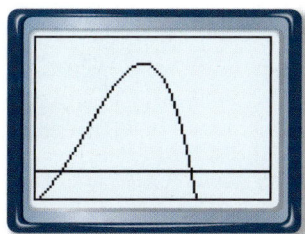

  d. Enlarge the window to $-100 \leq x \leq 100$ with x-scale of 10, and $-500 \leq y \leq 500$ with y-scale of 100. Copy this graph on your paper.

  e. Change to a window of $-3 \leq x \leq 3$ with x-scale of 0.5 and $-3 \leq y \leq 3$ with y-scale of 1. Copy this graph on your paper.

APPLYING THE MATHEMATICS

In 9 and 10, an equation is given.
a. Graph the equation in a standard window.
b. Label at least three points on the graph with their coordinates.

9. $y = 3x - 4$

10. $y = x^2 - 3$

Chapter 1

In 11 and 12, first graph the equation using the window $-10 \leq x \leq 10$ and $-10 \leq y \leq 10$. The result will be a line. Then adjust the window so the points where the line crosses both the x- and y-axes are visible. Describe your window with two inequalities.

11. $y = -4x - 32$

12. $y = \frac{1}{2}x - 12$

13. Test to see if $(6x - 14) - (x + 11)$ and $5x + 3$ are equivalent using a graphing calculator. Explain your conclusion.

14. Graph $y = \sqrt{x}$ on a calculator. Use the window $-10 \leq x \leq 10$ and $-10 \leq y \leq 10$.

REVIEW

15. Ayita's Gym charges new members a sign-up fee of $54, which they pay only once. For every month a person is a member, there is a fee of $26. Let $m =$ the number of months of membership. (Lessons 1-1, 1-4)
    a. Write an expression for the cost of membership for m months.
    b. Make a table of values and plot the graph.

In 16–18, choose the most reasonable domain for the variable. (Lesson 1-4)

   a. the set of whole numbers
   b. the set of integers
   c. the set of real numbers
   d. the set of positive real numbers

16. the number of students n in your math class

17. the time t it takes to drive to school

18. the temperature T of a location on Earth

19. An air conditioning unit with a high energy efficient ratio (EER) gives more cooling with less electricity. To find the EER of a unit, divide the BTU (British Thermal Unit) number by the number of watts the unit uses. The higher the EER, the more efficient the air conditioner. (Lesson 1-1)

$$\text{EER} = \frac{\text{BTU}}{\text{number of watts}}$$

   a. Find the EER to the nearest tenth for an air conditioner having 8,500 BTUs and 925 watts.
   b. Find the EER to the nearest tenth for an air conditioner having 12,700 BTUs and 1,500 watts.
   c. Which of the two air conditioners above is more efficient? Justify your answer.

Walking for 20–45 minutes four to five times per week at 3 miles per hour is a great way to stay fit.

40 Using Algebra to Describe

20. Evaluate the expression $\frac{a+b}{3} + 3(a-b)$ for the given values of a and b. (**Lesson 1-1**)

   a. $a = 11, b = 4$ b. $a = 7, b = 2$

In 21–26, compute in your head. (Previous Course)

21. $15 + -19 + 4$

22. $-10{,}000 - 20{,}000$

23. $\frac{-16 + 8}{2}$

24. $-6 - -10 + 3$

25. $-8 - 16 + 5$

26. $\frac{1}{3} + \frac{1}{2} - \frac{1}{5}$

EXPLORATION

27. Explain what the following features do on your graphing calculator.

    a. Zoom In
    b. Zoom Out
    c. ZBox

28. Graph $y = \frac{19x}{x^2 + 4}$ on your graphing calculator. Experiment with the values of the window until the graph looks like each of the following. Give the values for your window's Xmin, Xmax, Ymin, and Ymax.

    a. a horizontal line
    b. a diagonal crossing from the lower left corner to the upper right corner of the window
    c. a vertical line
    d. a diagonal crossing from the upper left corner to the lower right corner of the window

---

QUIZ YOURSELF ANSWERS

1. Each equation is represented in the graph.

2. The window $45 \leq x \leq 65$, and $-800 \leq y \leq -600$ displays the scatterplot.

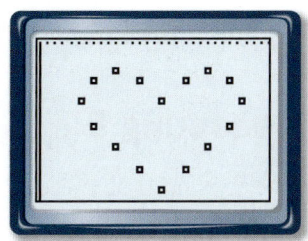

Chapter 1

Lesson 1-6
Absolute Value and Distance

Vocabulary

absolute value
origin

▶ **BIG IDEA** The absolute value of a number and the distance between two numbers on a number line are closely related.

Cameron and Maria took a vacation to Whites City, New Mexico, near the Guadalupe Mountains and the famous Carlsbad Caverns. Being avid hikers, they wanted to go mountain climbing and explore the caves. They found the following information on the Internet.

Guadalupe Peak is the highest point in Texas.
Source: National Park Service

Mental Math

a. How many words can you type in 10 minutes if you type 52 words per minute?

b. How many words can you type in 20 minutes if you type 52 words per minute?

c. How many words can you type in t minutes if you type 52 words per minute?

Whites City is at an elevation of 5,740 feet (above sea level). Guadalupe Peak in Guadalupe Mountains National Park, 35 miles away, is at 8,749 feet. The entrance to Carlsbad Caverns National Park is 7 miles away. The entrance to Carlsbad Caverns is at an elevation of 4,400 feet, and the King's Palace Cavern is 900 feet below the entrance.

🛑 See Quiz Yourself 1 at the right.

▶ **QUIZ YOURSELF 1**

What is the elevation of King's Palace Cavern?

Measuring Elevations from Whites City

Because they were starting from Whites City (W), Cameron and Maria wanted to know how much they would be going up and down from 5,740 feet. They subtracted the elevation of Whites City from the

42   Using Algebra to Describe

elevation of each destination. The difference is a positive number if they were going up and a negative number if they were going down.

Destination	Elevation (ft)	Difference from Whites City (ft)
Guadalupe Peak (G)	8,749	8,749 − 5,740 = 3,009
Carlsbad Caverns entrance (C)	4,400	4,400 − 5,740 = −1,340
King's Palace Cavern (K)	3,500	3,500 − 5,740 = −2,240

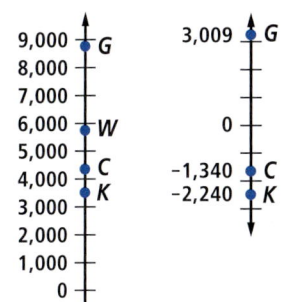

These elevations are graphed on the left vertical number line at the right.

The difference from Whites City's elevation to the other places is a *deviation* in altitude. A deviation can be positive or negative. The deviations are on the right vertical number line. However, sometimes we just want to know how much difference there is, and not the direction of the deviation. Then we take the *absolute value* of the deviation.

The symbol for absolute value is two vertical lines: $|\ |$.

If $x > 0$, then $|x| = x$.

If $x < 0$, then $|x| = -x$.

Since $3{,}009 > 0$, $|3{,}009| = 3{,}009$.

Since $-1{,}340 < 0$, $|-1{,}340| = -(-1{,}340) = 1{,}340$.

Then we say that the *absolute difference* in elevations between Whites City and the entrance to Carlsbad Caverns is 1,340 feet.

 See Quiz Yourself 2 at the right.

▶ **QUIZ YOURSELF 2**
a. What is $|-2{,}240|$?
b. What is the absolute difference in the elevations of Whites City and the King's Palace Cavern?

Absolute Value and Distance

From the right vertical number line drawn above, you can see the geometric interpretation of absolute value.

> **Absolute Value**
> The **absolute value** of a number is its distance from 0.

Because the absolute value of a number is a distance, it is never negative. Using the definition of absolute value, $-x$ looks like a negative number. But that is only the definition when x is negative, so $-x$ stands for a positive number. For example, when $x = 830$, $|x| = |830| = 830$, and when $x = -830$, $|x| = |-830| = -(-830) = 830$.

 See Quiz Yourself 3 at the right.

▶ **QUIZ YOURSELF 3**
Evaluate each expression.
a. $|3 - 24|$
b. $|3| - |24|$
c. $|-3| - |-24| - |18|$

Absolute Value and Distance   43

Chapter 1

Expressions with Absolute Value

In this chapter you have seen many tables where the values are small, positive integers. However, using only these numbers can give you misleading information. For example, consider the expressions $2x$ and $|2x|$. The table at the right seems to show that $2x$ and $|2x|$ are equivalent. When x is a positive number, both expressions double it. So both $2x$ and $|2x|$ are positive. On the graph below the table, points for $2x$ are shown with open circles (○) and points for $|2x|$ are marked with squares (□). They are the same points.

x	2x	|2x|
0	0	0
1	2	2
2	4	4
3	6	6
4	8	8

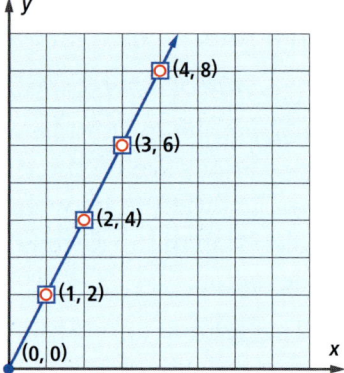

But this is misleading! No negative numbers were used in the table above. As soon as negative numbers are included, you can see that the expressions $2x$ and $|2x|$ have different values. In fact, for negative numbers the values are opposites. For negative values of x, $2x$ is also negative, so taking the absolute value changes its sign. This can be seen on the graph below by looking to the left of the point $(0, 0)$, called the **origin.** For each negative value of x, the point for $2x$ (marked ○) and the point for $|2x|$ (marked □) are on opposite sides of the x-axis.

x	2x	|2x|
−5	−10	10
−4	−8	8
−3	−6	6

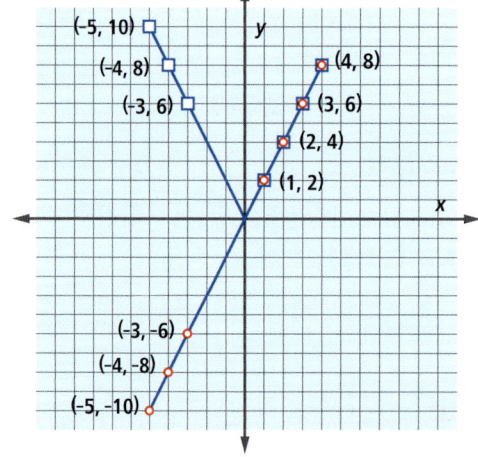

Activity

The graphing calculator or computer symbol for absolute value of a number x is `abs(x)`. In Lesson 1-5, you used technology to compare expressions. Now we use technology to compare $|x + 4|$ and $|x| + 4$.

Enter `Y1=abs(x+4)` and `Y2=abs(x)+4` in your graphing calculator with window $-10 \leq x \leq 10$ and $-10 \leq y \leq 10$. Sketch the graphs in your window. Do the expressions appear to be equivalent?

🛑 See Quiz Yourself 4 at the right.

> ▶ **QUIZ YOURSELF 4**
>
> Test whether $|x^2 − 6|$ and $−(x^2 − 6)$ are equivalent. Explain your reasoning.

44 Using Algebra to Describe

Questions

COVERING THE IDEAS

1. The deepest cavern in Carlsbad Caverns is 1,567 feet below its entrance, making it the deepest cave in the United States.
   a. What is the elevation of that deepest cavern?
   b. What is the difference in its elevation from Whites City?
   c. What is the absolute difference in its elevation from Whites City?

2. Suppose you begin the day at an elevation of E feet. After some hiking, you are at an elevation of H feet.
   a. What is the deviation of where you are now from where you began?
   b. What is the absolute deviation of where you are now from where you began?
   c. When will the answers to Parts a and b be different?

3. If $x = -40$ and $y = 35$, find each expression.
   a. $|x - y|$
   b. $|y - x|$
   c. $|x + y|$
   d. $|x| + |y|$

In 4–7, evaluate each expression.

4. $\left|\frac{3}{5}\right| + \left|\frac{3}{-5}\right|$

5. $|-1 - (-2)|$

6. $\text{abs}(2) - \text{abs}(-20)$

7. $\frac{\text{abs}(x)}{\text{abs}(-x)}$ when $x = 4.8673$

8. Graph $y = |x + 5|$.

9. Find a value of t for which $3t$ and $|-3t|$ are not equal.

Carlsbad Caverns in New Mexico has more than 100 caves.

Source: National Park Service

APPLYING THE MATHEMATICS

10. Test whether $|x + 4|$ and $|x| + 4$ are equivalent.

11. Graph $y = |x - 3| - 5$ on your calculator.
    a. Sketch the graph in a standard window.
    b. Sketch the graph in the window $-20 \leq x \leq 25$ with an x-scale of 5, and $-6 \leq y \leq 2$ with a y-scale of 1.
    c. Give Xmin, Xmax, Ymin, and Ymax for a window where only the left side of the angle is visible.

In 12–15, find all values of the variable that satisfy the equation.

12. $|x| = 15$
13. $|w - 5| = 0$
14. $|z| = -8.8$
15. $|A| = |A - 1|$

Absolute Value and Distance

16. Use the graph at the right to answer the following questions. (**Lesson 1-4**)

    a. Complete the following table of values.

x	1	2	3	4	5
y	?	?	?	?	?

    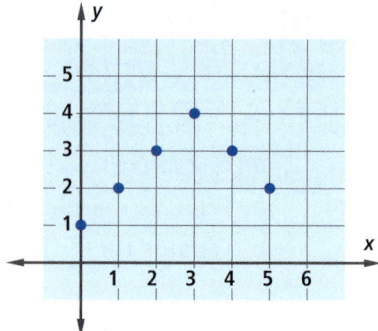

    b. What is the value of the expression when x equals 4?
    c. What is the x-value when the value of the expression is 4?
    d. **Multiple Choice** Which of the following equations best describes the graph?
       A $y = |x - 2|$
       B $y = |x - 3| + 4$
       C $y = -1 \cdot |x - 3| + 4$
       D $y = x + 1$

REVIEW

17. a. Graph $y = x^3 - 3x^2 + 3x - 1$ using the standard window.
    b. What are the coordinates of the x-intercept? (**Lesson 1-5**)

18. Make a table of values and a graph for $y = -x^2 + 4$ using the window $-5 \leq x \leq 5$, and $-30 \leq y \leq 30$. On your table, increase x by 2 for each row. (**Lesson 1-5**)

19. Express the set $\{x: x \text{ is an integer}, 5 \leq x \leq 10\}$ in roster form. (**Lesson 1-4**)

20. a. Give a counterexample to show that the following pattern is not true for all real numbers x and y. (**Lesson 1-3**)

    $$x - y = y - x$$

    b. State the result of Part a using the word *commutative*. (**Lesson 1-2**)

21. Evaluate the following expressions for the given value of x. (**Lesson 1-1**)

    a. $-(-x^2) + x \cdot x \div (-x)$ for $x = 4$
    b. $\left(\dfrac{2}{1 - (-x)}\right)^{x+1}$ for $x = 1$

EXPLORATION

22. Describe all integer values of x that satisfy $|x| < |x + 1|$.

QUIZ YOURSELF ANSWERS

1. 3,500 ft
2. a. 2,240
   b. 2,240 ft
3. a. 21
   b. -21
   c. -39
4. The expressions are not equivalent. For example, when $x = 3$: $|3^2 - 6| = |9 - 6| = 3$, and $-(3^2 - 6) = -(9 - 6) = -3$.

Lesson 1-7

Data and Spread

Vocabulary

range
mean absolute deviation
symmetric
skewed right
skewed left
uniform

▶ **BIG IDEA** Two measures of the spread of a data set are *range* and *mean absolute deviation*.

In this chapter you have investigated number sequences that fit a pattern exactly and are described by an expression. However, many collections of numbers are "messy" and do not have an exact algebraic description. Statistics can help analyze and summarize these kinds of data.

Mental Math

a. What is the mean of 2, 3, and 4?

b. What is the mean of −2, −3, and −4?

c. What is the mean of −1, 0, and 1?

d. What is the mean of $x - 1$, x, and $x + 1$?

Picturing the Mean of a Data Set

Suppose that a school board has received complaints that classrooms at Central School are small and crowded. To decide if this is the case, data about room size were gathered. Areas of 15 rooms were calculated in square feet, then rounded to the nearest 50 square feet. Here are the areas they found.

750; 700; 800; 750; 650; 800; 750; 750; 650; 750; 600; 900; 650; 600; 1,000

You can represent this data in a frequency table at the right.

Two other ways are in a *dot plot* (below at the left) and a *bar graph* (below at the right).

Room Area (ft²)	Frequency
600	2
650	3
700	1
750	5
800	2
900	1
1,000	1

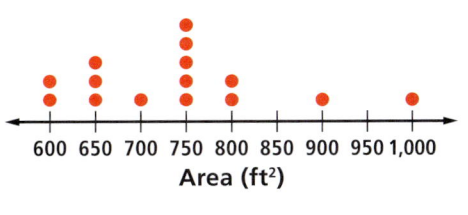

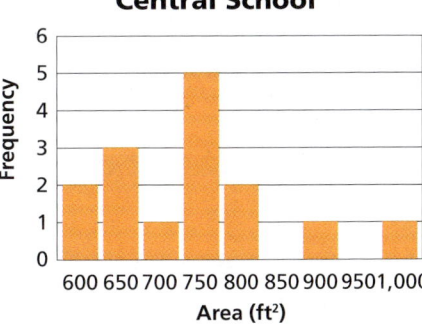

It is customary to use the Greek letter μ (*mu*, pronounced "mew") to stand for the mean. The mean classroom area is shown below.

$$\mu = \frac{2 \cdot 600 + 3 \cdot 650 + 1 \cdot 700 + 5 \cdot 750 + 2 \cdot 800 + 1 \cdot 900 + 1 \cdot 1{,}000}{15}$$
$$= 740 \text{ ft}^2$$

Data and Spread

Chapter 1

The school board at Central decided to compare the areas of classrooms at Central with those of Whittier School. Here is the list of classroom areas and a dot plot for Whittier.

750; 750; 750; 700; 800; 750; 700; 750; 800; 650; 700; 650; 750

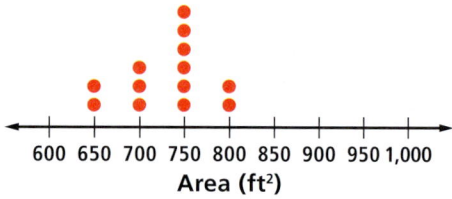

Areas of Classrooms at Whittier School

The Range of a Data Set

The mean of the classroom areas for Whittier is about 730 square feet, which is less than the mean of 740 square feet for Central. But notice that the room areas at Central are more spread out than those at Whittier. The data values for Whittier lie in the interval $650 \le A \le 800$, where A is the area of a room. For Central, A is in the interval $600 \le A \le 1{,}000$. The **range** r of a collection of data is the difference between the maximum value M and minimum value m, so $r = M - m$. The range of classroom areas for Whittier is $800 - 650$, or 150, square feet.

 See Quiz Yourself 1 at the right.

▶ **QUIZ YOURSELF 1**

What is the range of classroom areas for Central School?

The Mean Absolute Deviation of a Data Set

The range is a measure of the spread of a data set. Another measure of spread is the average difference between the areas of the rooms and the mean area. This is the **mean absolute deviation** of the data set.

Example

Use the information about the classroom areas at Central School on page 47 to find the mean absolute deviation.

Solution We show how to find the mean absolute deviation for the classroom areas at Central School using spreadsheets or lists, although the same computations can be done by hand or on a calculator without lists. See the table on the next page.

48 Using Algebra to Describe

Lesson 1-7

Step	Spreadsheet	Graphing Calculator
1. Organize the data.	After naming column A in cell A2, enter the 15 room areas in cells A3 through A17.	Enter the 15 room areas into L1, the first list on your calculator.
2. Calculate the mean.	As shown below, it is 740 square feet. Enter the mean in cells B3 through B17.	Keep the mean in mind.
3. Calculate the deviation of each area from the mean.	Enter "=A3−B3" in cell C3. Copy down to C17.	Enter L1−740 in place of L2 at the top of the screen. L2 will contain the deviations.
4. Calculate the mean of the deviations.	In cell C19 use the formula "=average(C3:C17)". The mean of the deviations should be 0 because the positive and negative differences balance.	Leave the LIST screen to calculate mean(L2).
5. Put the absolute value of each deviation in column D.	Enter "=abs(C3)" in cell D3 and copy the formula down the rest of the column.	Return to LIST. Enter abs(L2) for L3.
6. Calculate the mean absolute deviation.	In cell D19 enter "=average (D3:D17)". The mean of the absolute deviations is approximately 78.7.	Leave LIST again. Calculate mean(L3).

	A	B	C	D
1	Central School			
2	classroom area	mean area	Deviation (area − mean area)	Absolute Deviation (|area − mean area|)
3	750	740	10	10
4	700	740	−40	40
5	800	740	60	60
6	750	740	10	10
7	650	740	−90	90
8	800	740	60	60
9	750	740	10	10
10	750	740	10	10
11	650	740	−90	90
12	750	740	10	10
13	600	740	−140	140
14	900	740	160	160
15	650	740	−90	90
16	600	740	−140	140
17	1000	740	260	260
18				
19	740		0	78.66667
20				

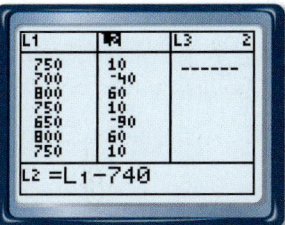

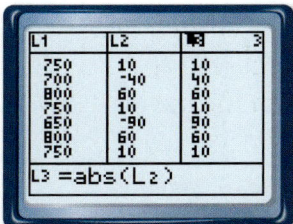

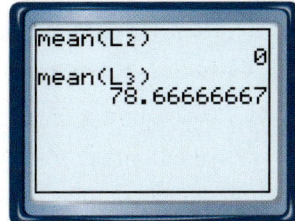

Data and Spread

Chapter 1

🛑 See Quiz Yourself 2 at the right.

An abbreviation for mean absolute deviation is m.a.d. You should have found that the m.a.d. for Whittier is much smaller than the m.a.d. for Central. The rooms at Central are larger, on average, than those at Whittier. The range of the rooms at Central is also larger than the range of the rooms at Whittier. Do you see what is probably causing the complaints that the classrooms at Central are too small?

> **QUIZ YOURSELF 2**
>
> Calculate the mean absolute deviation for the areas of the classrooms at Whittier on page 48.

An Algorithm for Finding the Mean Absolute Deviation

As its name indicates, the mean absolute deviation is found using the following three steps:

Step 1 Find the mean, μ, of the data values.

Step 2 For each data value v, find the absolute value of its deviation from the mean, which is represented by $|v - \mu|$.

Step 3 Take the mean of the absolute deviations.

The Shape of a Distribution

Here are dot plots of the number of runs scored by Westview's softball team and the number of goals scored by its soccer team.

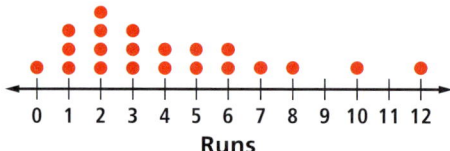

Softball Runs Westview Scored in Games

Soccer Goals Westview Scored in Games

There was a 67.3% increase in the number of female high school softball players from the 1980–1981 season to the 2000–2001 season.

Source: profastpitch.com

Some shapes of distributions are given special names. When most of the data are centered around one point and the values on the left and right sides are roughly mirror images, the distribution is called **symmetric.** The distribution of goals scored by the soccer team is symmetric. When the upper half of the values extends much farther to the right than the lower half, leaving a tail on the right, the shape is said to be **skewed right.** The softball team's run distribution has this kind of shape because most games are relatively low-scoring, but a few have high scores. Likewise, when the lower half of the data is much farther out, leaving a tail to the left, the shape is said to be **skewed left.** If the distribution has roughly the same height for all values it is called **uniform.**

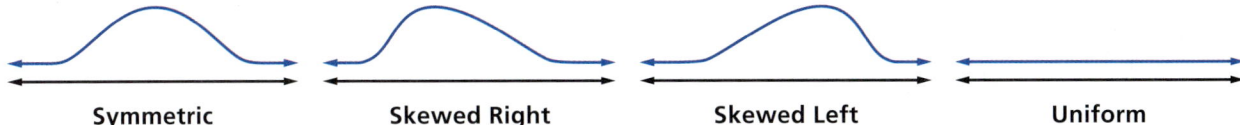

Symmetric   Skewed Right   Skewed Left   Uniform

Questions

COVERING THE IDEAS

1. Calculate the mean absolute deviation of the runs scored by Westview's softball team on page 50.
2. Calculate the m.a.d. of the goals scored by Westview's soccer team on page 50.
3. The table below gives the quiz scores of a group of students. Copy the table and fill in the columns. Then find the mean absolute deviation of the quiz scores.

Score	Mean	Deviation	Absolute Deviation
21	?	?	?
15	?	?	?
25	?	?	?
22	?	?	?
27	?	?	?

APPLYING THE MATHEMATICS

In 4 and 5, find the mean and mean absolute deviation in your head.

4. 7, 7, 7, 7, 7, 7, 7, 7, 7, 7, 7, 7
5. 2, 2, 2, 2, 2, 2, 4, 4, 4, 4, 4, 4

For 6–9, use the following sketches.

a.
b.
c.
d.

6. Which distribution is symmetric?
7. Which distribution is skewed left?
8. Which distribution is skewed right?
9. Which of the distributions most closely describes the scores of all baseball games of one team in a season?
10. In this lesson, a story is told in which people at Central School complain about the classrooms being too small. Whittier School has a mean classroom size that is lower than Central's, but there are no complaints at Whittier. What seems to be the cause of the complaints at Central?
11. These two data sets give the lengths of terms of presidents of the United States and the reigns of kings and queens of England.
    Years served by the first 42 American presidents: 8, 4, 8, 8, 8, 4, 8, 4, 0, 4, 4, 1, 3, 4, 4, 4, 4, 8, 4, 1, 3, 4, 4, 4, 5, 7, 4, 8, 2, 6, 4, 12, 8, 8, 3, 5, 6, 2, 4, 8, 4, 8
    Years reigned by 39 English rulers: 21, 13, 35, 19, 35, 10, 17, 56, 35, 20, 50, 22, 13, 9, 39, 22, 0, 24, 38, 6, 5, 44, 22, 24, 25, 3, 13, 6, 12, 13, 33, 59, 10, 7, 63, 9, 25, 1, 15

    a. Use a spreadsheet or graphing calculator to calculate the mean, mean absolute deviation, and range of each collection of data.
    b. Write a few sentences explaining why the two distributions have such different values for their means, mean absolute deviations, and ranges.

12. a. Construct a data set of 8 items that has a mean absolute deviation of 0.5 and for which $\mu = 8$.
    b. What are the M, m, and r for this data set?

The first Oval Office was built in 1909 in the center of the south side of the West Wing of the White House.

Source: www.whitehouse.gov

13. Construct a data set of 7 items that has a median of 9 and a mean of 8.

14. Construct a data set of 12 items that has a mode of 6 and for which $\mu = 9$.

15. The mean absolute deviation can be used to measure consistency. The more consistent data set is the one with the smaller m.a.d. Here are the mean maximum daily temperatures for each month in San Diego and Miami. In which city are the year's high temperatures more consistent?

Month	Miami (°C)	San Diego (°C)
January	24.0	18.8
February	24.7	19.2
March	26.2	19.1
April	28.0	20.2
May	29.6	20.6
June	30.9	22.0
July	31.7	24.6
August	31.7	25.4
September	31.0	25.1
October	29.2	23.7
November	26.9	21.1
December	24.8	18.9

REVIEW

16. Evaluate each expression. (**Lesson 1-6**)
    a. $|3 - |-3 + 9||$
    b. $|2x - 10|$ when $x = 3.5$
    c. $-4|x - y|$ when $x = 0.5$ and $y = 3$

17. Use the window at the right. The tick marks on the *x*-axis occur every 20 units. On the *y*-axis, tick marks occur every 0.25 unit. Find the values of each. (**Lesson 1-5**)

    x-min __?__   *x*-max __?__
    y-min __?__   *y*-max __?__

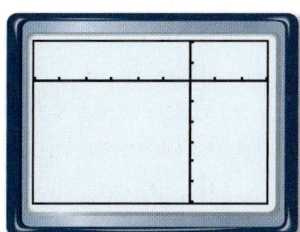

18. On a graphing calculator, graph $y = x^3 - 5x$ using the window $-10 \leq x \leq 10$ and $-10 \leq y \leq 10$. Sketch the graph that results. (**Lesson 1-5**)

19. Make a table and scatterplot for the ordered pair $(n, n^2 - n)$ using the following values of *n*: –3, –2, 0, 2, and 3. (**Lesson 1-4**)

20. Complete the table of values for each expression. Use the tables to determine whether the expressions appear to be equivalent. Explain your reasoning. (**Lesson 1-3**)

n	$(n+2)+7+(n+2)$
−5	?
−3	?
1	?
2	?
5	?

n	$3n+3-n+8$
−5	?
−3	?
1	?
2	?
5	?

21. Consider the sequence of triangles. Assume the designs are made with toothpicks. (**Lesson 1-2**)

1

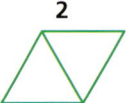
2

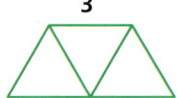
3

a. Draw a table describing the number of toothpicks needed for the first 5 terms.

b. Write an expression for the number of toothpicks in the nth term.

c. How many toothpicks will be needed to make the 100th term?

EXPLORATION

22. Choose the data set that would probably have the smaller m.a.d. and explain why.
    a. incomes of 10 union members from a factory; incomes of 10 management members from the same factory
    b. temperatures in one place on the moon for a month; temperatures in one place on Earth for a month
    c. recovery times for 6 people from an appendectomy; recovery times for 6 basketball players from knee operations

Unionized factory workers

QUIZ YOURSELF ANSWERS

1. $r = 1{,}000 - 600 = 400$ ft^2
2. 39

Using Algebra to Describe

Chapter 1 Projects

1 Examining Pi

Look up the first fifty digits of the decimal expansion of π. Calculate the frequency with which each digit 0 through 9 appears. Calculate the mean absolute deviation of these digits. Find two numbers: one whose first fifty digit decimal expansion has a smaller mean absolute deviation and one with a greater mean absolute deviation.

The constant π is used in the formulas to find the circumference and area of a circle.

2 Estimating Square Roots

Al-Karkhi (also known as al-Karaji) was an Arab mathematician who lived during the early 11th century. He found a formula for approximating the square root of a positive integer n. The formula is $\sqrt{n} \approx w + \frac{n + w^2}{2w + 1}$. Before using his formula to find the square root of n, you must first find w, the whole number part of $\sqrt{n}$. Investigate al-Karkhi's method by making a table with different values for n from 1 to 50 and using the formula to estimate $\sqrt{n}$ rounded to the nearest thousandth. Compare the results to what you obtain by finding $\sqrt{n}$ with a calculator. What conclusions can you make concerning al-Karkhi's method?

3 Formula 1

The game of Formula 1 is played in the following way: Two players secretly make up an expression with one variable and write it on a piece of paper. After both players have their formulas, they "race" them in the following way: Substitute the numbers 1 through 10 in the formula and take the sum of all of the results. The player with the higher total wins. Play this game with a friend several times, each time with a new formula. What changes did you make in order to increase your total? (You may want to use a spreadsheet to do the calculations.)

Formula One (F1) also deals with Grand Prix auto racing. The "Formula" in Formula One is a set of rules which all race teams must follow.

4 Figurate Numbers

Some numbers are called *figurate numbers* because they can easily be represented by geometric figures. Pictured below are the first four triangular numbers, the first four square numbers, and the first four pentagonal numbers.

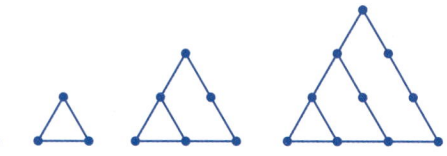

triangular numbers 1, 3, 6, 10

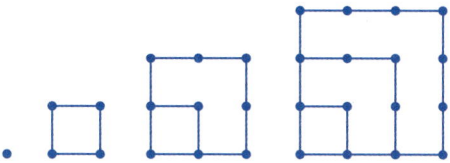

square numbers 1, 4, 9, 16

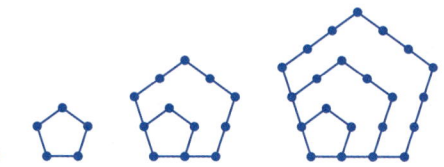

pentagonal numbers 1, 5, 12, 22

a. Draw a picture of the 5th triangular number, 5th square number, and 5th pentagonal number.

b. Based on the patterns in the number of dots on the side of each figure, find the 10th triangular number, the 10th square number, and the 10th pentagonal number.

c. Make a poster or write a report about figurate numbers for your classroom.

5 Encoding and Decoding Using Formulas

a. Formulas can be used to create codes. Consider the formula $27 - n$. Convert a short sentence to numbers using a = 1, b = 2, and so on. Use the formula to encode it. For example, the word "code" in numbers would be 3, 15, 4, and 5 with c = 3, o = 15, d = 4, and e = 5. Using the formula, we would get the encoded message 24, 12, 23, 22.

b. Have a friend give you their encoded sentence. Explain how you would use the formula to decode their message.

c. One famous method for decoding is called "frequency analysis." This method is based on the fact that some letters in English appear more commonly than others. For example, with our formula, the code for the letter "e" is 22. Because e is the most common letter in written English, the number 22 should be the most common one to appear. Look up the term "frequency analysis" on the Internet, and find a table describing how commonly each letter is used in English prose. Choose a paragraph from a book, and find the frequencies for the letters "e," "q," and "p" in the paragraph. Does this agree with the table you found? Explain how you would use frequency analysis to decode an encoded message.

This Enigma cypher machine was used to encode messages during World War II.

Chapter 1 Summary and Vocabulary

- Algebra is a powerful language for describing patterns and real-world situations. Its power comes from the use of **variables**. Variables are letters or other symbols that can be replaced by any number from a set, the **domain** of the variable.

- An **algebraic expression** is a sequence of symbols that contains variables, numbers, and operations. Rules for the **order of operations** ensure that different individuals evaluating the same expressions will get the same values. Scientific and graphing calculators usually follow the same rules of order of operations.

- Algebraic expressions can describe the nth term in a pattern. Two algebraic expressions are **equivalent** if they have the same value. Graphs and tables can be used to explore whether algebraic expressions are equivalent. For example, tables show that $(x - 1) + 3 + (x - 1)$ and $2x + 1$ each have the same value for any particular value of x. This suggests that the expressions are equivalent. The graphs of $y = (x - 1) + 3 + (x - 1)$ and $y = 2x + 1$ also suggest that the expressions are equivalent. On the other hand, the expression $2x + 3$ produces different values from those of the other two expressions, so it is not equivalent to the others.

- The **absolute value** of a number is its distance from 0 on a number line. The distance between two numbers on a number line with coordinates x and y is $|x - y|$. Statistics can be useful in describing data that do not fit an exact algebraic pattern. Two measures of spread are the **range** and the **mean absolute deviation (m.a.d.)**. The m.a.d. applies the idea of absolute value.

Theorems and Properties

Algebraic Definition of Subtraction (p. 7)
Algebraic Definition of Division (p. 7)
Associative Property of Multiplication (p. 9)
Associative Property of Addition (p. 10)
Transitive Property of Equality (p. 10)
Commutative Property of Addition (p. 16)
Commutative Property of Multiplication (p. 16)

Vocabulary

1-1
variable
algebraic expression
evaluating the expression

1-2
pattern
instance
define a variable
term
factor

1-3
sequence
term
equivalent expressions
counterexample

1-4
scatterplot
domain of a variable

1-5
window
Xmin, Xmax
Ymin, Ymax
Xscl, Yscl
standard window

1-6
absolute value
origin

1-7
range
mean absolute deviation (m.a.d.)
symmetric
skewed right
skewed left
uniform

Chapter 1 Self-Test

Take this test as you would take a test in class. You will need graph paper. Then use the Selected Answers section in the back of the book to check your work.

1. Consider the following pattern:

    $9 \cdot 4 + 9 \cdot 1 = 9 \cdot 5$
    $9 \cdot 4 + 9 \cdot 2 = 9 \cdot 6$
    $9 \cdot 4 + 9 \cdot 3 = 9 \cdot 7$
    $9 \cdot 4 + 9 \cdot 4 = 9 \cdot 8$
    $9 \cdot 4 + 9 \cdot 5 = 9 \cdot 9$

    a. Write the next three instances of the pattern.
    b. Describe the pattern using one variable.

2. Provide an example illustrating the Commutative Property of Multiplication.

3. Rewrite the division problem as a multiplication problem $3x \div 7y$.

4. At a sports store, authentic jerseys cost $179 and customized T-shirts cost $24. Let j = the number of jerseys purchased and let t = the number of T-shirts purchased. Write an expression describing the total cost of all clothing purchased at the sports store.

5. a. Complete the table of values for the provided expressions.

n	$\frac{6n-12}{3}$	$-4 + 2n$
-5	?	?
-3	?	?
0	?	?
2	?	?

    b. Do $\frac{6n-12}{3}$ and $-4 + 2n$ seem to be equivalent expressions?

6. Consider the sequence of square tile designs shown here.

    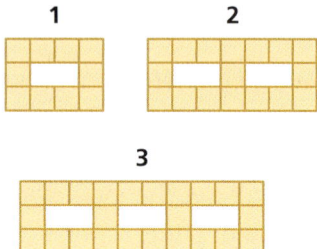

    a. Determine how many tiles would be required to make the next term.
    b. Complete the table describing the number of square tiles used for various patterns in the sequence.

n	1	2	3	4	5	6
Number of Tiles	10	17	?	?	?	?

    c. Write an expression for the number of square tiles in the nth term.
    d. Create a scatterplot from the table.

7. Consider the following expressions:

    Expression 1: $\frac{m}{2} + \frac{3}{2}$

    Expression 2: $\frac{3+m}{4}$

    a. Find a value for m that shows that Expressions 1 and 2 are not equivalent.
    b. Write an expression that is equivalent to Expression 1.

8. Consider the following expressions.

    Expression 1: $\frac{101x + 200}{100}$

    Expression 2: $x + 2$

    a. What does a standard window seem to suggest about equivalence of the expressions?

b. Are the expressions equivalent? Why or why not?

9. An expression is graphed below. Use the graph to answer the following questions.

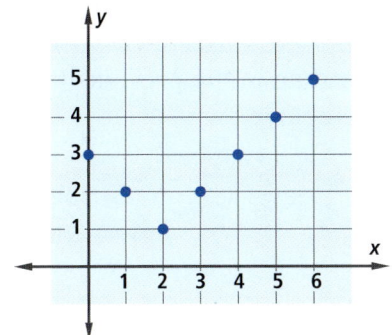

a. What is the value of the expression when $x = 4$?

b. Use the graph to complete the following table of values.

x	0	1	2	3	4	5
y	?	?	?	?	?	?

c. **Multiple Choice** Which of the following expressions best describes the values plotted on the graph?

A $|x + 2|$
B $|x - 2| + 1$
C $|x + 2| - 1$
D $3 - x$

10. Graph $y = x - 15$ using the window $-10 \le x \le 10$ and $-10 \le y \le 10$. The result will be a line. Adjust the window to reveal where the line crosses both the *x*- and the *y*-axes. Describe your new viewing window in the space provided.

Xmin: __?__   Xmax: __?__
Ymin: __?__   Ymax: __?__

In 11–13, use the following information. Over the past week, students in Mr. Cy Metric's class reported paying the following amounts for entrance to a movie. The results are graphed in a dot plot.

$6, $7, $9, $8, $10, $2, $7, $10, $10, $9, $8, $8, $9, $9, $10

11. Identify the shape of the dot plot as symmetric, skewed right, skewed left, or uniform.

Movie Price

12. Calculate the mean and mean absolute deviation for the data set. Round to the nearest tenth.

13. Do data have more spread or less spread than a set with a mean of 8.8 and a mean absolute deviation of 2.6? Explain your reasoning.

14. a. Graph $y = 0.5x^3 + x^2 - 5.5x - 5$ on a graphing calculator. View the graph with the standard window and sketch the results.

b. Change the values in the window to get the graph to look like the one below. What are the values of your window's Xmin, Xmax, Ymin, and Ymax?

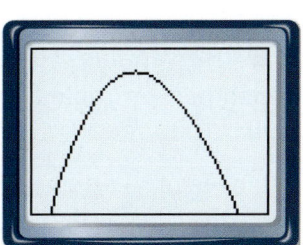

Chapter 1 Chapter Review

SKILLS **PROPERTIES** **USES** **REPRESENTATIONS**

SPUR stands for Skills, Properties, Uses, and Representations. The Chapter Review Questions are grouped according to the SPUR Objectives in this Chapter.

SKILLS Procedures used to get answers

OBJECTIVE A Evaluate numerical and algebraic expressions. (Lesson 1-1)

In 1–4, evaluate the numerical expression.

1. $20 \div 5 \div 5$
2. $70 - 4 \cdot (-3) \div 2$
3. $\frac{5 \cdot 0.43^2}{0.43}$
4. $\left(\frac{3}{8} + \frac{5}{6}\right) \cdot 4 + \frac{3}{5}$

In 5–10, evaluate the algebraic expression for the given variable.

5. $4x^2$ for $x = 13$
6. $-6q - q$ when $q = -3.64$
7. $\frac{-5x}{2} + \frac{3}{-x}$ if $x = 7$
8. $\left(\frac{n}{3}\right)^3$ for $n = 24$
9. $4(p - q)$ when $p = 13\frac{1}{2}$ and $q = 2\frac{3}{5}$
10. $y - x \div 4 \cdot (-2)$ when $x = 8$ and $y = -3$

OBJECTIVE B Use variables to describe patterns in instances or tables. (Lessons 1-2, 1-3)

11. Three instances of a pattern are given below. Describe the pattern using one variable.

    $4(2) + 3(2) = 7(2)$

    $4(-3.1) + 3(-3.1) = 7(-3.1)$

    $4\left(\frac{2}{5}\right) + 3\left(\frac{2}{5}\right) = 7\left(\frac{2}{5}\right)$

12. The number of zeros in each term is 2 greater than in the previous term in the table below.

Term Number	1	2	3	4
Term	0	000	00000	0000000

a. How many zeros are in the sixth term?

b. In words, explain how to use the term number to find the number of zeros in the nth term.

c. Use n to represent the term number. Write an algebraic expression for the number of zeros in the nth term.

13. Refer to the sequence below.

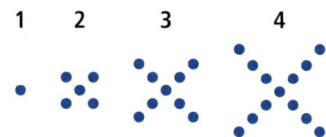

a. How many dots will be in the seventh term?

b. In words, describe a method to find the number of dots in the nth term.

c. Write an algebraic expression for the number of dots in the nth term.

OBJECTIVE C Determine if two expressions seem equivalent by substituting values or making a table. (Lesson 1-3)

14. Alyssa and Odell both looked at a pattern. Alyssa thought the nth term could be represented by the expression $3n - 6 + n$, and Odell came up with the expression $-5 + 4n - 1$. Substitute the numbers 3, 5, and 9 in for each expression to determine if they seem to be equivalent.

60 Using Algebra to Describe

15. To convert temperatures from Fahrenheit to Celsius, Kimi used the expression $\frac{5}{9}F - 32$ and Edward used the expression $\frac{5}{9}(F - 32)$. Substitute the numbers 9 and 18 for F in each expression to determine if they seem to be equivalent.

16. Fill in each table to determine if $5 + |3n|$ is equivalent to $5 + |-3n|$.

n	$5 + |3n|$
−3	?
−2	?
0	?
4	?
5	?

n	$5 + |-3n|$
−3	?
−2	?
0	?
4	?
5	?

OBJECTIVE D Evaluate expressions involving absolute value. (Lesson 1-6)

In 17 and 18, evaluate the numerical expressions.

17. $|(-3)| + |5 - 3|$
18. $|9 \cdot (-2)| + |(-9) \cdot 2|$

In 19–21, evaluate the algebraic expression for the given value of the variable.

19. $|x - 3| \cdot x$ for $x = -2$
20. $|x - |x||$ for $x = -1$
21. $||x| - |2x + 1||$ for $x = -3$

OBJECTIVE E Calculate the range and mean absolute deviation. (Lesson 1-7)

In 22–25, calculate the range and mean absolute deviation of each data set.

22. 12.6, 10.4, 3.8, 7.2, 5.9, 4.1, 1.5, 2.5
23. $\frac{3}{8}, \frac{3}{4}, \frac{5}{2}, \frac{1}{4}, 2\frac{1}{8}$
24. A student's test scores: 87, 94, 90, 73, 84, 83, 97, 72
25. The Miami Heat's total points in games played in January 2006: 97, 92, 93, 118, 110, 117, 100, 92, 94, 119, 94, 98, 91, 101, and 118

PROPERTIES Principles behind the mathematics

OBJECTIVE F Apply the Algebraic Definitions of Subtraction and Division. (Lesson 1-1)

In 26 and 27, rewrite each subtraction as an addition.

26. $x - y - z$
27. $-8 - y - 32$

28. **Multiple Choice** Which expression is not equivalent to the others?

    A $a - b$ B $a + -b$ C $-b + a$ D $b + -a$

29. **True or False** $\frac{x}{7} = \frac{1}{7}x$

In 30 and 31, rewrite the division problem as a multiplication problem.

30. $\frac{7d + 2}{4st}$
31. $6.21 \div 3.14$

OBJECTIVE G Identify and apply the associative, commutative, and transitive properties. (Lessons 1-1, 1-2)

32. $(45 + 23) + 77 = 45 + (23 + 77)$ is an example of what property?

33. What property is described by $x \cdot y = y \cdot x$?

34. Wesley was evaluating the expression $(2 - x) + 3$ and used the Associative Property to rewrite it as $2 - (x + 3)$ to make it easier to compute.

    a. Is this correct?
    b. If so, evaluate the expressions for values of x to show they are equal. If not, find and correct the mistake Wesley made.

35. Erin had the expression $x + 2 + 2x + 7$. Jamal had the expression $3x + 9$. Kelly's expression was $x + 9 + 2x$. Erin and Jamal discovered their expressions are equivalent. Erin and Kelly's expressions are also equivalent. What property makes Jamal and Kelly's expressions equivalent?

USES Applications of mathematics in real-world situations

OBJECTIVE H Create expressions to model real-world situations. (Lesson 1-2)

36. In a football board game, a person earns 6 points per receiving touchdown and 3 points per passing touchdown. The person will lose a point for interceptions thrown. Let $r=$ the number of receiving touchdowns, let $p=$ the number of passing touchdowns, and let $i=$ the number of interceptions thrown. Write an expression to represent the total points a person playing the football board game has.

37. Juan is a wedding photographer. He offers two package deals to his clients. Package A costs $1,225 and Package B costs $1,405. Let $a=$ the number of Package A deals he sells, and let $b=$ the number of Package B deals he sells. Write an expression to represent the total amount Juan makes selling his wedding package deals.

OBJECTIVE I Calculate and interpret the spread of a distribution using mean absolute deviation. (Lesson 1-7)

38. The costs of four video games at two stores are shown in the table below.

Game	Store 1	Store 2
Play Soccer	$35	$40
Adventure Trip	$45	$40
Be a Robot!	$48	$50
Catch the Dragon	$50	$60

a. Calculate the mean and the mean absolute deviation for the four videogames at each store.

b. What is the meaning of the mean absolute deviations calculated in Part a?

c. If you could go to only one store for your video game purchases, which would it be? Explain your reasoning.

d. Which store has less variation in the price of video games? Explain your reasoning.

REPRESENTATIONS Pictures, graphs, or objects that illustrate concepts

OBJECTIVE J Create a scatterplot from a table or expression. (Lesson 1-4)

39. Suppose Ben Inriver begins with $50 in a savings account and adds $20 per week. The table shows the number of weeks that he has saved and the total amount saved.

Week (w)	Total (t)
0	50
1	?
2	?
3	?
4	?

a. Complete the table.

b. Plot the five pairs (w, t).

40. The table below shows the wind chill index for various temperatures when there is a 10 mile per hour wind. Plot the data with temperature on the horizontal axis and wind chill on the vertical axis.

Actual Temperature (°F)	Wind Chill Index (°F)
30°	21
20°	9
10°	−4
0°	−16
−10°	−28
−20°	−41
−30°	−53

Source: NOAA's National Weather Service

Chapter Wrap-Up

OBJECTIVE K Graph ordered pairs from expressions. (Lessons 1-4, 1-5)

41. Use the table below.

n	$2n + (n + 2)$
1	5
2	?
3	?
4	?
5	?

  a. Complete the table of values.

  b. Graph the ordered pairs with n on the x-axis and $2n + (n + 2)$ on the y-axis.

  c. From the graph, predict the value of the expression when n is 11.

42. Suppose the number of seats in the nth row of an auditorium is $32 + 2n$.

  a. Evaluate the expression for the rows numbered 2, 3, 4, 5, 6, 7, and 8.

  b. Graph the ordered pairs (row number, number of seats) for these values.

OBJECTIVE L Use graphs to determine whether expressions seem to be equivalent. (Lessons 1-4, 1-5, 1-6)

43. Determine if the expressions $x + 2$ and $|x + 2|$ are equivalent by using a graph.

44. James and Luanda both looked at a pattern. James thought the nth term could be represented by the expression $n - 6 + n$, and Luanda came up with the expression $-6 + 2n$. Graph the data and determine if the expressions seem to be equivalent.

45. Use a graph to determine whether the expressions $(y - 1)^2$ and $y^2 + 1$ seem to be equivalent.

OBJECTIVE M Use graphs to find values, create tables, and select appropriate windows. (Lesson 1-5)

46. a. Graph $y = x^3 - 3x^2$ on your graphing calculator in the standard window and sketch your result.

  b. Change your window to match the calculator screen below. Record Xmin, Xmax, Ymin, and Ymax.

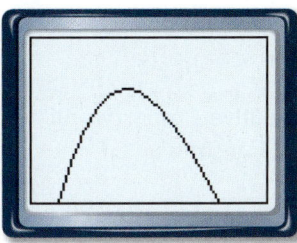

In 47 and 48, use the calculator screen below. Each tick mark represents one unit.

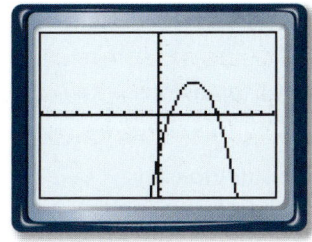

47. Make a table of values for x and y when $x = 0, 2, 3,$ and 6.

48. **Multiple Choice** Which of the following expressions best represents that of the values on the graph?

  A $|x - 3| + 4$

  B $4(x - 3)$

  C $-(x - 3)^2 + 4$

  D $(x - 3)^2 + 4$

Chapter 2

Using Algebra to Explain

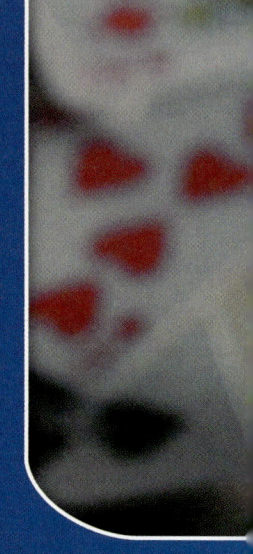

Contents

- **2-1** The Distributive Property and Removing Parentheses
- **2-2** The Distributive Property and Adding Like Terms
- **2-3** Explaining Number Puzzles
- **2-4** Opposites
- **2-5** Testing Equivalence
- **2-6** Equivalent Expressions with Technology
- **2-7** Explaining Addition and Subtraction Related Facts
- **2-8** Explaining Multiplication and Division Related Facts

Here is a card trick that can be explained using algebra. The trick is typically performed with two people: an illusionist and a spectator. The spectator should not read the directions while the illusionist is performing the trick.

Directions

Step 1 Before the trick begins, shuffle a deck of cards. Look at the card in the ninth position from the top of the deck. Without letting the spectator see, write what the card is on a piece of paper, fold the paper several times, and give it to the spectator. Tell the spectator not to open it.

Step 2 Ask the spectator to choose any number from 10 to 19. Count out that number of cards from the top of the deck. For example, if the spectator chooses 17, count out 17 cards from the top of the deck. Put them in a smaller, new pile faceup next to the deck of cards.

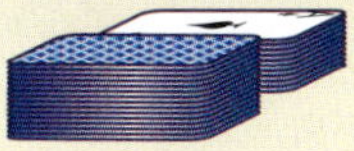

Step 3 Tell the spectator to add the digits of the number they choose. Count out that number of cards from the top of the smaller pile and place them back on top of the original deck. For example, the sum of the digits of 17 is 8. So place 8 cards from the smaller pile back on top of the original deck.

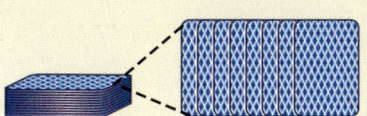

Step 4 Tell the spectator to open the piece of paper. The top card on the smaller deck should match the card written on the piece of paper.

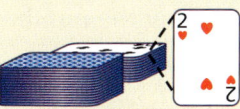

In this chapter, you will use the properties of algebra to explain relationships among numbers, show equivalence, and explain why number tricks, such as this one, work.

Chapter 2

Lesson 2-1

The Distributive Property and Removing Parentheses

> **BIG IDEA** By applying the Distributive Property, you can rewrite the product $a(b + c)$ as the sum $ab + ac$.

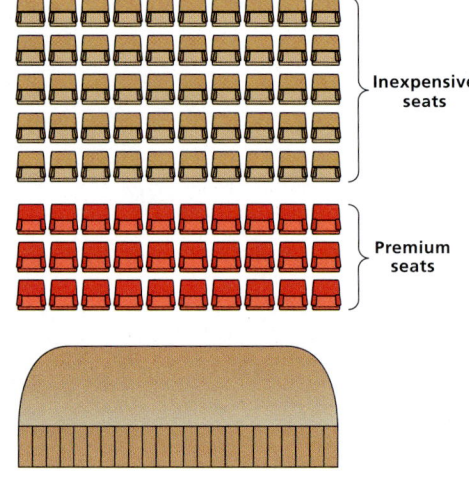

Mental Math

a. Order from least to greatest: $\frac{5}{32}, \frac{1}{4}, \frac{1}{8}, \frac{3}{16}$.

b. Order from least to greatest: $-\frac{5}{32}, -\frac{1}{4}, -\frac{1}{8}, -\frac{3}{16}$.

Suppose an auditorium has 8 rows of seats with 10 seats to a row. The tickets for the first 3 rows cost more than the tickets for the other 5 rows.

Two ways to count the number of seats in the auditorium illustrate a useful pattern. One way is to treat all the seats alike. Multiply the number of seats in each row, 10, by the total number of rows, $5 + 3$, or 8.

$$10(5 + 3) = 10 \cdot 8 = 80$$

The second way is to count the number of inexpensive seats and premium seats separately, then add the results.

$$10 \cdot 5 + 10 \cdot 3 = 50 + 30 = 80$$

These two ways of counting the number of seats yield the same result.

$$10(5 + 3) = 10 \cdot 5 + 10 \cdot 3$$

This is an example of the basic property that involves both addition and multiplication. It is called the *Distributive Property of Multiplication over Addition* since the multiplication by 10 in $10(5 + 3)$ is "distributed" to both terms in the parentheses.

The Distributive Property of Multiplication over Addition

For all real numbers a, b, and c, $c(a + b) = ca + cb$.

The name of this property is very long, so we just call it the *Distributive Property*. The Distributive Property can be used to rewrite the product $c(a + b)$ as the sum of terms $ca + cb$. This is called expanding the expression. Expanding a product has the effect of removing parentheses.

66 Using Algebra to Explain

Lesson 2-1

🛑 QY1

> QY1
>
> Use the Distributive Property to expand $6(x + 5)$.

Example 1

Jason earns money by mowing the lawns of three houses in his neighborhood each week. Since the lawns are of different sizes, he charges the neighbors different prices: *a*, *b*, and *c*. He mows the second neighbor's lawn twice a week, so each week Jason earns $a + 2b + c$ dollars. The neighbors pay monthly. Assuming 4 weeks per month, Jason earns $4(a + 2b + c)$ dollars each month. Use the Distributive Property to give an expression that is equivalent to $4(a + 2b + c)$.

Solution Distribute 4 over each term. Then simplify.

$$4(a + 2b + c) = 4 \cdot a + 4 \cdot (2b) + 4 \cdot c$$
$$= 4a + 8b + 4c$$

Check The expression $4a + 8b + 4c$ indicates that Jason has mowed the first lawn 4 times, the second lawn 8 times, and the third lawn 4 times during the month.

Recycling grass clippings into lawns lowers soil temperature, reduces water loss, and reduces yard waste going into landfills.

Source: Servicemaster

Explaining a Multiplication Shortcut

Suppose a motel room costs $59 a day, and a person will stay in that room for 8 days. To find the total cost, you can multiply $8 \cdot 60$, and then subtract $8 \cdot 1$. (Calculate as if the price for each of the eight days was $60. Then subtract $1 per day for 8 days.) The Distributive Property explains why this works.

$$8 \cdot 59 = 8(60 - 1) = 8 \cdot 60 - 8 \cdot 1, \text{ or } \$472$$

Here we have distributed the multiplication over a *subtraction*. Since $60 - 1 = 60 + (-1)$, the subtraction can be thought of adding the opposite. Some people like to think of this variant of the Distributive Property as a separate property.

> **The Distributive Property of Multiplication over Subtraction**
>
> For all real numbers *a*, *b*, and *c*, $c(a - b) = ca - cb$.

Thus, there are two forms of the Distributive Property used to expand expressions. You can use either of these versions to expand a subtraction expression, as shown in Example 2 on page 68.

The Distributive Property and Removing Parentheses

Example 2
Expand $-11(5 - 6w)$.

Solution Begin by rewriting the subtraction expression as an addition expression.

$$-11(5 - 6w) = -11(5 + (-6w))$$
$$= -11 \cdot 5 + -11 \cdot -6w$$
$$= -55 + 66w$$

Check Substitute the same value for w in both the given expression and the expanded expression. We use $w = 3$. Remember to follow the order of operations.

When $w = 3$, $-11(5 - 6w) = -11(5 - 6 \cdot 3) = -11(-13) = 143$.
When $w = 3$, $-55 + 66w = -55 + 66 \cdot 3 = -55 + 198 = 143$.
It checks.

GUIDED

Example 3
Expand $2x(5x - 3)$.

Solution

$2x(5x - 3) = 2x \cdot \underline{\ ?\ } - 2x \cdot \underline{\ ?\ }$ Distributive Property
$\qquad\qquad\quad = \underline{\ ?\ } - \underline{\ ?\ }$ Multiplication

The Distributive Property also works in the cases where the multiplier is on the right, as in $(a + b)c$, because multiplication is commutative. So, $(a + b)c = c(a + b) = ca + cb$.

Expanding a Fraction

Because every division can be converted to multiplication, the Distributive Property can also be used to rewrite expressions involving division. Suppose the sum $(a + b)$ is to be divided by c.

$\frac{a + b}{c} = \frac{1}{c}(a + b)$ Dividing by c is the same as multiplying by $\frac{1}{c}$.

$\qquad\ = \frac{1}{c} \cdot a + \frac{1}{c} \cdot b$ Distributive Property of Multiplication over Addition

$\qquad\ = \frac{a}{c} + \frac{b}{c}$ algebraic definition of division

In this way, a fraction with a sum in its numerator can be rewritten using $\frac{a + b}{c} = \frac{a}{c} + \frac{b}{c}$. This step may allow you to simplify an expression, as shown in Example 4.

Lesson 2-1

Example 4
Use the Distributive Property to write $\frac{36 + 3x}{18}$ as a sum of two fractions.

Solution
$$\frac{36 + 3x}{18} = \frac{36}{18} + \frac{3x}{18} = 2 + \frac{x}{6}$$

Check Let $x = 6$. (Do you see why we chose 6?)
Then $\frac{36 + 3x}{18} = \frac{36 + 3 \cdot 6}{18} = \frac{36 + 18}{18} = \frac{54}{18} = 3$.
Also $2 + \frac{x}{6} = 2 + \frac{6}{6} = 3$. It checks.

 QY2

> ▶ QY2
>
> **Multiple Choice** For all x, $\frac{20 - 8x}{4}$ equals which of the following?
> A $5 - 8x$
> B $3x$
> C $5 - 2x$
> D $3 - 8x$

Questions

COVERING THE IDEAS

1. Use the Distributive Property to find equivalent expressions.
   a. $n(k + w)$
   b. $g(d - e)$
   c. $\frac{n + p}{r}$

2. Write an expression for the area of the largest rectangle below in two ways.

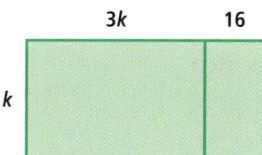

   a. as length times width
   b. as the sum of areas

3. A person buys 45 envelopes at $1.03 each. Explain how you can use the Distributive Property to calculate the total cost in your head.

4. Calculate in your head the total width of 5 windows that are each 39 inches wide.

In 5–10, expand the expression.

5. $(m + 4)5$
6. $-30x(x + 2 + 4n)$
7. $12(k - \frac{1}{6})$
8. $(2b + c)10b$
9. $6(3v - 8w + 9z^3)$
10. $-7a(a - b)$

11. Suppose the cost of a cell phone call is $0.12 per minute. Two calls are made. One lasts 17 minutes and the other lasts 6 minutes. Find the total cost of the calls in two different ways.

12. Rewrite $\frac{24 + 6x}{8}$ as the sum of two fractions. Check your answer.

The Distributive Property and Removing Parentheses

APPLYING THE MATHEMATICS

In 13 and 14, complete each sentence to show examples of the Distributive Property.

13. $24(k + m) = 24\underline{} + 24\underline{}$

14. $10a + 80 = 10(\underline{} + \underline{})$

15. For each hour of television, there is an average of $13\frac{1}{2}$ minutes of commercials. If you watch 9 hours of television in a week, how many minutes of commercials will you see? Explain how you can find the answer in your head.

16. You and your friend decide to start a dog-walking business and plan to charge $12 for each dog walked.
    a. You are scheduled to walk 13 dogs this week and your friend has scheduled 17 more appointments. Determine the total amount of money you will earn using two different methods.
    b. You are scheduled to walk 13 dogs this week. If your friend schedules x more appointments for the week, how much money will you earn?

20% of owned dogs were adopted from an animal shelter.
Source: www.thepetprofessor.com

REVIEW

17. The table below gives the number of films (up to 2006) in which the ten top-ranked actresses have starred. (**Lesson 1-6**)

Actress	Number of Films
Cate Blanchett	34
Patricia Clarkson	52
Toni Collette	32
Kirsten Dunst	52
Scarlett Johansson	27
Nicole Kidman	46
Julianne Moore	52
Samantha Morton	32
Michelle Pfeiffer	46
Kate Winslet	30

Source: Internet Movie Database

a. Find the mean number of films.
b. Find the mean absolute deviation. Explain what the mean absolute deviation means in this case.

18. Graph $y = (0.65x)^3$ and $y = 2^x$ on a graphing calculator with the following window settings: $0 \leq x \leq 60$ and $5 \leq y \leq 45$. (Lessons 1-5, 1-3)

   a. Create a table of values for both graphs with $x = -10, -5, 0, 5, 10,$ and 15.

   b. Are the expressions equivalent? Why or why not?

In 19 and 20, use the Associative and Commutative Properties of Multiplication to compute in your head. (Lesson 1-1)

19. $2 \cdot 11 \cdot 3 \cdot 1.5$ 20. $8 \cdot 2 \cdot 7 \cdot 5$

21. A raffle to raise money for charity sells tickets for $4 each. The winner of the raffle will receive half of all the money raised from selling the tickets, and the other half will be given to the charity. If 721 tickets are sold, how much money will be given to the charity? (Lesson 1-1)

In 22 and 23, which of the numbers –2, 5, and 8 makes the sentence true? (Lesson 1-1)

22. $12 + 3n - 4 < 7$ 23. $11 \geq -3y + \frac{15}{3}$

24. According to a study by the Kaiser Family Foundation, typical American teenagers spent an average of 12.5% of their days watching television in 2004. (Previous Course)

   a. How many hours does this represent in a day?

   b. How many hours does this represent in one week?

   c. How many days does this represent in one year?

Teens spend the majority of their TV-watching time between the hours of 8 P.M. and 11 P.M.

Source: Nielsen Media Research

EXPLORATION

25. Some people overgeneralize the Distributive Property. They think that because $6x + 3x = 9x$, both of the following must be true. Find a counterexample to each equation to show that it is false.

   a. $6x \cdot 3x = 18x$ b. $\frac{6x}{3x} = 2x$

QY ANSWERS

1. $6x + 30$

2. C

Chapter 2

Lesson 2-2
The Distributive Property and Adding Like Terms

Vocabulary

like terms
coefficient
factoring

> **BIG IDEA** By applying the Distributive Property, you can add or subtract like terms.

In Lesson 2-1, you learned that the Distributive Property can be used to remove parentheses by changing $c(a + b)$ into $ca + cb$. Now we will reverse the direction.

Adding Like Terms: From $ac + bc$ to $(a + b)c$

Algebraic expressions such as $x^2 + 3x + 5$ or $4a - 9b$ are made up of terms. The terms of $x^2 + 3x + 5$ are x^2, $3x$, and 5. Recall that a term is either a single number or a variable, or a product of numbers and variables. In an expression, addition separates terms. For instance, the terms of $4m^3 - 2m + 9.2m$ are $4m^3$, $-2m$, and $9.2m$. The terms of $-8k^2n + \frac{1}{3}k - 77$ are $-8k^2n$, $\frac{1}{3}k$, and -77.

The terms $-2m$ and $9.2m$ are called **like terms** because they contain the same variables raised to the same powers.

Mental Math

a. $-1(1 + -1)$
b. $-1(1 + -1(1 + -1))$
c. $-1(1 + -1(1 + -1(1 + -1)))$

Like Terms	Unlike Terms
$3t$ and $40t$	$5t$ and $16t^2$ (different powers)
y^2 and $-19y^2$	$37x^3$ and y^3 (different variables)
$200u^5c^3$ and $8u^5c^3$	$9u^5c^3$ and $4u^{10}c$ (different powers)

Reversing the sides of the Distributive Property in Lesson 2-1 shows how to add or subtract like terms. For any real numbers a, b, and c, $ac + bc = (a + b)c$ and $ac - bc = (a - b)c$.

Here are the sums of two of the three pairs of like terms from the table above.

$$3t + 40t = (3 + 40)t \qquad y^2 + (-19y^2) = 1y^2 + -19y^2$$
$$= 43t \qquad\qquad\qquad = (1 + -19)y^2$$
$$\qquad\qquad\qquad\qquad\qquad = -18y^2$$

 QY

▶ **QY**

Find the sum of $200u^5c^3$ and $8u^5c^3$.

72 Using Algebra to Explain

Notice that in combining $y^2 + -19y^2$, the first step was to rewrite y^2 as $1y^2$. Multiplying a number by 1 does not change its value. The following expression can be combined in a similar way.

$$5n - n = 5n - 1n$$
$$= (5 - 1)n$$
$$= 4n$$

When there are two or more collections of like terms in an expression, you can group the like terms together using the Commutative and Associative Properties of Addition.

Example 1

Write a simplified expression for the perimeter of the quadrilateral QUAD.

Solution The perimeter is the sum of the lengths of the sides.

$4u + (2t + u) + 7t + (5u - t)$

Group like terms, changing subtraction to addition.

$= (2t + 7t + -t) + (4u + u + 5u)$

Combine like terms using the Distributive Property. Notice that u is the same as $1 \cdot u$ and $-t$ is the same as $-1 \cdot t$.

$= (2t + 7t + -1t) + (4u + 1u + 5u)$

Add like terms.

$= 8t + 10u$

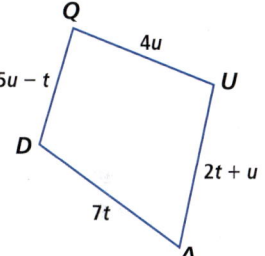

Check Pick values for t and u. We pick 10 for t and 25 for u because they are easy numbers to multiply. The values of the original and simplified expressions must be equal. Substitute 10 for t and 25 for u into the initial expression.

$4u + 2t + u + 7t + 5u - t = 4 \cdot 25 + 2 \cdot 10 + 25 + 7 \cdot 10 + 5 \cdot 25 - 10$
$= 100 + 20 + 25 + 70 + 125 - 10$
$= 330$

Substitute 10 for t and 25 for u into the simplified expression.

$8t + 10u = 8 \cdot 10 + 10 \cdot 25 = 80 + 250 = 330$

The values of the original and simplified expressions are equal, so it checks.

A number that is a factor in a term is called a **coefficient** of the other variables in the term. For instance in $-8k^2n$, -8 is the coefficient of k^2n. You can see from Example 1 that *like terms are combined by adding the coefficients*.

The Distributive Property and Adding Like Terms

It is very important to distinguish coefficients from exponents. For example, the Distributive Property applies to $3x + 11x$, but not to $x^3 + x^{11}$. To give a numerical example, $2 \cdot 10 + 3 \cdot 10 = (2 + 3)10 = 5 \cdot 10 = 50$, but $10^2 + 10^3 = 100 + 1{,}000$ or $1{,}100$ is not equal to 10^5 or $100{,}000$. *Different powers of the same number are not like terms.*

Sometimes the Distributive Property can be used twice to simplify an expression, first to remove parentheses and then to combine like terms.

> **GUIDED**
>
> **Example 2**
> Combine like terms for $(4f^2 + f + 9) + 10(12f - f^2 - 3)$.
>
> **Solution** First remove parentheses.
> $= 4f^2 + f + 9 + \underline{}f - \underline{}f^2 - \underline{}$
>
> Group like terms, changing subtractions to additions.
> $= [\underline{}f^2 + (\underline{}f^2)] + (\underline{}f + \underline{}f) + [\underline{} + (\underline{})]$
>
> Add like terms.
> $= \underline{}f^2 + \underline{}f + \underline{}$
>
> Be sure to check that the simplified expression is equivalent to the original expression by substituting a number for *f* into both expressions and making sure they are equal.

Explaining the Addition of Fractions

Adding fractions with the same denominator is an example of adding like terms. Below are two sums to consider: one is numeric and the other is algebraic. Dividing by a number (5 or *m*) is the same as multiplying by its reciprocal $\left(\frac{1}{5} \text{ or } \frac{1}{m}\right)$.

$$\frac{3}{5} + \frac{4}{5} = 3 \cdot \frac{1}{5} + 4 \cdot \frac{1}{5} \qquad \frac{k}{m} + \frac{2}{m} = k \cdot \frac{1}{m} + 2 \cdot \frac{1}{m}$$
$$= (3 + 4) \cdot \frac{1}{5} \qquad\qquad = (k + 2) \cdot \frac{1}{m}$$
$$= 7 \cdot \frac{1}{5} \qquad\qquad\qquad = \frac{k + 2}{m}$$
$$= \frac{7}{5}$$

If the fractions do not have a common denominator, they must be changed to equivalent fractions that have a common denominator before they can be considered as like terms.

Factoring

When $ac + bc$ is rewritten as $(a + b)c$, the original addition has been changed into a multiplication in which $(a + b)$ and c are factors. This process is called **factoring**. We say that c has been "factored out" of the expression. Notice that in this process, instead of removing parentheses, factoring introduces parentheses. For this reason, some people say that factoring "undoes" expansion.

Example 3

Factor 8 out of the expression $8x + 8y$.

Solution First write $8x + 8y = 8(\underline{\ ?\ } + \underline{\ ?\ })$. To get $8x$, there must be an x. To get $8y$, there must be a y.

$8x + 8y = 8(x + y)$

GUIDED

Example 4

Factor $3z^2$ out of the expression $3mz^2 - 6pz^2$.

Solution

$3mz^2 - 6pz^2 = 3z^2(\underline{\ ?\ } - \underline{\ ?\ })$

(*Hint:* Work backwards to see what $3z^2$ could be multiplied by to get each original term.)

Questions

COVERING THE IDEAS

1. Determine whether the term and $8m^2$ are like terms.
   a. $8m$ b. $8x^2$ c. m^2 d. $3m^2$

In 2–7, combine like terms.

2. $\frac{7}{11} + \frac{5}{4}$

3. $19x + -7x$

4. $8y^3 + y^3 + 5y$

5. $(6x + 2y) + (2x + y)$

6. $(9m^2p + 5mp^2 + 8) + (-4m^2 + 7mp^2)$

7. $\frac{x + 1}{3m} + \frac{4}{3m}$

8. Group the six terms below into three pairs of like terms.

   $5m^3n^2, -8.9m, 68m^2n, -m^2n, m, \frac{4}{9}m^3n^2$

9. Gregory measured his garden using the lengths h and f of his hands and feet. Write a simplified expression for the perimeter of his garden below.

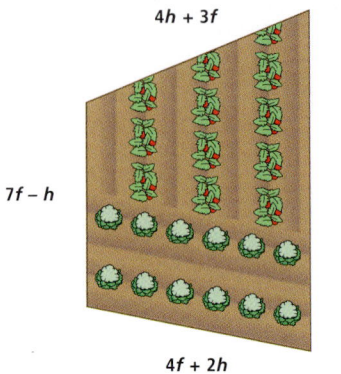

Fill in the Blanks In 10 and 11, factor the expression.

10. $16k - 16m = 16(\underline{\ ?\ } - \underline{\ ?\ })$

11. $2a^2 + 8b = 2(\underline{\ ?\ } + \underline{\ ?\ })$

In 12–14, simplify the expression.

12. $27 + 10(z + 3) + 8z$

13. $p(3p + 1) + 5p^2$

14. $2x + 5(x - 4 + 3x) + (2 - 2x)$

15. Factor 5 out of $15ab + 40c - 10$.

16. Factor $3y$ out of $9y^2 - 24xy$.

APPLYING THE MATHEMATICS

17. Frank, Susan, and Kazu collect stamps. Frank has f stamps, and Susan and Kazu each have 4 times as many stamps as Frank. Write an expression for the number of stamps they have altogether.

18. Some taxicab companies allow their drivers to keep $\frac{3}{10}$ of all the money they collect. The rest goes to the company. If a driver collects F dollars from the fares, write an expression for the company's share.

19. Find a counterexample to show that $6^x + 3^x = 9^x$ is not always true.

20. Does $6\sqrt{x} + 3\sqrt{x} = 9\sqrt{x}$ for all values of x? Why or why not?

Cabs contain a meter that indicates the fare based on the distance covered and other factors.

REVIEW

21. Timothy was asked to simplify $-4(-5 + 3n + -4m)$. His answer was $20 + 3n + -4m$. Write a note to Timothy explaining what he did wrong. (**Lesson 2-1**)

76 Using Algebra to Explain

22. Malia went out to dinner with two friends. They lost track of the number of sodas ordered. But before the bill came, they figured they owed $3x + 48$ dollars, where x represented the number of drinks. If they split the bill evenly, what expression represents how much Malia owes? (**Lesson 2-1**)

In 23 and 24, explain how to use the Distributive Property to calculate the cost in your head. (Lesson 2-1)

23. 8 footballs at $39.95 each

24. 20 bottles of water at $0.99 each

25. Suppose JI is 5 centimeters more than JM, and IM is 7 centimeters less than twice JM. Recall that JI represents the distance from J to I. (**Lesson 1-2**)

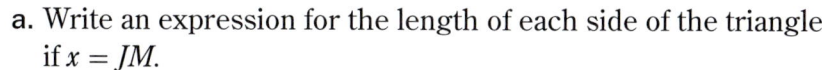

   a. Write an expression for the length of each side of the triangle if $x = JM$.
   b. Write a simplified expression for the perimeter of $\triangle JIM$.

26. a. Describe the pattern below with one variable.
$$7^2 < 7^3$$
$$6.2^2 < 6.2^3$$
$$\left(\tfrac{11}{8}\right)^2 < \left(\tfrac{11}{8}\right)^3$$

   b. Find an integer that is an instance of this pattern.
   c. Find an integer that is a counterexample to the pattern.
   d. Find a noninteger that is a counterexample to the pattern. (**Lesson 1-2**)

Matching In 27–30, match each algebraic expression to its English expression. (**Lesson 1-1**)

27. $2(x + 5)$ a. six less than a number
28. $2x + 5$ b. double the sum of five and a number
29. $x - 6$ c. five more than double a number
30. $6 - x$ d. take away a number from six

31. Suppose an apartment rents for $775 per month. Find the rent for the given time. (**Previous Course**)
   a. 8 months   b. 3 years
   c. 4.5 years   d. y years
   e. m months

In 32 and 33, use the following information. The cost c of painting the four walls of a room is given by $c = \frac{p(2h(\ell + w))}{400}$, where p is the price per gallon of paint and ℓ, w, and h are the length, width, and height of the room in feet. (Previous Course)

32. Find the cost of paint for the four walls of a great room that is 24 feet long, 18 feet, 6 inches wide, and 8 feet high, using paint that costs $29.95 per gallon.

33. At $18.99 per gallon, what is the cost of paint for the four walls of a high-school cafeteria that is 50 feet long, 35 feet wide, and 12 feet high?

EXPLORATION

34. Consider this sequence of sums of increasingly large multiples of x whose coefficients are alternately positive and negative.

Step 1	x
Step 2	$x + -2x$
Step 3	$x + -2x + 3x$
Step 4	$x + -2x + 3x + -4x$
Step 5	$x + -2x + 3x + -4x + 5x$
⋮	

    a. Simplify the five lines that are shown.
    b. What will be the simplified form of the 25th line?
    c. What will be the simplified form of the 100th line?
    d. What is the simplified form of the nth line when n is even?
    e. What is the simplified form of the nth line when n is odd?

In May 2004, the median hourly wage for house painters was $14.55.

Source: Bureau of Labor Statistics

QY ANSWER

$208u^5c^3$

Lesson 2-3

Explaining Number Puzzles

▶ **BIG IDEA** Algebra explains why many number puzzles work and how to invent them.

The number puzzle at the right was sent by e-mail. Try it yourself.

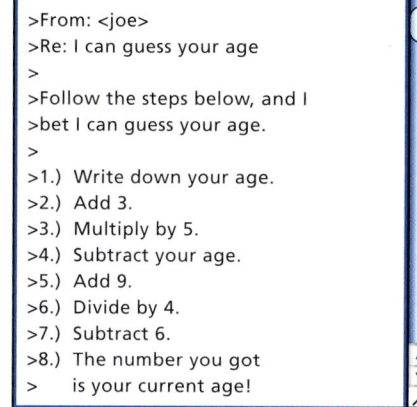

```
>From: <joe>
>Re: I can guess your age
>
>Follow the steps below, and I
>bet I can guess your age.
>
>1.) Write down your age.
>2.) Add 3.
>3.) Multiply by 5.
>4.) Subtract your age.
>5.) Add 9.
>6.) Divide by 4.
>7.) Subtract 6.
>8.) The number you got
>    is your current age!
```

Mental Math

a. In $\triangle ABC$, $m\angle A = 35°$ and $m\angle B = 105°$. What is $m\angle C$?

b. In $\triangle LMN$, $m\angle L = x$ and $m\angle M = 74°$. What is $m\angle N$?

c. In $\triangle WYZ$, $m\angle W = r$ and $m\angle Y = s$. What is $m\angle Z$?

The number puzzle raises some questions. Does it work for other ages? Will it work if you count your age in months? Will it work for your great-grandfather Odell? Will it work for the baby a mother is expecting in six months, who the family fondly refers to as $-\frac{1}{2}$ year old because she is half a year before her birth? Will it work for any age?

Explaining the Puzzle

The steps shown below are for a 16-year-old student.

Step 1 Write your age.	16
Step 2 Add 3.	$16 + 3 = 19$
Step 3 Multiply by 5.	$19(5) = 95$
Step 4 Subtract your age.	$95 - 16 = 79$
Step 5 Add 9.	$79 + 9 = 88$
Step 6 Divide by 4.	$88 \div 4 = 22$
Step 7 Subtract 6.	$22 - 6 = 16$
The final result is the age.	16

Explaining Number Puzzles 79

Using algebra to explain the puzzle on page 79, we let the variable A stand for the age used in the first step. Now we follow the given directions, simplifying as we go by using the Distributive Property.

Step 1 Write your age. $\quad A$

Step 2 Add 3. $\quad A + 3$

Step 3 Multiply by 5. $\quad 5(A + 3) = 5A + 5 \cdot 3 = 5A + 15$

Step 4 Subtract your age. $\quad 5A + 15 - A = 4A + 15$

Step 5 Add 9. $\quad 4A + 15 + 9 = 4A + 24$

Step 6 Divide by 4. $\quad \frac{4A + 24}{4} = \frac{4A}{4} + \frac{24}{4} = A + 6$

Step 7 Subtract 6. $\quad A + 6 - 6 = A$

The algebra shows that any age works. The "trick" is really just an algebraic process. No matter what the original age is, the process will end with the same age from which you started.

The "Seven Is Heaven" Puzzle

In the next puzzle, the result is not the original number, but it is surprising in a different way.

GUIDED

Example 1

Work the puzzle on the right with a few numbers. Then use a variable to create an expression to explain why the puzzle works.

Solution Follow the steps in the puzzle at the right.

Step 1 ___?___

Step 2 ___?___

Step 3 ___?___

Step 4 ___?___

Step 5 ___?___

Step 6 ___?___

Step 7 ___?___

Your answer should be 7.

Seven Is Heaven Puzzle

Step 1	Pick a number.
Step 2	Add 1.
Step 3	Multiply by 2.
Step 4	Multiply by 3.
Step 5	Subtract 4.
Step 6	Add 5.
Step 7	Subtract 6 times your original number.

Your answer should be 7.

We show the process beginning with any number n.

Step 1 Begin with a number.	Any number n
Step 2 Add 1.	$n + 1$
Step 3 Multiply by 2.	$2(n + 1) = 2n + 2$
Step 4 Multiply by 3.	$3(2n + 2) = 6n + 6$
Step 5 Subtract 4.	$6n + 6 - 4 = 6n + 2$
Step 6 Add 5.	$6n + 2 + 5 = 6n + 7$
Step 7 Subtract 6 times your original number.	$6n + 7 - 6n = 7$
Your answer should be 7.	7

Using Algebra to Create Number Puzzles

In Examples 2 and 3, we show how algebra can be used to create a number puzzle. This number puzzle will begin and end with the same number. Begin by choosing a variable to represent the starting number. A new expression is formed by performing an arithmetic operation on the existing expression. After several steps, carefully choose operations to return the expression to the variable.

Example 2

Create a number puzzle that begins and ends with the same number.

Solution We show the process and create one puzzle.

Changing

Begin with n.	n
Add 6.	$n + 6$
Multiply by 4.	$4(n + 6) = 4n + 24$
Divide by 2.	$\frac{4n + 24}{2} = \frac{4n}{2} + \frac{24}{2} = 2n + 12$
Subtract 19.	$2n + 12 - 19 = 2n - 7$

Returning

Add 7.	$2n - 7 + 7 = 2n$
Divide by 2.	$\frac{2n}{2} = n$
The answer equals n.	n

Explaining Number Puzzles

Chapter 2

GUIDED

Example 3
Create a number puzzle so that the answer always equals 6.

Solution

Step 1 Begin with n. n

Step 2 Subtract 4.   ___?___

Step 3 Multiply by 8. $8(n - 4) =$ ___?___

Step 4 Add $8n$.   ___?___ $=$ ___?___

Step 5 Divide by 16.   ___?___ $=$ ___?___

Step 6 Add 8.   ___?___ $=$ ___?___

Step 7 Subtract n.   ___?___ $=$ ___?___

One of algebra's greatest strengths is the use of a variable to create algebraic expressions that can represent *all* the possibilities of a number puzzle, or even a real-life puzzle. This helps to explain mysteries and many other things in mathematics and the world.

Questions

COVERING THE IDEAS

1. Why would you use a variable in a number puzzle?

In 2–5, complete the number puzzle at the beginning of this lesson for the individual.

2. a person who is 50 years old
3. great-grandfather Odell, who is 97 years old
4. a baby expected to be born in half a year
5. a person who is 192 months old

In 6 and 7, complete the "Seven Is Heaven" puzzle on page 80 for a number x in the given range.

6. $250 < x < 500$
7. $0 < x < 1$

8. Will the "Seven Is Heaven" puzzle work for negative numbers? Explain why or why not.

9. Complete the "We're Number One!" puzzle at the right by starting with each number shown.

    a. 7
    b. −2.9
    c. n

We're Number One! Puzzle	
Step 1	Pick a number.
Step 2	Subtract 8.
Step 3	Add 7.
Step 4	Multiply by 6.
Step 5	Add 5.
Step 6	Add 4.
Step 7	Divide by 3.
Step 8	Subtract 2 times your original number.
Step 9	Your answer should be 1.

82 Using Algebra to Explain

10. Complete the "Double Trouble" puzzle at the right by starting with each number shown.
    a. 17
    b. 0.4
    c. n

APPLYING THE MATHEMATICS

11. Create a number puzzle that begins and ends with the same number.

12. The "Seven Is Heaven" puzzle in Example 1 always ended with 7. Create an "Eight Is Great" puzzle that always ends with 8.

In 13 and 14, use the "magic square" at the right. In a magic square, the sums of any row, column, or diagonal are equal. In this magic square the sums are all 15.

13. Add 13 to every number in the magic square.
    a. Is the result still a magic square?
    b. By how much did the sum of the rows, columns, and diagonals change?
    c. Add k to each of the numbers in the magic square. Is the result still a magic square? Explain your answer using algebra.

14. Multiply every number in the magic square by 7.
    a. Is the result still a magic square?
    b. By how much did the sum of the rows, columns, and diagonals change?
    c. If you multiply each number in the magic square by m, will it still be a magic square? Explain your answer using algebra.

15. Refer to the card trick described on page 64. Make a table with 5 rows. In each row, give a number that the spectator might choose and the resulting number of cards in the small pile after Step 3 of the trick.

Double Trouble

Step 1	Pick a number.
Step 2	Subtract 11.
Step 3	Multiply by 3.
Step 4	Add 5 times your original number.
Step 5	Add 1.
Step 6	Divide by 4.
Step 7	Add 8.
Step 8	Your answer should be twice your original number.

6	7	2
1	5	9
8	3	4

REVIEW

In 16–19, determine whether the terms are like or unlike. (Lesson 2-2)

16. $k, \dfrac{k}{6}$
17. $-y, \dfrac{26}{y}$
18. $22n, 22n^2$
19. $5d^2, -d^{-2}$

20. The figure at the right consists of a big rectangle split into three rectangular parts. (Lesson 2-2)
    a. Write the area as length times width.
    b. Write the area as the sum of three areas.

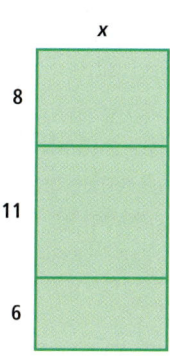

Chapter 2

21. Write each expression as a single fraction. (**Lesson 2-2**)
    a. $\frac{3}{5} + \frac{4}{3}$
    b. $\frac{x}{2} + \frac{2x}{7}$
    c. $\frac{2}{5y} + \frac{4}{5y}$

In 22 and 23, use the Distributive Property to simplify the expression. Check your answer by substituting 5 for the variable in both the original expression and your answer. (**Lesson 2-1**)

22. $4(3(2x + 7))$

23. $n(11(3(n - 4)))$

24. A local coffee shop sells "Coffee Club" cups for $1.99. With the cup, each additional refill of coffee only costs $1.25. Let $n =$ the number of refills purchased. (**Lesson 1-4**)
    a. Make a table for 0, 1, 2, 3, 4, 5, and 6 refills purchased.
    b. Graph your results from Part a.
    c. Write an expression describing the cost of n refills.

25. Write an expression for the amount of money each person has or owes after w weeks. (**Lesson 1-1**)
    a. Eddie is given $100 and spends $4 per week.
    b. Liseta owes $350 on a stereo and is paying it off at $5 per week.

In the United States, it is estimated that as many as 70% of 18–24 year olds drink coffee in the morning.

Source: National Coffee Association of U.S.A., Inc.

In 26–30, evaluate the expression. (Previous Course)

26. $-23 + 41$
27. $-42 - (-87)$
28. $(-8)(12)$
29. $(-13)(-6)$
30. $\frac{-24}{6}$

EXPLORATION

31. The puzzle at the right involves a square array of numbers. Pick a number in the first row, then pick a number from the second row that is in a different column than the first number. Then choose the number in the third row that is not in the same column as the first or second number. One example is 4, −24, and −15.

8	4	−20
−24	−12	60
6	3	−15

    a. Multiply the three numbers together. What value do you get for the product?
    b. Repeat the process. What do you notice?
    c. The numbers in the puzzle above were chosen on purpose. To see why, begin by replacing 3 with a and 4 with b. Then the other seven numbers can be expressed in terms of a and b. This has already been done for some cells in the diagram at the right. Fill in the remaining cells, and explain why the puzzle works.

?	b	−5a
?	−ab	?
2a	a	?

84 Using Algebra to Explain

Lesson 2-4

Opposites

Vocabulary

additive inverse

opposite

▶ **BIG IDEA** The opposite $-k$ of a number k has properties relating it to addition, multiplication, and subtraction: $-k + k = 0$; $-k = -1 \cdot k$; and $a - k = a + -k$.

The real numbers k and $-k$ are opposites. You have dealt with opposites in previous lessons. In this lesson, we look at some of the basic properties of opposites.

Opposites as Additive Inverses

Suppose you walk forward 10 steps and then walk backward 10 steps. We signal these opposite directions by calling the walk forward 10 and the walk backward −10. The result of doing these actions one after the other is to end in the same place from where you started. So $10 + -10 = 0$. When two numbers add to zero, they are called **additive inverses,** or **opposites.** Because they signal opposite actions, another name for additive inverse is opposite. So, if two real numbers x and y are opposites, then their sum is zero ($x + y = 0$). Reversing this, if you know that $x + y = 0$, then x and y must be opposites ($x = -y$ or $y = -x$).

🛑 **QY1**

The numbers k and $-k$ are opposites regardless of the value of k. If $k = 5$, then $-k = -5$. If $k = -10$, then $-k = -(-10) = 10$. *When k is negative, $-k$ is positive.*

What is the opposite of $-k$? One way to denote the opposite of $-k$ is as $-(-k)$. Yet we know that the opposite of $-k$ is k. This tells us that these two expressions are equivalent. We call this the *Opposite of Opposites Property*.

Opposite of Opposites Property

For any real number a, $-(-a) = a$.

Mental Math

a. How many games must a team win to win a best-of-5 playoff series?

b. How many games must a team win to win a best-of-7 playoff series?

c. How many games must a team win to win a best-of-n playoff series?

▶ **QY1**

What is the opposite of each number?

a. 3.5 **b.** $\frac{-17}{8}$

Opposites **85**

Taking the Opposite by Multiplying by –1

One way of changing a number to its opposite is to multiply it by –1. The Distributive Property explains why this works. If the sum of two numbers is 0, then they must be opposites. Add k to $-1 \cdot k$ and see if the result is 0.

$$k + (-1) \cdot k = 1 \cdot k + (-1) \cdot k$$
$$= (1 + (-1))k$$
$$= 0k$$
$$= 0$$

So k and $-1 \cdot k$ are opposites. In symbols, this can be written $-1 \cdot k = -k$. We call this the *Multiplication Property of –1*.

> **Multiplication Property of –1**
>
> For any real number a, $a \cdot -1 = -1 \cdot a = -a$.

The Opposite of Opposites Property and the Multiplication Property of –1 can be used to rewrite and simplify algebraic expressions.

An Example of the Opposite of a Sum

Suppose Marisol has $800 in her savings account. She withdraws a dollars from her account. Deciding that this is not enough, she makes another withdrawal of b dollars. The amount of money left in her savings account can be expressed in several different ways. One way is to think that Marisol withdrew a total of $(a + b)$ dollars. So she has $800 - (a + b)$, or $800 + -(a + b)$ dollars left.

Another way is to think that Marisol withdrew a dollars, then withdrew b dollars. So she has $800 - a - b$, or $800 + -a + -b$ dollars left.

The fact that $-(a + b)$ is the same as $-a - b$ is due in part to the Distributive Property.

$$-(a + b) = -1(a + b) \quad \text{Multiplication Property of -1}$$
$$= -1a + -1b \quad \text{Distributive Property}$$
$$= -a + -b \quad \text{Multiplication Property of -1}$$
$$= -a - b \quad \text{definition of subtraction}$$

The opposite of a sum is the sum of the opposites of its terms.

> **Opposite of a Sum Property**
>
> For all real numbers a and b, $-(a + b) = -a + (-b) = -a - b$.

For example, $-(15y + 3) = -15y - 3$.

The first automated cash dispenser in use in the U.S. was at Chemical Bank, Long Island, New York in 1969.

Source: www.history.com

QY2

▶ **QY2**

Simplify $-(a^2 + 2b)$.

Opposite of a Difference

Suppose the expression in parentheses involves subtraction rather than addition. How can you rewrite its opposite? Again, the Distributive Property can be used.

$-(a - b) = a + -b$ definition of subtraction
$ = -a + -(-b)$ Opposite of a Sum Property
$ = -a + b$ Opposite of Opposites Property

> **Opposite of a Difference Property**
>
> For all real numbers a and b, $-(a - b) = -a + b$.

Some problems involve subtracting an expression with two or more terms. Begin by rewriting the subtraction as adding the opposite.

Example 1

Simplify $4x - (3x + 7)$.

Solution Rewrite the subtraction as adding the opposite.

$4x - (3x + 7) = 4x + -(3x + 7)$ definition of subtraction
$ = 4x + -3x + (-7)$ Opposite of a Sum Property
$ = x + -7$ Add like terms.
$ = x - 7$ definition of subtraction

 QY3

▶ **QY3**

Graph $y = 4x - (3x + 7)$ and $y = x - 7$ in the same window. Do both equations have the same graph?

GUIDED

Example 2

Simplify $(x + 6) - 7(2x - 3)$.

Solution

$(x + 6) + \underline{}\,(2x - 3)$ definition of subtraction
$= x + 6 + \underline{} + \underline{}$ Distributive Property
$= \underline{} + \underline{}$ Add like terms.

Check When you are asked to simplify expressions, you can make a quick check to see if your answer is equivalent to the given expression by substituting a value for the variable. If the given expressions have one variable each, you can graph the expressions or generate a table of values to check that they are equivalent.

Chapter 2

Questions

COVERING THE IDEAS

1. A bottle of apple juice contains 48 fluid ounces. You pour f ounces into a glass and drink it. Then you pour n ounces more into the glass.
   a. Express the amount of juice left in the bottle in two different ways.
   b. Check that the two expressions in Part a are equal by letting $f = 12$ ounces and $n = 5$ ounces.

2. Simplify $-(-(-w))$.

In 3–5, find the opposite of the expression.

3. $-2n$
4. $6p - 8$
5. $-2a^2 + 28a - 15$

6. **Multiple Choice** The opposite of $-x$ is *not*
   A $-x$. B x. C $-(-x)$. D $-(-(-(x)))$

7. **Multiple Choice** Which of the following is equal to $-(P + 7)$?
   A $-P + 7$ B $P + -7$ C $-P + -7$ D $P - 7$

8. **Multiple Choice** Which expression does *not* equal $-(x - y)$?
   A $-x + y$ B $-1x + -1y$ C $-x - (-y)$ D $y - x$

In 9–16, write an equivalent expression without parentheses.

9. $-(x + 15)$
10. $-(4n - 3m)$
11. $x - (x + 2)$
12. $3y - 5(y + 1)$
13. $(3k^4 + 4) - (7k^4 - 9)$
14. $-(5 + k^3) + (k^3 - 18)$
15. $a^2 + b - c - (a^2 - b + c)$
16. $(-4a)(-a)$

Research suggests that nutrients in apples and apple juice improve memory and learning.

Source: www.applejuice.org

APPLYING THE MATHEMATICS

17. Theo and Rafael completed the toothpick activity from Lesson 1-3. The expression Theo wrote for the number of toothpicks in the nth term was $2n + n + 1$, and the expression Rafael wrote was $4n - (n - 1)$.
    a. Graph $y = 2n + n + 1$ and $y = 4n - (n - 1)$ on your calculator. Do the expressions appear to be equivalent?
    b. Use the Opposite of a Difference Property to show the expressions are equivalent.

18. Evaluate each of the following expressions.
    a. $(-1)^6$
    b. $(-1)^8$
    c. $(-1)^7$
    d. $(-1)^9$

19. **True or False** Justify your answer.
    a. $(-5)^3 = -5^3$
    b. $(-5)^4 = -5^4$
    c. $-(5^4) = -5^4$

88 Using Algebra to Explain

20. a. Which powers of –1 are positive?

 b. Which powers of –1 are negative?

 c. If –1 was changed to –3, would this change your answers to Parts a and b? Explain why or why not.

21. The command "about-face" in the military signals a soldier to rotate 180 degrees. Two commands of "about-face" result in the soldier facing in the original direction again. How does the table at the right relate to $(-1)^n$, where n is the number of about-faces?

Number of About-Faces	Facing Direction
1	Reverse
2	Forward
3	Reverse
4	Forward
.	.
.	.
.	.

22. Determine whether the number is positive, negative, or zero.

 a. $(-5)^{10}$
 b. $(-1)(-5)^{10}$
 c. $(-1)^{10}(-5)^{10}$
 d. $(5)^{10}(-5)^{10}$
 e. $[5 + (-5)]^{10}$
 f. $(-1)^{10}(-5)$

23. **Skill Sequence** Evaluate each expression.

 a. $-\frac{1}{2} \cdot -\frac{2}{3}$

 b. $-\frac{1}{2} \cdot -\frac{2}{3} \cdot -\frac{3}{4}$

 c. $-\frac{1}{2} \cdot -\frac{2}{3} \cdot -\frac{3}{4} \cdot -\frac{4}{5}$

 d. $-\frac{1}{2} \cdot -\frac{2}{3} \cdot -\frac{3}{4} \cdot \ldots \cdot -\frac{9}{10}$

REVIEW

24. a. Pick a number and complete the number puzzle below.

 Step 1 Subtract 1 from your number.

 Step 2 Multiply this by 8.

 Step 3 Add 20.

 Step 4 Divide by 4.

 Step 5 Subtract 5.

 Step 6 Divide by 2.

 Step 7 Add 1.

 The result will be your original number.

 b. Let n be your original number. Develop an algebraic expression for each step of the puzzle. (**Lesson 2-3**)

In 25–27, simplify the expression. (Lessons 2-2, 2-1)

25. $3(t + 5) + (4t - 7) + (-7t + 8)$

26. $2(2v - 11w) + 3(v + 7w)$

27. $\frac{7}{y} - \frac{4}{3y}$

In 28 and 29, use the following situation. A golf caddie who works at Pine Oaks Country Club makes $15 per golfer, but he has to pay the country club a $12 equipment fee each day.

28. Write an expression for the amount a golf caddie makes if he caddies for *g* golfers each day. (Lesson 1-1)

29. Write an expression for the amount a golf caddie makes after working 6 days, if he caddies for *g* golfers each day. (Lesson 2-2)

30. Consider the sequence made with dots below. Each term is 1 row and 2 columns larger than the previous term. (Lesson 1-2)

Caddies carry clubs, replace divots, rake sand traps or bunkers, look for lost balls, and clean the player's club after each time it is used.

Source: www.teachingkidbusiness.com

a. Complete the table describing the number of dots for the next three terms.

n	1	2	3	4	5	6
dots	1 · 2 = 2	2 · 4 = 8	3 · 6 = 18	?	?	?

b. Write an expression for the number of dots required to make the *n*th term.

31. a. Find two numbers, each greater than 1, with a product of 728.
    b. Find two numbers, each less than zero, with a product of 72.8.
    c. Find two numbers, each greater than 1, with a product of 7.28.
    (Previous Course)

EXPLORATION

32. The difference of two numbers is subtracted from their sum. What can be said about the answer? Explain how you explored this problem.

QY ANSWERS

1. $-3.5, \frac{17}{8}$

2. $-a^2 - 2b$

3. yes

90 Using Algebra to Explain

Lesson 2-5

Testing Equivalence

▶ **BIG IDEA** You can test whether algebraic expressions are equivalent using substitution, tables, graphs, or properties.

One of the most powerful aspects of algebra is that it offers many different approaches to solving problems. It may seem to you that learning different ways to do the same thing is a waste of time, but it is not. The knowledge of different approaches increases your ability to undertake new and different situations or problems on your own.

In Chapter 1, you began working with one of the most important ideas in algebra: *equivalence*. This idea will be examined and reexamined throughout your study of mathematics. Numbers, graphs, and algebraic properties give us three different approaches for testing equivalence.

In Lesson 1-3, you were asked to write expressions to describe the number of toothpicks used to form patterns, as shown below.

Mental Math

Estimate to the nearest integer.

a. $\frac{4}{5} + 2\frac{1}{10} + 7\frac{7}{8}$

b. $\frac{4}{5} - 2\frac{1}{10} + 7\frac{7}{8}$

c. $\frac{4}{5} - 2\frac{1}{10} - 7\frac{7}{8}$

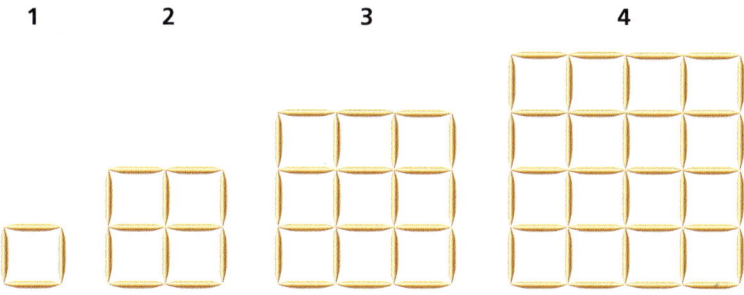

Suppose Yolanda and Suki created the following expressions to represent the number of toothpicks in the *n*th term.

Yolanda: $4n + 2n(n - 1)$
Suki: $n(n + 1 + n + 1)$

How can you tell if these expressions produce the same result for each value of *n*? You have seen a variety of methods used in this chapter. Some methods are more powerful than others.

Testing Equivalence

Chapter 2

Example 1

Test whether $4n + 2n(n - 1)$ and $n(n + 1 + n + 1)$ are equivalent.

Method 1: Use numbers to test for equivalence. Choose $n = 1$. Evaluate each expression.

If $n = 1$, then $4n + 2n(n - 1) = 4 \cdot 1 + 2 \cdot 1(1 - 1) = 4$.

If $n = 1$, then $n(n + 1 + n + 1) = 1(1 + 1 + 1 + 1) = 4$, the number of toothpicks in the 1st term.

There are 12 toothpicks in the second term. Do both expressions have the value 12 when $n = 2$?

If $n = 2$, then $4n + 2n(n - 1) = 4 \cdot 2 + 2 \cdot 2(2 - 1) = 12$.

If $n = 2$, then $n(n + 1 + n + 1) = 2(2 + 1 + 2 + 1) = 12$, which checks.

You can repeat this process to test many numbers.

Technology can allow you to test many numbers at once, as seen in Method 2.

Method 2: Use tables to test equivalence. A graphing calculator can quickly make a table. Substitute x for n when entering the expressions into the calculator.

For Yolanda's expression enter into Y1:
$4x + 2x(x - 1)$.
For Suki's expression enter into Y2:
$x(x + 1 + x + 1)$.
We set a table starting at 1 with increments of 1.

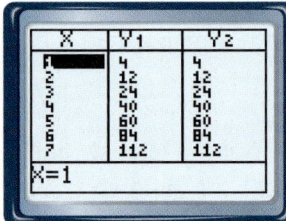

The seven numbers located in the left column are each substituted into the two expressions. The results are automatically calculated and displayed in the second and third columns. In each case, the values of the expressions are equal, and they equal the numbers of toothpicks.

You can scroll up or down the left column of the table to see more values. So for these values of n, the expressions $4n + 2n(n - 1)$ and $n(n + 1 + n + 1)$ have the same value.

 QY

Method 3: Use graphs to test for equivalence.
Use the formulas $Y1 = 4x + 2x(x - 1)$ and $Y2 = x(x + 1 + x + 1)$. The table helps you decide on a viewing window. One window that fits the ordered pairs listed is $-1 \leq x \leq 10, 0 \leq y \leq 120$.

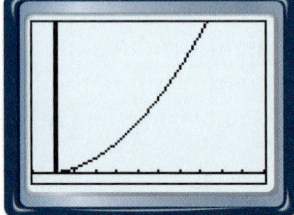

▶ **QY**

Two of the three expressions below are equivalent. Use the table feature on the graphing calculator to determine which are equivalent. Use at least three rows of your table to explain your answer. The screen below shows how the equations should be entered into the calculator.

a. $|x^2 + 1| \cdot (x - 3)$
b. $x^3 - 3x^2 + x - 3$
c. $(x^2 + 1) \cdot |x - 3|$

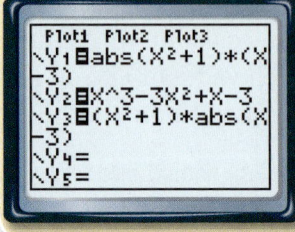

92   Using Algebra to Explain

Lesson 2-5

The graphs of the two expressions seem to be identical. But, even on a calculator, only a limited number of points are actually being graphed. How can you be sure the expressions are always equal? The answer is to use the properties of operations that are true for all real numbers.

Method 4: Use properties to test equivalence.

$4n + 2n(n - 1) = 4n + 2n^2 - 2n$ Expand the expression.
$ = 2n^2 + 2n$ Combine like terms.
$n(n + 1 + n + 1) = n(2n + 2)$ Combine like terms.
$ = 2n^2 + 2n$ Expand the expression.

In each case, the result is $2n^2 + 2n$. Since both expressions are equal to the same third expression, they are equal to each other (by the Transitive Property of Equality). Therefore, for any value of n, $4n + 2n(n - 1) = n(n + 1 + n + 1)$.

Properties of operations are powerful because they can show that a pattern is true for all real numbers. But the other methods are useful too. Testing specific numbers, either by hand or in a table, can help you decide if two expressions *seem* equivalent. These methods can often help you detect a counterexample. Testing numbers is also a good way to catch your own mistakes.

GUIDED

Example 2

A common error that some students make is to think that $4x - x$ is equivalent to 4 for all values of x. Here are three ways to show that these expressions are not equivalent.

Solution

Method 1: Substitute a value for x to show that $4x - x$ is not equal to 4.

Method 2: Graph $Y1 = 4x - x$ and $Y2 = 4$. Are the graphs identical?

Method 3: Create a table of values for $Y1 = 4x - x$ and $Y2 = 4$.

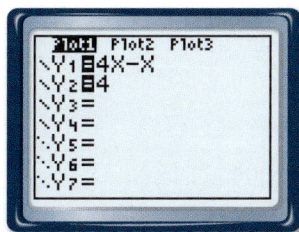

You should find that for almost all values of x, the expressions do not have the same values. Therefore, $4x - x$ is not equivalent to 4. You could also simplify $4x - x$ to $3x$, which clearly is not 4 for every value of x.

Testing Equivalence   93

Chapter 2

GUIDED

Example 3

Are $-x^2$ and $(-x)^2$ equivalent expressions? If so, explain. If not, provide a counterexample.

Solution Pick a value for x. Suppose you pick 6. Then $-x^2 = -6^2$ and $(-x)^2 = (-6)^2$.

For -6^2, follow the order of operations and square 6 *before* taking the opposite.

If $x = 6$, then $-x^2 =$ __?__   Substitute.
$\qquad\qquad\qquad = -(\underline{\ ?\ })(\underline{\ ?\ })$ Evaluate powers first.
$\qquad\qquad\qquad = \underline{\ ?\ }$ Simplify.

For $(-6)^2$, square -6.

If $x = 6$, then $(-x)^2 =$ __?__   Substitute.
$\qquad\qquad\qquad = (\underline{\ ?\ })(\underline{\ ?\ })$ Evaluate powers first.
$\qquad\qquad\qquad = \underline{\ ?\ }$ Simplify.

Because __?__ and __?__ are not equal, the expressions are not equivalent. Therefore, 6 is a counterexample.

Guided Example 3 used just one counterexample to show that $-x^2$ and $(-x)^2$ are not equivalent. Only one is necessary. However, to show that expressions *are* equivalent requires much more. You must show equality for *every* possible value of the variable. This is why using properties that apply to all numbers is so important.

Example 4

Use properties to show that $(2x^2 + 3) - 8(3x^2 + 4)$ is equivalent to $1 - 27x^2 - 30 + 5x^2$.

Solution Simplify each expression and check that the results are the same. Expand and combine like terms.

$(2x^2 + 3) - 8(3x^2 + 4) = 2x^2 + 3 - 24x^2 - 32$ Distributive Property
$\qquad\qquad\qquad\qquad\quad = -22x^2 - 29$ Combine like terms.

Now simplify $1 - 27x^2 - 30 + 5x^2$.

$1 - 27x^2 - 30 + 5x^2 = -22x^2 - 29$ Combine like terms.

Because both $(2x^2 + 3) - 8(3x^2 + 4)$ and $1 - 27x^2 - 30 + 5x^2$ equal $-22x^2 - 29$, we can say that

$(2x^2 + 3) - 8(3x^2 + 4) = 1 - 27x^2 - 30 + 5x^2$ by the Transitive Property of Equality.

Using Algebra to Explain

Questions

COVERING THE IDEAS

1. Refer to the sequence of toothpicks shown below and the table at the right.

   1   2   3

Term Number	Number of Toothpicks
1	6
2	11
3	16

   a. How many toothpicks would be used to make the 4th term?
   b. Dion and Ellis wrote the expressions below to give the number of toothpicks used to make the nth term. Test to see if these two expressions are equivalent for the values $n = 4, 5$, and 6.
   Dion: $1 + 3n + 2n$ Ellis: $6n - (n - 1)$
   c. Use graphs with $0 \leq n \leq 10$ to test whether the two expressions in Part b are equivalent.
   d. Simplify each expression in Part b to test whether they are equivalent.

In 2–4, test the two given expressions for equivalence by using a table or graph, or by simplifying the expressions.

2. $4n - 15$ and $4(n - 4) - 1$
3. $3x^2 + 6x(x + 2)$ and $3x^2 + 6x^2 + 2$
4. $(5 + x)^2$ and $25 + 10x + x^2$
5. Is $3x^2$ equivalent to $(3x)^2$? If so, explain. If not, provide a counterexample.

APPLYING THE MATHEMATICS

6. Consider the expressions $x \cdot x$ and $2x$.
   a. Copy and complete the table of values at the right.
   b. Give two values of x for which $x \cdot x = 2x$.
   c. Give two values of x for which $x \cdot x$ does *not* equal $2x$.
   d. Graph $y = x \cdot x$ and $y = 2x$. Circle the points that correspond to your answer from Part b.

x	$x \cdot x$	$2x$
−3	?	?
−2	?	?
−1	?	?
0	?	?
1	?	?
2	?	?
3	?	?

7. The perimeter of the square below can be written as $2x + 3 + 2x + 3 + 2x + 3 + 2x + 3$, or $4(2x + 3)$. Verify the two expressions are equivalent by using a table or graph, or simplifying each expression.

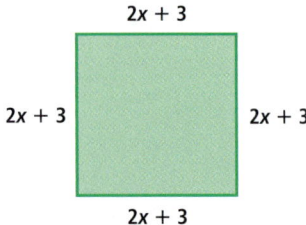

8. Write two equivalent expressions for the perimeter of the regular pentagon at the right.

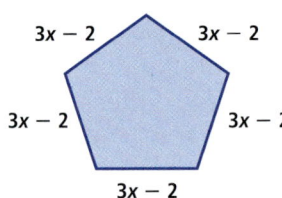

9. Manuel found the area of the "H" shape below to be $20(3x) - 2(8x)$, while Lina got $20x + 4x + 20x$. Are they equivalent?

    a. Use a table to find your answer.
    b. Use algebraic expressions to answer the question.
    c. Let $x = 5$. Find the area of the shape.

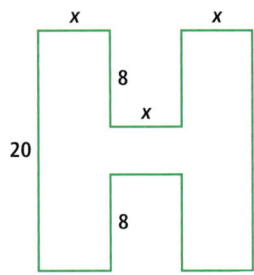

REVIEW

In 10–12, simplify the expression. (Lessons 2-4, 2-1)

10. $8b - 5(7b + 3)$

11. $-1.4(2v - 4) + 3.6v$

12. $\dfrac{12m + 2}{3} - \dfrac{6m - 1}{3}$

13. Determine whether each expression is equivalent to $-(18 - 5x)$. (Lesson 2-4)

    a. $-18 - (-5x)$ b. $-9(2 - 5x)$
    c. $-18 - 5x$ d. $-18 + 5x$

Using Algebra to Explain

14. a. Pick any number and do the following number puzzle.

   Step 1 Add 23 to your number.
   Step 2 Double the sum.
   Step 3 Subtract twice your original number.
   Step 4 Add 3.
   Step 5 Subtract 42.
   You will be left with 7.

   b. Let n be your original number. Develop an algebraic expression for each step of the puzzle. (**Lesson 2-3**)

15. **Skill Sequence** Simplify each expression. (**Lesson 2-2**)

   a. $p \cdot p \cdot p$
   b. $p + p + p$
   c. $2p \cdot 2p \cdot 2p$
   d. $2p + 2p + 2p$

16. Use the Distributive Property to find $5 \cdot 999{,}999$ in your head. (**Lesson 2-1**)

In 17 and 18, give a counterexample to show that the two expressions are *not* equivalent. (**Lesson 1-3**)

17. $6 + m$ and $2m - 3(m - 2)$

18. $\frac{y}{2} + \frac{3}{2}$ and $\frac{3y}{4}$

EXPLORATION

19. Two expressions used to calculate the area of the regular octagon below are $4ba$ and $\frac{1}{2}pa$, where p is the perimeter of the octagon. Explain how b and p are related and use this to demonstrate algebraically that the two expressions are equivalent.

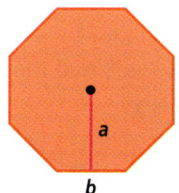

QY ANSWER

Assume
$Y1 = |x^2 + 1| \cdot (x - 3)$,
$Y2 = x^3 - 3x^2 + x - 3$,
and
$Y3 = (x^2 + 1) \cdot |x - 3|$.
From the table it appears that Y1 and Y2 are equivalent. We also see that neither Y1 nor Y2 is equivalent to Y3.

X	Y1	Y2	Y3
−1	−8	−8	8
0	−3	−3	3
1	−4	−4	4

Chapter 2

Lesson 2-6
Equivalent Expressions with Technology

> **BIG IDEA** A computer algebra system (CAS) uses properties of operations to create equivalent expressions.

In this lesson, you will use a computer algebra system (CAS) to test whether expressions are equivalent. Then you will use it to explore a variety of equivalent forms for a single expression. A CAS does algebra just like a calculator does arithmetic. Among the many things that a CAS can do is simplify expressions and solve equations. It has been programmed to use the same algebraic properties that you have been learning. The first computer software with CAS capabilities was created in 1968 at the Massachusetts Institute of Technology (MIT). Until then, mathematicians had to do even the most complicated algebraic calculations by hand.

Mental Math

a. 35 raffle tickets cost $10. What is the price per ticket?

b. 60 raffle tickets cost $20. What is the price per ticket?

c. n raffle tickets cost $15. What is the price per ticket?

d. y raffle tickets cost x dollars. What is the price per ticket?

Equivalent Formulas in Geometry

In geometry, it is not unusual to see different area formulas for the same type of figure. If the area is found using different formulas, will the results be the same? Where do different formulas come from?

Example 1
The picture below shows the area of a deck surrounding a rectangular swimming pool. The length of the entire region (pool and deck) is L and the width of the entire region is W. The distance across the deck is x.

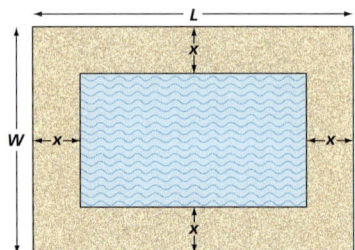

a. Find a formula for the area of the deck by splitting up the deck into 4 rectangles.
b. Split up the deck in another way to find its area.

98 Using Algebra to Explain

Solutions Sonia and Roxy both found formulas for the area of the deck, but they are different.

a. Sonia's Method Sonia broke the deck region into four smaller rectangles and then added them. The length of Areas 2 and 4 can be found by taking the total length L and subtracting two lengths of x, or $L - 2x$. She found the areas of Areas 1 through 4 and added them together.

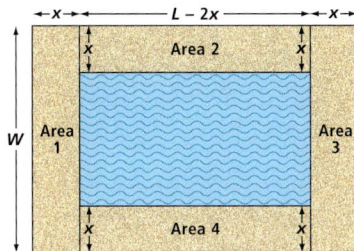

Area 1 = $W \cdot x = Wx$

Area 2 = $x \cdot (L - 2x) = x(L - 2x)$

Area 3 = $W \cdot x = Wx$

Area 4 = $x \cdot (L - 2x) = x(L - 2x)$

Sonia's Deck Area = $Wx + x(L - 2x) + Wx + x(L - 2x)$

b. Roxy's Method Roxy saw the area of the deck as the area of the pool and deck minus the area of the pool.

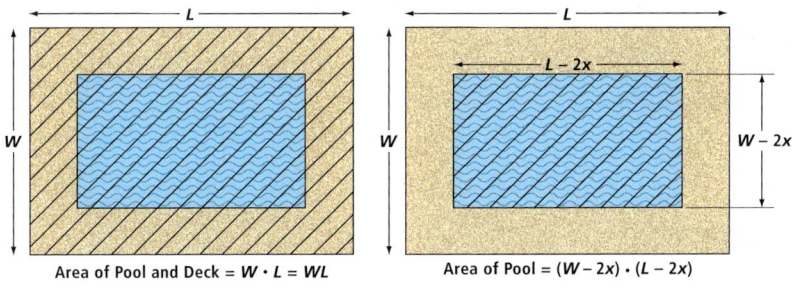

Area of Pool and Deck = $W \cdot L = WL$ Area of Pool = $(W - 2x) \cdot (L - 2x)$

Roxy's Deck Area = $WL - (W - 2x) \cdot (L - 2x)$

Sonia's and Roxy's methods lead to two expressions that appear to be different. But are they? A CAS can answer this question.

Is $Wx + x(L - 2x) + Wx + x(L - 2x)$ equivalent to
$WL - (W - 2x) \cdot (L - 2x)$?

The approach that we use is one you saw in the last lesson. We work with the two expressions to see if the result is the same third expression.

Equivalent Expressions with Technology

Chapter 2

Using a CAS to Test for Equivalence

Just as with any new technology, you need to learn the commands that your CAS uses. In this first activity, you will use the Distributive Property to expand multiplication expressions and to combine like terms. Most CAS make both of these changes to an expression by using the expand command. The expression that is to be changed appears in parentheses. For example, the command expand(3(2x+50)+11) does two operations. It first does the multiplication $3(2x + 50)$, which produces $6x + 150$. Then it adds $6x + 150 + 11$ to get $6x + 161$.

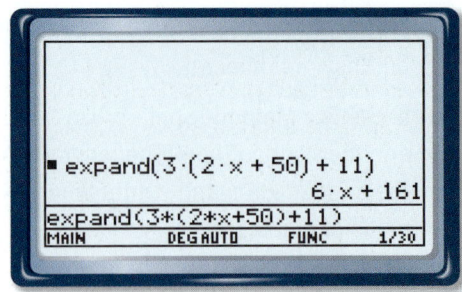

Activity 1

Use a CAS to expand the two expressions from Example 1 to determine whether they are equivalent. Caution: You may need to enter W • x into your CAS so that the CAS will recognize W and x as separate variables.

Step 1 Expand $Wx + x(L - 2x) + Wx + x(L - 2x)$, as shown below.

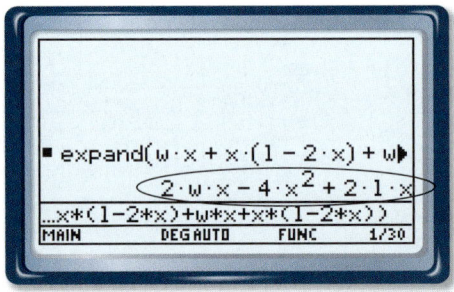

Step 2 Expand $WL - (W - 2x) \cdot (L - 2x)$, as shown below.

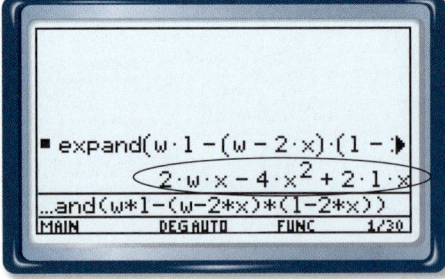

The CAS applied the properties of algebra and found that $Wx + x(L - 2x) + Wx + x(L - 2x)$ and $WL - (W - 2x) \cdot (L - 2x)$ are both equal to $2Wx - 4x^2 + 2Lx$. Therefore, they are equivalent to each other. Both Sonia's and Roxy's approaches will correctly find the area of the deck.

Lesson 2-6

Activity 2

Find three expressions equivalent to $2a^2 + 4b$. In each case, use a CAS to verify the equivalence.

To form equivalent expressions, use the properties you have learned, but in reverse. Instead of trying to make the expression $2a^2 + 4b$ simpler, you need to make it more complicated. There are many approaches to take.

Expression 1 To find a first expression, notice that 2 and 4 are both divisible by 2. Therefore, one possibility is to "undo" the Distributive Property: $2a^2 + 4b = 2(a^2 + 2b)$.
Is $2(a^2 + 2b)$ an equivalent expression? Check using the expand command with a CAS. With some problems, you may find using a CAS is slower than doing it yourself. But a good check is one that uses a method different from the one originally used to get the answer.

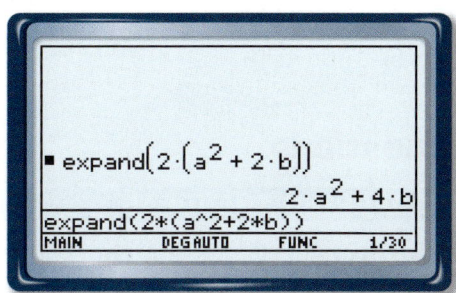

Using expand gives the expression $2a^2 + 4b$.

Thus, $2(a^2 + 2b)$ is equivalent to $2a^2 + 4b$.

Expression 2 To find a second expression, apply the Identity Property of Multiplication. Multiplying a number by 1 does not change its value. The number 1 can take several forms, including $\frac{3y}{3y}$. Therefore, multiplying an expression by $\frac{3y}{3y}$ will result in an equivalent form. Is $\frac{3y}{3y} \cdot 2a^2 + \frac{3y}{3y} \cdot 4b$ an equivalent expression? Using the expand command gives $2a^3 + 4b$.

Thus, $\frac{3y}{3y} \cdot 2a^2 + \frac{3y}{3y} \cdot 4b$ is equivalent to $2a^2 + 4b$.

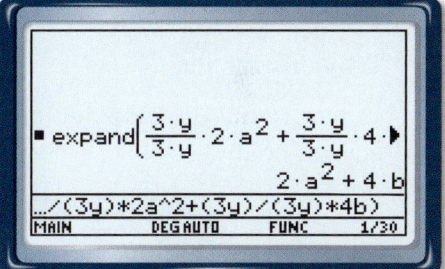

 **QY1**

▶ **QY1**

Multiply $2a^2 + 4b$ by a different form of 1 to create another equivalent expression.

Expression 3 To find a third expression, apply the Identity Property of Addition and then rearrange the terms. The idea behind this method is to add a "clever" form of zero to the expression. This does not change its value, but produces an expression that looks very different from the original. Here we use $3a^2 + (-3a^2)$.

$2a^2 + 4b + 3a^2 + -3a^2$ Add $3a^2 + -3a^2$, which is 0.
$= (2a^2 + 3a^2) + 4b - 3a^2$ Group two of the like terms.
$= 5a^2 + 4b - 3a^2$ Add the first two like terms.

Is the new expression $5a^2 + 4b - 3a^2$ equivalent to $2a^2 + 4b$? Check using the expand command with a CAS.

Using the expand command gives the answer $2a^2 + 4b$.
Thus $5a^2 + 4b - 3a^2$ is equivalent to $2a^2 + 4b$.

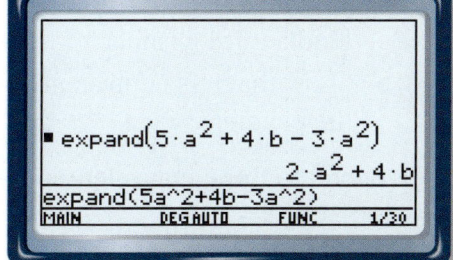

 **QY2**

▶ **QY2**

Create a new expression equivalent to $2a^2 + 4b$ by adding another form of zero.

Equivalent Expressions with Technology **101**

Activity 3

Find at least four equivalent expressions for each of the following expressions. Try to use more than one property on each expression. See how complicated-looking you can make your expression while maintaining equivalence. Use a CAS to verify that your expressions are equivalent.

1. $4(4a^2 - b^2)$
2. $-9m^2 + 12m - 8p$
3. $\frac{1}{2}m \cdot n + \frac{1}{4}$
4. $\frac{6y}{5x}$
5. $100p^2r^2$
6. $2\ell + 2w$

Questions

COVERING THE IDEAS

1. Write an expression equivalent to $5x(7x - 9y)$ by using the Distributive Property.
2. Are $6(x^3 - 5y) + 17x^2 + 12y$ and $6x^3 + 17x^2 - 18y$ equivalent? Why or why not?
3. Explain what the CAS did to simplify the expression on the screen below.

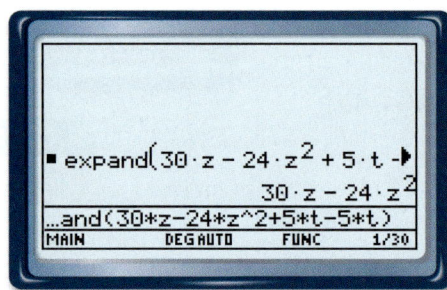

APPLYING THE MATHEMATICS

4. a. Find a third formula for the total area of the deck in Example 1.
   b. Show that it is equivalent to one of the other two formulas in the example.

In 5–8, create three equivalent expressions.

5. $9k^2 - 3k$
6. $13m^2 + 8m$
7. $-24y$
8. $16x^4 + 12xy + 20y$

9. Write a process you could use to convert the expression $5x - 7y$ into the equivalent expression $5(x - 2y) + 3y$.

10. Write an expression equivalent to $34wn + w^2$ using each property.
    a. Commutative Property of Addition
    b. Commutative Property of Multiplication

In **11** and **12**, test if the two expressions are equivalent by using a table. Then check your results by simplifying each expression.

11. $3n - 15, 3(n - 4) - 3$

12. $-2(a - 5), -2a - 5$

REVIEW

13. a. In your own words, state the Multiplication Property of –1.
    b. Give an example of this property. **(Lesson 2-4)**

14. **Multiple Choice** Which of the following must be negative? **(Lesson 2-4)**

    A $-(-z)$ B $-(-6)$ C $-x$ D $-(-(-\frac{2}{5}))$

15. Take a piece of paper. Fold it in half. The paper now has a thickness twice that of the original piece of paper. Fold the folded paper in half. The paper now has a thickness four times the original piece of paper. **(Lesson 1-2)**

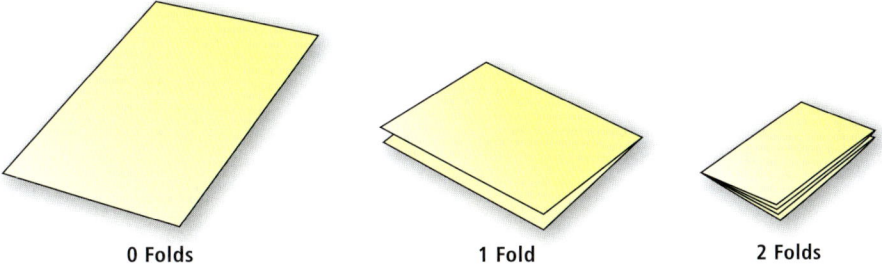

0 Folds   1 Fold   2 Folds

a. Complete the table for the thickness of a piece of paper that is folded 3 and 4 times.

Number of Folds	Thickness of Paper
0	1
1	1 · 2 = 2
2	1 · 2 · 2 = 4
3	?
4	?

b. If a piece of paper is folded six times, what is the thickness of the folded paper?

c. Write an expression to describe the thickness of paper for n folds.

Equivalent Expressions with Technology **103**

In 16 and 17, let $L =$ the length of a segment. Write an expression for the following. (**Lesson 1-1, Previous Course**)

16. one quarter the length
17. five and one half times the length

18. **Skill Sequence** Write each as a decimal. (**Previous Course**)
    a. 1 divided by 5
    b. 1 divided by 0.5
    c. 1 divided by 0.05
    d. 1 divided by 0.00005

19. Convert 0.325823224 to the nearest percent. (**Previous Course**)

EXPLORATION

20. Use a CAS to determine which of these expressions are equivalent.

    A $x^3 + y^3$
    B $(x + y)^3$
    C $(x - y)^3$
    D $x^3 - y^3$
    E $(x + y)(x^2 - xy + y^2)$
    F $(x - y)(x^2 + xy + y^2)$
    G $(x + y)(x + y)(x + y)$
    H $(x - y)(x - y)(x - y)$

QY ANSWERS

1. Answers vary. Sample answer: $\dfrac{2a^3b + 4ab^2}{ab}$

2. $2a^2 + 4b + 5c - 5c$

Lesson 2-7

Explaining Addition and Subtraction Related Facts

▶ **BIG IDEA** The Addition Property of Equality explains how addition and subtraction facts are related and helps solve equations of the form $a + x = b$.

A diagram called a *fact triangle* is shown at the right. Any pair of numbers in the triangle can be combined with addition or subtraction to produce the third number. The numbers 6, 8, and 14 produce the four related number facts listed below the triangle. You know that the first two facts are equivalent because $6 + 8 = 8 + 6$ by the Commutative Property of Addition.

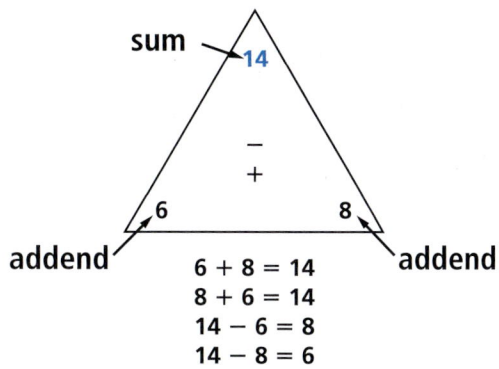

$6 + 8 = 14$
$8 + 6 = 14$
$14 - 6 = 8$
$14 - 8 = 6$

Mental Math

a. 1 km $\approx \frac{3}{5}$ mile, so 1 mile $\approx$ ___?___ km.

b. 1 in. $\approx$ 2.5 cm, so 1 cm $\approx$ ___?___ in.

Fact triangles can be used to show addition/subtraction related facts with *any* kinds of numbers, including fractions and negative numbers. Here are two examples below. The fact triangle on the left uses fractions and the fact triangle on the right uses negative numbers. Each fact triangle has two addition facts and two subtraction facts.

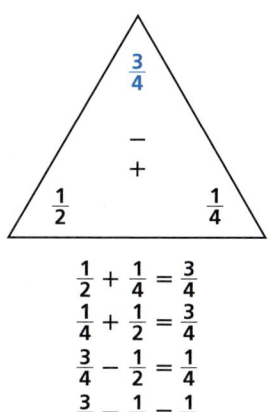

$\frac{1}{2} + \frac{1}{4} = \frac{3}{4}$
$\frac{1}{4} + \frac{1}{2} = \frac{3}{4}$
$\frac{3}{4} - \frac{1}{2} = \frac{1}{4}$
$\frac{3}{4} - \frac{1}{4} = \frac{1}{2}$

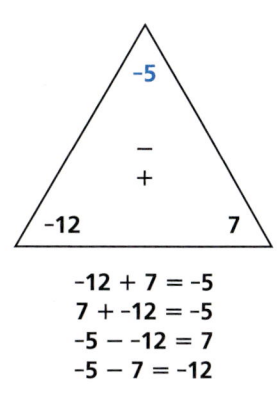

$-12 + 7 = -5$
$7 + -12 = -5$
$-5 - -12 = 7$
$-5 - 7 = -12$

STOP QY1

▶ **QY1**

a. What number goes in the empty corner of the fact triangle below?

b. Write the four related facts.

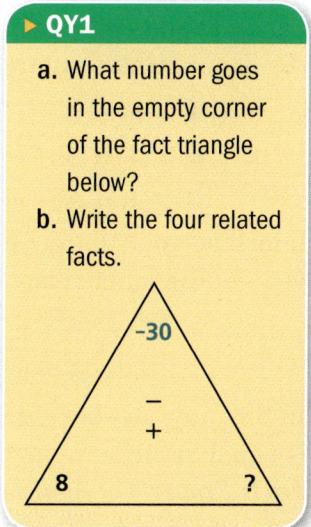

Chapter 2

The Addition Property of Equality

You know from arithmetic that if you start with equal quantities and add the same amount to each, the resulting quantities are still equal. For example, we know that $4 + 3 = 7$.

Adding 5 to each side gives $4 + 3 + 5 = 7 + 5$.

$$12 = 12$$

Note that the value of each side changes from 7 in the first equation to 12 in the second equation. But in each case, the two sides of the equation are equal. The idea that adding a number to both sides produces another true equation is called the *Addition Property of Equality*. It is a basic property of addition.

> **Addition Property of Equality**
>
> For all real numbers a, b, and c, if $a = b$, then $a + c = b + c$.

By the Definition of Subtraction, $a - c = a + -c$ for all real numbers. Every subtraction can be converted to an addition. So if you wanted to subtract a number, say 40, from both sides of an equation, you could add –40 instead. For this reason, the Addition Property of Equality means also that there is a *Subtraction Property of Equality*.

> **Subtraction Property of Equality**
>
> For all real numbers a, b, and c, if $a = b$, then $a - c = b - c$.

The Related Facts Property of Addition and Subtraction

These properties also explain why related facts work. Write down a general addition fact. The sum of two numbers is a third number. The result is a related subtraction fact.

$$h + m = S$$
$$h + m - m = S - m \quad \text{Subtract } m \text{ from both sides.}$$
$$h = S - m$$

If instead you subtract h from both sides of the original addition fact, you get the other related subtraction fact.

$$h + m - h = S - h$$
$$m = S - h$$

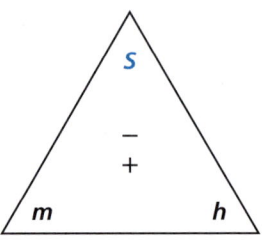

In this way, algebra explains why fact triangles work. Either addend is equal to the sum minus the other addend. We call this the *Related Facts Property of Addition and Subtraction*.

Lesson 2-7

Related Facts Property of Addition and Subtraction

For all real numbers a, b, and c,
if $a + b = c$, then $b + a = c$,
$c - b = a$, and $c - a = b$.

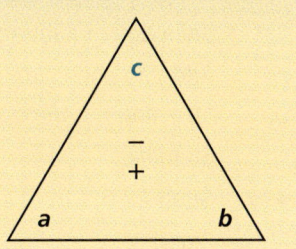

🛑 QY2

> ▶ QY2
>
> Write the two related subtraction facts for $5a + x = k$.

Example 1

Solve the equation $x + -8 = -30$
a. using a fact triangle (the Related Facts Property).
b. using the Subtraction Property of Equality.

Solutions

a. Draw a fact triangle. Place x and -8 at the lower corners. The sum at the top is -30, as shown at the right.

   The triangle shows that $x = -30 - (-8)$ is a related fact.
   $x = -30 - -8 = -30 + 8 = -22$

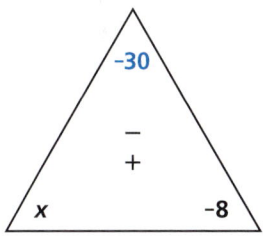

b. Subtract -8 from both sides.
   $$x + -8 = -30$$
   $$x + -8 - -8 = -30 - -8$$
   $$x + -8 + 8 = -30 + 8$$
   $$x = -22$$

Example 2

Solve the equation $-3.4 - y = 6.1$
a. using a fact triangle.   b. using properties.

Solutions

a. Because y is subtracted from -3.4, it must be that -3.4 is the sum. So put -3.4 in the upper corner and y in one of the lower corners. Put 6.1 in the third corner.
   $y = -3.4 - 6.1$
   $y = -3.4 + -6.1$
   $y = -9.5$

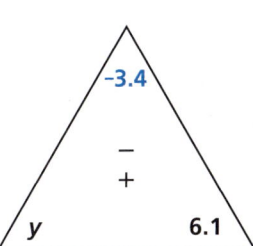

(continued on next page)

Explaining Addition and Subtraction Related Facts

b. $-3.4 - y = 6.1$ Write the equation.
    $-3.4 + -y = 6.1$ definition of subtraction
    $-3.4 + -y + y = 6.1 + y$ Add y to each side since we are looking for y and not the opposite of y.
    $-3.4 = 6.1 + y$ Simplify.
    $-3.4 - 6.1 = 6.1 - 6.1 + y$ Subtract 6.1 from each side.
    $-9.5 = y$ Simplify.

Check Substitute -9.5 for y in the original equation.
Does $-3.4 - -9.5 = 6.1$?
   $-3.4 + 9.5 = 6.1$
   $6.1 = 6.1$ Yes.

Special Numbers for Addition

Important properties can be seen from a fact triangle in which one of the numbers is zero. Two cases are possible.

Case 1: Zero is one of the addends.   **Case 2:** Zero is the sum.

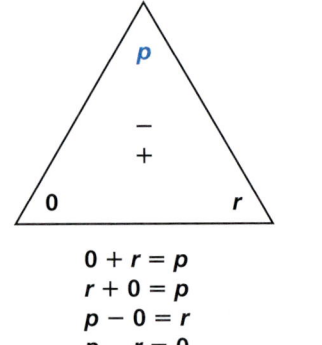
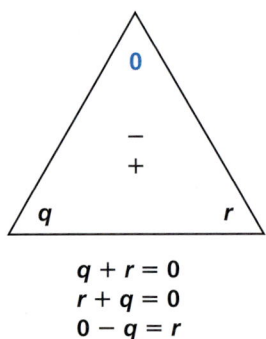

$0 + r = p$ $q + r = 0$
$r + 0 = p$ $r + q = 0$
$p - 0 = r$ $0 - q = r$
$p - r = 0$ $0 - r = q$

When 0 is an addend (Case 1), one of the related facts is $p - 0 = r$. This means that $p = r$. So the first related fact $0 + r = p$ can be rewritten as $0 + p = p$ or as $0 + r = r$. Adding 0 to a number keeps the *identity* of that number. So 0 is called the *additive identity*.

> **Additive Identity Property**
>
> For any real number a, $a + 0 = 0 + a = a$.

When 0 is the sum (Case 2), a related subtraction fact is $0 - q = r$.
$0 + (-q) = r$ definition of subtraction
$-q = r$ Additive Identity Property

108   Using Algebra to Explain

Since $-q = r$, $q + r = 0$ becomes $q + (-q) = 0$. As you know from Lesson 2-4, q and $-q$ are *additive inverses*, or *opposites*. When you add two inverses, the sum is 0, the additive identity. Every number, including 0, has exactly one additive inverse.

> **Additive Inverse Property**
>
> For any real number a, $a + -a = -a + a = 0$.

Questions

COVERING THE IDEAS

1. What are the related facts for the fact triangle shown at the right?

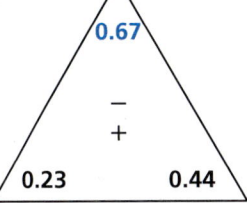

In 2–4, make a fact triangle to fit the equation. Then write the other three related facts.

2. $82 + -5 = 77$
3. $-\frac{1}{8} - 3\frac{7}{8} = -4$
4. $x + 5x = 6x$

5. Solve the equation $y + (-6) = 14$
    a. using a fact triangle (the Related Facts Property).
    b. using the Addition Property of Equality.

6. a. Write the other three related facts of $7 - b = -8$.
   b. What is the value of b?

7. **Fill in the Blank** $0 + -10 = \underline{}$

8. Why is zero called the additive identity?

9. Give an example of two numbers that are additive inverses.

10. What is another name for an additive inverse?

11. What is the additive inverse of $-x$?

APPLYING THE MATHEMATICS

12. Use $-7 + 7 = 0$ to describe a real situation.

13. Make a fact triangle where the sum is
    a. $\frac{5}{9}$.
    b. $2x + 9$.

14. Write a note to a friend explaining how to make a fact triangle for $x + 10 = -19$.

15. Show all possible fact triangles that can be made where two of the numbers are
    a. 3 and -3.
    b. n and $-n$.

Chapter 2

16. Consider this expression:
$$-0.3 + 1.7 - 14.2 - 1.7 + 0.3 - 2.8 + 14.2$$
Use the Associative and Commutative Properties of Addition to first group the additive inverses together. Then evaluate the expression.

17. At 11:00 A.M., the temperature was 15°F. By the time the football game started at 7:00 P.M., the temperature had dropped to −4°F. By how much did the temperature drop?

In the United States, there were a total of 1,071,775 high school football players during the 2005–2006 season.

Source: National Federation of State High School Associations

REVIEW

18. The figure below shows a rectangle with width $40 + a$ and height b. Find two different expressions that describe the area of the figure. Verify that the expressions are equivalent using properties. (Lessons 2-5, 2-2, 2-1)

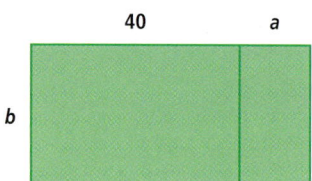

19. A farmer owns a piece of land that is 350 yards wide by 200 yards long, as shown below. She uses one part to harvest wheat and another to harvest corn. She does not use the brown part. Let W = the area of the wheat field and C = the area of the corn field. The farmer decides to sell the wheat and corn fields. Write an expression for the area of the land she will own after the sale. (Lesson 2-4)

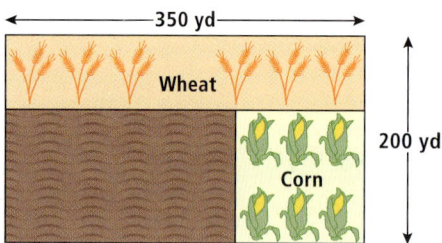

20. **Skill Sequence** Simplify without using a calculator. (Lesson 2-4, Previous Course)

  a. $23.7 + -23.7$ b. $-(-23.7 + 23.7)$ c. $-23.7(23.7 + -23.7)$

In 21 and 22, use the Distributive Property to write an equivalent expression. (Lessons 2-2, 2-1)

21. $-3p + 4p - 8$

22. $-2(3 - 4d)$

23. Two times a number is increased by 5. The resulting quantity is tripled. If m is the original number, write the final result without parentheses. (**Lesson 2-1**)

24. **Multiple Choice** Which formula was used to create the table at the right? (**Lessons 1-3, 1-1**)

    A $y = 4x$ **B** $y = x + 2$
    C $y = 2x + 2$ **D** $y = 2x + 1$

x	y
0	2
1	4
2	6
3	8
4	10

25. Evaluate the following expressions. (**Lesson 1-1**)
    a. $n^3 - \frac{(-1)^n}{2+n}$ for $n = -5$
    b. $a\left(a + \frac{1}{a}\right)$ for $a = 2$

EXPLORATION

26. A tetrahedron is a 3-dimensional solid whose four faces are equilateral triangles. A net to construct a tetrahedron is given below. A set of related facts has been formed within each triangle and at each vertex of the tetrahedron. Create a tetrahedron of your own with these properties. Describe the method you used to create it.

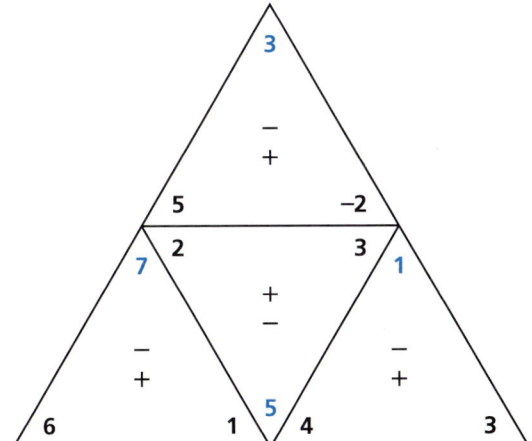

QY ANSWERS

1a. -38

1b. $8 + -38 = -30$,
$-38 + 8 = -30$,
$-30 - -38 = 8$,
$-30 - 8 = -38$

2. $k - x = 5a$,
$k - 5a = x$

Chapter 2

Lesson 2-8
Explaining Multiplication and Division Related Facts

▶ **BIG IDEA** The Multiplication Property of Equality explains how multiplication and division facts are related and helps to solve equations of the form $ax = b$.

Related Facts for Multiplication

Fact triangles can also be used to represent related facts in multiplication and division. In a multiplication fact triangle, the number at the top is the product of the other two.

Mental Math

Calculate.
a. $2 \cdot {-5} - 10 + 60$
b. $2 \cdot {-5} - (10 + 60)$
c. $2 \cdot ({-5} - 10 + 60)$

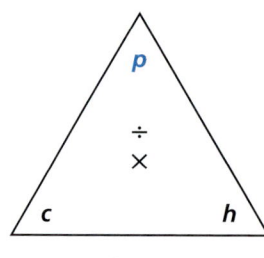

$6 \cdot 5 = 30$
$5 \cdot 6 = 30$
$30 \div 6 = 5$
$30 \div 5 = 6$

$c \cdot h = p$
$h \cdot c = p$
$p \div c = h$
$p \div h = c$

The related facts that can be obtained from these triangles are shown above. As with addition, the first two facts in each list are equivalent by the Commutative Property of Multiplication. Again, as with addition, the fact triangles work for any nonzero real numbers. Here are some other fact triangles and related facts with positive and negative numbers and with fractions.

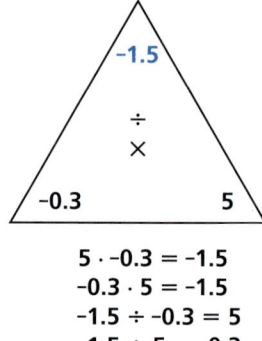

$5 \cdot {-0.3} = -1.5$
${-0.3} \cdot 5 = -1.5$
$-1.5 \div {-0.3} = 5$
$-1.5 \div 5 = -0.3$

$18 \cdot \frac{1}{2} = 9$
$\frac{1}{2} \cdot 18 = 9$
$9 \div \frac{1}{2} = 18$
$9 \div 18 = \frac{1}{2}$

112   Using Algebra to Explain

Just as you can add the same number to equal quantities and get equal quantities as a result, you can multiply (or divide) equal quantities by the same number and get equal quantities. For example, we know that $-0.3 \cdot 5 = -1.5$.

$-0.3 \cdot 5 = -1.5$

$10 \cdot -0.3 \cdot 5 = 10 \cdot -1.5$ Multiply each side by 10.

$\qquad -15 = -15$ Simplify.

This property is called the *Multiplication Property of Equality*.

> **Multiplication Property of Equality**
>
> For all real numbers a, b, and c, if $a = b$, then $ca = cb$.

Because $x \div c = x \cdot \frac{1}{c}$ for all values of x and c (provided c is not zero), every division can be converted to multiplication. So to divide both sides of an equation by c, you can multiply both sides by $\frac{1}{c}$. For this reason, the Multiplication Property of Equality means that there is also a *Division Property of Equality*.

> **Division Property of Equality**
>
> For all real numbers a, b, and all real nonzero numbers c, if $a = b$, then $\frac{a}{c} = \frac{b}{c}$.

The Related Facts Property of Multiplication and Division

The Multiplication Property of Equality also explains why related facts work. Write down a multiplication fact, as shown below.

$$8 \cdot 45 = 360$$

Now divide each side by 8. Do the computation on the left side only.

$$\frac{8 \cdot 45}{8} =$$

$$45 = \frac{360}{8}$$

The result is a related division fact. You can divide each side of the original fact by 45 to find the other related division fact.

$$8 = \frac{360}{45}$$

Related multiplication and division facts cannot be found if one of the numbers being multiplied is 0 because division of 0 is undefined. However, related facts can be found with nonzero numbers. We call this the *Related Facts Property of Multiplication and Division*.

Chapter 2

> **Related Facts Property of Multiplication and Division**
>
> For all nonzero real numbers a, b, and c, if $ab = c$, then $ba = c$, $\frac{c}{b} = a$, and $\frac{c}{a} = b$.

Multiplication equations can be solved by using related facts or by performing the same operation on each side of the equation.

Example 1
Solve $16x = 192$.

Solution 1 Use a fact triangle.

The fact triangle shows $16x = 192$.

$$16x = 192$$
$$192 \div 16 = x$$
$$12 = x$$

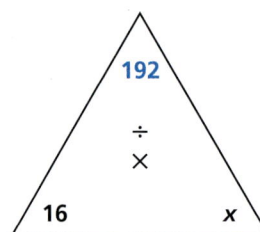

Solution 2 Divide each side by 16.

$$16x = 192$$
$$\frac{16x}{16} = \frac{192}{16}$$
$$x = 12$$

Solution 3 Multiply each side by $\frac{1}{16}$.

$$16x = 192$$
$$\frac{1}{16} \cdot 16x = \frac{1}{16} \cdot 192$$
$$x = 12$$

 QY1

> ▶ **QY1**
>
> Solve $\frac{3}{5}x = 60$ by using
> a. a fact triangle.
> b. the Multiplication Property of Equality.

The Role of Zero in Multiplication and Division

The operations of addition and multiplication often behave in similar ways. Both have commutative and associative properties. In both, fact triangles illustrate four related number facts. But there is one special case that arises for multiplication which has no parallel in addition. Zero is special in multiplication. You know that whenever 0 is multiplied by a number, the result is zero.

> ▶ **READING MATH**
>
> In some countries, zero is called the annihilator.

114 Using Algebra to Explain

Lesson 2-8

> **Multiplication Property of Zero**
>
> For any real number a, $a \cdot 0 = 0 \cdot a = 0$.

You have also learned that you cannot divide by 0. Another way of putting it is that a fraction cannot have zero in its denominator. This can be explained by using related facts.

Suppose you tried to divide 0 by 0. You write $\frac{0}{0} = b$. What is the value of b? Using related facts, you get $0 \cdot b = 0$. But any value of b would work to make the equation true. Since there is no unique value for $\frac{0}{0}$, we say it is undefined.

Now suppose you tried to divide some nonzero number a by 0. You write $\frac{a}{0} = b$. Then by related facts, you get $0 \cdot b = a$. The Multiplication Property of Zero says that a would have to be 0. But a was specifically indicated as being a nonzero number. So there is no value of b that makes the equation true.

Therefore, an attempt to divide a number by 0 can never give exactly one answer. And since operations must give a single answer and the same answer each time they are performed, division by zero is not allowed.

When zero is involved in multiplication or division, we do not draw a fact triangle. But zero can still be involved in sample multiplication equations. Three types of equations are possible.

Example 2

Solve each equation.

a. $0x = 0$
b. $0x = 4$
c. $13x = 0$

Solutions Each equation uses zero in a slightly different way.

a. Zero times x is zero. **All real numbers are solutions.**
b. Zero times x is a nonzero number. Since $0 \cdot x$ is always 0, it cannot be 4. **There is no solution.**
c. A nonzero number times x is zero. **There is exactly one solution, 0.**

Part c of Example 2 illustrates a simple fact.

> **Zero Product Property**
>
> If the product of two real numbers a and b is 0, then $a = 0$, $b = 0$, or both a and b equal 0.

Explaining Multiplication and Division Related Facts

STOP QY2

> ▶ QY2
> a. Find an equation (not in this lesson) that has no solution.
> b. Find an equation (not in this lesson) that is true for every real number.

Some Properties of the Number One

Another number that has special properties involving multiplication is 1. There are two cases to investigate.

Case 1: One is a factor.

Whatever number you choose for the second factor, the result is that same number.

Case 2: One is the product.

What pairs of numbers multiply to give 1 as a result? They must be reciprocals, like 3 and $\frac{1}{3}$, or $\frac{2}{11}$ and $\frac{22}{4}$, or -8 and -0.125.

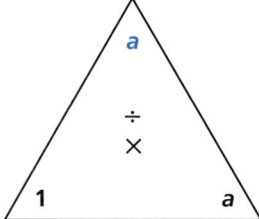
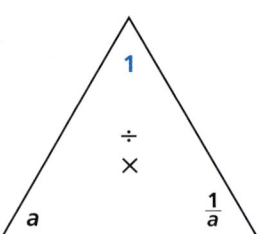

The two fact triangles above illustrate two properties of multiplication.

Multiplicative Identity Property

For any real number a, $a \cdot 1 = 1 \cdot a = a$.

The Multiplicative Identity Property is also true for zero, since $0 \cdot 1 = 1 \cdot 0 = 0$.

The second property involves the reciprocals, or *multiplicative inverses*, of a and $\frac{1}{a}$, whose product is 1.

Multiplicative Inverse Property

For any real number a, where $a \neq 0$, $a \cdot \frac{1}{a} = \frac{1}{a} \cdot a = 1$.

GUIDED

Example 3

Find the reciprocal of each number.

a. 1.5625   b. $\frac{4}{15}$ c. -34

Solutions

a. The reciprocal of 1.5625 is $\frac{1}{1.5625}$. To find the decimal for $\frac{1}{1.5625}$, you can divide 1 by 1.5625 or find 1.5625^{-1} on your graphing calculator.
$\frac{1}{1.5625} = \underline{}$

b. The reciprocal of $\frac{4}{15}$ is $\underline{} = 1 \cdot \underline{} = \frac{15}{4}$. So taking the reciprocal inverts a fraction.

c. The reciprocal of −34 is $\underline{}$.

Questions

COVERING THE IDEAS

1. Write the related facts for the fact triangle shown below.

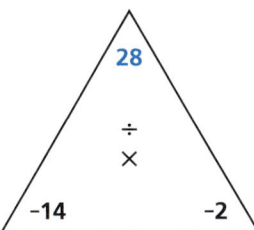

2. a. Make a fact triangle with the numbers 275, 25, and $\frac{1}{11}$.
   b. Write the four related facts.

3. If a, b, and c are not equal to zero and $a = bc$, find the other three related facts.

4. Use the equation $\frac{2}{3}p = 96$.
   a. Make a fact triangle for this equation.
   b. Use the fact triangle to solve the equation for p.
   c. Solve the equation using the Multiplication Property of Equality.
   d. Solve the equation using the Division Property of Equality.

5. **Fill in the Blank** If $a = b$, then $6a = \underline{}$.

In 6–8, find the reciprocal.

6. 0.8
7. −6
8. $\frac{4}{7}$

9. What number is the multiplicative identity?

10. a. If $ab = 1$, then what is true about a and b?
    b. If $ab = 0$, then what is true about a and b?

Explaining Multiplication and Division Related Facts

Chapter 2

11. Explain in your own words why 0 does not have a reciprocal.
12. What number(s) satisfy the following sentences?
    a. Zero times a number is eight.
    b. Zero times a number is zero.
    c. Seven times a number is zero.

In 13–16, solve the equation.

13. $4x = 0$
14. $0y = \frac{1}{2}$
15. $0 \cdot a = 0$
16. $3b - 3b = 0$

APPLYING THE MATHEMATICS

In 17 and 18, what property is shown by the statement?

17. All the books on the table are free, so a book is free.
18. Since two packages of batteries cost $7.98, one package costs $3.99.
19. Illustrate the rules for multiplying and dividing positive and negative numbers by drawing fact triangles. Label corners "pos" for positive number and "neg" for negative numbers. Draw all the possible triangles and give the rules that each triangle illustrates.
20. Explain why the fact triangle below is not possible.

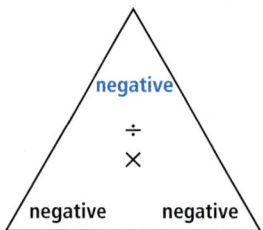

In 21 and 22, find the quotient by multiplying by the reciprocal.

21. $\dfrac{\frac{6}{7}}{\frac{3}{21}}$

22. $\dfrac{\frac{9}{3}}{\frac{3}{8}}$

23. A rectangle has a length of 50 centimeters and an area of 1 square centimeter.
    a. Is this rectangle possible?
    b. If so, what is the width of the rectangle? If not, why is it not possible?

24. Refer to the pattern in the table.

Quotient	$\frac{p}{10}$	$\frac{p}{1}$	$\frac{p}{0.1}$	$\frac{p}{0.01}$	$\frac{p}{0.001}$	$\frac{p}{0.0001}$
Equal Expression	0.1p	1p	10p	?	?	?

  a. Copy and complete the pattern in the table.
  b. Use the pattern to rewrite $\frac{p}{0.000000001}$ as a multiple of p.

25. Consider the formula $d = rt$, where d is distance, r is rate, and t is time.
  a. Write the related facts for the formula.
  b. Suppose Sam D. Yago is traveling from his home to Santa Clara, California, a distance of 160 miles. To the nearest mile, how long will the trip take him if he can average 42 miles per hour?

26. Meli went grocery shopping. Her least expensive purchase was a drink. She bought bread which cost twice as much as the drink, salad that was four times as much as the drink, and laundry detergent that was five times as much as the drink. Her bill came to $18. How much did each item cost Meli?

REVIEW

27. The formula $P = 2a + 2b$ gives the perimeter P of a rectangle with sides a and b. Create a fact triangle using this formula, and list the other three related facts. (**Lesson 2-7**)

28. The measure of ∠PQR equals 133°. Find y, the measure of ∠SQR. (**Lesson 2-7**)

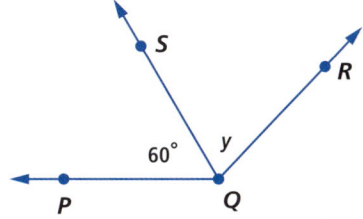

29. Evaluate the expressions $18p^2 + p$ and $9p^2 + p^3 + 19p$ when $p = 6$. Are the expressions equivalent? Why or why not? (**Lessons 2-6, 1-3**)

30. a. Evaluate $-(-194) + -(-(-194))$.
  b. Explain how you found your answer. (**Lesson 2-4**)

Chapter 2

In 31–34, suppose that *x* is positive and *y* is negative. Tell whether the value of the expression is *always positive*, *sometimes positive*, or *never positive*. (Lessons 2-4, 1-1)

31. $\frac{x}{y}$
32. $x \cdot y$
33. $x - y$
34. $x + y$

35. To estimate the number of bricks *N* needed in a wall, some bricklayers use the formula $N = 7LH$, where *L* and *H* are the length and height of the wall in feet. If a wall is to be 8.25 feet high and 27.5 feet long, about how many bricks would a bricklayer need? (Lesson 1-1)

36. The value of Birchmere stock went down 1.64 on March 30. On March 31 and April 1 it went up 0.88 of a point each day. Find the net change in Birchmere stock over this 3-day period. (Previous Course)

EXPLORATION

37. Find single values for *a*, *x*, *w*, and *z* that make all five of these equations true. (*Hint*: *a*, *x*, *w*, and *z* are all different numbers.)

    $a^2 = a$

    $xw = x$

    $z \cdot z = z + z$

    $x + w = w$

    $\frac{z}{a} = z$

Nearly 1 in 3 bricklayers are self-employed.

Source: Bureau of Labor Statistics

QY ANSWERS

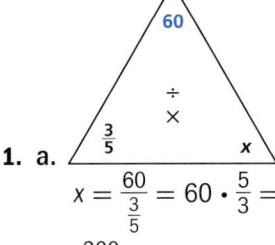

1. a. $x = \frac{60}{\frac{3}{5}} = 60 \cdot \frac{5}{3} = \frac{300}{3} = 100$

   b. $\frac{3}{5}x = 60$; Multiply both sides by $\frac{1}{\frac{3}{5}}$, or $\frac{5}{3}$; the Multiplication Property of Equality tells us $\frac{5}{3} \cdot \frac{3}{5}x = \frac{5}{3} \cdot 60$, so $x = \frac{300}{3} = 100$.

2. Answers vary. Sample answers:

   a. $0 \cdot x = 17$

   b. $0 \cdot 3 \cdot x = 2 \cdot 0$

120 Using Algebra to Explain

Chapter 2 Projects

1 Repeating Number Puzzle

An unusual number trick is performed as follows.

Your teacher asks Student A to jot down any three-digit number and then to repeat the digits in the same order to make a six-digit number (for example, 546,546). The teacher asks Student A to pass the sheet of paper to Student B, who is requested to divide the number by 7.

"Don't worry about the remainder," your teacher claims, "because there won't be any." Student B is surprised to discover that your teacher is correct (546,546 ÷ 7 = 78,078). Without telling the teacher the result, Student B passes the paper on to Student C, who is told to divide the quotient by 11. Once again, your teacher states that there will be no remainder. This, in fact, is the case (78,078 ÷ 11 = 7,098).

With no knowledge of the answers obtained by these computations, your teacher directs a fourth Student D to divide the last quotient by 13. Again, the quotient comes out even (7,098 ÷ 13 = 546). The final result is written on a slip of paper which is folded up.

Without opening it, the teacher passes the folded paper to Student A. The teacher says, "Open this and you will find your original three-digit number."

Prove that the trick works regardless of the digits chosen by the first student. Write out your explanation in several complete sentences. Include sufficient mathematical work to support your explanation. Make certain that you explain why the trick works in general (rather than showing it works for specific cases).

2 A New Operation

Addition and multiplication are both commutative and associative. Consider the operation # in which for all x and y, $x \# y = xy + 2$. Some examples are shown below.

$1 \# 2 = 4$ $2 \# 2 = 6$ $3 \# 1 = 5$ $5 \# 2 = 12$

a. Is the operation # associative? Is it commutative?

b. Give an example of an operation that is *not* commutative.

c. Is there a number that is an identity for the operation #? In other words, is there a number n, so that $x \# n$ always equals x?

Chapter 2

3 Pascal's Triangle

Use a CAS to expand the expressions $(x+y)^2$, $(x+y)^3$, $(x+y)^4$, $(x+y)^5$, and $(x+y)^6$. For each expression, write down in a row the coefficients the CAS gives. For example: $(x+y)^2 = x^2 + 2xy + y^2 = 1 \cdot x^2 + 2 \cdot xy + 1 \cdot y^2$, so you should write **1 2 1**.

a. For each of the expressions above, compute the sum of all the coefficients in the row. For example, for $(x+y)^2$ you would calculate $1 + 2 + 1$, or 4. What seems to be the formula for this sum? Count the coefficients in each row. What seems to be the formula for this number?

b. There are many relationships among the coefficients you wrote down in Part a. For example, in every row there are repeating numbers. Describe the pattern behind these repetitions.

c. Another famous relationship explains how to use the coefficients in a row to figure out the row that comes after it. This relationship is named after the French mathematician Blaise Pascal (1623–1662). Can you figure out this relationship? Use it to predict the coefficients for $(x+y)^7$.

4 Magic Squares

In Lesson 2-3, you encountered magic squares. There are many methods you can use to construct your own magic squares. For example, the following is a magic square for any real numbers a, b, and c.

$a + c$	$a + b - c$	$a - b$
$a - b - c$	a	$a + b + c$
$a + b$	$a - b + c$	$a - c$

a. Choose three real numbers a, b, and c and check that the above method does give a magic square.

b. Explain algebraically why this method always gives a magic square.

c. Using the Internet or the library, find a method of constructing magic squares with more cells. Make a poster illustrating the method you found.

5 Postal Rates

In 2006, the cost to mail a first-class letter was $0.39 for the first ounce and $0.24 for each additional ounce up to 13 ounces. So people who mailed many letters would buy many $0.39 and $0.24 stamps. If you buy x $0.39 stamps and y $0.24 stamps, then the postage you have paid for is $0.39x + 0.24y$. For example, if you use one $0.39 stamp and three $0.24 stamps for a letter, you have paid a total of $1(\$0.39) + 3(\$0.24) = \$1.11$ for postage.

a. Which weights can be sent first class for less than $3 using these stamps?

b. What is the maximum weight you can send with these stamps without going over $10 for postage?

c. Explore the problem and find the postage beyond which it is *always* possible to use these stamps.

Chapter 2 Summary and Vocabulary

- With algebra, a wide variety of relationships among numbers can be explained. An important tool in these explanations is the **Distributive Property.** This property says that for all real numbers a, b, and c, $c(a + b) = ca + cb$. Moving from $c(a + b)$ to $ca + cb$ is called expanding the expression. Moving from $ca + cb$ to $c(a + b)$ is called factoring the expression $ca + cb$.

- In an equation, the expression on the left side is equivalent to the expression on the right. Algebraic properties provide a way of determining whether expressions are equivalent and, at the same time, explaining why they are equivalent. This makes the use of algebraic properties a more powerful method than using tables or graphs, which can suggest that expressions are equivalent but cannot prove them equivalent. Yet tables and graphs can be useful when algebraic methods are not available. Also, by using algebraic properties, you can create your own equivalent forms of algebraic expressions.

- Among the earliest properties you ever learned were the **related facts of addition and subtraction, and of multiplication and division.** The related facts properties and the other algebraic properties can also help explain why various number puzzles work. Algebra is a language that helps you better understand the world.

Vocabulary

2-2
like terms
coefficient
factoring

2-4
additive inverse
opposite

Theorems and Properties

Distributive Property of Multiplication over Addition (p. 66)
Distributive Property of Multiplication over Subtraction (p. 67)
Opposite of Opposites Property (p. 85)
Multiplication Property of −1 (p. 86)
Opposite of a Sum Property (p. 86)
Opposite of a Difference Property (p. 87)
Addition Property of Equality (p. 106)
Subtraction Property of Equality (p. 106)
Related Facts Property of Addition and Subtraction (p. 107)
Additive Identity Property (p. 108)
Additive Inverse Property (p. 109)
Multiplication Property of Equality (p. 113)
Division Property of Equality (p. 113)
Related Facts Property of Multiplication and Division (p. 114)
Multiplication Property of Zero (p. 115)
Zero Product Property (p. 115)
Multiplicative Identity Property (p. 116)
Multiplicative Inverse Property (p. 116)

Chapter 2 Self-Test

Take this test as you would take a test in class. You will need a calculator. Then use the Selected Answers section in the back of the book to check your work.

In 1–7, simplify the expression.

1. $5w - 2w$
2. $\frac{6}{7}(3v + 78 + v)$
3. $-7h + 2(h - 4)$
4. $3(k + 10) - (11 - 4k)$
5. $-(-(-(r)))$
6. $\frac{5x + 7}{6} + \frac{2x}{6}$
7. $\frac{2}{3x} - \frac{1}{x}$

8. Carlos bought seven shirts at $19.98 each. Show how Carlos could use the Distributive Property to calculate the total cost of the shirts in his head.

9. Write two expressions to describe the total area of the figure below.

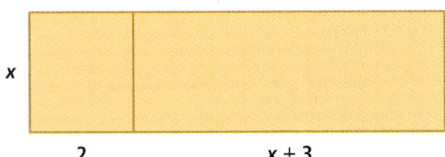

10. **Fill in the Blank** If $2 \cdot n = 4$, then $2 \cdot n \cdot 6 = \underline{\ ?\ }$.

11. Write the related facts for $C = \pi \cdot d$, where C is the circumference of a circle and d is its diameter.

12. If you know the circumference of a circle and want to find the diameter, which of the related facts from Question 11 should you use?

13. Make a fact triangle for $-12 = -4 + -8$, and write the related facts.

In 14 and 15, find the opposite of the expression.

14. $-3p$
15. $b + 2$

In 16 and 17, find the reciprocal of the expression.

16. -2.5
17. $\frac{4}{11d}$

18. **True or False** Is $(-2)^5 = 2^5$? Justify your response in one or two sentences.

19. Let n be any real number. Use algebra to show how the number puzzle below works.
    Step 1 Pick a number.
    Step 2 Multiply by 10.
    Step 3 Add 30.
    Step 4 Divide by 5.
    Step 5 Subtract your original number.
    Step 6 Add 1.
    Step 7 Subtract 7.
    You have your original number.

20. Use a graphing calculator to determine whether $w + (2w - 1) + 3(w + 6)$ and $6w - 5$ seem to be equivalent.

21. Use properties of operations to determine whether $w + (2w - 1) + 3(w + 6)$ and $6w - 5$ are equivalent.

22. Four siblings worked on a project. Darryl worked the shortest amount of time. Carol worked twice as long as Darryl. Beryl worked three times as long as Darryl, and Errol worked four times as long as Darryl. They decided to split up the $1,500 they earned from the project based on the amount of time each of them worked. How much did each sibling get?

23. Make a table to show that $2x + 1$ and $|-2x - 1|$ are not equivalent expressions.

Chapter 2 Chapter Review

SKILLS
PROPERTIES
USES
REPRESENTATIONS

SKILLS Procedures used to get answers

OBJECTIVE A Use the Distributive Property to expand and combine like terms. (Lessons 2-1, 2-2)

In 1–12, simplify the expression by distributing and/or combining like terms.

1. $3(x + 4)$
2. $(2a - 1.3)10$
3. $5(3x + 7) - 11(x + 6)$
4. $\frac{2}{5}(10 + -15w + 4w)$
5. $4(y - 13) + 6(3y - 3y)$
6. $1.5x + -4x + 17x$
7. $7x + -2x + 13x$
8. $-6m + 5m + -m$
9. $\frac{3x}{4} - \frac{3}{8} + 2x$
10. $\frac{2x}{5} + \frac{4z}{5} - \frac{2x}{5}$
11. $\frac{n+1}{3} + \frac{5}{3}$
12. $\frac{3}{2x} + \frac{1}{x}$

OBJECTIVE B Use the Opposite of Opposites Property, the Opposite of a Sum Property, and the Opposite of a Difference Property to simplify expressions. (Lesson 2-4)

In 13–20, remove the parentheses and then combine like terms, if possible.

13. $-(7a + 4)$
14. $-(3h - 7g + 8)$
15. $1 - (1 - z)$
16. $7x - (4x - 8)$
17. $1\frac{1}{2} - (\frac{3}{4} - y)$
18. $3(b - 2) - 5(3 + 2e)$
19. $(-2)^3$
20. $(-x)^4$

21. Evaluate each expression.
    a. $(-3)^4$
    b. -3^4
    c. $(-3)^5$
    d. -3^5

In 22–24, determine whether the expression is positive or negative. How do you know?

22. $-3(7.4)(-237)(-2)$
23. $(-1,135)4$
24. $x \cdot -x$

OBJECTIVE C Use related facts to solve sentences. (Lessons 2-7, 2-8)

In 25–27, identify all the real numbers that complete the sentence.

25. $0n = 3$
26. $-2 + k = 0$
27. $0g = 0$

In 28–31, find the related facts for the sentence.

28. $d = cg$
29. $\frac{1}{2} = \frac{1}{6} + \frac{1}{3}$
30. $a + b = 5$
31. $317.23 = 1 \cdot 317.23$

In 32–34, use related facts to find the value of the variable.

32. $2.6 = 13 + a$
33. $56 = x \cdot 0.8$
34. $0 = 1853.42b$

PROPERTIES The principles behind the mathematics

OBJECTIVE D Apply and recognize the following multiplication properties: Multiplicative Identity Property, Multiplicative Inverse Property, Multiplication Property of Zero, Multiplication Property of Equality, and the Zero Product Property. (Lessons 2-4, 2-8)

In 35–37, write the reciprocal of the number.

35. -5
36. 0.513
37. $\frac{1}{8x}$

38. Write the following statement in symbols: *The product of a number and its reciprocal is the multiplicative identity.*

39. Of what property is this an example? If $w = x$, then $w - 1.7 = x - 1.7$.

40. Fill in the Blank Multiplication by −1 changes a number to its ___?___.

In 41–43, evaluate the expression $(x + 5)(x + 4)(x + 3)$ for the given value of x.

41. $x = -3$ 42. $x = 3$ 43. $x = -2$

OBJECTIVE E Apply and recognize the following properties: Additive Identity Property, Additive Inverse Property, and Addition Property of Equality. (Lesson 2-7)

44. If $a + b = 0$, how are a and b related?

45. If $a + b = a$, what is the value of b?

46. **Fill in the Blank** Let $r = t$. Then $r + 2.576 = t +$ ___?___.

In 47–49, write the additive inverse of the number.

47. -7.536 48. 0 49. $-(-x)$

50. Write the following statement in symbols: *If two numbers are additive inverses, their sum is the additive identity.*

OBJECTIVE F Use and apply the Distributive Property to perform calculations in your head. (Lesson 2-1)

51. In the sentence $2(r + w) = 2r + 2w$, what property has been applied?

In 52–55, explain how the Distributive Property can be used to do the calculations in your head.

52. $7 \cdot \$5.95$

53. $103 \cdot 36$

54. $4 \cdot 59$

55. the cost of 11 shirts if each one costs $3.50

USES Applications of mathematics in real-world situations

OBJECTIVE G Use algebra to explain how number puzzles work. (Lesson 2-3)

In 56–58, let n be the number used to solve the number puzzle. Use algebra to explain the result.

56. Step 1 Pick a number.
    Step 2 Subtract 5.
    Step 3 Multiply by 4.
    Step 4 Add 4.
    Step 5 Divide by 2.
    Step 6 Add 10.
    Step 7 Subtract twice your original number.
    You will always end up with 2.

57. Step 1 Pick a number.
    Step 2 Multiply by 4.
    Step 3 Add 10.
    Step 4 Add 2 more.
    Step 5 Divide by 4.
    Step 6 Subtract 3.
    You will always end up with your original number.

58. Step 1 Pick a number.
    Step 2 Add 11.
    Step 3 Multiply by 6.
    Step 4 Subtract 12.
    Step 5 Divide by 2.
    Step 6 Subtract 30.
    Step 7 Add 3.
    You will always end up with 3 times your original number.

OBJECTIVE H Apply the Distributive Property in real-world situations. (Lessons 2-1, 2-2)

59. A $150,000 estate is to be split among 4 children, 2 grandchildren, and a charity. Each child gets the same amount, while the grandchildren get half as much. If the charity receives $5,000, how much will each child receive?

60. Two next-door neighbors' yards are pictured below. Find the total area of both yards.

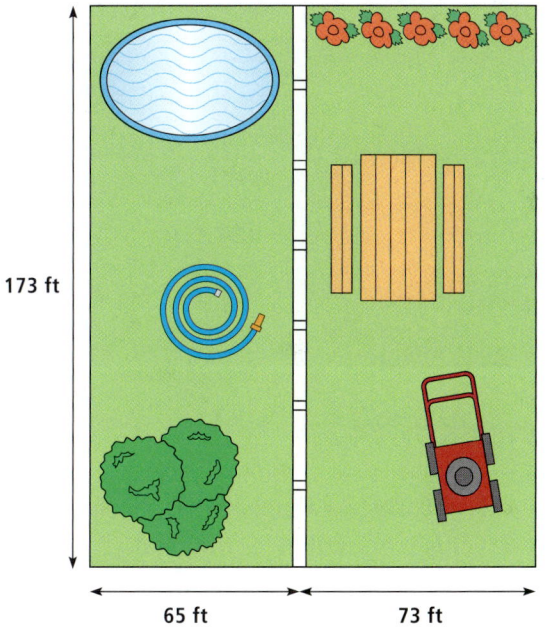

61. Suppose a taxicab driver is allowed to keep $\frac{2}{5}$ of all fares collected. The remaining fares go to the company. If a driver collects F dollars in fares, what is the driver's share to keep?

REPRESENTATIONS Pictures, graphs, or objects that illustrate concepts

OBJECTIVE I Use a spreadsheet or table to test the equivalence of expressions. (Lesson 2-5)

62. Make a table of values to show that $3(2x + 4)$ and $6x + x + 12 - x$ are equivalent expressions.

63. Make a table of values to show that x^2 and $2x$ are not equivalent expressions. Circle a counterexample in the table.

64. A table of values generated by two expressions is shown below. Do the expressions seem to be equivalent? Why or why not?

Expression 1		Expression 2	
x	y	x	y
−5	−4	−5	4
−4	−2	−4	2
−3	0	−3	0
−2	2	−2	2
−1	4	−1	4

OBJECTIVE J Use technology to test for equivalence of expressions. (Lessons 2-5, 2-6)

In 65–67, use a CAS or a graphing calculator to determine whether the expressions are equivalent.

65. $n - (1 - (2 - (3 - n)))$ and $2n + 2$

66. $x^2 - 4x + 4$ and $(x - 2)(x - 2)$

67. $(x + 1)(x)(x - 1)$ and x^3

68. Suppose two different students use the formulas $A = \pi r^2$ and $A = \frac{\pi d^2}{4}$ to find the area of a circle. Remember, the diameter d of a circle is twice as long as the radius r. Use a graphing calculator or CAS to determine if the two formulas are equivalent.

Chapter 3
Linear Equations and Inequalities

Contents

- 3-1 Graphing Linear Patterns
- 3-2 Solving Equations with Tables and Graphs
- 3-3 Solving Equations by Creating Equivalent Equations
- 3-4 Solving $ax + b = c$
- 3-5 Using the Distributive Property in Solving Equations
- 3-6 Inequalities and Multiplication
- 3-7 Solving $ax + b < c$
- 3-8 Solving Equations by Clearing Fractions

Stephen collects coins. He begins with 25 coins and each week is sent 10 new coins in the mail. After w weeks he will have $25 + 10w$ coins. Let t stand for the total number of coins he has at the end of w weeks. The size of Stephen's collection over these weeks can be described in a number of ways. Three of them are shown here.

Equation

$t = 25 + 10w$

Table

w	t
0	25
1	35
2	45
3	55
4	65
⋮	⋮

Graph

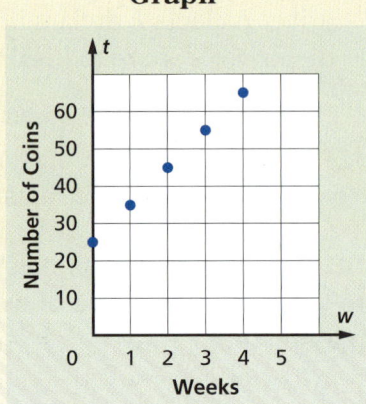

The table lists the ordered pairs (0, 25), (1, 35), (2, 45), (3, 55), and (4, 65). All these pairs make the equation $t = 25 + 10w$ true. The equation $t = 25 + 10w$ is called a *linear equation* because all the points of its graph lie on the same line. For the same reason, $25 + 10w$ is called a *linear expression*.

Linear equations are the backbone of relationships among variables. In Chapter 1 you connected points to make graphs of algebraic expressions. In this chapter and the next, you will see more of their many applications.

Chapter 3

Lesson 3-1
Graphing Linear Patterns

Vocabulary

constant-increase situation
collinear
constant-decrease situation

▶ **BIG IDEA** Constant-increase and constant-decrease situations lead to linear graphs and are represented by linear equations.

Constant-Increase Patterns

The size of Stephen's coin collection described on pages 128 and 129 increases by 10 coins each week. It provides an example of a **constant-increase situation** because his collection grows by the same amount each week. The graph of every constant-increase situation consists of points that lie on the same line. We call these points **collinear**.

Mental Math

Simplify.
a. $\frac{m}{x} + \frac{n}{x}$
b. $\frac{m}{x} - \frac{n}{kx}$
c. $\frac{5}{x} + \frac{3}{kx}$

But notice that the graph does not include every point on the line. Because Stephen receives coins at specific whole-number intervals, the domain of w is the set of whole numbers. It does not make sense to connect the points on the graph because numbers such as $2\frac{1}{2}$ are not in the domain of w.

Constant-Decrease Patterns

A **constant-decrease situation** involves a quantity that decreases at a constant rate. In Example 1 below, the variable can be any positive real number between two whole numbers. The graph is no longer a set of separate or *discrete* points; it is connected, or *continuous*.

Example 1

After a flash flood, the level of water in a river was 54 inches above normal and dropping at a rate of 1.5 inches per hour. Let x equal the number of hours since the water started dropping and y equal the height, in feet, above normal of the lake. We can model this situation with the equation $y = 54 - 1.5x$. Graph this relationship.

Solution Find the lake level at various times and make a table. A table for 0, 1, 2, 3, and 4 hours is shown on the next page.

Flash floods result from a large amount of rain within a short amount of time and usually occur within 6 hours of a storm.

130 Linear Equations and Inequalities

Time x (hr)	Height y (in.)	Ordered Pair (x, y)
0	54 − 1.5 • 0 = 54	(0, 54)
1	54 − 1.5 • 1 = 52.5	(1, 52.5)
2	54 − 1.5 • 2 = 51	(2, 51)
3	54 − 1.5 • 3 = 49.5	(3, 49.5)
4	54 − 1.5 • 4 = 48	(4, 48)

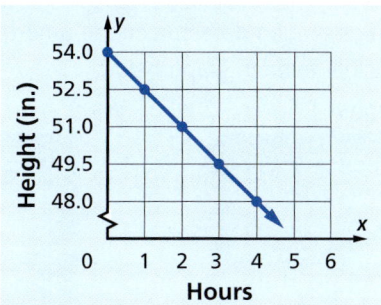

Plot the ordered pairs in the table and look for patterns. You should see that the five points lie on the same line. Time in hours can be any nonnegative real number, such as 1.75 or $3\frac{1}{2}$. This means that other points lie between the ones you have already plotted. So, draw the line through them for the domain $x \geq 0$.

Example 2

In Example 1, if the water level continues to drop at the same rate, how many hours will it take for the water level to fall to 3 feet above normal?

Solution Look at the graph from Example 1. The level of the water above the normal level is given by y. Find the point for 3 feet, or 36 inches, on the y-axis. The x-coordinate of this point is 12. This is shown by the arrows on the graph. **The water will be 3 feet above normal after 12 hours.**

🛑 QY

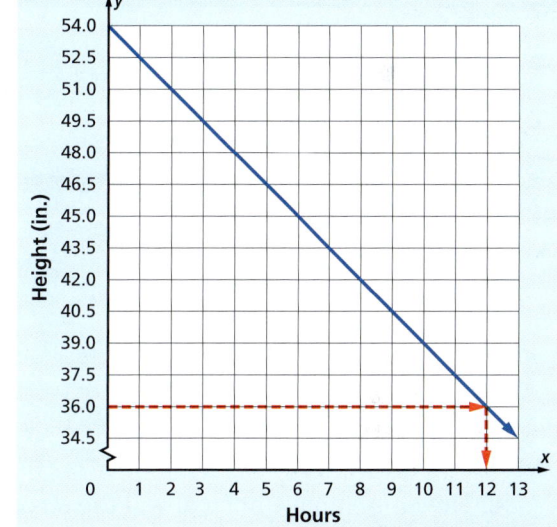

In Examples 1 and 2, notice that the ⌇ symbol appears on the graphs. This symbol indicates a break in the scale of the axis. It is often used so that patterns in graphs become more apparent.

Activity

You can create your own continuous graph using a motion detector that hooks up to a computer or your graphing calculator. These graphs are time-distance graphs. They plot the distance between the motion detector and a stationary solid object (like a wall) over time. If a person holds a motion detector and moves closer to or farther from a wall, those changes will be seen in the graph. An example of such a graph is shown at the right.

To create this graph, start about 5 feet away from the wall and move farther away from the wall at a constant speed.

(continued on next page)

▶ QY

Based on the graph in Example 2, when is the water level 42 inches above normal?

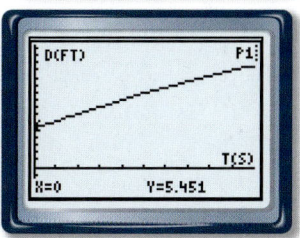

Graphing Linear Patterns

Chapter 3

Step 1 Using a motion detector, create time-distance graphs as similar as possible to each graph below.

Graph A

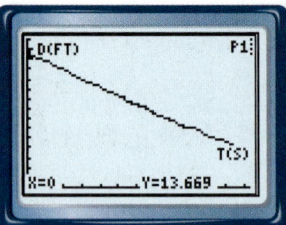

Graph B

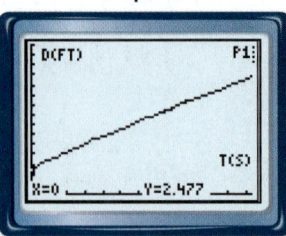

Graph C

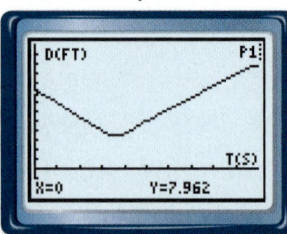

Step 2 Explain how you created each graph. Describe your starting point, direction, and speed while walking. Explain what quantity changes during the walk.

Graph D

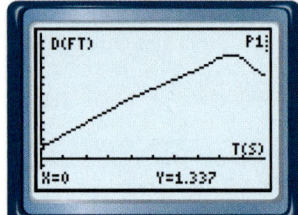

Questions

COVERING THE IDEAS

1. A flooded stream is 18 inches above its normal level. The water level is dropping 3 inches per hour. Its height y in inches above normal after x hours is given by the equation $y = 18 - 3x$. A table and a graph of the equation are shown at the right.

Time x (hr)	Height y (in.)
0	18
1	?
2	?
3	?
4	?

   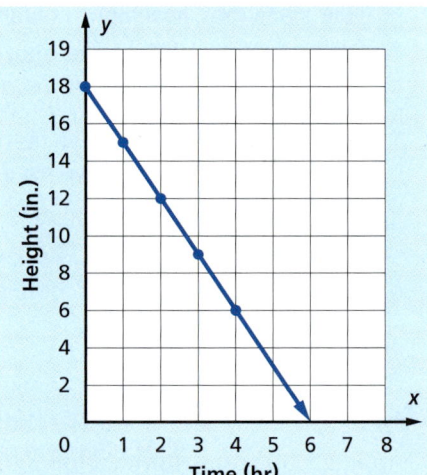

   a. Complete the table at the right.
   b. After how many hours will the stream be 12 inches above normal?
   c. How high above normal will the stream be after 3 hours?
   d. After how many hours will the stream be back to its normal level?

2. Suppose Miguel begins with $500 in an account and adds $20 per week.

Weeks (w)	Total (t)
0	?
1	?
2	?
3	?
4	?

   a. Complete the table at the right, showing t, the total amount Miguel will have at the end of w weeks.
   b. Graph the ordered pairs (w, t).
   c. Write an equation that represents t in terms of w.
   d. What is the domain of w?

3. A train consists of an engine that is 60 feet long and cars that are each 40 feet long. There is a distance of 2.5 feet between two cars and between the first car and the engine. Let T be the total length, in feet, of a train with c cars.

   a. What is the total length of a train with 1 car?
   b. What is the total length of a train with 2 cars?
   c. **Multiple Choice** How are T and c related?
      - **A** $T = 60 + 40c$
      - **B** $T = 62.5 + 40c$
      - **C** $T = 60 + 42.5c$
      - **D** $T = 62.5 + 42.5c$
   d. Graph the equation you found in Part c for values of c from 1 to 5.
   e. Find the length of the train if it has 12 cars.

John Stevens built and operated the first steam locomotive in the United States in 1825.

Source: *The World Almanac*

APPLYING THE MATHEMATICS

4. A tree has a trunk with a 12-centimeter radius. The radius increases by 0.5 centimeter per year. Its radius y, in centimeters, after x years is described by $y = 12 + 0.5x$.
   a. Make a table of values for this relationship.
   b. Draw a graph of this situation.
   c. After how many years will the radius equal 20 centimeters?

5. a. Draw the graph of $y = 4x$. Choose your own values for x.
   b. On the same grid from Part a, draw the graph of $y = -4x$.
   c. At which points do the graphs of $y = 4x$ and $y = -4x$ intersect?
   d. Describe any patterns you observe in these graphs.

6. A flooded stream is now 30 centimeters above its normal level. The water level is dropping at a rate of 3 centimeters per hour. Let x equal the number of hours from now and y equal the water level above normal after x hours.
   a. Suppose the water level continues to drop at the same rate. Write an equation for y in terms of x.
   b. Graph your equation from Part a.
   c. Use your graph to estimate when the stream is expected to drop to a level of 6 centimeters above normal.

In 7–10, graph the equation using a graphing calculator, and determine whether the graph is linear. You may have to adjust the window to view the graph.

7. $y = -0.02(x - 3)$

8. $y = \frac{1}{100}x^2$

9. $y = \frac{x-3}{5} - x$

10. $y = |x|$

Graphing Linear Patterns

Chapter 3

REVIEW

In 11–14, compute in your head. (Lesson 2-8)

11. $\frac{9}{11}h \cdot \left(\frac{9}{11}h \cdot \frac{11}{9h}\right)$

12. $95{,}620 \cdot 657 - 95{,}620 \cdot 656$

13. $85.69 \cdot 6.514 \cdot 0 \cdot 12$

14. $\frac{4}{13} \cdot 9.85 \cdot \frac{13}{40}$

In 15–17, use the table at the right and the formula $d = rt$ to calculate the following. (Lesson 2-8)

15. At top speed, how far can an ostrich run in 15 seconds?

16. a. At top speed, how long will it take a cheetah to run 0.78 mile?
    b. How long will it take a cheetah to run 100 yards at top speed?

17. If a greyhound runs at top speed for 8 seconds and a giraffe runs at top speed for 17 seconds, which animal has run farther?

18. Consider the equation $-0.6(5x) = 90$. (Lesson 2-8)
    a. Simplify the left side of the equation.
    b. Solve for x.
    c. Check your solution.

19. **Skill Sequence** Evaluate each expression. (Lesson 2-4)
    a. 6^2
    b. -6^2
    c. $(-6)^2$
    d. -6^3
    e. $(-6)^3$
    f. $-(-6)^4$

20. Is $x = -9$ a solution to $28 = -3x - 1$? Explain your reasoning. (Lesson 1-1)

Fast Land Animals

Animal	Top Speed (mi/hr)
Cheetah	70
Ostrich	40
Greyhound	39
Giraffe	32

Source: American Museum of Natural History

EXPLORATION

21. In Question 3, approximate lengths are given for an engine, train cars, and the distance between them.
    a. Find the lengths of an engine and of a car for an actual train. Identify the type of train and where you found the information.
    b. Give an equation relating the total length T of the train and the number of cars c.

QY ANSWER

after 8 hr

Lesson 3-2

Solving Equations with Tables and Graphs

Vocabulary

solution to an equation

▶ **BIG IDEA** Tables and graphs lead to exact or approximate solutions to equations.

Consider an equation that contains one variable, such as $2 + x = 6$. A **solution to an equation** is a value of the variable that makes the equation true. For $2 + x = 6$, 4 is a solution because $2 + 4 = 6$.

 QY1

Mental Math

Evaluate.

a. 50% of $120P$

b. 5% of $120P$

c. 50% of $12P$

d. 5% of $12P$

Solving with a Table and a Graph

In this lesson, you will learn two ways to solve equations: using a table and using a graph. We apply each way to solve $11 = 4x + 3$.

Example 1

Solve $11 = 4x + 3$ using a table.

Solution Replace y with 11 and make a table of values for $y = 4x + 3$. Look in the table to the right to see what value of x gives a value of 11 for y.

x	$4x + 3$	y
1	$4(1) + 3$	7
2	$4(2) + 3$	11
3	$4(3) + 3$	15
4	$4(4) + 3$	19
5	$4(5) + 3$	23

Looking at the table, you can see that when $x = 2$, $y = 11$. Based on the table, you can conclude that **the solution to the equation $11 = 4x + 3$ is $x = 2$.**

▶ **QY1**

Is 5 a solution to $|-12m + 19| = 41$?

Example 2

Solve $11 = 4x + 3$ using a graph.

Solution As in Example 1, replace y with 11. Then graph $y = 4x + 3$ and look for the value of x that corresponds with the value of 11 for y. To do this, start at 11 on the y-axis. Move right to the line, and then move down to the x-axis. As in Example 1, **$x = 2$ is the solution.**

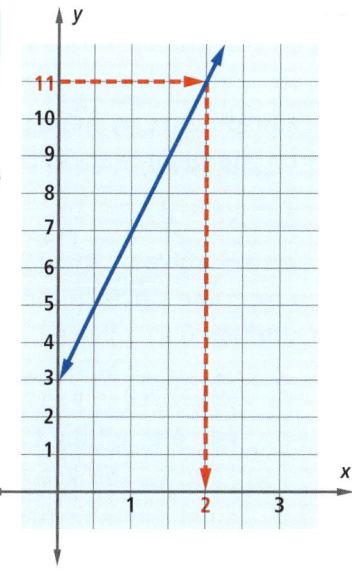

(continued on next page)

Solving Equations with Tables and Graphs

You can also see this on a graphing calculator. Most graphing calculators have a trace option. The trace option allows you to move a cursor along the graph. As the cursor moves, it lists the coordinates on the graph. The calculator screen at the right shows the cursor at (–0.8510638, –0.4042553). Move the cursor until the y-coordinate is as close as possible to 11. Remember that the x value is only an approximation for the solution if the y value is also an approximation.

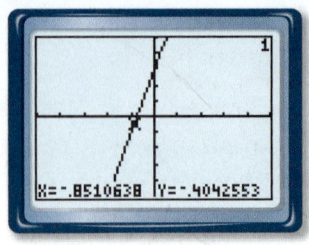

Situations Leading to Equations

Real-world situations can often be translated into relationships among numbers. If you can represent these relationships with an equation, then the methods of this and the next lesson can lead to a solution.

Example 3

Khalid lives in Alaska where there is no sales tax. He bought 3 tacos with a $5 bill and received $0.77 in change. Define the variable and represent the relationship with an equation.

Solution Let t represent the unknown cost of one taco. Because $5 minus the cost of 3 tacos equals $0.77 in change, $5 - 3t = 0.77$ is the equation.

🛑 QY2

In 2006, the states without a sales tax were Alaska, Delaware, Montana, New Hampshire, and Oregon.

Source: Federation of Tax Administrators

▶ **QY2**

Todd has $250 in his savings account. If he has a job that pays $11 per hour, how many working hours will it take for him to have enough money to buy a $481 surfboard?

Questions

COVERING THE IDEAS

1. Determine if the value of x is a solution for the equation.
   a. $11 - 9x = 47, x = 4$
   b. $3x^2 + 4x = 55, x = -5$
   c. $-x^3 = 64, x = -4$

In 2 and 3, solve by making a table for each equation and circling the row on the table that contains the solution.

2. $6 = 2x + 2$
3. $-5a + 7 = -23$

In 4 and 5, solve by making a graph for each equation and drawing lines from the y-axis to the graph, and then to the x-axis to indicate the solution.

4. $y = -3x - 4, y = 2$
5. $y = 5x - 18, y = -3$

6. a. With a graphing calculator, make a table for $y = 5(5x - 2)$ using the integers from –5 to 0.
   b. Adjust the table to solve $5(5x - 2) = -26$.

136 Linear Equations and Inequalities

In 7–9, define the variable and represent the relationship with an equation.

7. Solana lives where there is no sales tax. She bought five pieces of pizza for her friends with a $20 bill. She received $7.65 in change. What was the price of one piece of pizza?

8. Trevor is riding his bicycle across the United States. He started in a town on the East Coast and has already biked 630 miles. If he can ride 82 miles per day, how many days will it take him to complete the 3,210-mile journey?

9. The Coles are saving to send their child to college. They currently have $5,275 in the bank and are saving $950 per year. How many years will pass before they save $20,000?

APPLYING THE MATHEMATICS

10. Use a graph to solve $5 - 3t = 0.77$ from Example 3.

11. Use a table to solve $250 + 11h = 481$ from QY2.

12. Use a table on a graphing calculator to solve $3x + 5 = -4$.

13. The graph at the right shows the equation $y = 6.2 - 2.15x$.
    a. Draw dotted lines to show how to find an approximate solution to $3 = 6.2 - 2.15x$.
    b. What is the approximate solution?

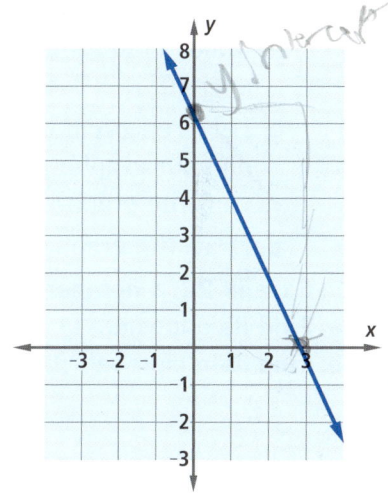

In 14 and 15, use the information provided to write an equation involving the variable. Then use any method to find the value of the variable.

14. The perimeter of the rectangle at the right is 46 feet.

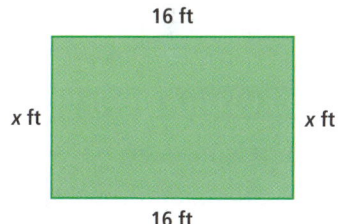

15. The perimeter of the triangle at the right is 97 meters.

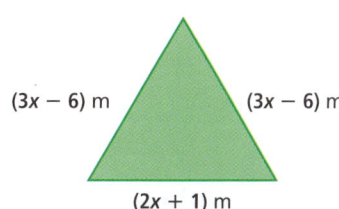

REVIEW

In 16–18, graph the equation on a calculator. Does the graph appear to be a line? You may have to adjust the window to view the graph. (Lesson 3-1)

16. $y = 0.402(x - 3) + 1.97x - 0.567$

17. $h = 7 - 2t^2 + t$

18. $y = 4x(5 + x)$

Chapter 3

19. Sierra redwood trees are known to be among the world's tallest trees, growing an average of 3.5 feet each year until they mature. If a redwood tree is now 28 feet tall, its height h after y years is described by $h = 28 + 3.5y$. (**Lesson 3-1**)

   a. Make a table of values for this relationship.
   b. Draw a graph of this situation.
   c. How tall will the tree be in 21 years?
   d. How many years will it take the tree to grow to a height of 259 feet?

20. *Internet World Stats* reported in the summer of 2005 that there were about 220 million Internet users in North America. That number was growing by an average of 23 million users per year. Suppose this rate continues between 2005 and 2010. (**Lessons 3-1, 1-5**)

   a. Make a table showing the total number of millions of Internet users 0, 1, 2, 3, 4, and 5 years after 2005.
   b. Let x represent the number of years after 2005. What window on your graphing calculator would be most appropriate to view the graph?
   c. Draw the graph using your graphing calculator.

Many Sierra redwoods are between 250 and 300 feet tall, the tallest being about 325 feet high.

Source: California Department of Parks & Recreation

In 21 and 22, a number is given.
a. Find its opposite.
b. Find its reciprocal. (**Lessons 2-8, 2-5**)

21. 8.3

22. $-\dfrac{26}{5}$

In 23 and 24, evaluate the expression $\dfrac{a+b}{3} + 3(a-b)$ for the given values of a and b. (**Lesson 1-1**)

23. $a = 11$ and $b = -4$

24. $a = 11x$ and $b = 55x$

EXPLORATION

25. Use the graph below to solve $11 = 7 + 2\sqrt{x+1}$. Draw dotted lines to show your method.

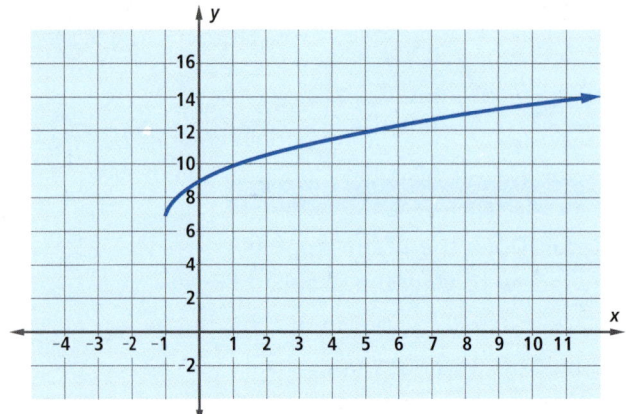

QY ANSWERS

1. $m = 5$ is a solution because
   $|-12 \cdot 5 + 19| =$
   $|-60 + 19| = |-41|$
   $= 41$.

2. 21 hr

Lesson 3-3
Solving Equations by Creating Equivalent Equations

Vocabulary

equivalent equations

▶ **BIG IDEA** Performing the same arithmetic operations on both sides of an equation can create an equation that is easier to solve.

In the previous lesson, you used a table and a graph to find the solution to an equation of the form $ax + b = c$. Those methods allow you to visualize what it means to solve an equation. But in practice, they are sometimes awkward to use. Other times they yield solutions that are not exact. In this lesson, we discuss a method that enables you to find exact solutions to many types of equations.

Mental Math

a. If $x = 4z - 9y$, what is $3x$?

b. If $x = 4z - 9y$, what is $-x$?

c. If $x = 4z - 9y$, what is $x + 9y$?

Solving with a Balance

A balance illustrates the meaning of the "=" sign in an equation by having equal weights placed on both sides. The scale will still balance as long as changes made to one side are also made to the other side. This is the idea behind solving an equation algebraically.

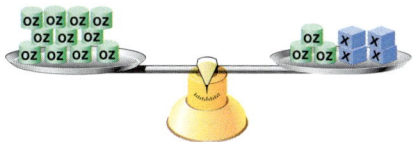

The equation $11 = 4x + 3$ is shown above with 11 identical one-ounce weights on the left side and 3 one-ounce weights with 4 unknown weights on the right side. You can find the weight x of one box in two steps. Each step keeps the scale balanced.

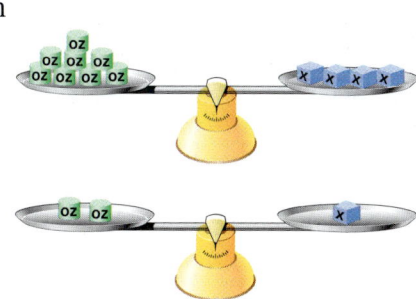

Step 1 Remove 3 one-ounce weights from each side of the scale.

Step 2 Leave $\frac{1}{4}$ of the contents on each side.

From the original equation two more equations were formed.

$$11 = 4x + 3$$
$$8 = 4x$$
$$2 = x$$

These three equations are called *equivalent equations* because 2 is the solution to each of them. **Equivalent equations** are equations with exactly the same solutions.

Solving Equations by Creating Equivalent Equations **139**

Chapter 3

Exploring $ax + b = c$ Equations with a Computer Algebra System (CAS)

The goal when solving equations of the form $ax + b = c$ is to add, subtract, multiply, or divide both sides of the equation to eventually get an equation of the form $x =$ a number. In the activity below, you will use a CAS to analyze the effect of performing various operations to the two sides of an equation.

Activity

Solve $8x - 12 = 4$.

Step 1 Enter $8x - 12 = 4$ into the calculator.

Step 2 Next enter $\boxed{+}12$ to instruct your CAS to add 12 to both sides of the equation. The result is $8x = 16$, which is simpler than $8x - 12 = 4$. So adding 12 was a good choice.

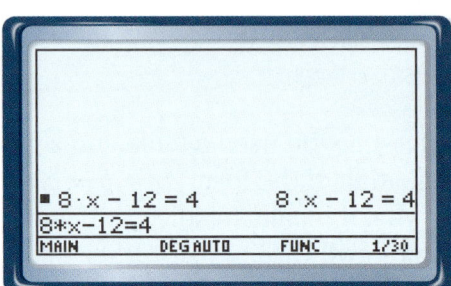

Step 3 Enter $\boxed{\div}8$ to divide each side by 8. The result is an equation that is as simple as possible. So $x = 2$.

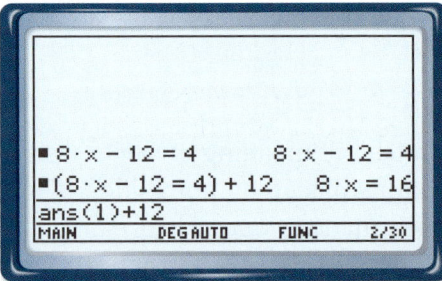

 QY

Solving a Linear Equation Using a CAS

Step 1 Enter your equation into your CAS.

Step 2 Decide which operation to perform on each side. For example, to add 9 to both sides enter $\boxed{+}9$.

Step 3 Do you get an equation that is simpler than the previous one? If the answer is "yes," continue until the equation is solved. If the answer is "no," re-enter the equation and try a different operation.

Step 4 You are finished solving when the equation has the form *variable = number*.

▶ QY

Solve each question using a CAS.

a. $\frac{4}{7}z + \frac{2}{3} = \frac{1}{7}$

b. $-3y + 14 = 98$

140 Linear Equations and Inequalities

Questions

COVERING THE IDEAS

1. Why are the equations below called equivalent equations?
   $$2x + 5 = 11$$
   $$2x = 6$$
   $$x = 3$$

2. a. What equation is represented by the diagram below?

   b. What two steps can be taken with the weights on each side of the scale to find the weight of a single box?
   c. How much does a single box weigh?

3. a. **Fill in the Blanks** When solving $5x - 27 = 13$, first ___?___ to each side, then ___?___ on both sides.
   b. Solve $5x - 27 = 13$ and check your result.

4. Consider the steps used in the solution of $82n - 51 = 441$ below.
   Given: $82n - 51 = 441$
   Step 1 $82n = 492$
   Step 2 $n = 6$

   a. What was done to go from the given equation to Step 1?
   b. What was done to go from Step 1 to Step 2?

In 5–8, a pair of equations is given. Determine what was done to each side of the first equation to arrive at the second equation.

5. $4x - 11 = 12$
   $4x = 23$

6. $72 - 18t = 864$
   $-18t = 792$

7. $\frac{3}{5}n = 30$
   $n = 50$

8. $0.004v = 1.2$
   $v = 300$

9. What might be done first to each side of $75x - 100 = 800$ to begin the process of solving the equation?

APPLYING THE MATHEMATICS

10. In solving $5x + 430 = 315$, Paula instructed her CAS to divide both sides of the equation by 5.

    a. What result did she get?
    b. Is this a reasonable first step to solve the equation? If so, use it to solve the equation. If not, explain why.

11. In solving $-3y + 14 = 98$, Paula instructed her CAS to divide both sides by -3.

    a. What result did she get?

    b. Is this a reasonable first step to solve the equation? If so, use it to solve the equation. If not, explain why.

12. A student showed the following work to solve $4x - 32 = 20$.

    $$4x - 32 = 20$$
    $$4x - 32 - 32 = 20 - 32$$
    $$4x = -12$$
    $$x = -3$$

    But substituting -3 in the original equation results in an equation that is not true. What did the student do wrong?

13. a. Solve $3x + 5y = c$ for x by hand or with a CAS.

    b. Check your answer to Part a by substituting numbers for y and c and solving that equation.

REVIEW

14. a. Using the equation $y = 7x - 2$, complete the table at the right. (**Lesson 3-2**)

    b. Use the table to find a solution to the equation $7x - 2 = -16$.

x	y
−2	?
−1	?
0	?
1	?
2	?

15. Jordan collects basketball cards. Last Sunday he had 200 cards. He then bought one pack of cards on each weekday, and two packs on Saturday. Jordan now has 284 cards. (**Lesson 3-2**)

    a. Define a variable and represent the relationship with an equation.

    b. Use a table or graph to find the number of cards in one pack.

16. Consider the following situation. A bathtub has 11 gallons of water in it. Rachana adjusts the faucet so it is now filling at approximately 4.5 gallons per minute. (**Lesson 3-2**)

    a. Create a table and a graph showing the amount of water in the tub in terms of the amount of time that has passed from when Rachana adjusted the faucet.

    b. How much water will be in the tub after 7 minutes?

    c. If the bathtub holds 56 gallons of water, how long does it take until the bathtub is full?

17. Death Valley, California once reached a record high of 134 degrees Fahrenheit. Suppose the temperature was 85 degrees at 9 A.M. and increased by 7 degrees every hour until 4 P.M., when it reached the record high. (**Lesson 3-1**)

    a. Use this information to complete the table at the right.
    b. Graph the data using the table.

18. Consider $\frac{3}{4}x = 16$. (**Lesson 2-8**)

    a. Solve this equation.
    b. Think of a real-world problem that can be solved with this equation.

19. Below is a dot plot of the number of mosquitoes caught in a trap over 15 days in Buckinghamshire, England. (**Lesson 1-7**)

    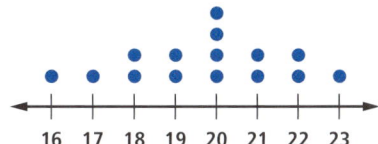

    a. Is the shape symmetric, skewed right, skewed left, or uniform?
    b. Calculate the mean and mean absolute deviation of the data.

Time of Day	Temperature (°F)
9:00 A.M.	?
10:00 A.M.	?
11:00 A.M.	?
12:00 P.M.	?
1:00 P.M.	?
2:00 P.M.	?
3:00 P.M.	?
4:00 P.M.	?

The average rainfall in Death Valley is about 1.65 inches per year.
Source: National Park Service

EXPLORATION

20. Enter $4(3x - 17) + 12 = 6(x + 2) + 6x - 68$ into your CAS as if you were going to solve the equation as in this section. Then press ENTER.

    a. Describe what the screen shows.
    b. Explain why. (*Hint:* Use algebra to simplify each side of the equation by hand.)
    c. Write down the other two equations that produce the same result when entered into a CAS.

QY ANSWERS

a. $z = \frac{-11}{12}$

b. $y = -28$

Solving Equations by Creating Equivalent Equations

Lesson 3-4: Solving $ax + b = c$

▶ **BIG IDEA** An equation of the form $ax + b = c$ can be solved in two major steps.

In 1637, the French philosopher and mathematician René Descartes started the practice of identifying known quantities by the letters a, b, and c from the beginning of the alphabet and unknown quantities by the letters x, y, and z from the end of the alphabet. Following the practice of Descartes, when we write "solving $ax + b = c$" we mean that a, b, and c are known numbers and x is unknown. For example, when $a = -\frac{3}{2}$, $b = -53$, and $c = 7$, we obtain the equation $-\frac{3}{2}x + -53 = 7$.

In general, any equation of the form $ax + b = c$, with a not equal to zero, can be solved in two steps. First add the opposite of b to both sides. Then multiply both sides by the reciprocal of a.

Mental Math

a. Which is greater, $\frac{1}{3}$ or 0.33?

b. Which is greater, 1.4 or $\frac{33}{22}$?

c. Which is greater, $-\frac{5}{4}$ or $-\frac{4}{5}$?

GUIDED

Example 1

Solve $-\frac{3}{2}x - 53 = 7$.

Solution

$-\frac{3}{2}x - 53 = 7$ Write the equation.

$-\frac{3}{2}x - 53 + \underline{\ ?\ } = 7 + \underline{\ ?\ }$ Add $\underline{\ ?\ }$ to each side.

$-\frac{3}{2}x = \underline{\ ?\ }$ Simplify.

$\underline{\ ?\ }\left(-\frac{3}{2}x\right) = \underline{\ ?\ }(60)$ Multiply each side by the reciprocal of $-\frac{3}{2}$.

$x = \underline{\ ?\ }$ Simplify.

Be sure to check your solution by substituting it back into the original equation.

Descartes was a French scientist, mathematician, and philosopher. His statement, "I think; therefore I am," is very famous.

Source: *Discourse on Method*

Equations That Require Simplifying First

Equations are often complicated, but they can be simplified into ones that you can solve.

Example 2

When Val works at the zoo on Saturday, she earns $10.80 per hour. She is also paid $8 for meals and $3 for transportation. Last Saturday she received $83.90. How many hours did she work?

Solution Let h = the number of hours Val worked.
In h hours she earned $10.80h$ dollars. So,
$10.80h + 8 + 3 = 83.90$.

Next, 8 and 3 are added. The resulting equation has the form $ax + b = c$. Solve for h.

$10.80h + 11 = 83.90$	Write the equation.
$10.80h + 11 + -11 = 83.90 + -11$	Addition Property of Equality
$10.80h = 72.90$	Simplify.
$\frac{1}{10.80} \cdot 10.80h = \frac{1}{10.80} \cdot 72.90$	Multiplication Property of Equality
$h = 6.75$	Simplify.

Val worked 6.75 hours.

Zookeepers take care of wild animals in zoos and animal parks. They feed the animals, clean their living spaces, work to keep them healthy, and keep them cool in the summer months.

Source: Bureau of Labor Statistics

Check If Val worked 6.75 hours at $10.80 per hour, she earned 6.75 · $10.80, or $72.90. Now add $8 for meals and $3 for transportation. The total comes to $83.90.

GUIDED

Example 3

The area of the largest rectangle is 94 square centimeters. What is the value of n?

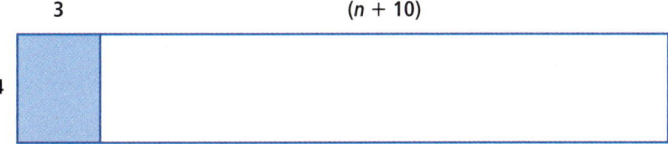

Solution Write an equation to represent the area. (*Hint:* You can use the sum of areas or the length and width of the big rectangle.) Then solve.

Area of left rectangle + Area of right rectangle = 94

$\underline{\ ?\ } \cdot \underline{\ ?\ } + \underline{\ ?\ } \cdot \underline{\ ?\ } = 94$	Write the equation.
$12 + \underline{\ ?\ } = 94$	Distributive Property
$\underline{\ ?\ }n + \underline{\ ?\ } = 94$	Simplify.
$\underline{\ ?\ }n + \underline{\ ?\ } + \underline{\ ?\ } = 94 + \underline{\ ?\ }$	Addition Property of Equality
$\underline{\ ?\ }(\underline{\ ?\ }) = \underline{\ ?\ }(\underline{\ ?\ })$	Multiplication Property of Equality
$n = \underline{\ ?\ }$	Simplify.

Be sure to check your solution.

Chapter 3

Variations of ax + b = c

If an equation has the variable on the right side, as in $c = ax + b$, the solution can still be obtained by adding the opposite of b, and multiplying by the reciprocal of a. The Commutative Property of Addition also implies that $ax + b = c$ is equivalent to $b + ax = c$. For example, the following equations can be solved with the same major steps.

$$7 = \frac{3}{2}x - 53 \qquad -53 + \frac{3}{2}x = 7$$

$$\frac{3}{2}x - 53 = 7 \qquad 7 = -53 + \frac{3}{2}x$$

Questions

COVERING THE IDEAS

1. a. **Fill in the Blanks** When solving $7t - 57 = 97$, first add __?__ to both sides. Then __?__ each side by __?__.
   b. Solve and check $7t - 57 = 97$.

2. Steps in solving $73y - 432 = 1{,}101$ are shown here.

   Given: $73y - 432 = 1{,}101$

   Step 1 $73y = 1{,}533$

   Step 2 $y = 21$

   a. What was done to arrive at Step 1?
   b. What was done to arrive at Step 2?

In 3 and 4, the equation is in the form $ax + b = c$. Find the values of a, b, and c.

3. $73y - 432 = 1{,}101$
4. $17 - 4y = 88$

5. **Multiple Choice** How do the solutions to $50x - 222 = 60$ and $60 = 50x - 222$ compare?
   A They are equal.
   B They are opposites.
   C They are reciprocals.
   D None of these are true.

In 6–13, solve and check the equation.

6. $6x + 42 = 126$
7. $31 = 11A - 24$
8. $-20y - 2 = 8$
9. $18 = 16 + 5B$
10. $7 + \frac{3}{5}d = -5$
11. $2.4n - 2.4 = 2.4$
12. $1.06P + 3.25 = 22.86$
13. $200 = 4 - \frac{7}{2}m$

Linear Equations and Inequalities

14. Write directions to teach a friend how to solve the equation given in Question 6 in a step-by-step process, but do not tell your friend the exact equations.

APPLYING THE MATHEMATICS

In 15–17, a situation is given.
a. Write an equation of the form $ax + b = c$ to describe the situation. Be sure to identify what the unknown variable represents.
b. Solve the equation and answer the question.

15. Ms. Toshio bought gas for $2.39 per gallon and a drink for $1.50. The total bill was $31.15 before sales tax. How many gallons of gas did she buy?

16. Bena lives in Delaware, where there is no sales or meals tax. She bought three chicken sandwiches with a $10 bill and received $1.45 change. What was the price of one sandwich?

17. Eighty students from a school went to the taping of a television program. They filled 7 rows of seats and there were 3 students in the eighth row. How many seats were in each row?

The average price per gallon of unleaded gasoline in 2004 in the United States was $1.92.
Source: U.S. Department of Energy

18. **Skill Sequence** Solve for n in each equation.
    a. $17n + 38 = 4$ b. $-2n + 15 = 7$ c. $3n + a = 9$

In 19–22, solve and check the equation.

19. $\frac{11}{17}x + \frac{17}{11} = 11\frac{11}{17}$ 20. $0.003 = 0.02y - 0.1$

21. $\frac{4}{3}\left(a - \frac{1}{2}\right) = 3\frac{1}{3}$ 22. $2.08 + 4.2n = 41.56$

REVIEW

23. Consider the balance below. (Lesson 3-3)

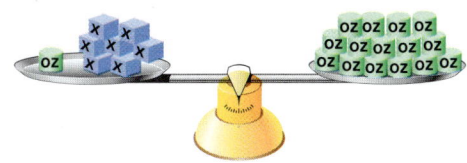

a. What equation is shown?
b. What two steps can be done with the weights on the balance to find the weight of a single box?
c. How much does a single box weigh?

24. **True or False** Determine whether the given equation is equivalent to $\frac{1}{2}x + 1 = 21$. (**Lesson 3-3**)

   a. $x + 2 = 42$
   b. $-x + 2 = -38$
   c. $x = 40$
   d. $\frac{1}{4}x + \frac{1}{4} = \frac{21}{4}$

25. a. Graph the following set of points: $\{(1, 2), (3, 4), (5, 6), (7, 8), (9, 10)\}$. (**Lessons 3-1, 1-4**)

   b. These points are collinear. The point $(99, 63a + 10)$ is on that line. What is a?

   c. **Fill in the Blank** The point $(m, \underline{\ ?\ })$ is also on this line.

In 26–28, apply the Distributive Property to rewrite the expression with fewer terms. (**Lessons 2-4, 2-2, 2-1**)

26. $7p + q - 7(q + 6p)$
27. $-(a - b) + 2(a + b)$
28. $3(-2n - 5) - 2(-3n + 5)$
29. Suppose $y = 11 - (9 - 14x)$. If x is 9, what is y? (**Lesson 1-1**)

EXPLORATION

30. Solve the general equation $ax + b = c$ for x.
31. Consider equations of the form $ax + b = c$, where $a \neq 0$.
   a. Write a program for a calculator or a computer that accepts values of a, b, and c as input and gives the value of x as output.
   b. Run your program with different values of a, b, and c that lead to both positive and negative solutions.

Lesson 3-5

Using the Distributive Property in Solving Equations

> **BIG IDEA** By collecting like terms or expanding expressions, you can transform many equations into the form $ax + b = c$.

Solving Equations by Collecting Like Terms

We need to solve $c + c + \frac{1}{2}c + \frac{1}{2}c + \frac{1}{2}c + 15{,}000 = 260{,}000$ in Example 1. It takes only one step more to solve this equation than to solve an equation in the form of $ax + b = c$. The first step is to simplify the left side of the expression. Simplifying sides of equations is common in solving equations.

Mental Math

a. How many cups is $\frac{1}{3}$ of 2 cups?

b. How many cups is $\frac{1}{3}$ of 4 cups?

c. How many cups is $\frac{1}{3}$ of n cups?

Example 1

A $260,000 estate is to be split among two children, three grandchildren, and a charity. Each child receives the same amount, while each grandchild receives half that amount. If the charity receives $15,000, how much will each child receive?

Solution First identify the unknown. Let c = the portion a child receives. Then $\frac{1}{2}c$ is a grandchild's portion of the estate.

Now translate the given information into an equation.

$$c + c + \tfrac{1}{2}c + \tfrac{1}{2}c + \tfrac{1}{2}c + 15{,}000 = 260{,}000$$

Step 1 Use the Distributive Property to add the like terms.

$$3\tfrac{1}{2}c + 15{,}000 = 260{,}000$$

Now the equation is in the form $ax + b = c$ and can be solved in two steps.

Step 2 Add −15,000 to each side. Rewrite $3\frac{1}{2}$ as 3.5 to make the computation easier.

$$3.5c + 15{,}000 + {-}15{,}000 = 260{,}000 + {-}15{,}000$$
$$3.5c = 245{,}000$$

Step 3 Multiply each side by $\frac{1}{3.5}$.

$$c = 70{,}000$$

Each child will receive $70,000.

(continued on next page)

Check Each grandchild receives half as much as a child. This amount is $35,000. So the two children, three grandchildren, and the charity will receive 2 · 70,000 + 3 · 35,000 + 15,000 dollars.
Does 2 · 70,000 + 3 · 35,000 + 15,000 = 260,000? Yes.

Notice that in Example 1 we check the result not by substituting into the original equation, but by checking whether the numbers work in the original statement of the problem.

Sometimes the use of the Distributive Property will result in an equation that requires only one step for its solution.

Example 2
Solve $9y - y = 40$.

Solution

Write the equation.

$9y - y = 40$

Many students like to rewrite y as $1 \cdot y$ to make the use of the Distributive Property more obvious.

$9y - 1y = 40$	$y = 1 \cdot y$
$(9 - 1)y = 40$	Distributive Property
$8y = 40$	Combine like terms.
$\frac{8y}{8} = \frac{40}{8}$	Divide each side by 8.
$y = 5$	Simplify.

Check

Substitute 5 for y in the original equation.

Does $9(5) - 5 = 40$?
$\quad\quad 45 - 5 = 40$?
$\quad\quad\quad\quad 40 = 40$? Yes.

Solving Equations by Removing Parentheses

You can use the Distributive Property to remove parentheses that appear in equations.

Example 3
Solve $2(x + 3) = 7$.

Linear Equations and Inequalities

Solution

$2(x + 3) = 7$ Write the equation.
$2x + 6 = 7$ Distributive Property
$2x + 6 - 6 = 7 - 6$ Subtraction Property of Equality
$2x = 1$ Simplify.
$\frac{2x}{2} = \frac{1}{2}$ Division Property of Equality
$x = \frac{1}{2}$ Simplify.

Check

Is $2(\frac{1}{2} + 3) = 7$?
$2(3\frac{1}{2}) = 7$
$7 = 7$ Yes.

Example 4

A company charges $29.95 each day to rent a truck, with the first 50 miles free. Subsequently, the cost is $0.60 per mile. Therefore, if the truck is driven m miles (where $m \geq 50$), the total rental cost is $29.95 + 0.60(m - 50)$ dollars. Isabel rented a truck for one day and paid $100.15. How many miles did Isabel drive?

Solution

Let the total cost equal $100.15 to determine how far the truck was driven.

$29.95 + 0.60(m - 50) = 100.15$
$29.95 + 0.60m - 0.60 \cdot 50 = 100.15$
$29.95 + 0.60m - 30 = 100.15$
$0.60m - 0.05 = 100.15$
$0.60m = 100.20$
$m = 167$

The truck was driven 167 miles.

 QY

In the U.S., one in five households, or 20% of the population, moves every year.

Source: www.ourtownamerica.com

> **QY**
>
> In Example 4, how much does it cost to drive the truck 225 miles?

GUIDED

Example 5

Solve $2(3k + 4) - (9k - 7) = 6$.

(continued on next page)

Chapter 3

Solution

$2(3k + 4) - (9k - 7) = 6$ Write the equation.
$\underline{}\,? \underline{} = 6$ Distributive Property
$\underline{}\,?\,\underline{}\,k + 15 = 6$ Add like terms.
$\underline{}\,?\,\underline{}\,k + 15 - 15 = 6 - 15$ $\underline{}\,?\,\underline{}$
$\underline{}\,?\,\underline{}\,k = \underline{}\,?\,\underline{}$ Add like terms.
$\underline{}\,?\,\underline{} = \underline{}\,?\,\underline{}$ Divide each side by -3.
$k = \underline{}\,?\,\underline{}$ Multiplication Property of Equality

Check Substitute 3 for k in the original equation.
Does $2(3 \cdot \underline{}\,?\,\underline{} + 4) - (9 \cdot \underline{}\,?\,\underline{} - 7) = 6$?

Questions

COVERING THE IDEAS

In 1–10, solve the equation.

1. $3r + r = 8$
2. $2x + 3x - 7 = 23$
3. $14 = 3(x + 2)$
4. $2(4x - 9) = -2$
5. $9 = 2(x + 2) + 2$
6. $5y - 3(5 - 2y) = -15$
7. $42 = t - 7t$
8. $3x - (x + 4) = 22$
9. $13(t + 1) - 3(2 + 4t) = 38$
10. $6(\frac{1}{2}b + 3) + 4(12 - b) = \frac{5}{6}$

11. The winner of a car race received a prize of $150,000. Ten percent went to the driver and the rest was split among the 4 owners and the head mechanic, with the head mechanic getting half as much as the owners.
    a. Let E be the amount each owner received. Write an equation that can be solved to determine E.
    b. Find E and the amount the head mechanic received.

12. In 2005, the federal income tax T for a single person whose taxable income I was between $7,300 and $29,700 was given by $T = 730 + 0.15(I - 7{,}300)$. This can be translated as $730 plus 15% of the amount over $7,300. If a single person paid $2,150 in income tax, what was the person's taxable income to the nearest dollar?

Race cars can reach speeds in excess of 230 mph.

Source: National Aeronautics and Space Administration

Linear Equations and Inequalities

Lesson 3-5

APPLYING THE MATHEMATICS

13. The area A of a trapezoid with parallel bases b_1 and b_2 and height h is given by the formula $A = 0.5h(b_1 + b_2)$, as shown at the right.
    a. If $h = 6$ in., $b_2 = 4$ in., and $b_1 = 8$ in., calculate A.
    b. If $A = 26$ in^2, $h = 4$ in., and $b_2 = 5$ in., calculate b_1.
    c. If $A = 48$ in^2, $h = 4$ in., and $b_1 = 8$ in., calculate b_2.

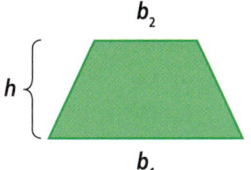

14. If the perimeter of the rhombus at the right is 940 centimeters, find y.

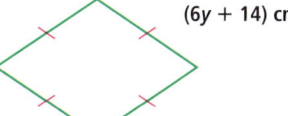

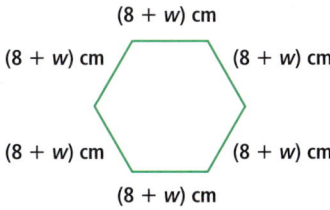

15. Find the value of w in the regular hexagon at the right if each side has length $(8 + w)$cm and the perimeter is 186 centimeters.

16. Find the value of z if the area of the rectangle at the right is 96 square miles.

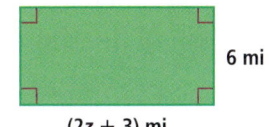

17. Suppose you have $100,000 to invest. You decide to put part of the money in a certificate of deposit (CD) that pays 6% annual interest and the rest in a savings account that pays 4% per year. If d dollars are invested in the CD, then $E = 0.06d + 0.04(100,000 - d)$ gives the interest earned E in one year. How much should you put in each place to earn $4,800 in the first year?

REVIEW

In 18–20, solve and check the equation. (Lesson 3-4)

18. $\frac{4}{7}d + 9 = -11$ 19. $6.21 = 3.4 + -c$ 20. $\frac{1}{2}t + 4 = -2$

21. There were already 5 inches of snow on the ground when it started snowing at midnight. Snow continued to fall throughout the night, accumulating at three-quarters of an inch per hour. (Lesson 3-4)
    a. How much snow was on the ground at 4 A.M.?
    b. Winterville Junior High cancels school when there are more than 10 inches of snow on the ground. At what time was school canceled?

In the U.S., snow depth is usually reported to the nearest 1 inch. 24-hour snowfall is reported to the nearest 0.1 inch.

Source: National Oceanic & Atmospheric Administration

Using the Distributive Property in Solving Equations

Chapter 3

22. Sareeta is planning a party and she wants to serve sandwiches to her guests. Each sandwich costs $4.50. (**Lessons 3-4, 3-1**)

   a. Labeling the *x*-axis *Number of Sandwiches* and the *y*-axis *Total Cost*, draw a graph to represent possible costs for the sandwiches. Use $0 \leq x \leq 20$.

   b. Suppose Sareeta paid $50 for the sandwiches and received $0.50 in change. How many sandwiches did she buy? (Assume there is no sales tax.)

23. Sherman was twice as late arriving home today as he was yesterday. If he is supposed to be home at 5:00 P.M. each day and didn't arrive until 7:30 P.M. today, at what time did he get home yesterday? (**Lesson 3-3**)

24. The perimeter of the triangle below is 54 yards. Find *x* and the length of the shortest side. (**Lesson 3-2**)

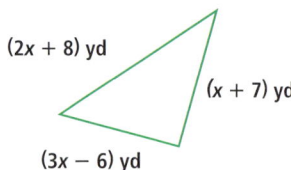

25. The equation $3\frac{1}{2}c = 245,000$ in Example 1 was solved by changing $3\frac{1}{2}$ to 3.5 then multiplying both sides by $\frac{1}{3.5}$. The equation can also be solved by multiplying by the reciprocal of $3\frac{1}{2}$. What is the reciprocal of $3\frac{1}{2}$? (**Lesson 2-8**)

EXPLORATION

26. a. Solve the equation $mx + (m + 1)x = 8mx + 4$ for *x*.

   b. Check your solution by substituting a number for *m* and solving the resulting equation.

27. You have $64 and you want to buy a pair of jeans and a $20 T-shirt. There is a 7% sales tax. If *x* represents the cost of the jeans, then the equation $x + 20 + 0.07(x + 20) = 64$ shows how much you can spend on jeans. What is the price of the most expensive jeans you can afford?

Consumers in the United States spent about $329 billion on clothing and shoes in 2004.
Source: *The World Almanac and Book of Facts*

QY ANSWER

$134.95

Lesson 3-6
Inequalities and Multiplication

Vocabulary

inequality
boundary point
interval
endpoint

▶ **BIG IDEA** Multiplying each side of an inequality by a positive number keeps the direction of the inequality; multiplying each side by a negative number reverses the direction of the inequality.

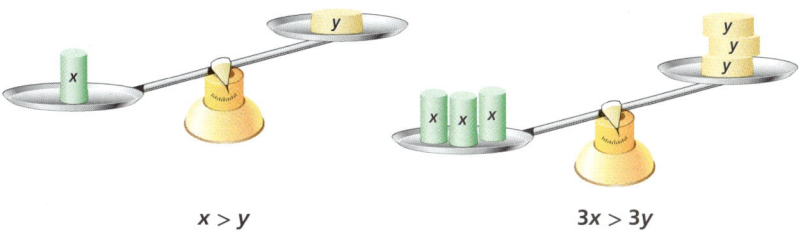

$x > y$ $3x > 3y$

Mental Math

Estimate to the nearest dollar.

a. 20% tip on a $49.56 bill

b. 20% tip on a $149.56 bill

c. 20% tip on a $249.56 bill

Graphs and Inequalities on a Number Line

An **inequality** is a mathematical sentence with one of the verbs $<$ ("is less than"), $>$ ("is greater than"), $\leq$ ("is less than or equal to"), or $\geq$ ("is greater than or equal to").

Even though they look similar, equations and inequalities are different in important ways. Consider $x = 3$, $x > 3$, and $x \geq 3$. The equation $x = 3$ has just one solution, while the inequalities $x > 3$ and $x \geq 3$ have infinitely many solutions. Their graphs at the right show these differences.

Exactly One Solution

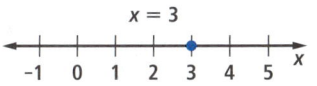

In the graph of $x > 3$ on a number line, the 3 is marked with an open circle because 3 does not make the sentence $x > 3$ true. The number 3 is the **boundary point** between the values that satisfy $x > 3$ and those that do not. Numbers just a little larger than 3 such as 3.01 and $3\frac{34}{10,000}$ are solutions, as are larger numbers like 1 million. The graph of $x \geq 3$ does include 3 because the sentence $3 \geq 3$ is true. Another way to write $x \geq 3$ is to use set-builder notation. It is written as $\{x: x \geq 3\}$ and is read as "the set of all x such that x is greater than or equal to 3."

Infinitely Many Solutions

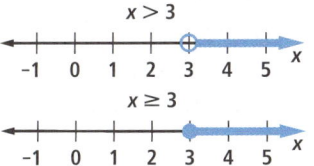

When solving real-world problems, you must often decide what domain makes sense for the situation. Inequalities are common in the world. For example, let $x =$ a Fahrenheit temperature at which water is in solid form (ice). Then, since water freezes below 32°F, $x < 32$, as graphed at the right. The open circle at 32 shows that all numbers to the left of 32 are graphed, but not 32 itself.

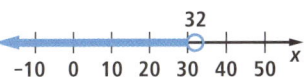

Inequalities and Multiplication 155

Chapter 3

 QY

Some situations lead to *double inequalities*. Let E = the elevation of a place in the United States. Elevations in the United States range from 86 meters below sea level (in Death Valley) to 6,194 meters above sea level (at the top of Denali, also known as Mt. McKinley). So, $-86 \leq x \leq 6{,}194$. This double inequality is graphed below.

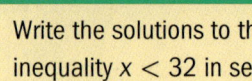

▶ **QY**

Write the solutions to the inequality $x < 32$ in set-builder notation.

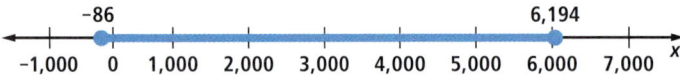

Draw dots on –86 and 6,194 to show that those numbers are included in the solution.

The graph of $-86 \leq x \leq 6{,}194$ is an *interval*. An **interval** is a set of numbers between two numbers a and b, which are called the **endpoints** of the interval. When the endpoints are included in the interval, it is described by $a \leq x \leq b$. If the endpoints are not included, the $\leq$ is replaced by $<$, giving $a < x < b$.

The height of Denali (formerly Mt. McKinley) is 20,320 feet.
Source: U.S. Geological Survey

Activity

The inequalities in 1–7 below are very similar. Use a CAS to solve each one. Record the operation you do to both sides and the inequality that results. A CAS screen for the first problem is shown.

1. $2x < 8$
2. $2x < -8$
3. $2x \leq 8$
4. $-2x < 8$
5. $-2x \leq -8$
6. $2x > 8$
7. $-2x \geq 8$

In 8 and 9, give what you expect the solution to be. Use a CAS to check your answer.

8. $4m > -20$
9. $-10w \leq 62$

10. The inequalities you solved in 1–9 can be grouped into two categories whose solution processes are somewhat different. What are the two categories?

The Multiplication Property of Inequality

Here are some numbers in increasing order. Because the numbers are in order, you can put the inequality sign ($<$) between any two of them.

$$-10 < -6 < 0 < 5 < 15$$

156   Linear Equations and Inequalities

Now multiply these numbers by some fixed *positive* number, say 2.

$$-20 \quad -12 \quad 0 \quad 10 \quad 30$$

The order stays the same, as shown in the number line at the right.

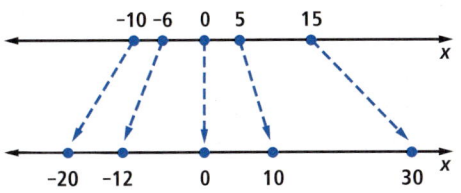

You could still put a < sign between any two of the numbers. This illustrates that if $x < y$, then $2x < 2y$. In general, multiplication by a positive number maintains the order of a pair or a list of numbers.

> **Multiplication Property of Inequality (Part 1)**
>
> If $x < y$ and a is *positive*, then $ax < ay$.

Here is the same list of numbers we used earlier.

$$-10 < -6 < 0 < 5 < 15$$

Now we multiply these numbers by –2 and something different happens.

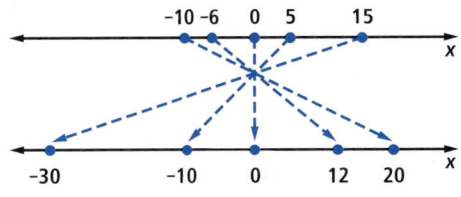

$$20 \quad 12 \quad 0 \quad -10 \quad -30$$

Notice that the numbers in the original list are in *increasing* order, while the numbers in the second list are in *decreasing* order. The order has been reversed.

If you multiply both sides of an inequality by a *negative* number, you must change the direction of the inequality. This idea can be generalized.

> **Multiplication Property of Inequality (Part 2)**
>
> If $x < y$ and a is *negative*, then $ax > ay$.

Changing from < to >, or from ≤ to ≥, or vice-versa, is called *changing the sense of the inequality*. You have to change the sense of an inequality when you are multiplying both sides by a negative number. Otherwise, solving $ax < b$ is similar to solving $ax = b$.

Solving Inequalities

To solve an inequality of the form $ax < b$, you must isolate the variable on one side (just like solving an equation). Be careful to notice whether you multiply or divide each side by a positive number or a negative number.

Chapter 3

Example 1

Solve $-7x \geq 126$ and check.

Solution Multiply both sides by $-\frac{1}{7}$, the reciprocal of -7. Since $-\frac{1}{7}$ is a negative number, Part 2 of the Multiplication Property of Inequality tells you to change the sense of the inequality sign from $\geq$ to $\leq$.

$$-7x \geq 126 \quad \text{Write the inequality.}$$
$$-\tfrac{1}{7} \cdot -7x \leq -\tfrac{1}{7} \cdot 126 \quad \text{Multiply each side by } -\tfrac{1}{7}.$$
$$x \leq -\tfrac{126}{7} \quad \text{Simplify.}$$
$$x \leq -18 \quad \text{Simplify.}$$

Check

Step 1 Substitute -18 for x. Does $-7 \cdot -18 = 126$? Yes.

Step 2 Try a number satisfying $x < -18$. We use -20. Is $-7 \cdot -20 \geq 126$? Yes, $140 \geq 126$.

> **READING MATH**
>
> Inequalities with variables are open sentences. When the variable is replaced with a number, the inequality is either true or false.

The two-step check of an inequality is important. The first step checks the boundary point in the solution. The second step checks the sense of the inequality.

GUIDED

Example 2

a. Solve $20 \geq 4x$.

b. Graph the solution.

c. Check your answer.

Solutions

a.
$$20 \geq 4x \quad \text{Write the inequality.}$$
$$\tfrac{1}{4} \cdot 20 \underline{\ ?\ } \tfrac{1}{4} \cdot 4x \quad \text{Multiply each side by } \tfrac{1}{4}.$$
$$5 \underline{\ ?\ } x \quad \text{Simplify.}$$

This can be rewritten as $x \underline{\ ?\ } 5$.

b. Graph the solution on the number line.

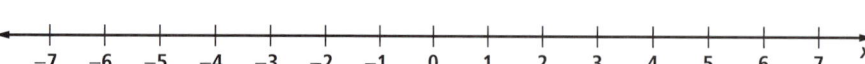

c. **Step 1** Check the boundary point by substituting 5 for x.
Does $\underline{\ ?\ } = 4(\underline{\ ?\ })$?

158 Linear Equations and Inequalities

Step 2 Check whether the sense of the inequality is correct. Pick some number from the shaded region in Part b. This number should also work in the original inequality. We choose 0.

Is __?__ ≥ 4(__?__)? Yes, __?__ ≥ __?__.

Since both steps worked, the solution to $20 \geq 4x$ can be described by the sentence __?__.

Examples 1 and 2 showed solutions of two types of inequalities.

> **Multiplication Property of Inequality (in words)**
>
> You may multiply both sides of an inequality by the same *positive* number without affecting the set of solutions to the sentence. You can also multiply both sides by a *negative* number, but then you must also *change the sense of the inequality*.

Questions

COVERING THE IDEAS

In 1 and 2, consider the situation. a. Write an inequality that describes the situation. b. Identify the boundary point for the inequality. c. Graph the solution set of the inequality.

1. To successfully leave Earth's orbit, a satellite must be launched at a velocity of at least 11.2 kilometers per second. Let v be the launch velocity of a satellite in kilometers per second.

2. To obtain a passing grade on the Spanish exam, Marlene needs to answer at least 60 percent of the test items correctly. Let p be the percentage of items answered correctly.

3. Wakana's dog w weighs over 120 pounds, but not more than 130 pounds. Graph the possible values of w on a number line.

In 4 and 5, graph the inequality on a number line.

4. $x < 2\frac{1}{5}$
5. $y \leq -3$

6. Consider the inequality $20 < 30$. What true inequality results if you multiply both sides of the inequality by the given number?
   a. 6
   b. $\frac{1}{2}$
   c. -4

In 7–12, solve the inequality. Then check your answer.

7. $15x > 5$
8. $-32 < 2n$
9. $\frac{4}{5}y \leq 20$
10. $-\frac{3}{2}z \leq 1$
11. $-3a \leq -6$
12. $7b \geq 6$

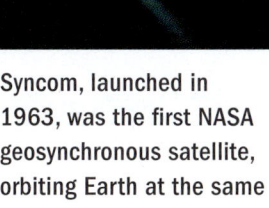

Syncom, launched in 1963, was the first NASA geosynchronous satellite, orbiting Earth at the same speed as Earth's rotation.

Source: NASA

13. Consider the following list of numbers. $\frac{1}{2}, \frac{1}{3}, \frac{1}{4}, \frac{1}{5}, \frac{1}{6}$
    a. Place an appropriate symbol between each pair of consecutive numbers ($<, \leq, \geq,$ or $>$).
    b. Multiply each number in the list by -8, then place an appropriate symbol between each pair of consecutive numbers ($<, \leq, \geq,$ or $>$) in the new list.
    c. Did the direction of the symbols change in Part b? Why or why not?

APPLYING THE MATHEMATICS

14. Consider the following list of numbers. 10, 20, 30, 40, 50
    a. Place an appropriate symbol between each pair of consecutive numbers ($<, \leq, \geq,$ or $>$).
    b. Create a new list by taking the reciprocal of each number, then place an appropriate symbol between each pair of consecutive numbers ($<, \leq, \geq,$ or $>$) in the new list.
    c. Did the direction of the symbols change in Part b? Why or why not?

15. **Fill in the Blanks** If $6 \leq m \leq 6.2$, then $\underline{} \leq 5m \leq \underline{}$.

In 16 and 17, read the problem situation. a. Write an inequality describing the situation. b. Solve the inequality.

16. An engineer is designing a rectangular parking lot that is to be 75 feet wide. According to the city building code, the area of the lot can be at most 6,250 square feet. What is the allowable length of the lot?

17. A concert hall is being designed to seat at least 2,200 people. Each row will have 80 seats. How many rows of seats will the hall have?

REVIEW

18. The boxes are of unknown equal weight W. (Lesson 3-5)

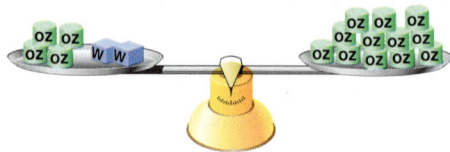

    a. What equation is pictured by the balance above?
    b. What two steps can be done with the weights on the balance to find the weight of a single box?
    c. How much does each box weigh?

In 19 and 20, solve the equation. (Lesson 3-5)

19. $-4(3x - 1.5) + 7 = 39$

20. $2(a + 5) - 3(5 + \frac{1}{2}a) = -19$

21. Grafton went to the store to buy bottles of soda and bags of chips for a party. He bought bottles of soda for $1.99 each and bags of chips for $2.99 each. He bought twice as many bags of chips as bottles of soda. After paying with two twenty-dollar bills, he received $0.15 in change. (Lessons 3-5, 3-3)
    a. Define a variable and write an equation describing the situation.
    b. How many bottles of soda and bags of chips did Grafton buy?

22. **Multiple Choice** How do the solutions to $2x - 111 = 35$ and $-35 = 2x + 111$ compare? (Lessons 3-4, 2-8, 2-4)
    A They are equal.
    B They are opposites.
    C They are reciprocals.
    D none of the above

In 23 and 24, a fact triangle is given. Write the related facts and determine the value of x. (Lesson 2-7)

23.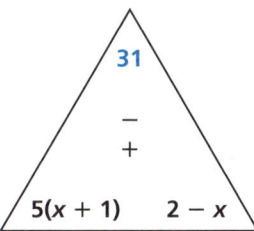

24.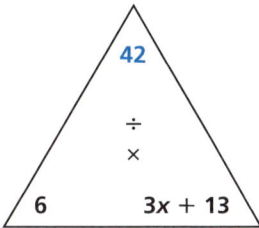

25. Tomás drove at an average speed of 61 miles per hour for $3\frac{1}{4}$ hours. About how many miles did he travel? (**Previous Course**)

EXPLORATION

26. A rectangle is 12 units by w units.
    a. Find the values of w that would make the area of the rectangle greater than 84 square units.
    b. Find the values of w that would make the area of the rectangle less than or equal to 216 square units.
    c. Write a sentence to explain what the inequality $60 \leq 12w < 108$ means in relation to the rectangle.
    d. Solve the inequality in Part c and explain its meaning.

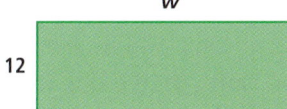

QY ANSWER

$\{x: x < 32\}$

Chapter 3

Lesson 3-7
Solving $ax + b < c$

> **BIG IDEA** Inequalities of the form $ax + b < c$ can be solved in two steps, similar to those used in solving $ax + b = c$.

If you are x years old and an older friend's age is y, then you can write the inequality $x < y$ to compare the ages.

Six years from now you will still be younger than your friend. The inequality that compares your ages then is $x + 6 < y + 6$.

In general, H years from now you will still be younger than your friend. This is written as $x + H < y + H$.

These examples illustrate the *Addition Property of Inequality*.

Mental Math

$n \parallel p$ and $m\angle 1 = 140°$.

a. What is $m\angle 2$?

b. What is $m\angle 3$?

Addition Property of Inequality

For all real numbers a, b, and c, if $a < b$, then $a + c < b + c$.

The Addition Property of Inequality can be represented with a balance, as shown below. Suppose a and b represent the weights of two packages where $a < b$.

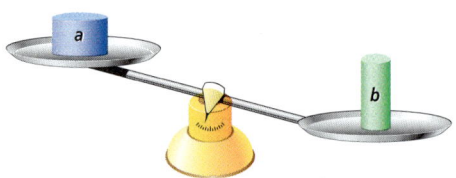

If the same weight c is added to each side of the balance at the right, then $a + c < b + c$.

Inequalities containing $>$, $\geq$, or $\leq$ signs work in the same way. Thus, sentences with $=$, $<$, $>$, $\leq$, or $\geq$ can all be solved in the same way.

Addition Property of Inequality (in words)

You may add the same number to both sides of an inequality or equation without affecting the set of solutions to the sentence.

Linear Equations and Inequalities

Remember that $b - a = b + -a$, so you can convert any subtraction expression to an addition expression. This means you may subtract the same number from each side of an inequality or equation without affecting its solutions.

Solving Inequalities with Positive Coefficients

The process of solving inequalities of the form $ax + b < c$ is quite similar to that of solving equations.

Example 1

A crate weighs 7 kilograms when empty. An orange weighs about 0.3 kilogram. For shipping, the crate and oranges must weigh at least 70 kilograms. How many oranges should be put in the crate?

Solution Let n be the number of oranges. Then the weight of n oranges is $0.3n$. The weight of the crate with n oranges is $0.3n + 7$, so the question can be answered by solving the inequality $0.3n + 7 \geq 70$.

This is of the form $ax + b \geq c$ and is solved the same way as $ax + b = c$.

$0.3n + 7 + -7 \geq 70 + -7$ Add -7 to both sides.
$\quad\quad 0.3n \geq 63$ Simplify.
$\quad\quad \dfrac{0.3n}{0.3} \geq \dfrac{63}{0.3}$ Divide each side by 0.3.
$\quad\quad n \geq 210$ Simplify.

At least 210 oranges should be put in the crate.

The United States produced almost 295 million boxes of oranges in 2004.

Source: U.S. Department of Agriculture

Check

Step 1 Check the boundary of $n \geq 210$.

$\quad$ Is $0.3(210) + 7 = 70$?
$\quad\quad 70 = 70$ Yes.

Therefore, 210 is the boundary point.

Step 2 Pick a number that satisfies the inequality $n \geq 210$. We choose 250.

$\quad$ Is $0.3(250) + 7 \geq 70$?
$\quad\quad 82 \geq 70$ Yes.

Since both Steps 1 and 2 produce true statements, $n \geq 210$ describes the solutions to the original inequality.

Chapter 3

Solving Inequalities with Negative Coefficients

Remember the Multiplication Property of Inequality when solving inequalities with negative coefficients. Example 2 is about drought, a natural phenomenon that affects many communities.

Example 2

Suppose that during a particularly dry summer, the level of water in a local reservoir decreased 1.5 feet each week. At the start of summer, the water level was 60 feet. When the water level is 30 feet or less, a "water emergency" is put into effect, with various water conservation measures enacted. When will a water emergency be in effect?

Solution First, express the problem as an inequality. Let x represent the number of weeks from the start of summer. Since the water level is initially 60 feet and the water level decreases 1.5 feet each week, the water level in week x is $60 - 1.5x$. A water emergency is put into effect when the water level is 30 feet or less. The problem is solved with the inequality $60 - 1.5x \leq 30$. Solve this inequality.

$60 - 1.5x + {-60} \leq 30 + {-60}$ Add -60 to each side.

$\phantom{60 - 1.5x + {-60} }-1.5x \leq -30$ Simplify.

$-1.5x \cdot -\frac{1}{1.5} \geq -30 \cdot -\frac{1}{1.5}$ Multiply each side by $-\frac{1}{1.5}$.
Reverse the inequality sign.

$\phantom{-1.5x \cdot -\frac{1}{1.5} }x \geq 20$ Simplify.

Here is a graph of the solutions.

 QY

The worst drought in 50 years affected at least 35 states during the long hot summer of 1988. Rainfall totals over the Midwest, Northern Plains, and the Rockies were 50–85% below normal.

Source: National Weather Service

▶ **QY**

Check that $\{x: x \geq 20\}$ is the set of solutions to Example 2 using the two-step check in Example 1.

Questions

COVERING THE IDEAS

1. Use the symbol $>$ to state the Addition Property of Inequality.
2. a. What inequality is suggested at the right?
   b. What is the solution to the inequality?

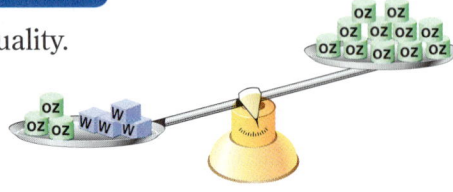

164   Linear Equations and Inequalities

Lesson 3-7

In 3–6, solve, graph, and check the inequality.

3. $3x + 4 < 19$
4. $6 \leq 4b + 10$
5. $5 \leq -3n + 2 - 7n + 203$
6. $-101 - 102x < 103$

7. T-shirts can be ordered from the Sports Central catalog for $7.50 each, with a shipping fee of $4 for the order. A team wants to spend $300 on shirts to give away to fans. How many shirts can they buy?

APPLYING THE MATHEMATICS

In 8–13, solve the inequality.

8. $6(-4x - 23) - 73 < 77$
9. $-0.2y + \frac{1}{2} \geq 0.48$
10. $15 \geq 12 + \frac{1}{3}a$
11. $\frac{-5d}{6} + 30 < 120$
12. $11 - 5(16 - 7p) \geq 49$
13. $594 > -2(q - 9) + 3(7 - 8q) - 17$

During sporting events, T-shirts are often launched into crowds.

14. Using the graph of $y = 13x - 20$ at the right, find each of the following for $13x - 20 \leq 45$.
    a. the boundary point
    b. a value of x in the solution set
    c. a value of x *not* in the solution set
    d. an inequality using x that describes all solutions

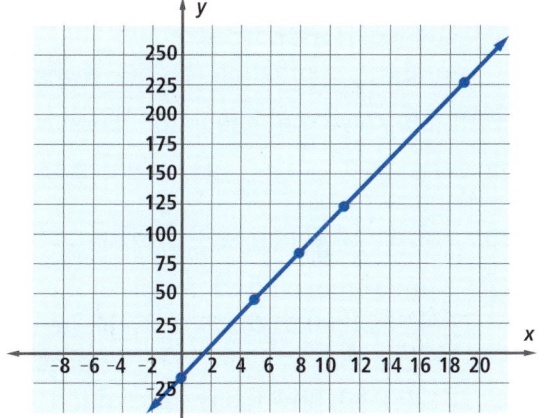

15. Using the table of values for $y = -9.6x - 41.4$ below, find each of the following for $-9.6x - 41.4 > -108.6$.

x	–3	–1	1	3	5	7	9	11	13
y	–12.6	–31.8	–51	–70.2	–89.4	–108.6	–127.8	–147	–166.2

   a. the boundary point
   b. a value of x in the solution set
   c. a value of x *not* in the solution set
   d. an inequality using x that describes all solutions

16. Use the formula $C = \frac{5}{9}(F - 32)$ to determine which Celsius temperatures are below 68°F.

17. A membership to Jenny's Gym costs $22 per month plus an initial $49.95 fee to join. If Arturo has $300 budgeted to join and pay monthly fees, for how many months can he be a member?

18. You are allowed to fly with a suitcase that weighs no more than 50 pounds. Suppose your suitcase and clothes total 35.2 pounds, and your shoes weigh 3.7 pounds per pair. How many pairs of shoes can you bring on the flight?

19. Compare and contrast the solutions to $-7w - 1 \leq -15$ and $7w + 1 \leq 15$.

20. a. Solve the sentence $ax + b < c$ for x when $a > 0$.
    b. How does the result from Part a change if $a < 0$?

Some people recommend that a person participate in an aerobic activity at least 3–5 times per week for 20–30 minutes per session as part of a healthy lifestyle.

Source: Aerobics and Fitness Association of America

REVIEW

21. Consider the inequality $8 > -2$. What inequality (if any) results if you multiply both sides of the inequality by each number? (Lesson 3-6)

    a. 100   b. -5 c. $\frac{1}{2}$ d. 0

22. The area of the foundation of a rectangular building is not to exceed 15,000 square feet. The width of the foundation is to be 150 feet. Write an inequality that should be solved to find how long the foundation can be. (Lesson 3-6)

In 23–25, solve the inequality. (Lesson 3-6)

23. $5j > 17$ 24. $4.2 < 8.4k$ 25. $\frac{5}{6}x \leq 12$

26. Solve and check $3x + (-5x) + 12(x - 15) = -4$. (Lesson 3-5)

27. Nomar gets paid $2 per pizza that he delivers in his car. He also gets paid $0.31 for every mile he drives while making deliveries. Last month he delivered 512 pizzas. If his monthly paycheck was $1,482.80, how many miles did he drive last month delivering pizzas? (Lesson 3-4)

28. **Skill Sequence** Simplify the expression. (Previous Course)

    a. $\frac{5}{6} + \frac{1}{9}$ b. $\frac{5}{6}c + \frac{1}{9}c$ c. $\frac{5c}{6} + \frac{c}{9}$

EXPLORATION

29. Create five inequalities of the form $ax + b < c$ equivalent to $x < 24$.

30. Create five inequalities of the form $ax + b < c$ equivalent to $x > 4$.

QY ANSWER

$x = 20$: $30 = 30$;
$x > 20$ ($x = 30$): $15 < 30$

Lesson 3-8

Solving Equations by Clearing Fractions

> **BIG IDEA** Equations with fractions can be transformed into equivalent equations without fractions.

Choosing a Multiplier to Clear Fractions

With the techniques you have learned, you can solve any linear equation. However, when you want to solve an equation containing fractions, for example $\frac{t}{3} + \frac{t}{2} - 350 = 270$, you may want to *clear the fractions* before you do anything else. The Multiplication Property of Equality allows you to do this. If you make a wise choice of a number by which to multiply both sides, the result will be an equation with no fractions.

We will examine the results of different multipliers for the equation $\frac{t}{3} + \frac{t}{2} - 350 = 270$, as shown below. For example, to tell a CAS to multiply both sides of the equation by 2, type (t/3+t/2−350 = 270)∗2 ENTER.

Mental Math

Let *n* be any real number. Determine if the statement is *always*, *sometimes but not always*, or *never* true.

a. $\frac{n}{3}$ is greater than *n*.

b. *n* is greater than −*n*.

c. 5*n* is equal to −5*n*.

Multiply by 2.	Multiply by 3.	Multiply by 6.
$\left[\frac{t}{3} + \frac{t}{2} - 350 = 270\right] \cdot 2$ $\frac{5 \cdot (t - 420)}{3} = 540$	$\left[\frac{t}{3} + \frac{t}{2} - 350 = 270\right] \cdot 3$ $\frac{5 \cdot (t - 420)}{2} = 810$	$\left[\frac{t}{3} + \frac{t}{2} - 350 = 270\right] \cdot 6$ $5 \cdot (t - 420) = 1620$
The CAS transformed the equation into $\frac{5(t-420)}{3} = 540$. But there is still a fraction in the equation. So 2 is not a useful multiplier.	Again, the result is an equation that has fractions. So 3 is also not a good multiplier.	Success! When 6 is a multiplier, the result is $5(t - 420) = 1,620$, an equation that has no fractions and is equivalent to $\frac{t}{3} + \frac{t}{2} - 350 = 270$.

Solving Equations by Clearing Fractions **167**

Chapter 3

On some CAS machines you must use the `expand` or `simplify` command to cause the multiplication to be carried out. You may need to type `expand((t/2 + t/3 − 350 = 270)*6)` or `simplify((t/2 + t/3 − 350 = 270)*6)` to multiply both sides by 6. So one multiplier that clears fractions in $\frac{t}{3} + \frac{t}{2} - 350 = 270$ is 6. But there are others, as you will see in the following activity.

Activity

Step 1 The table below shows the effect of three different multipliers on the equation $\frac{t}{3} + \frac{t}{2} - 350 = 270$. Experiment to find three more multipliers that clear the fractions. Record your results in the table.

Multiplier	Resulting Equation	Fractions Cleared?
2	$\frac{5(t - 420)}{3} = 540$	No
3	$\frac{5(t - 420)}{2} = 810$	No
6	$5(t - 420) = 1{,}620$	Yes
?	?	?
?	?	?
?	?	?

Step 2 Consider the multipliers you tried in Step 1. Describe the relationship between the multipliers that eliminate fractions and the original equation $\frac{t}{3} + \frac{t}{2} - 350 = 270$.

Step 3 Use what you have learned about multipliers in Steps 1 and 2 to find an equation equivalent to $\frac{5n}{6} + \frac{n}{4} + \frac{2n}{3} = 42$ that contains no fractions.

  a. Predict a value by which you could multiply each side of the equation to clear the fractions.

  b. Test your prediction using a CAS. Multiply each side of the equation $\frac{5n}{6} + \frac{n}{4} + \frac{2n}{3} = 42$ by the value and write down the results.

Lesson 3-8

Clearing Fractions in Equations

In the preceding activity you saw how to clear fractions in an equation. The idea is to multiply both sides of the equation by a common multiple of the denominators. The result is an equation in which all of the coefficients are integers.

Example 1

In 2004 the Washington Redskins and the Cleveland Browns had the highest earnings in the National Football League (NFL). The Redskins accounted for $\frac{1}{12}$ of the league's income and the Browns accounted for $\frac{1}{15}$ of the league's income. Their combined income was $129 million. What was the total league income for 2004?

a. Write an equation to describe the situation.
b. Solve by clearing the fractions.

Solution 1

a. Let T be the NFL's total earnings, in millions of dollars.
Redskins' earnings + Browns' earnings = 129
$$\frac{1}{12}T + \frac{1}{15}T = 129$$

b. Multiply each side by a common multiple of 12 and 15. We use 60.

$60\left(\frac{1}{12}T + \frac{1}{15}T\right) = 60 \cdot 129$ Multiply each side by 60.
$5T + 4T = 7{,}740$ Distributive Property
$9T = 7{,}740$ Combine like terms.
$T = 860$ Divide each side by 9.

The total income for the NFL in 2004 was $860 million.

Solution 2

b. Add the fraction coefficients.

$\frac{1}{12}T + \frac{1}{15}T = 129$ Write the equation.
$\frac{5}{60}T + \frac{4}{60}T = 129$ Find equivalent fractions with the same denominator.
$\frac{9}{60}T = 129$ Add like terms.
$60 \cdot \frac{9}{60}T = 60 \cdot 129$ Multiply each side by 60 to clear the fractions.
$9T = 7{,}740$ Simplify.
$T = 860$ Divide each side by 9.

In 2005, the average NFL team was worth $733 million.
Source: *Forbes*

 QY1

▶ **QY1**

Use a CAS to solve
$\frac{1}{12}T + \frac{1}{15}T = 129$.

Solving Equations by Clearing Fractions 169

Chapter 3

Clearing Fractions in Inequalities

When solving an inequality, you can multiply to clear fractions just like you do when solving an equation.

> **GUIDED**
>
> ### Example 2
> Solve $\frac{x}{4} - 8 > \frac{1}{6}$.
>
> **Solution** The two denominators in the sentence are 4 and 6. The least common denominator is __?__. So multiply each side of the inequality by __?__.
>
> $$\underline{\;?\;}\left(\frac{x}{4} - 8\right) > \underline{\;?\;} \cdot \frac{1}{6}$$
> $$\underline{\;?\;} \cdot \frac{x}{4} + \underline{\;?\;} \cdot 8 > 2$$
> $$\underline{\;?\;}\,x - \underline{\;?\;} > 2$$
> $$\underline{\;?\;}\,x > \underline{\;?\;}$$
> $$x > \underline{\;?\;}$$

> **To Clear Fractions in an Equation or Inequality**
>
> 1. Choose a common multiple of all of the denominators in the sentence.
> 2. Multiply each side of the sentence by that number.

 QY2

> ▶ **QY2**
>
> To clear the fractions, what number could you use to multiply each side of $\frac{m}{6} - \frac{3}{8}m \leq 5$?

Clearing Decimals

Like fractions, decimals can be cleared from an equation to give a simpler equation with integer coefficients. A decimal can be thought of as a fraction whose denominator is a power of 10. For example, 0.4 can be written as $\frac{4}{10}$, so the "denominator" of 0.4 is 10. Similarly, the "hidden denominators" of 9.38 (or $9\frac{38}{100}$) and 6.022 (or $6\frac{22}{1,000}$) are 100 and 1,000, respectively.

> ### Example 3
> Solve $5.85n - 9 = 2.7$.
>
> **Solution** The equation involves two decimals: 5.85 and 2.7. Their "hidden denominators" are 100 and 10. Since 100 is divisible by both 100 and 10, multiply each side of the equation by 100.

170   Linear Equations and Inequalities

$$5.85n - 9 = 2.7 \qquad \text{Write the equation.}$$
$$100(5.85n - 9) = 100 \cdot 2.7 \qquad \text{Multiply each side by 100.}$$
$$585n - 900 = 270 \qquad \text{Simplify.}$$
$$585n - 900 + 900 = 270 + 900 \qquad \text{Add 900 to each side.}$$
$$585n = 1{,}170 \qquad \text{Simplify.}$$
$$\frac{585n}{585} = \frac{1{,}170}{585} \qquad \text{Divide each side by 585.}$$
$$n = 2 \qquad \text{Simplify.}$$

Questions

COVERING THE IDEAS

1. Suppose $\frac{3}{5}w + 2 = 26$.
   a. Multiply each side of the equation by 5.
   b. Solve the resulting equation.
   c. Check your answer.

2. Consider the equation $\frac{m}{9} + \frac{m}{3} = 16$.
   a. Multiply each side by 9 and solve the resulting equation.
   b. Multiply each side by 27 and solve the resulting equation.
   c. What conclusions can you make from your work in Parts a and b?

3. Consider the equation $0.152 = 0.3m - 0.43$.
   a. What are the "hidden denominators" in the equation?
   b. Multiply each side by 100 and solve the resulting equation.
   c. Convert the decimals in the equation to fractions and solve the resulting equation.
   d. What conclusions can you make from your work in Parts b and c?

In 4 and 5, an inequality and a number are given.
a. Write the inequality that results if both sides of the inequality are multiplied by the given number.
b. Solve the inequality.
c. Graph the solution set on a number line.
d. Check your work.

4. $\frac{3n}{2} - \frac{n}{4} < 6$; 4
5. $\frac{2}{3}a + \frac{a}{5} \geq 21$; 15

Solving Equations by Clearing Fractions

Chapter 3

In 6–13, solve and check the sentence.

6. $\frac{3}{7}x + 2 = \frac{2}{5}$

7. $\frac{3}{4}y - \frac{1}{3} = 5$

8. $0.05n + 3.75 = 22.50$

9. $40{,}000 = 138{,}000 - 2{,}000c$

10. $\frac{d}{3} + \frac{3d}{5} < \frac{3}{4}$

11. $1 - \frac{n}{10} \geq -\frac{4}{5}$

12. $5 - \frac{t}{3} = -7$

13. $\frac{m}{5} - \frac{1}{13} \geq \frac{3}{22}$

14. Philo Dendrun owns $\frac{3}{8}$ of the stock in Blossom Industries and his wife Rhoda Dendrun owns $\frac{1}{4}$ of it. This means that they receive $\frac{3}{8}$ and $\frac{1}{4}$, respectively, of the dividends paid to the stockholders.

   a. Last year the Dendruns together earned $25,400 from the stock. What was the total amount of dividends paid to the stockholders?

   b. How much did stockholders other than Philo and Rhoda receive in dividends?

Nine of the top ten greatest one-day point gains on the Dow Jones Industrial Average occurred in 2000 or later.

Source: Dow Jones & Company, Inc.

APPLYING THE MATHEMATICS

15. When solving $4{,}000 = 8{,}000 - 2{,}000x$, a student first multiplies both sides by $\frac{1}{1{,}000}$. Is this a good idea? Why or why not?

In 16 and 17, solve the sentence.

16. $\frac{1}{6}\left(\frac{17}{3} - \frac{y}{4}\right) < -7$

17. $\frac{x}{6} + \frac{17}{36} - \frac{x}{4} = -7$

REVIEW

18. What inequality is suggested by the balance below? What is the solution to the inequality? (**Lesson 3-7**)

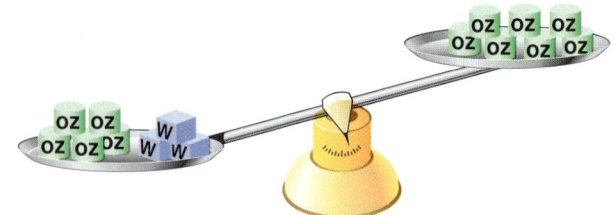

In 19–21, solve and check the inequality. (**Lesson 3-7**)

19. $5t - 3(7t + 1) < 93$

20. $77 \leq -3n + 29$

21. $5y - 16 > 49$

22. Use the formula $C = \frac{5}{9}(F - 32)$ to determine which Fahrenheit temperatures are between $50°C$ and $70°C$. (**Lessons 3-7, 1-1**)

23. Find the value of w in the pentagon below if the perimeter is 105 meters. (Lessons 3-5, 1-1)

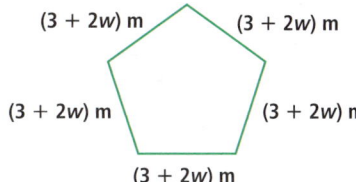

24. Felipe has been trying to lower his cell phone bill by limiting the length of his calls to an average of 2.5 minutes. His calls on November 30th were 2, 3, 6, 1, 1, 2, 1, 3, 4, 1, and 7 minutes long. (Lessons 3-4, 1-7)

    a. Was his average call less than 2.5 minutes long?

    b. What is the mean absolute deviation for the calls?

In 25–27, combine like terms. (Lesson 2-2)

25. $\dfrac{5}{t} + \dfrac{-4}{7t}$

26. $\dfrac{3x + y}{3} - \dfrac{2z + 8y}{5}$

27. $\dfrac{4x^2}{9} - \dfrac{7x^2}{18}$

EXPLORATION

28. Diophantus, a Greek mathematician who lived in the third century, was the first known person to use variables to stand for unknown numbers. About 200 years after his death, an algebraic riddle was written to honor him. Here is one version of that riddle, written as a rhyming poem. Decipher the riddle to find an equation and solve the equation to determine how long Diophantus lived.

    "Here lies Diophantus." The wonder behold
    Through art algebraic, the stone tells how old.
    "God gave him his boyhood one-sixth of his life,
    One-twelfth more as youth while whiskers grew rife;
    And then yet one-seventh ere marriage begun;
    In five years there came a bounding new son.
    Alas, the dear child of master and sage
    Met fate at just half his dad's final age.
    Four years yet his studies gave solace from grief,
    Then leaving scenes earthly he, too, found relief."

In 2006, about 22% of people in the world owned a cellular phone.
Source: International Telecommunication Union

QY ANSWERS

1. $T = 860$

2. Answers vary. Sample answer: 24

Chapter 3 Projects

1. Modeling Buying and Selling

Sue visits the CD store to sell some of her old CDs and buy used ones. The store will give her $2 for every CD she sells to them, and each CD costs $4 to buy. Sue also has $20 to buy used CDs in addition to any money she gets from selling her old ones.

a. Find an equation relating x, the number of CDs she buys; y, the number of CDs she sells; and z, the amount of money she has left afterwards.

b. Use three-dimensional graphing software or a graphing calculator to graph this equation. Describe the shape of the graph.

c. If Sue doesn't sell any CDs but still buys x CDs, your equation from Part a can be reduced to a linear equation. Graph this equation. Is this a constant-increase or constant-decrease situation? What inequality must be satisfied by x?

d. Repeat Part c for if Sue sells 5 CDs. What do you notice about the lines from Parts c and d? Explain.

2. Appliances and Energy

Visit a store or Web site where appliances are sold. New major appliances like refrigerators are tested for their expected energy consumption per year. That information is available to consumers to aid them in their decisions. Pick a range of sizes of refrigerators (for example, 16 to 21 cubic feet) and find the price and expected yearly energy cost for at least four different models. For each model, develop an equation that describes its total cost over its lifespan. Graph these equations. Is the least expensive model always the most economical? Which model is the best value if it is kept for 10 years?

3 Integer Solutions to Linear Equations

You have seen that equations of the form $ax + b = c$ always have one solution when a is nonzero. But even if a, b, and c are integers, the solution may not be an integer. Write a program for a graphing calculator or computer to solve $ax + b = c$ when a, b, and c are entered. Roll a die three times to determine values for a, b, and c, and use your program to solve for x. Record these values in a table and repeat the process until you have solved 20 equations. What percentage of solutions are integers? Is this result surprising? (Note that you can broaden this project by writing a program on your computer which generates and solves the equations.)

4 Combining Linear Equations

If a and d are nonzero, then the equations $ax + b = c$ and $dx + e = f$ always have one solution for x. What happens if you add these two equations together, that is, add the left sides together and the right sides together? Do you get a linear equation? When does it have a solution? What if you subtract one equation from the other? Using a CAS, multiply the two equations together. Do you get a linear equation? Explain.

5 The Strength of Spaghetti

Support a strand of uncooked spaghetti on each end, leaving a gap of about 7 inches. Unfold a paper clip. Hook one end over the spaghetti. Hook the other end through the rip of a small paper cup. Drop pennies into the cup one by one until the spaghetti "bridge" breaks. Record the number of pennies it takes to break the bridge. Repeat this process for bridges made with 2, 3, 4, 5, 6, and 7 strands of spaghetti. Each time, record the number of strands and the number of pennies that broke the bridge. Draw a line that approximates your data. Using your graph, predict how many pennies it will take to break a bridge made of 12 strands.

Chapter 3

Summary and Vocabulary

- **Constant-increase** and **constant-decrease situations** can be described by algebraic expressions of the form $ax + b$ (where x is changing) and equations of the form $ax + b = c$. Since the points (x, y) that satisfy the equation $y = ax + b$ lie on a line, we call the expressions *linear expressions* and the equations *linear equations*.

- There are many ways to solve an equation of the form $ax + b = c$. A **table of values** of the expression $ax + b$ may give a value that equals c. A **graph** of $y = ax + b$ may reach the value c at a point whose x-coordinate can be determined. For example, to solve $3x - 46 = 17$, you can make a table of values of $3x - 46$ for various values of x and check if 17 appears as a value of y. You can graph $y = 3x - 46$ and check if it crosses the line $y = 17$. Graphs and tables may be created by hand, but they are easily produced with the aid of technology.

- Tables and graphs can picture how a quantity is changing, but they are not reliable methods for finding an exact solution to a linear equation. A sure method is to use the **Addition and Multiplication Properties of Equality** to change the given equation into a simpler equation. For example, to solve $3x - 46 = 17$, you might add 46 to each side, resulting in $3x = 63$. Then multiply each side by $\frac{1}{3}$, resulting in $x = 21$. The solution 21 checks in the original equation.

- **Linear inequalities** can be solved by finding equivalent inequalities in much the same way that equivalent equations can be found to solve linear equations. For example, $-3x - 46 < 17$ is solved by adding 46 to each side and then dividing each side by -3 to get $x > 21$. Notice that when multiplying or dividing by a negative number, the inequality sign is reversed.

- To remove fractions from an equation, you can find a common multiple of the denominators and then multiply each side by that multiple.

Vocabulary

3-1
constant-increase situation
collinear
constant-decrease situation

3-2
solution to an equation

3-3
equivalent equations

3-6
inequality
boundary point
interval
endpoint

Theorems and Properties

Multiplication Property of Inequality (Parts 1 and 2) (p. 157)

Addition Property of Inequality (p. 162)

Chapter 3 Self-Test

Take this test as you would take a test in class. You will need a calculator. Then use the Selected Answers section in the back of the book to check your work.

In 1–3, solve and check the equation.

1. $4t - 5 = 13$
2. $5(2 + t) = -10$
3. $101 = 13f - 2(4 + 3f)$

In 4–6, solve the inequality.

4. $3(x - 4) \geq 12$
5. $5x + 14 < -26$
6. $26 - 2x > 10$

7. Match the solutions to Questions 4–6 with the graphs below.

   a.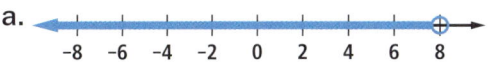

   b.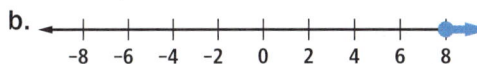

   c.

8. Solve $\frac{n}{4} - \frac{n}{8} = 3$ by clearing fractions.

9. A school held a fund-raising raffle that had three winners and one grand-prize winner. The value of the grand prize was twice the value of the other prizes combined. Together the prizes had a value of $3,500. What was the value of the grand prize?

10. Write down the steps you should take to solve the equation $-15 = -2x + 7$ for x. Solve the equation and check your answer.

In 11–13, use the following situation. Allison has $350 in her checking account and she withdraws $20 each month to pay for her school lunch ticket. After m months she has $t = 350 - 20m$ dollars in the account.

11. Make a table of values for the relationship.

12. Make a graph from the table.

13. When will Allison have $190?

14. Make a table to solve the inequality $30 < 2x - 6$.

15. Which commands would you enter on a CAS to find solutions to the inequality $5 - 3x > 17$? What are the solutions?

16. Toni is collecting leaves for a school science project. She needs to have 37 different types of leaves for the project. Toni already has 9 leaves and she plans on collecting 7 more each weekend. When will Toni have enough leaves to complete her project?

17. In 2006, the United States Postal Service charged $0.39 for the first ounce and $0.24 for each additional ounce for first-class mail.

    a. Write an equation for the price P of a first class letter that weighs w ounces.

    b. Use your equation from Part a to find the weight of a package that costs $3.27 to ship.

Chapter 3

Chapter Review

SKILLS
PROPERTIES
USES
REPRESENTATIONS

SKILLS Procedures used to find answers

OBJECTIVE A Solve and check linear equations of the form $ax + b = c$. (Lessons 3-4, 3-5)

In 1–12, solve and check the equation.
1. $4t + 3 = 15$
2. $5x + -3x + 6 = 12$
3. $(4 + n) + -10 = -4 + 5$
4. $-470 + 2r = 1{,}100$
5. $0.9y + 11.2 + 1.7y = 131.2$
6. $5(s + 4) = 85$
7. $4{,}000W - 8{,}000 = 12{,}000$
8. $21 = 2x + 3(2 + x)$
9. $\frac{2}{3}z + 14 = 4$
10. $16 = \frac{3}{4}x + 22$
11. $3(w + 4) - 4(2w - 2) = 7$
12. $\frac{n}{5} - \frac{2n}{11} = 6$

OBJECTIVE B Solve and check linear inequalities of the form $ax + b < c$. (Lessons 3-7, 3-8)

In 13–16, solve and check the inequality.
13. $2x + 11 < 199$
14. $-3 + d + 6 < 4$
15. $-28 \leq 18 - 3y - 7$
16. $4 < -16t + 7t + 5$

PROPERTIES The principles behind the mathematics

OBJECTIVE C Apply the Addition and Multiplication Properties of Equality and Inequality. (Lessons 3-3, 3-6, 3-8)

In 17 and 18, explain what has been done to both sides of the first equation to get the second equation.
17. If $17d - 17 = 22$, then $17d = 39$.
18. If $\frac{11}{12}b = \frac{2}{3}$, then $11b = 8$.

In 19 and 20, write a command you would enter in a CAS to complete the next step in solving the equation. Then predict the output of the CAS.

19.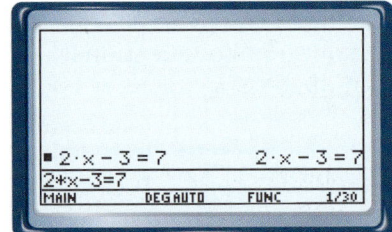

20.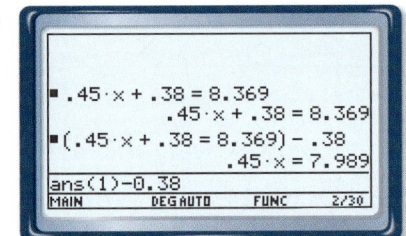

21. Given the inequality $-x \geq 4$, Desiree writes $x \leq -4$ as the next step. Is she correct? Why or why not?

22. To solve $5y + 38 < 50$, Kaya subtracts 38 from both sides. What inequality should she get?

USES Applications of mathematics in real-world situations

178 Linear Equations and Inequalities

OBJECTIVE D Use linear equations and inequalities of the form $ax + b = c$ or $ax + b < c$ to solve real-world problems. (Lessons 3-2, 3-4, 3-5, 3-7, 3-8)

23. If the temperature is $-12°C$, by how much must it increase to become hotter than $14°C$?

24. Monica has $250 in the bank and has a job that pays her $9 per hour. She deposits all the money she earns into a savings account. How long will it take her to save a total of $439?

25. Bo Constrictor earns $7.80 per hour at the zoo. He also receives weekly a $25 meal allowance, $15 for transportation, and $7.50 for dry cleaning. Last week he was paid a total of $297.10. How many hours did he work last week?

26. A $98,100 estate is to be split among four children and a grandchild after $5,000 in estate expenses are paid. Each child gets the same amount and the grandchild gets half that amount. How much will each receive?

27. Saudi Arabia has about one-fourth of the world's oil reserves. The rest of the Middle East has approximately two-fifths of the world's oil reserves. Together they have about 660 billion barrels. How many barrels are estimated to be in the world's oil reserve?

28. A small cup is 2 inches high. When stacked, each cup adds $\frac{3}{32}$ inch to the height of the stack. How many cups are in a stack that is $5\frac{3}{8}$ inches high?

REPRESENTATIONS Pictures, graphs, or objects that illustrate concepts

OBJECTIVE E Solve problems involving equations of the form $y = ax + b$ using tables or graphs. (Lessons 3-1, 3-2)

29. A tree now has a trunk with a radius of 12 centimeters. The radius is increasing by 0.5 centimeter per year. Its radius y after x years is described by $y = 12 + 0.5x$.

    a. Make a table of values for this relationship.
    b. Use the table to draw a graph.
    c. How many years will it take for the tree trunk to reach a radius of 15 centimeters?
    d. What will the radius be in 8 years?

30. In the summer of 2005, the Chicago area had its worst drought on record. On July 3, the level of the Fox River was 6.1 feet and dropping 0.2 foot per week. Let y be the level of the river after x weeks.

    a. Suppose the river continues to drop at the same rate. Write an equation for y in terms of x.
    b. Make a table of values for this relationship.
    c. Use the table to draw a graph.

31. Darnell has $55 in the bank and is adding $20 every week. Let b be his balance after w weeks.

    a. Write an equation for b in terms of w.
    b. Make a table of values for this relationship.
    c. Make a graph from the table.
    d. When will he have $175 in the bank? Explain how you found your answer.

OBJECTIVE F Graph all the solutions to a linear inequality. (Lessons 3-6, 3-7)

In 32–35, graph all solutions to each inequality on a number line.

32. $x \geq -4.3$

33. $d < 3\frac{1}{2}$

34. $2 < 5 - a$

35. $4m + 6 > -2$

Chapter Review

Chapter 4
More Linear Equations and Inequalities

Contents

- 4-1 Solving Percent Problems Using Equations
- 4-2 Horizontal and Vertical Lines
- 4-3 Using Tables and Graphs to Solve
- 4-4 Solving $ax + b = cx + d$
- 4-5 Solving $ax + b < cx + d$
- 4-6 Situations That Always or Never Happen
- 4-7 Equivalent Formulas
- 4-8 Compound Inequalities, *And* and *Or*
- 4-9 Solving Absolute Value Equations and Inequalities

A lightbulb manufacturer produces two kinds of bulbs: regular bulbs and new compact fluorescent (CF) bulbs. The CF bulbs are more energy efficient than regular bulbs, since they produce the same amount of light but use less electricity. However, CF bulbs cost more than regular bulbs. Would the money a person saves in electricity make up for the higher initial cost of the CF bulb? This question and other related questions can be answered by using tables, by drawing graphs, and by solving linear equations and inequalities.

$$a + bx > c + dx$$
$$a + bx < c + dx$$
$$a + bx = c + dx$$

Hours (h)	CF Bulb Cost	Regular Bulb Cost
0	$1.99	$0.52
100	$2.15	$1.00
200	$2.31	$1.48
300	$2.47	$1.96
400	$2.63	$2.44
500	$2.79	$2.92
600	$2.95	$3.40
700	$3.11	$3.88
800	$3.27	$4.36
900	$3.43	$4.84
1,000	$3.59	$5.32

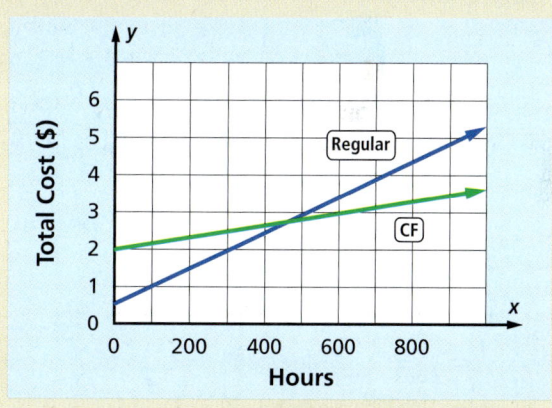

Equations and Inequalities

	CF Bulb Cost	Regular Bulb Cost
When is the regular bulb cheaper?	$1.99 + 0.16h > 0.52 + 0.48h$	
When is the CF bulb cheaper?	$1.99 + 0.16h < 0.52 + 0.48h$	
When are the costs the same?	$1.99 + 0.16h = 0.52 + 0.48h$	

In this chapter, you will use all these ways to solve linear sentences and study how they are related to each other.

Chapter 4

Lesson 4-1

Solving Percent Problems Using Equations

Vocabulary

percent

> **BIG IDEA** Many types of questions involving percents can be answered by solving an equation of the form $ax = b$.

The word **percent** (often written as the two words per cent) comes from the Latin words "per centum," meaning "per 100." So 7% literally means 7 per 100, the ratio $\frac{7}{100}$, or the decimal 0.07. The symbol % for percent is only a little more than 100 years old.

Mental Math

Evaluate.

a. -2^2

b. $-(-2^2)^2$

c. $-((-2^2)^2)^2$

Putting Percent Problems into the Form $p \cdot q = r$

Recall that to find a percent p of a given quantity q you simply multiply q by p. For example, to find 12% of 85, multiply $0.12 \cdot 85 = 10.2$. To find $5\frac{1}{4}\%$ of $3,000, calculate $0.0525 \cdot 3,000 = \$157.50$. This gives a straightforward method for solving many percent problems. Just translate the words into an equation of the form $p \cdot q = r$, where p is the percent in decimal or fraction form, q is the initial quantity, and r is the resulting quantity.

Example 1

7% of what number is 91?

Solution

7% of what number is 91?

$0.07 \cdot q = 91$ Change 7% to 0.07.

$\frac{0.07q}{0.07} = \frac{91}{0.07}$ Divide each side by 0.07.

$q = 1{,}300$ Simplify.

Check 7% of $1{,}300 = 0.07 \cdot 1{,}300 = 91$. It checks.

182 More Linear Equations and Inequalities

Example 2

What percent of 160 is 38.4?

Solution

What percent of 160 is 38.4?

$p \cdot 160 = 38.4$ Translate into an equation.

$\dfrac{p \cdot 160}{160} = \dfrac{38.4}{160}$ Divide each side by 160.

$p = 0.24$ Simplify. This is the solution to the equation.

$p = 24\%$ Rewrite the solution as a percent.

Check 24% of $160 = 0.24 \cdot 160 = 38.4$. It checks.

Percent Add-Ons and Discounts

Percents are very common in business, science, statistics, and even everyday shopping. In the next example, the Distributive Property gives a useful approach to a common situation.

Example 3

Ian, Cassady, Quincy, and Dylan ate at a local restaurant. The meal cost $36.50 without a tip. The bill states, "An 18% gratuity (tip) will be added for parties of 4 or more people." Find the total cost of the meal and tip.

An average of one out of five meals consumed by Americans—4.2 meals per week—is prepared in a commercial setting.

Source: Meal Consumption Behavior

Solution 1 Let M be the cost of the meal. We know $M = \$36.50$, but we ignore that for a while. Note that total cost = cost of meal + cost of tip.

total cost	$= 100\% \cdot M + 18\% \cdot M$	18% of $36.50
	$= 118\% \cdot M$	Distributive Property
	$= 1.18 \cdot 36.50$	Rewrite 118% as 1.18.
	$= 43.07$	Multiply.

So the total cost was $43.07.

Solution 2 Find the amount of the tip and then add it to the $36.50 meal cost.

tip	$= 18\% \cdot 36.50$	18% of $36.50
	$= 0.18(36.50)$	$18\% = 0.18$
	$= 6.57$	Simplify.

total cost $= 36.50 + 6.57 = 43.07$

Again, the total cost was $43.07.

Solving Percent Problems Using Equations

Chapter 4

 QY

The use of the Distributive Property in Solution 1 of Example 3 enables the total cost to be found with just one calculation. This idea is used in Example 4 to quickly calculate the price for an item on sale.

▶ **QY**

If a meal costs $21.16 with a 15% tip, what was the cost before the tip?

GUIDED

Example 4

A dishwasher normally costs $320, but it is on sale for 15% off. What is the sale price?

Solution Let P be the regular price. Then the discount is $15\% \cdot P$, or $0.15P$.

Sale price = regular price − discount

= __?__ − __?__ Translate into an equation.

= __?__ − __?__ Multiplicative Identity Property

= __?__ Combine like terms.

We know $P = \$320$. So the sale price is __?__ (__?__) = $__?__.

Another way to think about the sale price is that when 15% of the price is removed, 100% − 15%, or 85% of the price remains.

Markups and Discounts

If an item is discounted $x\%$, you pay $(100 - x)\%$ of the original or listed price.

If an item is marked up or taxed $x\%$, you pay $(100 + x)\%$ of the original or listed price.

Questions

COVERING THE IDEAS

1. 123% of 780 is what number?
2. 40% of what number is 440?
3. What percent of 4.7 is 0.94?
4. What number is 62% of 980?
5. Suppose a shirt is on sale for 10% off its original price of $23.50.
   a. What percent of the original price does the customer pay?
   b. How much does the customer pay before tax?

184 More Linear Equations and Inequalities

6. Suppose a dinner costs D dollars and you wish to give a 20% tip.
   a. What is the amount of the tip?
   b. What is the total cost of the meal with tip?

7. A table is being sold at "40% off." If the price of the table before the sale was T dollars, what is the sale price?

8. An electronics store owner buys a television from the manufacturer, and then adds 45% of that cost to get the price the customer pays. If a TV sells for $499, what was the price the store owner paid?

9. The total cost of a digital camera including an 8.5% sales tax is $215. How much tax was paid on this purchase?

10. The Cupertinos bought a new car. The total amount they paid was $28,250.75 including the 8% sales tax. What was the price before the sales tax was added?

Digital-camera sales surpassed $4 billion in 2004.
Source: *Twice*

11. Consider the sales receipt at the right. Determine the sales-tax rate as a percent.

12. Clearwater High School expects a 4% increase in enrollment next year. There are 1,850 students enrolled this year.
    a. How many students will the school gain?
    b. What is the expected enrollment next year?

13. According to the 2000 census, 73% of the 221 million U.S. residents age 15 and older had been married at least once. How many U.S. residents over 15 had *never* been married?

> **Ian's Computer Store**
> 2217 Smithville Road
> Middleboro, Ohio 42155
>
> – CASH RECEIPT –
> 9/30/2004 1:29:07 PM
>
> Ref ID: 4727-1324
>
> AMOUNT   15.85
> TAX   1.15
> ----------
> TOTAL:
> CASH:
>
> CHANGE:
>
> THANK YOU
> To Reorder Call: (714) 449–8211
> MADE IN THE USA

APPLYING THE MATHEMATICS

In 14 and 15, use the following information. Sucrose, or common table sugar, is composed of carbon, hydrogen, and oxygen. Suppose an experiment calls for 68.4 grams of sucrose.

14. If 4.2% of the weight of sucrose is carbon, how many grams of carbon are in the 68.4 grams?

15. If 35.2 grams of the 68.4 grams are oxygen, what percent of the weight of sucrose is oxygen?

Chapter 4

16. On a mathematics test there were eight A's, twelve B's, ten C's, two D's, and zero F's. What percent of the students earned A's?

17. Consider the following situation. Jorge works at a clothing store in the local mall. As an employee, he receives a 20% discount on clothes. He spent $118.50 at the store.

    a. What was the regular price of the clothes Jorge bought before the discount was figured?

    b. Suppose a student answers Part a in the manner shown below. Show why the student's answer is not correct.

    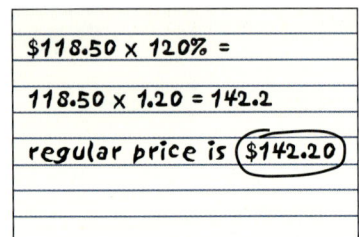

18. A DVD player originally cost $150. It is on sale for $72.50. What is the percent of the discount, rounded to the nearest tenth of a percent?

In 2002, about 40 million households in the United States had DVD players.

Source: Consumer Electronics Association

REVIEW

In 19–21, solve and check the equation. (Lesson 3-8)

19. $\frac{k}{7} + 18 = \frac{19}{2}$

20. $\frac{u}{12} - \frac{u}{2} = 35$

21. $\frac{a+3}{4} + 3 = \frac{1}{2}$

22. Penelope Nichols spends half her monthly income on housing and food, and budgets the other half as follows: $\frac{1}{4}$ on clothes, $\frac{1}{3}$ on entertainment, and $\frac{1}{4}$ on transportation. She saves the remaining $40. What is her monthly income? (Lesson 3-8)

23. Rebecca works at a clothing store that pays her $7.25 per hour plus 9% of the cost of the clothes that she sells. If she works for 8 hours, what is the cost of the clothing she must sell if she wants to earn at least $87? (Lesson 3-7)

24. **Multiple Choice** Which graph below shows the solutions of $6 < -4n + 10$? (Lesson 3-7)

A    B

C    D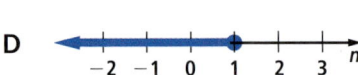

25. Consider the sequence made with cubes below.
(Lessons 3-2, 1-4)

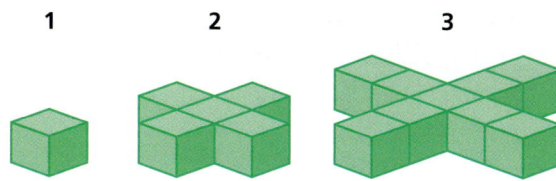

   a. How many cubes will be needed to make the 4th term?
   b. Write an equation for the number c of cubes that are in the nth term.
   c. Using your equation from Part b, find the term number that contains 33 cubes.

26. Trina timed her commute to school every morning for two weeks. Her times, in minutes, were 15, 22, 17, 12, 14, 16, 20, 21, 21, and 19. Compute the m.a.d. of Trina's commute. (Lesson 1-7)

27. Evaluate $(w + 4.7)(2.6 - w)(w + 7.1)$ when $w = -4.7$.
(Lesson 1-1)

28. A recipe calls for $\frac{1}{4}$ tablespoon of vanilla. If Sally wants to make $\frac{2}{3}$ of the recipe, how much vanilla will she use? (Previous Course)

EXPLORATION

29. Jerome noticed that 40% of 50 is equal to 50% of 40 and concluded a% of b is always equal to b% of a. Explain why this works for any positive values of a and b.

QY ANSWER

$18.40

Chapter 4

Lesson 4-2

Horizontal and Vertical Lines

Vocabulary

horizontal line, $y = k$

vertical line, $x = h$

▶ **BIG IDEA** Every horizontal line has an equation of the form $y = k$; every vertical line has the form $x = h$.

Equations for Horizontal Lines

In Lesson 2-3, you saw several number puzzles in which you began with a number and then performed a series of calculations to get a final answer. The puzzling part came from the surprising relationship between the starting number and the final result. Below you see ordered pairs for two of these puzzles in a table and on a graph. For each puzzle, the points lie on a line. In the "I Can Guess Your Age Puzzle," the result is always the same as the starting number. The line that is graphed has equation $y = x$.

Mental Math

△ABC is equilateral.

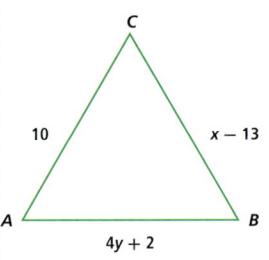

a. What is x?

b. What is y?

I Can Guess Your Age Puzzle	
Starting Number (x)	Result (y)
3	3
10	10
−8	−8

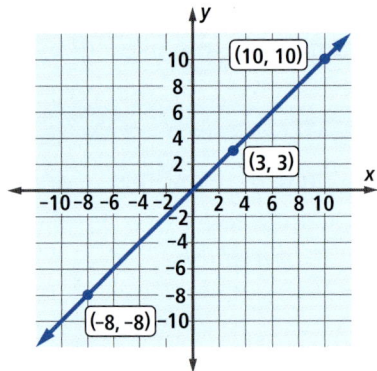

But the "Seven Is Heaven" graph is different.

Seven Is Heaven Puzzle	
Starting Number (x)	Result (y)
3	7
10	7
−1	7

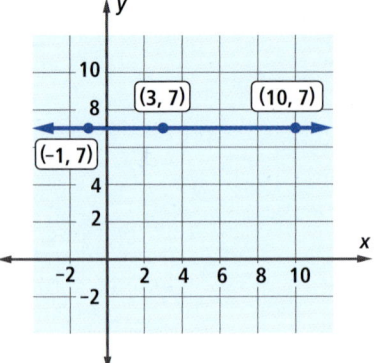

The points lie on a *horizontal line*. For all of these points, the y-coordinate equals 7. In short, $y = 7$. Every *horizontal line* has an equation of this type.

188   More Linear Equations and Inequalities

> **Equation of a Horizontal Line**
>
> Every **horizontal line** has an equation of the form **y = k**, where k is a real number.

Example 1

Find an equation describing all points on the line graphed at the right.

Solution The points are on a horizontal line that crosses the y-axis at 2. An equation for the line is $y = 2$.

Check Two of the points on the line have coordinates (1, 2) and (−3, 2). These numbers satisfy the equation $y = 2$.

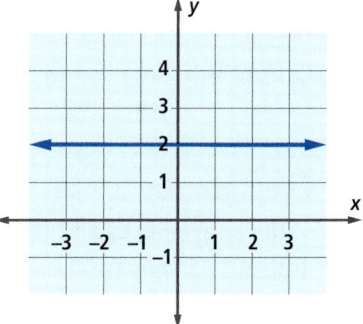

Equations for Vertical Lines

A vertical line is drawn at the right.

Notice that the x-coordinate of the ordered pairs is 4.5 regardless of the y-coordinate. Thus an equation for the line is $x = 4.5$. This means x is fixed at 4.5, but y can be any number.

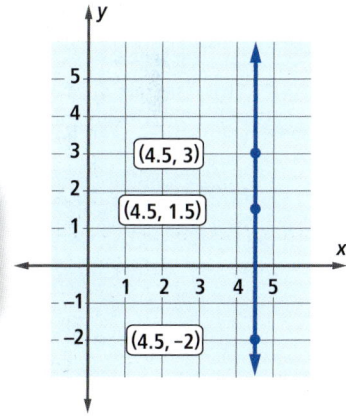

> **Equation of a Vertical Line**
>
> Every **vertical line** has an equation of the form **x = h**, where h is a real number.

An equation with only one variable, such as $x = -4.5$ or $y = 7$, can be graphed on a number line (in which case its graph is a point), or on a coordinate plane (in which case its graph is a line). The directions or the context of the problem will usually tell you which type of graph to draw.

Example 2

a. Graph $x = -20$ on a number line.
b. Graph $x = -20$ on a coordinate plane.

Solutions

a. Draw a number line. Mark the point with coordinate equal to −20, as shown below. The graph of $x = -20$ on a number line is the single point with coordinate −20.

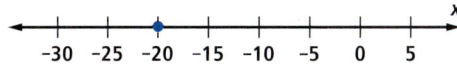

(continued on next page)

Horizontal and Vertical Lines

b. Draw a coordinate grid. Plot points whose x-coordinate is −20. Some points are (−20, 10), (−20, 4), and (−20, −8). Draw the line through these points.

The graph of $x = -20$ in a coordinate plane is the vertical line that crosses the x-axis at −20.

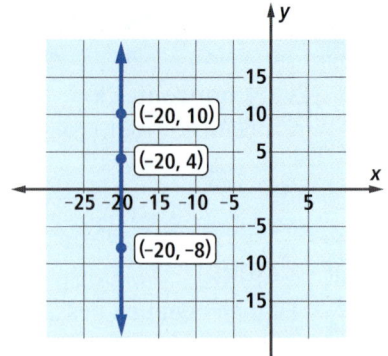

Horizontal Lines and Linear Patterns

Horizontal lines can be helpful in solving equations or inequalities.

Example 3

At the beginning of a vacation, Matt has $900 in his savings account. As long as he has at least $300 in his account, he does not have to pay a service fee. Each morning he withdraws $50 from an ATM for spending money. For how many days can he withdraw $50 per day without paying a service fee?

Solution 1 Let x = the number of days after Matt's vacation began. Let y = the amount of money in his bank account. Then $y = 900 - 50x$.

Day (x)	Balance (y)
0	900
4	700
8	500
12	300
16	100

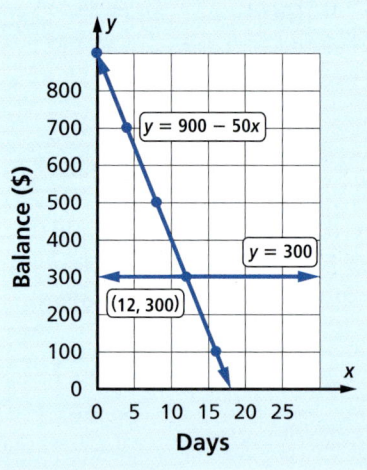

The first modern day ATM in the United States was introduced to consumers in 1971 by Chemical Bank.

Source: Cash Technologies, Inc.

The table of (x, y) values can be used to graph $y = 900 - 50x$. Also graphed is $y = 300$ to represent the balance that he must keep to avoid a service fee. As long as the point for Matt's balance is at or above the $y = 300$ line he does not have to pay a service fee. The lines appear to intersect at $x = 12$, so for the first 12 days, Matt's balance is high enough to avoid paying a service fee.

Solution 2 Translate "Matt's balance is at least $300" into an inequality and solve. Let x = the number of days Matt has been on vacation. In x days, he will have withdrawn 50x dollars, so we want to know when $900 - 50x \geq 300$.

$$900 - 50x - 900 \geq 300 - 900 \quad \text{Subtract 900 from each side.}$$
$$-50x \geq -600 \quad \text{Collect like terms.}$$
$$\frac{-50x}{-50} \leq \frac{-600}{-50} \quad \text{Divide each side by } -50.$$
$$\text{Change the sense of the inequality.}$$
$$x \leq 12 \quad \text{Simplify.}$$

For the first 12 days, Matt does not have to pay a service charge.

Deviation from the Mean

For statistical data in a scatterplot, a horizontal line at the mean can help to show how the data relate to the average value. The hourly temperatures in Flagstaff, Arizona, on a June day are shown in the graph below.

Related to each temperature is its deviation from the mean, which is the difference between the actual temperature and the mean temperature.

For example, when $h = 15$ (3 P.M.), the temperature was 82.9°F, giving a deviation of $82.9 - 68.1 = 14.8$°F from the mean. At $h = 3$ (3 A.M.), the deviation was $52 - 68.1 = -16.1$°F.

Activity

Use the table and graph of the temperatures at Flagstaff on a June day. Notice that the values on the vertical axis begin at 45. The interval $0 < y < 45$ is compressed on the graph since there are no data in that interval.

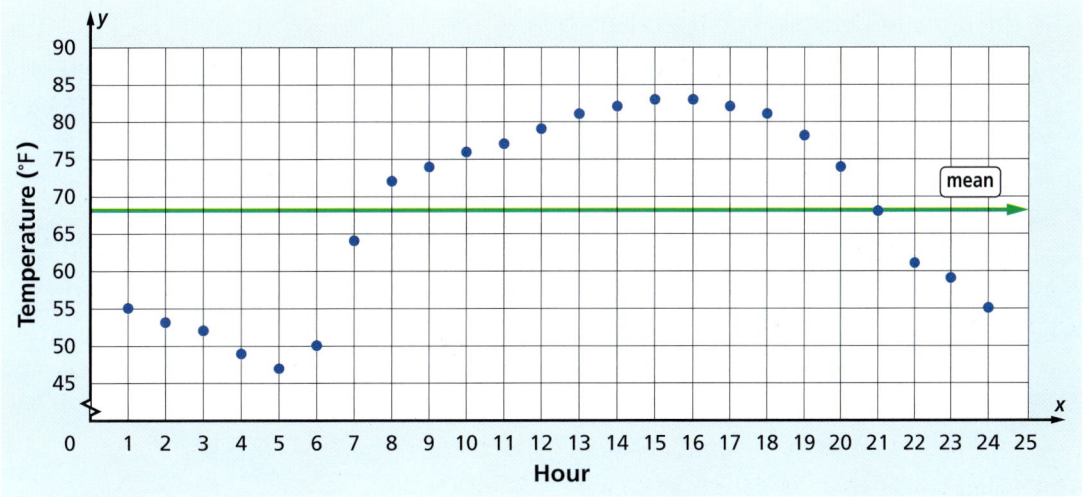

(continued on next page)

Chapter 4

1. What is the equation of the horizontal line that is graphed?
2. Give the deviation from the mean at 7 A.M. and 8 A.M.
3. For which two hours was the deviation +5.8°F?
4. The temperatures t are described by the interval $\underline{?} \leq t \leq \underline{?}$.
5. The deviations d are described by the interval $\underline{?} \leq d \leq \underline{?}$.
6. The deviation is positive when h, the number of hours since midnight, is in the interval $\underline{?} \leq h \leq \underline{?}$.
7. The deviation is negative when $h \leq \underline{?}$ or $h \geq \underline{?}$.
8. The maximum deviation and the maximum absolute deviation are $\underline{?}$ and $\underline{?}$ respectively. Why is the minimum deviation not equal to the minimum absolute deviation?

Hour (h)	Temperature (°F)
1	55
2	53.1
3	52
4	48.9
5	46.9
6	50
7	64
8	72
9	73.9
10	75.9
11	77
12	79
13	81
14	82
15	82.9
16	82.9
17	82
18	81
19	78.1
20	73.9
21	68
22	61
23	59
24	55

Sometimes deviations are taken from a number other than the mean. For instance, it is common in golf to give the player's results not by the score, but by how the score deviates from par. (*Par* is the expected number of strokes a golfer should take on the hole.) For example, in the 2006 U.S. Open, par was 70 strokes for 18 holes. Some scores in the last round were; Jim Furyk 0 (his score was 70), Ryuji Imada +4 (score: 74), Jeff Sluman –1 (score: 69), and Vijay Singh +3 (score: 73). In other sports, the deviation of a team's score from its opponent's score is described with statements like "We're down by 3 points!"

Questions

COVERING THE IDEAS

1. a. List three ordered pairs whose *y*-coordinate is 4.
   b. Graph your points from Part a and draw a line through them.
   c. Write the equation of your line from Part b.

In 2 and 3, write an equation for each graph.

2.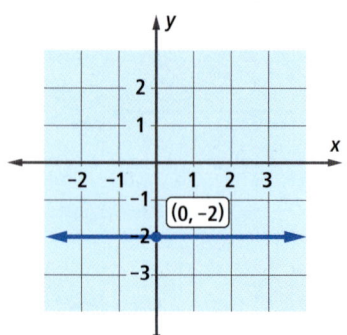

3.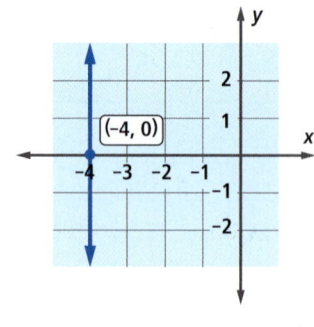

192   More Linear Equations and Inequalities

4. **Fill in the Blank** All points on a horizontal line have the same __?__-coordinate.

5. **Fill in the Blank** All points on a vertical line have the same __?__-coordinate.

In 6 and 7, an equation is given.
  a. Graph all points on a number line that satisfy the equation.
  b. Graph all points in the coordinate plane that satisfy the equation.

6. $x = 2$

7. $y = 1.5$

8. At the beginning of the year, Kylie has $580 in her savings account. As long as she has $200 in her account she does not have to pay a service fee. How long can she withdraw $20 per week without paying a service fee?

In 9–11, the graph shows the annual snowfall on Mt. Hood near Portland, Oregon. The mean snowfall is 98.5 inches per year. This information is important for the water supply in the region, and also for people who like to ski on Mt. Hood.

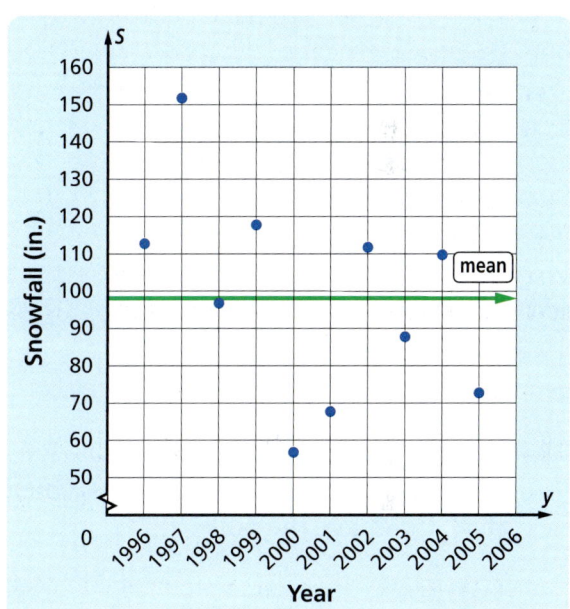

9. a. Let S be the snowfall in year y. What is an equation of the line that is graphed?
   b. Which year had the greatest absolute deviation from the mean?

10. a. Which year's snowfall was closest to the mean?
    b. What does that tell about its absolute deviation from the mean?

11. a. In which two consecutive years was the deviation negative?
    b. Was this good or bad news for Portland residents? Explain your answer.

In 12 and 13, write an equation for the line containing the given points.

12. (−9, 12), (−4, 12), (0.04, 12)

13. (−6, −3), (−6, 4), (−6, 22)

14. a. Write an equation for the horizontal line through (7, −13).
    b. Write an equation for the vertical line through (7, −13).

15. **Fill in the Blanks** Horizontal lines are parallel to the __?__-axis and perpendicular to the __?__-axis.

Chapter 4

16. **Matching** Match each table with the appropriate graph.

a.
x	y
−3	1
−2	0
−1	−1
0	−2
1	−3
2	−4

b.
x	y
−3	1
−3	2
−3	3
−3	4
−3	5
−3	6

c.
x	y
−4	−5
−2	−5
0	−5
2	−5
4	−5
6	−5

i.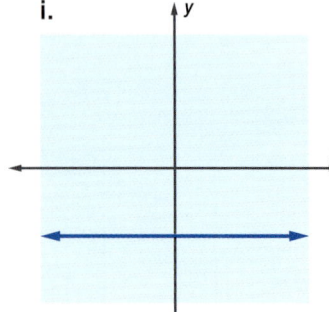

ii.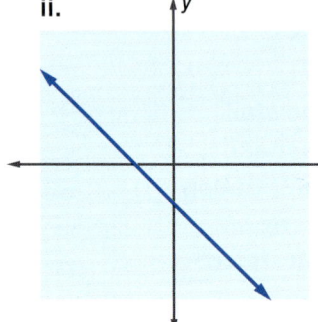

iii.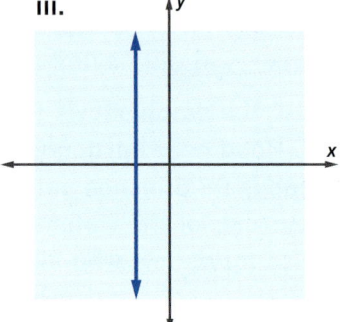

APPLYING THE MATHEMATICS

17. Predict where the lines $y = 5$ and $x = 7$ will intersect without graphing.

18. The table at the right gives information about a set of data values that has a mean of 53. Complete the table.

Value	Deviation
45	?
42	?
?	−1
?	13
?	7

19. The temperature in Acapulco, Mexico does not vary greatly whereas in a desert location like Reno, Nevada, the temperature can change greatly over the course of a day. Suppose that at 7 A.M. the temperature in Acapulco was 80°F and in Reno it was 65°F. If the Acapulco temperature stayed steady all morning, and the Reno temperature rose 5°F per hour, when were the temperatures in the cities the same?

20. Write an equation for a line through the points $(4, m)$, $(0, m)$, $(6, m)$, and $(−8, m)$.

Lesson 4-2

21. A furnace repair service charges $35 for travel time plus $70 per hour to repair furnaces. Lloyd's Smoothie Emporium needs its furnace repaired, but Lloyd is willing to spend no more than $315 on the repair.
    a. Write an equation that relates the repair cost y to the time spent x.
    b. Draw a graph of the equation that you wrote in Part a.
    c. On the same coordinate axes as in Part b, draw the line $y = 315$ to represent the money that Lloyd is willing to spend on the repair.
    d. Use your graph to determine the maximum number of hours that the furnace repairperson could work and still keep Lloyd's bill under $315.
    e. Check your answer to Part d by solving an inequality.

REVIEW

22. In 2006, the seating capacity in Fenway Park was 38,805 people. If 28,486 people show up for a baseball game, what percent of the stadium is full? Round your answer to the nearest percent. (**Lesson 4-1**)

23. After a 6% sales tax, it cost the Clarences $17,000 for a car. What was the amount before the tax was applied? (**Lesson 4-1**)

24. The Sholitons bought a farm, but they did not know the capacity of the heating-oil storage tank. At one point, the tank was $\frac{1}{8}$ full. After a delivery of 450 gallons, the tank's gauge showed that it was $\frac{7}{8}$ full. How many gallons of oil does the tank hold? (**Lesson 3-8**)

25. Last year the Guptas paid $18,550 in income taxes, more than one fourth of their earned income. (**Lessons 3-7, 3-6**)
    a. Let I = the Guptas earned income last year. Write a sentence that describes the situation above.
    b. Solve the sentence.

26. Solve $2(x + 1) + 7 = 5x$. (**Lesson 3-3**)

27. Graph $y = -5x - (1 + x)$. (**Lesson 3-1**)

Major League Baseball's oldest ballpark, Fenway Park, is located in Boston, Massachusetts.

Source: Major League Baseball

EXPLORATION

28. Find a set of data with numbers that lend themselves to finding the mean. Find the mean. Then find the deviations. Describe the advantages and disadvantages of using deviations rather than the actual values.

Lesson 4-3

Using Tables and Graphs to Solve

▶ **BIG IDEA** Tables and graphs are often useful for comparing values from two or more situations.

GUIDED

Example

You are the manager of an office. Your company needs to lease a copy machine. You must choose between two office supply firms. Acme Copiers offers a copier for $250 per month with an additional charge of $0.01 per copy.

Best Printers offers the same machine for $70 per month with a per-copy charge of $0.03.

Describe the *break-even point*, the situation for which the costs are the same.

Solution 1 Use a table. The cost of the machine depends upon the number of copies made. To compare, you must look at prices for many situations.

1. Let $x =$ the number of copies made per month.

   Price for x copies from Acme Copiers $= 250 + 0.01x$

   Price for x copies from Best Printers $= 70 + 0.03x$

   Use your calculator to make a table showing the costs for 0; 2,000; 4,000; 6,000; …; 20,000 copies per month. Use the table to draw conclusions about which company's pricing plan best suits your needs.

Number of Copies (x)	Acme Copiers Price $250 + 0.01x$	Best Printers Price $70 + 0.03x$
0	?	?
2,000	?	?
4,000	?	?
6,000	?	?
8,000	?	?
10,000	?	?
12,000	?	?
14,000	?	?
16,000	?	?
18,000	?	?
20,000	?	?

Mental Math

a. Half a serving of soup is 150 mL. How many mL are 2 servings?

b. How many mL are 3 servings?

c. How many mL are 4 servings?

Lesson 4-3

2. When $x \geq$ __?__, Acme's price is lower. When $x \leq$ __?__, Best's price is lower.

3. From this table you cannot tell the exact break-even point. However, it seems to occur between __?__ copies and __?__ copies.

4. Another table with x between the values in Step 3 above may yield a solution. Complete this table.

Number of Copies (x)	Acme Copiers Price $250 + 0.01x$	Best Printers Price $70 + 0.03x$
8,000	?	?
8,500	?	?
9,000	?	?
9,500	?	?
10,000	?	?

5. The break-even point occurs when $x =$ __?__ and the price is __?__.

6. It is important to describe the two variables for the break-even point.
   a. The x value: How many copies are made when Best and Acme charge the same amount? __?__
   b. The y value: What is the amount charged by each company when the prices are the same amount? __?__

A graph can help you interpret the information in a table and can help answer many questions.

Solution 2 Use a graph.

1. Let the prices be as in Step 1 of Solution 1.

   Graph $y_1 = 250 + 0.01x$ on the window $0 \leq x \leq 20{,}000, 0 \leq y \leq 500$.

   Graph $y_2 = 70 + 0.03x$ on the same window.

2. Use the INTERSECT command to determine the point at which the graphs intersect. (__?__ , __?__)

3. What do the coordinates of the point of intersection mean?

   The x-coordinate, __?__, means __?__.

   The y-coordinate, __?__, means __?__.

One of the challenges of finding solutions using tables or graphs is that the result is a coordinate point with both an x and a y value. You need to pay close attention to the question you are being asked to determine whether the x- or the y-coordinate is the final answer.

Using Tables and Graphs to Solve

Chapter 4

Graphs, tables, and algebraic sentences each have advantages. Graphs can display a great deal of information and are useful for comparing values, but may be time-consuming to make. Tables might also be time-consuming to make. Graphing calculators and spreadsheets can make both tables and graphs, but still may not always give exact solutions. Solving equations and inequalities using algebraic properties is often preferred because they are efficient tools, and the results are precise. In the next two lessons, you will use properties to solve equations with variables on each side of the equal sign.

Questions

COVERING THE IDEAS

1. The table below lists the charges for color copies of digital photos at two different camera shops.

Number of Copies	Cost at Shop A	Cost at Shop B
1	$1.30	$1.20
2	$1.55	$1.50
3	$1.80	$1.80
4	$2.05	$2.10
5	$2.30	$2.40
6	$2.55	$2.70

   a. For what number of copies is Shop A's cost less?
   b. For what number of copies is Shop B's cost less?
   c. Graph the costs of the copies. Which graph is higher for 2 copies? Which graph is higher for 7 copies?
   d. If y is the cost of x copies, what are the (x, y) coordinates of the break-even point?

2. A bakery keeps a supply of flour and sugar. The graph at the right shows how many pounds of each are in the bakery storeroom over a period of days.
   a. Estimate when the bakery has the same amount of flour as sugar. How many pounds of each are in the storeroom?
   b. Give an example of a day for which there is more flour than sugar.
   c. Use an inequality to describe when there is more sugar than flour.

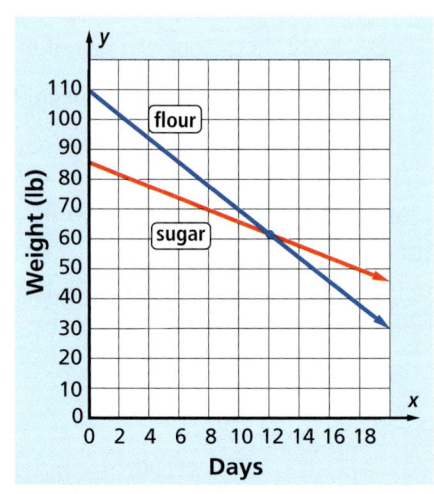

198   More Linear Equations and Inequalities

3. Rental-car companies sometimes charge a set fee plus an amount for each mile that the car is driven. Suppose Extra Value Cars charges $18.32 plus $0.32 per mile and Rhodes Rental charges $26.24 plus $0.20 per mile.

   a. Use the graph to approximate the point of intersection of the two lines.
   b. What does the x-coordinate of the point of intersection represent in the problem? What does the y-coordinate represent?
   c. Using the graph, determine which company is less expensive if 80 miles are driven. Explain how you know this from the graph.

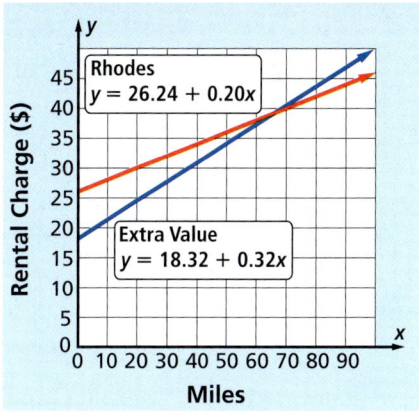

4. The population of Coolsville is currently 25,000 and is growing at a rate of 600 people per year. Across the river is the town Dulle, which currently has a population of 34,900 and is decreasing by 300 people per year. Use the graph and the table to answer the following questions.

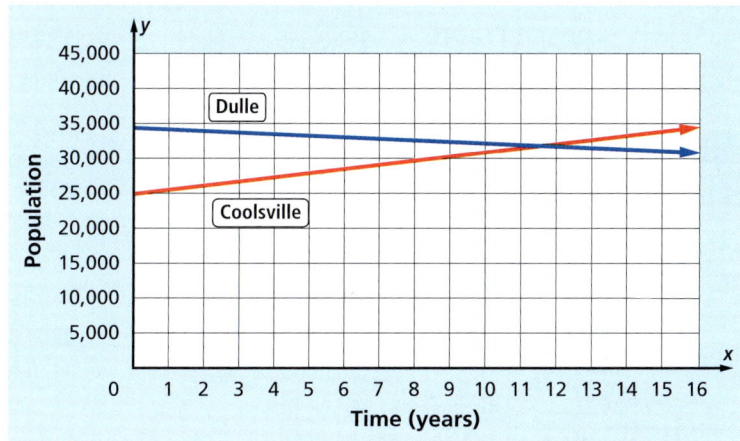

Time in Years (x)	Coolsville Population (y_1)	Dulle Population (y_2)
0	25,000	34,900
1	25,600	34,600
2	26,200	34,300
3	26,800	34,000
4	27,400	33,700
5	28,000	33,400
6	28,600	33,100
7	29,200	32,800
8	29,800	32,500
9	30,400	32,200
10	31,000	31,900
11	31,600	31,600
12	32,200	31,300
13	32,800	31,000
14	33,400	30,700
15	34,000	30,400
16	34,600	30,100

   a. Write an equation for y, the population of Coolsville after x years. Write a similar equation for the population of Dulle after x years.
   b. Give the approximate coordinates of the intersection or break-even point. Explain what the two coordinates of this point represent in the problem.
   c. **Fill in the Blanks** Until __?__ years, __?__ had the larger population. After __?__, __?__ had the larger population. At __?__ years, Coolsville and Dulle had the same population of __?__.
   d. Write an inequality that represents the values of x for which the population of Coolsville is greater, and an inequality that represents the values of x for which the population of Dulle is greater.

APPLYING THE MATHEMATICS

5. Theo has $30 and is *saving* at a rate of $6 per week. Michelle has $150 and is *spending* at a rate of $5 per week.
   a. Write an expression for the amount Theo has after w weeks.
   b. Write an expression for the amount Michelle has after w weeks.
   c. Make a graph. Use it to determine when Theo and Michelle will have the same amount.

6. Alicia is offered two sales positions. With Company Q, she would earn $800 per month plus 5% commission on sales. (This means that 5% of the money her customers spend is added to her $800 salary.) With Company P, she would earn $600 per month plus a 6% commission on sales.
   a. If Alicia expects sales of about $20,000, which company would pay her more monthly?
   b. Complete the table at the right.
   c. Determine how much Alicia must sell to be paid more at Company P than at Company Q.

Sales (S)	Earnings at Company Q	Earnings at Company P
$12,000	$1,400	$1,320
$14,000	$1,500	$1,440
$16,000	?	?
$18,000	?	?
$20,000	?	?
$22,000	?	?
$24,000	?	?
$26,000	?	?
$28,000	$2,200	$2,280
$30,000	$2,300	$2,400

REVIEW

In 7 and 8, write the equation for the line pictured in the graph. (Lesson 4-2)

7.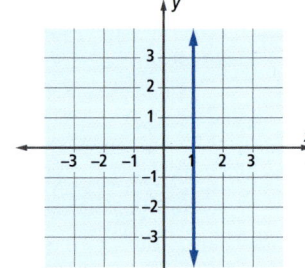

8.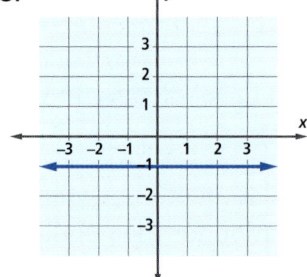

9. On a coordinate grid, graph the following three lines. (Lessons 4-2, 3-1)
   line ℓ: $y = 7$ line m: $x = 2$ line n: $y = 2x - 3$
   a. At what point do lines ℓ and m intersect?
   b. At what point do lines ℓ and n intersect?
   c. Find the area of the triangle formed by the three lines.

10. Ms. Chang invested $8,450 in stocks. After 1 year, the value of her stocks had fallen and her investment was now worth $7,625. By what percent did her investment fall? (Lesson 4-1)

In 11 and 12, compute in your head. (Lesson 4-1)

11. What is 25% of 60?

12. 110 is what percent of 100?

13. Mary has a collection of foreign coins. $\frac{5}{12}$ of her coins are from China and $\frac{1}{4}$ are from Japan. If she has 248 Japanese and Chinese coins altogether, how many coins does Mary have in all? (Lesson 3-8)

14. **Skill Sequence** Solve each sentence. (Lessons 3-7, 3-3)
    a. $0n = 8$ b. $8n = 0$ c. $8n > 0$ d. $-8n > 0$

15. a. Determine whether the equations $(2x + 2)(-9) + 6x = 10$ and $2(-6x - 9) = 10$ are equivalent.
    b. If the equations in Part a are equivalent, explain why they are equivalent. (Lesson 3-3)

16. Mika and June traveled from Denver, Colorado to San Francisco, California, a distance of about 1,250 miles. Mika drove in a car, while June left later and flew by airplane. The graph shows the distance y each had traveled x hours after Mika began his trip. (Previous Course)
    a. Who arrived in San Francisco first? How can you tell?
    b. How long did June's trip take?

Here are various Japanese coins, ranging in value from 1 to 500 yen.

EXPLORATION

17. There are three kittens Tic, Tac, and Toe. They each have different colored fur: orange, gray, and yellow. They also live in different homes: a hotel, a condo, and a house. Using the following information, find the fur color of each kitten and where each lives.
    (1) Tic does not live in the condo.
    (2) The cat that lives in the house does not have orange fur.
    (3) Tac lives in the hotel.
    (4) Tic's favorite color is gray, but that is not her fur color.
    (5) The cat that lives in the hotel has orange fur.

(*Hint:* It can help to record what you know in a table like the one shown below. You can eliminate an incorrect pairing by placing an X in the cell. Place an O in the cell for a correct pairing.)

	Orange	Gray	Yellow	Hotel	Condo	House
Tic	?	?	?	?	?	?
Tac	?	?	?	?	?	?
Toe	?	?	?	?	?	?

Chapter 4

Lesson 4-4

Solving $ax + b = cx + d$

Vocabulary

general linear equation, $ax + b = cx + d$

▶ **BIG IDEA** An equation of the form $ax + b = cx + d$ can be solved with just one more step than solving one of the form $ax + b = c$.

Mental Math

Estimate to the nearest integer.

a. $4.1 \cdot 3.9$

b. $-6.2 \cdot 3.8$

c. $-0.5 \cdot -5.05$

Solving $ax + b = cx + d$ with a CAS

In this lesson, you will see how to solve the **general linear equation, $ax + b = cx + d$.** First we explore solving them with a CAS.

Example 1

Solve $5x - 12 = 16 + 9x$ using a CAS.

Solution

Step 1 Begin by entering the equation in a CAS.

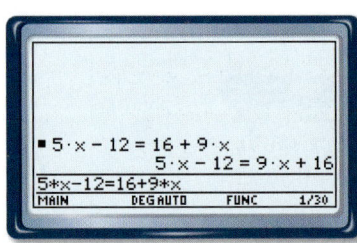

Step 2 Each side has a variable term, $5x$ on the left and $9x$ on the right. Eliminate one of these by subtracting it from each side. We choose to subtract $5x$. On some CAS, you can just type in $-5x$. On other CAS, put the equation inside parentheses and then type $-5x$.

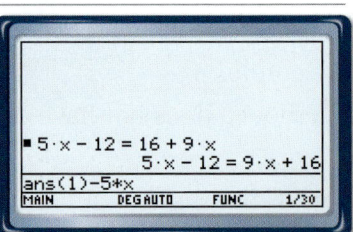

Step 3 Once you press ENTER, the CAS will subtract $5x$ from both sides of the equation. The result is $-12 = 4x + 16$. This is an equation of the type you solved in Chapter 3.

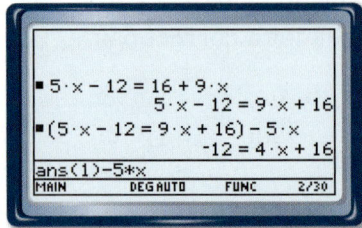

202 More Linear Equations and Inequalities

Step 4 Use the CAS to carry out the familiar steps to complete the solution. Subtract 16 from each side.

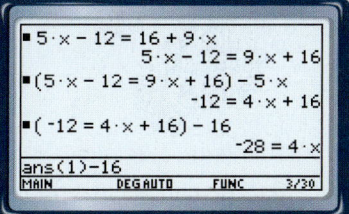 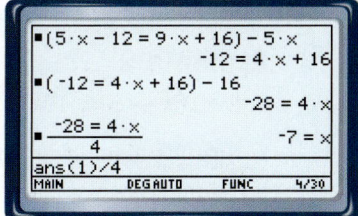

Step 5 Divide each side by 4. The equation is solved, $x = -7$.

As you can see, the methods you learned to solve simpler equations are also used to solve equations when the variable is on both sides. As before, sometimes the first step is to simplify one side of the equation. Some CAS will automatically simplify expressions. On other CAS, you must use the EXPAND command. The first step for one CAS is shown at the right. The original equation is $3z - 4 = 2 - 5(z + 5)$. The EXPAND command simplifies it to $3z - 4 = -5z - 23$. The next step would be to eliminate one of the variable terms, either $3z$ or $-5z$.

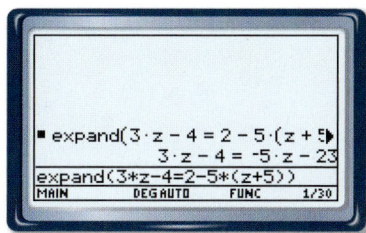

Activity

Use a CAS to solve the following equations. Remember, a CAS can help you make good decisions on the process of solving equations. Write down what you do to both sides of the equation as you go through the process.

1. $3z - 4 = 5z - 23$
2. $21 - 6x = 22 - {-8x}$
3. $-p = 13p - 42$
4. $33n - 102 + 4n = -23n - 252$
5. $9(9 - k) = 6k - 1$
6. $\frac{2}{3}b - \frac{6}{7} = \frac{5}{9}b + \frac{11}{3}$
7. $\frac{3}{4}(24 - 8y) = 2(5y + 1)$
8. $12.3 - (3.4w - 4.5) = 5.6(6.7 - 7.8w) - 8.9$
9. Write a description of the process you can use to solve $4x + 3 = 6x - 8$.
10. Use the process described in Question 9 to solve $4x + 3 = 6x - z$ for x without a CAS. Then check your solution with a CAS.
11. Solve $4x + t = 6x - z$ for x without a CAS. Then check your solution using a CAS.

Solving $ax + b = cx + d$ with Algebraic Processes

An important problem-solving strategy in mathematics and other fields is to turn a problem into a simpler one that you already know how to solve. In the Activity, you saw that to solve an equation in which the unknown is on both sides, you should turn it into an equation that has the variable on just *one* side.

Chapter 4

Example 2
Solve $16x - 5 = 10x + 19$.

Solution Each side has a variable term, $16x$ and $10x$. Subtract one of these from each side to eliminate it. We choose to subtract $10x$.

$16x - 5 = 10x + 19$	Write the equation.
$16x - 5 - 10x = 10x + 19 - 10x$	Subtract $10x$ from each side.
$6x - 5 = 19$	Combine like terms.

The result is a simpler equation that you know how to solve.

$6x - 5 + 5 = 19 + 5$	Add 5 to each side.
$6x = 24$	Simplify.
$\frac{6x}{6} = \frac{24}{6}$	Divide each side by 6.
$x = 4$	Simplify.

Check Substitute 4 for x in the original equation.
Does $16(4) - 5 = 10(4) + 19$?
$64 - 5 = 40 + 19$
$59 = 59$ Yes.

But a question remains. In the first step, should you subtract $10x$ or $16x$ first? In solving $16x - 5 = 10x + 19$, we subtracted $10x$. We could have subtracted $16x$ instead. Either way works.

 QY

> ▶ QY
>
> Solve $16x - 5 = 10x + 19$ by first subtracting $16x$ from both sides.

Equations that Require Simplifying First

Now consider an equation that is more complicated. Again, we work to turn this problem into a simpler one.

GUIDED
Example 3
Solve $7 - (x + 5) = 4(x + 11)$.

Solution First, simplify each side. On the left side rewrite the subtraction in terms of addition.

$7 + -(x + 5) = 4(x + 11)$	Rewrite the equation.
$7 + \underline{} = \underline{}$	Distributive Property to remove parentheses
$-x + 2 = 4x + 44$	Simplify.

Now the equation has the form $ax + b = cx + d$. It is similar to Examples 1 and 2.

204 More Linear Equations and Inequalities

$-x + 2 + x = 4x + 44 + x$ Add x to both sides.
   $\underline{}$ = $\underline{}$ Add like terms.
   $\underline{}$ = $\underline{}$ Subtract 44 from each side.
   $\underline{}$ = x Divide each side by 5.

Check

$7 - (\underline{} + 5) = 4(\underline{} + 11)$ Substitute -8.4 for x in the original equation.

$7 - \underline{} = 4(\underline{})$ Remember to use the order of operations.

$10.4 = 10.4$ Yes, it checks.

Example 4

A dog breeder raises two kinds of dogs. At birth, the average puppy of breed A weighs 14.8 ounces and gains weight at a rate of 0.5 ounce per week. Breed B puppies are smaller at birth, weighing about 11.6 ounces. But they gain weight faster, at 0.9 ounce per week. How many weeks will it be before the puppies are the same weight?

Solution Let w = number of weeks that have passed. The weight of a breed A puppy will be $14.8 + 0.5w$. The weight of a breed B puppy will be $11.6 + 0.9w$. Set the expressions equal to each other to indicate that the weights are the same.

The American Kennel Club (AKC) officially recognizes more than 150 breeds of dogs.

Source: American Kennel Club

$14.8 + 0.5w = 11.6 + 0.9w$ Write the equation.
$14.8 + 0.5w - 0.5w = 11.6 + 0.9w - 0.5w$ Subtract $0.5w$ from each side.
$14.8 = 11.6 + 0.4w$ Add like terms.
$14.8 - 11.6 = 11.6 + 0.4w - 11.6$ Subtract 11.6 from each side.
$3.2 = 0.4w$ Simplify.
$\dfrac{3.2}{0.4} = \dfrac{0.4w}{0.4}$ Divide each side by 0.4.
$8 = w$ Simplify.

After 8 weeks, the weights of breed A and breed B puppies will be the same.

Check 1 Substitute 8 for w in the original equation.

Does $14.8 + 0.5(8) = 11.6 + 0.9(8)$?

$14.8 + 4.0 = 11.6 + 7.2$

$18.8 = 18.8$ Yes, it checks.

Check 2 Use a table and graph. Let x = the number of weeks that have passed, let Y_1 = weight of a breed A puppy, and Y_2 = weight of a breed B puppy. Enter Y1=14.8+0.5x and Y2=11.6+0.9x.

(continued on next page)

Solving $ax + b = cx + d$

Chapter 4

The row where $x = 8$ shows both breeds weigh 18.8 ounces. The INTERSECT command shows the two lines intersect at (8, 18.8).

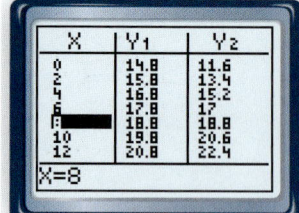 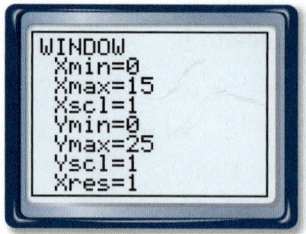 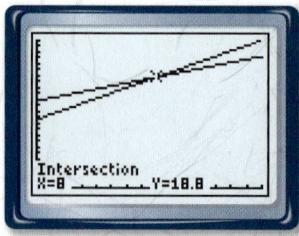

No matter how long and complex the two sides of a linear equation are to begin with, the equation can never be more complicated than $ax + b = cx + d$ after each side is simplified.

Questions

COVERING THE IDEAS

1. The equation $9.3m - 4 = 11 + 2m$ is an equation of the form $ax + b = cx + d$. What are a, b, c, and d?

2. **Fill in the Blanks** An equation is solved below. Fill in the blanks to explain the steps of the solution.
   a. $x - 4 = -3x - 7$ Add __?__ to each side.
   b. $4x - 4 = -7$ Add __?__ to each side.
   c. $4x = -3$ __?__ each side by __?__.
   d. $x = -\frac{3}{4}$ Simplify.

3. a. To solve $10t + 5 = 4t + 7$, what can you add to both sides so t is on only one side of the equation?
   b. Solve the equation in Part a.

4. a. Solve $3x + 18 = 5x - 22$ by first adding $-5x$ to each side.
   b. Solve $3x + 18 = 5x - 22$ by first adding $-3x$ to each side.

5. Solve $250 + 0.01n = 70 + 0.03n$ by subtracting $0.03n$ from each side.

6. a. In solving $7m - \frac{23}{4} = -5m$, what advantage does adding $-7m$ to each side of the equations have over adding $5m$?
   b. Solve $7m - \frac{23}{4} = -5m$.

In 7–14, solve the equation and check your solution.

7. $4k + 39 = 7k + 6$
8. $9n = -6n + 5$
9. $14y + 5 = 8y - 1$
10. $46 - 8p = 19 + p$
11. $7 - m = 8 - 3m$
12. $1.55t - 2.85 = 8.4t + 10.85$
13. $3d + 4d + 5 = 6d + 7d + 8$
14. $4(x - 4) = 5(x - 3)$

206 More Linear Equations and Inequalities

15. Nebraska's population had been increasing at a rate of 13,300 people per year and reached 1,711,000 in 2000. West Virginia's population had been increasing at a rate of 1,400 people per year and reached 1,808,000 in 2000. If the rates of increase do not change in the future, when will the populations be equal?

16. The 2000 population of Dallas, Texas was 1,189,000. It has been increasing at a rate of 21,500 people each year. The 2000 population of Philadelphia, Pennsylvania was 1,518,000 and has been decreasing at a rate of 6,500 people each year. Assuming the rates do not change in the future, when will the populations of Dallas and Philadelphia be the same?

APPLYING THE MATHEMATICS

17. The boxes on the balance are of equal weight. Each cylinder represents 1 ounce.

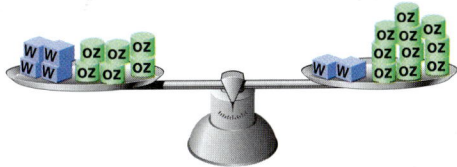

   a. What equation is represented by the balance?
   b. Describe the steps you could use to find the weight of a box using a balance.
   c. What is the weight of one box?
   d. When you use a balance to represent solving $ax + b = cx + d$, why is there only one sensible choice of a variable term to eliminate from each side?

18. Five more than three times a number is three more than five times the number. What is the number?

19. In 2004, the women's Olympic winning time for the 100-meter freestyle in swimming was 53.84 seconds. The winning time had been decreasing at an average rate of 0.32 second per year. The men's winning time was 48.17 seconds and had been decreasing by an average of 0.18 second a year. Assume that these rates continue in the future.
   a. What will the women's 100-meter winning time be x years after 2004?
   b. What will the men's 100-meter winning time be x years after 2004?
   c. After how many years will the winning times be the same?

Ranomi Kromowidjojo of the Netherlands won the gold medal in the 100-meter freestyle at the 2012 Olympic Games in London.

Source: International Olympic Committee

Chapter 4

20. Refer to the figures below. The perimeter of the triangle is equal to the perimeter of the square. Find the length of a side of the square.

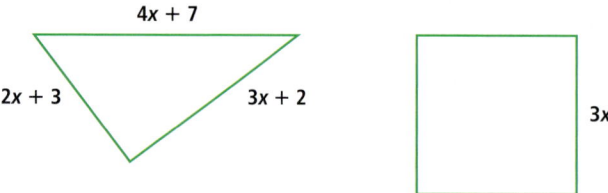

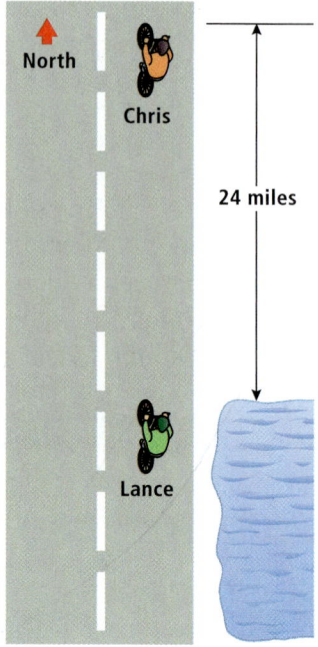

21. Lance and Chris are biking along the same road. Chris travels at a speed of 9 miles per hour, while Lance is faster and goes 13 miles per hour. Right now, Lance is next to a lake and Chris is 24 miles north of the lake.
    a. Make a table showing how far each cyclist is from the lake at hours 0, 1, 2, 3, 4, 5, and 6.
    b. Use the table to find how long it takes Lance to catch up with Chris.
    c. Solve an equation to find how long it takes Lance to catch up to Chris.

(not drawn to scale)

REVIEW

22. The table at the right shows the cost of making copies of photos at two different photo shops. (**Lesson 4-3**)
    a. For how many copies does it cost less to get photos copied at Ruby's than at Paula's?
    b. For how many copies does it cost less to get photos copied at Paula's than at Ruby's?

Number of Copies	Ruby's Photos	Paula's Prints
1	$0.75	$0.80
2	$1.25	$1.40
3	$2.00	$2.00
4	$2.75	$2.60
5	$3.50	$3.20

23. Write an equation for the line containing the points $(8, d)$, $(0, d)$, and $(-4, d)$. (**Lesson 4-2**)

24. During the 2005–2006 National Basketball Association (NBA) season, Steve Nash scored approximately 17% of all the points that the Phoenix Suns scored, and Shawn Marion scored approximately $\frac{2}{10}$ of all the Suns' points. Nash and Marion combined to score 3,258 points that season. Approximately how many points did the team score in all? (**Lesson 3-8**)

25. Solve $9 - \frac{5c}{8} < \frac{21}{4}$. (**Lessons 3-8, 3-7, 3-6**)

26. Alvin wants to spend no more than $18 at the state fair for admission and rides. If admission to the fair is $8 and each ride costs $1.25, how many rides can Alvin take? (**Lesson 3-6**)

27. The reciprocal of 6 is added to the opposite of 6. What is the result? (**Lessons 2-7, 2-4**)

28. Rewrite $18(3 - x)$ using each property. (**Lessons 2-1, 1-2**)
    a. Commutative Property of Addition
    b. Commutative Property of Multiplication
    c. Distributive Property

In 29 and 30, estimate in your head to the nearest whole number. (Previous Course)

29. $\frac{9}{10} + \frac{12}{13} + \frac{8}{7}$

30. $9.92 \cdot 23$

EXPLORATION

31. This puzzle is from the book *Cyclopedia of Puzzles,* written by Sam Loyd, Jr. in 1914. If a bottle and a glass balance a pitcher, a bottle balances a glass and a plate, and two pitchers balance three plates, how many glasses will balance with a bottle?

QY ANSWER

$16x - 5 - 16x = 10x + 19 - 16x$

$-5 = 19 - 6x$

$-5 - 19 = 19 - 6x - 19$

$-24 = -6x$

$\frac{-24}{-6} = \frac{-6x}{-6}$

$4 = x$

Chapter 4

Lesson 4-5

Solving $ax + b < cx + d$

> **BIG IDEA** An inequality of the form $ax + b < cx + d$ can be solved by applying the same operations to both sides as in solving the equation $ax + b = cx + d$.

Just as an equation can have the variable on both sides of the equal sign, an inequality can have the variable on both sides of the inequality sign. The methods from the last lesson can also be used to solve linear inequalities of the form $ax + b < cx + d$.

Consider the following situation. Years ago, the Moseley family planted a 12-foot beech tree and a 4-foot maple tree. Beech trees grow at a rate of about $\frac{1}{2}$ foot per year. Maple trees grow about 1 foot per year. So t years after planting, the height of the beech tree is $12 + 0.5t$ feet and the height of the maple tree is $4 + t$ feet.

Mental Math

a. How long is a 9-inning baseball game that averages 20 minutes per inning?

b. How long is a 4-quarter basketball game that averages 40 minutes per quarter?

c. How long is a 3-period hockey game that averages 50 minutes per period?

Example 1

Mrs. Moseley looked at an old photograph of the two trees. In it, the beech tree was taller than the maple tree. She wondered when the photo was taken.

Solution 1 Use an algebraic solution. Write an inequality relating the heights. The beech tree was taller than the maple tree, so the heights satisfy the inequality $12 + 0.5t > 4 + t$.

Solve this inequality as you would solve an equation.

$12 + 0.5t > 4 + t$	Write the equation.
$12 + 0.5t + (-0.5t) > 4 + t + (-0.5t)$	Add $-0.5t$ to each side.
$12 > 4 + 0.5t$	Add like terms.
$12 + (-4) > 4 + (-4) + 0.5t$	Add -4 to each side.
$8 > 0.5t$	Simplify.
$\frac{8}{0.5} > \frac{0.5t}{0.5}$	Divide each side by 0.5.
$16 > t$	Simplify.

So $t < 16$. The photo was taken less than 16 years after the Moseleys planted the trees.

Solution 2 Use a table as shown on the next page. You can see the beech tree was taller when $t < 16$. This verifies that the photo was taken less than 16 years after the Moseleys planted the trees.

National Arbor Day was founded by J. Sterling Morton in Nebraska in 1872.

Source: National Arbor Day Foundation

210 More Linear Equations and Inequalities

Solution 3 Use a graph. Write an equation describing each tree height. For the beech tree, $h = 12 + 0.5t$. For the maple tree, $h = 4 + t$. Graph each equation and find the point of intersection.

t Number of Years	Height h (ft) Beech Tree	Maple Tree
0	12	4
4	14	8
8	16	12
12	18	16
16	20	20
20	22	24
24	24	28

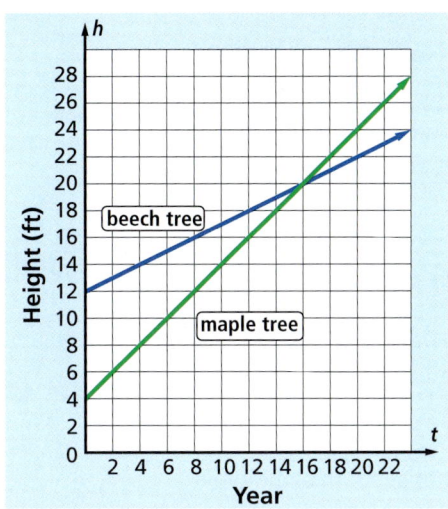

The lines intersect at the point (16, 20). So sixteen years after they were planted, the trees were both 20 feet tall. We are looking for the times when the beech tree was taller than the maple tree. So look for values of t where the beech's line is *above* that for the maple. These times lie to the left of the intersection point (16, 20). This is where $t < 16$. The photo was taken less than 16 years after the trees were planted.

The Addition and Multiplication Properties of Inequality can also be used to solve any inequality of the form $ax + b < cx + d$. In Guided Example 2, two algebraic solutions are given. Solution 2 involves dividing the inequality by a negative number. Recall that multiplying or dividing an inequality by a negative number reverses the inequality sign.

GUIDED
Example 2
Solve $7 - 11x \geq 4x + 12$.

Solution 1

$\begin{aligned} 7 - 11x &\geq 4x + 12 & &\text{Write the inequality.} \\ 7 - 11x + \underline{} &\geq 4x + 12 + \underline{} & &\text{Add } \underline{} \text{ to each side.} \\ 7 &\geq 15x + 12 & &\text{Add like terms.} \\ 7 - \underline{} &\geq 15x + 12 - \underline{} & &\text{Subtract } \underline{} \text{ from each side.} \\ -5 &\geq 15x & &\text{Simplify.} \\ \underline{} &\geq \underline{} & &\underline{} \text{ each side by 15.} \\ \underline{} &\geq x & &\text{Simplify.} \end{aligned}$

(continued on next page)

Solution 2

$7 - 11x \geq 4x + 12$	Write the inequality.
$7 - 11x - \underline{} \geq 4x + 12 - \underline{}$	Subtract __?__ from each side.
$7 - 15x \geq 12$	Add like terms.
$7 - 15x - \underline{} \geq 12 - \underline{}$	Subtract __?__ from each side.
$-15x \geq 5$	Simplify.
$\frac{-15x}{-15} \underline{} \frac{5}{-15}$	Divide each side by -15. Be sure to __?__ the inequality sign.
$x \underline{} -\frac{1}{3}$	Simplify.

Check Recall that checking an inequality requires two steps.

Step 1 Try the boundary value of x. Check that $x = \underline{}$ makes both sides of the original sentence equal.

$$7 - 11(\underline{}) \geq 4(\underline{}) + 12$$

Step 2 Try a number that satisfies $x \leq \underline{}$. Test to see if this number makes the original inequality true.

$$7 - 11(\underline{}) \geq 4(\underline{}) + 12$$

Example 3

Five times a number is less than three times the same number. Find the number.

Solution It may seem that there is no such number. But let's work it out and see. Let n be such a number. Then n must be a solution to $5n < 3n$. Solve this as you would any other linear inequality.

$5n < 3n$	Write the inequality.
$5n - 3n < 3n - 3n$	Add $-3n$ to each side.
$2n < 0$	Combine like terms.
$\frac{2n}{2} < \frac{0}{2}$	Divide each side by 2.
$n < 0$	Simplify.

So n must be less than zero. Any negative number will work.

Check

Step 1 Check the boundary value. If $n = 0$, $5 \cdot 0 = 3 \cdot 0$, and $0 = 0$. It checks.

Step 2 Pick $n < 0$ to check that the inequality is true. We let $n = -4$.

$$\text{Is } 5(-4) < 3(-4)?$$
$$-20 < -12? \quad \text{Yes, it checks.}$$

5 times n will be less than 3 times n exactly when n is a negative number.

Questions

COVERING THE IDEAS

1. Two hot air balloons are descending at a constant rate from their cruising altitude to the ground. Both balloons begin descending at the same time, but from different elevations. The green balloon starts at 700 feet and the purple balloon starts at 550 feet as shown in the graph below.

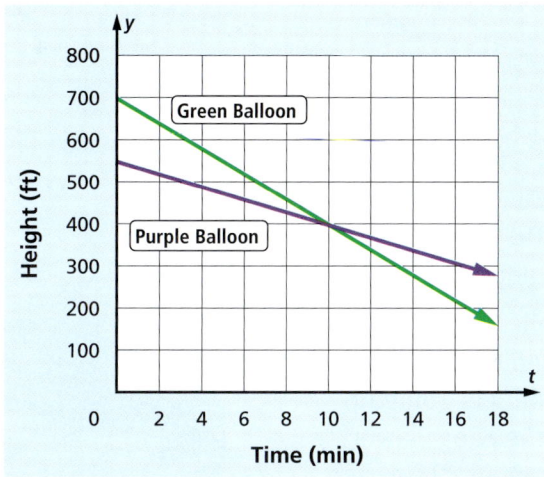

   a. Give three values of t for which the green balloon is higher.
   b. Write an inequality that describes when the green balloon is higher.
   c. Write an inequality that describes when the purple balloon is higher.

2. In a typical year, a willow tree grows 3.5 feet, while a Chitalpa tree grows 2 feet. Suppose a 6-foot willow tree and a 13-foot Chitalpa tree were planted at the same time.
   a. Which tree is taller 2 years after they were planted?
   b. Which tree is taller 10 years after they were planted?
   c. Write an inequality to describe when the willow tree is taller.
   d. Solve the inequality you wrote in Part c.

The Chitalpa tree is a hybrid created in Uzbekistan by Nikolai Rusanov in 1964 and was first introduced into the United States in 1977.
Source: Mid-Columbia Community Forestry Council

APPLYING THE MATHEMATICS

3. a. Solve $5m + 4 > 8m + 14$ by first adding $-5m$ to each side.
   b. Solve $5m + 4 > 8m + 14$ by first adding $-8m$ to each side.
   c. Should you get the same answers for Parts a and b? Why or why not?
   d. Describe how the steps in the solutions to Parts a and b are different.

Solving $ax + b < cx + d$ 213

Chapter 4

In 4–7, a. solve the inequality, and
   b. check the inequality.

4. $7a + 4 \geq 3a + 28$

5. $-53 + 15x \leq -8 + 20x$

6. $2y + 13 < -5y - 8$

7. $14 - 4n > 29 + 6n$

8. Three times a number is less than two times the same number. Find such a number.

In 9 and 10, solve the inequality.

9. $9 - 4(x + 2) < 10x$ 10. $-5y \geq -y + 8 + 7y$

11. In solving the inequality $12x - 16 > 20x + 15$, you must decide which term to eliminate first, $12x$ or $20x$. Explain why this decision matters more for solving an inequality than an equation.

12. Suppose two skyscrapers are being built. The Edwards Building was already 155 feet high when work began on the King Tower. The Edwards Building is going up at an average of 24 feet per day, while the King Tower construction is progressing at an average of 29 feet per day. Let t represent the number of days since construction started on the King Tower.
    a. How tall will each building be when $t = 10$?
    b. Write an inequality for the following question: *When will the King Tower be taller than the Edwards Building?*
    c. Solve your inequality in Part b.
    d. When each building is completed, the graph of its height stops increasing and becomes horizontal. Suppose that the King Tower will be 1,230 feet tall and the Edwards Building will be taller than the King Tower. Draw a graph to show when the Edwards Tower will again be taller than the King Tower.

A typical tower crane can lift up to 18 metric tons of material at one time.
Source: howstuffworks.com

13. Angelina Wright has a Web site. She allows two music companies to advertise on it, Elevator Tunes and Sleepy Songs. Elevator Tunes pays her $5.00 plus 6¢ each time their ad gets a hit. Sleepy Songs pays her $3.50 plus 10¢ each time their ad gets a hit. (A hit is when someone clicks on the ad from Angelina's Web site.) For how many hits are the Elevator Tunes' ads more profitable than the Sleepy Songs' ads? Justify your answer.

REVIEW

In 14–17, solve the equation. (Lesson 4-4)

14. $2t + 38 = 5(t + 1)$

15. $\frac{7}{3} - \frac{1}{4}z = 7 + \frac{2}{5}z$

16. $3(x - 6) + 12(2x + 5) = 10(2x + 7)$

17. $2.83 - 0.4r = 9.02 - 4.2r$

18. According to the Census Bureau, in 2005, Delaware had a population of approximately 840,000, which was increasing at a rate of about 12,000 people a year. Montana had a population of approximately 935,000, increasing at a rate of about 6,700 per year. If these rates continue in the future, in how many years after 2005 will the populations be equal? **(Lesson 4-4)**

19. a. On a coordinate grid, graph $x = -5$ and $y = 9$. **(Lesson 4-2)**
    b. Give the coordinates of the point the two lines have in common.

20. Suppose a sweater on sale costs $35 and a pair of jeans on sale costs $27. If they originally cost a total of $86, what is the percent of discount, rounded to the nearest percent? **(Lesson 4-1)**

21. **Multiple Choice** Suppose $k > m$. What is true about $k - m$? **(Lesson 3-6)**
    A $k - m$ is always negative.
    B $k - m$ is always positive.
    C $k - m$ can be either positive or negative.

EXPLORATION

22. The square of a number is less than the product of one less than the number and two greater than the number.
    a. Find one such number that makes the statement true.
    b. Find one such number that makes the statement false.
    c. Find all such numbers that make the statement true.

Chapter 4

Lesson 4-6
Situations That Always or Never Happen

> **BIG IDEA** Some linear sentences have no solution; others are true for all real numbers.

Comparing Situations

Which job would you take?

Job 1
starting salary $30,000
yearly raises of $5,000

Job 2
starting salary $28,000
yearly raises of $5,000

Mental Math

a. How many cups are in 2 pints?

b. How many cups are in 3 quarts?

c. How many cups are in 4 gallons?

Of course, the answer is obvious. Job 1 will always pay more than Job 2. Looking at the pay in a table of values or on a graph supports this conclusion.

Years Worked (n)	Job 1	Job 2
0	$30,000	$28,000
1	$35,000	$33,000
2	$40,000	$38,000
3	$45,000	$43,000
4	$50,000	$48,000
5	$55,000	$53,000
6	$60,000	$58,000

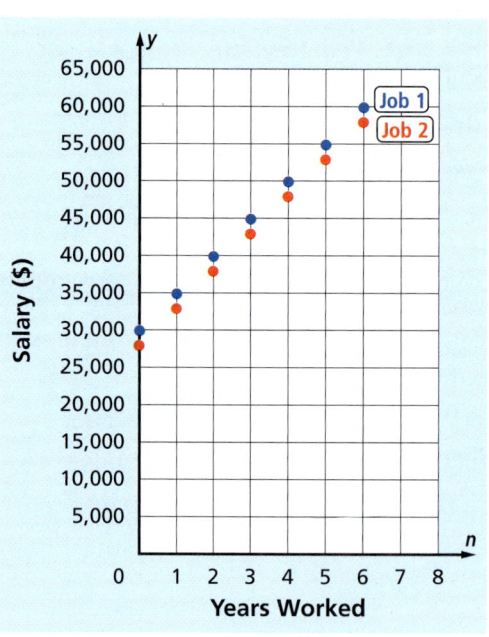

As you compare the salaries using the table, you can see that the money earned at Job 1 is always $2,000 greater than the money earned at Job 2. When the points are graphed, notice that they lie on parallel lines. For any year you pick, the pay for Job 1 is greater than for Job 2. But what happens when this is solved algebraically? Let n = number of years worked.

Salary in Job 1
$30{,}000 + 5{,}000n$

Salary in Job 2
$28{,}000 + 5{,}000n$

When is the pay in Job 1 better than the pay in Job 2? You must solve
$$30{,}000 + 5{,}000n > 28{,}000 + 5{,}000n.$$

Subtract $5{,}000n$ from each side.
$$30{,}000 + 5{,}000n - 5{,}000n > 28{,}000 + 5{,}000n - 5{,}000n$$

Now collect like terms.
$$30{,}000 > 28{,}000$$

The variable has disappeared! Since $30{,}000 > 28{,}000$ is always true, the disappearance of n signals that n can be any real number. Job 1 will always pay a better salary than Job 2, as expected. For any equation or inequality the following generalization is true.

> **Sentences That Are Always True**
>
> When solving a sentence, if you get a sentence that is *always* true, then the original sentence is always true.

When does Job 1 pay *less* than Job 2? Looking at the table or graph on page 216, this never appears to be true. To answer this algebraically, you could solve $30{,}000 + 5{,}000n < 28{,}000 + 5{,}000n$.

$$30{,}000 + 5{,}000n < 28{,}000 + 5{,}000n$$
$$30{,}000 + 5{,}000n - 5{,}000n < 28{,}000 + 5{,}000n - 5{,}000n$$
$$30{,}000 < 28{,}000$$

It is never true that $30{,}000$ is less than $28{,}000$. So Job 1 never pays less than Job 2, something which was obvious from the pay rates. The following generalization is also true.

> **Sentences That Are Never True**
>
> When solving a sentence, if you get a sentence that is never true, then the original sentence is never true.

True and False Sentences

Suppose the sentence you are solving has only one variable and the variable disappears. The chart below summarizes the possibilities.

Solving leads to a	Possible Examples	Solutions
True statement	$0 = 0$ or $92 \leq 92$	any real number
False statement	$0 = 3$ or $-9 > 14$	no real number

Example 1

Solve $18m - 20 + 3m = 3(7m + 2)$.

Solution

$18m - 20 + 3m = 3(7m + 2)$	Write the equation.
$21m - 20 = 21m + 6$	Simplify each side.
$21m - 21m - 20 = 21m - 21m + 6$	Subtract $21m$ from each side.
$-20 = 6$	Simplify.

Since -20 does not equal 6, this statement is not true, so the original equation has no solution.

GUIDED

Example 2

Solve $42k < 80k + 6 - 38k$.

Solution

$42k < 80k + 6 - 38k$	Write the inequality.
$42k < 42k + 6$	Combine like terms.
__?__ < __?__	Subtract $42k$ from each side.
$0 < 6$	__?__

Since $0 < 6$ is always true, __?__ is true for any real number k.

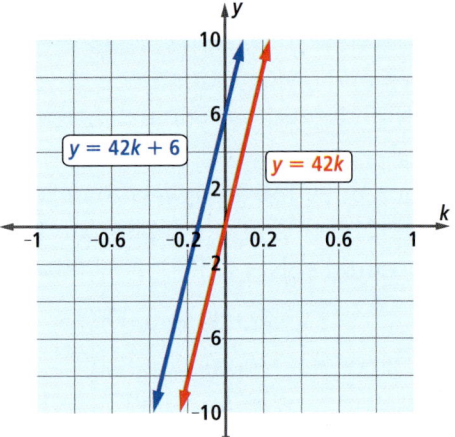

Check Graph both sides of the inequality, as shown at the right. The lines appear to be parallel and never intersect. The graph of $42k$ is always below the graph of $80k + 6 - 38k$. This supports the conclusion that $42k < 80k + 6 - 38k$ is true for any real number k.

Questions

COVERING THE IDEAS

1. Hamburger Heaven pays a starting salary of $6.90 an hour and each year increases it by $1 an hour. Video King starts at $7.50 an hour and also increases it by $1 an hour per year.
   a. Create a table for the data.
   b. Write an inequality to represent this situation.
   c. When does Hamburger Heaven pay more?
2. Add $-12y$ to both sides of the inequality $12y + 11 > 12y - 1$.
   a. What inequality results?
   b. Describe the solutions to this inequality.

In 2006, the federal minimum wage for covered, nonexempt employees was $5.15 per hour.

Source: U.S. Department of Labor

3. Add $2g$ to both sides of the equation $40 - 2g = 11 + 29 - 2g$.
   a. What equation results?
   b. Describe the solution(s) to this equation.
   c. Check your solution(s) using a graph.

4. Check Guided Example 2 by substituting any real number for k.

In 5–8, solve the sentence.

5. $-6 + 4n > 4(n + 11)$
6. $65w = 11 + 65w$
7. $12x + 33 = 33 + 12x$
8. $-3(-9d + -2) > 30d - 3d$

9. The population of Yorkville is about 40,000 and growing at about 800 people a year. Newburgh has a population of about 200,000 and is growing at about 800 people a year.
   a. What will be the population of Yorkville in y years?
   b. What will be the population of Newburgh in y years?
   c. When will their populations be the same?

APPLYING THE MATHEMATICS

10. Apartment A rents for $810 per month including utilities. Apartment B rents for $700 per month but the renter must pay $110 per month for utilities and a one-time $50 fee for a credit check.
    a. What sentence could you solve to find out when apartment A is cheaper?
    b. Solve this inequality.
    c. If you wanted to rent one of these apartments for two years, which one would be cheaper?

Nationally, the median rent for a one-bedroom apartment in 2003 was $550.
Source: National Multi-Housing Council

In 11 and 12, create an example of an equation different from those in this lesson with the given solution.

11. There is no real solution.
12. The equation is true for all real numbers.

REVIEW

In 13 and 14, solve the inequality. (Lessons 4-5, 3-7)

13. $-3(5p + 2) > 24$
14. $2n \geq -7 - (n + 4)$

15. Sending a package by Wedropem shipping service costs $3.50 plus $0.25 per ounce. Brokefast Company charges $4.75 plus $0.10 per ounce. If your package weighs 16 ounces, which service is cheaper? Explain your reasoning. (Lesson 4-5)

Chapter 4

16. Mama's Pizza charges $12 for a large cheese pizza and $0.90 for each additional topping. Pizza Palace charges $14 for a large cheese pizza but only $0.40 for each additional topping. (**Lessons 4-4, 4-3**)
    a. Write an equation for each pizza shop describing the cost c of a large pizza in terms of the number of toppings t.
    b. For how many toppings are the pizza prices at the two shops equal?

17. In the 2000 United States presidential election, 105,396,641 people voted, according to the *World Almanac and Book of Facts*. In the 2004 presidential election, 122,293,332 people voted. By what percent did the number of voters in the election increase from 2000 to 2004? (**Lesson 4-1**)

18. Four instances of a general pattern are given below. (**Lesson 1-2**)
    $$-(43 - 20) = 20 - 43$$
    $$-(5.21 - 8.49) = 8.49 - 5.21$$
    $$-\left(\frac{6}{5} - \frac{3}{7}\right) = \frac{3}{7} - \frac{6}{5}$$
    $$-(w - 16) = 16 - w$$
    a. Write the pattern using the variables x and y.
    b. Is the pattern true for all real number values of x and y? Justify your answer.

EXPLORATION

19. When solving the equation $ax + b = cx + d$ for x, there may be no solution, exactly one solution, or infinitely many solutions. What must be true about a, b, c, and d to guarantee each of the following?
    a. There is exactly one solution.
    b. There are no solutions.
    c. There are infinitely many solutions.

Lesson 4-7

Equivalent Formulas

Vocabulary
Celsius scale
centigrade scale
Fahrenheit scale
input
output
equivalent formulas

▶ **BIG IDEA** If a formula gives a first variable in terms of a second variable, it may be possible to transform the formula to give the second variable in terms of the first.

Two different temperature scales are in common use throughout the world. The scale used wherever people use the metric system is the **Celsius scale,** named after Anders Celsius (1701–1744), a Swedish astronomer and physicist. It is also sometimes called the **centigrade scale** because of the 100-degree interval between 0°C and 100°C, the freezing and boiling points of water. The other scale is the **Fahrenheit scale,** named after the German physicist Gabriel Fahrenheit (1686–1736). In the Fahrenheit scale, which is now used only in the United States and a few other countries, water's freezing and boiling points are 32° and 212°.

Mental Math

Evaluate $x^2 - 2xy + y^2$ if
a. $x = 5$ and $y = 5$.
b. $x = -5$ and $y = 5$.
c. $x = 5$ and $y = -5$.

People outside the United States seldom use the Fahrenheit scale. An exception is when a visitor from the U.S. asks about the weather forecast. They have to change the Celsius temperature from their forecast into Fahrenheit using the formula $F = 1.8C + 32$. This formula uses the Celsius temperature C as the **input** and produces the Fahrenheit temperature F as the **output.** We say that it gives F in terms of C. However, visitors to the United States have to make the reverse conversion. They must convert Fahrenheit temperatures from the U.S. forecast into Celsius. If you are going to convert from Fahrenheit to Celsius often, it is convenient to have a formula that begins with F as input and gives C as output. To find this formula, we write 1.8 as the fraction $\frac{9}{5}$.

The lowest recorded temperature in the U.S. occurred in Prospect Creek Camp, Alaska, on January 23, 1971. The temperature fell to −80°F.

Source: *The World Almanac and Book of Facts*

GUIDED

Example 1
Solve $F = \frac{9}{5}C + 32$ for C.

Solution Solve like any linear equation. Isolate C on the right side.

$F = \frac{9}{5}C + 32$ Write the formula.

____?____ = ____?____   Add −32 to each side.

$\frac{5}{9}(\underline{\ ?\ }) = \frac{5}{9}(\underline{\ ?\ })$ Multiply each side by $\frac{5}{9}$.

____?____ = C Simplify.

(continued on next page)

Equivalent Formulas 221

Check In each formula, substitute 212 and 100 for the appropriate variables.

Using $F = 1.8C + 32$
$\underline{\ ?\ } = 1.8 \cdot \underline{\ ?\ } + 32$
$\underline{\ ?\ } = \underline{\ ?\ } + 32$
$\underline{\ ?\ } = \underline{\ ?\ }$

Using $C = \frac{5}{9}(F - 32)$
$\underline{\ ?\ } = \frac{5}{9}(\underline{\ ?\ } - 32)$
$\underline{\ ?\ } = \frac{5}{9}(\underline{\ ?\ })$
$\underline{\ ?\ } = \underline{\ ?\ }$

The numbers $F = 212$ and $C = 100$ satisfy both equations.

The formulas $F = 1.8C + 32$ and $C = \frac{5}{9}(F - 32)$ are called **equivalent formulas** because every pair of values of F and C that satisfies one equation also satisfies the other.

GUIDED

Example 2

A formula for the volume of a cone is $V = \frac{1}{3}\pi r^2 h$, where V is the volume, r is the radius of the base, and h is the height. Octavio solved this formula for h. His work is shown below. Explain what he did to get each equation.

$$V = \frac{1}{3}\pi r^2 h$$

a. $3V = \pi r^2 h$ $\underline{\ ?\ }$
b. $\frac{3V}{\pi} = r^2 h$ $\underline{\ ?\ }$
c. $\frac{3V}{\pi r^2} = h$ $\underline{\ ?\ }$

Activity

In 1–5, use a CAS to solve for the given variable. Tell what is done to both sides in each step.

1. $-6x + 3y = 24$, for y

2. $x - y = -14$, for y

3. $S = \frac{C - 1}{3}$, for C (formula for cap size S when $C =$ circumference of the head in inches)

4. $E = \pi a b$, for b (formula for the area E of an ellipse when a is one-half the length of the major axis and b is one-half the length of the minor axis)

5. $G = \frac{s + 2d + 3t + 4h}{a}$, for h (formula for baseball slugging average G where $s =$ number of singles, $d =$ number of doubles, $t =$ number of triples, $h =$ number of home runs, and $a =$ number of at-bats)

Whether working with a CAS or by hand, different people may write formulas that look quite different. However, they might be equivalent. For example, here is what Alf and Beth wrote to solve $p = 2\ell + 2w$ for w.

Alf's Solution

$p = 2\ell + 2w$	Write the formula.
$p - 2\ell = 2w$	Subtract 2ℓ.
$\dfrac{p - 2\ell}{2} = w$	Divide by 2.
$w = \dfrac{p - 2\ell}{2}$	

Beth's Solution

$p = 2\ell + 2w$	Write the formula.
$p - 2\ell = 2w$	Subtract 2ℓ.
$\frac{1}{2}(p - 2\ell) = w$	Multiply by $\frac{1}{2}$.
$w = \frac{1}{2}(p - 2\ell)$	

> ▶ **QY**
> Explain how you know that $w = \dfrac{p - 2\ell}{2}$ and $w = \frac{1}{2}(p - 2\ell)$ are equivalent.

 QY

Using a Graphing Calculator

One important use of equivalent formulas arises when using graphing calculators. Often formulas that are entered must give y in terms of x.

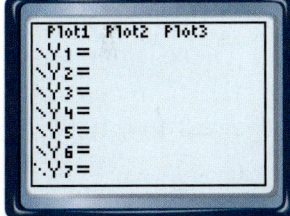

Example 3

Use a graphing calculator to graph $5x - 2y = 100$.

Solution Solve for y.

$5x - 2y = 100$	Write the equation.
$-5x + 5x - 2y = -5x + 100$	Add $-5x$ to each side.
$-2y = -5x + 100$	Combine like terms.
$\dfrac{-2y}{-2} = \dfrac{-5x + 100}{-2}$	Divide each side by -2.
$y = \dfrac{-5}{-2}x + \dfrac{100}{-2}$	Expand the fraction.
$y = 2.5x - 50$	Simplify.

Now enter the equation Y1 = 2.5x − 50 into the calculator. A window of $-20 \leq x \leq 30$ and $-60 \leq y \leq 60$ is shown below.

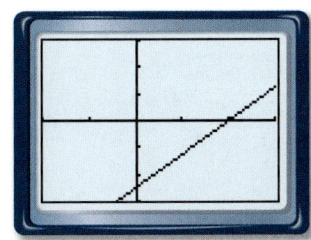

(continued on next page)

Equivalent Formulas **223**

Check 1 Use the TRACE feature to read the coordinates of some points on the line. Check that these satisfy the original equation. For example, our TRACE showed the point with $x \approx 10.5$, $y \approx -23.7$ on the graph.

Does $5(10.5) - 2(-23.7) = 100$?

$$99.9 \approx 100 \text{ Yes. It checks.}$$

The point $(10.5, -23.7)$ is very close to the graph of $5x - 2y = 100$.

Check 2 Compute the coordinates of a point on the line. For example, when $x = 0$, $5(0) - 2y = 100$.

$$5(0) - 2y = 100$$
$$-2y = 100$$
$$y = -50$$

The TRACE on our calculator shows that the point $(0, -50)$ is on the line.

Questions

COVERING THE IDEAS

1. There is one temperature at which Celsius and Fahrenheit thermometers give the same reading: $-40°$. Verify that $C = -40$, $F = -40$ satisfies both $F = 1.8C + 32$ and $C = \frac{5}{9}(F - 32)$.

2. A person with a head circumference of 23.5 inches wears a size $7\frac{1}{2}$ baseball cap. Verify that $C = 23.5$ and $S = 7\frac{1}{2}$ satisfy the formula $S = \frac{C - 1}{3}$.

3. a. Solve $p = 2\ell + 2w$ for ℓ.
   b. **Fill in the Blanks** In Part a you are asked to find a formula for __?__ in terms of __?__ and __?__.
   c. Check your solution to Part a by substituting values for ℓ, w, and p.

In 4 and 5, solve the equation for y.

4. $8x + y = 20$

5. $4x - 8y = -40$

In 6–9, solve the formula for the indicated variable.

6. $r = \frac{d}{t}$ for d

7. $S = 180n - 360$ for n

8. $F = m \cdot a$ for a

9. $A = \frac{1}{2}(b_1 + b_2)h$ for h

APPLYING THE MATHEMATICS

10. The formula $S = 3F - 24$ gives the size of a person's shoe in terms of F, the length of a person's foot in inches.
    a. Solve this formula for F.
    b. Estimate the length of a person's foot if the person wears a size 9 shoe.

11. Jocelyn and Alma were asked to solve the equation $5x - 2y = 100$ for y. When solving the equation, they got the following answers.

Jocelyn	Alma
$y = \dfrac{100 - 5x}{-2}$	$y = 2.5x - 50$

    Is the work of either student correct? Explain how you know.

12. a. Solve the following equations for y. $2x + 6y = 15$ and $x = \dfrac{6y - 15}{-2}$
    b. Graph each equation and $y = 2.5 - \dfrac{x}{3}$ on a calculator.
    c. Which of these equations appear to be equivalent? Provide evidence to support your answer.

13. A formula for the circumference C of a circle is $C = \pi d$, where d is the diameter.
    a. Solve this formula for π.
    b. How could you use the formula to find a value of π?
    c. Use your answer to Part b to estimate π from the measurements of some circular object you have.

14. The formula $S = 2\pi r^2 + 2\pi rh$ gives the total surface area of a cylindrical solid shown below with radius r and height h.

    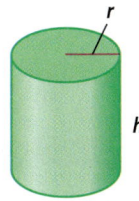

    a. Solve this formula for h.
    b. Find the height to the nearest hundredth if the radius is 10 centimeters and the surface area is 2,000 square centimeters.

15. Solve for y and use a graphing calculator to graph the equations $5x + 2y = 8$ and $5x + 2y = 12$. What is true about these graphs?

REVIEW

In 16 and 17, solve the sentence. (Lessons 4-6, 4-4)

16. $6(y - 4) = 2(y - 4) - 8(2 - y)$

Chapter 4

17. $52v < 22v - 7 + 30v$

18. Five more than twice a number is three more than four times the number. What is the number? (**Lesson 4-4**)

19. a. Write an equation for the horizontal line through (5, –3). (**Lesson 4-2**)
    b. Write an equation for the vertical line through (5, –3).

20. According to the *World Almanac and Book of Facts*, the Middle East is reported to have approximately 65% of the world's oil reserves. All together, the Middle East's crude oil reserves are estimated to total 686 billion barrels. How many barrels are estimated to be in the world's total crude oil reserves? (**Lesson 4-1**)

In 2004, the United States imported over 10,000,000 barrels of oil per day from foreign nations.
Source: *The World Almanac and Book of Facts*

21. a. A pentagon has two sides of length $2x + 22$ and three sides of length $x - 1$. Its perimeter is 55. Solve for x. (**Lesson 3-5**)
    b. Suppose the pentagon had two sides of length $x - 1$, three sides of length $2x + 22$, and still had a perimeter of 55. Why is this impossible?

In 22 and 23, write the related facts and determine the value of x for each fact triangle. (**Lessons 3-5, 2-7**)

22.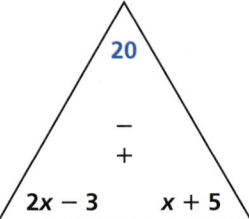

23.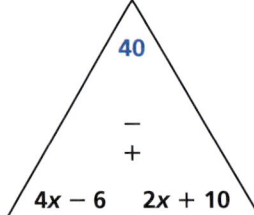

EXPLORATION

24. Ask a friend or relative for a formula used in his or her job. Explain what the variables represent and show an example of how it is used. Solve the formula for one of its other variables.

QY ANSWER

The definition of division says that dividing by 2 is the same as multiplying by $\frac{1}{2}$.

Lesson 4-8

Compound Inequalities, *And* and *Or*

Vocabulary

compound sentence
intersection
union
± notation

▶ **BIG IDEA** Placing AND between two sentences means that you want the intersection of their solutions; placing OR between them means that you want the union of their solutions.

The Language of Compound Sentences

A car's water temperature gauge is an indicator of its engine cooling system. This gauge is marked to highlight normal temperatures as shown. Temperatures below 40°C occur when the car is warming up. Temperatures above 120°C indicate that the engine is in jeopardy of breaking down. These temperatures can be graphed. Using t to represent temperature, each graph can be described with a pair of inequalities.

Mental Math

a. Katie's temperature is 4 degrees away from 98.6. What are her possible temperatures?

b. Jeremy was 7 points away from getting 90 points on a test. What are his possible scores?

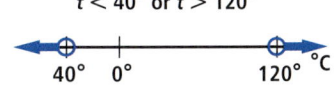

Normal Temperatures	Abnormal Temperatures
$t \geq 40°$ and $t \leq 120°$	$t < 40°$ or $t > 120°$
The graph is an interval.	The graph is two rays without their endpoints.

A **compound sentence** is a single sentence consisting of two or more sentences linked by the words *and* or *or*. The above graphs are described by *compound inequalities*. The compound inequality at the left can be written as the *double inequality* $40° \leq t \leq 120°$.

Intersection and Union of Sets

The graphs of the compound inequalities above come from the *intersection* and *union* of two sets. On the left, the interval showing normal engine temperatures is the set of points shared by the graphs of the two simple inequalities. It consists of the points where the two rays $t \geq 40°$ and $t \leq 120°$ overlap.

> **Intersection of Sets**
>
> The **intersection** of sets A and B, written $A \cap B$, is the set of elements that are in both A and B.

Compound Inequalities, *And* and *Or* **227**

The graph of abnormal engine temperatures also begins with two rays without their endpoints, which in this case show those temperatures below 40° and above 120°. However, we do not look for the overlap. Instead, we take the union of the two rays. The meaning of *union* in mathematics is similar to its meaning in other contexts. For example, the Preamble to the Constitution of the United States of America reads:

> We the People of the United States, in Order to form a more perfect Union, establish Justice, insure domestic Tranquility, provide for the common defense, promote the general Welfare, and secure the Blessings of Liberty to ourselves and our Posterity, do ordain and establish this Constitution for the United States of America.

Here the word union describes a new set (the United States) formed by joining together component sets (the thirteen colonies).

Union of Sets

The **union** of sets A and B, written A ∪ B, is the set of elements in either A or B or in both.

 QY1

Example 1

Let A = the set of numbers for which $x > 8$.
Let B = the set of numbers for which $x \leq -3$.
Graph the set A ∪ B. Describe the set with an *and* or an *or* statement.

Solution Draw the graph of $x > 8$. The open circle at 8 indicates that 8 is not a solution.

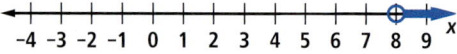

Draw the graph of $x \leq -3$. The closed circle at −3 shows that −3 is a solution.

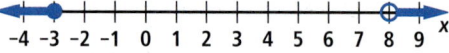

The union includes all points that satisfy either sentence or both sentences. The graph of A ∪ B is shown below. This is the set "$x > 8$ or $x \leq -3$."

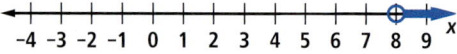

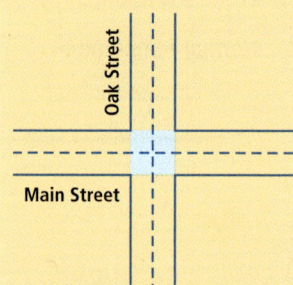

▶ QY1

Consider this situation: A police officer is directing traffic at the corner of Main and Oak Streets. Choose the correct word to complete each sentence.

The police officer was in the (union/intersection) of Main Street and Oak Street. This means that the officer was on Main Street (and/or) the officer was on Oak Street.
The officer was directing cars that were in the (union/intersection) of Main Street and Oak Street. This means that each car was on Main Street (and/or) on Oak Street.

228 More Linear Equations and Inequalities

Example 2

A family purchased some neon tetras to put in their new fish tank. They looked on the Internet to determine at what temperature to set the tank water. One site wrote to keep the water temperature from 72° to 78°F. A second site wrote 68° to 74° and a third site wrote 73° to 81°. To be safe, at what temperature should the family keep the tank?

A neon tetra can live 10 years or more with the proper conditions.

Source: animal-world.com

Solution Let t represent the water temperature. Appropriate temperatures for tetras are those satisfying the following.

$$72° \leq t \leq 78° \text{ according to Site 1}$$
$$68° \leq t \leq 74° \text{ according to Site 2}$$
$$73° \leq t \leq 81° \text{ according to Site 3}$$

The family wondered if any temperatures could satisfy all three conditions. The graphs show the three intervals separately.

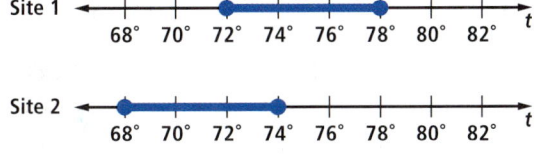

The best temperatures for the tetras are those that lie in all three intervals. This is the intersection in which the three graphs overlap. **The tank should have a temperature satisfying 73° < t < 74°.**

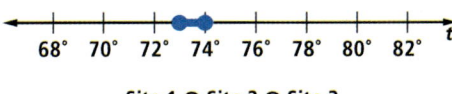

Site 1 ∩ Site 2 ∩ Site 3

Describing the Intervals $a \leq x \leq b$

Most people would say that the average body temperature is 98.6°F. This figure was arrived at in the 19th century. Recent medical research has established that the mean temperature for healthy people is 98.2°F. However, there is some variability among healthy people. According to the new standard, the normal range varies above or below 98.2° by 1.5°. This means that the normal body temperatures t of healthy people range from $98.2 + 1.5 = 99.7$ to $98.2 - 1.5 = 96.7$. So $96.7 \leq t \leq 99.7$.

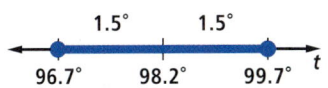

You can combine $98.2 + 1.5$ and $98.2 - 1.5$ into one expression using **± notation.** Then the interval of normal temperatures is written $98.2°F \pm 1.5°F$. The graph on the next page shows this interval and the temperatures of 129 men and women.

Compound Inequalities, *And* and *Or*

Chapter 4

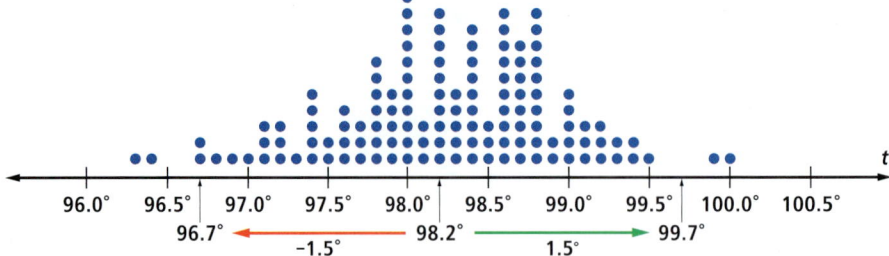

STOP QY2

Solving Inequalities with *And* and *Or*

If the variable is not isolated in an inequality, its solutions are not evident. The first task then is to use the Addition and Multiplication Properties of Inequality to isolate the variable.

▶ **QY2**

Write an expression of the form $a \pm d$ to represent the interval $40\,°C \leq t \leq 120\,°C$ for normal car engine temperatures from the beginning of this lesson.

Example 3

Solve and graph $-8 \leq 5m + 4 < 19$.

Solution 1 $-8 \leq 5m + 4 < 19$ can be rewritten as $-8 \leq 5m + 4$ and $5m + 4 < 19$. Each inequality can be solved separately and then we take the intersection of their solutions.

$$-8 \leq 5m + 4 \qquad \text{and} \qquad 5m + 4 < 19$$
$$-8 + {-4} \leq 5m + 4 + {-4} \qquad 5m + 4 + {-4} < 19 + {-4}$$
$$-12 \leq 5m \qquad 5m < 15$$
$$\frac{-12}{5} \leq \frac{5m}{5} \qquad \frac{5m}{5} < \frac{15}{5}$$
$$-2.4 \leq m \qquad \text{and} \qquad m < 3$$

Now combine these two statements to describe the interval.

$$-2.4 \leq m < 3$$

Solution 2 Notice that after breaking the interval into two inequalities, the steps for solving each inequality are the same. In the future, the inequality does not need to be split apart at all. You perform the same operations to all three parts, as shown below.

$-8 \leq$	$5m + 4$	< 19		Write the equation.
$-8 + {-4} \leq$	$5m + 4 + {-4}$	$< 19 + {-4}$		Add -4 to each part.
$-12 \leq$	$5m$	< 15		Simplify.
$\frac{-12}{5} \leq$	$\frac{5m}{5}$	$< \frac{15}{5}$		Divide by 5.
$-2.4 \leq$	m	< 3		Simplify.

230 More Linear Equations and Inequalities

Lesson 4-8

GUIDED

Example 4

Solve $13y + 86 > y + 2$ or $2y + 9 < 5y$.

Solution Solve each inequality separately.

$13y + 86 > y + 2$ or $2y + 9 < 5y$
$12y + 86 > \underline{}$ or $-3y + 9 < \underline{}$
$12y > \underline{}$ or $-3y < \underline{}$
$y > \underline{}$ or $y \underline{} \underline{}$

The intervals overlap and their union is described by $y > -7$.

Questions

COVERING THE IDEAS

1. **Fill in the Blank** Fill in the blank with *and* or *or*.

   $-8.2 < x \le 4.75$ means $-8.2 < x \underline{} x \le 4.75$.

Matching In 2–5, match the inequality with its graph.

2. $x < -2$ or $5 < x$

   a.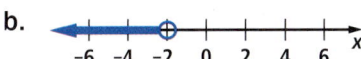

3. $-2 < x < 5$

   b.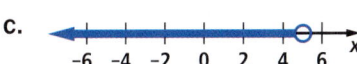

4. $x < -2$ or $x < 5$

   c.

5. $x < -2$ and $x < 5$

   d.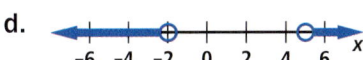

6. During the summer months, Mr. and Mrs. Boller have to agree on a temperature at which to set the air conditioner. Mr. Boller prefers the room temperature to be between 68° and 72°. Mrs. Boller likes temperature between 70° and 75°. Write an inequality to show the temperatures when Mr. and Mrs. Boller will both be comfortable.

7. During thyroid surgery, doctors make a 1.5 inch ± 0.5 inch incision. Write a double inequality showing the possible lengths of the incision. Then graph the sentence.

8. A movie theater gives a discount price to children under 3 years old and to senior citizens over 65 years old.
   a. Write a compound inequality to show the ages that receive the discounted price.
   b. Graph the ages that receive this discounted price.

Compound Inequalities, *And* and *Or*   **231**

Chapter 4

9. Temperatures in space vary greatly. Astronauts on the International Space Station have to be able to endure outside temperatures that are $-18°C \pm 139°C$.
   a. Write an inequality to represent these temperatures in space.
   b. Graph the inequality.

In 10–13, solve the inequality and graph all solutions.

10. $2x + 9 > 17$ or $8 - 5x \leq 13$
11. $\frac{1}{4}x - 7 < -2$ and $6x + 3.8 \geq 9.2$
12. $-4 \leq 2.5x - 9 < 15$
13. $20c + 5 \geq 30c - 15$ or $9 - c > 4 - 2c$

The International Space Station (ISS) weighs 206,043.3 kilograms and has a habitable volume of 420 cubic meters.
Source: NASA

APPLYING THE MATHEMATICS

14. Marcos and Lydia want to hire a band for their wedding. The band charges a flat fee of $950 and an additional $275 per hour. They are willing to spend at least $1,200 on the band, but not more than $2,300. For how many hours can they have the band play for their wedding?

Fill in the Blanks For 15 and 16, fill in the blanks to describe the interval shown on the graph with $\pm$ notation.

15. $\underline{} \pm \underline{}$

   $-14\ -7\ \ 0\ \ 7\ \ 14\ \ 21\ \ 28$

16. $\underline{} \pm \underline{}$

   $-3.7 \qquad\qquad 1.6$
   $-4\ -3\ -2\ -1\ \ 0\ \ 1\ \ 2$

17. Make a table and graph of $y = -2x + 8$. Highlight the x-coordinates on the table and the part of the graph where $-2x + 8 > 12$ or $-2x + 8 < 1$.

18. Make a table and graph of $y = 3 + 5x$. Highlight the x-coordinates on the table and the part of the graph where $-2 \leq 3 + 5x \leq 23$.

REVIEW

19. a. Solve the formula $d = rt$ for r, where d is distance, r is rate, and t is time.
    b. What was the average speed of a truck that traveled 245 miles in 4 hours? (**Lesson 4-7**)

In 20 and 21, solve the sentence. (Lesson 4-6)

20. $7(k - 2) - 11 > (7k - 2) - 11$

21. $20p - 4(p + 2) = (6p - 8) + 10p$

22. Solve $-5 - 2y > 14 - 6y$ and graph the set of solutions. (Lesson 4-5)

23. Jennifer compared prices for the same pair of shoes at three different stores. The first store sold them at $75 less a discount of 15% due to a storewide sale. Jennifer had a $10 gift certificate for the second store, where the shoes cost $72. She would have to pay 8% sales tax at these two stores. The shoes at the third store cost $67 including tax. At which store can Jennifer buy the shoes for the cheapest price? (Lesson 4-1)

24. Evaluate $|-4xy + y - x|$ for each situation. (Lesson 1-6)

    a. $x = 3, y = 4$
    b. $x = -1, y = 2$
    c. $x = -10, y = -5$

EXPLORATION

25. In mathematics, the difference between the words *and* and *or* is determined by union and intersection, but the English usage of these words is not always as clear.

    a. Consider this statement:

       In case of an emergency, women and children go first.

       Write how a mathematician would view the statement and then explain whether there is a difference in what was probably meant by the statement.

    b. Find at least two other situations in which the English usage of the words *and* and *or* varies from the mathematical usage.

QY ANSWERS

1. intersection; and; union; or
2. $80°C \pm 40°C$

Chapter 4

Lesson 4-9

Solving Absolute Value Equations and Inequalities

> **BIG IDEA** Inequalities with $|ax + b|$ on one side and a positive number c on the other side can be solved by using the fact that $ax + b$ must equal either c or $-c$.

A school carnival had a "Guess the Number" booth featuring a jar full of pennies. Whoever guessed closest to the actual number of pennies would win a prize. Only the principal knew there were 672 pennies in the jar. When the prize was announced, the winner was off by 9 pennies. How many pennies did the winner guess?

The winning guess deviated from the actual number of pennies by 9. This does not say whether the guess was too high or too low. It could have been either $672 + 9 = 681$ or $672 - 9 = 663$. These two possibilities are shown on the number line at the right.

The two numbers 663 and 681 are the solutions of $|n - 672| = 9$. The expression $|n - 672|$ is the absolute deviation of the guess n, from the actual number of pennies, 672. In this case, $|n - 672| = 9$ and the solutions to the equation are 663 and 681.

The equation $|n - 672| = 9$ is of the form $|ax + b| = c$, with $a = 1$, n in place of x, $b = -672$, and $c = 9$. All equations of this form can be solved using what you know about linear equations and compound sentences.

Mental Math

When $a > 0$, determine whether the following are positive or negative.

a. $\left(-\dfrac{a}{2}\right)^2$

b. $(-5a)^3$

c. $-(-0.9a)^2$

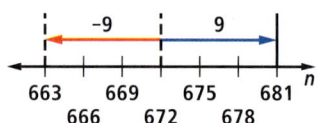

Solving $|x| = a$

Remember that $|x|$ is the distance of x from 0 on a number line. At the right is a table of some pairs of values of x and $|x|$ and the graph of $y = |x|$. Also, the line $y = 3$ is graphed.

x	$\lvert x \rvert$
−4	4
−3	3
−2	2
−1	1
0	0
1	1
2	2
3	3
4	4

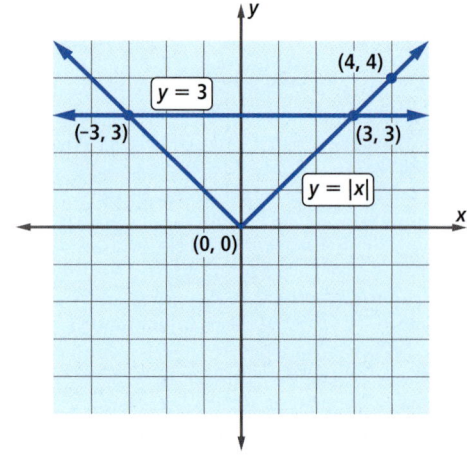

234 More Linear Equations and Inequalities

The following conclusions about the equation $|x| = c$ can be made from the table and graph.

- $|x|$ is never negative, so there are no solutions to $|x| = -10$ or to any other equation of the form $|x| = c$ when c is negative.
- $|x| = 0$ only when $x = 0$.
- $|x| = 3$ has two solutions, 3 and −3. The two solutions can be seen in the table in the rows where $|x| = 3$. Also, the graph of the horizontal line $y = 3$ intersects $y = |x|$ in two points, where $x = 3$ and $x = -3$. Therefore, there are always two solutions to $|x| = c$ when c is positive.

Solutions to $|x| = c$

- When c is positive, then there are two solutions to $|x| = c$, namely c and $-c$.
- When c is negative, then there are no solutions to $|x| = c$.
- When c is zero, then $|x| = 0$ and there is one solution: $x = 0$.

 QY1

▶ QY1

Find all the solutions to each equation.
a. $|t| = 15$
b. $|u| = -88.2$
c. $|v| = 0$

Solving $|ax + b| = c$

The above ideas apply to any equation where there is an absolute value of an expression on one side and a number on the other.

GUIDED

Example 1

Consider the graph, which shows $y_1 = |6 - 2x|$ and $y_2 = 18$. Use the graph and use algebraic properties to solve $|6 - 2x| = 18$.

Solution 1 Use the graph.

The points of intersection of the two graphs are __?__ and __?__. The x-coordinates of these points of intersection are __?__ and __?__.
The solutions to $|6 - 2x| = 18$ are __?__ and __?__.

Solution 2 Use algebraic properties.

Ask yourself: What numbers have absolute value 18?
$6 - 2x = $ __?__ or $6 - 2x = $ __?__
Solve this compound sentence as you did in Lesson 4-8.

$6 = 2x +$ __?__   or $6 = 2x +$ __?__
__?__ $= 2x$ or   __?__ $= 2x$
__?__ $= x$ or   __?__ $= x$

You should have the same answers from both Solutions 1 and 2.

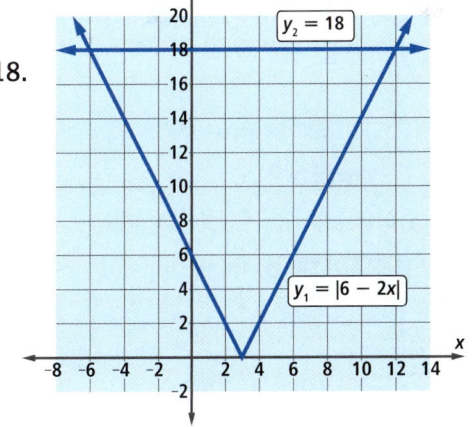

Solving Absolute Value Equations and Inequalities

Solving $|x| < c$ and $|x| > c$

When c is positive, the two solutions to $|x| = c$ can be represented on a number line. They are the points at a distance c from 0.

$|x| = c$ if and only if $x = c$ or $x = -c$.

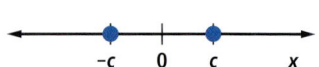

The points closer to 0 than c are the solutions to the inequality $|x| < c$. They can be described by the interval $-c < x < c$. For example, the solutions to $|x| < 3$ can be described by the double inequality $-3 < x < 3$.

$|x| < c$ if and only if $-c < x < c$.

The solutions to the inequality $|x| > 3$ are the points whose distance from 0 is greater than 3. The graph of these points has two parts and is described by the compound inequality $x < -3$ or $x > 3$.

$|x| > c$ if and only if $x < -c$ or $x > c$.

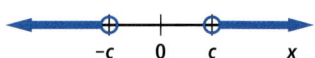

> **Solutions to $|x| < c$ when c is positive**
>
> $|x| < c$ if and only if $-c < x < c$.
>
> **Solutions to $|x| > c$ when c is positive**
>
> $|x| > c$ if and only if $x < -c$ or $x > c$.

🛑 **QY2**

▶ **QY2**
Describe all solutions to $|d| > 0.9$ using a compound inequality without the absolute value symbol.

Solving $|ax + b| < c, |ax + b| > c$

To solve these inequalities, think of the simpler inequalities $|x| > c$ and $|x| < c$.

> **GUIDED**
>
> ### Example 2
> Solve $|8x + 24| \leq 40$.
>
> **Solution 1** Think: $|x| < a$ means $-a < x < a$.
>
> | $\|8x + 24\| \leq 40$ | Write the inequality. |
> |---|---|
> | $-40 \leq 8x + 24 \leq 40$ | $\|x\| < a$ means $-a < x < a$. |
> | $\underline{} \leq 8x \leq \underline{}$ | Add -24 to each side. |
> | $\underline{} \leq x \leq \underline{}$ | Divide both sides by 8. |

236   More Linear Equations and Inequalities

Solution 2 Use a graph or table.

Ask yourself: When is $|8x + 24|$ below or at 40 on the graph? When is $|8x + 24|$ less than or equal to 40 in the table?

x	$|8x + 24|$
-10	?
-8	?
-6	?
-4	?
-2	?
0	?
2	?
4	?
6	?

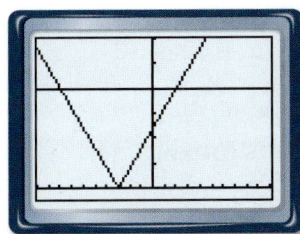

$-10 \leq x \leq 10$; x scl = 1
$0 \leq y \leq 60$; y scl = 10

So the solution to $|8x + 24| \leq 40$ is ___?___ $\leq x \leq$ ___?___.

Questions

COVERING THE IDEAS

1. **Multiple Choice** You are asked on a test for the year of the Emancipation Proclamation in the United States. The correct answer is 1863. You guessed g and you were off by 4 years. What equation's solution gives the possible values of g?

   A $|1863 - 4| = g$
   B $|g| = 1863 - 4$
   C $|g - 1863| = 4$
   D $|g - 4| = 1863$

2. Determine whether the number is a solution to the equation $60 = |n - 90|$.

   a. 30 b. -30 c. 150 d. -150

In 3–6, find all solutions in your head.

3. $|A| = 6$ 4. $|B| = -600$ 5. $|C| = 0$ 6. $5|D| = 40$

7. Use the table at the right to solve each sentence.

   a. $|2x - 3| = 7$
   b. $|2x - 3| < 7$
   c. $|2x - 3| > 7$

x	$Y_1 = |2x - 3|$	$Y_2 = 7$
-5	13	7
-4	11	7
-2	7	7
0	3	7
1	1	7
2	1	7
5	7	7
6	9	7

In September 1862, Abraham Lincoln called on the seceded states to return to the Union or have their slaves declared free. When no state returned, he issued the proclamation on January 1, 1863.

Source: Britannica

Chapter 4

8. **Fill in the Blanks** The sentence $|ax + b| = 15$ is equivalent to the compound sentence ___?___ or ___?___.

9. **Multiple Choice** The green portion of the graph can be used to find the solution to which of the following?

   A $|x + 5| = 10$ B $|x + 5| \leq 10$
   C $|x + 5| \geq 10$ D $|x + 5| > 10$

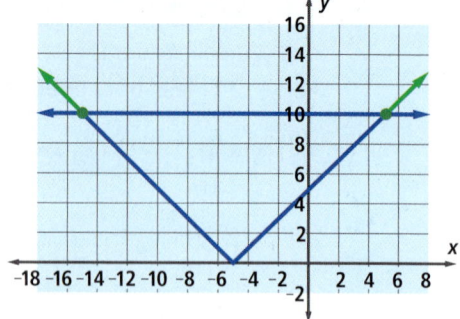

10. **Fill in the Blanks** $|x - 4| = 2.3$ means the distance between ___?___ and ___?___ on a number line is ___?___.

In 11–16, solve the sentence.

11. $|a + 10| = 12$
12. $|42 - 3b| = 45$
13. $|8 - c| < 9$
14. $|10d + 0.3| > 5.6$
15. $\frac{1}{2} \leq \left|\frac{5}{8}g + \frac{3}{4}\right|$
16. $|h + 11| - 3 \leq 0$

17. Use the graph of $y = |7 - 2x|$ at the right to solve the sentence.

   a. $|7 - 2x| = 9$ b. $|7 - 2x| = 0$
   c. $|7 - 2x| = -2$ d. $|7 - 2x| > 5$

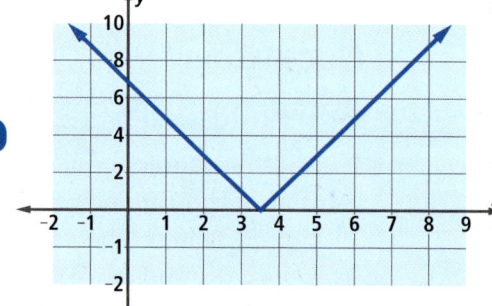

APPLYING THE MATHEMATICS

18. Let $|5x + 2| = m$. Find a value of m so that the absolute value equation has the given number of solutions.

    a. two solutions   b. one solution   c. no solutions

19. A box of Wheat-Os breakfast cereal says that it contains 24 ounces. However, because the machinery that fills the boxes cannot be exactly precise, they can be from $\frac{1}{8}$ ounce below to 1 ounce above this weight.

    a. Graph the possible number of ounces of Wheat-Os.
    b. Write an absolute value inequality to show this amount.

20. It is recommended that teenagers get 8.5 ± 0.7 hours of sleep.

    a. Write a double inequality to express the recommended amount of sleep.
    b. Write an absolute value inequality for the recommended amount of sleep.

21. Write a single inequality for the graph below.

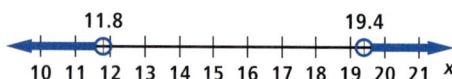

238   More Linear Equations and Inequalities

22. At the right is the graph that Zoey used to solve $|3x - 5| - 8 < x + 1$. Use her graph and intersection points to give the solution.

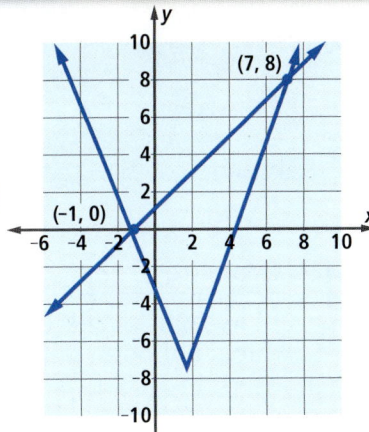

REVIEW

23. Because of traffic, Nina's drive to school takes anywhere from 12 to 20 minutes. (**Lesson 4-8**)
    a. Write an inequality expressing the time it takes Nina to get to school.
    b. Write an expression of the form $a \pm d$ to represent the interval in Part a.

In 24 and 25, solve the compound inequality and graph the solution. (**Lesson 4-8**)

24. $4a - 7 \leq 17$ and $14 - a > -5a + 3$

25. $2(m - 2) \geq 7m + 6$ or $4m - 7 > 30$

26. Consider the equations $y = 5x + 9$ and $3x - y = a - 2x$. Find a value of a so that the two lines do not intersect. (**Lesson 4-7**)

27. Write equations for the horizontal and vertical lines that go through the point $\left(-4.25, \frac{9}{11}\right)$. (**Lesson 4-2**)

28. Gail had $11.50 to spend on snacks for herself and her friends. She wanted to buy as many 85 cent energy bars as she could, in addition to 3 boxes of popcorn at 75 cents each and 3 juice bottles at $1.50 each. Set up an inequality and solve by clearing decimals to find the number of energy bars she can buy. (**Lesson 3-8**)

EXPLORATION

29. You learned that absolute value equations can have 0, 1, or 2 solutions.
    a. Write an absolute value inequality whose solution is all real numbers.
    b. Write an absolute value inequality that has no solutions.

30. Solve $|x - 2| = |x| - 2$.

QY ANSWERS

1a. $t = 15$ or $t = -15$

b. no solution

c. $v = 0$

2. $d < -0.9$ or $d > 0.9$

Chapter 4 Projects

1 Growing Trees

In Lesson 4-5 you read about a maple tree which, although initially smaller, grew faster than a beech tree and overtook it in height. Give three other real-world examples whose graphs would result in intersecting lines. Make sure that at least one example is meaningful for negative values of the independent variable.

2 Hybrid Cars

Suppose car A costs $20,000 and gets 25 miles per gallon of gas. Suppose a hybrid car B costs $25,000 but gets 39 miles per gallon. Suppose gas costs an average of $2.75 per gallon and you drive 12,000 miles per year.

a. Plot the cost of owning and operating each car on the same graph, with years of ownership along the x-axis.

b. Use an inequality to determine how long it will take for the cost of owning car A to overtake the cost of owning car B. Check your answer with your graph from Part a.

3 Planning Your Trip

Suppose you want to take a trip to Washington, D.C. You can either drive or fly. If you fly, the airline requires you to be at the airport two hours before your departure time and it will take you half an hour to pick up your luggage once you arrive at your destination. Let W = the distance from your home to Washington. The plane averages 500 miles per hour, and you drive at an

average speed of 60 miles per hour. Write expressions in terms of W for your travel time by car and your travel time (including waiting time) by plane. What is the farthest possible distance that you could live from Washington and still get there faster by car? Where could you live for this to be the case? Illustrate on a map similar to the one above.

4 Absolute Value Equations

a. Solve the equation $|5x + 3| = 2x - 8$ by any method.

b. Generalize Part a by solving the equation $|mx + n| = 2x - 8$ for x.

c. Find values of m and n for which the equation of Part b has infinitely many solutions. Describe those solutions.

d. Find values of m and n for which the equation of Part b has no solution.

5 Getting Closer and Closer to Zero

Begin with the expression $\frac{1}{x}$. Find the value of the expression when $x = 1, 0.1, 0.01, 0.001, 0.0001,$ and 0.00001.

a. **Fill in the Blank** As x gets closer to zero, the value of the expression $\frac{1}{x}$ gets ___?___.

b. Represent as a ray the values that the expression can take when x is in the interval $0 < x \leq 1$.

c. Find the value of the expression when $x = -1, -0.1, -0.01, -0.001, -0.0001,$ and -0.00001. Represent as a pair of rays the values that the expression can take when x is in the interval $-1 \leq x < 0$ or $0 < x \leq 1$.

Chapter 4

Summary and Vocabulary

- Equations involving percents may be written in the form $p \cdot q = r$, where p is the decimal form of the percent, q is the initial quantity, and r is the resulting quantity. The equation can then be solved as you would solve any other equation.

- Linear sentences are equivalent to sentences of the forms $ax + b = cx + d$ and $ax + b < cx + d$. Graphs, tables, CAS, and algebraic processes all provide ways of solving these sentences. The same algebraic processes and a CAS enable you to solve many formulas for one of their variables.

- Some sentences like $12 - 30x = -3(10x - 4)$ are true for all real numbers. When solving them, the same number appears on both sides of the equal sign. Other sentences like $2y + 5 = 2y - 3$ have no solution. When solving them, an equation with different numbers on both sides of the equal sign will result.

- The union and intersection of sets help to describe situations with *or* and *and*, respectively. They can be used to solve compound inequalities.

- Graphs and algebraic processes are also used to find the solution set for absolute value equations and absolute value inequalities.

- All of this sentence-solving is for a purpose. Many real-world situations lead to linear equations or inequalities, and many common formulas involve linear expressions.

Vocabulary

4-1
percent

4-2
horizontal line, $y = k$
vertical line, $x = h$

4-4
general linear equation,
$ax + b = cx + d$

4-7
Celsius scale
centigrade scale
Fahrenheit scale
input
output
equivalent formulas

4-8
compound sentence
intersection
union
± notation

Chapter 4 Self-Test

Take this test as you would take a test in class. You will need a calculator. Then use the Selected Answers section in the back of the book to check your work.

1. Gloria found a prom dress at a local shop. The price tag said $262, but Gloria would have to pay 9% sales tax. She found the same dress for $285 online with no tax. There was no charge for shipping. Find out whether the dress costs less in the shop or online. Explain your process.

2. **Fill in the Blank** To solve the equation $-8t + 73 = -49 + 4t$, an effective first step is to add __?__ to each side.

3. By what number can each side of $\frac{7}{8}x - \frac{2}{5} = 6 - \frac{1}{8}x$ be multiplied to clear the fractions?

In 4–10, solve each sentence. Show your work.

4. $-5m + 21 = 6m - 56$
5. $0.73v + 37.9 = 16 - v$
6. $\frac{x}{4} + \frac{3}{5} = \frac{x}{2}$
7. $-2(3 - d) > 3(2d - 3)$
8. $-29 < 7.5p - 44 \leq 28$
9. $|2x - 3| = 21$
10. $8(4 - 2n) = -16n + 16$

11. Suppose you have graphed Y_1 and Y_2 and the result is the graph below.

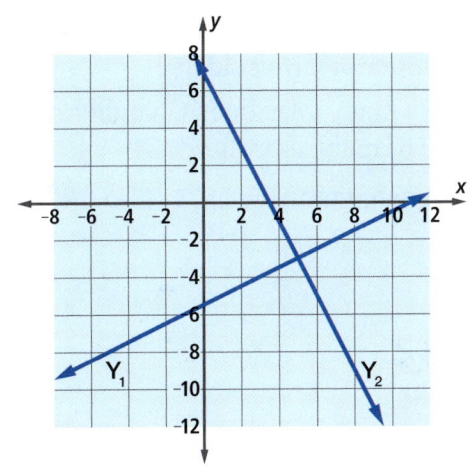

a. Estimate the x-value that is the solution to $Y_1 = Y_2$.
b. Write an inequality using x whose solutions tell when $Y_1 \geq Y_2$.

12. Silvia is offered two jobs selling air time for a local cable television station. From TurboTV she could earn a $30,000 annual salary plus 5% of the amount of her sales. Sparkle Cable offered Silvia a $25,000 annual salary plus 8% of her sales.

a. Complete the table.

Sales	Total Salary from TurboTV	Total Salary from Sparkle Cable
$25,000	$31,250	?
$50,000	?	$29,000
$75,000	?	?
$100,000	?	?
$125,000	?	?
$150,000	?	?
$175,000	$38,750	?

b. Use the table to find for what sales amount Silvia will earn more money working for Sparkle Cable. Write the answer as an inequality.

c. In her current job, Silvia's sales average $83,000 per year. If this amount of sales were to continue, which job do you think Silvia should take? Explain your reasoning.

13. Graph in the coordinate plane the solution to $4 - |x + 1| > -2$.

14. Give an equation for each line in the graph.

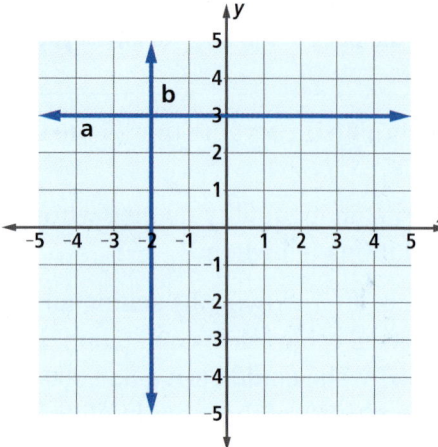

15. Solve $-6x + 7y = -84$ for x.

16. Use a graphing calculator to determine whether there is a solution to $-7.4(x - 3) = 1.2x + 8.1 - 2(4.3x - 8.05)$. Explain how the graph supports your conclusion.

17. The graph below gives the annual consumption of ice cream in the United States from 1990 to 2002 in millions of dollars. The horizontal line shows the mean ice cream consumption, which was approximately 1,290 million gallons.

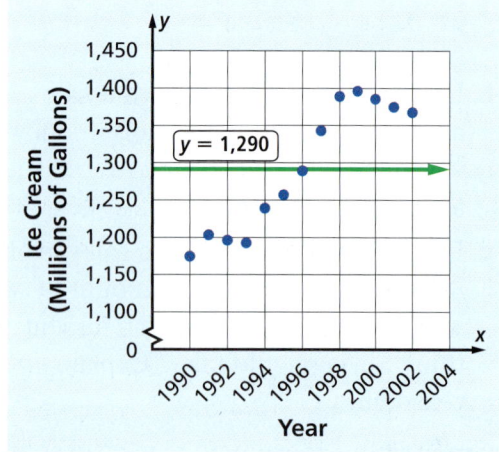

Source: U.S. Dept. of Agriculture Economic Research Service

a. Which year had the greatest absolute deviation from the mean?

b. Which year's ice cream consumption was closest to the mean? What does that tell you about the deviation for that year?

c. In the last four years of the data, the consumption levels have been getting closer to the mean. Is this situation good or bad for ice cream producers? Why or why not?

18. At Las Sendas Golf Club, Joe's handicap is 9 while Elena's is 18. To level the playing field, the players subtract their handicaps from their stroke totals at the end of the round. Suppose Joe took $\frac{9}{10}$ as many shots as Elena but they ended up with the same score after subtracting their handicaps.

a. How many strokes did Elena take?

b. How many strokes did Joe take?

c. What were their scores with the handicaps?

19. Chandler has a company that produces and sells CDs. He paid $452.54 in equipment fees, and it costs him $5.25 to produce each CD. He charges customers $17.99 per CD.

a. Write an expression describing Chandler's expenditure based on the number of CDs produced.

b. Write an expression describing Chandler's revenue based on the number of CDs sold.

c. How many CDs does Chandler need to sell to make a profit?

20. 30% of what number is 7?

Chapter 4 Chapter Review

SKILLS
PROPERTIES
USES
REPRESENTATIONS

SKILLS Procedures used to get answers

OBJECTIVE A Solve and check equations of the form $ax + b = cx + d$. (Lesson 4-4)

In 1–8, solve the equation and check your solution.

1. $14 + 8A = 4A - 10$
2. $-5a + 9 = 5a$
3. $3n = n + 7 + 9n$
4. $8x - 21.25 = -0.5x$
5. $4f = 2f - 7(5 - 6f)$
6. $7(2 - y) = 3(y + 2)$
7. $\frac{1}{3}a - 1 = 3a$
8. $\frac{x}{2} + \frac{x}{5} + 20 = x$

OBJECTIVE B Solve and check compound inequalities of the form $ax + b < cx + d$. (Lessons 4-5, 4-8)

In 9 and 10, solve and check the inequality.

9. $14w + 64 \geq 17w - 323$
10. $4(5 + 2t) < 9(2 + t)$

In 11–16, solve and graph all solutions on a number line.

11. $\frac{7}{10}n + \frac{2}{5} > -\frac{1}{2}n - \frac{12}{5}$
12. $0.42h + 3 \leq 0.6h - 0.78$
13. $3p + 4 < 31$ and $8p - 12 \geq 7p - 11$
14. $5s - 7 > 2.5$ or $8 \leq s + 10$
15. $-1 < 4x + 7 < 23$
16. $22 \geq 2d - 40 > -55$

OBJECTIVE C Find equivalent forms of formulas and equations. (Lesson 4-7)

In 17–22, solve the equation for the stated variable.

17. $A = \frac{1}{2}ap$ for p
18. $V = \ell wh$ for h
19. $S = 2\pi r^2 + 2\pi rh$ for h
20. $k = \frac{s}{w}$ for w
21. $6x + 3y = 21$ for y
22. $5y - 7x = 25$ for y

PROPERTIES The principles behind the mathematics

OBJECTIVE D Solve percent problems. (Lesson 4-1)

23. How much tax is there on a $32 item if the tax rate is 7.5%?
24. According to the Census Bureau about 2% of people in the United States are age 85 or older. In a town of 35,000, about how many people would be expected to be at least 85 years old?
25. To the nearest percent, 10 is what percent of 23?
26. 47 is what percent of 30?
27. 85% of what number is 170?
28. To the nearest whole number, 6.3% of what number is 7?

OBJECTIVE E Solve absolute value equations and inequalities involving linear expressions. (Lesson 4-9)

In 29–31, suppose $|x - 14| = n$.

29. Find all solutions when $n = 2$.
30. Find all solutions when $n = 0$.
31. Find all solutions when $n = -5$.

In 32–34, translate the given English sentence into a mathematical sentence. Then graph the solutions on a number line.

32. The distance between x and 4 is less than 20.
33. Liseli's dog will only go outside when the temperature is within 15 degrees of 60 degrees Fahrenheit.
34. In a math competition, Howard missed the winning score of 180 points by more than 5 points.

Chapter 4

OBJECTIVE F Apply and recognize Addition and Multiplication Properties of Equality and Inequality when solving linear sentences. (Lessons 4-4, 4-5)

35. Consider the equation $3x + 2 = 5x + 12$.
    a. Solve the equation by first adding $-5x$ to each side.
    b. Solve by first adding $-3x$ to each side.
    c. Compare your answers to Parts a and b.

36. Consider the inequality $2a + 3 < 5a - 6$.
    a. Solve the inequality by first adding $-5a$ to each side.
    b. Solve by first adding $-2a$ to each side.
    c. How are your answers to Parts a and b related?

In 37 and 38, a sentence is solved. State the property that justifies each step.

37. $\frac{2}{3}x - 4 = \frac{1}{6}x + 2$
    a. $6\left(\frac{2}{3}x - 4\right) = 6\left(\frac{1}{6}x + 2\right)$
    b. $6 \cdot \frac{2}{3}x - 6 \cdot 4 = 6 \cdot \frac{1}{6}x + 6 \cdot 2$
       $4x - 24 = x + 12$
    c. $3x - 24 = 12$
    d. $3x = 36$
    e. $x = 12$

38. $-3y + 7 \geq 8y - 5$
    a. $-3y + 12 \geq 8y$
    b. $12 \geq 11y$
    c. $\frac{12}{11} \geq y$

39. Alexis is trying to solve the equation $100n + 10 = -90n + 4$. After the first step the resulting equation was $10n + 10 = 4$.
    a. Identify the mistake Alexis made.
    b. Correct the mistake and solve the equation.

OBJECTIVE G Recognize when sentences have no solution or every real number as a solution. (Lesson 4-6)

40. Explain why any real number is a solution to the equation $8y - 30 = 4(2y - 7.5)$.

41. Find all the solutions to the equation $t - t = 1$.

42. Find all solutions to the equation $t - t = 0$.

43. Explain why no real number is a solution to $x + 5 > x + 6$.

USES Applications of mathematics in real-world situations

OBJECTIVE H Use linear equations and inequalities of the form $ax + b = cx + d$ or $ax + b < cx + d$ to solve real-world problems. (Lessons 4-4, 4-5)

44. Sam has $1,850 in her savings account and adds $25 each month. Diego has $2,000 in his account and adds $20 each month.
    a. How much will be in each account after n months?
    b. After how many months of saving will they have the same amount in their accounts?

45. Taxi, Inc., charges a fee of $5 for each ride and an additional $0.75 for each mile you travel. Calling Cabs charges an initial fee of $3.50 and an additional $0.85 per mile.
    a. If you are going to take a cab 10 miles, which company should you use?
    b. If you are going to take a cab 15 miles, which company should you use?
    c. What is the break-even point?

246 More Linear Equations and Inequalities

46. Kim has $15 and is saving $9 per week. Alberto has $100 and is spending $8 per week.
    a. Let x = the number of weeks that have passed. Write a linear inequality that can be used to find out when Kim will have more money than Alberto.
    b. Solve the inequality.

47. A sign-making company, Signs-R-We, charges a set-up art fee of $50 plus an additional $1.50 for each sign printed. Their competitor Sign-Me-Up waives the art fee but charges $3.50 per sign. What is the largest number of signs you could print so that Sign-Me-Up would be a better deal than Signs-R-We?

OBJECTIVE I Use tables and graphs to solve real-world problems involving linear situations. (Lessons 4-2, 4-3)

48. Two music downloading Web sites offer music discounts. Site 1 has a $19.95 membership fee and charges $0.99 per download. Site 2 charges $14.95 to join and $1.03 per download.
    a. Copy and complete the table below.

Number of Downloads	Charges	
	Site 1	Site 2
2	?	?
4	?	?
6	?	?
8	?	?
10	?	?

    b. How many downloads must you buy for the charges of the two sites to be equal?
    c. When is the price of Site 1 better?
    d. When is the price of Site 2 better?

49. Sherita was investigating the number of baby carrots that came pre-packaged. She bought ten bags and recorded her data in a table.

Bag Number	Number of Carrots
1	30
2	35
3	27
4	32
5	39
6	29
7	31
8	35
9	32
10	29

    a. Make a dot plot of these data.
    b. Determine the mean number of carrots per bag and draw the horizontal line representing the mean on your plot.
    c. Which bag(s) is closest to the mean?
    d. Which bag has the greatest absolute deviation from the mean?

50. Michael is considering two different sales positions. Rent-A-Vehicle would pay a total salary of $1,100 per month plus a 6% sales commission. Borrow-Our-Car would pay a total salary of $900 plus an 8% commission.
    a. Copy and complete the table below.

Sales	Rent-A-Vehicle	Borrow-Our-Car
$0	$1,100	$900
$5,000	?	?
$10,000	?	?
$15,000	?	?
$20,000	?	?
$25,000	?	$2,900
$30,000	$2,900	?

    b. For what amounts of sales will Borrow-Our-Car pay a greater total salary?

Chapter Review

Chapter 4

OBJECTIVE J Solve real-world problems involving percents. (Lesson 4-1)

51. According to the *World Almanac and Book of Facts*, there were approximately 217,000 women serving in the Armed Forces of the United States in 2004, accounting for about 15% of total military personnel. In all, about how many persons are serving in the Armed Forces?

52. According to the *Pew Internet & American Life Project*, 17% of United States households used online banking in 2000. In 2005, 35% of U.S. households used online banking. If there were 98,000,000 U.S. households in both years, how many more households used online banking in 2005 than in 2000?

53. After a 30% discount, a mattress sold for $896. What was its price before the discount?

REPRESENTATIONS Pictures, graphs, or objects that illustrate concepts

OBJECTIVE K Graph horizontal and vertical lines. (Lesson 4-2)

In 54 and 55, graph the points in the coordinate plane satisfying each equation.

54. $x = -3$ 55. $y = 6$

56. **True or False** The graph of all points in the coordinate plane satisfying $y = 23$ is a horizontal line.

57. Write an equation for the line containing the points $(5, 12)$, $(5, -3)$, and $(5, -15)$.

58. Write an equation for the line in the graph below.

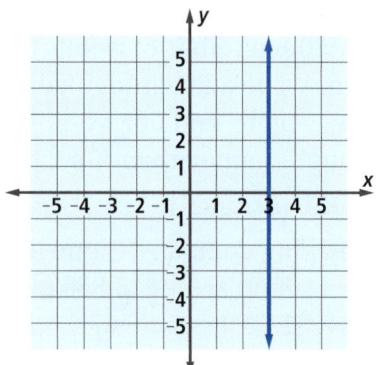

OBJECTIVE L Use graphs to solve problems involving linear equations. (Lesson 4-3)

59. An airplane is cruising at 35,000 feet when it begins its descent into an airport at 1,750 feet per minute.
    a. Write an equation that relates the plane's altitude a in feet and the time m in minutes since it started to descend.
    b. Graph your equation from Part a.
    c. Use the graph to determine how many minutes it will take to land.

In 60 and 61, lines ℓ, m, and n are graphed below. Line ℓ has equation $y = -\frac{8}{7}x + 8$, line m has equation $y = -5 - x$, and line n has equation $y = 2x - 5$.

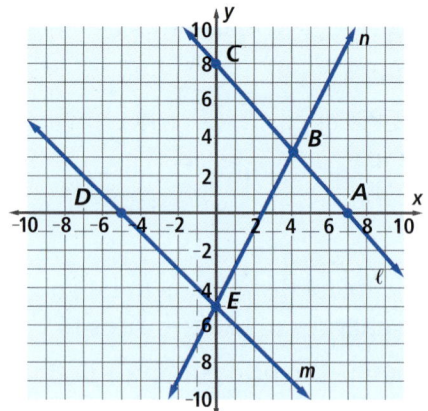

248 More Linear Equations and Inequalities

60. a. Fill in the blank, then answer the question. The solution to the equation $2x - 5 = -\frac{8}{7}x + 8$ is the __?__-coordinate of which named point?
   b. Estimate a solution to the equation in Part a.

61. Suppose $E = (0, -5)$. Use this information to solve the inequality $2x - 5 \leq -5 - x$.

62. Picture Perfect will develop a roll of film for $0.35 per photo with no developing charge. You Oughta Be In Pictures charges $0.10 per photo plus a $3.50 developing charge. Make a graph to determine the break-even point.

OBJECTIVE M Use graphs to model sentences that have no solution or every real number as a solution. (Lesson 4-6)

63. Nate starts July with $80 and decides to mow lawns for $10 per lawn to earn money. His friend Owen starts with $65 and also decides to mow lawns for $10 per lawn.
   a. How much money will Nathan have after mowing n lawns?
   b. How much money will Owen have after mowing n lawns?
   c. Let y equal each of your expressions in Parts a and b. Graph the two equations.
   d. Using your graph, when will Owen have as much money as Nate?

In 64–66, solve by letting y equal each side of the sentence and by using a graph.

64. $4(h - 7) + 6h < 2(5h - 4)$

65. $-7g + 4 = 5 - 7g$

66. $2.5(4.6p - 4) \geq 10p + \frac{3}{2}(p + 20)$

OBJECTIVE N Use graphs to solve absolute value inequalities of the form $|ax + b| < c$ or $|ax + b| > c$. (Lesson 4-9)

67. The air (heat or cooling) in Peggy's house turns on when the temperature varies 5 degrees or more from 70°F.
   a. On a number line, graph the temperatures corresponding to when the air is on.
   b. Using t for temperature, write inequalities describing the temperatures graphed in Part a.

68. Use the graph of $y = |2x - 7|$ and $y = 4$ below to solve each of the following.

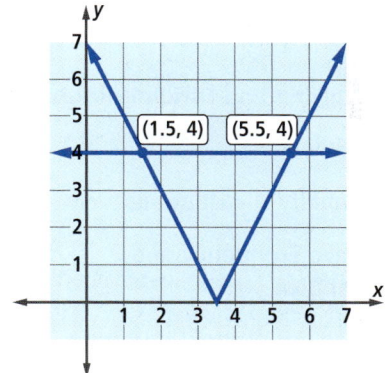

   a. $|2x - 7| = 4$
   b. $|2x - 7| > 4$
   c. $|2x - 7| \leq 4$

Chapter 5
Division and Proportions in Algebra

Contents

- 5-1 Multiplication of Algebraic Fractions
- 5-2 Division of Algebraic Fractions
- 5-3 Rates
- 5-4 Multiplying and Dividing Rates
- 5-5 Ratios
- 5-6 Probability Distributions
- 5-7 Relative Frequency and Percentiles
- 5-8 Probability Without Counting
- 5-9 Proportions
- 5-10 Similar Figures

In Chapter 1, we noted that division is related to multiplication by its algebraic definition: dividing by b is the same as multiplying by the reciprocal of b, or $a \div b = a \cdot \frac{1}{b}$.

In Chapter 2, you saw that division is also related to multiplication by related facts.

If a and b are not zero and $ab = c$, then $a = \frac{c}{b}$ and $b = \frac{c}{a}$.

Division is also an important operation in its own right. Three kinds of situations lead directly to division.

Splitting Up

If a quantity *a* is split into *b* equal parts, then each part has measure $\frac{a}{b}$. For this reason, every fraction can be viewed as a division $\left(\frac{a}{b} = a \div b\right)$.

The value of the fraction is the quotient of *a* and *b*.

Rate

If *a* and *b* are quantities with different units, then the quotient is a *rate*. For example, dividing 50 miles by 2 hours yields the rate 25 miles per hour: $\frac{50 \text{ mi}}{2 \text{ hr}} = \frac{25 \text{ mi}}{\text{hr}}$.

Ratio

If *a* and *b* have the same kind of units, then the quotient is a *ratio*. For example, if one doll weighs 36 grams and another weighs 4.5 grams, then the quotient $\frac{36 \text{ g}}{4.5 \text{ g}}$ equals 8, the ratio of the first weight to the second weight.

If two rates or ratios are equal, then the result is a *proportion*. For example, dividing 75 miles by 3 hours yields the same rate as dividing 50 miles by 2 hours. The equation $\frac{50 \text{ mi}}{2 \text{ hr}} = \frac{75 \text{ mi}}{3 \text{ hr}}$ is a proportion. In this chapter, you will study these and related topics.

Chapter 5

Lesson
5-1

Multiplication of Algebraic Fractions

Vocabulary

algebraic fraction

> **BIG IDEA** Algebraic fractions are multiplied in the same way you multiply numeric fractions.

In algebra, a division is represented by a fraction. An **algebraic fraction** is a fraction with a variable in the numerator, in the denominator, or in both. Here are some algebraic fractions.

$$\frac{7t}{2} \qquad \frac{-a}{6.4bc} \qquad \frac{3m+4}{4m+3} \qquad \frac{\frac{2}{3}+\frac{4}{5}}{x^2} \qquad \frac{x-y}{\sqrt{x^2+y^2}}$$

Mental Math

Evaluate.
a. $0.5 \cdot 4$
b. $5 \cdot 0.4$
c. $0.5 \cdot 0.4$

Multiplying Algebraic Fractions

Algebraic fractions are multiplied just as you multiply numeric fractions. Below is a way to picture the product of the fractions $\frac{a}{b}$ and $\frac{c}{d}$. First draw a unit square as shown below. Split one side into b parts and the other side into d parts, and draw lines creating bd small rectangles. Then find $\frac{a}{b}$ of one side and $\frac{c}{d}$ of an adjacent side.

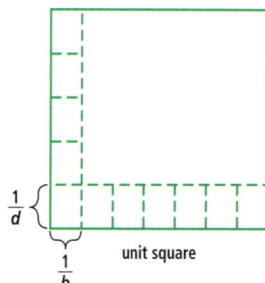 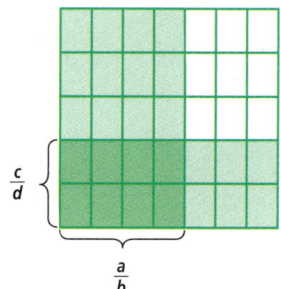

There are ac shaded rectangles out of bd small rectangles in the unit area. So the area is $\frac{ac}{bd}$. This describes the common rule for multiplying fractions, which applies to all algebraic fractions.

Multiplying Fractions Property

For all real numbers a, b, c, and d, with $b \neq 0$ and $d \neq 0$,
$\frac{a}{b} \cdot \frac{c}{d} = \frac{ac}{bd}$.

Division and Proportions in Algebra

Example 1

Multiply $\frac{L}{4} \cdot \frac{W}{3}$.

Solution Use the Multiplying Fractions Property.

$$\frac{L}{4} \cdot \frac{W}{3} = \frac{LW}{12}.$$

Check 1 Substitute values for L and W, say $L = 20$ and $W = 8$.

Does $\frac{20}{4} \cdot \frac{8}{3} = \frac{20 \cdot 8}{12}$?

Yes, the left side is $5 \cdot \frac{8}{3}$, or $\frac{40}{3}$. The right side is $\frac{160}{12}$, or $\frac{40}{3}$.

Check 2 Use an area model. Draw a rectangle with length L and width W. Divide the length into fourths and the width into thirds. Shade a smaller rectangle with dimensions $\frac{L}{4}$ and $\frac{W}{3}$. Since the L by W rectangle with area LW is divided into twelfths, the shaded rectangle has area $\frac{1}{12} \cdot LW$, or $\frac{LW}{12}$.

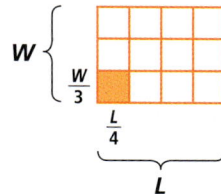

The fraction $\frac{LW}{12}$ is another way of writing $\frac{1}{12}LW$. Example 2 below shows how the Multiplying Fractions Property can be used to explain why many different expressions with algebraic fractions are equivalent.

Example 2

Show that each of the following expressions equals $\frac{5x}{3}$.

a. $\frac{5}{3}x$ b. $\frac{1}{3} \cdot 5x$ c. $5 \cdot \frac{x}{3}$

Solutions Notice how each part below uses the property that $x = \frac{x}{1}$ and the Multiplying Fractions Property.

a. $\frac{5}{3}x = \frac{5}{3} \cdot \frac{x}{1} = \frac{5x}{3}$

b. $\frac{1}{3} \cdot 5x = \frac{1}{3} \cdot \frac{5x}{1} = \frac{5x}{3}$

c. $5 \cdot \frac{x}{3} = \frac{5}{1} \cdot \frac{x}{3} = \frac{5x}{3}$

This shows that $\frac{5x}{3}$, $\frac{1}{3} \cdot 5x$, and $5 \cdot \frac{x}{3}$ are all equal to each other.

Equal Fractions

As you know, every numerical fraction is equal to many other fractions. For example, $\frac{3}{5} = \frac{30}{50}$ and $\frac{30}{50} = \frac{9}{15}$. These equalities are examples of the *Equal Fractions Property*.

> **Equal Fractions Property**
>
> For all real numbers a, b, and k, if $b \neq 0$ and $k \neq 0$, then $\frac{a}{b} = \frac{ak}{bk}$.

The Equal Fractions Property holds for all fractions $\frac{a}{b}$ and values of k as long as the denominator is not zero. In $\frac{3}{5} = \frac{30}{50}$, $a = 3$, $b = 5$, and $k = 10$. That is, $\frac{3}{5} = \frac{3 \cdot 10}{5 \cdot 10} = \frac{30}{50}$. In $\frac{30}{50} = \frac{9}{15}$, $a = 9$, $b = 15$, and $k = \frac{10}{3}$. The Equal Fractions Property is true because of the Multiplying Fractions Property and the Multiplicative Identity Property. Here is how.

$$\frac{a}{b} = \frac{a}{b} \cdot 1 \quad \text{Multiplicative Identity Property}$$

$$= \frac{a}{b} \cdot \frac{k}{k} \quad \frac{k}{k} = 1, k \neq 0$$

$$= \frac{ak}{bk} \quad \text{Multiplying Fractions Property}$$

Algebraic fractions, like numeric fractions, can sometimes be written in simpler form. To use the Equal Fractions Property to simplify algebraic fractions, find common factors in the numerator and denominator of the fraction.

Example 3

Simplify $\frac{112ab}{7a}$.

Solution 1 $7a$ is a common factor of the numerator and denominator.

$$\frac{112ab}{7a} = \frac{7a \cdot 16b}{7a \cdot 1} \quad \text{Multiplying Fractions Property}$$

$$= \frac{16b}{1} \quad \text{Equal Fractions Property}$$

$$= 16b \quad x = \frac{x}{1} \text{ for all } x.$$

Solution 2 People often skip steps. They sometimes show division of the common factors with slashes.

$$\frac{\overset{16}{\cancel{112}}\overset{1}{\cancel{a}}b}{\underset{1}{\cancel{7}}\underset{1}{\cancel{a}}} = \frac{16b}{1} = 16b$$

 QY1

▶ **QY1**

Show that $\frac{25m}{30n}$ and $\frac{5m^2}{6mn}$ equal the same algebraic expression.

Activity

Use a CAS to simplify the following algebraic fractions. Check your work with a CAS or by substitution.

1. $\frac{5ab}{10b}$
2. $\frac{27x^2}{9x^2}$
3. $\frac{2y}{2yz}$
4. $\frac{48cd}{6ac}$
5. $\frac{12m^2}{18m}$
6. $\frac{8\pi x}{6x^2}$

254 Division and Proportions in Algebra

Lesson 5-1

Example 4
Assume $x \neq 0$ and $y \neq 0$. Multiply $\frac{4x}{27y}$ by $\frac{3y}{2x^2}$ and simplify the product.

Solution 1 Here we show all the major steps.

$\frac{4x}{27y} \cdot \frac{3y}{2x^2} = \frac{4x \cdot 3y}{27y \cdot 2x^2}$ Multiplying Fractions Property

$= \frac{12xy}{54x^2y}$ Multiply.

$= \frac{6 \cdot 2 \cdot x \cdot y}{6 \cdot 9 \cdot x \cdot x \cdot y}$ Factor each expression.

$= \frac{6}{6} \cdot \frac{2}{9} \cdot \frac{x}{x} \cdot \frac{1}{x} \cdot \frac{y}{y}$ Multiplying Fractions Property

$= \frac{2}{9} \cdot \frac{1}{x}$ $\frac{k}{k} = 1$ if $k \neq 0$; Identity Property

$= \frac{2}{9x}$ Multiplying Fractions Property

Solution 2 Look for common factors in the numerator and denominator.

$\frac{4x}{27y} \cdot \frac{3y}{2x^2} = \frac{\overset{2}{\cancel{4}} \cdot \overset{1}{\cancel{x}} \cdot \overset{1}{\cancel{3}} \cdot \overset{1}{\cancel{y}}}{\underset{9}{\cancel{27}} \cdot \underset{1}{\cancel{y}} \cdot \underset{1}{\cancel{2}} \cdot \underset{1}{\cancel{x}} \cdot x} = \frac{2}{9x}$

 QY2

> **READING MATH**
> The Equal Fractions Property is a property related to multiplication. It does not work when the same terms are *added* to the numerator and denominator.

> **QY2**
> Multiply $\frac{-5a^2}{12b} \cdot \frac{2b^2}{6a}$ and simplify the product.

Questions

COVERING THE IDEAS

1. State the Multiplying Fractions Property.

2. The rectangle at the right has base *b* and height *h*.
   a. If all the small rectangles have the same dimensions, what is the area of the shaded region?
   b. What product of algebraic fractions is represented by the area of the shaded region?

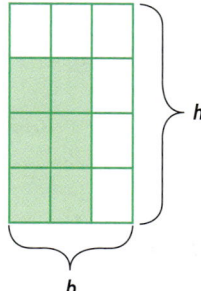

In 3 and 4, multiply the fractions.

3. $\frac{a}{7} \cdot \frac{b}{2}$ 4. $\frac{x}{30} \cdot \frac{3y}{z^2}$

5. Determine whether $\frac{1}{5}n = \frac{n}{5}$ is *always*, *sometimes but not always*, or *never* true.

6. Explain why $\frac{3}{8}x$ is equal to $\frac{3x}{8}$.

7. **Multiple Choice** Which expression does *not* equal the others?

   A $\frac{5n}{8}$ B $\frac{5}{8}n$ C $5n \cdot \frac{1}{8}$ D $\frac{5}{n} \cdot 8$

Multiplication of Algebraic Fractions   255

In 8–10, use the Equal Fractions Property to simplify the fraction.

8. $\dfrac{1{,}875}{225}$
9. $\dfrac{-4n}{24n^2}$
10. $\dfrac{10mn}{15np}$

In 11–14, multiply and simplify the result.

11. $\dfrac{1.2m}{n} \cdot \dfrac{1.2n}{m}$
12. $\dfrac{7v}{x^2} \cdot \dfrac{x^2}{7v}$
13. $\dfrac{4abc}{27c} \cdot \dfrac{3}{2a^2 b^3}$
14. $\dfrac{1}{4} \cdot 2n \cdot \dfrac{3n}{6}$

15. a. One rectangle is half as wide and one-fourth as long as another rectangle. How do their areas compare?
    b. Draw a figure to illustrate your answer.

16. a. Show that $\dfrac{30+x}{10+x}$ and $\dfrac{30}{10}$ are *not* equivalent by letting $x = 7$.
    b. Show that $\dfrac{30+x}{10+x}$ and $\dfrac{3+x}{1+x}$ are *not* equivalent by letting $x = -4$.
    c. Why can't the Equal Fractions Property be applied in Parts a and b?

17. The Brock and Pease families have rectangular vegetable gardens. The length of the Brocks' garden is $\dfrac{2}{3}$ the length and $\dfrac{3}{4}$ the width of the Peases' garden.
    a. How do the areas of the gardens compare?
    b. Check your answer by using a specific length and width for the Peases' garden.

APPLYING THE MATHEMATICS

18. **Skill Sequence** Compute in your head.
    a. $\dfrac{5}{3} \cdot 3b$
    b. $\dfrac{9}{x} \cdot xc$
    c. $\dfrac{a}{b} \cdot bd$
    d. $n^2 \cdot \dfrac{a}{n^2}$

19. Combine and simplify $\dfrac{4n-5}{n} + \dfrac{5}{n}$.

In 20–22, multiply and simplify where possible.

20. $\dfrac{a}{b} \cdot \dfrac{b}{c} \cdot \dfrac{c}{a}$
21. $\dfrac{-30r^3}{7s} \cdot \dfrac{-28s}{120r^4}$
22. $\dfrac{-2y}{3} \cdot \dfrac{5y}{6} \cdot z$

23. Find two *algebraic* fractions whose product is $\dfrac{36a^2}{5x}$.

24. **Multiple Choice** Find the fraction that is *not* equal to the other three.
    A $\dfrac{110}{130}$ B $\dfrac{121}{143}$ C $\dfrac{121}{169}$ D $\dfrac{550}{650}$

Recent surveys show the average size of a garden is between 500 and 1,000 square feet.

Source: www.oldhouseweb.com

25. a. Find the volume of the brick at the right.
    b. Check your answer by letting $L = 12$.
    c. Think of a cube with sides of length L. How many of these bricks would fit into the cube? How can you tell?

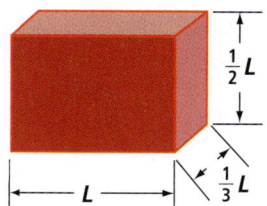

REVIEW

26. If 5% of a number is 12, what is 12% of that number? (**Lesson 4-1**)

27. Solve $150x + 200x = 14{,}000$. (**Lessons 3-4, 2-2**)

28. *Skill Sequence* State the reciprocal of each number. (**Lesson 2-8**)
    a. 5
    b. $\frac{1}{100}$
    c. $\frac{-2}{3}$

29. A single-story house is to be built on a lot 75 feet wide by 100 feet deep. The shorter side of the lot faces the street. The house must be set back from the street at least 25 feet. It must be 20 feet from the back lot line, and 10 feet from each side lot line. What is the maximum area the house can have? (**Lesson 2-1**)

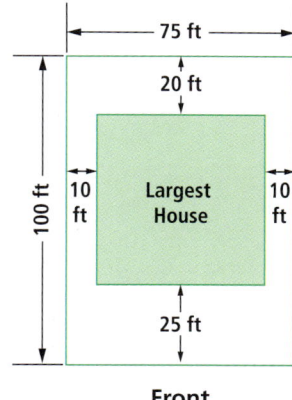

30. Consider $\frac{3}{4} \div \frac{9}{32}$.
    a. Rewrite the problem as the multiplication of two fractions.
    b. What is the answer in lowest terms?
    c. What is the answer as a decimal? (**Previous Course**)

EXPLORATION

31. a. Calculate the products at the right.
    b. Write a sentence or two describing the patterns you observed in Part a.
    c. Predict the following products.
    $\frac{2}{1} \cdot \frac{3}{2} \cdot \frac{4}{3} \cdot \frac{5}{4} \cdot \ldots \cdot \frac{2{,}010}{2{,}009}$
    $\frac{2}{1} \cdot \frac{3}{2} \cdot \frac{4}{3} \cdot \frac{5}{4} \cdot \ldots \cdot \frac{n+1}{n}$

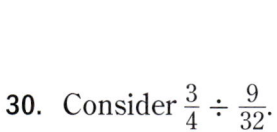

QY ANSWERS

1. $\frac{25m}{30n} = \frac{5 \cdot 5m}{5 \cdot 6n} = \frac{5m}{6n}$ and $\frac{5m^2}{6mn} = \frac{m \cdot 5m}{m \cdot 6n} = \frac{5m}{6n}$.

2. $\frac{-5ab}{36}$

Lesson 5-2

Division of Algebraic Fractions

Vocabulary

complex fraction

▶ **BIG IDEA** Algebraic fractions are divided the same way you divide numeric fractions.

Remember that two numbers are *reciprocals* if the product of the numbers is 1. The reciprocal of the number $\frac{c}{d}$ is $\frac{d}{c}$, with $c \neq 0$ and $d \neq 0$, because $\frac{c}{d} \cdot \frac{d}{c} = \frac{c \cdot d}{d \cdot c} = 1$.

When 32 ounces of orange juice are shared equally among 5 people, each person gets $\frac{1}{5}$ of a quart. This is an example of the Algebraic Definition of Division: dividing by a number is the same as multiplying by its reciprocal.

Mental Math

Use the triangle below.

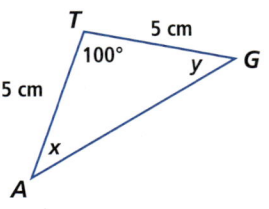

a. What is x?

b. What is y?

Dividing Algebraic Fractions

Consider the division of fractions $\frac{a}{b} \div \frac{c}{d}$. Since dividing by a number is the same as multiplying by its reciprocal, dividing by $\frac{c}{d}$ gives the same result as multiplying by $\frac{d}{c}$.

> **Dividing Fractions Property**
>
> For all real numbers a, b, c, and d, with b, c, and $d \neq 0$,
> $$\frac{a}{b} \div \frac{c}{d} = \frac{a}{b} \cdot \frac{d}{c}.$$

For example, $\frac{5}{3} \div \frac{7}{11} = \frac{5}{3} \cdot \frac{11}{7} = \frac{55}{21}$. The Dividing Fractions Property is sometimes called the "invert and multiply" rule.

258 Division and Proportions in Algebra

Lesson 5-2

Example 1
Simplify $\frac{x}{4} \div \frac{3}{5}$.

Solution Dividing by $\frac{3}{5}$ is the same as multiplying by $\frac{5}{3}$.

$$\frac{x}{4} \div \frac{3}{5} = \frac{x}{4} \cdot \frac{5}{3} = \frac{5x}{12}$$

Check Substitute some value for x. Use this number to evaluate the original expression and your answer. Suppose $x = 2$. Does $\frac{2}{4} \div \frac{3}{5} = \frac{5 \cdot 2}{12}$? To determine this, change each fraction to a decimal. Does $0.5 \div 0.6 = \frac{10}{12}$? Yes, each side equals $0.8\overline{3}$.

Simplifying Complex Fractions

Recall that a horizontal fraction bar indicates division. The division $\frac{a}{b} \div \frac{c}{d}$ can be written as $\frac{\frac{a}{b}}{\frac{c}{d}}$. Fractions of the form $\frac{\frac{a}{b}}{\frac{c}{d}}$ are called *complex fractions*. A **complex fraction** consists of three fractions; One is the numerator and the second is the denominator of a third "bigger" fraction.

$$\begin{array}{l}\text{fraction in numerator} \\ \text{fraction in denominator}\end{array} \quad \frac{\frac{a}{b}}{\frac{c}{d}} \text{ "big fraction"}$$

Since a fraction is a division, one way to simplify a complex fraction is as follows: $\frac{\frac{a}{b}}{\frac{c}{d}} = \frac{a}{b} \div \frac{c}{d}$.

GUIDED

Example 2
Simplify $\frac{\frac{6x}{5}}{\frac{9x}{10}}$.

Solution Rewrite the fraction as a division.

$$\frac{\frac{6x}{5}}{\frac{9x}{10}} = \underline{} \div \underline{}$$

$$\frac{6x}{5} \div \underline{} = \frac{6x}{5} \cdot \underline{} \quad \text{Dividing Fractions Property}$$

$$= \frac{\underline{} x}{\underline{} x} \quad \text{Multiply the fractions.}$$

$$= \frac{4}{3} \quad \text{Simplify.}$$

Check Let $x = 20$. Then $\frac{6x}{5} = \underline{}$ and $\frac{9x}{10} = \underline{}$.

Does $\frac{6x}{5} \div \frac{9x}{10} = \frac{4}{3}$? $\underline{}$

Division of Algebraic Fractions

Chapter 5

When entering complex fractions into a calculator, be sure to group the numerator fraction in parentheses and the denominator fraction in parentheses. Otherwise, the calculator will follow the order of operations and the result will be incorrect.

For $\dfrac{\frac{2}{3}}{\frac{5}{12}}$, enter (2/3)/(5/12). If you enter 2/3/5/12, you will obtain $\dfrac{1}{90}$, which is incorrect. The correct quotient is $\dfrac{8}{5}$, or 1.6.

 QY

▶ QY

Write $\dfrac{\frac{3}{8}}{\frac{3}{11}}$ as a decimal.

Questions

COVERING THE IDEAS

1. In this lesson, we note that $\dfrac{5}{3} \div \dfrac{7}{11} = \dfrac{55}{21}$. Check this result by approximating all three fractions by decimals.

2. State the Algebraic Definition of Division.

In 3 and 4, fill in the blanks.

3. a. $\dfrac{m}{n} = m \div$ ___?___   b. $\dfrac{m}{n} = m \cdot$ ___?___

4. a. $\dfrac{\frac{p}{q}}{\frac{r}{s}} = \dfrac{p}{q} \div$ ___?___   b. $\dfrac{p}{q} \div \dfrac{r}{s} = \dfrac{p}{q} \cdot$ ___?___

In 5–11, simplify the expression.

5. $\dfrac{\frac{4}{5}}{\frac{5}{6}}$ 6. $\dfrac{\frac{4}{x}}{\frac{x}{y}}$ 7. $\dfrac{\frac{3a}{2}}{\frac{a}{2}}$ 8. $\dfrac{1}{2} \div x$

9. $\dfrac{m}{30} \div \dfrac{n}{84}$ 10. $\dfrac{8v}{5} \div \dfrac{2v}{25}$ 11. $\dfrac{\frac{3\pi}{5}}{6\pi}$

APPLYING THE MATHEMATICS

12. Cody and Troy solved $\dfrac{3}{8}x = 15$ using different methods.
    Explain why Cody and Troy got the same solution.

    Cody's Method
    $\dfrac{\frac{3}{8}x}{\frac{3}{8}} = \dfrac{15}{\frac{3}{8}}$
    $x = 15 \div \dfrac{3}{8}$
    $x = 15 \cdot \dfrac{8}{3}$
    $x = 40$

    Troy's Method
    $\dfrac{8}{3} \cdot \dfrac{3}{8}x = 15 \cdot \dfrac{8}{3}$
    $x = 40$

13. The area of a rectangle with side lengths of m and $\dfrac{4}{23}$ is 16. Find the value of m.

260 Division and Proportions in Algebra

14. Half of a pizza was divided equally among 3 people. How much of the original pizza did each person receive?

15. Le Parfum Company produces perfume in 200-ounce batches and bottles it in quarter-ounce bottles.
    a. Write a division problem that will tell you how many bottles will be filled by one batch.
    b. Find the answer.

16. A dozen bagels are bought for a group of *x* people. On average, how many bagels are there per person?

In 17–19, simplify the expression.

17. $b \div \frac{1}{b}$
18. $\frac{xy}{21} \div \frac{x}{47}$
19. $\frac{\frac{12m}{5}}{\frac{mn}{20}}$

20. a. Evaluate $x \div y$ and $y \div x$ for each of the following.
    i. $x = 12$ and $y = 2$
    ii. $x = 20$ and $y = -5$
    iii. $x = \frac{2}{3}$ and $y = \frac{4}{5}$
    b. Do your answers in Part a indicate that division is commutative? Explain your answer.
    c. Describe how $x \div y$ and $y \div x$ are related in general.

REVIEW

21. Multiply and simplify $\frac{5}{13}d \cdot \left(\frac{d}{5} \cdot \frac{5}{13}d\right)$. (**Lesson 5-1**)

22. The graph at the right compares the values of two computers A and B over time. (**Lesson 4-3**)
    a. Which computer is decreasing in value faster?
    b. About how much does the value of the computer you found in Part a change each year?
    c. After about how many years do the computers have the same value?
    d. Suppose you buy these two computers and you wish to sell one of them after 3 years. For which computer will you get more money? About how much more will you get for it?

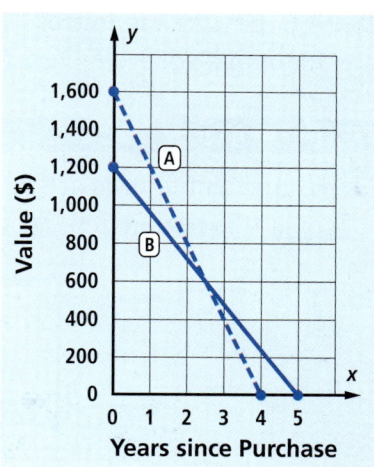

23. a. Solve $V + 0.06V - 100 = 14{,}289.16$.
    b. **Fill in the Blanks** The equation in Part a could arise from this situation. After a discount of \$100 and with a __?__ tax, the car cost __?__. Find V, the cost of __?__. (**Lessons 4-1, 3-5**)

Chapter 5

24. Let y = the depth of a point in Lake Baikal in Siberia, the deepest lake in the world. (**Lesson 3-6**)
    a. Give a reasonable domain for y.
    b. It is known that the deepest point in the Lake Baikal is 1,940 meters below the surface. What inequality does y satisfy?
    c. Graph the solution set to Part b.

25. Use the picture of the balance below. The boxes are equal in weight and the other objects are one-kilogram weights. (**Lesson 3-3**)

    a. Write an equation describing the situation, with B representing the weight of one box.
    b. What is the weight of one box?

26. A circle has a radius of 1.2 meters. Find its area to the nearest tenth of a square meter. (**Previous Course**)

Lake Baikal is situated nearly in the center of Asia in a huge stone bowl set 445 meters above sea level.

Source: www.irkutsk.org

EXPLORATION

27. Congruent figures are figures with the same size and shape. Split this region into 6 congruent pieces.

    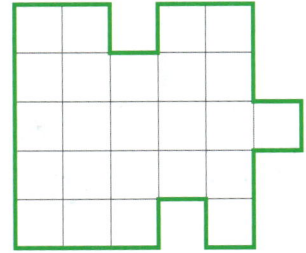

QY ANSWER

1.375

262   Division and Proportions in Algebra

Lesson 5-3
Rates

Vocabulary

rate

reciprocal rates

▶ **BIG IDEA** The quotient of two measures with different units is a rate.

What Is a Rate?

In Lesson 5-2, 32 ounces of orange juice were split evenly among 5 people. By doing the division with the units left in, we see that the answer is a *rate*.

$$\frac{32 \text{ oz}}{5 \text{ people}} = \frac{6.4 \text{ oz}}{\text{person}} = 6.4 \text{ oz/person} = 6.4 \text{ ounces per person}$$

Every rate consists of a number and a *rate unit*. You may see rate units expressed using a slash "/" or a horizontal bar "——". The slash and the bar are read "per" or "for each." The rate unit $\frac{\text{oz}}{\text{person}}$ is read "ounces per person."

In general, a **rate** is the quotient of two quantities with different units.

Mental Math

Given $3x + 4 = 12$, evaluate:

a. $3x + 5$.

b. $3x + 12$.

c. $6x + 8$.

How Are Rates Calculated?

Since rates are quotients, rates are calculated by dividing.

Example 1

Tanya and Gary drove 400 miles in 8 hours during a trip. What was their average speed?

Solution 1 Divide the distance in miles by the time in hours.

$$\frac{400 \text{ mi}}{8 \text{ hr}}$$

Separate the measurement units from the numerical parts.

$$\frac{400 \text{ mi}}{8 \text{ hr}} = \frac{400}{8} \frac{\text{mi}}{\text{hr}} = 50 \text{ miles per hour}$$

They were traveling at an average speed of 50 miles per hour.

Solution 2 You could also divide the time by the distance.

$$\frac{8 \text{ hours}}{400 \text{ miles}} = \frac{8}{400} \frac{\text{hr}}{\text{mi}} = \frac{1}{50} \text{ hour per mile}$$

This means that on the average, it took them $\frac{1}{50}$ of an hour to travel each mile.

In Example 1, the first solution gives the rate in *miles per hour*. The second solution gives the rate in *hours per mile*, or how long it takes to travel one mile. These are **reciprocal rates.** Notice that $\frac{1}{50}$ of an hour is $\frac{1}{50} \cdot 60$ min, or 1.2 minutes. In other words, it takes a little over a minute to go one mile. Either rate is correct. The one to use depends on the situation in which you plan to use it.

Rates and Negative Numbers

Rates can be positive or negative quantities.

Example 2
a. If the temperature rises from 70 to 85 degrees in 2 hours, what is the change in temperature in degrees per hour?

b. If the temperature goes down from 44 to 32 degrees in 5 hours, what is the rate of temperature change in degrees per hour?

Solutions To find the rate, divide the number of degrees changed by the number of hours.

a. rate of temperature change $= \dfrac{(85 - 70) \text{ degrees}}{2 \text{ hours}}$

$= \dfrac{15 \text{ degrees}}{2 \text{ hours}}$

$=$ rise of 7.5 degrees per hour

b. rate of temperature change $= \dfrac{(32 - 44) \text{ degrees}}{5 \text{ hours}}$

$= \dfrac{-12 \text{ degrees}}{5 \text{ hours}}$

$=$ drop of 2.4 degrees per hour

or −2.4 degrees per hour

> ▶ QY1
>
> Find the rate of change of the number of people at a restaurant if it decreased from 161 to 98 people in 45 minutes. (Pay attention to positives and negatives.)

 QY1

In Part b of Example 2, the rate −2.4 degrees per hour came from dividing a negative number (−12 degrees) by a positive one (5 hours). The same negative rate can be found by dividing 12 by −5, which would describe the 12-degree rise in temperature that would come from moving backward in time 5 hours.

So $\dfrac{-12}{5}$ degrees per hour $= \dfrac{-12 \text{ degrees}}{5 \text{ hr}} = \dfrac{12 \text{ degrees}}{-5 \text{ hr}}$.

Here is a way to think of this situation. If you change the numerator or denominator of a fraction to its opposite, then the value of the fraction also changes to its opposite.

> **Negative Fractions**
>
> In general, for all a and b, and $b \neq 0$, $-\frac{a}{b} = \frac{-a}{b} = \frac{a}{-b}$.

When *both* the numerator and denominator are changed to their opposites, the value of the fraction is unchanged.

$\frac{-a}{-b} = \frac{-1 \cdot a}{-1 \cdot b}$ Multiplication Property of –1

$\phantom{\frac{-a}{-b}} = \frac{a}{b}$ Equal Fractions Property

Fractions with negative numbers are common, as you will see in the next chapter.

Division by Zero and Rates

Consider the rate $\frac{0 \text{ meters}}{10 \text{ seconds}}$, which has 0 in the numerator. This means that you do not travel at all in 10 seconds. So your rate is 0 meters per second. This reinforces that $\frac{0}{10} = 0$. In contrast, try to imagine a rate such as $\frac{10 \text{ meters}}{0 \text{ seconds}}$, which has 0 in the denominator. This would mean you travel 10 meters in 0 seconds. For this to occur, you would have to be in two places at the same time! That is impossible. Rates show that the denominator of a fraction can never be zero.

When a fraction has a variable in its denominator, then the expression is said to be *undefined* for any value of the variable that would make the denominator zero.

> **GUIDED**
>
> ### Example 3
> a. For what value(s) of x is $\frac{x}{x+4}$ undefined?
> b. For what value(s) of x does $\frac{x}{x+4} = 0$?
>
> **Solutions**
> a. A fraction is undefined whenever its ___?___ is zero.
> So $\frac{x}{x+4}$ is undefined when ___?___ = 0.
> Solve the equation on the preceding line for x. $x =$ ___?___.
> Therefore, $\frac{x}{x+4}$ is undefined when $x =$ ___?___.
>
> *(continued on next page)*

Rates

Chapter 5

b. A fraction equals zero whenever its ___?___ is zero.
So $\frac{x}{x+4} = 0$ when ___?___ = 0.
Therefore, $\frac{x}{x+4} = 0$ when $x = $ ___?___.

STOP QY2

▶ **QY2**
a. For what values of k is $\frac{2k-6}{k-55}$ undefined?
b. For what values of k does $\frac{2k-6}{k-55} = 0$?

Questions

COVERING THE IDEAS

1. Name all of the rates (including the rate units) in the following paragraph.

    The Indianapolis 500 is one of the most famous auto races in the United States, with hundreds of thousands of people attending annually. Attendees in 2006 paid from $40 to $150 per ticket plus $20 to $50 per hour to park. Drivers reached speeds of more than 230 mph (354 km/hr) as they raced the 2.5-mile oval track.

2. Give an example of a rate with a rate unit that is not mentioned in the reading of this lesson.

An IRL IndyCar Series car accelerates from 0 to 100 mph in less than three seconds.

Source: www.indy500.com

In 3–5, calculate a rate suggested by the given information.

3. Danielle walked her dog 6 blocks in t minutes.

4. In the last seven days, Salali slept 6.5 hours one night, 7 hours two nights, 7.5 hours two nights, 8 hours one night, and 9.5 hours one night.

5. In 2004, 2.3 billion books were sold in the United States, which had a population of about 296 million people.

6. In playing a video game 4 times, Bailey scored a points, b points, c points, and d points.
    a. Give an expression for her average score.
    b. Bailey's average is a rate. What is the rate unit?

7. Translate the change in time and temperature into positive and negative quantities. Then calculate the rate.
    a. 8 hours ago it was 5 degrees warmer than it is now.
    b. If this rate continues, then 8 hours from now it will be 5 degrees colder.

8. **Multiple Choice** Which of these numbers is *not* equal to the others?

    A $\frac{-153x}{82}$ B $-\frac{153x}{82}$ C $\frac{-153x}{-82}$ D $\frac{153x}{-82}$ E $\frac{-(-153)x}{-82}$

266   Division and Proportions in Algebra

In 9–11, an expression is given.

a. For what values of the variable is the expression equal to zero?
b. For what value of the variable is the expression undefined?

9. $\dfrac{w-12}{w+5}$
10. $\dfrac{17}{m-4}$
11. $\dfrac{2+y}{15}$

APPLYING THE MATHEMATICS

12. When you buy something in quantity, the cost of one item is the *unit cost*. Find the unit cost for each of the following.
    a. frozen juice at 3 cans for $5
    b. 500 sheets of notebook paper for $2.49
    c. x paper clips for $0.69

13. Let $y = \dfrac{1+x}{2-x}$. Complete the table at the right.

x	y
−3	?
−2	?
−1	?
0	?
1	?
2	?
3	?

14. A very fast runner can run a half-mile in 2 minutes. Express the average rate in each of these units.
    a. miles per minute   b. minutes per mile   c. miles per hour

15. For each state below, find the number of people per square mile to the nearest tenth. This is the state's *population density* for 2005.
    a. New Jersey: population = 8.7 million; area = 8,700 square miles
    b. Montana: population = 0.9 million; area = 147,000 square miles

16. **Multiple Choice** In t minutes, a copy machine made n copies. At this rate, how many copies per second does the machine make?

    A $\dfrac{n}{60t}$ B $\dfrac{60t}{n}$ C $\dfrac{60n}{t}$ D $\dfrac{t}{60n}$

17. The Talkalot cell phone company sells a pay-as-you-go phone with 700 minutes for $70.
    a. What is the rate per minute?
    b. What is the rate per hour?
    c. Elizabeth buys a phone and talks for m minutes. What is the value of the phone now?

REVIEW

In 18–20, simplify the expression. (Lessons 5-2, 5-1, 2-2)

18. $\dfrac{4xy}{-3y^2} \cdot \dfrac{-6y}{5x^2}$

19. $\dfrac{ab}{21} \div \dfrac{a}{4b}$

20. a. $\dfrac{-8n}{3} \cdot \dfrac{8n}{3}$ b. $\dfrac{-8n}{3} + \dfrac{8n}{3}$

21. Alice has n pounds of bologna. If she uses $\dfrac{1}{8}$ pound of bologna to make one bologna sandwich, how many bologna sandwiches can Alice make? (Lesson 5-2)

In 22–25, simplify the expression. (Lesson 5-2)

22. $\dfrac{\frac{9}{4}}{5}$

23. $\dfrac{-6}{7} \div \dfrac{-6}{5}$

24. $\dfrac{a}{4b} \div \dfrac{b}{6}$

25. $\dfrac{2d}{3} \div 5d$

26. a. Draw a square with side of length x.
    b. Shade or color the diagram to show $\dfrac{x}{2} \cdot \dfrac{3x}{4}$.
    c. What is the result of the multiplication? (Lesson 5-1)

27. **Skill Sequence** Solve each inequality. (Lessons 4-5, 3-7)
    a. $-y > 10$
    b. $-5x < 10$
    c. $-2A + 3 \leq 10$
    d. $-9B + 7 \geq 10 + 3B$

28. a. How many seconds are in one day?
    b. A second is what fraction of a day? (Previous Course)

EXPLORATION

29. Use the Internet, an almanac, or some other source to find the estimated current U.S. national debt and an estimate of the U.S. population. Then calculate the average debt per capita. (The phrase *per capita* means "per person.")

QY ANSWERS

1. a decrease of 1.4 people/min

2. a. $k = 55$
   b. $k = 3$

Lesson 5-4
Multiplying and Dividing Rates

Vocabulary

conversion rate

▶ **BIG IDEA** When rates are multiplied or divided, the unit of the answer follows rules of arithmetic on the units of the original rates.

Rates can be multiplied or divided by other quantities. The units are multiplied as if they were numeric fractions. Keeping the units in the calculations can help you understand the meaning.

If you bought 2.5 pounds of fish at $8.49 a pound, it would cost $2.5 \text{ lb} \cdot 8.49 \frac{\text{dollars}}{\text{lb}} = 21.225$ dollars, or $21.23.
Notice how the unit "lb" in "2.5 lb" cancels the unit "lb" in the denominator of $\frac{\text{dollars}}{\text{lb}}$, so that the result is in dollars.

Mental Math

Find a pair of numbers x and y that satisfy the equation.

a. $2x + 3y = 600$
b. $2x - 3y = 600$
c. $20x + 30y = 600$

Conversion Rates

A **conversion rate** is a rate determined from an equality between two quantities with different units. For example, since 1 hour = 60 minutes, dividing one unit by the other equals 1.

$$\frac{1 \text{ hour}}{60 \text{ min}} = \frac{60 \text{ min}}{1 \text{ hour}} = 1$$

Both $\frac{1 \text{ hour}}{60 \text{ min}}$ and $\frac{60 \text{ min}}{1 \text{ hour}}$ are conversion rates. Multiplying a quantity by a conversion rate does not change its value.

Example 1
On average, an adult human heart beats about 70 times per minute. At this rate, how many times does a heart beat in a 365-day year?

Solution We begin with the given information and multiply by conversion rates until one minute is converted into one year.

$$70 \frac{\text{beats}}{\text{minute}} = 70 \frac{\text{beats}}{\text{minute}} \cdot \frac{60 \text{ minutes}}{1 \text{ hour}} \cdot \frac{24 \text{ hours}}{1 \text{ day}} \cdot \frac{365 \text{ days}}{1 \text{ year}}$$

$$= 70 \cdot 60 \cdot 24 \cdot 365 \frac{\text{beats}}{\text{year}}$$

$$= 36{,}792{,}000 \frac{\text{beats}}{\text{year}}$$

The Pike Place Market in Seattle, Washington attracts 10 million visitors per year.

Source: Pike Place Market

Rates are used in many formulas. One of the more important formulas is $d = rt$, which gives the distance d traveled by an object moving at a constant rate r during a time t. (r is often called the speed.)

> **QY**
>
> The International Sports Medicine Institute has a formula for daily water intake. A nonactive person should consume $\frac{1}{2}$ ounce per pound of body weight. An athletic person should consume $\frac{2}{3}$ ounce per pound of body weight. (Note: 1 cup = 8 ounces.) If a person weighs p pounds, how many cups of water should the person drink each day if he or she is:
>
> a. nonactive?
>
> b. athletic?

GUIDED

Example 2

Explain why a train traveling 68 $\frac{mi}{hr}$ is traveling about 100 $\frac{ft}{sec}$.

Solution The goal is to convert miles to __?__ and an hour to __?__. Use conversion rates so that the units cancel in the numerator and denominator just like common factors.

$$\frac{68 \text{ mi}}{\text{hr}} = \frac{68 \text{ mi}}{\text{hr}} \cdot \frac{1 \text{ hr}}{\underline{}} \cdot \frac{\underline{}}{60 \text{ sec}} \cdot \frac{\underline{}}{1 \text{ mi}}$$

$$= \frac{68 \cdot \underline{}}{60 \cdot 60} \frac{\text{ft}}{\underline{}}$$

$$= \underline{}$$

So 68 $\frac{mi}{hr}$ is about 100 $\frac{ft}{sec}$.

STOP QY

Using Reciprocal Rates

Sometimes a rate does not help to simplify the computation but its reciprocal does. Guided Example 2 used $\frac{1 \text{ hr}}{60 \text{ min}}$, while Example 1 used $\frac{60 \text{ min}}{1 \text{ hr}}$.

Remember the algebraic definition of division: $\frac{a}{b} = a \cdot \frac{1}{b}$. So to divide by a rate, you can multiply by its reciprocal. For example, if a book has 672 pages and Marta reads an average of 35 pages per day, you can use division.

$$\frac{672 \text{ pages}}{\frac{35 \text{ pages}}{\text{day}}} = 672 \text{ pages} \cdot \frac{1}{35} \frac{\text{day}}{\text{pages}} = 19.2 \text{ days}$$

So it will take her a little more than 19 days to finish reading the book.

Not every reciprocal rate is meaningful in every situation. For example, in a school with an average of 23 $\frac{\text{students}}{\text{class}}$, the equivalent reciprocal rate of $\frac{1}{23} \frac{\text{class}}{\text{students}}$ does not have much meaning. Still, it might be useful in a computation.

There are approximately 8,000 to 10,000 professional online book dealers, selling books with an average price range of $10–$19 per book.

Source: www.bookologist.com

270 Division and Proportions in Algebra

Questions

COVERING THE IDEAS

1. a. If a car travels 30 minutes at an average rate of 44 miles per hour, how far will it have gone?
   b. If a car travels m minutes at an average rate of 44 miles per hour, how far will it have gone?

2. Suppose ground beef is $2.59 per pound.
   a. How much will it cost you to purchase P pounds?
   b. How many pounds can you purchase for x dollars?

3. One version of an English translation of the novel *War and Peace* by Leo Tolstoy has 1,370 pages. If you read 12 pages per day, show how to calculate how many months it will take you to read the entire novel. Use the conversion 1 month ≈ 30 days.

In 4 and 5, a rate is given.
  a. Write a sentence using the rate to describe a situation.
  b. Name the reciprocal rate.
  c. Write a sentence using the reciprocal rate to describe a situation.

4. 12 feet per second

5. $4 per kilogram

In 6 and 7, give the two conversion rates that result from the given equation.

6. 1,000 m = 1 km

7. 36 in. = 1 yd

8. A penny has a thickness of about 0.08 inch. How high, to the nearest 0.1 mile, would a stack of 1 million pennies be?

9. Convert 30 miles per hour into kilometers per hour using the conversion 1 mile ≈ 1.6 kilometers.

APPLYING THE MATHEMATICS

10. In 2005, Danica Patrick drove the fastest lap ever at the time during trials for the Indianapolis 500 auto race. In a South African newspaper, her speed for the lap was given as 360.955 km/hr.
    a. What is this speed to the nearest hundredth of a mile per hour? Use the conversion 1 mi = 1.609344 km.
    b. To the nearest hundredth of a minute, how long did it take her to drive one lap of the 2.5-mile track?

Each penny costs 1.23 cents to make, but the U.S. Mint collects only one cent for it.
Source: U.S. Department of Treasury

In 11 and 12, write a rate multiplication problem whose answer is the given quantity.

11. 200 inches

12. $4.2 \frac{\text{min}}{\text{page}}$

13. Kenji can wash *k* dishes per minute. His sister Suna is twice as fast.
    a. How many dishes can Suna wash per minute?
    b. How many minutes does Suna spend per dish?

14. Suppose Chip is baking cookies for a school bake sale. His oven bakes 36 cookies every 10 minutes. He wants to bake *d* dozen cookies for the sale. How long will it take him to bake all of the cookies?

15. In 2006, Gordy Savela ran 300 miles across northern Minnesota in 12 days.
    a. What distance did Gordy average each day?
    b. If a marathon is 26.2 miles, how many marathons did Gordy run during his journey?

Bake sales have long been one of the most popular ways of raising funds for schools, social clubs, and other organizations.

16. People sometimes go for a walk after a big meal to burn off calories. To lose 1 pound, a person must burn 3,500 calories. Suppose a person burns about 300 calories per hour by walking.
    a. If a person walks for 2.5 hours, how many pounds will the person lose?
    b. If a person walks for 2.5 hours, how many ounces will the person lose?
    c. If a person walks for *h* hours, how many ounces will the person lose?
    d. If a person walks for *m* minutes, how many ounces will the person lose?

REVIEW

17. Over the past five days, Mykia has run 3.6 miles one day, 2.9 miles each of two days, 4.2 miles one day, and 1.9 miles one day. Calculate the average number of miles run per day. Write your answer as a rate. (**Lesson 5-3**)

18. For what value(s) of *n* is $\frac{19 - 2n}{19 + 2n}$ undefined? (**Lesson 5-3**)

19. A box is $\frac{1}{3}$ as long, $\frac{1}{3}$ as wide, and $\frac{1}{3}$ as high as a crate. How many of these boxes will fit in the crate? (**Lesson 5-1**)

In 20–22, use the formula $H = \frac{M + F + 5}{2}$. It predicts the adult height *H* of a boy based on his mother's height *M* and his father's height *F*, where all measurements are given in inches. The formula $H = \frac{M + F - 5}{2}$ applies to the adult height of girls. (**Lesson 4-7**)

20. Solve each formula for *M*.

21. Booker's father's height is 73 inches, and his mother's height is 68 inches. How tall does the formula predict Booker will be when he is an adult?

22. Predict your adult height using this formula.

Multiple Choice In 23–25, decide whether each equation has
   A no solution.
   B one solution.
   C more than one solution. (Lesson 4-5)

23. $\frac{x}{5} + 5 = 5$

24. $\frac{y}{4} + 4 = \frac{y}{4}$

25. $\frac{z}{3} + 3 = 3\left(\frac{z}{3} + 3\right)$

26. Solve $8(-3d - 4) \leq 13(2 - d) - (7d - 2)$. (Lessons 4-4, 2-1)

EXPLORATION

27. a. Here is a problem found in many puzzle books. If a hen and a half can lay an egg and a half in a day and a half, how long will it take 24 hens to lay 24 eggs? Explain how you got your answer.

   b. Generalize Part a. If a hen and a half can lay an egg and a half in a day and a half, how long will it take h hens to lay e eggs?

QY ANSWERS

a. $\frac{p}{16}$ c

b. $\frac{p}{12}$ c

Chapter 5

Lesson
5-5 Ratios

Vocabulary

ratio
tax rate
discount rate

▶ **BIG IDEA** The quotient of two measures with the same type of units is a ratio.

What Is a ratio?

A *ratio* describes how many times larger one number is compared to another. For example, dogs come in a huge variety of shapes and sizes. Consider the three types pictured below, which are very different from each other. To compare the three dog breeds, you could use corresponding lengths on their bodies to form ratios.

Mental Math

Give the coordinates of a point on

a. the *x*-axis.

b. the *y*-axis.

c. both the *x*- and *y*-axes.

Dachshund    Labrador Retriever    Greyhound

Activity

Measure the pictures to answer each question.

1. Which dog has the greatest $\frac{\text{head length}}{\text{front leg length}}$ ratio?

2. Which dog has the least $\frac{\text{head length}}{\text{body length}}$ ratio?

3. Which dog has the least $\frac{\text{front leg length}}{\text{body length}}$ ratio?

4. Which dog has the $\frac{\text{front leg length}}{\text{body length}}$ ratio that is closest to 1?

The direction of the comparison is important. The dachshund's $\frac{\text{head length}}{\text{front leg length}}$ ratio is much greater than 1 because its head is much longer than its front legs. But the reciprocal ratio $\frac{\text{front leg length}}{\text{head length}}$ is much less than 1.

274 Division and Proportions in Algebra

A **ratio** is a comparison of two quantities with the same type of units. For example, a to b is written $\frac{a}{b}$. Similarly, the ratio $\frac{b}{a}$ compares b to a. Notice the difference between a rate and a ratio. In the rate a per b, the units for a and b are different. In a ratio, the units are the same. $\frac{40 \text{ km}}{2 \text{ hr}} = 20 \frac{\text{km}}{\text{hr}}$ is a rate. $\frac{40 \text{ km}}{8 \text{ km}} = 5$ is a ratio.

Example 1

It takes Max $\frac{1}{4}$ of an hour to ride his bike to school, and it takes Riley 21 minutes to walk to school. Write a ratio comparing Max's time to Riley's time.

Solution The units of measure for 21 minutes and for $\frac{1}{4}$ of an hour are not the same, so we need to change hours to minutes. Since Max's time is to be compared to Riley's, his time is in the numerator.

$$\frac{\text{Max's Time}}{\text{Riley's Time}} = \frac{\frac{1}{4} \text{ hr}}{21 \text{ min}} = \frac{\frac{1}{4} \text{ hr} \cdot \frac{60 \text{ min}}{\text{hr}}}{21 \text{ min}} = \frac{15 \text{ min}}{21 \text{ min}} = \frac{5}{7}$$

This means that it takes Max $\frac{5}{7}$ of the time to ride his bike as it takes Riley to walk to school.

Ratios and Percents

Ratios can be expressed as fractions, decimals, or percents. In Example 1, $\frac{5}{7} \approx 0.\overline{714285}$, so you could say that it takes Max about 71% or $\frac{71}{100}$ of the time it takes Riley to go to school.

While in this book we distinguish rates from ratios, in the real world some ratios are called rates. In money matters, the **tax rate** is the ratio of the tax amount to the selling price. The **discount rate** is the ratio of the discount to the original price.

About 20 million bicycles were sold in the United States in 2005.

Source: National Bicycle Dealers Association

Example 2

A TV that normally sells for $400 is on sale for $340. The tax on the reduced price is $23.80, so the total cost with tax is $363.80.

a. What is the discount rate?
b. What is the tax rate?
c. Including tax, how much would a customer save by buying the TV on sale?

Solutions

a. discount rate $= \frac{\text{amount of discount}}{\text{original price}} = \frac{\$400 - \$340}{\$400} = \frac{\$60}{\$400} = 0.15 = 15\%$

(continued on next page)

Ratios 275

b. tax rate = $\frac{\text{amount of tax}}{\text{discounted price}} = \frac{\$23.80}{\$340} = 0.07 = 7\%$

c. Paying full price, the 7% tax is paid on $400. Recall that price + 7% of price = 1.07 • price. So at full price, the customer would pay 1.07(400) = $428. Buying the TV on sale saves the customer $428 − $363.80 = $64.20.

Using Ratios to Set Up Equations

When two numbers are in the ratio a to b, then they are also in the ratio ka to kb because $\frac{a}{b} = \frac{ka}{kb}$ for any nonzero value of k. For example, if the ratio of boys to girls in a band is $\frac{7}{6}$, the ratio is also $\frac{14}{12}$, or $\frac{21}{18}$.

If you know the original ratio, then you can multiply the numerator and denominator by the same constant and never change the value of the ratio.

Directions for mixing foods or chemicals are frequently given as ratios. You can use equations to determine actual quantities from the ratios.

Example 3

One Internet site warns painters: "Always • always • ALWAYS! make a note of the paint ratios when you mix paints." In painting a pink wall, Bethany mixed 1 part red paint with 3 parts white paint. How much of each color will be needed for a wall that will use about 5 gallons of paint in all?

Solution The red paint and white paint are in the ratio of 1 to 3. So their ratio is $\frac{1}{3}$. Now we use the fact that $\frac{1}{3} = \frac{1k}{3k}$ for any nonzero k. So let k be the number of gallons of red paint and $3k$ be the number of gallons of white paint. The total paint needed is 5 gallons, so you can set up the following equation and solve for k.

$$1k + 3k = 5$$
$$4k = 5$$
$$k = \frac{5}{4} = 1.25$$

So $3k = 3 \cdot 1.25 = 3.75$.

She will need 1.25 gallons of red paint and 3.75 gallons of white paint.

Check $\frac{1.25}{3.75} = \frac{1}{3}$ and 1.25 gallons + 3.75 gallons = 5 gallons. It checks.

One gallon of paint generally covers 400 square feet of wall space.

Source: Paint and Decorating Retailers Association

▶ **QY**

To paint a house light green, you want a ratio of 1 part green to 4 parts white. Find out how much of each color you must buy if you need 12 gallons of paint.

STOP QY

Changing Ratios

It is natural to ask how a ratio will change if the numerator or denominator (or both) change. For example, right now Myron is 6 years old and his little sister Ella is 2. The ratio of Myron's age to Ella's is $\frac{6 \text{ years}}{2 \text{ years}} = \frac{3}{1}$. Now, Myron is 3 times as old as his sister. But when they get 2 years older, the ratio will have changed. In two years, $\frac{6 + 2 \text{ years}}{2 + 2 \text{ years}} = \frac{8}{4} = \frac{2}{1}$, so Myron will be twice as old as Ella.

Example 4

Suppose a team has won 15 of its first 38 games. How many games must it win in a row to bring its winning percentage to at least 0.500?

Solution If the team now wins r games in a row, then the team will have played $38 + r$ games and have won $15 + r$ of them. So we want to know when $\frac{\text{number of wins}}{\text{number of games played}} = \frac{15 + r}{38 + r} \geq 0.500$.

To clear this inequality of fractions, multiply both sides by $38 + r$. This is a positive number so the sense of the inequality remains the same.

$15 + r \geq 0.500(38 + r)$	Multiply each side by $(38 + r)$.
$15 + r \geq 19 + 0.5r$	Distributive Property
$r \geq 4 + 0.5r$	Add -15 to both sides.
$0.5r \geq 4$	Add $-0.5r$ to both sides.
$r \geq 8$	Divide each side by 0.5.

The team must win 8 games in a row to bring its winning percentage up to at least .500.

Questions

COVERING THE IDEAS

1. Grass snakes and rattlesnakes are found in the Great Plains. Grass snakes grow up to 20 inches in length, and rattlesnakes grow to 45 inches. Express the ratio of rattlesnake length to grass snake length
    a. as a fraction.
    b. as a percent.

2. An item is on sale for $16. It originally cost $28.
    a. What is the discount rate?
    b. If the tax on the sale price is $1.04, what is the tax rate?

Rattlesnakes are usually between 7 and 15 inches long at birth.

Source: San Diego Zoo

3. A recipe for fruit-juice punch calls for 1 part tropical fruit-punch and 2 parts ginger ale. If a person wants to make 5 gallons (20 quarts) of fruit-juice punch, how many quarts of each ingredient are needed?

4. A paint mixture calls for 7 parts of linseed oil, 5 parts of solvent, and 1 part of pigment. How much of each ingredient is needed to make 60 gallons of paint?

5. A team has won 11 of 17 games.

   a. If the team wins its next 2 games, what will its winning percentage be? Round your answer to three decimal places.

   b. If the team loses its next n games, what will its winning percentage be?

   c. How many games could the team lose and still have a winning percentage above .600?

APPLYING THE MATHEMATICS

6. At a local concert, $\frac{1}{6}$ of the attendees are teenagers and $\frac{5}{8}$ of the people are between the ages of 20 and 40. The rest of the people are older. Write the ratio (in lowest terms) of the number of people over 40 to the number of people between 20 and 40.

7. The ratio of adults to children at a concert is expected to be 2 to 5. If 200 people attend, how many are expected to be children?

8. The circles at the right have radii 2 centimeters (Circle A) and 5 centimeters (Circle B).

   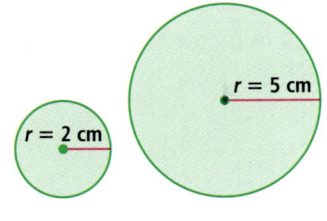

   Circle A   Circle B

   a. Find the circumference and area of each circle using $C = \pi d$ and $A = \pi r^2$.

   b. Find the ratio of the diameter of Circle A to the diameter of Circle B in lowest terms.

   c. Give the ratio of the area of Circle A to the area of Circle B in lowest terms.

9. a. In 4 minutes, Liana can type 140 words. If she has an essay that is 700 words long, about how long will it take Liana to type it?

   b. In 5 minutes, Elan can type 160 words. About how many words can he type in 12 minutes?

10. Recall that two angles are called *supplementary angles* if the sum of their measures is 180°. In the diagram at the right, $\angle ABC$ and $\angle CBD$ are supplementary angles whose measures have a ratio of 7 to 3. Find the measure of each angle.

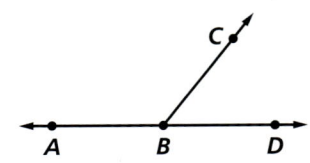

Lesson 5-5

REVIEW

11. An airplane is scheduled to fly from LaGuardia Airport in New York to Los Angeles International Airport in 6 hours and 18 minutes. If the distance between the two airports is 2,779 miles, what is the average speed of the airplane? (Lesson 5-3)

12. For what values of c is $\frac{16 - 4c}{2c + 9}$ undefined? (Lesson 5-3)

13. Tell whether each fraction equals $\frac{99}{100}$. (Lessons 5-3, 5-2)

    a. $\frac{999}{1,000}$ b. $-\frac{-99}{-100}$ c. $\frac{1}{1 - \frac{1}{100}}$ d. $\frac{\frac{99}{7}}{\frac{7}{100}}$

14. **Skill Sequence** Simplify each expression. (Lessons 5-2, 5-1, 2-1)

    a. $\frac{x}{4} + \frac{x}{5}$ b. $\frac{x}{4} - \frac{x}{5}$ c. $\frac{x}{4} \cdot \frac{x}{5}$ d. $\frac{x}{4} \div \frac{x}{5}$

15. The volume of a box is to be less than 20 cubic meters. If the base has dimensions 185 cm by 250 cm, what inequality describes possible heights of the box? (Lesson 3-7)

16. What fraction of a complete turn is a rotation of 75°? (Previous Course)

In 2014, Los Angeles International Airport (LAX) was the fourth busiest airport in the world in terms of the number of take-offs and landings.

Source: *The New York Times*

EXPLORATION

17. The right triangle shown here has side lengths of 7, 24, and 25. One of its angles has a measure of about 74°. The lengths of the three sides can form six different ratios.

    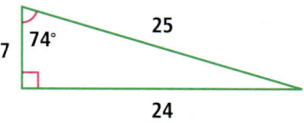

    a. Write the values of all six of these ratios.
    b. These ratios have special names. One of these is called the *sine*. For this triangle it is written sin 74°. Put your calculator in degree mode and compute sin 74° using the key sequence 74 [SIN] or [SIN] 74 (depending on your calculator), and determine which of the six ratios in Part a is the sine.
    c. Another special ratio is called the *cosine*. Compute cos 74° on your calculator and determine which of the six ratios in Part a is the cosine.
    d. A third special ratio is called the *tangent*. Compute tan 74° on your calculator and determine which of the six ratios in Part a is the tangent.
    e. Which of the three ratios in Parts b, c, and d is a ratio of the other two?

QY ANSWER

2.4 gal of green paint, 9.6 gal of white paint

Chapter 5

Lesson 5-6
Probability Distributions

Vocabulary

outcome
event
probability
P(x)
probability distribution
unbiased
fair
conditional probability
complement
odds of an event

▶ **BIG IDEA** The distribution of the probabilities of all possible outcomes from a situation can be displayed in tables and graphs.

In a situation such as asking a question or flipping a coin, the possibilities are called **outcomes**. No two outcomes can occur at the same time. A set of outcomes is an **event**. Recall from your earlier courses that the **probability** of an event is a number from 0 to 1 that measures how likely it is that the event will happen. The sum of the probabilities of all possible outcomes must equal 1.

If an outcome or event is identified as x, then the symbol **P(x)** stands for the probability of the outcome or event.

How Are Probabilities Determined?

There are three common ways in which people determine probabilities.

1. Pick a probability close to the relative frequency with which the outcome or event has occurred in the past.
2. Deduce a probability from assumptions about the situation.
3. Give a best guess.

The gender of a newborn baby is a situation where there are two outcomes: B = a boy is born and G = a girl is born. Any one of the three common ways might be used to identify P(B), the probability that a boy is born.

Mental Math

Find the value of the variable.

a. The temperature T is 30° cooler than 82°.

b. A book's price p is 7 times as much as a $2.25 pen.

c. A plane departs 45 minutes later than its 6:55 A.M. departure time, at d.

1. According to the National Center for Health Statistics, there were 4,089,950 children born in the United States in 2003. The relative frequency of boys to births is found by dividing the number of boys born by the total number of births.

relative frequency = $\frac{\text{number of baby boys born}}{\text{total number of births}} = \frac{2{,}093{,}535}{4{,}089{,}950} \approx 0.512 = 51.2\%$

Using relative frequency, we might say that P(B) is about 51.2%.

2. However, some people would rather assume that the probabilities of a boy and a girl are equal. If the two probabilities are equal, then because the probabilities must add to 1, P(B) = P(G) = $\frac{1}{2}$ = 50%. More generally, if there are n outcomes in a situation and each has equal probability, then the probability of each is $\frac{1}{n}$.

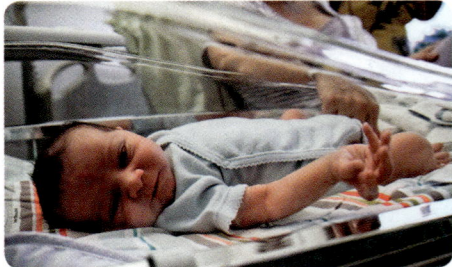

The birthrate gives the number of live births per 1,000 of population. The U.S. birthrate declined from 23.7 in 1960 to a record low of 13.9 in 2002.

Source: U.S. National Center for Health Statistics

280 Division and Proportions in Algebra

3. In a family where more boys than girls have been born, some people might think that boys are much more likely than 50% or 51.2% to be born and guess that the probability that a boy will be born is much greater. But usually we only guess when the event is a one-time event and there are no data from past experience. For example, if a new drug has been developed to cure a disease, at the start researchers may be able to only guess at the probability that the drug will actually work.

Probability Distributions

Some situations have many outcomes. A **probability distribution** is the set of ordered pairs of outcomes and their probabilities. For example, in many board games, two dice are thrown and the sum of the numbers that appear is used to make a move. Since the outcome of the game depends on landing or not landing on particular spaces, it is helpful to know the probability of obtaining each sum. The following diagram is helpful. It shows the 36 possible outcomes when two dice are thrown.

We call a situation **unbiased,** or **fair,** if each outcome has the same probability. If the dice are fair, then each of the 36 outcomes has a probability of $\frac{1}{36}$.

Let P(x) = the probability of getting a sum of x. In this case, x can only be a whole number from 2 to 12. There are 36 outcomes but only 11 possible sums. For example, the event "getting a sum of 7" has 6 possible outcomes: 1 and 6, 2 and 5, 3 and 4, 4 and 3, 5 and 2, and 6 and 1. So P(7) = $6 \cdot \frac{1}{36} = \frac{6}{36}$, or $\frac{1}{6}$. The probability distribution is shown on page 282 in the table and graph. A sum of 7 is the most likely outcome, so P(7) is plotted as the highest point on the graph.

Chapter 5

x	P(x)
2	$\frac{1}{36} = 0.02\overline{7}$
3	$\frac{2}{36} = 0.0\overline{5}$
4	$\frac{3}{36} = 0.08\overline{3}$
5	$\frac{4}{36} = 0.\overline{1}$
6	$\frac{5}{36} = 0.13\overline{8}$
7	$\frac{6}{36} = 0.1\overline{6}$
8	$\frac{5}{36} = 0.13\overline{8}$
9	$\frac{4}{36} = 0.\overline{1}$
10	$\frac{3}{36} = 0.08\overline{3}$
11	$\frac{2}{36} = 0.0\overline{5}$
12	$\frac{1}{36} = 0.02\overline{7}$

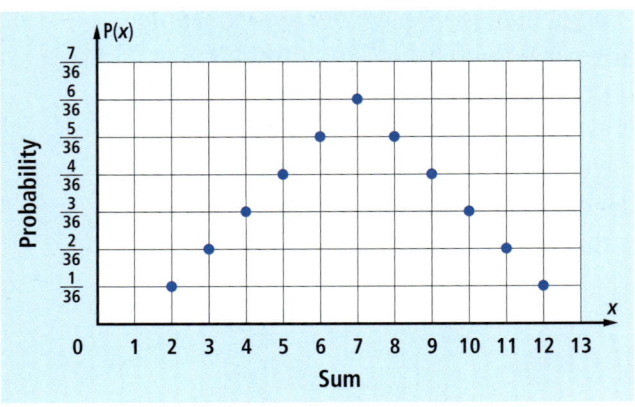

Notice that the sum of all the probabilities adds to $\frac{36}{36}$, or 1. This is because these 11 events include each outcome exactly once.

STOP QY1

You may have answered QY1 by adding P(9), P(10), P(11), and P(12). Another way to answer QY1 is to count the outcomes that give a sum of 9, 10, 11, and 12, and then divide that number by 36. This is because probabilities of events with equally likely outcomes satisfy the following property.

> **Probability of an Event with Equally Likely Outcomes**
>
> If a situation has a total of *n* equally likely outcomes and *E* is an event, then $P(E) = \frac{\text{number of outcomes in } E}{n}$.

▶ **QY1**

In tossing two fair dice, what is the probability of tossing a sum greater than 8?

A Probability Distribution from Relative Frequencies

Your blood and the blood of all humans is one of eight types: O, A, B, or AB and either positive (+) or negative (−) for each type depending on the existence of a Rh antigen. Overall in the United States, the approximate percents of people with these types are shown in the table on page 283. Note: The percents do not add to 100% due to rounding.

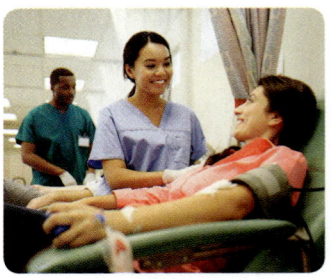

Blood makes up about 7% of a person's body weight.

Source: www.bloodbook.com

282 Division and Proportions in Algebra

We can take these relative frequencies as probabilities that a person at random in the United States has one of these types.

Example 1

a. What is the probability that a person in the United States has type-O blood?

b. What is the probability that a person with type-O blood is O−?

Type	Percent
O+	38.4
O−	7.7
A+	32.3
A−	6.5
B+	9.4
B−	1.7
AB+	3.2
AB−	0.7

Source: www.bloodbook.com

Solutions

a. A person has type-O blood if he or she is either O+ or O−.
$$P(O) = P(O+) + P(O−)$$
$$= 38.4\% + 7.7\%$$
$$= 46.1\%$$

b. Since 46.1% of the population has type-O blood, and 7.7% of the population is O−, the probability that a person with type-O blood is O− is $\frac{P(O-)}{P(O)}$, which is $\frac{7.7\%}{46.1\%} \approx 0.167$, or about 16.7%.

The probability calculated in Example 1b is called a **conditional probability.** It is the probability that a person has blood type O− given the condition that they are already known to be type O. You can write: P(O− given O) = 16.7%. In general, the conditional probability of event B given event A, $P(B \text{ given } A) = \frac{P(A \text{ and } B)}{P(A)}$. That is, it is the probability that both events occur divided by the probability of the first event.

Complementary Events

If the probability of a tornado occurring on a weekend is $\frac{2}{7}$, then the probability of a tornado occurring on a weekday is $\frac{5}{7}$ because there are five weekdays in a week. The events "occurring on a weekend" and "occurring on a weekday" are called *complements* of each other. Two events are **complements** if they have no elements in common and together they contain all possible outcomes.

The sum of the probability of an event and the probability of its complement is 1. The same goes for relative frequencies. Thus the probability or relative frequency of the complement of an event is found by subtracting the probability or relative frequency of the original event from 1.

The greatest incidence of tornadoes around the world is in North America.

Source: Oklahoma Climatological Survey

Chapter 5

 QY2

The probabilities that weather forecasters use are found by using mathematical models. These models are created from a study of what happened in the past under conditions like those when the forecast is made.

> ▶ **QY2**
>
> A weather forecaster reports that the probability of rain tomorrow is 80%. What is the probability that it does not rain tomorrow?

Odds

Probabilities and complements of probabilities are used to compute *odds*. Odds are stated and used in different ways that are not always the same. One meaning is that the **odds of an event** occurring is the ratio of the probability that the event *will occur* to the probability that the event *will not occur*.

$$\text{odds of } E \text{ occurring} = \frac{P(E)}{P(\text{complement of } E)} = \frac{P(E)}{1 - P(E)}$$

For example, if you think that the odds of your being selected for a particular honor are 2 to 1, then you mean that you will be selected 2 out of 3 times, and that the probability of the event is $\frac{2}{3}$. This shows how to calculate a probability from odds. If the odds for the event are m to n, then the probability for the event is $\frac{m}{m+n}$ and the probability the event will not occur is $\frac{n}{m+n}$.

> ▶ **QY3**
>
> Cameron is managing a large project at work that is behind schedule. He estimates that there is a $\frac{1}{4}$ chance it will be completed on time. What are the odds it will be done on time?

 QY3

Questions

COVERING THE IDEAS

1. If E is an event, what does $P(E)$ stand for?

2. Use the data for births in the United States found on page 280.
   a. How many girls were born in the United States in 2003?
   b. What was the relative frequency of female births in the United States in 2003?
   c. Let G represent a girl being born. Using relative frequency, what is the value of $P(G)$?
   d. Let G represent a girl being born. If a baby has an equal probability of being a boy or a girl, what is $P(G)$?
   e. Suppose someone chooses one of the children in your family at random. What is the probability that the person chosen is a girl?

284 Division and Proportions in Algebra

3. Suppose a multiple-choice question has 5 choices: A, B, C, D, and E. Jasmine guesses each answer randomly.
   a. What is the probability that Jasmine will get a particular question correct?
   b. What is the probability that Jasmine will miss the question?

4. Suppose you pick a number from 1 to 25 randomly out of a hat.
   a. How many outcomes are possible?
   b. What is the probability that you will pick the number 17?
   c. Let $E =$ you pick an even number. What is $P(E)$?
   d. Let $D =$ you pick an odd number. What is $P(D)$?
   e. **Fill in the Blank** D and E are called ___?___ events.

5. Examine the probability distribution in this lesson for the sum of the numbers on two fair dice when they are tossed.
   a. What is $P(2)$?
   b. What is $P(13)$?
   c. What is P(a number less than 5)?

In 6 and 7, use the information on blood types in the United States found on page 283.

6. a. What blood type is the least common?
   b. What is the most common blood type?

7. a. What is the probability that a person has type-B blood?
   b. What is the probability that a person with type-B blood is B+?
   c. What is the probability that a person with type-B blood is B−?
   d. **Fill in the Blank** The probability in Part c is called the ___?___ probability that a person with type-B blood is B−.

8. When two equally matched teams play a best-of-5 series, the odds that one team will win in three games is 1 to 3. From this information, what is the probability that one team will win in three games?

9. Suppose the probability that an event will occur is $\frac{5}{12}$.
   a. What are the odds in favor of the event occurring?
   b. What are the odds against the event occurring?

10. **Fill in the Blank** Use *always*, *sometimes but not always*, or *never*. If p is the probability of an event and q is the probability of its complement, then the value of $p + q$ ___?___ equals 1.

In **11** and **12**, find the complement of the event.

11. A heart is chosen from a standard deck of playing cards. A standard deck has 52 cards. Each card is one of four suits (clubs ♣, diamonds ♦, hearts ♥, and spades ♠) and has one of 13 values (ace, 2, 3, 4, 5, 6, 7, 8, 9, 10, jack, queen, and king).

12. You were born on a weekday.

APPLYING THE MATHEMATICS

In **13–15**, a card is picked randomly from a standard deck of playing cards.

13. Determine the probability of selecting each card.
    a. the ace of spades
    b. a 5
    c. a 5 or a 9

14. Determine the probability of selecting each card.
    a. a club
    b. a club or a heart
    c. a club or a 5 (This is a tricky one.)

15. If you know the card you have selected is a face card (jack, queen, or king), what is the probability that it is a king?

16. Detectives investigating a crime have narrowed the search for the criminal to five suspects. The table below lists each suspect's personal features.

Features	Suspect				
	1	2	3	4	5
Height	6'2"	5'8"	5'6"	6'3"	6'0"
Eye Color	Green	Blue	Green	Brown	Green
Gender	Male	Female	Female	Male	Female
Handedness	Right	Left	Right	Right	Right

   a. If chosen at random, what is the probability that Suspect 3 is the criminal?
   b. Suppose the detectives receive evidence that the criminal is female. If this new information is true, what is the probability that Suspect 3 is the criminal?
   c. Suppose the detectives also receive evidence that the female criminal is between 5'5" and 5'9" tall. Given the evidence from Part b and this new information, what is the probability that Suspect 3 is the criminal?

In 17 and 18, suppose that slips of paper containing the integers from 1 to 200 are put in a hat. A number x is drawn.

17. Determine each probability.
    a. $x = 135$
    b. $x > 99$
    c. $x < 1$

18. Determine P($x = 135$) given each circumstance.
    a. The number x is odd.
    b. The ones digit of x is 5.
    c. The hundreds digit of x is 1.
    d. The tens digit of x is 8.

19. A person buys a raffle ticket. The person says, "The probability of winning the raffle is $\frac{1}{2}$ since either I will win or I won't." What is wrong with this argument?

REVIEW

20. A television station has scheduled n hours of news, c hours of comedy, d hours of drama, s hours of sports, and x hours of other programs during the week. (Lesson 5-5)
    a. What is the ratio of hours of news to hours of drama?
    b. What is the ratio of hours of sports to total number of hours of programs during the week?

21. The Jones family earned $48,735 last year on their 95-acre farm.
    a. What is their income per acre?
    b. Is the income per acre a ratio or a rate?
    (Lessons 5-5, 5-3)

22. The list price of a car is c dollars. Find the selling price according to the following conditions.
    a. You pay a 7% sales tax and there is no discount.
    b. You get a 20% discount and there is no sales tax.
    c. You pay a 7% sales tax and get a 20% discount.
    (Lessons 5-5, 4-1, Previous Course)

In 2004, 74 million acres of soybeans were harvested in the United States, a 31% increase since 1990.

Source: U.S. Department of Agriculture

In 23 and 24, solve the equation. (Lessons 3-5, 3-4)

23. $D - 8.5 - 0.25D = 7.5$

24. $\frac{4}{3}w + 72 = 8$

25. Use the Distributive Property to compute $0.50 times 299 in your head. (**Lesson 2-1**)

26. Two circles with radii 6 cm and 4 cm are shown below. Let A = area of Circle 1 and B = area of Circle 2. Calculate each expression. (**Previous Course**)

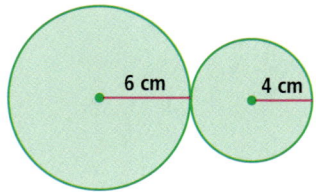

Circle 1   Circle 2

a. $A - B$ b. $\frac{B}{A}$ c. $\frac{A - B}{A}$

EXPLORATION

27. a. Pick a letter of the alphabet. Estimate what percent of words used in the English language begin with that letter.
    b. Pick a reading selection that has more than 200 words. Determine the relative frequency that a word in your reading selection begins with your letter.
    c. Having done the experiment, decide whether you should change the probability you guessed in Part a. Explain your decision.

QY ANSWERS

1. $\frac{10}{36} = 0.2\overline{7}$
2. 20%
3. 1 to 3

Lesson 5-7

Relative Frequency and Percentiles

Vocabulary
relative frequency
pth percentile

▶ **BIG IDEA** The distribution of the relative frequencies of the outcomes from a situation can be displayed in tables and graphs.

Suppose a particular event has occurred with a frequency of F times in a total of T opportunities for it to happen. Then the **relative frequency** of the event is $\frac{F}{T}$. Like other ratios, relative frequencies may be written as fractions, decimals, or percents.

The Differences between Probability and Relative Frequency

There are similarities and differences between probability and relative frequency. When a single die is tossed, we often think that it is equally likely that the die will land with each side up. If we think this way, then each side will land up $\frac{1}{6}$ of the time. So $P(1) = P(2) = ... = P(6) = \frac{1}{6}$. But in an actual experiment, it is rare that outcomes occur exactly the same number of times even if their probabilities of occurring are the same. A person simulated the random tossing of a die 100 times and recorded the results in the table below.

Outcome	1	2	3	4	5	6
Probability	$\frac{1}{6} = 0.1\overline{6}$	$\frac{1}{6} = 0.1\overline{6}$	$\frac{1}{6} = 0.1\overline{6}$	$\frac{1}{6} = 0.1\overline{6}$	$\frac{1}{6} = 0.1\overline{6}$	$\frac{1}{6} = 0.1\overline{6}$
Relative Frequency	$\frac{24}{100} = 0.24$	$\frac{16}{100} = 0.16$	$\frac{13}{100} = 0.13$	$\frac{14}{100} = 0.14$	$\frac{16}{100} = 0.16$	$\frac{17}{100} = 0.17$

Notice that the sum of the relative frequencies is 1, just as the sum of the probabilities of these events is 1. The probability of an event is what we would expect the relative frequency of the event to be close to "in the long run," after a large number of experiments.

The table on page 290 summarizes some of the important similarities and differences between relative frequencies and probabilities.

Mental Math

Use the graph to answer the questions.

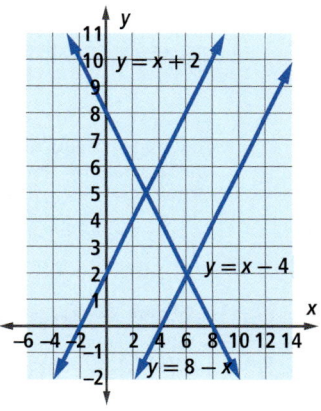

a. Where do $y = 8 - x$ and $y = x + 2$ intersect?

b. Where do $y = 8 - x$ and $y = x - 4$ intersect?

c. Where do $y = x - 4$ and $y = x + 2$ intersect?

Chapter 5

Relative Frequency	Probability
1. Calculated from data	1. Deduced from assumptions (like randomness) or assumed to be close to some relative frequency
2. The ratio of the number of times an event has occurred to the number of times it could occur	2. If outcomes are equally likely, the ratio of the number of outcomes in an event to the total number of possible outcomes
3. 0 means that an event did not occur. 1 means that the event occurred every time it could.	3. 0 means that an event is impossible. 1 means that an event is sure to happen.
4. The more often an event occurs relative to the number of times it could occur, the closer its relative frequency is to 1.	4. The more likely an event is, the closer its probability is to 1.
5. The sum of the relative frequencies of all outcomes in an experiment is 1.	5. The sum of the probabilities of all outcomes in an experiment is 1.
6. If the relative frequency of an event is r, then the relative frequency of its complement is $1 - r$.	6. If the probability of an event is p, then the probability of its complement is $1 - p$.

Relative Frequency Distributions

Just as there are probability distributions, there are distributions of relative frequency. Relative frequency distributions can be useful in answering questions about data when there are many possible answers. For example, how tall are professional basketball players?

GUIDED

Example 1

The dot plot on the next page shows the heights of 198 players in the National Basketball Association (NBA) during the 2005-06 season. Each player's height is represented by a dot. The information is summarized in the table at the right.

In the NBA, what is the relative frequency of a player

a. being shorter than 6 feet tall?
b. being at least 7 feet tall?
c. being at most 6 feet 7 inches tall?
d. having a height h given by $72 \leq h \leq 78$?

Height (in.)	Frequency
87	1
86	1
85	3
84	16
83	18
82	20
81	22
80	22
79	23
78	15
77	9
76	11
75	15
74	7
73	10
72	3
71	1
67	1

Source: NBA

Lesson 5-7

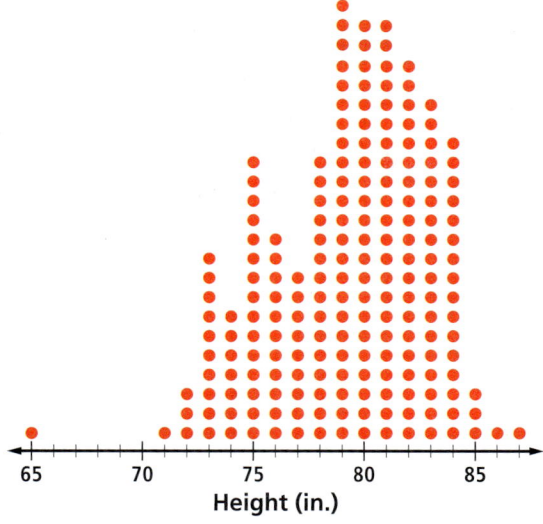

Solutions

a. Two players are under 6 feet. So the relative frequency is
$\frac{2}{?} \approx 0.01 = 1\%$.

b. Include all the players 7 ft or taller.
relative frequency = $\frac{?}{198} \approx 0.106 = 10.6\%$.

c. Include players 6 ft 7 in. and shorter.
relative frequency = $\frac{23 + 15 + 9 + 11 + 15 + 7 + 10 + 3 + 1 + 1}{198}$
= $\frac{?}{?} \approx \underline{\ ?\ } \approx 48\%$

d. These heights include __?__ players. So the relative frequency is __?__ %.

How tall are NBA players? Part c of the Example shows that in the NBA, almost half of the players (48%) are 6 ft 7 in. or shorter. Since 48% is very close to 50%, we know that 6 ft 7 in. is close to the median. Recall that the median is the middle number when a data set is arranged in order. For the NBA heights, the median is 6 ft 8 in. Another name for median is *50th percentile*. Fifty percent of the data are at or below the 50th percentile.

Percentiles

Sometimes it is interesting to describe the relative position of a person within a distribution. You have received standardized test reports in the past that report your score as a percentile. What does it mean if the report says you were at the 70th percentile? It means that 70 percent of the people that took the test had scores that were less than or equal to your score.

Chapter 5

Percentile

The **pth percentile** of a data set is the smallest data value that is greater than or equal to *p* percent of the data values.

 QY

▶ QY

Your school counselor tells you that your class rank or grade point average (GPA) is the 78th percentile. What does this mean?

Percentiles cut the data set into 100 equal-size parts. This is similar to the way data are cut into four equal-size parts by the median and quartiles.

The 25th percentile is the same as the 1st quartile.
The 50th percentile is the same as the median.
The 75th percentile is the same as the 3rd quartile.

The dot plot of the NBA heights shown below has these three percentiles marked, along with the 10th percentile and the 90th percentile. A player who is 84 in. tall (7 ft) is at the 90th percentile. About 90% of the players are 7 ft or shorter and about 10% are taller.

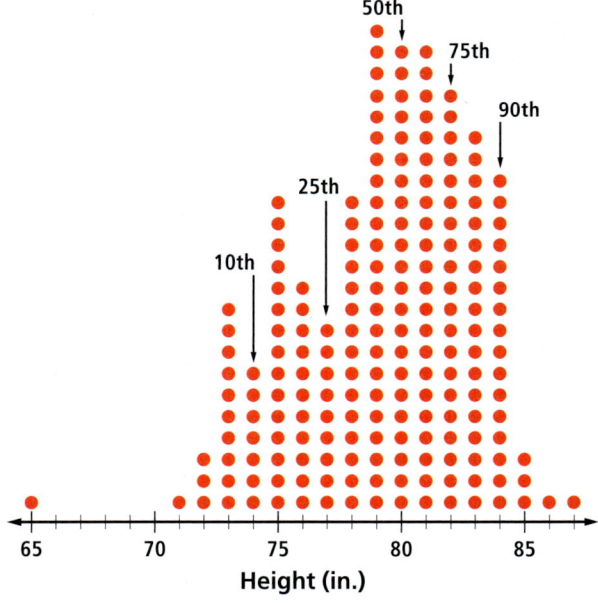

Example 2

Find the 10th percentile in the NBA data and explain its meaning.

Solution The 10th percentile is the smallest data value that is greater than or equal to 10% of the data values. There are 198 values, and 10% of 198 = 19.8, so look at the 20th height from the bottom of the list. This height is 74 in. So 74 in. is at the 10th percentile of NBA heights.

Questions

COVERING THE IDEAS

1. What is the meaning of a relative frequency of 0?
2. What is the meaning of a relative frequency of 1?
3. What is the meaning when the probability of an event equals 0?
4. What is the meaning when the probability of an event equals 1?
5. A letter is picked randomly from the English alphabet.
   a. What is the probability of picking a vowel (A, E, I, O, or U)?
   b. Count the number of vowels and consonants in the following sentence.
      The quick brown fox jumped over the lazy dog.
   c. What is the relative frequency that a letter randomly picked in the sentence is a vowel?
   d. Why aren't the answers to Parts a and c equal?
6. A 2006 estimate was that 116 million households in the United States would have televisions in 2007. Of these homes, 82 million were thought to have TVs with digital capability.
   a. What was the relative frequency of households with a TV having digital capability?
   b. What was the relative frequency of households with a TV not having digital capability?
7. What is meant by a test score that is at the 46th percentile?

In 8 and 9, refer to Example 2.

8. What height is at the 50th percentile of NBA heights? Explain how you got your answer.
9. What height is at the 80th percentile of NBA heights?

APPLYING THE MATHEMATICS

In 10 and 11, use the table which gives percentiles for weights of 3-year-old boys. Doctors use weight percentiles to see whether children are growing properly.

10. 75% of the boys are heavier than Fernando. How much does he weigh?
11. Find possible values of a and b if 80% of the boys have weights in the interval $a \leq$ weight $\leq b$.

Weight (lb)	Percentile
37	90th
34	75th
32	50th
29	25th
28	10th

Chapter 5

12. In 1983 the American Veterinary Medical Association surveyed 20,000 households and found that 5,680 had a cat as a pet. In 2001, a similar survey of 54,000 households found that 31.6% had a cat. Was cat ownership becoming more popular?

13. **Multiple Choice** An event occurred c times out of t possible occurrences. The relative frequency of the event was 50%. Which equation is true?

    A $\frac{c}{t} = 0.5$ **B** $\frac{t}{c} = 0.5$ **C** $1 - t = 0.3$ **D** $1 - c = 0.3$

14. Of the people surveyed, $\frac{3}{5}$ thought the American League team would win the World Series. If $3n$ people were surveyed, how many thought the American League team would win?

In 15 and 16, use the following data, which show the level of education of U.S. adults through 2004. The numbers in the cells are in thousands.

Education Level	Age					Sum
	25–34	35–44	45–54	55–64	> 64	
Did not complete high school	5,072	5,232	4,251	3,856	9,339	27,750
Completed high school	11,244	13,739	12,910	9,436	12,482	59,811
1–3 years of college	7,583	7,420	7,210	4,824	4,771	31,808
4 or more years of college	15,304	17,183	16,699	10,258	8,067	67,511
Sum	39,203	43,574	41,070	28,374	34,659	186,880

Source: U.S. Census Bureau

15. What is the probability that an individual older than 64 completed at least one year of college?

16. What is the probability that a person between the ages of 25 and 44 did not graduate from high school?

REVIEW

17. The end-of-the-year raffle at Lincoln Elementary is the school's biggest fund-raiser. Last year, the school sold 578 tickets. The grand prize was a $1,000 gift certificate to an electronics store. In addition, there were 3 prizes of computers, 5 bikes, and 10 sweatshirts with the school seal on them. (Lesson 5-6)

    a. Given that Rufus won a prize, what is the probability it was a computer?

    b. If Randall bought one ticket, what is the probability that he won the grand prize?

    c. How many times as likely is it to win a sweatshirt as a bike?

18. Joanie joined a bowling league to improve her game. Her scores over 10 weeks were: 52, 65, 59, 72, 70, 92, 85, 100, 95, and 93. **(Lesson 5-6)**
    a. What is the relative frequency of Joanie bowling over 70?
    b. What is the relative frequency of Joanie bowling 70 or under?

19. In planning her trip from Toronto, Ontario to Montreal, Quebec Bianca looked at the legend of a map that says $\frac{3}{4}$ in. = 50 km. If the two cities are approximately 542 km apart, how far should the distance be between them on the map? **(Lesson 5-4)**

20. Simplify each expression. **(Lessons 5-3, 5-2, 5-1)**
    a. $\frac{63}{34} \div \frac{9}{2}$
    b. 76 mi/hr · 2.15 hr
    c. $\frac{7m^3(8n)}{4m^2n^2}$

21. Graph $y = |6 - 3x|$ and use your graph to solve $|6 - 3x| = 15$. **(Lesson 4-9)**

EXPLORATION

22. a. Create an experiment in which randomness indicates certain probabilities that outcomes will occur.
    b. Conduct the experiment a large number of times. Record the appropriate data.
    c. Compare the relative frequencies you get with the probabilities predicted by randomness.
    d. Do you think the outcomes occurred randomly? Why or why not?

QY ANSWER

78 percent of students in your class have GPAs less than or equal to your GPA.

Chapter 5

Lesson 5-8
Probability Without Counting

▶ **BIG IDEA** The probability that a point lands in a particular region can be calculated by taking the ratios of measures of regions.

When a situation has equally likely outcomes, the probability of an event is the ratio of the number of outcomes in the event to the total number of outcomes. But sometimes the number of outcomes is infinite and not countable. In such cases, probabilities may still be found by division.

Mental Math

Compare using $>$, $=$, or $<$.

a. $-50 + 74$ and $-74 + 50$
b. -5 and $(-5)^2$
c. $y + 7$ and $y + 6$
d. $|x|$ and -4

Probabilities from Areas

Example 1

Suppose a dart is thrown at a 24-inch square board containing a target circle of radius 3 inches, as shown at the right. Assuming that the dart hits the board and that it is equally likely to land on any point on the board, what is the probability that the dart lands in the circle?

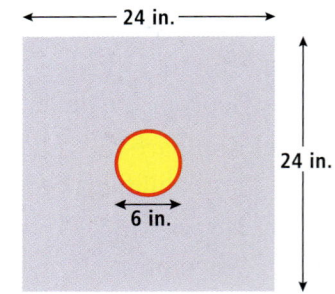

Solution Recall that the area of a circle with radius r is πr^2. Compare the area of the circle to the area of the square.

Probability the dart lands in the circle $= \dfrac{\text{area of circle}}{\text{area of square}}$

$= \dfrac{\pi \cdot 3^2}{24 \cdot 24}$

$= \dfrac{9\pi}{576}$

≈ 0.049, or about 5%

So, the probability of the dart landing in the circle is about 5%.

Example 1 illustrates the Probability Formula for Geometric Regions.

> **Probability Formula for Geometric Regions**
>
> Suppose points are selected at random in a region and part of that region's points represent an event E of interest. The probability P of the event is given by
>
> $\dfrac{\text{measurement of region in the event}}{\text{measure of entire region}}$.

296 Division and Proportions in Algebra

QY1

Probabilities from Lengths

Example 2
Points A, B, C, D, and E below represent exits on an interstate highway.

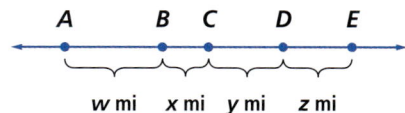

If accidents occur at random along the highway between exits A and E, what is the probability that when an accident occurs, it happens between exit C and exit D?

Solution First find the length of the entire segment.

Length of $\overline{AE} = w + x + y + z$

Probability the accident is in $\overline{CD} = \dfrac{y}{w + x + y + z}$

Traffic safety engineers might compare the probabilities in Example 2 with the actual relative frequency of accidents. If the relative frequency along one stretch of the highway is greater than predicted, then that part of the highway might be a candidate for repair or new safety features.

 QY2

Probabilities can also be determined by finding ratios of angle measures.

Activity
A basic spinner used in many games is shown here. Suppose the spinner is equally likely to point in any direction. There is a 50% probability the spinner lands in the red region. Draw a different spinner that still has a 50% probability of landing in a red region.

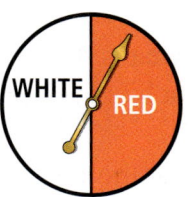

Sometimes the calculation of the measures needed to compute a probability requires you to do some addition or subtraction first.

▶ **QY1**

A target consists of two concentric circles as shown below. The smaller circle (called the "bull's eye"), has a radius of 4 cm and the larger circle has a radius of 6 cm. If a point is selected at random from inside the target, what is the probability it *misses* the bull's eye?

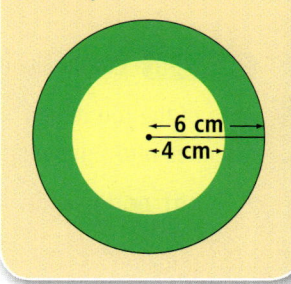

▶ **QY2**

What is the probability that an accident occurring between exits A and E happens between exits C and E from Example 2?

Chapter 5

Example 3
A target consisting of three evenly spaced concentric circles is shown below. If a point is selected at random from inside the circular target, what is the probability that it lies in the red region?

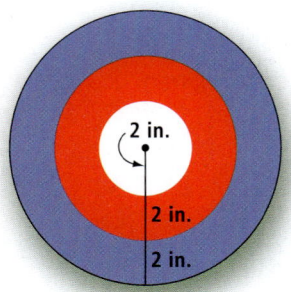

Solution Probability of a point in the red region = $\frac{\text{area of red region}}{\text{area of largest circle}}$

Area of the red region = the difference in the areas of the circles with radii 4 inches and 2 inches

Area of red region = $\pi(4)^2 - \pi(2)^2$
$= 16\pi - 4\pi$
$= 12\pi$ in^2

The radius of the largest circle is 6 inches.

Area of largest circle = $\pi(6)^2 = 36\pi$ in^2

Thus the probability of choosing a point in the red region is $\frac{12\pi}{36\pi} = \frac{1}{3}$.

Questions

COVERING THE IDEAS

1. Consider the square archery target board at the right.
   a. What is the area of the bull's eye?
   b. What is the area of the entire target board?
   c. To the nearest percent, what is the probability that an arrow shot at random that hits the board will land in the bull's eye?
   d. What is the probability that the arrow hitting the board will land on the target outside the bull's eye?

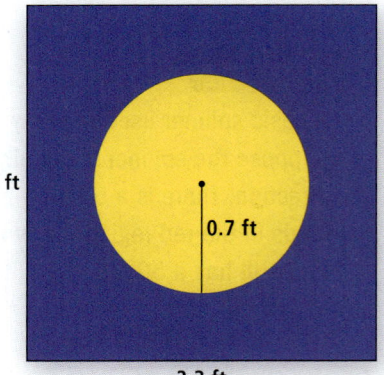

2. Draw three different spinners that have a $\frac{2}{3}$ probability of landing in a blue region.

3. An electric clock with a continuously-moving second hand is stopped by a power failure. What is the probability that the second hand stopped between the following two numbers?
   a. 12 and 2
   b. 5 and 6
   c. 7 and 11

298   Division and Proportions in Algebra

Lesson 5-8

In 4 and 5, use the following scenario and diagram. A student from the University of Chicago wanted to ride her bike north to Loyola University. Along the bike trek, she planned on making stops at Navy Pier and North Avenue Beach. If the student has a flat tire on the trip, what is the probability it occurs between each pair of locations?

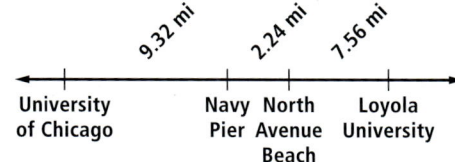

4. Navy Pier and North Avenue Beach

5. University of Chicago and North Avenue Beach

In 6–8, refer to the target at the right. Suppose a point on the target is chosen at random.

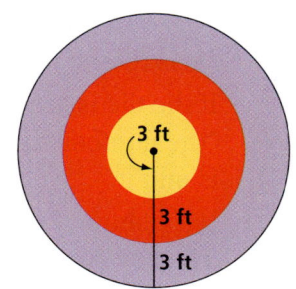

6. What is the probability that it lies inside the bull's eye?

7. What is the probability that it lies in the outermost ring?

8. What is the probability that it lies in the middle ring?

APPLYING THE MATHEMATICS

9. The land area of Earth is about 57,510,000 square miles, and the water surface area is about 139,440,000 square miles. Give the probability that a meteor hitting the surface of Earth will
   a. fall on land.   b. fall on water.

Ocean water covers nearly 71% of Earth's surface, whereas fresh water in lakes and rivers covers less than 1%.
Source: NASA

10. In a rectangular yard of dimensions q by p, there is a rectangular garden of dimensions b by a. If a newspaper is thrown randomly into the yard, what is the probability that it lands on a point in the garden?

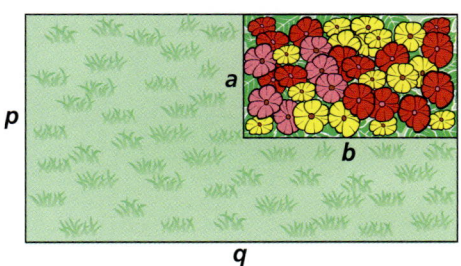
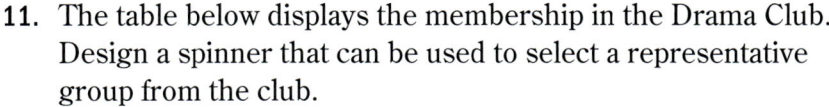

11. The table below displays the membership in the Drama Club. Design a spinner that can be used to select a representative group from the club.

Grade	Members
9	5
10	15
11	17
12	23

Probability Without Counting   299

12. One student, seeing that the answer to Example 3 is $\frac{1}{3}$, said that if there were five concentric circles instead of 3, then the middle ring would have $\frac{1}{5}$ the area of the largest circle. Is the student correct? Why or why not?

13. What is the probability that a point selected from the region within the red rectangle at the right is also inside the circle?

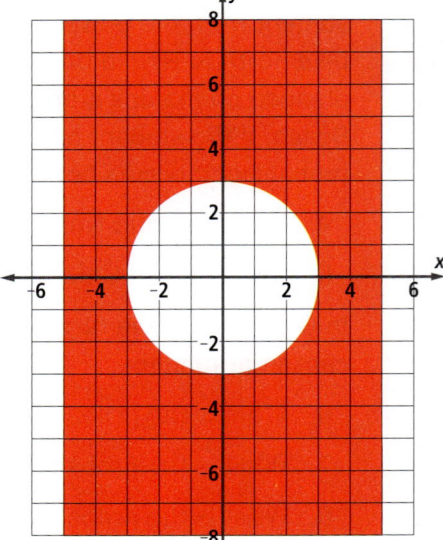

REVIEW

14. A die is tossed once. (Lessons 5-7, 5-6)
    a. If the die is assumed to be fair, which event is more likely, "the number showing is less than 3" or "the number showing is odd"?
    b. If a 4 showed on the die, what was the relative frequency that a 2 showed?

15. A 14-foot-long metal rod is cut so that the two pieces formed have lengths in a ratio of $\frac{7}{3}$. How long is each piece? (Lesson 5-5)

16. Write an equation of a line that passes through the points (1, −8) and (1, 1). (Lesson 4-2)

17. a. If $42n = 0$, then $\frac{2n}{5} = \underline{\ ?\ }$.
    b. What property was used to answer Part a? (Lesson 2-8)

In 18–20, rewrite the fraction in lowest terms. (Previous Course)

18. $\frac{20}{25}$ 19. $\frac{42}{54}$ 20. $\frac{112}{28}$

EXPLORATION

21. a. A circle with radius of 10 units is drawn inside a square with sides of 20 units, as shown below. What part of the area inside the square is outside the circle?

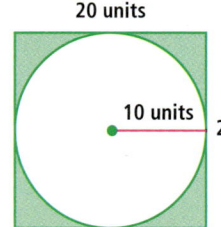 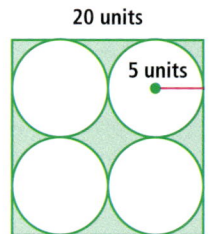

b. Four circles with radii of 5 units are drawn inside a square with sides of 20 units, as shown at the right. What part of the area inside of the square lies outside the four circles?

c. Generalize Parts a and b. Explain why you believe your generalization to be true.

QY ANSWERS

1. $0.5\overline{5}$ or about 56%

2. $\frac{y + z}{w + x + y + z}$

Lesson 5-9

Proportions

Vocabulary
proportion
extremes
means
population
sample
randomly
capture-recapture method

▶ **BIG IDEA** Proportions can be solved algebraically using the Means-Extremes Property.

Mental Math

Write in lowest terms.
a. $\frac{35}{7,000}$
b. $-\frac{27}{72}$
c. $\frac{18}{-78}$

In his 1859 autobiography, Abraham Lincoln wrote about his childhood in Indiana. "There were some schools, so called; but no qualification was ever required of a teacher beyond 'readin, writin, and cipherin' to the Rule of Three. . . . Of course when I came of age I did not know much. Still somehow, I could read, write, and cipher to the Rule of Three; but that was all." After his short stint at school, Lincoln went on to teach himself algebra and geometry from books.

The "Rule of Three" refers to a method of solving a *proportion*, a sentence such as "4 is to 6 as 3 is to __?__." It is equivalent to solving $\frac{4}{6} = \frac{3}{x}$. In a simple proportion such as this, whenever you know three out of the four numbers, you can determine the fourth. The Rule of Three is a method that dates from ancient times and is not usually taught today. Today we use algebra, and so are able to solve this and more complicated proportions.

Solving Proportions

A **proportion** is a statement that two ratios are equal. Thus any equation of the form $\frac{a}{b} = \frac{c}{d}$ is a proportion. This equation is sometimes written $a:b = c:d$. Because a and d are at the two ends of this statement, a and d are called the **extremes**. Because b and c are in the middle, b and c are called the **means**.

Example 1

Complete the sentence: "4 is to 6 as 3 is to __?__."

Solution Let x be the unknown number.

$\frac{4}{6} = \frac{3}{x}$ Write the sentence as a proportion.

$\frac{4}{6} \cdot 6x = \frac{3}{x} \cdot 6x$ Multiply each side of the equation by $6x$.

$\frac{4}{\cancel{6}} \cdot \cancel{6}x = \frac{3}{\cancel{x}} \cdot 6\cancel{x}$ Simplify each side.

$4x = 18$ Simplify.

$x = \frac{18}{4} = 4.5$ Divide each side by 4 and simplify.

(continued on next page)

The Rail Splitter, a painting by J. L. G. Ferris of a young Abraham Lincoln splitting logs

Check Substitute 4.5 for x in the proportion $\frac{4}{6} = \frac{3}{x}$. Does $\frac{4}{6} = \frac{3}{4.5}$? Yes, in lowest terms the left side equals $\frac{2}{3}$, and the right side equals $\frac{30}{45}$ or, in lowest terms, $\frac{2}{3}$.

The Means-Extremes Property

Look at Step 4 in the solution to Example 1. The left side, $4x$, is the product of the extremes, 4 and x, of the original proportion. The right side, 18, is the product of the means 6 and 3. The general pattern is that the product of the means is equal to the product of the extremes. Algebra explains why this is true. Consider any proportion $\frac{a}{b} = \frac{c}{d}$.

$\frac{a}{b} \cdot bd = \frac{c}{d} \cdot bd$ Multiply both sides of the equation by bd to clear the fractions.

$\frac{a}{\cancel{b}} \cdot \cancel{b}d = \frac{c}{\cancel{d}} \cdot b\cancel{d}$ Simplify both sides of the equation.

$ad = bc$ Simplify.

> **Means-Extremes Property**
>
> For all real numbers a, b, c, and d (with b and d not zero), if $\frac{a}{b} = \frac{c}{d}$ then $ad = bc$.

Proportions and Statistics

In statistics, the set of individuals or objects you want to study is called the **population** for that study. If you cannot collect data from the entire population, the part studied is called a **sample.** When samples are taken **randomly,** every member of the population has an equal chance of being chosen. Therefore, data from the sample can be used to estimate information about the population.

Activity

Step 1 Fill a large bowl with pennies, beans, popcorn kernels, or something similar. Do not count how many there are.

Step 2 Take out at least 30 pieces. Mark or tag them and then return the pieces to the bowl.

Step 3 Mix the contents of the bowl. Draw out a handful of the pieces. Be sure to get at least one tagged piece.

How many pieces are there in your handful? __?__

How many pieces are tagged? __?__

Step 4 Set up a proportion to estimate how many pieces are in the bowl and solve it.

Step 5 Return all pieces to the bowl and repeat Steps 3 and 4 at least two more times.

Step 6 Compare and contrast your totals. Based on your random samples, do you believe you have a good estimate of how many pieces are in the bowl? Explain why or why not.

In this activity, the method you used to find the total pieces in the bowl is called the **capture-recapture method.** This method has been used to estimate the number of fish in a lake or deer in a forest. In Step 2, you captured pieces and tagged them. Step 3 is where you recaptured the pieces. If the sample in the recapture is chosen randomly, the ratio of tagged pieces to the number of pieces is nearly the same as the percentage of tagged pieces in the entire bowl.

Surveys give information about a part of a population. Proportions are used to extend that information to a larger group. When we do this, we assume that the ratios are the same for the survey group and the larger population.

A biologist uses a computer while colleagues capture spring-summer Chinook salmon smolts for counting and tagging on the Idaho Salmon River.

Example 2

A survey of 454 undergraduates at the University of Texas in 2003 found that 409 of them used their own computers at home. The actual number of undergraduates at the school was 37,409. Based on the survey, how many undergraduates were expected to use computers at home?

Solution 1 Let x be the number of 2003 undergraduates using a personal computer. Set up a proportion with equal ratios that compares undergraduates with computers to the total number of undergraduates.

$$\frac{\text{number of undergraduates with computers in survey}}{\text{total number of undergraduates surveyed}} = \frac{\text{number of total undergraduates with computers}}{\text{total number of undergraduates}}$$

$\frac{409}{454} = \frac{x}{37,409}$

$454x = 409 \cdot 37,409$ Means-Extremes Property

$454x = 15,300,281$ Simplify.

$x \approx 33,701$ Multiply each side by $\frac{1}{454}$.

About 33,701 undergraduates would be expected to have computers at home in 2003.

(continued on next page)

Solution 2 Let x be the number of 2003 undergraduates using their own computer. Use ratios that compare undergraduates with computers to the total number of undergraduates.

$$\frac{\text{number of undergraduates with computers in survey}}{\text{total number of undergraduates with computers}} = \frac{\text{total undergraduates surveyed}}{\text{total number of undergraduates}}$$

$$\frac{409}{x} = \frac{454}{37{,}409}$$

$409 \cdot 37{,}409 = 454x$ Means-Extremes Property

$15{,}300{,}281 = 454x$ Simplify.

$33{,}701 \approx x$ Multiply each side by $\frac{1}{454}$.

About 33,701 undergraduates were expected to have computers at home in 2003.

The equations that were written for the two solutions are different. But notice that when the Means-Extremes Property is used, the resulting equations are the same.

 QY

Some proportions contain algebraic expressions.

Example 3

Two candidates, A and B, ran in a village election. Candidate A received 450 more votes than Candidate B. Their vote counts were in the ratio of 8 to 3. How many votes did each candidate get?

Solution Let x = number of votes Candidate B received. Because Candidate A received 450 more votes than Candidate B, the expression $x + 450$ is the number of votes Candidate A received.

$$\frac{\text{number of votes for A}}{\text{number of votes for B}} = \frac{x + 450}{x} = \frac{8}{3}$$

$3(x + 450) = x \cdot 8$ Means-Extremes Property

$3x + 1{,}350 = 8x$ Distribute.

$1{,}350 = 5x$ Add $-3x$ to both sides.

$270 = x$ Divide both sides by 5.

Because $x = 270$, Candidate B got 270 votes. Candidate A got $x + 450$, or $270 + 450 = 720$ votes.

Check Is the ratio of A's votes to B's votes equal to $\frac{8}{3}$?

Is $\frac{720}{270} = \frac{8}{3}$?

$\frac{8}{3} = \frac{8}{3}$ Yes, it checks.

▶ **QY**

Emilio knows he can do 11 pushups in 15 seconds.

a. If his wrestling coach times him for a minute, at this rate how many pushups could he do?

b. At this rate, how long will it take Emilio to do 74 pushups?

c. Why is the answer to Part b likely an underestimate?

Lesson 5-9

Questions

COVERING THE IDEAS

In 1 and 2, complete the sentence.

1. 4 is to 12 as 18 is to __?__.
2. 5 is to 13 as 17 is to __?__.
3. What is a proportion?

In 4 and 5, a proportion is given.

   a. Use the Means-Extremes Property to solve the proportion.
   b. Check your work.

4. $\frac{n}{4} = \frac{20}{48}$

5. $\frac{-28}{21} = \frac{64}{p}$

6. On a map of Spain, 3 centimeters represents 200 kilometers.
   a. Seville and Madrid are approximately 417 kilometers apart. How far apart would they be on the map?
   b. If Barcelona and Madrid are 9.4 centimeters apart on a map, about how many kilometers apart are they?

7. After soccer's World Cup was held in Germany in 2006, the national tourist board surveyed about 1,300 foreign visitors who attended. Of them, approximately 1,200 responded that the World Cup had been a great event.
   a. Based on the survey, about how many of the 2 million foreign visitors felt the World Cup was a great event?
   b. How many foreign visitors felt the World Cup was *not* a great event?

Holland tackles Romania during the 2006 World Cup in Stuttgart, Germany.

8. In a capture-recapture study, suppose 60 deer in a forest are tagged. On the recapture, 52 deer are caught, of which 10 are found to have been tagged. Estimate the number of deer in the forest.

9. In 1995, scientists began restoring gray wolves to Yellowstone National Park. The recovery plan called for introducing 10 breeding pairs of gray wolves each year for three years. Suppose that in 2000, the scientists recaptured 14 wolves and 3 had tags. What was the estimated population of gray wolves in Yellowstone in 2000?

10. Two numbers are in the ratio of 9:5. One number is 76 greater than the other. What are the numbers?

In 11 and 12, solve the proportion.

11. $\frac{4m-1}{7} = \frac{m+2}{2}$

12. $\frac{5}{12} = \frac{2p-3}{3p+5}$

Proportions 305

APPLYING THE MATHEMATICS

In 13–15, solve the proportion for the indicated variable. No variables equal 0.

13. $\frac{2}{3} = \frac{b}{c}$ for c

14. $\frac{4x}{w} = \frac{3}{m}$ for w

15. $\frac{a}{b} = \frac{x}{y}$ for a

16. A baseball team plays 2 innings in 25 minutes. At this rate, how many minutes will a 9-inning game take?

17. During the first 7 days of November, Gabby used her cell phone for 133 minutes. At this rate, how many minutes will she talk during the entire month?

18. Kauai, Hawaii, is considered the rainiest place on Earth. In an average week, 9.1 inches of rain falls on the island. If you are on Kauai for 2 days, how much rain would you expect to fall during your stay?

19. The target heart rates for 22-year-old females exercising in the "fitness zone" is 122–143 beats per minute. Annie, a 22-year-old female, regularly checks her pulse rate while exercising. She found that her heart beats 19 times in 10 seconds.

    a. At this rate, how many times does Annie's heart beat in 60 seconds?

    b. Is Annie in her target heart rate zone?

20. The Havalot family bought a 26-inch and a 50-inch plasma TV. The total cost of the two televisions was $4,600. If the ratio of the prices was 6:17, how much did each TV cost?

21. A useful baseball statistic is a pitcher's earned run average (ERA), which is a measure of the average number of runs a pitcher allows during 9 innings. Suppose a pitcher has an ERA of 3.33 and has pitched 150 innings. How many earned runs has he allowed during those innings?

Mt. Waialeale on the island of Kauai in Hawaii (3,000 feet high) is the wettest spot on Earth, averaging about 460 inches of rain per year.

Source: www.infoplease.com

REVIEW

22. A raft that is a rectangle 8 feet by 12 feet is in a circular pool that is 40 feet in diameter. If a watch is at the bottom of the pool, what is the probability it is under the raft? (**Lesson 5-8**)

23. Consider the following situation. A bowl contains 8 green beads, 4 red beads, 11 blue beads, and 6 black beads. One bead from the bowl is then chosen at random. (**Lesson 5-6**)

    a. Find the probability of choosing a red bead.

    b. Find the probability of choosing a black or blue bead.

    c. Find the probability of choosing a bead that is *not* green.

24. Square I has sides of length a and Square II has sides of length $3a$, as shown at the right. (**Lesson 5-5**)
    a. Find the ratio of a side of Square II to a side of Square I.
    b. Find the areas of Square I and Square II.
    c. Find the ratio of the area of Square II to the area of Square I.

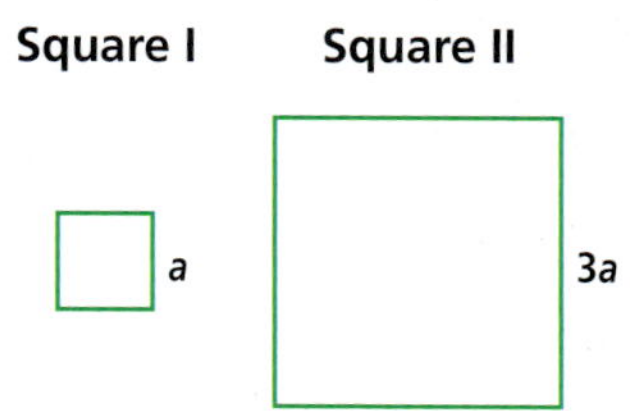

25. A formula to find the sum S of the measures of the interior angles of a polygon is $S = 180(n - 2)$, where n is the number of sides of the polygon. (**Lesson 4-7**)
    a. Solve this formula for n.
    b. If the sum of the measures of the interior angles of a polygon is 1,260°, find the number of sides of the polygon.

26. **Skill Sequence** Find each reciprocal. (**Lesson 2-8**)
    a. $\frac{5}{9}$
    b. $\frac{5x}{9}$
    c. $\frac{-5x}{9}$

27. a. The following number puzzle deals with your seven-digit phone number, not including the area code.

Step 1	Write down the first 3 digits of your phone number.
Step 2	Multiply this by 80.
Step 3	Add 1.
Step 4	Multiply this by 250.
Step 5	Add the last four digits of your phone number.
Step 6	Add the last four digits of your phone number again.
Step 7	Subtract 250.
Step 8	Divide by 2.

    What is your result?

    b. Use algebra to explain your answer to Part a.
    (**Lesson 2-3**)

EXPLORATION

28. The tallest person ever measured was Robert Wadlow, who was 8 feet 11.1 inches tall. At 13 years of age, he was 7 feet 4 inches tall. Because schools are constructed for a much shorter person, many things were too small for him. Assume that schools are designed for a person who is up to 6 feet tall. Measure the dimensions of five things that you see in school every day. How big would these objects be if they were made proportionally to fit Robert Wadlow when he was 13 years old?

QY ANSWERS

a. 44

b. about 101 seconds

c. As Emilio does more push-ups, he probably gets tired and slows down, making his rate change and the time it takes to do pushups longer.

Chapter 5

Lesson 5-10
Similar Figures

Vocabulary

ratio of similitude

> **BIG IDEA** Ratios of lengths of similar geometric figures are equal, giving rise to many applications of proportions.

Model airplanes, architect's drawings, models of buildings, and photographs are all pictures of objects that have the same shape as the originals but not necessarily the same size. Mathematically, the original object and the model are *similar*. Blow-ups of photographs are also similar.

Mental Math

If you can bike to a friend's house in 15 minutes, averaging 10 miles an hour, how long will it take if you average 20 miles an hour?

In some species of animals, babies are shaped much like their parents, like the elephants shown below. However, for humans this is not the case. Infants have very different shapes from adults. This drawing shows a baby and an adult, with each one's height divided into 8 equal parts. The divisions allow us to form ratios to compare the shapes of the infant and adult.

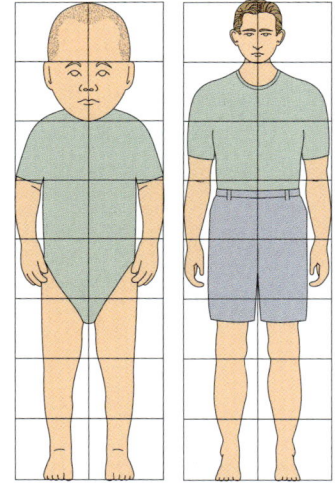

The largest land animal is the African bush elephant, standing 13 feet high and weighing 8 tons.

Source: *The World Almanac for Kids*

	Infant	Adult
Ratio of head length to height	$\frac{2 \text{ parts}}{8 \text{ parts}} = \frac{1}{4}$	$\frac{1 \text{ part}}{8 \text{ parts}} = \frac{1}{8}$
Ratio of trunk length to leg length	$\frac{3 \text{ parts}}{3 \text{ parts}} = \frac{1}{1}$	$\frac{3 \text{ parts}}{4 \text{ parts}} = \frac{3}{4}$

Notice that the ratios in an infant's body are quite different from an adult's body. An infant is not a scaled-down version of an adult.

308 Division and Proportions in Algebra

Lesson 5-10

Activity

Pictured at the right is a coffeemaker that is 12.5 inches tall and 5.625 inches wide.

Step 1 Measure AB and CD to find the height and width of the coffeemaker in the picture.

Step 2 Calculate these ratios to the nearest tenth.

a. $\dfrac{AB}{\text{actual height of coffeemaker}}$

b. $\dfrac{CD}{\text{actual width of coffeemaker}}$

You should find that these ratios are about equal.

Step 3 Measure EF, the height of the coffeepot.

Step 4 Solve a proportion to find the height of the actual coffeepot.

You also should have found that the dimensions of the picture are $\frac{1}{5}$ of the length of the corresponding dimensions of the coffeemaker. This illustrates a basic property of similar figures.

Fundamental Property of Similar Figures

If two polygons are similar, then ratios of corresponding lengths are equal and corresponding angles have the same measure.

The ratio of the lengths of corresponding sides of two similar figures is called a **ratio of similitude.** In the activity the ratio of similitude is $\frac{1}{5}$ because $\dfrac{AB}{\text{actual height}} = \dfrac{CD}{\text{actual width}} = \frac{1}{5}$.

Finding Lengths in Similar Figures

When two figures are similar, a true proportion can be written using corresponding lengths. If three of the four lengths in the proportion are known, the fourth can be found by solving an equation.

Similar Figures   309

Example 1

An adult African elephant can be 30 feet long and 11 feet high at the shoulder. Estimate the length of a baby elephant that is 3 feet high at the shoulder.

Solution Compare lengths on the adult with the corresponding lengths on the baby. Set up a proportion by forming two equal ratios. Let x be the length of the baby. Since the elephants are similar, the ratios are equal.

$$\frac{\text{height of adult}}{\text{height of baby}} = \frac{\text{length of adult}}{\text{length of baby}}$$

$$\frac{11}{3} = \frac{30}{x}$$

$$11x = 90$$

$$x = \frac{90}{11} \approx 8.2$$

We estimate that the baby elephant is slightly over 8 feet long.

GUIDED

Example 2

The two quadrilaterals at the right are similar, with corresponding sides parallel. Find x, the length of $\overline{CD}$.

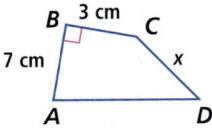

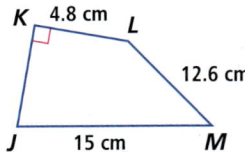

Solution The side corresponding to the unknown length $\overline{CD}$ is __?__. There is a pair of corresponding sides whose lengths are both known. These are __?__ and __?__. Because the figures are similar, the ratios of lengths of these corresponding sides are equal.

$\dfrac{CD}{?} = \dfrac{?}{?}$ Write the proportion.

$\dfrac{x}{?} = \dfrac{?}{?}$ Substitute the known lengths.

$\underline{}\cdot x = \underline{} \cdot \underline{}$ Means-Extremes Property

$x = \underline{}$ cm Divide by __?__ and simplify.

Using Similar Figures to Find Lengths without Measuring

Similar figures have many uses. For example, you can use similar triangles to find the height of an object you cannot measure easily. Suppose you want to find the height h of a flagpole. Here is how you can do it. Holding a yardstick parallel to the flagpole, measure the length of the yardstick's shadow. Then measure the length of the shadow of the flagpole. The picture on the next page illustrates one possible set of measurements.

Lesson 5-10

Example 3

Use the measurements at the right to find the height *h* of the flagpole.

Solution Two similar right triangles are formed. Now, use ratios of corresponding sides to find *h*.

$\frac{3}{h} = \frac{5}{72}$

$5h = 72 \cdot 3$

$5h = 216$

$h = 43.2$

The flagpole is about 43 feet tall.

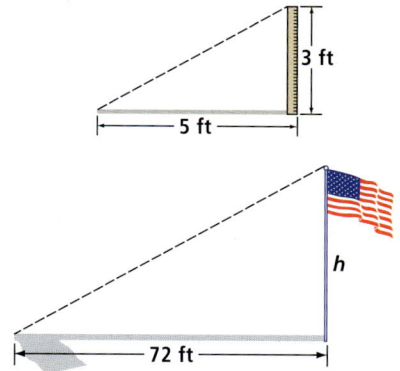

Questions

COVERING THE IDEAS

1. What is the fundamental property of similar figures?

2. An adult male African elephant is about 11 feet tall, with ears that measure 5 feet from top to bottom. If a baby elephant is 3 feet tall, find out how big its ears are.

In 3–5, refer to the two similar triangles below. Corresponding sides are parallel.

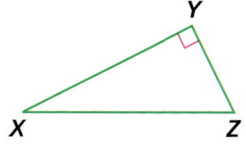 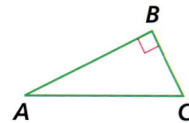

3. Which side of △XYZ corresponds to the given side of △ABC?
   a. $\overline{AC}$
   b. $\overline{BC}$
   c. $\overline{AB}$

4. Find two ratios equal to $\frac{XY}{AB}$.

5. Suppose $AB = 12$, $BC = 5$, $AC = 13$, and $XY = 18$. Find
   a. YZ.
   b. XZ.

6. The quadrilaterals shown below are similar.

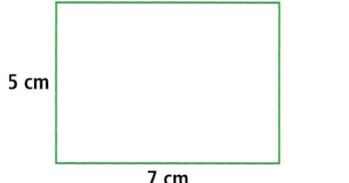 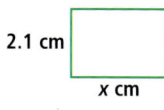

   a. Find *x*.
   b. Write two possible ratios of similitude.

Similar Figures **311**

7. A bookcase is pictured at the right. The actual bookcase is 36 in. wide.

   a. Measure the width of the bookcase in the picture.
   b. What is the ratio of the similitude comparing the picture's width to the actual width?
   c. Measure the height of the bookcase in the picture.
   d. Use your answers to Parts b and c to determine the height of the actual bookcase.

8. Suppose a 3-foot yardstick casts a 4-foot shadow. A nearby building casts a shadow of 56 feet at the same time. What is the height of the building?

APPLYING THE MATHEMATICS

9. A person who is 160 cm tall is photographed. On the photo, the image of the person is 12 cm tall. What is the ratio of similitude?

10. The Crazy Horse Memorial in the Black Hills of South Dakota will be the world's largest mountain carving. From the chin to the top of the head is 87.5 feet. Use the picture, a ruler, and your knowledge of similar figures to approximate the length of the outstretched arm in the carving.

    When completed, the Crazy Horse mountain carving (shown in the background) will be 641 feet long by 563 feet high. Crazy Horse's completed head is 87 feet 6 inches high. The horse's head, currently the focus of work on the mountain, is 219 feet, or 22 stories, high.

11. The two rectangles below are similar.

    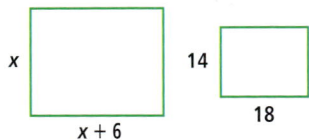

    a. Use a proportion to find the value of x.
    b. Find the perimeter of the larger rectangle.

12. The quadrilaterals at the right are similar. Corresponding sides are parallel.

    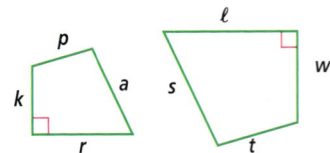

    a. Write a true proportion involving $\frac{s}{a}$.
    b. **Fill in the Blank** Complete $\frac{k}{w} = \frac{?}{t}$ and solve for k.

13. At a certain time on a sunny day, Shadrack, who is 6 feet tall, casts a shadow that is 9 feet long. A nearby building that is t feet tall casts a shadow that is 24 feet long.
    a. Draw a diagram of this situation and label the lengths.
    b. Write a proportion that describes the situation.
    c. How tall is the building?

14. For this question, you need to use a ruler and properties of similar figures. A scale drawing of a house, as seen from its front, is shown below. The actual width (across the front, not including the roof) of the house is 12 meters.

   a. Write a ratio comparing the width of the house in the drawing to the actual width of the house.
   b. Write a proportion you could use to find the actual distance from the ground to the peak of the roof.
   c. Solve the proportion in Part b.

REVIEW

In 15 and 16, solve the proportion. (Lesson 5-9)

15. $\dfrac{64}{3} = \dfrac{4x}{9}$

16. $\dfrac{2}{1-a} = \dfrac{4}{a-3}$

In 17 and 18, use the fact that a *karat* is a measure of fineness used for gold and other precious materials. Pure gold is 24 karats. Gold of 18-karat fineness is 18 parts pure gold and 6 parts other metals, giving 24 parts in all. (Lessons 5-9, 4-1)

17. A ring is 18-karat gold. What percent gold is this?

18. A necklace weighing 6 ounces is 14-karat gold. How many ounces of pure gold are in the necklace?

Central banks of nations hold an estimated 32,000 tons of gold as official stock, and about 96,000 tons is privately held in bullion, coin, and jewelry.

19. In a 3-ounce serving of beef, there are about 26 grams of protein. About how many grams of protein are in an 8-ounce steak? (**Lesson 5-9**)

20. The scale of a map for Yellowstone National Park is 1.75 in. = 10 miles. If the distance between Old Faithful and Mammoth Hot Springs on the map is about 8.5 inches, what is the approximate distance between these two places in miles? (**Lesson 5-9**)

21. When rolling two 6-sided dice and recording their sum, there are two ways to get a 3—rolling a 1 on the first die and a 2 on the second die, or rolling a 2 on the first die and a 1 on the second die. (**Lesson 5-6**)
    a. How many ways are there to roll a 4?
    b. Find the probability of rolling 4 if the dice are fair.
    c. When rolling two 6-sided fair dice, a sum of seven is the most likely outcome. Explain why this is true.

A system of small fissures carries water upward and creates about 50 hot springs in Mammoth Hot Springs.

Source: National Park Service

22. a. Graph $y = 6 + x$ and $y = 2 + 3x$ on the same set of axes.
    b. According to the graph, for what value(s) of x is $6 + x = 2 + 3x$? (**Lesson 4-3**)

In 23 and 24, consider the table at the right that shows the land area of three of the five largest countries in the world in area. (**Lessons 4-1, 3-4**)

Country	Square Miles
Russia	6,592,735
Canada	?
United States	3,717,792
China	?
Brazil	3,286,470

Source: infoplease.com

23. If the land area of Russia is 399,121 square miles less than the sum of the areas of China and Brazil, find the land area of China.

24. If the area of Canada is 3.6% larger than the area of the United States, estimate the area of Canada.

EXPLORATION

25. Find the highest point of a tree, a building, or some other object, using the shadow method described in this lesson. Draw a diagram to illustrate your method.

Chapter 5 Projects

1 Buffon's Needle

The French naturalist George Buffon (1707–1788) discovered the following method of approximating π.

Step 1 Measure the length of a needle.

Step 2 On a piece of paper, draw many horizontal lines, where the distance between the lines is the same as the length of the needle.

Step 3 Drop the needle randomly on the paper, and check if it crossed any of the lines. Repeat this process at least twenty times. Keep a tally of the number of times you dropped the needle and the number of times it crossed one of the lines.

Step 4 As you repeat the process in Step 3, the relative frequency of the times the needle crosses one of the horizontal lines should get closer and closer to $\frac{2}{\pi}$.

a. Calculate the relative frequency of the number of times that the needle crossed one of the lines to the number of times that you dropped the needle. Calculate the reciprocal of this relative frequency, and multiply it by 2. The number you get should be close to π.

b. Suppose Trevor performed this experiment 50 times, and the needle crossed a horizontal line every time. Does this mean that $\frac{2}{\pi} = 1$? Explain how your answer shows the difference between probability and relative frequency.

2 Population Densities

A *population density* is a type of rate, defined as the number of people living in a region divided by the area of that region.

a. Find the most dense and least dense countries in the world.

b. The area of the United States is 3,531,838 square miles. What would the population of the United States be if it had the same population density as the most dense country in the world? What would the population of the United States be if it had the same population density as the least dense country in the world?

c. Which would you expect to have a higher population density: the United States or Singapore? Explain your answer. Give some examples of geographic and cultural features that affect a country's population density.

d. Suppose a country has population density d. What is the meaning of the number $\frac{1}{d}$?

Projects 315

3 Copy Machine Puzzle

Most copy machines have an enlargement feature that creates a figure similar to the one being copied.

a. Use the enlarge feature on a copy machine to enlarge the figure below on the left so that the copy is exactly the same size as the figure below on the right. You may have to enlarge more than once. Record how much you enlarged the figure each time.

 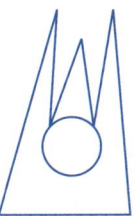

b. Many copiers can copy in a range of sizes that goes from 100% to 141%. Why is 141% used in this case?

4 Calculating Density

In science, the *density* of an object is determined by dividing the mass of the object (weight in grams) by its volume (in cubic centimeters). For example, the density of cool tap water is one gram per cubic centimeter. You already know how to measure the weight of an object. But how do you measure its volume? The Greek mathematician Archimedes (287 B.C.E.–212 B.C.E.) discovered that when an object is placed in a full container of water, the volume of the water that spills out is equal to the volume of the object that was placed in the water.

a. Use Archimedes' discovery and the density of water to explain how you could calculate the volume of an object.

b. Explain how you could use the result from Part a to calculate the density of an object.

c. Look up the story of Archimedes' discovery. How was density used in this story?

5 Converting Rates

Write a computer or calculator program that converts measurements in $\frac{ft}{sec}$ to measurements in $\frac{km}{hr}$. Write a short paragraph explaining how your program works.

Chapter 5 Summary and Vocabulary

- Fractions in algebra are generalizations of fractions in arithmetic. Every fraction can be treated as a division, and division by 0 is undefined (the denominator cannot be 0). For all real numbers a and b and $b \neq 0$, $\frac{a}{b} = a \div b = a \cdot \frac{1}{b}$.

- Algebraic fractions are multiplied and divided just like fractions in arithmetic. For all real numbers a, b, c, and d, with b, c, and $d \neq 0$, $\frac{a}{b} \cdot \frac{c}{d} = \frac{ac}{bd}$ and $\frac{a}{b} \div \frac{c}{d} = \frac{\frac{a}{b}}{\frac{c}{d}} = \frac{a}{b} \cdot \frac{d}{c} = \frac{ad}{bc}$.

- Because fractions represent division, all the applications of division in arithmetic lead to applications of fractions in algebra. When quantities with different units are divided, the result is a rate. Rates are often signaled by the word "per" as in students per class, miles per hour, and people per square mile. Rates have rate units, and these units are multiplied and divided just as if they were fractions. Rate units are useful for converting from one unit to another.

- When quantities with the same type of units are divided, the result is a ratio, a number without a unit. Ratios may be represented as fractions, percents, or decimals. The relative frequency of an event is the ratio of the number of times an event has occurred to the number of times it could have occurred. A probability is a number that is the expectation of what the relative frequency ratio would be in the long run. Both relative frequencies and probabilities are numbers from 0 to 1.

- When two fractions, rates, or ratios are equal, the result is a proportion, an equation of the form $\frac{a}{b} = \frac{c}{d}$. So proportions are found wherever there are fractions. Proportions can be solved by applying the Multiplication Property of Equality or by using the Means-Extremes Property. In similar figures, the ratios of corresponding lengths are equal to a ratio of similitude k. Proportions are everywhere when there are similar figures.

Theorems and Properties

Multiplying Fractions Property (p. 252)
Equal Fractions Property (p. 253)
Dividing Fractions Property (p. 258)
Means-Extremes Property (p. 302)
Fundamental Property of Similar Figures (p. 309)

Vocabulary

5-1
algebraic fraction

5-2
complex fraction

5-3
rate
reciprocal rates

5-4
conversion rate

5-5
ratio
tax rate
discount rate

5-6
outcome
event
probability
$P(x)$
probability distribution
unbiased
fair
conditional probability
complement
odds of an event

5-7
relative frequency
pth percentile

5-9
proportion
extremes
means
population
sample
randomly
capture-recapture method

5-10
ratio of similitude

Chapter 5 Self-Test

Take this test as you would take a test in class. You will need a calculator. Then use the Selected Answers section in the back of the book to check your work.

In 1–3, simplify the expression.

1. $\dfrac{75c}{8p} \cdot \dfrac{2}{15c}$
2. $\dfrac{5}{a} \div \dfrac{9}{3a^2}$
3. $\dfrac{\frac{2x}{5}}{\frac{x}{5}}$

4. **Multiple Choice** Which of the following is *not* equal to $\dfrac{a}{b}$?

   A $\dfrac{1}{b} \cdot \dfrac{a}{1}$
   B $6 \div \dfrac{6a}{b}$
   C $\dfrac{a}{b} \cdot \dfrac{c}{c}$
   D $\dfrac{1}{2b} \div \dfrac{1}{2a}$

In 5–7, solve the proportion. Show your work.

5. $\dfrac{h}{17} = \dfrac{6}{101}$
6. $\dfrac{9}{5} = \dfrac{4u}{3}$
7. $\dfrac{x}{8} = \dfrac{2x-3}{24}$

8. Suppose $\dfrac{7}{x} = \dfrac{8}{15}$.
   a. What does the Means-Extremes Property tell you?
   b. Solve for x.

9. Mrs. Wright bought six boxes of pencils, with each box containing 10 pencils to be split among her four children. Writing your answer as a rate, find how many pencils each child received.

10. The Colorado Department of Public Health and Environment reported that for the years 2001–2005, 1,563 bats were examined for rabies. Of that number, 221 actually had rabies. Use this information to estimate the probability that if you see a bat in Colorado, it has rabies.

11. Teresa can run one mile in seven and a half minutes. Using the approximation 1.6 km ≈ 1 mi, how many seconds does it take her to run one kilometer?

12. The two rectangles below are similar with corresponding sides parallel. Solve for x.

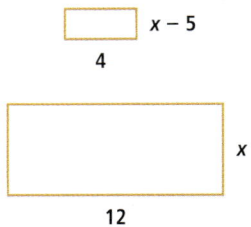

13. If the electricity goes out and a clock stops, what is the probability that the second hand stops between 3 and 4?

14. A DVD costs $18 after a discount of $4.50. What is the percent of discount?

15. In 2003, Markus Riese set a world record by bicycling backwards a distance of 50 kilometers in 1 hour and 47 minutes.
    a. To the nearest kilometer, what was his speed in kilometers per hour?
    b. Write the reciprocal rate to your answer from Part a.

In 16–18, consider the table below that displays the results of a survey of 50 people asked about their types of allergies.

Allergy	Number of People
Only Peanuts	7
Only Bees	16
Both Peanuts and Bees	2

16. What is the relative frequency of people who are allergic to peanuts?

17. What is the relative frequency of people who are allergic to neither peanuts nor bees?

18. What is the relative frequency of people who are allergic to either peanuts or bees, but not both?

19. In the expression $\frac{12k}{k-2}$, k cannot be what value? Explain.

20. A pet store has f goldfish, s snakes, c cats, and d dogs. What is the ratio of the number of cats to the total number of these animals?

21. If a bus travels 350 miles on 20 gallons of gas, about how far can it travel on 35 gallons of gas?

22. A circle with radius 21 inches is contained inside a rectangle that is 4 feet by 7 feet. If a point within the rectangle is chosen at random, what is the probability that it lies inside the circle?

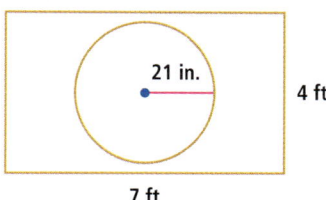

23. The Canadian National (CN) Tower in Toronto is one of the tallest towers in the world. Suppose the tower casts a shadow that is 1,210 feet long, and at the same time a 6-foot-tall man standing next to the tower casts a shadow that is 4 feet long. About how tall is the CN Tower?

24. Suppose an 8th grade gym class recorded the number of sit-ups each student could do in 1 minute. Janice was in the 75th percentile and only 6 students did more sit-ups than she did. How many people are in the gym class?

Chapter 5 Chapter Review

SKILLS
PROPERTIES
USES
REPRESENTATIONS

SKILLS Procedures used to get answers

OBJECTIVE A Multiply and simplify algebraic fractions. (Lesson 5-1)

In 1–4, multiply the fractions. Simplify if possible.

1. $\frac{7a}{2} \cdot \frac{4}{5}$
2. $\frac{12}{5x} \cdot \frac{2x}{3}$
3. $\frac{6}{5p} \cdot 10$
4. $\frac{2ax}{3} \cdot \frac{9x}{a}$

In 5 and 6, simplify the fraction.

5. $\frac{121bcd}{11cd}$
6. $\frac{-24x^3y}{32x^2y^2}$

OBJECTIVE B Divide algebraic fractions. (Lessons 5-2, 5-3)

In 7–10, find the quotient.

7. $\frac{c}{6} \div \frac{9}{2}$
8. $\frac{x}{y} \div \frac{x}{z}$
9. $\frac{\frac{4a}{15}}{-\frac{20a}{9}}$
10. $\frac{\frac{-625x}{50x}}{\frac{50x}{8}}$

In 11 and 12, what value(s) can the variable not have?

11. $\frac{10 + x}{8 + x}$
12. $\frac{28v}{4v - 2.4}$

OBJECTIVE C Solve proportions. (Lesson 5-9)

In 13–16, solve the equation.

13. $-\frac{28k}{5} = \frac{14}{3}$
14. $\frac{6}{y - 4} = \frac{2}{5}$
15. $\frac{3(t - 5)}{4} = \frac{9t}{2}$
16. $\frac{a + 12}{a - 3} = 4$

17. **Fill in the Blank** 6 is to 54 as 54 is to __?__.

PROPERTIES Principles behind the mathematics

OBJECTIVE D Use the language of proportions and the Means-Extremes Property. (Lesson 5-9)

18. Consider the proportion $\frac{7}{8} = \frac{28}{32}$.
    a. Which numbers are the means?
    b. Which numbers are the extremes?

19. If $\frac{6}{15} = \frac{x}{8}$, what does $\frac{8}{x}$ equal?

20. **Fill in the Blanks** If $\frac{a}{b} = \frac{x}{y}$, then by the Means-Extremes Property __?__ = __?__.

21. **Fill in the Blank** If $\frac{m}{n} = \frac{u}{v}$, then $\frac{v}{n} =$ __?__.

USES Applications of mathematics in real-world situations

OBJECTIVE E Use rates in real situations. (Lesson 5-3)

22. Suppose a 225-mile train ride took 3 hours.
    a. What was the rate in miles per hour?
    b. What was the rate in hours per mile?

23. A 16-oz box of pasta costs $1.20 and a 32-oz box of pasta costs $1.80.
    a. Find the unit cost (cost per ounce) for the 16-oz box.
    b. Find the unit cost of the larger box of pasta.
    c. Which is the better buy?

24. It took Jamila 45 minutes to answer 32 questions on her algebra test. On average, how much time did it take her to answer 1 question?

25. Which is faster, typing w words in $2m$ minutes or typing $4w$ words in $6m$ minutes? Explain your answer.

In 26 and 27, calculate a rate for the given situation.

26. In 22 almonds there are about 160 calories.

27. The red oak tree grew 12 feet in 8 years.

OBJECTIVE F Convert units and use reciprocal rates in real situations. (Lesson 5-4)

28. During a meteor shower, some meteors approach Earth's atmosphere at speeds of 95 kilometers per second. Using the fact that 1 mile ≈ 1.6 km, convert this rate into miles per hour.

29. The average human adult at rest takes 16 breaths per minute.
    a. At this rate, how many breaths would a human take in a week?
    b. If a cat takes 1,500 breaths per hour, does the cat or human breathe at a faster rate? Explain your answer.

30. Sliced turkey costs $6.50 per pound and there are 20 slices per pound. How many slices of turkey can you buy for $2.60?

31. It takes Clara 1 min to stuff $4n$ envelopes with letters. Melanie is half as fast.
    a. How many envelopes can Melanie stuff per minute?
    b. How many minutes does Melanie spend per envelope?

32. A halogen bulb can be used exactly six hours per day for a year before burning out. How many hours can the halogen bulb be used?

OBJECTIVE G Use ratios to compare two quantities. (Lesson 5-5)

33. **Multiple Choice** Which of the following is *not* equal to the ratio of 12 to 7?

    A $\frac{12x}{7x}$ B 60:35   C $\frac{24 \text{ ft}}{14 \text{ ft}}$ D 700 to 1,200

In 34 and 35, consider the table below that lists the types and number of televisions in stock at Eli's Electronics Store.

Type of TV	Number of TVs
High-Definition (HD)	12
Flat-Screen	36
Projection	8
Cathode-Ray Tube (CRT)	64

34. What is the ratio of HD televisions to all televisions?

35. What is the ratio of CRT televisions to projection televisions?

36. A pair of shoes that originally cost $53 is on sale for $42.40.
    a. What is the discount rate?
    b. Find the total amount saved, including tax, if the tax rate is 6.25%.

37. To make a certain shade of green paint, a painter mixes 5 parts blue paint with 3 parts yellow paint. If he needs 20 gallons of green paint, how many gallons of each paint color are needed in the mixture?

In 38 and 39, consider the following information. A baseball player's batting average can be viewed as the ratio of total number of hits divided by total number of at-bats. In 2005, Vladimir Guerrero got 20 hits in his first 57 at-bats for a batting average of .351.

38. If Guerrero got only 2 hits in his next 10 at-bats, what would have been his new batting average?

39. How many hits in a row would Guerrero have needed to raise his batting average to at least .400?

Chapter 5

OBJECTIVE H Calculate relative frequencies and probabilities in situations with a finite number of equally likely outcomes. (Lessons 5-6, 5-7)

40. A number is selected randomly from the integers {−1, 0, 1, ..., 8}. What is the probability that the number is greater than 1?

41. A fair die is thrown once. Find the probability of getting an even number greater than 2.

42. If the probability of winning a raffle is $\frac{1}{25,000}$, what is the probability of *not* winning?

43. A study shows that the relative frequency of people who eat cold cereal for breakfast in the United States is 31%. What is the relative frequency of people in the United States who do *not* eat cold cereal for breakfast?

44. Event X has a probability of 42%, event B has a probability of $\frac{5}{12}$, and event C has a probability of 0.45.

    a. Which event is most likely to happen?

    b. Which event is least likely to happen?

OBJECTIVE I Find probabilities involving geometric regions. (Lesson 5-8)

45. A 5-cm square inside a 6-cm square is shown below. If a point is selected at random from the figure, what is the probability that it lies in the shaded region?

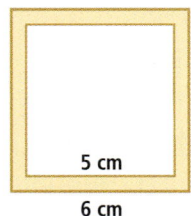

46. Tate drives to work every morning and follows the same route each day. The map below shows his path. One morning Tate runs out of gas while on the way to work. If each point on the map is equally likely, what is the probability that Tate ran out of gas on the highway?

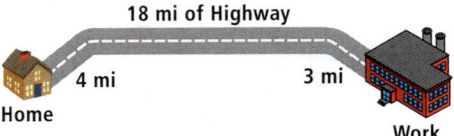

47. A target consists of a set of 4 concentric circles with radii of 4 inches, 8 inches, 12 inches, and 16 inches. The largest circle is inscribed in a square. A person with a bow and arrow randomly shoots at the target so that all points inside the square are equally likely to be hit. The arrow hits somewhere inside the square.

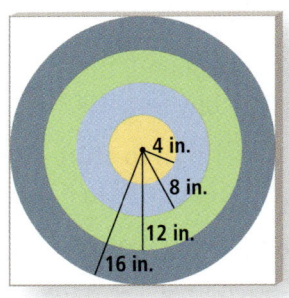

    a. What is the probability that the arrow hits the bull's eye?

    b. What is the probability that the arrow hits within one of the two middle rings (but not within the bull's eye)?

OBJECTIVE J Solve problems involving proportions in real situations. (Lessons 5-9, 5-10)

48. If $\frac{1}{2}$ cup of brown sugar equals 24 teaspoons of brown sugar, how many teaspoons are there in $2\frac{1}{3}$ cups of brown sugar?

49. A school donating money to a charity decides that for every student donation of $5, the school will donate $12. If the total student donation amount is $490, how much money will the school donate?

50. On September 27, 2005, you could buy 10.89 pesos (the currency in Mexico) for one U.S. dollar. If a sombrero cost 290 pesos then, what was its cost in U.S. dollars, rounded to the nearest cent?

51. Suppose a ranger caught, tagged, and released 28 moose in a state park. Two months later, the ranger caught 20 moose, and 14 of these had tags. Based on these findings, estimate the total number of moose in the park.

OBJECTIVE K Interpret the meaning of percentile for benchmarks of 10th, 25th, 50th, 75th, and 90th percentiles. (Lesson 5-7)

In 52–55, a class of 20 students at O'Sullivan High School received the following SAT scores.

2330	2200	1900	1870	2050
1680	1790	1950	2110	2020
1880	1790	2230	2000	1970
2050	1680	1550	1780	1910

52. What is the median of the students' scores?

53. Nolan scored a 1790. At what percentile is he in his class?

54. Tia hopes that her score is at least at the 90th percentile of her classmates. What score must she have for this to be true?

55. How many scores are in each of the following percentiles?

    a. 10th   b. 25th   c. 75th

REPRESENTATIONS Pictures, graphs, or objects that illustrate concepts

OBJECTIVE L Find lengths and ratios of similitude in similar figures. (Lesson 5-10)

56. One rectangular field has dimensions 900 m by 1,200 m; another rectangular field has dimensions 800 m by 1,100 m. Are the fields similar in shape? Explain your reasoning.

57. The quadrilaterals below are similar. Corresponding sides are parallel.

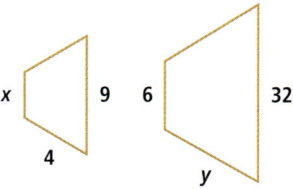

    a. Give the two possible ratios of similitude.
    b. Solve for y.   c. Solve for x.

58. A building casts a shadow that is 480 feet long. A yardstick casts a shadow n feet long at the same time. How tall is the building?

    a. Draw a sketch of the situation.
    b. Show how a proportion can be used to solve the problem.

59. Pentagons $PQRST$ and $VWXYZ$ are similar with ratio of similitude $\frac{4}{5}$. If $VWXYZ$ has area 60 square units, what is the area of $PQRST$?

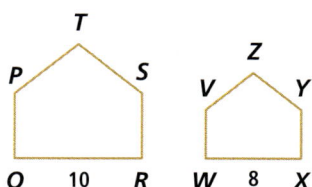

Chapter 6

Slopes and Lines

Contents

- **6-1** Rate of Change
- **6-2** The Slope of a Line
- **6-3** Properties of Slope
- **6-4** Slope-Intercept Equations for Lines
- **6-5** Equations for Lines with a Given Point and Slope
- **6-6** Equations for Lines through Two Points
- **6-7** Fitting a Line to Data
- **6-8** Standard Form of the Equation of a Line
- **6-9** Graphing Linear Inequalities

On the next page is a table of the population of Chicago from 1830 to 2000 according to the United States census. The ordered pairs (year, population that year) are also graphed.

The slopes of the line segments connecting the points indicate how fast the population increased or decreased in each decade. In this chapter, you will study many examples of lines and slopes.

Year	Population
1830	100
1840	4,470
1850	29,963
1860	112,172
1870	298,977
1880	503,185
1890	1,099,850
1900	1,698,575
1910	2,185,283
1920	2,701,705
1930	3,376,438
1940	3,396,808
1950	3,620,962
1960	3,550,404
1970	3,369,357
1980	3,005,072
1990	2,783,726
2000	2,896,016

Source: U.S. Census Bureau

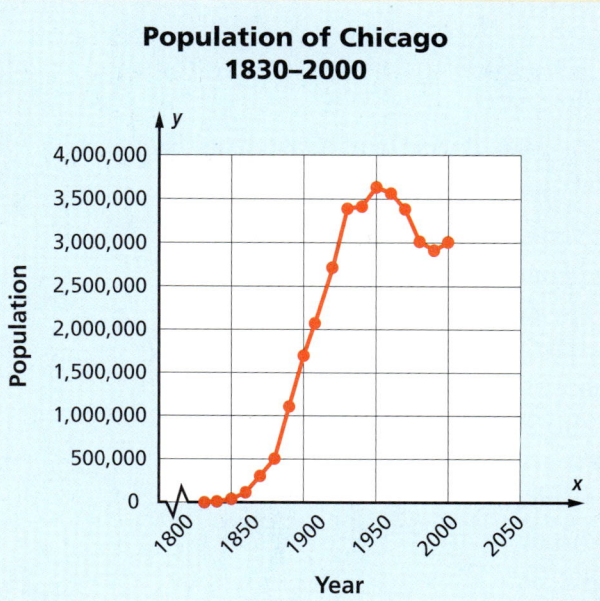

Population of Chicago 1830–2000

Chapter 6

Lesson 6-1
Rate of Change

Vocabulary

rate of change

rate unit

▶ **BIG IDEA** The rate at which a quantity changes can be determined either by computation or by looking at a graph.

What Is a Rate of Change?

Mr. and Mrs. Overjoyed had a healthy baby girl named Joy who weighed 7.5 pounds at birth. At the end of 4 months, the baby weighed 13.5 pounds. How fast did her weight change from birth to 4 months? To answer this question, we calculate the *rate of change* of Joy's weight, that is, how much she gained per month.

From 0 to 4 months

$$\frac{\text{change in weight}}{\text{change in age}} = \frac{13.5 \text{ lb} - 7.5 \text{ lb}}{4 \text{ mo} - 0 \text{ mo}} = \frac{6 \text{ lb}}{4 \text{ mo}} = 1.5 \frac{\text{lb}}{\text{mo}}$$

At age 6 months, Joy weighed 15.75 pounds. How fast did she grow from 4 months to 6 months? Use the same method to calculate the rate of change in her weight per month, from 4 months to 6 months.

From 4 to 6 months

$$\frac{\text{change in weight}}{\text{change in age}} = \frac{15.75 \text{ lb} - 13.5 \text{ lb}}{6 \text{ mo} - 4 \text{ mo}} = \frac{2.25 \text{ lb}}{2 \text{ mo}} = 1.125 \frac{\text{lb}}{\text{mo}}$$

Because 1.5 > 1.125, Joy gained weight at a faster rate from 0 to 4 months than from 4 to 6 months.

These data points have been plotted at the right and connected to make a *line graph*. The **rate of change** measures how fast the graph goes up or down when reading from left to right. Since Joy's rate of change was greater for the 0 to 4 month period than for the 4 to 6 month period, the line segment connecting (0, 7.5) to (4, 13.5) is steeper than the line segment connecting (4, 13.5) to (6, 15.75).

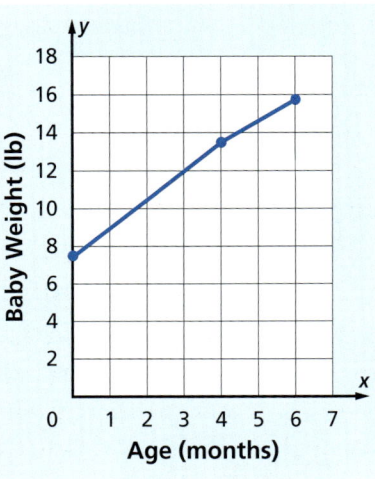

Mental Math

A square chessboard has 64 squares, alternating black and white.

a. How many black squares are there?

b. How many white squares are there?

c. How many squares are on the edge of the board?

In 2003, there were 4,089,950 births in the United States.

Source: National Center for Health Statistics

326 Slopes and Lines

Lesson 6-1

Negative Rates of Change

Joy was gaining weight, so the rate of change was positive. But a rate of change can be negative.

Example 1

The population of Chicago from 1830 to 2000 is shown in the table and graph on page 325. Find the rate of change of the population of Chicago (in people per year) during the given time period, and describe how the rate is pictured on the graph.

a. 1890 to 1900
b. 1970 to 1980

Solutions

a. the rate of change, in people per year, from 1890 to 1900:

$$\frac{\text{change in population}}{\text{change in years}} = \frac{1{,}698{,}575 - 1{,}099{,}850}{1900 - 1890}$$

$$= \frac{598{,}725 \text{ people}}{10 \text{ yr}}$$

$$= 59{,}872.5 \, \frac{\text{people}}{\text{yr}}$$

Between 1890 and 1900 the population increased, so the rate of change is positive. The graph slants upward as you read the graph from left to right.

b. the rate of change between 1970 and 1980:

$$\frac{\text{change in population}}{\text{change in years}} = \frac{3{,}005{,}072 - 3{,}369{,}357}{1980 - 1970}$$

$$= \frac{-364{,}285 \text{ people}}{10 \text{ yr}}$$

$$= -36{,}428.5 \, \frac{\text{people}}{\text{yr}}$$

Between 1970 and 1980 the population decreased, so the rate of change is negative. In that interval, the graph slants downward as you read the graph from left to right.

When you read graphs in algebra, read them from left to right just as you would read a line in a book. A *positive rate of change* indicates that the graph slants *upward*. A *negative rate of change* indicates that the graph slants *downward*.

In both Joy's weight and Chicago's population, the *changes* are found by subtraction. The *rates* are found by division. So, you can calculate the rate of change between two points by dividing the difference in the y-coordinates by the difference in the x-coordinates. We use the subscripts $_1$ and $_2$ to identify the coordinates of the two points. For example, x_1 is read "x one" or "x sub one." The point (x_1, y_1) simply means the *first point*, while (x_2, y_2) means the *second point*.

Chapter 6

Here is the general formula.

> **Rate of Change**
>
> The rate of change between points (x_1, y_1) and (x_2, y_2) is $\dfrac{y_2 - y_1}{x_2 - x_1}$.

Because every rate of change comes from division, the unit of a rate of change is a **rate unit.** In Example 1, a number of people is divided by a number of years. So, the unit of the rate of change is *people per year*, written as $\dfrac{\text{people}}{\text{year}}$ or people/year.

Using a Spreadsheet to Calculate Rate of Change

Spreadsheets and other table generators can be used to calculate rates of change. The spreadsheet below shows the years from 1830 to 2000 in column A, the population of Chicago in column B, and the rate of change of population for the previous decade in column C.

	A	B	C
1	Year	Population	Rate of change for previous decade
2	1830	100	
3	1840	4,470	437
4	1850	29,963	2549.3
5	1860	112,172	8220.9
6	1870	298,977	18680.5
7	1880	503,185	20420.8
8	1890	1,099,850	59666.5
9	1900	1,698,575	59872.5
10	1910	2,185,283	48670.8
11	1920	2,701,705	51642.2
12	1930	3,376,438	674773.3
13	1940	3,396,808	2037
14	1950	3,620,962	22415.4
15	1960	3,550,404	-7055.8
16	1970	3,369,357	-18104.7
17	1980	3,005,072	-36428.5
18	1990	2,783,726	-22134.6
19	2000	2,896,016	11229

10 years

| 1840 | 4,470 |
| 1850 | 29,963 |

25,493 people

Each rate of change is calculated using years and populations from two rows of the spreadsheet. Each formula in column C involves two subtractions, one to find the change in population and one to find the change in years. Then the population change is divided by the change in years. For example, C4 describes the change from 1840 to 1850.

The rate of change is $\frac{25{,}493 \text{ people}}{10 \text{ yr}} = 2{,}549.3 \frac{\text{people}}{\text{yr}}$. The formula to calculate this for C4 is = (B4–B3) / (A4–A3). Notice that cell C2 is empty because there is no population previous to 1830 in column B.

 QY

> ▶ **QY**
>
> Chicago's population in 2005 was 2,842,518.
>
> a. Find the rate of change of Chicago's population from 2000 to 2005.
>
> b. The answer to Part a is negative. Explain what that means in the context of the problem.

A Zero Rate of Change

Sometimes quantities do not change over a certain interval. Then the rate of change is zero.

Example 2

The table below shows estimated attendance by the hour at a professional baseball game that went into extra innings.

a. During what time interval did the attendance not change?

b. Find the rate of change of attendance during that time.

Time	1 P.M.	2 P.M.	3 P.M.	4 P.M.	5 P.M.	6 P.M.	7 P.M.
Attendance	1,200	18,400	23,200	23,200	23,200	20,100	2,000

AT&T Park in San Francisco

Solutions

a. From 3:00 pm to 5:00 pm the attendance did not change. Notice this is associated with a horizontal line segment on the graph at the right.

b. Use the coordinates (3, 23,200) and (5, 23,200) to calculate the rate of change.

rate of change = $\frac{23{,}200 - 23{,}200}{5 - 3} = \frac{0 \text{ people}}{2 \text{ hours}} = 0 \frac{\text{people}}{\text{hour}}$

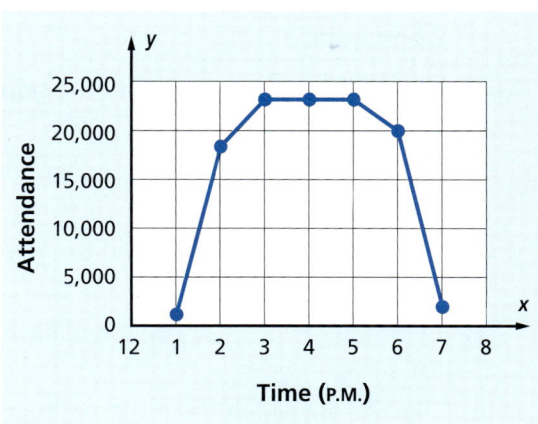

In general, a rate of change of 0 corresponds to a horizontal segment on the graph. The table on page 330 summarizes the relationship between rate of change and the graph.

Chapter 6

Situation	Constant Increase	Constant Decrease	No Change
Rate of Change	positive	negative	zero
Slant (left to right)	upward	downward	horizontal
Sketch of Graph	(graph with line slanting upward)	(graph with line slanting downward)	(graph with horizontal line)

Questions

COVERING THE IDEAS

1. **Fill in the Blanks** A rate of change is given in dollars per hour. It is found by dividing change in ___?___ by change in ___?___.

2. The table at the right shows Jack's weight from birth to age 2.
   a. Make a line graph for this data.
   b. Which is steeper, the segment joining (0, 8) to (4, 15) or the segment joining (6, 19) to (12, 24)?
   c. Which has the greater rate of change, the segment joining (0, 8) to (4, 15) or the segment joining (6, 19) to (12, 24)?
   d. Explain why we would not expect the rate of change of Jack's weight to be negative during his childhood.
   e. Was the rate of change in his weight greater from ages 6 to 12 months or from 12 to 18 months?
   f. What is the rate of change in his weight from birth to 24 months?

Age	Weight (lb)
birth	8
4 mo	15
6 mo	19
12 mo	24
18 mo	27
24 mo	29

3. Refer to the graph and table of the population of Chicago on page 325.
   a. In which decade did Chicago's population grow the fastest?
   b. In which 20-year period did the population of Chicago grow the fastest?
   c. In which decade did the population of Chicago decline the most?
   d. Use the spreadsheet of the population of Chicago on page 328 to find the formula that created cell C10.

330   Slopes and Lines

In 4–8, refer to the graph at the right of attendance at a football game.

4. Identify all one-hour time periods where the rate of change is negative.

5. Did attendance increase or decrease between 2:00 P.M. and 3:00 P.M.?

6. Calculate the rate of change in attendance from 5:00 P.M. to 6:00 P.M.

7. Did the attendance increase, decrease, or stay the same from 6:00 P.M. to 7:00 P.M.?

8. When do you think the game started and ended? Explain your answer.

9. Evaluate the expression $\frac{y_2 - y_1}{x_2 - x_1}$ for the points $(-6, 9)$ and $(-3, 7)$.

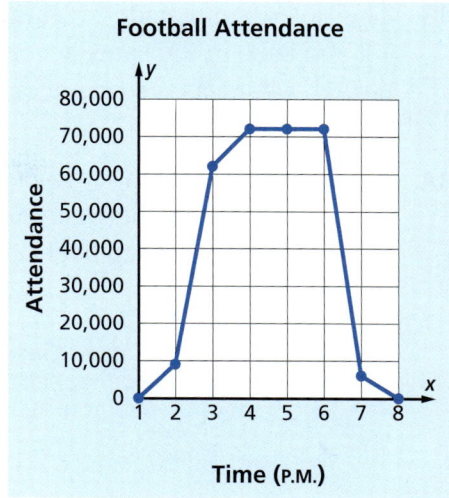

APPLYING THE MATHEMATICS

10. Find the rate of change from point A to point B on the graph at the right.

11. Refer to the table below.

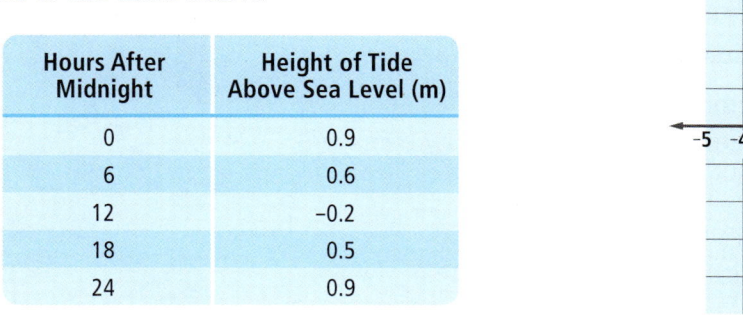

Hours After Midnight	Height of Tide Above Sea Level (m)
0	0.9
6	0.6
12	−0.2
18	0.5
24	0.9

   a. Calculate the rate of change from 12 hours to 18 hours after midnight.

   b. Is the rate of change between 6 and 12 hours after midnight positive, negative, or zero?

12. Refer to the table at the right of the population of Hong Kong, one of the most densely populated places in the world.

   a. In which time period was the rate of population growth per year the greatest?

   b. Hong Kong has an area of about 425 square miles. In 1999, Hong Kong had about how many people per square mile?

Year	Population
2006	6,940,432
1999	6,840,600
1991	5,647,114
1986	5,395,997
1981	4,986,560
1976	4,439,250

Source: *Advertising Age*

13. The Beier family had $2,324 in a checking account on December 1st and $490 in the same account two weeks later. In this time period, what was the rate of change of the amount of money in their account per day?

14. A bulldozer is working on a highway. The graph at the right shows the distance between the bulldozer and a fixed point on the highway measured over time. Use the graph to complete the table.

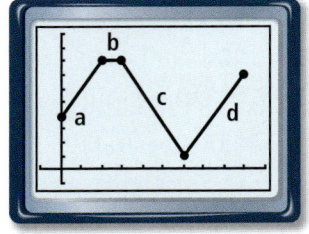

Section of Graph	Rate of Change (positive, negative, or zero)	Bulldozer's Movement (towards point, away from point, standing still)
a	?	away from point
b	?	?
c	?	?
d	positive	?

REVIEW

15. **True or False** The rate *9 apple pies for 15 people* is equal to the rate $\frac{3}{5}$ *apple pie for 1 person*. Explain your answer. (Lesson 5-1)

16. a. Graph $y = 2x + 5$ and $y = -2x + 5$ on the same set of axes.
    b. Where do the two lines intersect? (Lessons 4-3, 3-1)

17. A numismatist (coin collector) has a collection of 850 coins. If the collection grows at 48 coins per year, how many coins will the numismatist have in x years? (Lesson 1-2)

18. Suppose $2x + 5y = 10$. (Lesson 1-1)
    a. Find the value of y when $x = 0$.
    b. Find the value of y when $x = 3$.
    c. Find the value of x when $y = 0$.
    d. Find the value of x when $y = -4$.

The collecting of coins is one of the oldest hobbies in the world.

Source: *Encyclopædia Britannica*

EXPLORATION

19. Find the population of the town you live in for the years 1940, 1950, 1960, 1970, 1980, 1990, and 2000. Put this information into a spreadsheet and calculate the rate of change of population for each decade.
    a. In which 10-year period was the rate of growth greatest?
    b. In which decades was there negative growth? Positive growth? Zero growth?
    c. In which 20-year period was the rate of growth greatest?

QY ANSWERS

a. $-10,699.6 \frac{\text{people}}{\text{yr}}$

b. Chicago's population decreased from 2000 to 2005.

Lesson 6-2
The Slope of a Line

Vocabulary

slope

▶ **BIG IDEA** A line has a constant slope equal to the rate of change between any two of its points.

Constant-Decrease Situation

During a fire drill in a skyscraper with 50 floors, people were asked to move swiftly down the stairwell. To see how much time it would take to empty the building in a real emergency, the evacuation of the people on the top floor was monitored closely. They walked down the stairs at a rate of 5 floors every 2 minutes, or down $2\frac{1}{2}$ floors each minute.

This is a constant-decrease situation. The floor number decreases at a rate of $2\frac{1}{2}$ floors per minute. You can see the constant decrease by graphing the floor of the people after 0, 1, 2, 3, 4, … minutes of walking. Below is a table of ordered pairs (time, floor) charting their progress.

Mental Math

Simplify.

a. $\frac{1}{2}x + \frac{1}{2}x$

b. $\frac{1}{2}x \cdot \frac{1}{2}x$

c. $\dfrac{\frac{1}{2}x}{\frac{1}{2}x}$

Time (min)	Floor
0	50
1	$47\frac{1}{2}$
2	45
3	$42\frac{1}{2}$
4	40
5	$37\frac{1}{2}$
6	35
7	$32\frac{1}{2}$

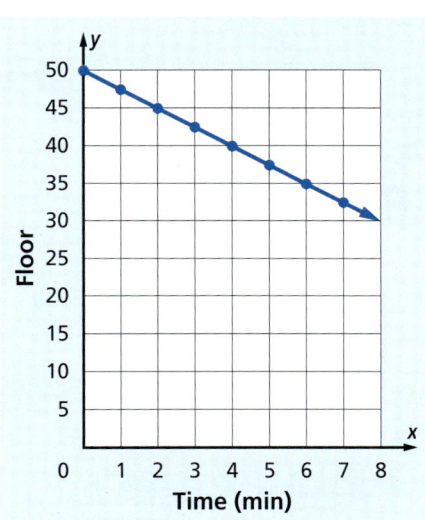

Pick any two points on this line, say (4, 40) and (6, 35). The rate of change between them is shown below.

$$\frac{\text{change in floor}}{\text{change in time}} = \frac{35\text{th floor} - 40\text{th floor}}{6\text{ min} - 4\text{ min}} = \frac{-5 \text{ floors}}{2 \text{ min}} = -2\frac{1}{2}\frac{\text{floors}}{\text{min}}$$

🛑 **QY**

▶ **QY**

Pick two other nonconsecutive points on the line and calculate the rate of change of the floor.

The Slope of a Line

Chapter 6

The Constant Slope of a Line

Notice that all the points on the graph on page 333 lie on the same line. In *any* situation in which there is a constant rate of change between points, the points lie on the same line. This constant rate of change is called the **slope** of the line.

> **Slope**
>
> The slope of the line through (x_1, y_1) and (x_2, y_2) is $\frac{y_2 - y_1}{x_2 - x_1}$.

In contrast, consider the graph at the right. On this graph, no three points lie on the same line, and the rate of change between each pair of the points is different. This was also the case for the graph of the population of Chicago on page 325.

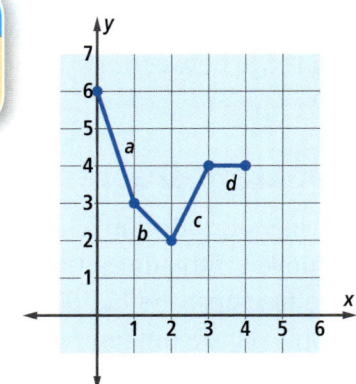

slope of $a = \frac{-3}{1} = -3$ slope of $b = \frac{-1}{1} = -1$

slope of $c = \frac{2}{1} = 2$ slope of $d = \frac{0}{1} = 0$

Example 1
Find the slope of line ℓ below.

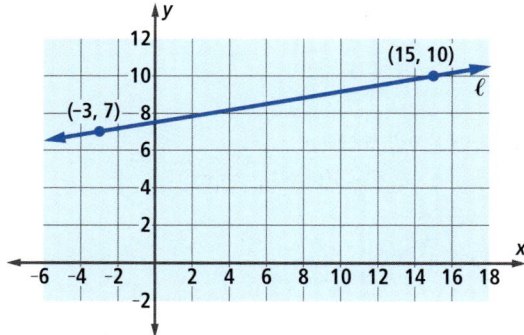

Solution 1 You can choose either point to be (x_1, y_1). The other will be (x_2, y_2). We pick $(x_1, y_1) = (-3, 7)$, and so $(x_2, y_2) = (15, 10)$. Apply the formula for slope.

$$\text{slope} = \frac{y_2 - y_1}{x_2 - x_1} = \frac{10 - 7}{15 - -3} = \frac{3}{18} = \frac{1}{6}$$

Solution 2 Instead, let $(x_1, y_1) = (15, 10)$ and $(x_2, y_2) = (-3, 7)$. Apply the formula for slope.

$$\text{slope} = \frac{y_2 - y_1}{x_2 - x_1} = \frac{7 - 10}{-3 - 15} = \frac{-3}{-18} = \frac{1}{6}$$

Check Because the line is going up to the right, the graph is expected to have a positive slope.

Example 1 shows that when you calculate a slope, it does not matter which of the two points you consider as (x_1, y_1) and which as (x_2, y_2). The slope is the same.

Calculating Slope from a Graph

The subtractions in the slope formula allow you to find the vertical and horizontal change between any two points. This can also be done from a graph simply by counting the units of change in each direction.

Example 2
In the coordinate grid below, each square of the grid is one unit. Find the slope of the line.
a. $\overleftrightarrow{AC}$
b. $\overleftrightarrow{DF}$

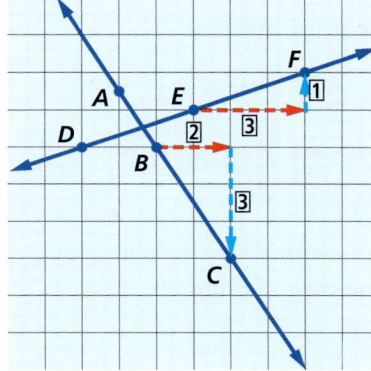

Solutions

a. Pick two points on $\overleftrightarrow{AC}$. We pick points B and C. The change in x-coordinates reading from left to right (B to C) is 2 units. The change in the y-coordinates is −3 units.

$$\text{slope of } \overleftrightarrow{AC} = \frac{-3}{2} = -\frac{3}{2}$$

b. We pick points E and F. The change in the x-coordinates reading from left to right (E to F) is 3 units. The change in the corresponding y-coordinates is 1 unit.

$$\text{slope of } \overleftrightarrow{DF} = \frac{1}{3}$$

Lines with negative slopes, such as $\overleftrightarrow{AC}$ in Example 2, go downward as they move to the right. Lines with positive slopes, such as $\overleftrightarrow{DF}$ in Example 2, go upward as they move to the right. If a line is horizontal, it is neither increasing or decreasing, and likewise the slope is zero.

If the rate of change, or slope, is the same for a set of points, then all the points lie on the same line. If the rate of change is different for different parts of a graph, then the graph is *not* a line.

Chapter 6

> **GUIDED**
>
> ### Example 3
> a. Show that (−2, 5), (2, 3), and (−10, 9) lie on the same line.
>
> b. Give the slope of that line.
>
> **Solutions**
>
> a. Pick pairs of points and calculate the slope between them.
>
> The slope between (−2, 5) and (2, 3) is $\frac{?-5}{2-?} = \underline{}$.
>
> The slope between (2, 3) and (−10, 9) is $\frac{9-?}{-10-?} = \underline{}$.
>
> The slope between (−10, 9) and (−2, 5) is $\frac{?-?}{?-?} = \underline{}$.
>
> Because the slope between any pair of the given points is $\underline{}$, the points lie on the same line.
>
> b. The slope of the line is the constant rate of change, which is $\underline{}$.

Finding the Slope of a Line from an Equation for the Line

You can find the slope of any line given its equation. Use the equation to find two points on the line and calculate the rate of change between them.

> ### Example 4
> Find the slope of the line with equation $2x - 5y = 4$.
>
> **Solution** First, find two points that satisfy the equation. Pick a value for x or y and then substitute it into $2x - 5y = 4$ to find the value of the other variable. To make the calculations easier, we let $x = 0$ for the first point and $y = 0$ for the second point.
>
> | Let $x = 0$. | Let $y = 0$. |
> |---|---|
> | $2 \cdot 0 - 5y = 4$ | $2x - 5 \cdot 0 = 4$ |
> | $0 - 5y = 4$ | $2x - 0 = 4$ |
> | $\frac{-5y}{-5} = \frac{4}{-5}$ | $\frac{2x}{2} = \frac{4}{2}$ |
> | $y = -\frac{4}{5}$ | $x = 2$ |
> | The point $(0, -\frac{4}{5})$ is on the line. | The point $(2, 0)$ is on the line. |
>
> Use the points $(0, -\frac{4}{5})$ and $(2, 0)$ in the slope formula.
>
> $$\text{slope} = \frac{0 - \left(-\frac{4}{5}\right)}{2 - 0} = \frac{\frac{4}{5}}{2} = \frac{4}{5} \cdot \frac{1}{2} = \frac{2}{5}$$

Check Find a third point on the line $2x - 5y = 4$. For example, we let $y = 2$ and find that $x = 7$. Thus the point $(7, 2)$ is on the line, as shown on the graph at the right. Now calculate the slope determined by $(7, 2)$ and one of the points you used to find the original slope, say $(2, 0)$. This gives $\frac{2-0}{7-2}$, which equals $\frac{2}{5}$. So all three points give the same slope. It checks.

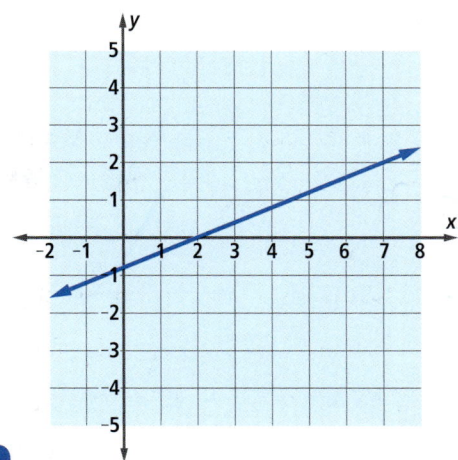

Questions

COVERING THE IDEAS

1. **Fill in the Blank** In a constant-increase or constant-decrease situation, all the points lie on the same ___?___.

2. What is the constant rate of change between any two points on a line called?

3. The table at the right lists the height of a burning birthday candle in centimeters based on time in seconds. The table shows the ordered pairs (time, height), and the relationship is linear.
    a. Find two more ordered pairs.
    b. Find the rate of change between any two points in the table.
    c. Explain the real-world meaning of your answer to Part b.

Time (sec)	Height (cm)
0	5.7
1	5.4
2	5.1
3	4.8
4	4.5
5	4.2

4. The table at the right gives coordinates of points on a line.
    a. If the pattern in the table is continued, what ordered pair should be entered in the next row?
    b. What is the slope of the line?

x	y
6	23
8	29
10	35
12	41
14	47
?	?

In 5 and 6, calculate the slope of the line through the given pair of points.

5. $(0, 4)$ and $(3, 19)$
6. $(4, 3)$ and $(-6, 8)$

7. a. Calculate the slope of the line through $(2, 1)$ and $(5, 11)$.
   b. Calculate the slope of the line through $(5, 11)$ and $(-4, -18)$.
   c. Do the points $(2, 1)$, $(5, 11)$, and $(-4, -18)$ lie on the same line? How can you tell?

The Slope of a Line   337

Chapter 6

In 8 and 9, refer to the graph below. Find the slope of the line.

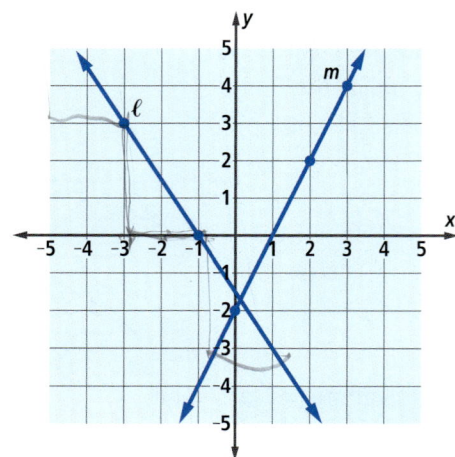

8. line ℓ 9. line m

In 10 and 11, an equation for a line is given.

a. Find two points on the line.
b. Find the slope of the line.
c. Check your work by graphing the line.

10. $y = -\frac{4}{3}x + 4$ 11. $6x - 5y = 30$

In 12 and 13, a graph is given.

a. Find the slope of the line.
b. Describe its real-world meaning.

12.
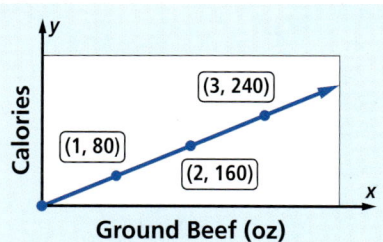

13.
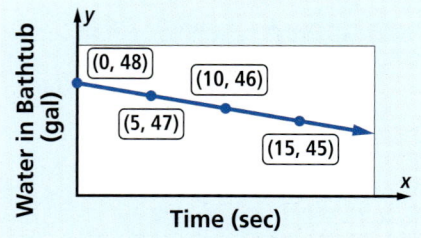

338   Slopes and Lines

APPLYING THE MATHEMATICS

14. When the fire drill described on page 333 was over, the people returned to the top floor of the building, but this time via the elevator. From the 26th floor, the elevator went up 1 floor every 3 seconds. (Notice the elevator's time is given in seconds, not minutes as with the fire drill.)
    a. Make a table and graph the people's progress for the first 6 seconds of the elevator ride from the 26th floor up using ordered pairs (seconds ridden, floor number).
    b. Find the rate of change between any two points on the graph.
    c. How many seconds will it take for them to reach the 50th floor after leaving the 26th floor?

15. Suppose you graphed the data below.

Time (x)	Temperature (y)
8 A.M.	62°F
9 A.M.	63°F
10 A.M.	67°F
11 A.M.	71°F
12 P.M.	73°F

    a. What is the unit of the rate of change?
    b. Describe the real-world meaning of the rate of change.

16. Use the figure at the right. Which line could have the indicated slope?
    a. slope $\frac{3}{5}$
    b. slope $-\frac{3}{5}$
    c. slope 0

17. The points (5, 7) and (4, y) are on a line with slope –4. Find the value of y.

18. Coordinates of points are given in columns A and B in the spreadsheet at the right with a partially blank column for the rate of change between consecutive points.
    a. Complete the rate of change column, leaving C2 blank.
    b. Do the points lie on a line? Explain how you know.

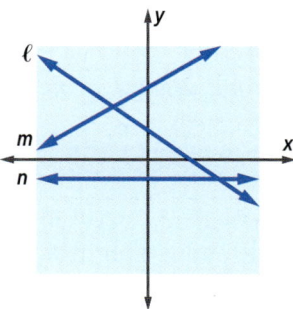

	A	B	C
1	x	y	rate of change
2	-4	22	
3	0	10	$\frac{10 - 22}{0 - -4} = \frac{-12}{4} = -3$
4	5	-5	
5	11	-23	

Chapter 6

19. In Lesson 1-4, the following sequence of toothpick designs was examined. A graph of ordered pairs (term number, number of toothpicks) is shown at the right.

   1st   2nd   3rd   4th

   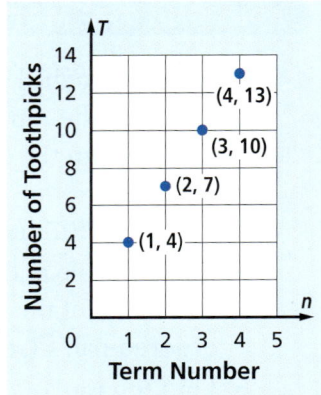

   a. What is the rate of change of the line through these points? (Remember to include the rate unit.)
   b. Let n be the term number and T be the number of toothpicks needed. Then $T = 1 + 3n$. Verify that the slope of the line with equation $T = 1 + 3n$ is the same as your answer to Part a.
   c. What does the slope mean in this situation?

REVIEW

20. A high-rise building was built at a constant rate. The builders completed the 18th floor on the 24th day of construction, and they finished the 81st floor on the 108th day of construction. Calculate the rate of construction in terms of floors per day. **(Lesson 5-3)**

21. Solve $-5 + 8p < 3(2p - 2) + 2p$. **(Lesson 4-5)**

22. Compute in your head. **(Lesson 4-1)**
    a. 16 is what percent of 64?   b. 75% of 80 is what number?

23. **Fill in the Blank** If $5a = \frac{1}{2}b$, then $20a =$ __?__. **(Lesson 2-8)**

24. Determine whether $(x + y)(x + y)(x + y) = x^3 + y^3$ is *always*, *sometimes but not always*, or *never* true. Explain. **(Lesson 1-3)**

EXPLORATION

25. Find a record of your height at some time over a year ago. Compare it with your height now. How fast (in inches or centimeters per year) has your height been changing from then until now?

A woman measures her daughter's height.

QY ANSWER

Answers vary. Sample answer: (5, 37.5) and (3, 42.5):
$\frac{-5 \text{ floors}}{2 \text{ min}} = -2.5 \frac{\text{floors}}{\text{min}}$

Lesson 6-3

Properties of Slope

▶ **BIG IDEA** The slope of a line can be determined by examining the graph; only one line through a point has a particular slope.

Most houses have slanted, or pitched, roofs. The *pitch* of a roof is its slope. In the United States, the pitch is usually measured in 12ths. Why are 12ths used? Because there are 12 inches in a foot. The number of 12ths tells you the number of inches the roof rises for each foot that the roof goes across. In areas where there is a lot of snow, roofs often have higher pitches. The steepest roof a carpenter can safely walk on without the aid of ropes has a pitch of $\frac{8}{12}$.

Mental Math

Consider the data set {3, 5, 9, 1, 7, 10, 16, 25, 24, 3, 18}. Find the

a. range.

b. mode.

c. median.

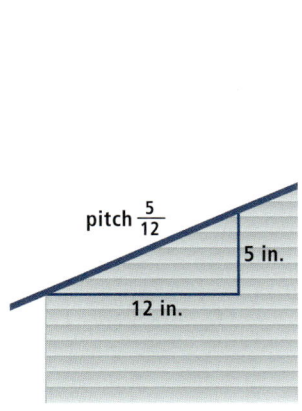
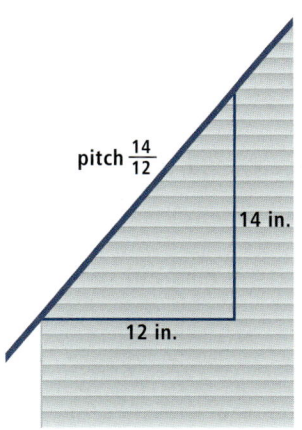

A roof with a pitch of $\frac{5}{12}$ goes up 5 inches for each 12 inches the roof goes across. But this ratio can also be expressed a different way. If the two inch measurements are converted to feet, the result is shown below.

$$\frac{5 \text{ in.}}{12 \text{ in.}} = \frac{5 \text{ in.} \cdot \frac{1 \text{ ft}}{12 \text{ in.}}}{12 \text{ in.} \cdot \frac{1 \text{ ft}}{12 \text{ in.}}} = \frac{\frac{5}{12} \text{ ft}}{1 \text{ ft}} = \frac{5 \text{ in.}}{1 \text{ ft}}$$

So a roof with a pitch of $\frac{5}{12}$ goes up 5 inches ($\frac{5}{12}$ of a foot) for every foot the roof goes across.

The situation with roofs illustrates an important property of the slope of a line. The slope of a line is the amount of change in the height of the line for every change of one unit to the right.

Properties of Slope 341

Chapter 6

For example, if a line has slope 2, as you move one unit to the right, the line goes up 2 units. So, if the line contains the point (a, b), it will also contain the point $(a + 1, b + 2)$.

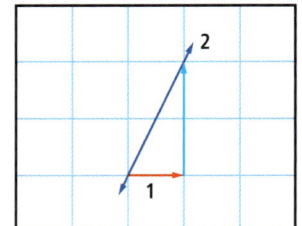 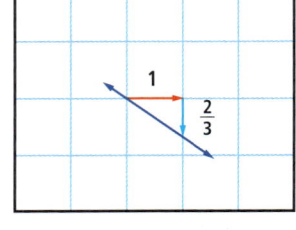

If the slope is $-\frac{2}{3}$, for every change of 1 unit to the right, the line goes down $\frac{2}{3}$ of a unit. So, if the line contains the point (a, b), it will contain the point $(a + 1, b - \frac{2}{3})$.

🛑 QY1

> ▶ QY1
>
> Find another point on the line through (20, 80) that has a slope of $\frac{3}{4}$.

Graphing a Line by Using Its Slope

If you know one point on a line, you can find a second point using its slope. With the two points, you can graph the line.

Example 1
a. Graph the line through (−3, 1) with slope $\frac{1}{4}$.
b. Name another point on the line with integer coefficients.

Solutions

a. Plot (−3, 1), then move right 1 unit and up $\frac{1}{4}$ unit. Plot the resulting point $(-2, 1\frac{1}{4})$, and draw the line through the two points.

b. One point with integer coordinates (−3, 1) is given. Continue plotting points by moving right 1 unit and up $\frac{1}{4}$ unit until you reach another point with integer coordinates. As shown on the graph, the point (1, 2) is also on the line.

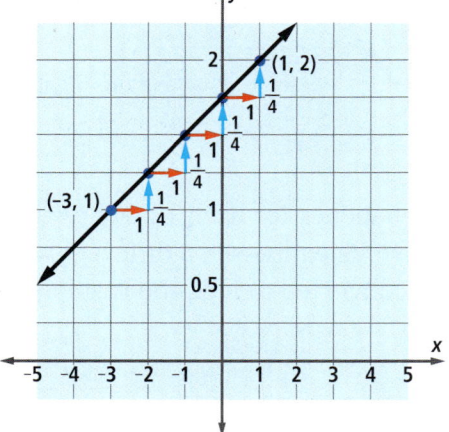

🛑 QY2

> ▶ QY2
>
> Find another point with integer coordinates on the line that contains (2, 6) and has a slope of $\frac{1}{2}$.

342 Slopes and Lines

The idea of Examples 1 and 2 helps in drawing graphs of constant-increase or constant-decrease situations.

Example 2

A certain stacking chair is 31 inches tall. In a stack, each chair adds 4 inches to the height of the stack. Graph the relationship of the number of chairs to the height of the stack.

Solution The first point is (1, 31). Because the height increases 4 inches for each chair stacked, the slope is 4 inches per chair. The points are on a line because the situation is a constant-increase situation.

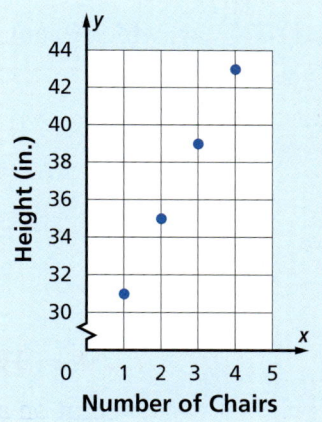

An equation for the height h of n stacking chairs is $h = 31 + 4(n - 1)$. This is an equation for the line that contains the points graphed in Example 2. Notice that the given numbers 31 and 4 both appear in this equation. In the next lesson, you will learn how the slope can be easily determined from an equation of a line.

Zero Slope and Undefined Slopes

Lines with slope 0 can be drawn using the method of the previous two examples.

Example 3

Draw a line through (2, 4) with slope 0.

Solution Plot the point (2, 4). From this point move one unit right and 0 units up. Place a point there and draw a line through the points. You should get a horizontal line.

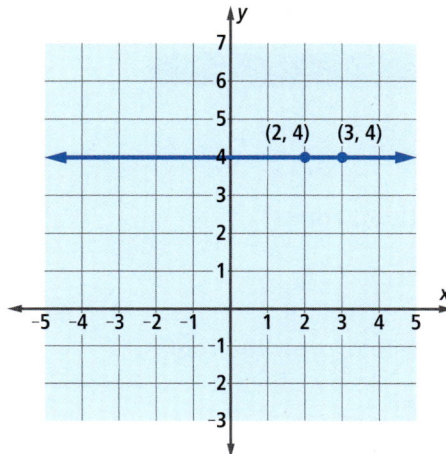

Properties of Slope

Chapter 6

There is one type of line for which the methods used in Examples 1–3 do not work. If you have points on a vertical line, you cannot move one unit to the right and stay on the line. Also, if you try to calculate the slope using the formula $\frac{y_2 - y_1}{x_2 - x_1}$, the denominator will be 0. For example, the slope of a line through (3, 4) and (3, 6) would be $\frac{6 - 4}{3 - 3} = \frac{2}{0}$, which is not defined. Thus the slope of a vertical line is *undefined*.

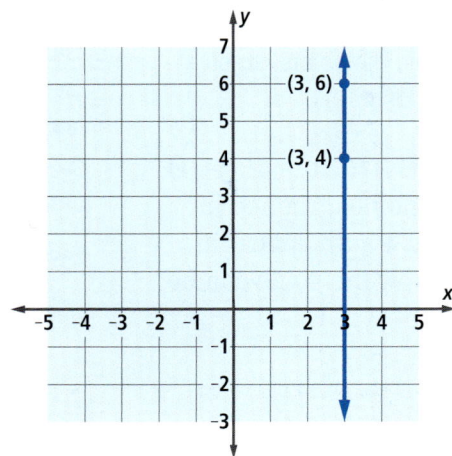

Notice the important difference between slopes of horizontal and vertical lines.

Slope of Horizontal and Vertical Lines

1. The slope of every horizontal line is 0.
2. The slope of every vertical line is undefined.

Questions

COVERING THE IDEAS

1. Suppose the pitch of a roof is $\frac{9}{12}$.
   a. Draw such a roof.
   b. Is this roof safe for a roofer to walk on without ropes? Why or why not?
   c. What is the slope of this roof?

In 2 and 3, a roof pitch is given. If the pitch is possible, draw the roof. If it is not possible, explain why.

2. pitch $\frac{0}{12}$
3. pitch $\frac{12}{0}$

4. **Fill in the Blanks** The slope of a line is the amount of change in the __?__ of the line for every change of __?__ to the right.

344 Slopes and Lines

In 5 and 6, find the slope of the line.

5.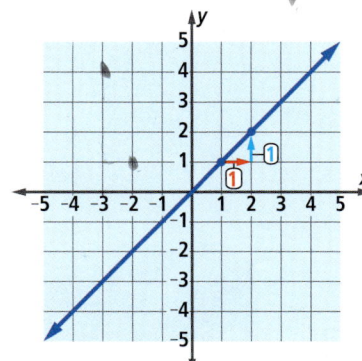

6.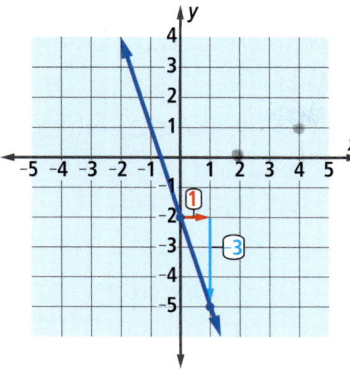

In 7 and 8, a line's slope and a point on it are given.
a. Graph the line.
b. Find one other point on this line.
c. Check that your second point is on the line using the slope formula.

7. point (2, 2), slope 4

8. point (4, −2), slope $-\frac{2}{3}$

In 9 and 10, find another point on the line with the given slope and containing the given point.

9. point (−3, 4), slope −3

10. point (4, 1), slope 0

11. Explain why the slope of a vertical line is not defined. Refer to a specific example that is not in this lesson.

12. Graph the line through (−1, 4) with an undefined slope.

13. A small plastic bathroom drinking cup is about 5.70 cm tall. These cups stack tightly and each cup adds only about 0.28 cm to the stack. Graph the relationship between the number of cups and the height of the stack.

APPLYING THE MATHEMATICS

In 14–16, the slope of a line is given, as are some coordinates of points on the line. Complete the remaining entries in the table.

14. slope $\frac{1}{5}$

x	y
4	8
5	?
6	?
7	?

15. slope −4

x	y
2	0
3	?
4	?
5	?

16. Slope is undefined.

x	y
−3	4
−2	?
−1	?
0	?

Properties of Slope

In 17 and 18, the question relates to the slope of a road. In the United States, the slope of a road is often called its *grade*. The grade is often given in percent. In the British Commonwealth, the word for slope is *gradient*.

17. The grade of Canton Avenue in Pittsburgh, Pennsylvania between Coast and Hampshire Streets is 37%. How much does that street rise for every 100 feet horizontally?

18. Baldwin Street in Dunedin, New Zealand has a maximum gradient of 1 in 2.86. This means that for every horizontal change of 2.86 units, the road goes 1 unit up. What is the slope of Baldwin Street? (*Caution:* The slope is not 2.86.)

A house sits on the steepest residential street in the world in Dunedin, New Zealand.

19. On a piece of graph paper, plot the point (0, 0). Draw the lines through (0, 0) with the following slopes on the same coordinate grid.

    line *a*: slope 3   line *b*: slope −3
    line *c*: slope $\frac{1}{3}$ line *d*: slope $-\frac{1}{3}$

    a. Which line(s) are slanted upward as you read from left to right?
    b. Which line(s) are slanted downward as you read from left to right?
    c. Can you relate the slant of the line to the sign of the slope?
    d. Which line(s) are steepest?
    e. Give the value of a slope that is steeper than the slope of any of the four lines on the graph.

20. In Tom Clancy's best selling novel, *The Hunt for Red October*, the American attack submarine *Dallas* is chasing the renegade Russian submarine *Red October*. The captain of the *Dallas* gave orders to descend to 1,200 feet below sea level. Engineer Butler "watched the depth gauge go below 600 feet. The diving officer would wait until they got to 900 feet before starting to level off, the object being to *zero the dive* out at exactly 1,200 feet." Explain, in terms of slope, what is meant by the italicized phrase.

Operators of a submarine control its buoyancy, allowing it to sink and surface at will.
Source: science.howstuffworks.com

REVIEW

21. Find the slope of the line with equation $12x - 8y = 4$. (**Lesson 6-2**)

22. The cost to rent a movie from Marcus's Movies is $2.99 for two nights with a late fee of $0.99 per day late. After *x* days late, the total cost will be *y* dollars. (**Lessons 6-2, 3-1, 1-2**)
    a. Write an equation relating *x* and *y*.
    b. Graph the line.
    c. Find the rate of change between any two points on the graph.

In 23 and 24, use the table and graph below. They show the unemployment rate, the percent of the United States labor force that was unemployed, for men and women. (Lesson 6-1)

Unemployment Rate		
Year	Men	Women
2000	3.9	4.1
2001	4.8	4.7
2002	5.9	5.6
2003	6.3	5.7
2004	5.6	5.4

Source: *Statistical Abstract of the United States*

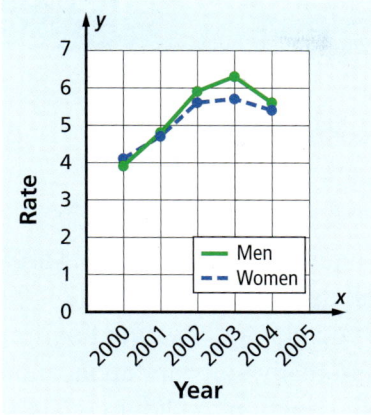

23. Between which two years was there the greatest increase in unemployment rate for
    a. men?
    b. women?

24. What was the rate of change from 2001 to 2003 for
    a. men?
    b. women?

25. Find b if $y = mx + b$, $y = -10$, $m = \frac{3}{4}$, and $x = 5$.
    (Lessons 4-7, 1-1)

26. If the quadrilateral at the right has perimeter 70, what is n?
    (Lessons 3-4, 2-2)

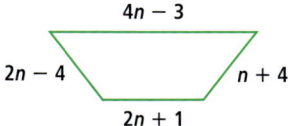

EXPLORATION

27. Canton Avenue and Baldwin Street, the roads in Questions 17 and 18, are among the steepest streets in the world. Search the Internet to find the grades of two other steep streets. Then make an accurate drawing indicating steepness of these roads.

28. Sam Saw wants to build a shed with a $\frac{5}{12}$ roof and the base of the roof being 12 feet long. To support the roof he needs to place a vertical support every 16 inches along the base. One way to do this is to climb the ladder, measure the length needed, go down the ladder and cut the board, then go back up the ladder to nail the board in place. Use your knowledge of slope to calculate the length of all the vertical supports so Sam does not need to go up and down the ladder constantly.

QY ANSWERS

1. Answers vary.
Sample answer: $(21, 80\frac{3}{4})$

2. Answers vary.
Sample answer: $(4, 7)$

Chapter 6

Lesson

6-4 Slope-Intercept Equations for Lines

Vocabulary

y-intercept
slope-intercept form
direct variation

▶ **BIG IDEA** The line that contains the point (0, b) and slope m has equation $y = mx + b$.

You have already worked with equations whose graphs are lines. In the following activity, you will use *dynamic graphing software* to experiment with equations. As you change the equation, watch how its graph changes in response. In many software applications, *sliders* allow you to change values in an equation. A slider consists of a portion of a number line that is used to control numeric values for a specific variable. As you drag a point along a slider, the value of a corresponding variable changes automatically.

Mental Math

Express as a mixed number.

a. 175 cm in meters

b. 35 ounces in cups

c. 5 feet, 7 inches in feet

Activity

Step 1 Create a slider for the variable *m*. Set the software so that *m* can vary in the interval $-5 \leq m \leq 5$. Move the slider so that $m = 2$.

Step 2 Enter the equation $y = mx$. Since $m = 2$, you have created the graph of $y = 2x$.

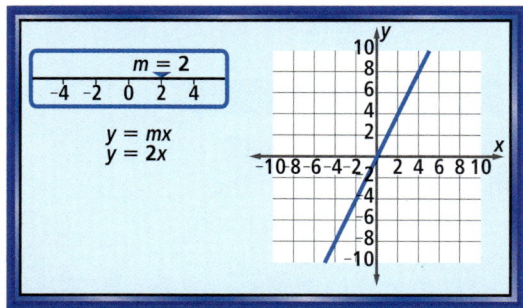

Step 3 Slowly move the slider to increase the value of *m*. What happens to the graph as *m* increases?

Step 4 Now move the slider so $m = 0$. What is the graph like when $m = 0$? Write an equation for this line.

Step 5 Move the *m* slider to the left of zero (into the negatives). What happens to the graph when *m* is negative?
What happens to the graph as *m* moves more and more to the left ("farther" into the negative values)?

348　Slopes and Lines

What appears to be true about the line when m is

a. positive?

b. negative?

c. zero (which is neither positive nor negative)?

Step 6 Move the slider so that $m = 2$. Enter the equation $y = mx + b$ and create a second slider for the variable b, using the interval $-5 \leq m \leq 5$. Move this slider so $b = 1$. Also, graph the point $(0, b)$.

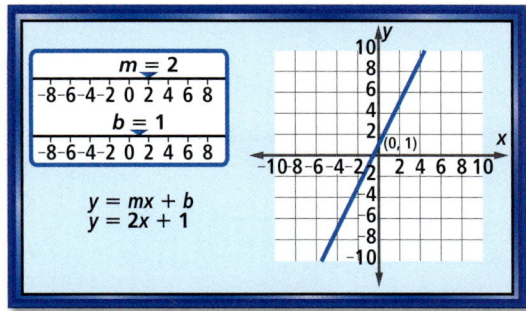

Step 7 Slowly move the b slider to the right (toward greater values of b). What happens to the graph as b increases?

Step 8 Now move the b slider so $b = 0$. Describe the graph when $b = 0$ and write the equation for this line.

Step 9 Slide the b slider to the left of zero (into the negatives). What happens to the graph when b is negative? What happens to the graph as b moves farther to the left? What appears to be true about the line when b is

a. positive?

b. negative?

c. zero (which is neither positive nor negative)?

Step 10 Refer to the graphs below. Move both sliders until you have a graph that resembles the line pictured. Give the values of m and b. Write the equation for the line that is shown by using your slider values.

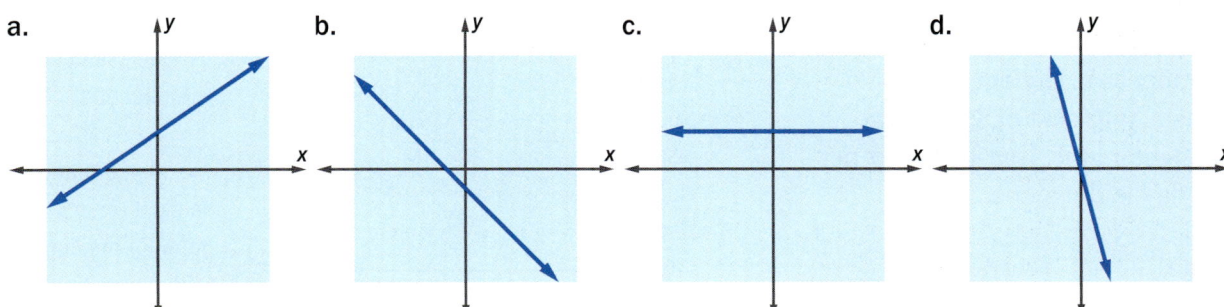

Chapter 6

In the Activity, you should have seen that the graph of a line whose equation is of the form $y = mx + b$ is determined by the values of m and b. In Steps 1–5, as you changed the value of m, the slope of the line changed.

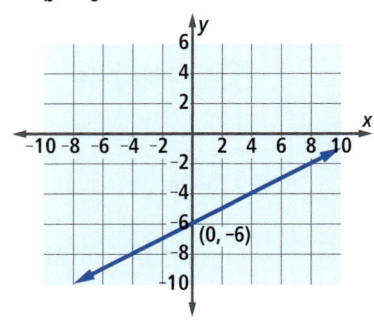
$b = -6$

When $m = 2$ and you varied the value of b in Steps 6–10, the line shifted and the point at which it crossed the y-axis changed. That number is the *y-intercept* of the line. In general, when a graph intersects the y-axis at the point $(0, b)$, the number b is a **y-intercept** for the graph. Each line at the right has slope $\frac{1}{2}$, but the y-intercepts are different.

So the equation $y = mx + b$ shows the slope of the line and its y-intercept. For this reason, $y = mx + b$ is called the **slope-intercept form** for an equation for a line.

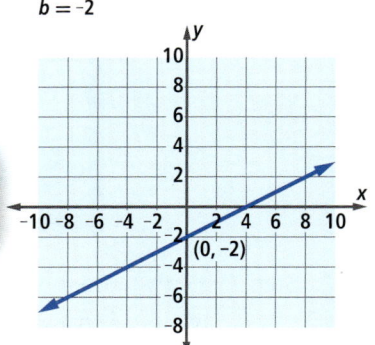
$b = -2$

> **Slope-Intercept Equation of a Line**
>
> The line with equation $y = mx + b$ has slope m and y-intercept b.

 QY1

Using the Slope-Intercept Form of the Equation of a Line

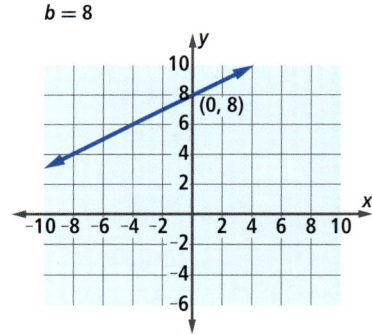
$b = 8$

Slopes and y-intercepts can be seen in equations of real-world situations. Suppose Jody is going on a vacation to Istanbul. The airfare is $940 and she expects to spend $300 per day for hotel and expenses. Then after x days, the total cost y of her trip is given by $y = 940 + 300x$. With the Commutative Property of Addition, $y = 940 + 300x$ becomes $y = 300x + 940$, which fits the slope-intercept form $y = mx + b$.

Notice how the two key numbers in Jody's expenses appear in the equation. Her starting cost of $940 is the y-intercept. Her $300 cost per day is the slope.

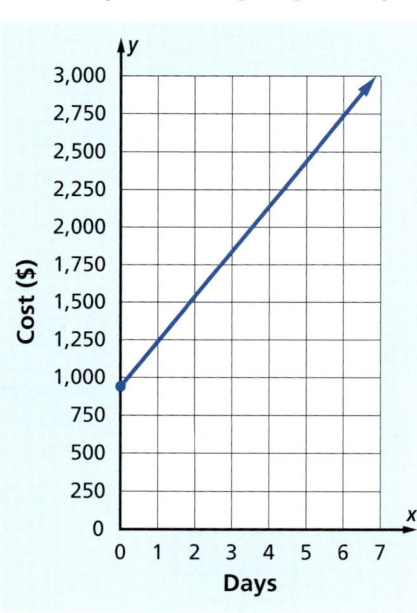

> ▶ **QY1**
>
> a. Find the slope and y-intercept of $y = \frac{1}{4}x + 11$.
>
> b. Write an equation in slope-intercept form for the line with slope 3 and y-intercept of −7.

350 Slopes and Lines

The same line can have many equations. For example, $3x + y = 7$, $30x + 10y = 70$, and $y = -3x + 7$ are all equations for the same line.

 QY2

> ▶ QY2
>
> Give the slope and y-intercept of $y = -2 - 6x$.

Slope-intercept equations allow us to get information about a line quickly. For this reason, it is often helpful to convert other equations for lines into slope-intercept form.

Example 1

Write the equation $3x + 7y = 9$ in slope-intercept form. Give the slope and y-intercept of the graph.

Solution Solve $3x + 7y = 9$ for y.

$$3x + 7y = 9$$
$$7y = -3x + 9$$
$$\frac{7y}{7} = \frac{-3x + 9}{7}$$
$$y = -\frac{3}{7}x + \frac{9}{7}$$

The slope is $-\frac{3}{7}$. The y-intercept is $\frac{9}{7}$, or $1\frac{2}{7}$.

Recall that every vertical line has an equation of the form $x = h$, where h is a fixed number. Equations of this form clearly cannot be solved for y. Thus, equations of vertical lines cannot be written in slope-intercept form (and they cannot be graphed on many graphing calculators). This confirms that the slope of vertical lines cannot be defined.

Writing an Equation for a Line from Its Graph

The graph of a line gives information that can be used to write its equation.

Example 2

Find the equation of the line graphed at the right.

Solution The graph crosses the y-axis at -25, so -25 is the y-intercept. The line contains $(0, -25)$ and $(10, -10)$. Its slope is found below.

$$m = \frac{-25 - -10}{0 - 10} = \frac{-15}{-10} = 1.5$$

The slope-intercept equation of the line is $y = 1.5x - 25$.

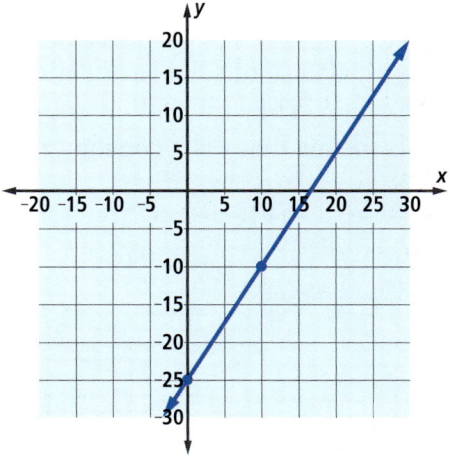

(continued on next page)

Slope-Intercept Equations for Lines

Check Do the coordinates of the two known points (0, −25) and (10, −10) satisfy the equation?

Does −25 = 1.5(0) − 25?   Does −10 = 1.5(10) − 25?
   −25 = 0 − 25   −10 = 15 − 25

Yes, both points check.

Every constant-increase or constant-decrease situation can be described by an equation whose graph is a line. The *y*-intercept of that line can be interpreted as the starting amount. The slope of that line is the amount of increase or decrease per unit.

Example 3

Assume that a skydiver opens a parachute and falls at a speed of 10 feet per second from 5,000 feet above the ground.

a. Find a slope-intercept equation for the height *h* of the skydiver after *x* seconds.

b. How high is the skydiver after 5 minutes (300 seconds)?

Solutions

a. The skydiver falls at 10 feet per second, so the slope is −10. The *y*-intercept is the starting height, 5,000 feet.

$y = 5{,}000 - 10x$ Write the equation.
$y = -10x + 5{,}000$ Rewrite in slope-intercept form.

b. Use the equation and substitute 300 for *x* because 5 minutes equals 300 seconds.

$y = 5{,}000 - 10(300)$
$y = 2{,}000$

After 5 minutes, the height of the skydiver is 2,000 feet.

Direct Variation

A special case of a linear equation occurs when the *y*-intercept is at the origin. Then the *y*-intercept is 0 and $y = mx + 0$ becomes $y = mx$. This means that *y* is a constant multiple of *x*, as shown below.

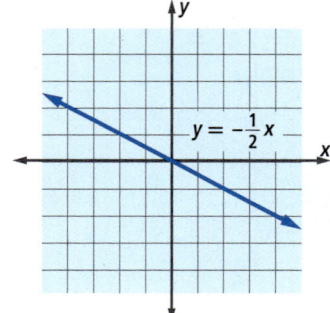

x	y
−4	2
0	0
1	$-\frac{1}{2}$
10	−5

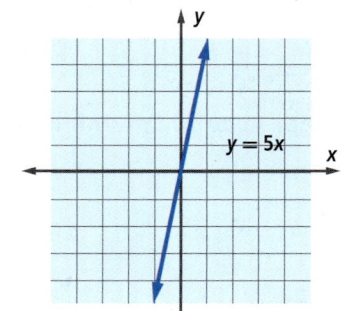

x	y
−2	−10
0	0
3	15
7	35

352 Slopes and Lines

When *y* is a constant multiple of *x*, it is said that *y* varies directly as *x*. This situation is called **direct variation.** The amount you multiply each *x* by to get *y* is the slope, also called the *constant of variation.* Direct variation equations arise in large numbers of real-world situations. For example, the distance driven at a constant speed varies directly as the time driven. The circumference of a circle varies directly as the circle's radius.

 QY3

> ▶ QY3
>
> Determine which equation is an example of direct variation and give its constant of variation.
>
> a. $y = -4x + 1$
>
> b. $y = \frac{3}{4}x$
>
> c. $y = 8$

Questions

COVERING THE IDEAS

1. **Fill in the Blanks** The set of ordered pairs (x, y) that satisfy $y = mx + b$ is a line with slope __?__ and *y*-intercept __?__.

2. A family is taking a car trip. They expect to pay $150 for someone to look after their pet dog while they are gone. Then they think it will cost $200 per day for a room and meals and $80 per day for other expenses.
   a. What is the expected cost *y* for *x* days of the trip?
   b. If the equation in Part a is graphed, what are the slope and *y*-intercept of the graph?

In 3 and 4, an equation of a line in slope-intercept form is given.
a. Determine the slope.
b. Determine the *y*-intercept.
c. Graph the line.

3. $y = 2x + 3$

4. $y = \frac{1}{4}x - 2$

In 5 and 6, an equation of a line is given.
a. Rewrite the equation in slope-intercept form.
b. Determine the slope of the line.
c. Determine the *y*-intercept of the line.

5. $y = 5.6 - 1.3x$

6. $x + 5y = 7$

7. A hot air balloon begins 3 feet above the ground. It then climbs at a constant rate of 2 feet per second.
   a. Determine an equation for the height *h* of the balloon at time *t*.
   b. Draw a graph of the equation in Part a.
   c. What are the slope and *y*-intercept of the graph in Part b?
   d. How high is the balloon after 60 seconds?

8. a. Find the equation in slope-intercept form using the data in the table at the right.
   b. Is this an example of direct variation? Explain.

Hot air balloons were invented in France in 1783.

Source: www.hotairballoons.com

x	y
1	6.5
2	11
3	15.5
4	20
5	24.5

Slope-Intercept Equations for Lines

Chapter 6

In 9 and 10, write an equation of a line in slope-intercept form with the following characteristics.

9. slope 4, *y*-intercept 3
10. slope 0, *y*-intercept −2

11. Write the equation of the line graphed at the right in slope-intercept form.

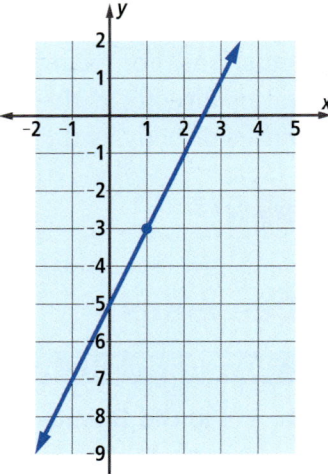

APPLYING THE MATHEMATICS

12. The equation $y = \frac{1}{2}x + 2$ was used to make the following table and graph. Find the slope and *y*-intercept. Explain how they are seen in the table and in the graph.

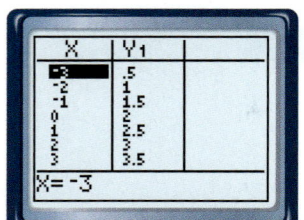 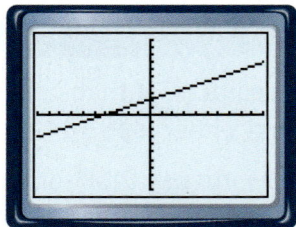

In 13–15, match the equation and graph with the situation.

Situation

13. Spencer has $80 in the bank and is spending $6 per week.

14. Owen borrowed $80 from his uncle and is paying him back $6 per week.

15. Savina has $80 in the bank and is adding $6 per week to the account.

Equation

a. $y = 6x - 80$
b. $y = 6x + 80$
c. $y = -6x + 80$

Graph

i.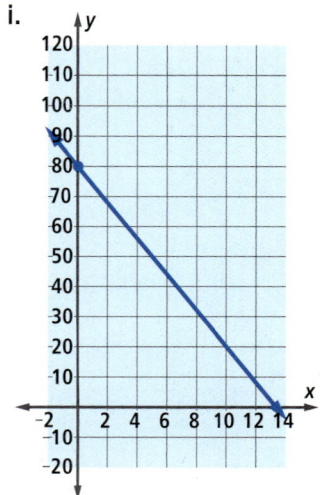

ii.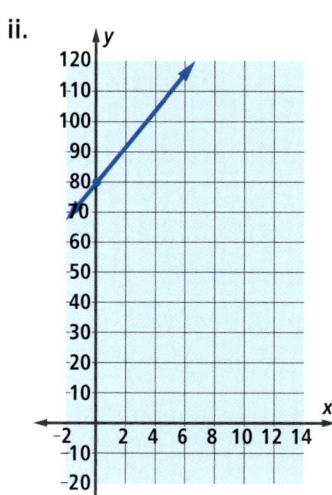

iii.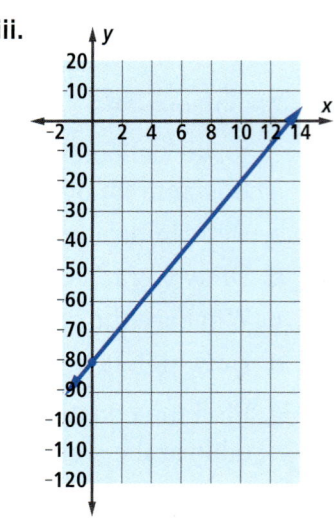

16. Consider the line through the points (3, 4) and (3, 7).
    a. Why does this line *not* have an equation in $y = mx + b$ form?
    b. What is an equation for this line?

17. a. Graph the equation $y = 3x + 4$ on a piece of graph paper.
    b. Draw a line through (2, 2) that is parallel to $y = 3x + 4$.
    c. Write an equation of the parallel line in slope-intercept form.
    d. How are the two equations the same?

18. A line has the equation $y = -3x + b$. For what value of b does the line pass through exactly 2 quadrants? (The axes are not considered to be in any quadrant.)

REVIEW

19. The Mount Washington Cog Railway in New Hampshire is one of the steepest mountain-climbing trains in the world with a gradient of 1 in 2.7. Find the slope of the path made as the train goes up the incline. (**Lesson 6-3**)

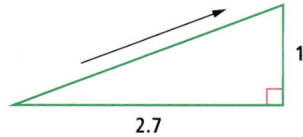

20. Determine if the following statement is *always*, *sometimes but not always*, or *never true* and explain your reasoning. *Vertical lines have a slope of zero.* (**Lesson 6-3**)

In 21 and 22, solve the equation. (**Lesson 5-9**)

21. $\dfrac{6x - 4}{3 + 2x} = \dfrac{5}{6}$

22. $\dfrac{w + 3}{3} = 3w - 1$

23. The sum of two consecutive integers is 8 less than three times the difference of the two numbers. Find the two numbers. (**Lesson 4-4**)

24. What is the area of the rectangle formed by the lines $y = 6$, $y = 2$, $x = -3$, and $x = 4$? (**Lesson 4-2**)

EXPLORATION

25. A line t has slope 4 and y-intercept 3.
    a. What is an equation for line t?
    b. By experimenting with a graphing calculator, find an equation for the line that has y-intercept 3 and seems to be perpendicular to line t.

The Mount Washington Cog Railway is the world's first mountain-climbing cog railway.

Source: Mount Washington Cog Railway

QY ANSWERS

1a. slope = $\dfrac{1}{4}$, y-intercept = 11

b. $y = 3x - 7$

2. slope = -6, y-intercept = -2

3. b; $\dfrac{3}{4}$

Chapter 6

Lesson 6-5

Equations for Lines with a Given Point and Slope

Vocabulary

point-slope form

x-intercept

> **BIG IDEA** An equation for the line through any given point with any given slope can be determined with the slope-intercept form for the equation.

In Lesson 6-4, you saw that it is easy to find an equation of a line if you know its slope m and y-intercept b. This means that if you know the line's slope and the point $(0, b)$ where the line intersects the y-axis, then you can write an equation for the line. But what if you know the slope and some other point on the line, not necessarily the y-intercept? This is often the situation, as shown in Example 1.

Mental Math

Evaluate.

a. $3 \cdot 2\frac{2}{7}$

b. $\frac{15}{6} \cdot \frac{4}{5}$

c. $-1\frac{2}{3} \cdot 5$

Example 1

In 2006, the population of Alaska was 664,000. It was increasing by 7,800 people each year. Assuming this rate of increase remains steady, find an equation relating the population of Alaska y to the year x.

Solution 1 This is a constant-increase situation, so it can be described by a line with equation in the form $y = mx + b$. The rate 7,800 people/year is the slope, so $m = 7,800$. The population of 664,000 in 2006 is described by the point (2006, 664,000). You now have the slope and one point $(x, y) = (2006, 664,000)$. Follow these three steps.

Step 1 Substitute for m, x, and y. $664{,}000 = 7{,}800 \cdot 2006 + b$

Step 2 Solve for b. $664{,}000 = 15{,}646{,}800 + b$

$-14{,}982{,}800 = b$

Step 3 Substitute the values for m and b in $y = mx + b$.

$$y = 7{,}800x - 14{,}982{,}800$$

Check After 3 years, the population should have increased by $7{,}800 + 7{,}800 + 7{,}800 = 23{,}400$ people to be 687,400.

When $x = 2009$, does $y = 687{,}400$?

Does $7{,}800 \cdot 2009 - 14{,}982{,}800 = 687{,}400$?

Yes, so it checks.

The solution to Example 1 begins with the slope-intercept form $y = mx + b$. Another method begins with the definition of slope.

Alaska's commercial fishing industry is the number one private sector employer in Alaska, providing more jobs than oil, gas, timber, and tourism.

Source: Alaska Department of Fish and Game

356 Slopes and Lines

Solution 2 This is a constant-increase situation. The amount of increase in a unit time, 7,800 $\frac{people}{year}$, is the slope, so $m = 7,800$. The given information means that the point (2006, 664,000) is on the line. Now picture the graph of the line, as shown at the right. If (x, y) is any point on the line, then

$$\frac{y - 664,000}{x - 2006} = 7,800.$$

Multiplying both sides by $x - 2006$, we obtain

$$y - 664,000 = 7,800(x - 2006).$$

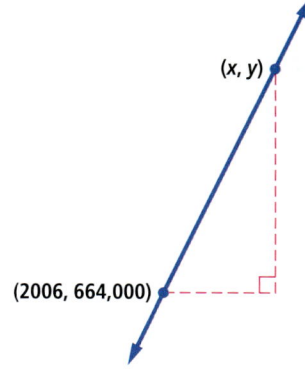

The equation found for the line using the method of the second solution has the advantage of showing all the given information. It displays the slope and the given point on the line. The general form is called the **point-slope form** of an equation of a line.

> **Point-Slope Equation of a Line**
>
> The line through the point (h, k) with slope m has equation $y - k = m(x - h)$.

Either the slope-intercept method or the point-slope method of solving a problem like Example 1 enables you to find an equation for any nonvertical line if you know its slope and the coordinates of one point on it.

In Example 2, we use the point-slope equation method to find an equation for a line whose slope and *x-intercept* are known. An **x-intercept** of a graph is the *x*-coordinate of a point where the graph intersects the *x*-axis.

GUIDED

Example 2

A line has slope −5, and its *x*-intercept is 22. Find an equation for the line.

Solution Because the *x*-intercept is 22, the point (22, 0) is on the line. So $(h, k) = (22, 0)$. Since the slope is −5, $m = -5$. Use the point-slope form $y - \underline{\ ?\ } = m(x - \underline{\ ?\ })$.

Substitute for *m*, *h*, and *k*. $y - \underline{\ ?\ } = \underline{\ ?\ }(x - \underline{\ ?\ })$.

When you find the equation of a line from a point and a slope, you can use a graphing calculator to check your answer. In Example 2, the slope of −5 and the point (22, 0) are indicated by the problem. Enter the equation you found into your [Y=] menu. To see if the point (22, 0) lies on the line, set up a table whose first *x*-coordinate is 22. The table shows that indeed the line passes through the point.

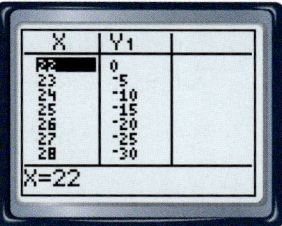

Chapter 6

Questions

COVERING THE IDEAS

1. In 2000, Vancouver, Washington, had a population of 143,560 and was growing at a rate of about 9,700 people per year. Suppose that this rate of increase stays steady.
   a. Find an equation relating y, the population of Vancouver, to the year x, where x is the number of years since 2000.
   b. Predict the population of Vancouver in the year 2015.
   c. Predict which year the population of Vancouver will exceed 400,000.

In 2–5, find an equation of the line given the slope and one point on the line.

2. point (h, k); slope m
3. point $(-8, 6)$; slope -2
4. point $(-2, 0)$; slope $\frac{1}{3}$
5. point $(5, -\frac{1}{2})$; slope 0

6. The slope of a line is -5 and the x-intercept is 4. Find an equation for the line.
7. Write the equation of the line graphed at the right.
8. What is an equation of the horizontal line through the point $(3, 9)$?
9. What is an equation for the line with slope -4.3 that contains the point $(6.8, -3.0)$?

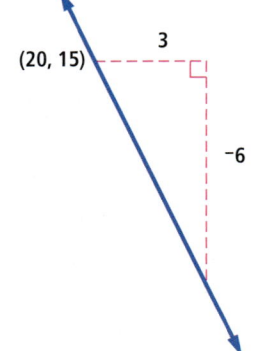

APPLYING THE MATHEMATICS

10. A mountain climber leaves his camp and hikes up the side of a mountain. His altitude increases at an average rate of 200 feet per hour. After 3 hours his altitude is 4,100 feet. Write an equation to find his altitude y after he has been hiking for x hours.

11. A newborn koala (age 0 months) is 2 cm long. Until maturity, it grows at an average rate of 1.5 cm per month. Koalas mature at about 4 years of age.
    a. From the given information, find an equation estimating the length y of a koala at age x months.
    b. About how long are mature koalas?

12. The slopes of two lines are reciprocals.
    a. An equation of one of the lines is $y = -8x - 9$. What is the slope of the second line?
    b. Find the equation of the second line if it passes through the point $(4, 6)$.

Koala bears are not bears. They are marsupials, like kangaroos. Baby koalas, called joeys, are the size of a jellybean at birth. A joey remains in the mother's pouch for about 22 weeks before emerging.

Source: *National Geographic*

358 Slopes and Lines

13. Suppose a 5-week-old baby weighs 12 pounds and is gaining weight at the rate of 0.3 pound per week.
    a. The given information represents the slope m, age x, and weight y. Find the values of m, x, and y.
    b. Determine an equation for the baby's weight y at age x weeks.
    c. Determine the baby's weight at 12 weeks.

14. The following sequence, seen in Lesson 1-3, is made up of yellow and green hexagonal tiles. Find a formula for the number of tiles y in terms of n, and find the term of the sequence.

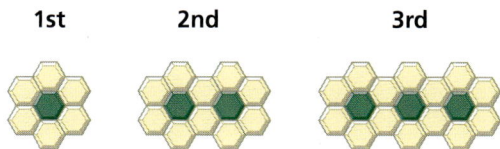

15. Determine an equation for the line that contains (0, 3) and is parallel to the line with equation $y = 4x - 9.89$.

In 16 and 17, some information is given.
a. Write the slope and an ordered pair described by the information.
b. Write an equation relating x and y.

16. Kathy is on vacation and is spending $75 per day. After 4 days she has $320 left.

17. To buy a pass for the city bus system, you must pay an initial fee and then pay $1.75 per ride. Dante paid $29.25 for 15 rides.

REVIEW

18. Match each of the lines k, m, and n at the right with its equation. (**Lesson 6-4**)
    a. $y = -\frac{1}{3}x$
    b. $y = -\frac{1}{3}x - 3$
    c. $y = 2 - \frac{1}{3}x$

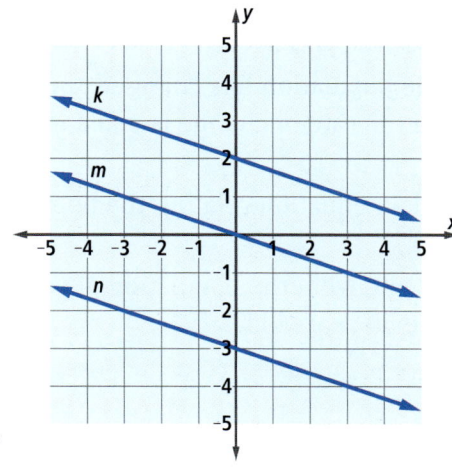

19. Graph the line with equation $y = -x$. (**Lesson 6-4**)

20. Do the points (−4, 3), (0, 2), and (4, −7) lie on the same line? Justify your answer. (**Lesson 6-2**)

Chapter 6

21. The following two points give information about prices of corn: (4 bushels, $9.24), (12 bushels, $27.72). Calculate the rate of change and describe what it stands for. (**Lesson 6-1**)

22. Which section(s) of the graph below shows the
    a. fastest increase?
    b. slowest decrease? (**Lesson 6-1**)

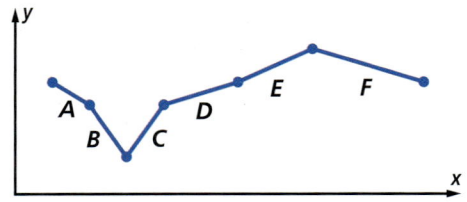

23. **Skill Sequence** Simplify each expression. (**Lesson 5-2**)
    a. $\frac{12}{5} \div \frac{12}{6}$
    b. $\frac{x}{5} \div \frac{x}{6}$
    c. $\frac{x}{y} \div \frac{x}{z}$

24. Phone Company A charges $4.99 per month for a special 7¢ per minute rate. Company B charges 10¢ per minute with no monthly fee. Suppose you talk for T minutes in a particular month.
    a. What will it cost you if you have signed up with Company A?
    b. What will it cost you if you have signed up with Company B?
    c. How many minutes must you talk before you pay less with Company A's rate than with Company B's rate? (**Lessons 4-5, 3-4, 1-2**)

EXPLORATION

25. Use the Internet or another resource to find the population at the last census of the state where you live.
    a. Find how much your state population has changed since the previous census. Assume the rate of change is constant and that it continues.
    b. Using your estimate, find an equation relating the population y to the year x.
    c. Use this equation to estimate what the population of your state will be when you are 50 years old.

Lesson 6-6
Equations for Lines through Two Points

> **BIG IDEA** An equation for the line through any two given points can be determined by first calculating the slope of the line.

A powerful approach to problem solving is to change the problem you are given into a simpler problem that you know how to solve. In Lesson 6-5, you learned how to find an equation of a line if you know its slope and one point on it. Now, consider a different problem. You are given two points and want to find the equation of the line that passes through them. With just a little work, you can turn this problem into a problem like one in the last lesson.

Example 1

Find an equation for the line that passes through (10, −4) and (18, 20).

Solution Use the coordinates of the given points to find the slope.

$$m = \frac{20 - -4}{18 - 10} = \frac{24}{8} = 3$$

Now you have more information than you need. You know two points and need to use only one of them to find an equation. Either point can be used. We choose to use (18, 20). Follow the algorithm from Lesson 6-5.

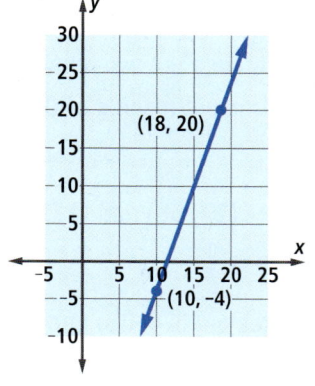

$y - k = m(x - h)$ Write the equation.
$y - 20 = 3(x - 18)$ Substitute $h = 18$, $k = 20$, and $m = 3$.

An equation of the line is $y - 20 = 3(x - 18)$.

Check Substitute the coordinates of the two points in the equation $y - 20 = 3(x - 18)$ to check if they produce true statements.

Using (10, −4), does $-4 - 20 = 3(10 - 18)$?
$$-24 = 30 - 54$$ Yes, the equation checks.

Using (18, 20), does $20 - 20 = 3(18 - 18)$?
$$0 = 3 \cdot 0$$ Yes, the equation checks.

Mental Math

Find the measure of the angle between spokes if the spokes are equally spaced around the circle and there are:

a. 3 spokes.

b. 10 spokes (pictured).

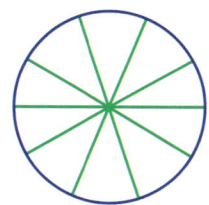

c. 15 spokes.

d. n spokes.

Relationships that can be described by an equation of a line occur in many places. Here is a relationship that may surprise you.

Example 2

Biologists have found that the number of chirps field crickets make per minute is related to the outdoor temperature. The relationship is very close to being linear. When field crickets chirp 124 times per minute, it is about 68°F. When they chirp 172 times per minute, it is about 80°F. Below is a graph of this situation.

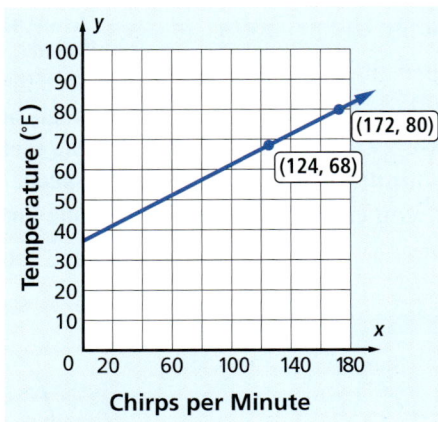

Field crickets are the crickets everyone sees and hears in late summer and fall.

a. Find an equation for the line through the two points.
b. About how warm is it if an instrument records 150 chirps in a minute?

Solutions

a. Find the slope using the points (124, 68) and (172, 80).

$$\text{slope} = \frac{80 - 68}{172 - 124} = \frac{12}{48} = \frac{1}{4}$$

Substitute $\frac{1}{4}$ and the coordinates of (124, 68) into $y = mx + b$.

$$68 = \frac{1}{4}(124) + b$$

Solve for b.

$$68 = 31 + b$$
$$37 = b$$

Substitute for m and b.

An equation is $y = \frac{1}{4}x + 37$.

b. Substitute 150 for x in the equation $y = \frac{1}{4}x + 37$.

$$y = \frac{1}{4} \cdot 150 + 37$$
$$= 37.5 + 37$$
$$= 74.5$$

It is about 75°F when you hear 150 chirps in one minute.

STOP QY

▶ **QY**

Check Example 2 by showing that (124, 68) is on the line with equation $y = \frac{1}{4}x + 37$.

The equation in Example 2 enables the temperature to be found for any number of chirps. By solving for x in terms of y, you could get a formula for the number of chirps expected at a given temperature. Formulas like these seldom work for values far from the data points. Field crickets tend not to chirp at all below 50°F, yet the formula $y = \frac{1}{4}x + 37$ predicts about 50 chirps per minute at 50°F.

GUIDED

Example 3
Suppose when using a calling card to call Iceland, you were charged $2.23 for a 10-minute call and $4.12 for a 20-minute call. Assume that this is a constant-increase situation.
a. Express the given information as two ordered pairs (minutes, cost).
b. Find an equation that expresses the cost y of the call in terms of x, the length of the call in minutes.
c. What does the y-intercept b represent in this situation?
d. What does the slope m represent in this situation?
e. How long could you talk for $25?

Solutions
a. Two points the line passes through are (__?__ , __?__) and (__?__ , __?__).
b. First, find the __?__ , m.
  $m = \frac{?-?}{?-?} = 0.189$
  Substitute 0.189 and the coordinates of one of the points into $y = mx + b$.
  __?__ $= 0.189($ __?__ $) + b$
  Solve for b.
  Substitute the values found for m and b into the equation $y = mx + b$.
  $y = $ __?__ $x + $ __?__
c. The y-intercept represents the cost of calling for __?__ minutes. This is usually a charge that is automatic no matter how long the call lasts. In this case, the automatic charge is $0.34.
d. The slope represents the __?__ .
e. Substitute $25 for __?__ in the equation $y = 0.189x + 0.34$ and solve for __?__ . Solving this equation shows that you could talk __?__ minutes for $25.

The equation $y = 0.189x + 0.34$ in Example 3 gives the cost for talking any number of minutes. If the equation were solved for x in terms of y, the result would be a formula for the number of minutes you could talk at a given cost.

Chapter 6

Questions

COVERING THE IDEAS

1. A katydid, or long-horned grasshopper, chirps about 70 times per minute at 70°F and 124 times per minute at 95°F.
   a. Find an equation relating x, the number of chirps per minute, and y, the temperature in degrees Fahrenheit.
   b. Use your equation from Part a to estimate the temperature when a katydid cricket is chirping 90 times a minute.

2. Suppose a 15-minute call costs $5.26, and a 30-minute call costs $10.24. Find a formula relating time (in minutes) and cost (in dollars).

3. Find an equation for the line shown at the right.

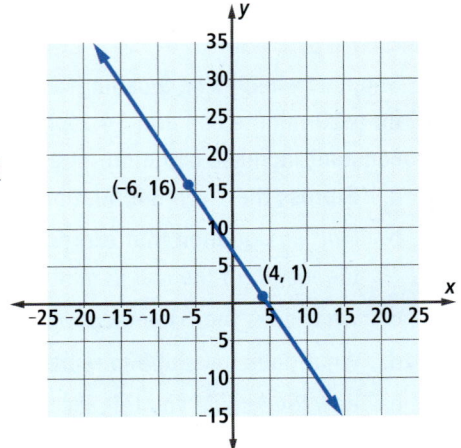

In 4 and 5, find an equation for the line through the two given points. Check your answer.

4. $(3, 0), (7, 16)$
5. $(4, 11), (10, 5)$

APPLYING THE MATHEMATICS

6. Raini ran a marathon, which is 26.2 miles long. It took him 2 hours and 6 minutes to reach mile 13 on the race course and he crossed the finish line in 4 hours and 18 minutes. Raini tends to run at a constant speed.
   a. Express the given information as two ordered pairs (time, distance).
   b. Write an equation to find the distance Raini ran in terms of the time.
   c. What does the slope represent in this situation?
   d. How long did it take Raini to reach mile 7?

7. Write an equation for the line that produced the table of values shown below.

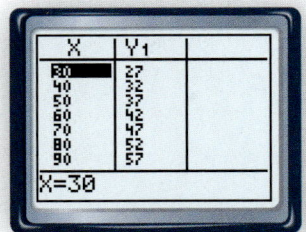

8. Find an equation for the line with x-intercept 7 and y-intercept 3.

364   Slopes and Lines

9. Penicillin was discovered in 1928. However, the medicine was not mass produced until the 1940s. In 1943, the price of a dose of penicillin was $20. By 1946, the price per dose was $0.55.
   a. Assuming the price of a dose of penicillin decreased at a constant rate, find an equation relating the year and price.
   b. According to the equation, what was the 1957 price of a dose of penicillin?
   c. Do you believe this equation gives good estimates for the price of penicillin? Why or why not?

10. a. Use the graph at the right to estimate the Italian shoe size that corresponds to a women's size 9 in the United States.
    b. Write an equation to relate women's shoe size in Italy to women's shoe size in the United States.
    c. Use your equation from Part b to find the Italian shoe size of a women's size 9 in the United States.

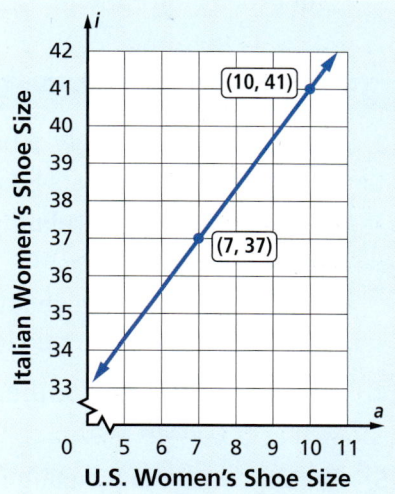

11. Old Faithful is a geyser in Yellowstone National Park that erupts often. The National Park Service has studied the length of the eruption and the amount of time between eruptions for many years. They have found that the relationship between how long an eruption lasts x and the amount of time until the next eruption y is approximately linear. One eruption lasted 2.1 minutes and the time before the next eruption was 59.24 minutes. Another eruption lasted 3.7 minutes and the time before the following eruption was 70.08 minutes.
    a. Write the data as two ordered pairs.
    b. Find the slope of this linear relationship.
    c. Explain what the slope represents in this situation.
    d. Find an equation for y in terms of x.
    e. If the park rangers posted a sign at the end of an eruption that said 94 minutes until the next eruption of Old Faithful, approximately how long had the previous eruption lasted?

An eruption at Old Faithful expels between 3,700 and 8,400 gallons of boiling water.

Source: National Park Service

12. The graph at the right shows the linear relationship between Fahrenheit and Celsius temperatures. The freezing point of water is 32°F and 0°C. The boiling point of water is 212°F and 100°C.
    a. Find an equation that relates Celsius temperatures C and Fahrenheit temperatures F.
    b. When it is 155°F, what is the temperature in degrees Celsius?
    c. When it is –10°C, what is the temperature in degrees Fahrenheit?

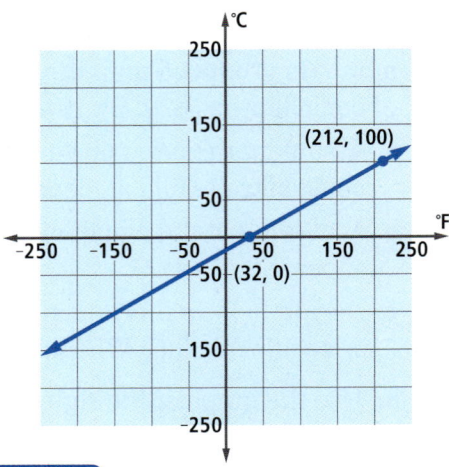

REVIEW

13. A cab company charges a base rate of $2.50 plus $1.60 per mile. A 15-mile cab ride costs $26.50.
    a. Write an equation relating the number of miles driven to the cost of the cab ride.
    b. Make up a question about this cab company that you can answer using your equation from Part a. **(Lesson 6-5)**

14. What is an equation for the line through the point (5, 5) with slope $-\frac{3}{5}$? **(Lesson 6-5)**

15. Graph $y = 3x + 9$ by using its slope and y-intercept. **(Lesson 6-4)**

16. The points (3, b) and (–2, 4) are on a line with slope $\frac{1}{2}$. Find b. **(Lesson 6-2)**

17. Interstate 70 leads from Denver, Colorado into the mountains. There are signs that post the elevation. **(Lesson 6-1)**

Miles from Denver	Elevation (ft)
0	5,260
33	7,524
45	8,512
47	9,100
72	9,042

Source: Colorado Tourism Office

View from Rocky Mountain National Park in Colorado

   a. Between which two signs is the rate of change of elevation negative?
   b. Calculate the rate of change of elevation for the entire distance listed in the table.

18. Simplify $\frac{3a^2b}{c^2} \div \frac{12a}{c^4}$. (**Lesson 5-2**)

19. A rectangle is 3 cm longer than twice its width w. Its perimeter is 42 cm. (**Lesson 3-4, Previous Course**)

   a. Write an expression for the length ℓ in terms of w.
   b. Find its width and length.
   c. What is its area?

In 20–22, simplify the expression. (Lessons 2-2, 2-1)

20. $4(-7x) - 9(x - 1)$

21. $\frac{3}{4}(16g - h) + 4(2h)$

22. $-4p - 3(p - 2.5)$

EXPLORATION

23. In many places, a taxi ride costs a fixed number of dollars plus a constant charge per mile.

   a. Find a rate for taxi rides in your community or in a nearby place.
   b. Find an equation relating distance traveled and the cost of a ride.

QY ANSWER

Does $68 = \frac{1}{4} \cdot 124 + 37$?
$68 = 31 + 37$?
Yes, it checks.

Chapter 6

Lesson 6-7
Fitting a Line to Data

Vocabulary

line of best fit
linear regression
least squares line

> **BIG IDEA** When points lie nearly on a line, it is useful to determine an equation for a line that lies on or comes close to the points.

Mental Math

Give the conversion factor for converting

a. $\frac{\text{inches}}{\text{year}}$ to $\frac{\text{inches}}{\text{month}}$.

b. $\frac{\text{meters}}{\text{pound}}$ to $\frac{\text{millimeters}}{\text{pound}}$.

c. $\frac{\text{feet}}{\text{second}}$ to $\frac{\text{inches}}{\text{minute}}$.

If data points are not all on one line but are close to being linear, you can often use an equation for a line to describe trends in the data. For example, the table and graph below show the life expectancy of people in the United States at birth in ten-year intervals from 1930 to 2000. Notice that the life expectancy has been increasing each decade.

It is natural to wonder what the life expectancy will be in 2020, 2030, or 2050. Of course no one knows, but we can make educated guesses by using algebra.

Notice that the change in life expectancy each decade is not constant. So the points do not lie on the same line. Still, the points seem to be reasonably close to a line. There are different ways of estimating a line that comes close to the data. This is called *fitting a line to the data*.

Year	Life Expectancy (yr)
1930	59.7
1940	62.9
1950	68.2
1960	69.7
1970	70.8
1980	73.7
1990	75.4
2000	77.0

Source: National Center for Health Statistics

Eyeballing a Line of Fit

Activity 1

Step 1 After carefully graphing the data from the table, take a ruler and draw a line that seems close to all the points. This is called "fitting a line by eye" or "eyeballing." One such line is graphed at the right.

Step 2 Find two points on the line. The line we drew happens to not pass through any of the original data points. Our line contains (1930, 61) and (1990, 76).

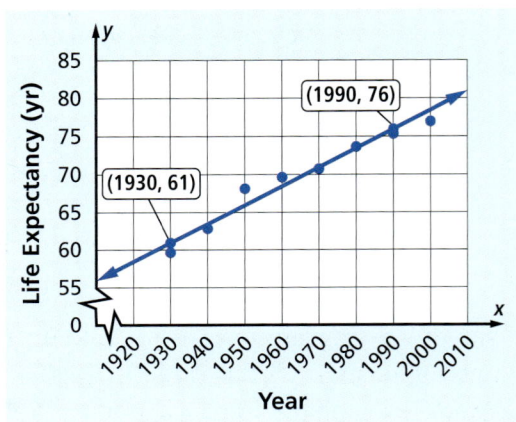

368 Slopes and Lines

Step 3 Find an equation for the line through the two points. We follow the algorithm in Lesson 6-6 for finding an equation of a line given two points. First we use these two points to find the slope of the line.

$$\text{slope} = \frac{76 - 61}{1990 - 1930} = \frac{15}{60} = 0.25$$

Now substitute the slope and the coordinates of one of the points into $y = mx + b$ and solve. We use (1930, 61).

$61 = 0.25 \cdot 1930 + b$

$61 = 482.5 + b$

$-421.5 = b$

An equation for the line is $y = 0.25x - 421.5$.

With this method, an estimate for the life expectancy for someone born in 2020 is $0.25 \cdot 2020 - 421.5$, or about 83.5 years.

Eyeballing is a simple method but it has a weakness in that two different people will likely eyeball two different lines.

Linear Regression

Most graphing calculators have a feature that will give you what is known as the **line of best fit.** The method they use is called **linear regression.** This is the most common way of finding a line to fit data. In Activity 2, we show only how to use a calculator to find an equation for this line.

Activity 2

Step 1 Enter the eight ordered pairs for the data into two lists, one for the x-coordinate and the other for the corresponding y-coordinate. On some calculators, these lists are called **L1** and **L2**.

Step 2 Have the calculator automatically calculate the line of best fit. One calculator showed the screen at the right. The letter a indicates slope.

Step 3 Round a and b to reasonable accuracy. Here we need four decimal places for a because the x values for the years are so large. Substitute the rounded values for a and b into the equation $y = ax + b$. So, by this method, a line of best fit is $y = 0.2395x - 400.9893$.

Using linear regression, an estimate for the life expectancy for someone born in 2020 is $0.2395 \cdot 2020 - 400.9893$, or about 82.8 years. This is a little lower than what was predicted by eyeballing.

Fitting a Line to Data **369**

Both an eyeballed line and the line of best fit can be considered as models of life expectancy in the United States from 1930 to 2000. Recall that the difference between the actual amount and the amount predicted by a model is called the deviation.

The table below shows the actual life expectancy, the expectancies predicted by these models, and the deviations for each model. It also shows the predicted life expectancies for 2010 and 2020.

Year	Life Expectancy	Eyeball a Line of Fit ($y = 0.25x - 421.5$)	Line of Best Fit ($y = 0.2395x - 400.9893$)	Eyeball Deviation	Best Fit Deviation
1930	59.7	61.0	61.2	1.3	1.5
1940	62.9	63.5	63.6	0.6	0.7
1950	68.2	66.0	66.0	−2.2	−2.2
1960	69.7	68.5	68.4	−1.2	−1.3
1970	70.8	71.0	70.8	0.2	0
1980	73.7	73.5	73.2	−0.2	−0.5
1990	75.4	76.0	75.6	0.6	0.2
2000	77.0	78.5	78.0	1.5	1.0
2010		81.0	80.4		
2020		83.5	82.8		

The line of best fit has the following property: The sum of the squares of the deviations of its values from the actual values is the least of all lines. For this reason, it is called the **least squares line.**

GUIDED

Example

Show that the sum of the squares of the eyeball line deviations for the life expectancies is greater than the sum of the squares of the best fit deviations.

Solution For the eyeball line, the sum of the squares of the deviations is $1.3^2 + 0.6^2 + (-2.2)^2 + (-1.2)^2 + 0.2^2 + (-0.2)^2 + 0.6^2 + 1.5^2 =$ __?__.

For the least squares line, the sum of the squares of the deviations is __?__.

Questions

COVERING THE IDEAS

1. If three people were to use the indicated method for fitting a line to a particular set of data, would their answers necessarily be the same? (Assume they made no errors in calculations.)
   a. eyeballing
   b. linear regression

In 2 and 3, use the table of life expectancies at the right for people in the United States.

2. a. Construct a scatterplot of the ordered pairs (year, female life expectancy).
   b. Eyeball a line to fit the data and find its equation.
   c. What female life expectancies does your equation predict for the years 1930–2000?
   d. Calculate the sum of the squares of the deviations of the predicted values in Part c from the actual values.
   e. Use your equation to predict the female life expectancy in the U.S. in the year 2020.

Year	All	Females	Males
1930	59.7	61.6	58.1
1940	62.9	65.2	60.8
1950	68.2	71.1	65.6
1960	69.7	73.1	66.6
1970	70.8	74.7	67.1
1980	73.7	77.4	70.0
1990	75.4	78.8	71.8
2000	77.0	79.7	74.3

Source: U.S. Census Bureau

3. Follow the directions for Question 2, but use linear regression.

4. Why is the line found using linear regression called the *least squares* line?

APPLYING THE MATHEMATICS

In 5 and 6, the table shows women's 800-meter freestyle swimming long course (50-meter pool) world records between 1971 and 1978.

Person and Country	Year	Time (min)
Shane Gould, Australia	1971	8.97
Keena Rothhammer, USA	1972	8.88
Novella Calligaris, Italy	1973	8.87
Jo Ann Harshbarger, USA	1974	8.79
Jennifer Turrall, Australia	1975	8.72
Petra Thumer, East Germany	1976	8.67
Petra Thumer, East Germany	1977	8.58
Tracey Wickham, Australia	1978	8.40

Source: USA Swimming

5. a. Construct a scatterplot of the ordered pairs (year, time).
   b. Use linear regression to predict the world record in 1989.
   c. In 1989, Janet Evans set the most recent world record in the long course women's 800-meter. Her time was 8 minutes, 16.22 seconds, or about 8.27 minutes. Calculate the deviation from the linear regression prediction.
   d. The first women's 800-meter freestyle world record was in 1919. The record was set by Gertrude Ederle in a time of 13.32 minutes. This time deviates from the linear regression equation's predicted time by how much?
   e. Is the linear regression line a good model for predicting world record times in the women's 800-meter freestyle before or after the 1970s? Explain why or why not.

Katie Ledecky (USA), 2015 world-record swimmer in the 800-meter freestyle, speaks to students at an elementary school in North Dakota.

6. Add the 1919 and 1989 world record times from Question 5 to the table. How does the linear regression equation change? Do you think it is more or less accurate for predicting world records in the women's 800-meter freestyle?

7. Refer to the data below.

City	Latitude (°North)	January Mean Low Temperature (°F)
Lagos, Nigeria	6	74
San Juan, Puerto Rico	18	70
Calcutta, India	23	55
Cairo, Egypt	30	47
Tokyo, Japan	35	31
Rome, Italy	42	39
Belgrade, Serbia	45	28
London, England	52	35
Copenhagen, Denmark	56	29
Moscow, Russia	56	9

Source: infoplease.com

a. Make a scatterplot showing a point for each city.
b. Use linear regression to find an equation.
c. **Fill in the Blank** As you go one degree north, the January low temperature tends to ___?___.
d. Which city's January mean low temperature deviates most from that predicted by the equation?
e. Predict the January mean low temperature for the North Pole.
f. The January mean low temperature for Acapulco, Mexico, which is at 17° north latitude, deviates from the equation by +8.3°F. Find the actual January mean low temperature in Acapulco.

The domes and minarets of the Sultan Hassan Madrasa stand on the eastern edge of Cairo.

In 8–11, tell whether fitting a line to the data points would be appropriate.

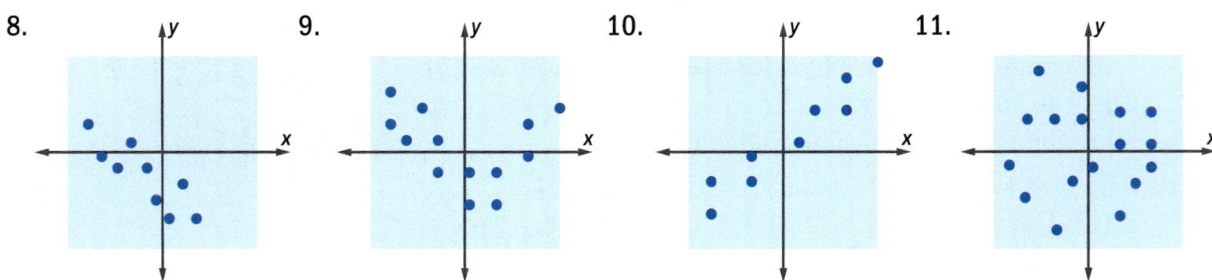

8.   9.   10.   11.

REVIEW

12. Find an equation for the line with y-intercept -2 and x-intercept 4. (**Lesson 6-6**)

13. a. Find an equation for the line through $(6, 4)$ with a slope of $\frac{5}{3}$.
    b. Give the coordinates of one other point on this line. (**Lesson 6-5**)

14. a. Give the slope and y-intercept of the line $4x - 9y = 3$.
    b. Graph the line. (**Lesson 6-4**)

In 15 and 16, find the slope and y-intercept of the line. (**Lesson 6-4**)

15. $y = -\frac{x}{6}$ 16. $y = 4$

17. a. Draw a line that has no y-intercept.
    b. What is the slope of your line? (**Lesson 6-4**)

18. The Tigers scored 15 points in the first 8 minutes of a basketball game. At that rate, how many points would the team score in a 48-minute game? (**Lesson 5-8**)

19. In a game of chess, each player begins the game with 8 pawns, 2 knights, 2 bishops, 2 rooks, 1 queen, and 1 king. What is the ratio of bishops to the total number of pieces? (**Lesson 5-5**)

Prior to the founding of the World Chess Federation in 1924, chess had existed as a sport played at a competitive level for centuries.

Source: World Chess Federation

20. Solve for y. (**Lesson 4-4**)
    a. $y - 3 = 6y + 2$ b. $y - a = by + c$

21. Describe all solutions to the following sentences. (**Lesson 2-8**)
    a. Six times a number is zero.
    b. Zero times a number is zero.
    c. Zero times a number is six.

22. If a is the cost of an adult's ticket to an amusement park and c is the cost of a child's ticket, what is the cost of tickets for 2 adults and 5 children? (**Lesson 1-2**)

EXPLORATION

23. The method of least squares was independently discovered by two mathematicians, the German Carl Friedrich Gauss and the Frenchman Adrien-Marie Legendre, in the years 1795–1809. Research to find the problem that led Gauss to this discovery.

Lesson 6-8

Standard Form of an Equation of a Line

Vocabulary

linear combination

standard form of an equation for a line

oblique

> **BIG IDEA** Every line has an equation of the form $Ax + By = C$.

Mental Math

Find the average speed of a car going

a. 300 miles in 6 hours.

b. 50 mph for an hour, then 70 mph for an hour.

c. 55 mph for 2 hours, 45 mph for an hour, then 60 mph for 2 hours.

In this chapter, lines have been used to describe situations involving a constant increase or decrease. Lines have also been used to model data in a scatterplot. The slope-intercept form $y = mx + b$ arises naturally from these applications. However, other situations can also lead to linear relations.

Linear Combination Situations

Lourdes had ordered party favors from an online store. When the 30 neon bouncing balls and the 12 glow-in-the-dark necklaces came, however, she had forgotten the price of each item. The bill said that the total cost was $24. What was the cost of one neon ball? What was the cost of one glow-in-the-dark necklace?

We can describe this situation as "The cost of the neon balls plus the cost of the necklaces is $24." Let $x = $ the cost of one neon ball and $y = $ the cost of one glow-in-the-dark necklace. Then $30x + 12y = 24$.

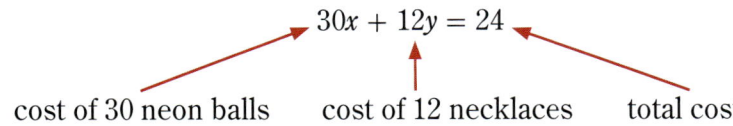

cost of 30 neon balls   cost of 12 necklaces   total cost

This equation can be quickly graphed by finding its x- and y-intercepts.

To find the x-intercept, find x when $y = 0$.

$$30x + 12 \cdot 0 = 24$$
$$30x = 24$$
$$x = 0.80$$

The point is $(0.80, 0)$.

To find the y-intercept, find y when $x = 0$.

$$30 \cdot 0 + 12y = 24$$
$$12y = 24$$
$$y = 2$$

The point is $(0, 2)$.

Plot the two intercepts and draw the line. Negative numbers do not make sense for x or y, so we ignore the part of the line in Quadrants II or IV.

Each pair of possible prices (cost of one ball, cost of one necklace) corresponds to a point. As shown at the right, the point (0.50, 0.75) lies on the graph. If a ball costs $0.50 and a necklace costs $0.75, then 30 balls and 12 necklaces cost $24. Other possible (x, y) pairs can be found algebraically by first changing the equation of the line into slope-intercept form.

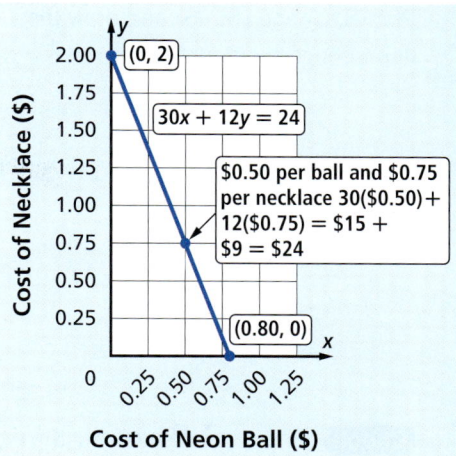

$$30x + 12y = 24$$
$$12y = -30x + 24$$
$$y = -2.5x + 2$$

This is a formula that gives y, the cost of a necklace, in terms of x, the cost of a neon ball. Suppose a neon ball costs $0.20. Then the cost of a necklace could be found by $y = -2.5(0.20) + 2 = -0.50 + 2 = \$1.50$.

 QY1

An expression of the form $Ax + By$, where A and B are fixed numbers, is called a **linear combination** of x and y. The name *linear combination* is appropriate because when $Ax + By$ has a constant value, the graph of all ordered pairs (x, y) lies on a line.

▶ QY1

Use $y = -2.5x + 2$ to find another possible pair of costs for the neon balls and necklaces that Lourdes ordered.

The Standard Form of an Equation for a Line

The equation $3x - 4y = 24$ has the form $Ax + By = C$, where $A = 3$, $B = -4$, and $C = 24$. The variables x and y are on one side of the equation and the constant term C is on the other. The equation $Ax + By = C$, where A, B, and C are constants, is the **standard form of an equation for a line**. Linear combination situations naturally lead to equations of lines in standard form.

To graph a line whose equation is in standard form, you do not need to rewrite the equation in slope-intercept form. Instead, you can find the intercepts and draw the line that contains the intercepts.

Example 1
Graph $3x - 4y = 24$.

Solution
Find the x-intercept. Let $y = 0$.

$$3x - 4(0) = 24$$
$$x = 8$$

The x-intercept is 8, so the point (8, 0) is on the line.

Find the y-intercept. Let $x = 0$.

$$3(0) - 4y = 24$$
$$y = -6$$

The y-intercept is −6, so the point (0, −6) is on the line.

(continued on next page)

Standard Form of the Equation of a Line 375

Plot (8, 0) and (0, −6) and draw the line through them, as shown.

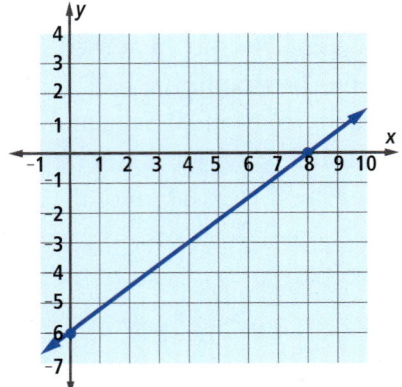

Check Find a point satisfying the equation, and check that it is on the graph of the line. The point (4, −3) is on the graph so it satisfies $3x - 4y = 24$.

 QY2

> ▶ **QY2**
>
> Give the equation in standard form that fits the following situation and draw a graph. A fruit grower ships peaches and berries to grocery stores. A carton of peaches weighs 5 pounds and a carton of berries weighs 4 pounds. The shipment to one store weighed 90 pounds. Let $x =$ the number of cartons of peaches and $y =$ the number of cartons of berries.

Rewriting Equations in Slope-Intercept and Standard Form

The idea of equivalent equations is useful in dealing with lines. In Example 1, the standard form equation $3x - 4y = 24$ is convenient to use to find the intercepts. But the slope-intercept equation is useful if you are graphing with a calculator, or if you need to know the slope. You should be able to quickly change an equation of a line into either of these forms. In standard form, the equation is usually written with A, B, and C as integers.

Example 2

Rewrite $y = \frac{4}{7}x + \frac{9}{7}$ in standard form with integer coefficients. Find the values of A, B, and C.

Solution

$7y = 7 \cdot \frac{4}{7}x + 7 \cdot \frac{9}{7}$ Multiply each side of the given equation by 7 to clear the fractions.

$7y = 4x + 9$ Simplify.

$-4x + 7y = 9$ Add $-4x$ to both sides so the x and y terms are both on the left side of the equation. Write the x term first.

This is written in standard form with $A = -4$, $B = 7$, and $C = 9$. Some people prefer that A be positive. Then multiply both sides of the equation by -1 to obtain $4x - 7y = -9$.

Lesson 6-8

Example 3
Rewrite $y = -0.75x$ in standard form with integer values of A, B, and C.

Solution

$y = -0.75x$
$100y = -75x$ Multiply each side by 100 to clear the decimal.
$75x + 100y = 0$ Add $75x$ to both sides so the x and y are both on the left side. Write the x term first.

Here $A = 75$, $B = 100$, and $C = 0$.

In the Activity below, you will practice changing the form of equations. An equivalent form of the equation of a line describes the same set of points.

Activity

Five equations are given below. Fill in the table so that each equation is written in both slope-intercept form and in standard form.

	Equation	Slope-Intercept Form	Standard Form
1.	$20x - 5y = 35$	?	?
2.	$y - 5 = 8(x - 1)$	?	?
3.	$2x + 4(y + x) = 5y + 15$	?	?
4.	$y = \frac{2}{3}x + 11$	?	?
5.	$14x + 4y = 20$	?	?

The lines graphed so far in this lesson are **oblique,** meaning they are neither horizontal nor vertical. In Lesson 4-2, you saw very short equations for horizontal and vertical lines. If a line is vertical, then its equation has the form $x = h$. If a line is horizontal, it has an equation $y = k$. These short equations are in standard form, but only one variable shows because one coefficient is zero. For example, $x = 4$ is equivalent to $x + 0y = 4$, where $A = 1$, $B = 0$, $C = 4$. The equation $y = -3$ is equivalent to $0x + y = -3$, where $A = 0$, $B = 1$, $C = -3$.

Thus, *every line* has an equation in the standard form $Ax + By = C$.

Questions

COVERING THE IDEAS

1. What is the form $Ax + By = C$ called?

Chapter 6

2. Refer to the following situation. The Hawkins family bought 3 sandwiches and 4 salads. They spent $36. Let $x =$ the cost of each sandwich and $y =$ the cost of each salad.

   a. Write an equation to describe the possible combinations of costs for the sandwiches and salads.
   b. Graph the equation from Part a.
   c. If the salads cost $6.30 each, how much did each sandwich cost?
   d. Give the coordinates of the point on the graph corresponding to your answer to Part c.
   e. Give another pair of possible costs for the sandwich and salad.

In 3 and 4, the equation is in standard form. Give the values of A, B, and C.

3. $5x - 3y = 9$
4. $8x + y = 2.4$

In 5 and 6, an equation for a line is given.
a. Find the x- and y-intercepts of the graph of the line.
b. Graph the line.

5. $2x + 5y = 20$
6. $3x - 2y = 12$

APPLYING THE MATHEMATICS

7. Refer to the following situation. Cheryl scored a total of 27 points in her basketball game last night. None of the points came from free throws. All of her points came from 2- or 3-point shots.

   a. Find three different combinations of 2- and 3-point shots that Cheryl may have had last night.
   b. Write an equation in standard form to describe all the different possible combinations of 2-point shots and 3-point shots Cheryl may have made.
   c. What is the greatest number of 3-point shots she may have made?
   d. What is the greatest number of 2-point shots she may have made?
   e. Graph the solutions to the equation in Part b.

In 8–10, an equation in slope-intercept form is given.
a. Find an equivalent equation in standard form with integer coefficients.
b. Give the values of A, B, and C.

8. $y = -\frac{4}{5}x + 10$
9. $y = \frac{9}{8}x$
10. $y = -x - 12$

Basketball is the most popular high school sport for girls, with 452,929 participants in 2006.

Source: National Federation of State High School Associations

In **11** and **12**, an equation in standard form is given. Find the equivalent equation in slope-intercept form.

11. $4x - 7y = 308$

12. $x + y = 0$

13. On many multiple-choice tests, 1 point is given for each correct answer and 0.25 point is taken away for every wrong answer. (This is to discourage guessing.) Answers that are left blank do not affect the score. Gloria scored 62 on a test with 100 questions. Let C be the number of questions Gloria correctly answered and let W be the number of wrong answers she had.

  a. Give three possible pairs of values of C and W.

  b. Write an equation that describes all possible solutions.

  c. Graph all possible solutions.

14. Suppose $Ax + By = C$, with $B \neq 0$.

  a. Solve this equation for y.

  b. Identify the slope and the y-intercept of this line.

REVIEW

15. The time of the winning runner in the men's 100-meter race has decreased since the first Olympics in 1896. The scatterplot below shows the time of the winning race in seconds in each Olympic race from 1900 through 2004. **(Lesson 6-7)**

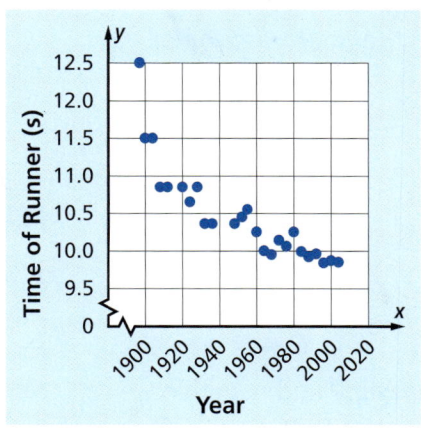

Source: International Olympic Committee

  a. Trace the graph and fit a line to the data.

  b. Find an equation for the line in Part a by approximating the coordinates of two points it goes through.

  c. Use your equation to predict the time of the winning runner in the year 2012. Why might this prediction be incorrect?

16. Find an equation for the line through the points $(-3, 4)$ and $(6, 1)$. **(Lesson 6-6)**

17. The water in a 6-foot-deep pool is 8 inches deep. A hose is being used to fill the pool at the rate of 4 inches per hour. Let x be the number of hours passed and y be the depth (in inches) of the water in the pool. (**Lessons 6-5, 6-2**)

   a. Write an equation for a line which relates x and y.
   b. Give the slope of the line.
   c. Use your answer to Part a to find how long it will take to fill the 6-foot pool.

18. A robot is moving along a floor that has a coordinate grid. The robot starts at the point (10, –4) and moves along a line with slope 3. Does the robot pass through the given point? Justify your answer. (**Lessons 6-5, 6-2**)

   a. (12, 0)   b. (17, 17)

19. What is the probability that you will randomly select a letter that is *not* U in the word ALBUQUERQUE? (**Lesson 5-6**)

In 20 and 21, tell whether (0, 0) is a solution to the sentence. (Lesson 3-7)

20. $x - 2y < 3$

21. $5x > 4 + \frac{2}{3}y$

EXPLORATION

22. Explain why an equation for the line with x-intercept a and y-intercept b is $\frac{x}{a} + \frac{y}{b} = 1$. (This is called the *intercept form* of the equation for a line.)

23. a. On a graphing calculator, graph the lines with the following equations.

   $3x - 2y = 6$
   $3x - 2y = 12$
   $3x - 2y = 18$

   b. What happens to the graph of $3x - 2y = C$ as C gets larger?
   c. Try values of C that are smaller, including 0 and negative values. What can you say about the graphs of $3x - 2y = C$ in these cases?

QY ANSWERS

1. Answers vary. Sample answer: cost of a neon ball = $0.30, cost of a necklace = $1.25
$30(0.30) + 12(1.25) = 24$.

2. $5x + 4y = 90$

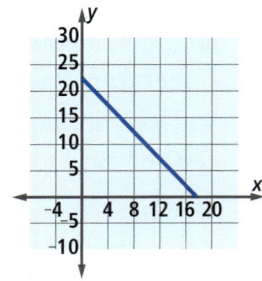

Lesson 6-9

Lesson 6-9
Graphing Linear Inequalities

Vocabulary

boundary line
half-planes
linear inequalities

> **BIG IDEA** The two sides (half-planes) of the line with equation $Ax + By = C$ can be described by the inequalities $Ax + By < C$ and $Ax + By > C$.

In Chapter 3, you graphed solutions to inequalities on a number line. Recall that to graph an inequality such as $x < 2$ you first find the point where $x = 2$. Next, decide which part of the line contains the solution to the inequality. The sentence $x < 2$ states that we want values less than 2, so we shade the points to the left of 2. An open circle is placed on the boundary point 2, because 2 is not a solution to $x < 2$ (2 is not less than 2).

Mental Math

Find a single rule that describes each sequence. Then use your rule to find the next term in the sequence.

a. 2, 5, 8, 11, …

b. 1, 2, 4, 8, 16, …

c. 101, 103, 105, 107, 109, …

These ideas can be extended to graphs of inequalities in two dimensions. In this case, the boundary is a line instead of a point. We call this line the **boundary line.**

Inequalities Involving Horizontal or Vertical Lines

Example 1

Graph $y < 4$ on the coordinate plane.

Solution Graph the line $y = 4$. This horizontal line is the boundary line of the solution. The line is *dashed* to show that the points having a y-coordinate of 4 are not part of the solution set. The solution set consists of all points that have a y-coordinate less than 4. This is the region below the boundary line, so this region is shaded purple.

Check Pick a point in the purple shaded region. We choose $(1, 1)$. Do the coordinates of this point satisfy the inequality $y < 4$? Yes, the y-coordinate is 1 and $1 < 4$.

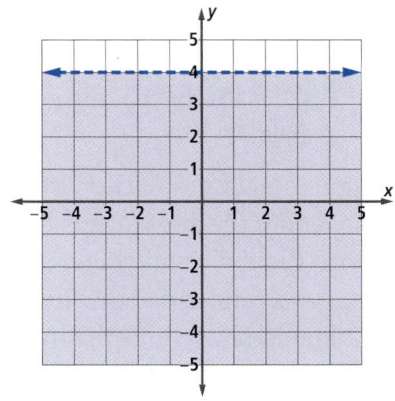

The regions on either side of a line in a plane are called **half-planes.** The boundary line is the edge of the half-plane. In Example 1, the line $y = 4$ is the edge of the half-plane $y < 4$. If you were asked to graph $y \leq 4$, then the boundary line $y = 4$ would be included and shown as a solid line.

Graphing Linear Inequalities **381**

Chapter 6

Example 2

Write an inequality that describes the set of points in the shaded region.

Solution The boundary line is solid, which indicates that the edge $x = 2$ should be included. All points to the right of the line $x = 2$ are shaded purple, meaning every point in the half-plane has an x-coordinate greater than 2. So the sentence describing the region is $x \geq 2$. This region is the union of a half-plane and its edge.

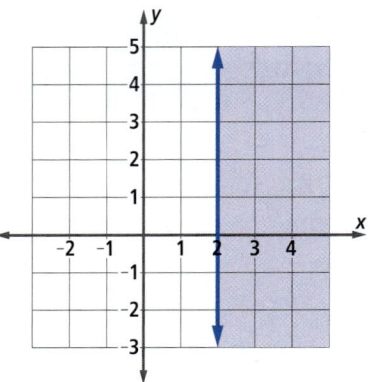

To distinguish $x > 2$ in the coordinate plane from $x > 2$ on a number line, we use set-builder notation. $\{(x, y): x \geq 2\}$ denotes a set of ordered pairs, so its graph is on the coordinate plane. $\{x: x \geq 2\}$ is a set of numbers, so its graph is on a number line.

Inequalities Involving Oblique Lines

Every line with an equation of the form $y = mx + b$ is the boundary line of the two half-planes described by $y < mx + b$ and $y > mx + b$.

Example 3

Draw the graph of $y \leq -2x + 3$.

Solution Begin by graphing the boundary line $y = -2x + 3$. The $\leq$ sign indicates the points on the line should be included in the solutions, so make the line solid. We want the points whose y-coordinates are less than or equal to the y values that satisfy $y = -2x + 3$. Since y values decrease as x increases, shade the region below the line.

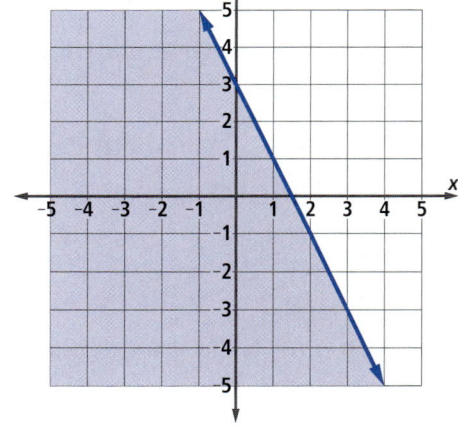

Check Pick a point in the shaded region. We choose (0, 0). Does it satisfy the inequality $y \leq -2x + 3$? Is $0 \leq 2(0) + 3$? Yes, 0 is less than 3, so the correct side of the line has been shaded.

If an inequality is of the form $y < mx + b$, then you shade below the line, and if the sentence is of the form $y > mx + b$, then you shade above the line. But when an inequality is in standard form $Ax + By < C$, you cannot use the inequality sign to determine which side of the line to shade. For this type of inequality, the method of testing a point is usually used. The point (0, 0) can be chosen if it is not on the boundary line. If (0, 0) is a solution to the inequality, then the half-plane that contains (0, 0) is shaded. If (0, 0) does not satisfy the inequality, shade the half-plane on the other side of the boundary line.

382 Slopes and Lines

Lesson 6-9

Example 4

Graph $2x - 3y > 12$.

Solution First graph the boundary line $2x - 3y = 12$. The line is dashed to show that the boundary line is not part of the solution. To determine which side of the line to shade, substitute $(0, 0)$ into the original inequality. Is $2(0) - 3(0) > 12$? No, $0 \not> 12$. Since $(0, 0)$ is in the upper half-plane and is *not* a solution, we shade the half plane that does not contain $(0, 0)$.

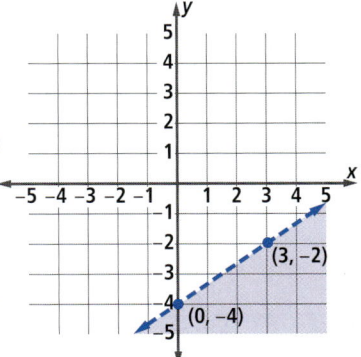

Sentences equivalent to $Ax + By < C$ or $Ax + By \leq C$ are called **linear inequalities.** The preceding examples show that there are two steps to graphing linear inequalities.

Graphing Linear Inequalities

Step 1 Graph the corresponding linear equation. Make this boundary line dashed (for < or >) or solid (for ≤ or ≥).

Step 2 Shade the half-plane that makes the inequality true. (You may have to test a point. If possible, use $(0, 0)$.)

Sometimes the graph of all solutions is not an entire half plane.

Example 5

An elevator has a capacity of 2,500 pounds. If an average adult weighs 150 pounds and an average child weighs 80 pounds, how many adults and children can the elevator hold?

Solution Let $A = $ the number of adults on the elevator. Let $C = $ the number of children on the elevator. Then the total weight of the adults and children is $150 \cdot A + 80 \cdot C$ pounds. So the elevator can hold A adults and C children as long as $150A + 80C \leq 2,500$.

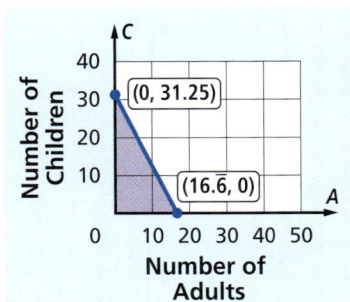

Elisha Graves Otis invented the first safety brake for elevators in 1852, thus starting the elevator industry.

Source: Elevator World, Inc.

We can make either variable be first. We choose A to be the horizontal coordinate and C to be the vertical coordinate. Notice that the domain of both A and C is the set of nonnegative integers. So the graph will have no points in Quadrants II, III, or IV. To graph the boundary line $150A + 80C = 2,500$, we find its intercepts.

When $A = 0$, $C = \frac{2,500}{80} = 31.25$.

When $C = 0$, $A = \frac{2,500}{150} = 16.\overline{6}$.

(continued on next page)

Graphing Linear Inequalities 383

Chapter 6

Because the elevator can hold 0 adults and 0 children, (0, 0) is a solution. So the points will be those on or below the line $150A + 80C = 2{,}500$ in which both coordinates are nonnegative integers. Although the graph is shaded for all rational numbers in the region, integers are the only possible values for A and C.

STOP QY

> ▶ **QY**
>
> If no children are on the elevator in Example 5, use the graph to find the maximum number of adults it can hold.

Questions

COVERING THE IDEAS

1. a. Graph $\{x: x > -7\}$
   b. Graph $\{(x, y): x > -7\}$

2. a. Graph $\{y: y < 0.5\}$
   b. Graph $\{(x, y): y < 0.5\}$

In 3 and 4, write an inequality describing the graph.

3.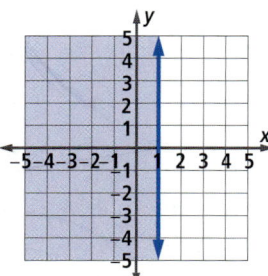

4.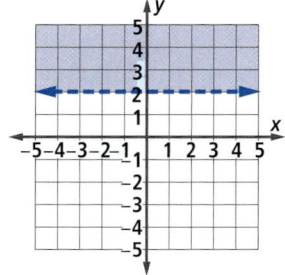

5. a. **Fill in the Blank** A line separates a plane into two distinct regions called   ?   .
   b. **Fill in the Blank** The line in Part a is called a   ?   line.

6. Match the inequality with its graph.

   a.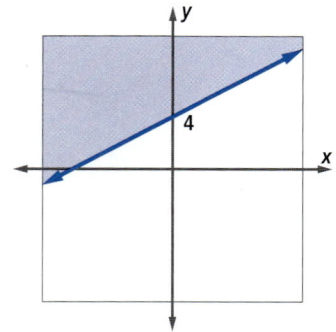
   b.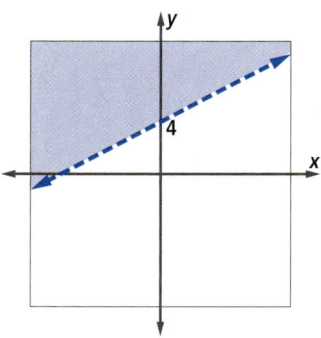
   c.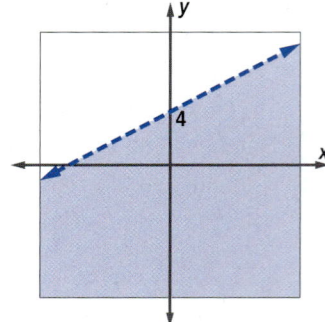

   i. $y < \frac{1}{2}x + 4$
   ii. $y \leq \frac{1}{2}x + 4$
   iii. $y > \frac{1}{2}x + 4$
   iv. $y \geq \frac{1}{2}x + 4$

7. What is the difference between the graphs of $y < -2x + 6$ and $y \leq -2x + 6$?

8. **True or False** The ordered pair (1, 3) is a solution to $y < -2x + 6$.

In 9 and 10, graph all points (x, y) that satisfy the inequality.

9. $x + y < 3$

10. $y \geq 2x - 5$

11. A person in another country read Example 5 and felt the weights should be in kilograms rather than pounds. So here is a similar situation, but using kilograms. An elevator has a capacity of 1,100 kg. If an average adult weighs 65 kg and an average child weighs 35 kg, how many adults and children can the elevator hold? Answer this question with an appropriate graph.

APPLYING THE MATHEMATICS

12. The Strikers volleyball team is selling spirit items to raise money. Pompons (p) cost $7 each and "Go Team!" buttons (b) cost $2.50 each. The team needs to make at least $400. Graph the set of points (p, b) that satisfies these conditions.

13. The scatterplot at the right shows data from the 75 top-ranked players in NCAA Division I men's basketball in 2006. Each point shows the number of field goals a player attempted and the number that the player actually made.

    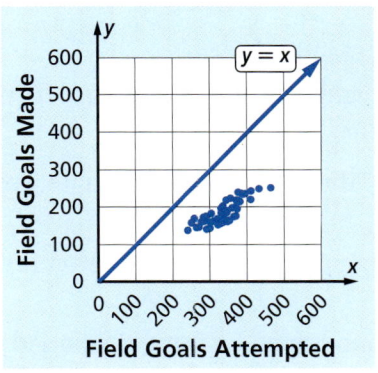

    Source: NCAA

    a. Explain why there are no points above the line $y = x$.
    b. Write an inequality that represents the half-plane bounded by the line $y = x$ and contains the data points.
    c. Suppose a similar scatterplot of pairs (field goals attempted, field goals actually made) is made for a group of 8-year-old players. How would you expect the graph to look compared to the graph for the Division I players?

Approximately 14,578 high schools in the U.S. participate in girls' volleyball.

Source: National Federation of State High School Associations

14. Suppose a person has less than $4.00 in nickels and dimes. Let n = the number of nickels and d = the number of dimes.
    a. Write an inequality to describe this situation.
    b. Give one example of a combination of nickels and dimes that satisfies the inequality.
    c. Graph the number of possible combinations of nickels and dimes.

15. Find a point that satisfies $y \leq x - 6$ but does not satisfy $y < x - 6$.

Graphing Linear Inequalities 385

Chapter 6

16. Refer to the graph at the right.
    a. Find an equation of the boundary line.
    b. Determine the inequality that is graphed.

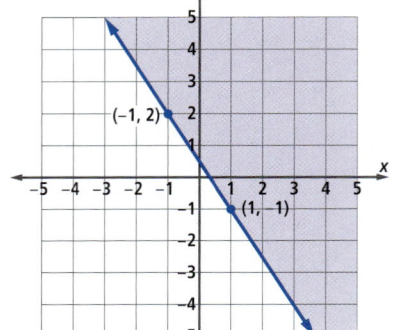

REVIEW

17. Anna and Dion are selling cakes for a bake sale. Some cost $8 and some cost $12. They forgot to keep track of the number of each type of cake they sold, but when the bake sale was over they had collected $288. Let x = the number of $8 cakes, and y = the number of $12 cakes. (**Lessons 6-8, 6-6**)
    a. Write an equation in standard form describing the relationship between x and y.
    b. Find the number of $12 cakes sold if only $12 cakes were sold.
    c. Find the number of $12 cakes sold if fifteen $8 cakes were sold.

18. a. Rewrite $5x - 36 = 3y$ in standard form.
    b. Give the values of A, B, and C from the standard form of an equation. (**Lesson 6-8**)

19. The weight of an object on the moon is one-sixth its weight on Earth. (**Lesson 3-4**)
    a. Write an equation relating an object's weight y on the moon to its weight x on Earth.
    b. How much will a 171-pound man carrying a 9-pound camera weigh on the moon?
    c. If a backpack weighs 6.2 lb on the moon, how much does it weigh on Earth?

20. Evaluate each of the following. (**Lesson 1-1**)
    a. 3^4
    b. $(-4)^3$
    c. $(-2)^8$

21. Rewrite each expression in decimal form. (**Previous Course**)
    a. $1 \cdot 10^{-2}$
    b. $6 \cdot 10^{-3}$
    c. $3 \cdot 4 \cdot 10^{-6}$

EXPLORATION

22. In previous lessons you graphed equations like $y = |x - 3|$ using tables. Graph $y = |x - 3|$; then shade the appropriate region to represent the inequality $y > |x - 3|$. Check a point to verify that your shading is correct.

QY ANSWER

16 adults

386 Slopes and Lines

Chapter 6 Projects

1. Marriage Age

Below are some data on the ages at which men and women first married in the second half of the 20th century.

Median Age at First Marriage

Year	Men	Women
1950	22.8	20.3
1960	22.8	20.3
1970	23.2	20.8
1980	24.7	22.0
1990	26.1	23.9
2000	26.8	25.1

Source: U.S. Census Bureau

a. Plot all the data on one graph. What seems to be the trend shown by these data?

b. Calculate the lines of best fit (one for men and one for women). What do these lines predict about the ages at first marriage for the year 2050? For the year 3000? For the year 1800?

c. Look up the data for the years 1900–1940 at 10-year intervals. Plot these data on a new graph and make two more lines of best fit. What is your prediction for the year 2050? Which of the two sets of lines of best fit do you think will better predict the correct ages for 2050, and why?

2. Paper Towels: Price vs. Absorbency

Are expensive paper towels more absorbent than less expensive ones? Get samples of about six different kinds of paper towels, record their prices, and calculate the price paid per towel. Perform the following experiment to measure absorbency. Fold one piece of towel in half vertically, and then in half again horizontally.

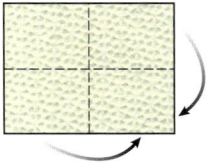

Fill an eyedropper with a fixed amount of water and drop it on the corner with the folds.

Open the towel and measure and record the diameter of the circular area that is wet. Greater diameter denotes greater absorbency.

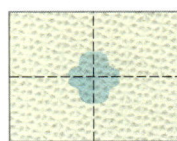

Repeat for each type of towel you have, making sure you use the same amount of water each time.

a. Plot your data with unit price on the horizontal axis and diameter on the vertical axis. Is the relation linear? If so, find a line to describe your data.

b. Plot your data with unit price on the horizontal axis and the area of the wet region on the vertical axis. If the relation is linear, describe it with a line.

c. What advice would you give someone shopping for paper towels?

3 The Diagonal Game

Choose a point (x, y), where both x and y are integers. The goal of the diagonal game is to start at the point $(0, 0)$ and end at the point (x, y), while following these two rules.

Rule 1 You may travel only along lines whose slope is 1 or –1.

Rule 2 At any point (m, n) where both m and n are integers, you may leave the line you were traveling along for a different line (with slope 1 or –1) that passes through that point. For example, suppose $(x, y) = (3, 1)$. You could travel along the line $y = -x$ until you hit the point $(1, -1)$, and then follow the line $y = x - 2$ until you hit the point $(3, 1)$.

a. Play the diagonal game in order to reach the points $(1, 3)$, $(4, -2)$, and $(-5, -1)$. Describe the lines you traveled along and the points at which you switched. At the point or points (m, n) where you switched, calculate $m + n$.

b. How many different lines did you have to travel along to reach each point? Could it have been done with fewer lines?

c. Some points cannot be reached by playing the diagonal game. One such point is $(0, 1)$. Explain why this point cannot be reached. (*Hint:* What do all the numbers $m + n$ that you calculated in Part a have in common?)

d. Suppose that Rule 1 of the game changed to say that you could travel along lines whose slope is 1, –1, or a. Give an example of a value of a that will allow you to reach the point $(1, 0)$.

4 Slopes of Perpendicular Lines

a. For each row of the table below, find an equation for a line with the indicated slope. Then, by experimenting on a graph, find an equation for a line that is perpendicular to the line you have found. Finally, give the slope of the perpendicular line. Record your information in a table like the one shown below.

Slope m	Equation of Line	Equation of Line Perpendicular to Line with Slope m	Slope of Line Perpendicular to Line with Slope m
2			
5			
$\frac{1}{4}$			
–3			
$-\frac{2}{3}$			

b. If a line ℓ has a positive slope, what sign is the slope of a line that is perpendicular to ℓ? If a line ℓ has a negative slope, what sign is the slope of a line that is perpendicular to ℓ?

c. Find a formula relating the slope of a line you started with to the slope of a line perpendicular to it. That is, find a relationship between the numbers in the far left and far right columns of the table.

d. Use the relationship you have found in Part c to determine an equation for the line that is perpendicular to the line with equation $y = -4x + 5$ and containing the point $(1, 3)$.

Chapter 6 Summary and Vocabulary

- The rate of change between two points (x_1, y_1) and (x_2, y_2) is $\frac{y_2 - y_1}{x_2 - x_1}$.

- When points all lie on the same line, the rate of change between them is constant and is called the slope of the line. The slope tells how much the line rises or falls for every move of one unit to the right. When the slope is positive, the line goes up and to the right. When the slope is negative, the line goes down and to the right. When the slope is 0, the line is horizontal. The slope of vertical lines is undefined.

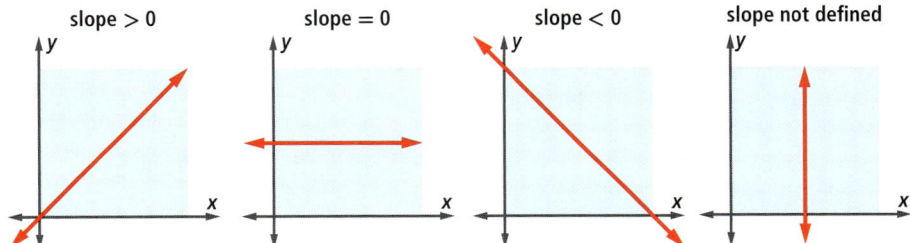

- Constant-increase or constant-decrease situations lead naturally to linear equations of the form $y = mx + b$. The graph of the set of points (x, y) satisfying this equation is a line with slope m and y-intercept b. Linear-combination situations lead naturally to linear equations in the standard form $Ax + By = C$. When the = sign in equations of either form is replaced by < or >, the graph of the resulting linear inequality is a half-plane, the set of points on one side of a line. If the inequality is ≤ or ≥, the boundary line is included in the graph.

- A line is determined by any point on it and its slope or by two points on the line. Its equation can be found from this information. If data are roughly linear, they can be modeled by lines. One model, the line of best fit or least-squares line, is easily found using a calculator or computer.

Vocabulary

6-1
rate of change
rate unit

6-2
slope

6-4
y-intercept
slope-intercept form
direct variation

6-5
point-slope form
x-intercept

6-7
line of best fit
linear regression
least squares line

6-8
linear combination
standard form of an equation for a line
oblique

6-9
boundary line
half-planes
linear inequalities

Theorems and Properties

Slope-Intercept Equation of a Line (p. 350)
Point-Slope Equation of a Line (p. 357)

Chapter 6 Self-Test

Take this test as you would take a test in class. You will need a calculator. Then use the Selected Answers section in the back of the book to check your work.

1. Refer to the line graphed below.

   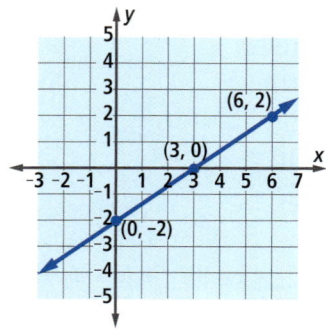

   a. Find its *y*-intercept.
   b. Find its *x*-intercept.
   c. Find its slope.

2. Do the points (8, 2), (18, –6), and (12, –1) lie on the same line? Justify your answer.

3. Find an equation of the line with slope $\frac{8}{9}$ and *y*-intercept –12.

4. The temperature at 8 A.M. is 58°F and is expected to rise 3°F each hour until 4 P.M. Write an equation in slope-intercept form to find the temperature *T* for each hour *h* from 8 A.M. to 4 P.M.

5. Describe the slope of every vertical line.

6. Rewrite the equation $8 - 3x = y$ in standard form $Ax + By = C$ and give the values of *A*, *B*, and *C*.

7. If the slope of a line is $-\frac{1}{2}$, how does the *y*-coordinate change as you go one unit to the right?

8. Find the slope and *y*-intercept of the line with equation $4x - 8y = 2$.

9. Lenny has $6,400 to spend at a music store buying trumpets and trombones for a jazz band. Suppose the cost of each trumpet *p* is $300 and the cost of each trombone *b* is $400.

   a. Write an inequality that describes all possible values of *p* and *b*.
   b. Graph these values.

In 10 and 11, use the following data of average yearly attendance at Pennsylvania State University football games.

Year	Average Attendance
1999	96,500
2000	95,543
2001	107,576
2002	107,239
2003	105,629
2004	103,111

Source: NCAA

10. Between which two consecutive years was there the greatest decrease in average yearly attendance?

11. What is the rate of change in average yearly attendance from 1999 to 2004?

In 12 and 13, graph the sentence.

12. $y = \frac{3}{8}x - \frac{5}{8}$

13. $y \leq 6 - 4x$

14. After driving for 22 miles, the gas tank in Masao's car had 14 gallons of gas. After 132 miles, the gas tank had 9 gallons of gas. Find a linear equation relating the amount of gas *y* in Masao's gas tank to the miles he has driven *x*.

15. Find another point on the line through the point (−4, 0) with slope $\frac{5}{4}$.

In 16 and 17, consider the data table and scatterplot that show the annual salary of employees of a company and the number of years they have worked for the company.

Years of Employment	Average Annual Salary (in thousands)
2	$35
3	$40
4	$39
5	$45
6	$60
7	$55
8	$62
9	$64
10	$68
12	$63
13	$75
15	$78
16	$80
17	$88
19	$82

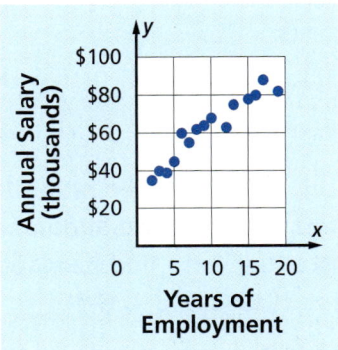

16. a. Find the coordinates of two points on a line that you estimate would fit these data.
    b. Write an equation for your line in slope-intercept form.
    c. Use the equation from Part b to estimate the annual salary of an employee who has worked at the company for 25 years.

17. Find an equation for the line of best fit for the data.

In 18 and 19, use the graph of lines a, b, c, and d below.

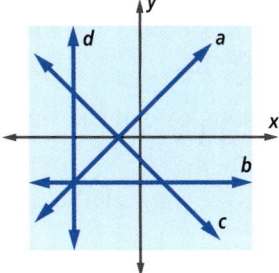

18. a. Name the line that has a negative slope.
    b. Name the line that has a slope of 0.

19. a. Which line could represent money in an account over time where money is added to the account at a constant rate?
    b. Name the line that has an x-intercept but does not have a y-intercept.

20. Felipe has only x ten-dollar bills and y five-dollar bills in his wallet. Graph all possible values of x and y if the total amount of money in his wallet is less than $100.

21. Patrick left his apartment and headed down the building stairs. He descended at a constant rate of 3 flights per minute. After 4 minutes, he was on the 10th floor.
    a. Write an equation in point-slope form relating the floor Patrick was on, y, to the number of minutes, x, it has been since he left.
    b. Use your equation to find the floor of Patrick's apartment.

Chapter 6 Chapter Review

SKILLS
PROPERTIES
USES
REPRESENTATIONS

SKILLS Procedures used to get answers

OBJECTIVE A Find the slope of the line through two given points. (Lesson 6-2)

1. Calculate the slope of the line containing (2, 9) and (7, 15).

2. Calculate the slope of the line through (−5, 1) and (−8, −8).

3. Find the slope of line ℓ below.

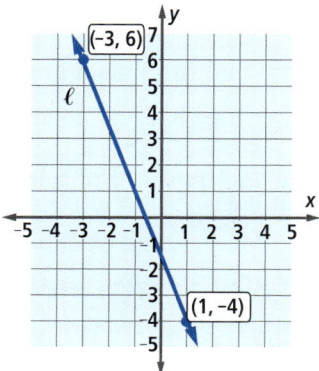

4. Using two different pairs of points, show that the slope of line m below is $\frac{2}{3}$.

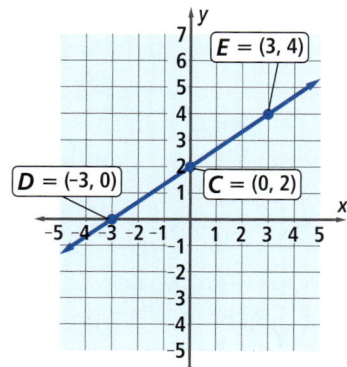

5. The points (3, −1) and (5, y) are on a line with slope 2. What is y?

6. The points (x, −6) and (2, 12) are on a line with slope $-\frac{3}{4}$. What is x?

OBJECTIVE B Find an equation for a line given either its slope and any point or two points on it. (Lessons 6-4, 6-5, 6-6)

7. Give an equation for the line with slope $2\frac{3}{5}$ and y-intercept 7.

8. What is an equation for the line with slope s and y-intercept t?

In 9–11, find an equation for the line through the given point with slope m.

9. (−5, 20), $m = -0.5$

10. (18, −16), $m = 0$

11. (30, 0.25), $m = 3$

12. What is an equation for the line through (19, −11) with undefined slope?

In 13–16, find an equation for the line through the two given points.

13. (3, 4), (1, 1) 14. (−16, −18), (0, 0)
15. (−3, −8), (−3, 8) 16. (1, 1), (2, 1)

OBJECTIVE C Write an equation for a line in standard form or slope-intercept form, and using either form find its slope and y-intercept. (Lessons 6-4, 6-8)

In 17 and 18, write the equation in the form $Ax + By = C$. Then give the values of A, B, and C.

17. $x + 14 = 2y$ 18. $y = \frac{7}{16}x + 14$

In 19 and 20, rewrite the equation in slope-intercept form.

19. $8x + 4y = 10$ 20. $30x - 90y = 270$

In 21–24, find the slope and y-intercept of the line.

21. $y = 12x - 6.2$
22. $3x - 6y = 2$
23. $x + y = 0$
24. $x = 3y + 500$

PROPERTIES The principles behind the mathematics

OBJECTIVE D Use the definition and properties of slope. (Lessons 6-2, 6-3)

25. **Fill in the Blanks** The slope determined by two points is the change in the __?__ coordinates divided by the __?__ in the x-coordinates.

26. **Fill in the Blanks** Slope is the amount of change in the __?__ of the graph for every change of one unit to the __?__.

In 27 and 28, five lines are shown below, including the x-axis and y-axis.

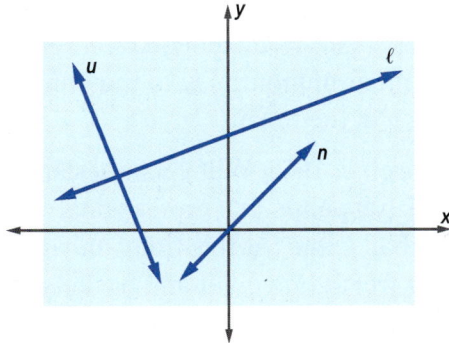

27. Which line or lines have negative slope?
28. Which line or lines have positive slope?
29. What is the slope of the line with equation $y = 5$?
30. What is true about the slope of the line with equation $x = 2$?
31. How can you use slope to show that three points do not all lie on the same line?

USES Applications of mathematics in real-world situations

OBJECTIVE E Calculate rates of change from real data and describe their real-world meanings. (Lessons 6-1, 6-3)

32. Consider the following ordered pairs of hours worked and pay received: (12, $93.60), (15, $117.00), (17, $132.60).

    a. Calculate the slope of the line through these points.
    b. What does the slope represent in this situation?

33. The picture below represents a ski slope. What is the slope of this ski slope?

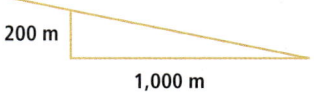

In 34–36, use the table below that shows the number of subscribers to a magazine.

Year	Subscribers
1985	1,145
1990	1,358
1995	1,601
2000	1,886
2005	2,023

34. Find the rate of change of the number of subscribers from 1985 to 1990.
35. Find the rate of change of the number of subscribers from 2000 to 2005.
36. a. According to these data, in which ten-year period did the number of subscribers increase the fastest?
    b. What is this rate of change?

In 37–39, use the table below of mean Fahrenheit temperatures each month in Fairbanks, Alaska, based on records from 1971–2000.

Month	Month Number	Mean (°F)
Jan.	1	−10
Feb.	2	−4
Mar.	3	11
Apr.	4	32
May	5	49
Jun.	6	60
Jul.	7	62
Aug.	8	56
Sept.	9	45
Oct.	10	24
Nov.	11	2
Dec.	12	−6

Source: National Climatic Data Center

37. Find the average rate of change of temperature per month from July to December.

38. Between which two months is the rate of change of temperature per month the greatest?

39. Between which two months is the rate of change of temperature per month the least (it will be negative)?

OBJECTIVE F Use equations for lines to describe real situations. (Lessons 6-4, 6-5, 6-6, 6-8)

In 40 and 41, each situation can be represented by a straight line. Give the slope and y-intercept of the line describing this situation.

40. Julie rents a truck. She pays an initial fee of $29.95 and then $0.49 per mile driven. Let y be the cost of driving x miles.

41. Nestor is given $1,000 to spend on a vacation. He decides to spend at most $100 a day. Let y be the minimal amount Nestor has left after x days.

42. The 28th Summer Olympic games were in 2004. The 27th summer Olympic games were 4 years earlier. Let y be the year of the nth summer Olympic games. Give a linear equation which relates n and y.

In 43 and 44, each situation leads to an equation of the form $y = mx + b$. Find that equation.

43. A stack of 25 small paper cups is 8 in. high. Each additional cup adds $\frac{1}{4}$ inch to the stack. Let y be the height of the stack when there are x cups.

44. A plane loses altitude at the rate of 5 m/sec. It begins at an altitude of 8,000 m. Let y be its altitude after x seconds.

45. Each month, about 50 new people move into a town. After 5 months, the town has 25,600 people. Write an equation relating the number of months m to the number of people p in the town.

46. On March 2, 1962, Wilt Chamberlain scored 100 points in a professional basketball game. At this time, only free throws worth one point and baskets worth two points were possible. (There were no 3-point shots.) Let F be the number of free throws and B be the number of baskets a team might make to score a total of 100 points. What equation do F and B satisfy?

47. Roberto baby-sat for $7 an hour and worked in a store for $8 an hour. He earned a total of $820. Write an equation that describes the possible hours B of baby-sitting and hours S of store work that he could have worked at those jobs.

OBJECTIVE G Given data whose graph is approximately linear, find a linear equation to fit the graph and make predictions about data values. (Lesson 6-7)

48. Beef production in the United States has tended to increase since 1970. Here is the production in selected years.

Year	Beef Production (millions of pounds)
1970	21,684
1980	21,643
1990	22,743
1995	25,525
2000	27,338
2003	26,339

Source: U.S. Dept. of Agriculture

a. Graph the data and draw a line of fit.
b. Find an equation for your line.
c. Use the equation to predict the amount of U.S. beef production in 2010.
d. Calculate the line of best fit for these data. Use that line to predict the amount of U.S. beef production in 2010.

49. The table below shows the number of fish, crustaceans, and mollusks caught in the Philippines from 1997 to 2002.

Year	Weight (thousands of tons)
1997	327
1998	313
1999	353
2000	394
2001	435
2002	443

Source: Food and Agricultural Organization of the United Nations

a. Graph the data and eyeball a line to fit the data.
b. Find the slope of the fitted line.
c. Explain what the slope tells you about the trend in the data.
d. Find an equation for the eyeballed line.
e. Find an equation for the line of best fit for these data.
f. By how much do your line and the line of best fit differ in their predictions for the number of tons that would be caught in 2010?

REPRESENTATIONS Pictures, graphs, or objects that illustrate concepts

OBJECTIVE H Graph a line given its equation, or given a point and its slope. (Lessons 6-3, 6-4, 6-8)

In 50–53, graph the line with the given equation.
50. $y = 3x + 5$
51. $y = -\frac{1}{2}x + 8$
52. $7x - 5y = 70$
53. $x + 2y = -6$

In 54–57, graph the line satisfying the given condition.
54. passes through $(0, 1)$ with a slope of 0.4
55. passes through $(4, 7)$ with a slope of -17
56. slope 8 and y-intercept -8
57. slope $-2\frac{1}{2}$ and y-intercept 10

OBJECTIVE I Graph linear inequalities. (Lesson 6-9)

58. Choose the correct words. The graph of a linear inequality with a $<$ sign is a (line, plane, half-plane) that (does, does not) contain its boundary line.

59. If you have only n nickels and q quarters and a total of less than $\$1.00$, graph all possible values of n and q.

In 60–65, graph the inequality on the coordinate plane.
60. $x \leq 12$
61. $y \geq 4.5$
62. $y \geq 2x + 1$
63. $y < -x + 5$
64. $2x - 5y < 10$
65. $x + 4y \geq 0$

Chapter 7
Using Algebra to Describe Patterns of Change

Contents

- **7-1** Compound Interest
- **7-2** Exponential Growth
- **7-3** Exponential Decay
- **7-4** Modeling Exponential Growth and Decay
- **7-5** The Language of Functions
- **7-6** Function Notation
- **7-7** Comparing Linear Increase and Exponential Growth

A small city of 100,000 people has been growing. School planners want to know how many classrooms the city might need during the next 50 years. They consider three possibilities.

1. The population stays the same.
2. The population increases by 3,000 people per year (increasing by a constant amount).
3. The population grows by 2% a year (increasing at a constant growth rate).

The graph on the next page shows what would happen under the three possibilities. P is the population x years from now.

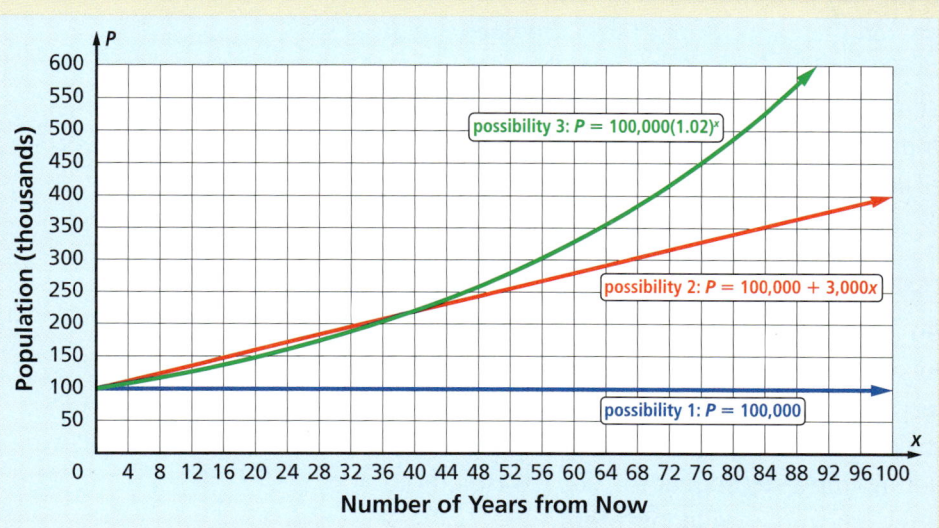

Possibility 3 is often considered the most reasonable. Under this assumption, $P = 100{,}000(1.02)^x$. Because the variable x is an exponent, this equation is said to represent *exponential growth*. This chapter discusses the important applications of exponential growth and compares them with the constant-increase and constant-decrease situations you studied in Chapter 6.

Chapter 7

Lesson 7-1

Compound Interest

Vocabulary

power, *n*th power
base
exponent
principal
interest
annual yield
compound interest

▶ **BIG IDEA** Compound interest is the way most banks and other savings institutions pay savers who put their money into their accounts.

Mental Math

Write as a fraction in lowest terms.

a. 0.02

b. 1.75

c. –3.57

Powers and Repeated Multiplication

A number having the form x^n is called a **power.** When n is a positive integer, x^n describes repeated multiplication. For example, $10^3 = 10 \cdot 10 \cdot 10 = 1{,}000$ and $3^5 = 3 \cdot 3 \cdot 3 \cdot 3 \cdot 3 = 243$. These are examples of the following property.

> **Repeated Multiplication Property of Powers**
>
> When n is a positive integer, $x^n = \underbrace{x \cdot x \cdot \ldots \cdot x}_{n \text{ factors}}$.

The number x^n is called the **nth power** of x and is read "x to the nth power" or just "x to the n." In the expression x^n, x is the **base** and n is the **exponent.** Thus, 3^5 is read "3 to the 5th power," or "3 to the 5th," where 3 is the base and 5 is the exponent. In the expression $100{,}000(1.02)^x$ found on page 397, 1.02 is the base and x is the exponent. The number 100,000 is the coefficient of the power 1.02^x.

How Is Interest Calculated?

An important application of exponents and powers occurs with savings accounts. When you save money, you can choose where to put it. Of course, you can keep it at home, but banks, savings and loan associations, and credit unions will pay you to let them hold your money for you. The amount you give them at the start is called the **principal.** The amount they pay you is called **interest.**

Interest is always a percent of the principal. The percent that the money earns per year is called the **annual yield.**

Lesson 7-1

Example 1
Suppose you deposit *P* dollars in a savings account on which the bank pays an annual yield of 4%. If the account is left alone, how much money will be in it at the end of a year?

Solution

Total = principal + interest (4% of principal)
= $P + 0.04P$
= $(1 + 0.04)P = 1.04P$

You will have 1.04P, or 104%, of the principal.

 QY1

> ▶ QY1
>
> If you deposited $1,000 in a savings account with an annual yield of 4%, what would you have at the end of a year?

Compound Interest and How It Is Calculated

When the year is up, the account will have extra money in it because of the interest it earned. If that money is left in the account, then at the end of second year, the bank will pay interest on all the money that is now in the account (the original principal and the first year's interest). This leads to **compound interest**, which means that the interest earns interest.

Example 2
Suppose you deposit $100 in a savings account on which the bank pays an annual yield of 4%. Assume the account is left alone in Parts a and b.
a. How much money will be in the account at the end of 4 years?
b. How much interest would you earn in the 4 years?

Solution

a. Refer to Example 1. Each year the amount in the bank is multiplied by $1 + 0.04 = 1.04$.

End of first year: $100(1.04) = 100(1.04)^1 = 104.00$

End of second year: $100(1.04)(1.04) = 100(1.04)^2 = 108.16$

End of third year:
$100(1.04)(1.04)(1.04) = 100(1.04)^3 = 112.4864 \approx 112.48$

End of fourth year:
$100(1.04)(1.04)(1.04)(1.04) = 100(1.04)^4 \approx 116.9858 \approx 116.98$

At the end of 4 years there will be $116.98 in the account.

b. Because you started with $100, you earned $116.98 − $100 = $16.98 in the 4 years.

Compound Interest **399**

Examine the pattern in the solution to Example 2. At the end of t years there will be $100(1.04)^t$ dollars in the account. By replacing 100 by P for principal, and 0.04 by r for the *annual yield*, we obtain a general formula for compound interest.

> **Compound Interest Formula**
>
> If a principal P earns an annual yield of r, then after t years there will be a total amount A, where $A = P(1 + r)^t$.

The compound interest formula is read "A equals P times the quantity 1 plus r, that quantity to the tth power."

GUIDED

Example 3

When Jewel was born, her parents put $2,000 into an account for college. What will be the total amount of money in the account after 18 years at an annual yield of 5.4%?

Solution Here $P = \$2{,}000$, $r = 5.4\%$, and $t = 18$. Substitute the values into the Compound Interest Formula. Use $5.4\% = 0.054$.

$A = P(1 + r)^t$
$= \underline{}(1 + \underline{})^{\underline{}}$

To evaluate this expression, use a calculator key sequence such as the following.

$\underline{}$ ⨯ $\underline{}$ ^ $\underline{}$ ENTER

Your display shows $\underline{}$, which rounded down to the nearest cent is $\underline{}$.

In 18 years, at an annual yield of 5.4%, $2,000 will increase to $\underline{}$.

 QY2

Eighteen years may seem like a long time, but it is not an unusually long amount of time for money to be in college accounts or retirement accounts.

Why Do You Receive Interest on Savings?

Banks and other savings institutions pay you interest because they want money to lend to other people. The bank earns money by charging a higher rate of interest on the money they lend than the rate they pay customers who deposit money.

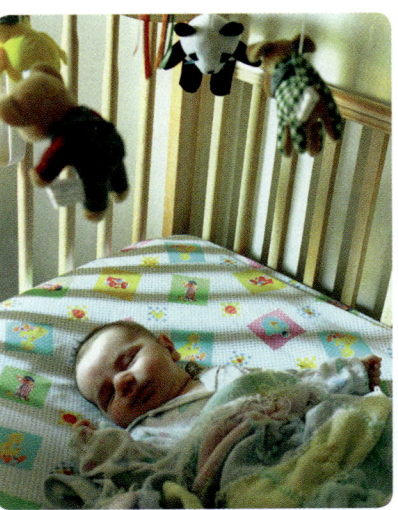

Tuition fees at public four-year colleges increased 35% between 2001 and 2006.

Source: The College Board

▶ **QY2**

Suppose you invest $6,240 in an account at 6.3% annual yield for 10 years. How much will be in the account at the end of the 10 years?

Thus, if the bank could loan the $1,000 you deposited at 4% (perhaps to someone buying a car) at 12% a year, the bank would receive 0.12($1,000), or $120 from that person. So the bank would earn $120 − $40 = $80 in that year on your money. Part of that $80 goes for salaries to the people who work at the bank, part for other bank costs, and part for profit to the owners of the bank.

Questions

COVERING THE IDEAS

1. How is the expression 4^{10} read?
2. Consider the expression $10x^9$. Name each of the following.
   a. base
   b. power
   c. exponent
   d. coefficient
3. a. Calculate 7^3 without a calculator.
   b. Calculate 7^3 with a calculator. Show your key sequence.

In 4–6, rewrite the following expressions using exponents.

4. $\underbrace{\frac{5}{9} \cdot \frac{5}{9} \cdot \ldots \cdot \frac{5}{9}}_{t \text{ times}}$

5. $18 \cdot -3 \cdot -3 \cdot -3 \cdot -3$

6. $21 \cdot x \cdot x \cdot x \cdot x \cdot x \cdot x \cdot x \cdot x \cdot x$

7. On page 397, three possibilities are offered for the growth of a population. What is the predicted population in 50 years using the indicated possibility?
   a. Possibility 1
   b. Possibility 2
   c. Possibility 3

8. **Matching** Match each term with its description.
   a. money you deposit   i. annual yield
   b. interest paid on interest   ii. compound interest
   c. yearly percentage paid   iii. principal

In 9 and 10, write an expression for the amount in the bank after 1 year if P dollars are in an account with the annual yield given.

9. 2%
10. 3.25%

11. a. Write the Compound Interest Formula.
    b. What does A represent?
    c. What does P stand for?
    d. What is r?
    e. What does t represent?

Chapter 7

In 12–14, assume the interest is compounded annually.

12. Suppose you deposit $300 in a new savings account paying an annual yield of 2.5%. If no deposits or withdrawals are made, how much money will be in the account at the end of 5 years?

13. A bank advertises an annual yield of 4.81% on a 5-year CD (certificate of deposit). If the CD's original amount was $2,000, how much will it be worth after 5 years?

14. How much interest will be earned in 7 years on a principal of $1,000 at an annual yield of 5.125%?

APPLYING THE MATHEMATICS

In 15 and 16, assume the interest is compounded annually.

15. Susana invests $250 at an annual yield of 4%. Jake invests $250 at an annual yield of 8%. They leave the money in the bank for 2 years.
    a. How much interest does each person earn?
    b. Jake's interest rate is twice Susana's. Does Jake earn twice the interest that Susana does? Why or why not?

16. Which yields more money, (a) an amount invested for 6 years at an annual yield of 5%, or (b) the same amount invested for 3 years at an annual yield of 10%? Explain your answer.

In 17–19 on the next page, use the following: Danica invested $100 in an account that earns an annual yield of 10%. On the same day, Todd deposited $200 in an account earning 5% annually. Below are a graph and a spreadsheet that compare the amount in Danica's and Todd's accounts.

	A	B	C
1	Time since Investment (yr)	Danica's Account	Todd's Account
2	0	$100.00	$200.00
3	2	$121.00	$220.50
4	4	$146.41	$243.10
5	6	$177.15	$268.01
6	8	$214.37	$295.49
7	10	$259.37	$325.78
8	12	$313.84	$359.17
9	14		
10	16		
11	18		
12	20		

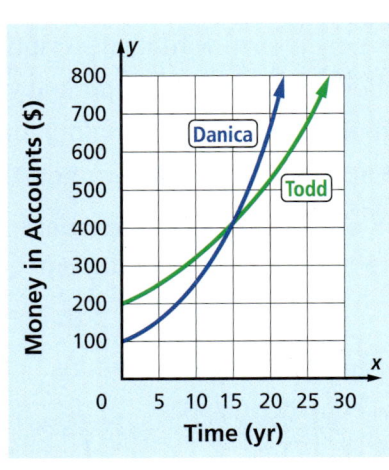

402   Using Algebra to Describe Patterns of Change

17. a. What formula will show the amount in Danica's account after t years?

    b. Complete the column indicating the amounts in Danica's account.

18. Repeat Question 17 for Todd's account.

19. In what year will Danica and Todd have the same amount in their accounts?

20. Use your calculator to make a table. If a principal of $1,000 is saved at an annual yield of 5% compounded annually and nothing is withdrawn from the account, in how many years will it double in value?

REVIEW

21. In World Cup Soccer, a team gets 3 points for a win and 1 point for a tie. Let W be the number of wins and T the number of ties. (Lessons 6-9, 3-7)

    a. If a team has more than 3 points, what inequality must W and T satisfy?

    b. Graph all possible pairs (W, T) for a team that has played 3 games and has more than 3 points.

Javier Mascherano of Argentina and Mario Götze of Germany compete for the ball during the 2014 World Cup Final in Brazil.

22. Miho puts $6.00 into her piggy bank. Each week thereafter she puts in $2.50. (The piggy bank pays no interest.)

    a. Write an equation showing the total amount of dollars Y after X weeks.

    b. Graph the equation. (Lesson 6-2)

23. Find the probability of getting a number that is a factor of 12 in one toss of a fair die. (Lesson 5-6)

In 24 and 25, solve the sentence. (Lessons 4-5, 4-4)

24. $38c - 14 = 6(c - 3) + 4$

25. $8(2 + \frac{1}{8}u) > 2u + 1 - u$

26. **Multiple Choice** Which formula describes the numbers in the table at the right? (Lesson 1-2)

    A $y = x + (x + 1)$
    B $y = 2x$
    C $y = 2^x$
    D $y = x^2$

x	0	2	4	6	8
y	1	4	16	64	256

27. Find t if $2^t = 32$. (Previous Course)

EXPLORATION

28. Find out the yield for a savings account in a bank or other savings institution near where you live. (Often these yields are in newspaper ads.)

QY ANSWERS

1. $1,040

2. $11,495.20

Chapter 7

Lesson 7-2
Exponential Growth

Vocabulary

exponential growth
growth factor
exponential growth equation

> **BIG IDEA** Growth at a constant percentage rate can be described by an expression of the form bg^x, where $g > 1$ and the variable is in the exponent.

Powering and Population Growth

An important application of powers is in population growth situations. As an example, consider rabbit populations, which can grow quickly. In 1859, 24 rabbits were imported to Australia from Europe as a new source of food. Rabbits are not native to Australia, but conditions there were ideal for rabbits and so they flourished. Soon, there were so many rabbits that they damaged grazing land. By 1887, the government was offering a reward for a way to control the rabbit population. How many rabbits might there have been in 1887? Example 1 provides an estimate.

Mental Math

Solve each inequality.
a. $2x < 5$
b. $-4m + 3 > 14$
c. $9 + 3b \leq 3 - b$

Example 1

Twenty-six rabbits are introduced to another area. Assume that the rabbit population doubles every year. How many rabbits would there be after 28 years?

Solution Since the population doubles every year, in 28 years it will double 28 times. The number of rabbits will be

$$26 \cdot \underbrace{2 \cdot 2 \cdot 2 \cdot \ldots \cdot 2}_{28 \text{ factors}}$$

To evaluate this expression on a calculator, rewrite it as $26 \cdot 2^{28}$. Use the y^x or $\wedge$ key. There would be 6,979,321,856, or about 7 billion rabbits after 28 years.

A pet rabbit's diet should be made up of good quality pellets, fresh hay (alfalfa, timothy, or oat), water, and fresh vegetables.

Source: House Rabbit Society

What Is Exponential Growth?

The rabbit population in Example 1 is said to grow exponentially. In **exponential growth**, the original amount is repeatedly *multiplied* by a positive number called the **growth factor**.

404 Using Algebra to Describe Patterns of Change

> **Growth Model for Powering**
>
> If a quantity is multiplied by a positive number g (the growth factor) in each of x time periods, then, after the x periods, the quantity will be multiplied by g^x.

In Example 1, the population doubles (is multiplied by 2) every year, so $g = 2$. There are 28 time periods, so $x = 28$. The original number of rabbits, 26, is multiplied by g^x, or 2^{28}. So the population y of rabbits after x years is given by the formula $y = 26 \cdot 2^x$.

In general, if the amount at the beginning of the growth period is b, the growth factor is g, and y is the amount after x time periods, then $y = b \cdot g^x$. We call this the **exponential growth equation.**

The compound interest formula $A = P(1 + r)^t$ is another example of an exponential growth equation. Suppose $5,000 is invested at an annual yield of 4%. Using the variable names of the exponential growth equation, $b = 5,000$ and $g = 1.04$. So y, the value of the investment after x years, is given by $y = 5,000 \cdot 1.04^x$.

If the money is kept invested for 11 years, then $x = 11$, and the total amount will be $5,000 \cdot 1.04^{11}$, which is $7,697.27.

What Happens If the Exponent Is Zero?

In the exponential growth equation $y = b \cdot g^x$, x can be any real number. Consider the situation when $x = 0$. In 0 time periods no time has elapsed. The starting amount b has not grown at all and so it remains the same. It can remain the same only if it is multiplied by 1. This means that $g^0 = 1$, regardless of the value of the growth factor g. This property applies also when g is a negative number.

> **Zero Exponent Property**
>
> If x is any nonzero real number, then $x^0 = 1$.

In words, the zero power of any nonzero number equals 1. For example, $4^0 = 1$, $(-2)^0 = 1$, and $\left(\frac{5}{7}\right)^0 = 1$. The zero power of 0, which would be written 0^0, is undefined.

What Does a Graph of Exponential Growth Look Like?

An equation of the form $y = b \cdot g^x$, where g is a number greater than 1, can describe exponential growth. Graphs of such equations are not lines. They are *exponential growth curves*.

Example 2

Graph the equation $y = 1.5 \cdot 2^x$, when x is 0, 1, 2, 3, and 4.

Solution Substitute $x = 0, 1, 2, 3,$ and 4 into the formula $y = 1.5 \cdot 2^x$. Below we show the computation and the results listed as (x, y) pairs.

Computation	(x, y)
$1.5 \cdot 2^0 = 1.5 \cdot 1 = 1.5$	(0, 1.5)
$1.5 \cdot 2^1 = 1.5 \cdot 2 = 3$	(1, 3)
$1.5 \cdot 2^2 = 1.5 \cdot 4 = 6$	(2, 6)
$1.5 \cdot 2^3 = 1.5 \cdot 8 = 12$	(3, 12)
$1.5 \cdot 2^4 = 1.5 \cdot 16 = 24$	(4, 24)

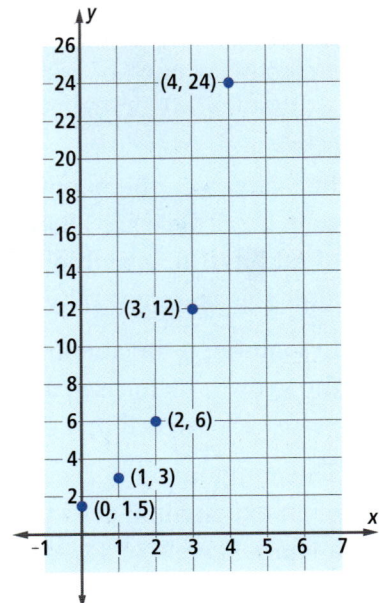

Notice that the *y*-intercept of the graph in Example 2 is 1.5. Also, the graph does not have a constant rate of change. When something grows exponentially, its rate of change is continually increasing.

Now/Next Method

Example 2 showed how to compute *y*-values. In an exponential growth equation, $y = b \cdot g^x$, you can compute the *y*-values by substituting for *x*. However, that method might not always be the fastest.

Numbers in an exponential growth pattern can be displayed on the homescreen of a graphing calculator using the Now/Next method.

Step 1 Type the starting value 48 (the Now) and press [ENTER].

Step 2 Multiply by the growth factor 1.5. The calculator represents the Now by Ans (for "answer"). The calculator at the right is now programmed for Now/Next. By pressing [ENTER], it automatically performs Now $\cdot$ 1.5 to get the Next term.

Try this on a calculator. Notice how quickly the table values can be displayed by repeatedly pressing [ENTER].

 QY

Exponential Population Growth

Other than money calculated using compound interest, few things in the real world grow exactly exponentially. However, exponential growth curves can be used to approximate changes in population. For example, consider the population of California from 1930 to 2000 shown in the table on the next page.

▶ **QY**

A Now/Next table is shown below.

x	y	
0	4	
1	?	$\cdot 2.5$
2	?	$\cdot 2.5$
3	?	$\cdot 2.5$

a. What are the three numbers that fill in the missing cells if the growth factor is 2.5?

b. Write an equation of the form $y = b \cdot g^x$ for the ordered pairs (x, y).

California's population is graphed below. The curve is the graph of the exponential growth model $y = 5.68(1.29)^x$, where y is the population (in millions) x decades after 1930. The growth factor 1.29 was chosen because it is an "average" growth factor for California for the six decades shown. The points lie quite close to the curve, indicating that California's population since 1930 has grown about 29% per decade.

Year	Population
1930	5,677,251
1940	6,907,387
1950	10,586,223
1960	15,717,204
1970	19,971,069
1980	23,667,764
1990	29,760,021
2000	33,871,648

Source: U.S. Census Bureau

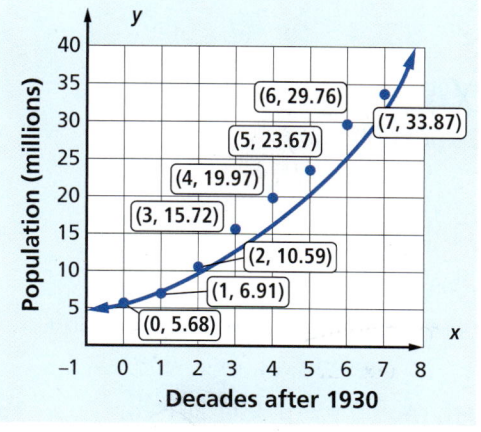

$y = 5.68(1.29)^6$

Questions

COVERING THE IDEAS

1. A round goby is a bottom-dwelling fish native to Eastern Europe. In 1995, it was found in the Great Lakes, where it is expected to be harmful to already existing habitats. The round goby is known to spawn several times during the summer, and biologists are tracking the growth of its population. Suppose 11 round gobies were in the Great Lakes in 1995 and that their population triples in size each year.
   a. Write an exponential growth equation to describe this situation.
   b. How many round gobies were in the Great Lakes after 2 years?
   c. How many round gobies were in the Great Lakes in 2000?
   d. How many round gobies will there be in the Great Lakes in 2025?

Round gobies have a well-developed sensory system that allows them to feed in complete darkness.

Source: University of Wisconsin Sea Grant

2. Copy the table at the right and complete it to make a Now/Next table for $y = 50 \cdot 1.2^x$.

3. Suppose that $3,000 is invested in an account with a 4.5% annual yield.
   a. What is the growth factor?
   b. What will be the value of the account after two years?

x	y
0	?
1	?
2	?
3	?

Exponential Growth

Chapter 7

4. **True or False** An amount is multiplied by 10 in each of 12 time periods. After the 12 time periods, the original amount will be multiplied by 120.

In 5 and 6, evaluate the expression when $x = 17$, $y = 1.05$, and $z = 1.04$.

5. a. x^0 b. $x \cdot x^0$ c. $(x + x)^0$
6. a. y^0 b. $y^0 - z^0$ c. $(z - y)^0$

7. Explain why $(-5)^0$ and -5^0 are not equal.

8. Explain how $g^0 = 1$ applies to exponential growth.

9. Let $y = 0.5 \cdot 2^x$.
   a. Make a table for $x = 0, 1, 2, 3, 4$.
   b. Graph the values from Part a.

10. Write an equation of the form $y = b \cdot g^x$ to fit the numbers in the calculator display at the right.

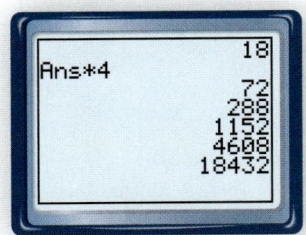

11. Consider the exponential growth model $y = 5.68(1.29)^x$ for California's population on page 407.
    a. How much does the model's value for 1990 deviate from the actual population?
    b. What does the model predict for the population of California in 2020?

APPLYING THE MATHEMATICS

12. Alexander Fleming discovered penicillin by observing mold growing on petri dishes. Suppose you are biochemist studying a type of mold that has grown from 3,000 spores to 192,000 spores in one hour. You record the following information.

Time Intervals from Now	Time (min)	Number of Mold Spores
0	0	3,000
1	20	12,000
2	40	48,000
3	60	192,000

    a. What is the growth factor as the time interval increases by 1?
    b. How many mold spores would there be after 2 hours? Explain how you found your answer.
    c. If this growth rate continues, how many mold spores will there be after x hours have passed?

Before tossing away some old petri dishes in 1928, Alexander Fleming accidentally discovered a blue mold growing on the culture of a harmful bacteria.

Source: San Jose State University

408 Using Algebra to Describe Patterns of Change

13. The equation $y = 34{,}277 \cdot 1.04^x$ can be used to model the population y of Colorado x years after 1860. This graph shows Colorado's actual population.

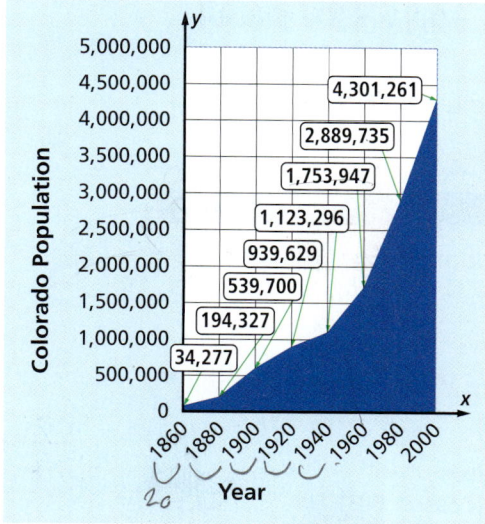

Source: Bureau of the Census

   a. What do 1.04 and 34,277 in the equation represent?
   b. What does the equation predict for the population of Colorado in 1960? By how much did the actual population deviate from the prediction?
   c. Use the model to predict the population of Colorado in 2030.

14. Gossip can be spread quickly in a school. Suppose one person begins spreading the gossip by telling 2 friends. Each friend then tells 2 of his or her different friends. Each person who hears the gossip continues to tell 2 more different friends.

   a. Complete the table below showing the number of new friends and total number of people who have heard the gossip if the pattern continues.

two friends sharing a secret

Stage of Gossip	0	1	2	3	4	5	6	7	8	9	10
New Friends Informed	1	2	4	?	?	?	?	?	?	?	?
Total Number of Friends Informed	1	3	7	?	?	?	?	?	?	?	?

   b. Make ordered pairs (x, y) from the first two rows of the table and plot the points on a graph.
   c. Make ordered pairs (x, y) from the first and third rows of the table and plot the points on a graph.
   d. How many stages of gossip will be needed before 800 people in all hear the gossip?

Chapter 7

15. a. Graph $y = 2.5^x$ for $x = 0, 1, 2, 3, 4,$ and 5.
    b. Calculate the rate of change on the graph from $x = 0$ to $x = 1$.
    c. Calculate the rate of change on the graph from $x = 4$ to $x = 5$.
    d. What do the answers to Parts b and c tell you about this graph?

REVIEW

16. Ashley deposits $3,400 in a savings account with an annual yield of 5%. What will be the total amount of money in the account after 8 years? **(Lesson 7-1)**

17. a. Suppose a person's birthday is in July. What is the probability that it is on July 4th?
    b. Suppose a person's birthday is in March. What is the probability that it is before the 10th? **(Lesson 5-6)**

18. At one time, the exchange rate for Swiss francs per U.S. dollar was 1.305, meaning that 1 dollar would buy 1.305 francs. With this exchange rate, how many U.S. dollars would 1 Swiss franc buy? **(Lesson 5-4)**

19. In the triangle at the right, side $\overline{AC}$ is 20% longer than side $\overline{AB}$ and side $\overline{BC}$ is 45% shorter than side $\overline{AC}$. If $AB = 9$, find the perimeter of the triangle. **(Lesson 4-1)**

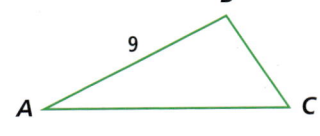

20. **Multiple Choice** Which expression does *not* equal $-(3x - 3y)$? **(Lessons 2-4, 2-1)**
    A $3(y - x)$ B $3y + 3x$ C $-3x + 3y$ D $-3(x - y)$

EXPLORATION

21. This exploration will help to explain why 0^0 is undefined. You will examine values of x^0 and 0^x when x is close to 0.
    a. Use your calculator to give values of x^0 for $x = 1, 0.1, 0.01, 0.001,$ and so on. What does this suggest for the value of 0^0?
    b. Use your calculator to give values of 0^x for $x = 1, 0.1, 0.01, 0.001,$ and so on. What does this suggest for the value of 0^0?
    c. What does your calculator display when you try to evaluate 0^0? Why do you think it gives that display?

QY ANSWERS

a. 10; 25; 62.5

b. $y = 4 \cdot 2.5^x$

Lesson 7-3

Exponential Decay

Vocabulary

exponential decay
half-life

> **BIG IDEA** Decay at a constant percentage rate can be described by an expression of the form bg^x, where $0 < g < 1$ and the variable is in the exponent.

The Growth Factor and the Type of Exponential Change

Three children were arguing. Tammy said, "If you multiply 5 by a positive number, the answer is always greater than 5." Nancy said, "No, you're wrong! I can multiply 5 by something and get an answer less than 5." Leon said, "I can multiply 5 by something and get 5 for an answer."

In the children's arguments, what matters is how the multiplier compares to 1. If Nancy chooses a multiplier between 0 and 1, multiplying by 5 gives a result that is less than 5. For example, $5 \cdot \frac{1}{10} = \frac{1}{2}$. Of course, Leon can do $5 \cdot 1$ and get 5 for an answer.

This relates to exponential equations of the form $y = b \cdot g^x$ because the growth factor g can be greater than, equal to, or less than 1. In the last lesson, you saw only situations in which the growth factor was greater than 1, so there was an increase over time in each case. While a growth factor always has to be positive, it can be less than 1. When this is true, there is a decrease over time. This happens in situations of **exponential decay.**

Mental Math

Find the number.

a. 7 less than 4 times the number is 13.

b. –10 times the number, plus 80, is 10.

c. 14 minus 3 times the number is 121.

Examples of Exponential Decay

Psychologists use exponential decay models to describe learning and memory loss. In Example 1, the growth factor is less than 1 so the amount remembered decreases.

> **Example 1**
>
> Assume that each day after cramming, a student forgets 20% of the vocabulary words learned the day before. A student crams for a French test on Friday by learning 100 vocabulary words Thursday night. But the test is delayed from Friday to Monday. If the student does not study over the weekend, how many words is he or she likely to remember on Monday?
>
> *(continued on next page)*

Exponential Decay

Solution If 20% of the words are forgotten each day, 80% are remembered.

Day	Day Number	Number of Words Remembered
Thursday	0	100
Friday	1	100(0.80) = 80
Saturday	2	100(0.80)(0.80) = 100(0.80)^2 = 64
Sunday	3	100(0.80)(0.80)(0.80) = 100(0.80)^3 = 51.2 ≈ 51
Monday	4	100(0.80)(0.80)(0.80)(0.80) = 100(0.80)^4 = 40.96 ≈ 41

On Monday the student is likely to remember about 41 vocabulary words.

 QY1

As they get old, cars and other manufactured items often wear out. Therefore, their value decreases over time. This decrease, called *depreciation*, is often described by giving the percent of the value that is lost each year. If the item is worth r percent less each year, then it keeps $(1 - r)$ percent of its previous value. This is the growth factor. (The word "growth" is used even though the value is shrinking.)

▶ **QY1**

If the student does not study between now (Thursday) and the test, how many words will be remembered x days from Thursday?

Example 2

In 1998, a new car cost $21,000. Suppose its value depreciates 15% each year.

a. Find an equation that gives the car's value y when it is x years old.
b. What was the predicted value of the car in 2005? How close is this to the actual price of a 1998 car, which was $7,050 in 2005?
c. Graph the car's value for the interval $0 \leq x \leq 8$.

About 35% of the cost of owning and operating a car comes from depreciation.
Source: Federal Highway Administration

Solutions

a. If the car loses 15% of its value each year, it keeps 85%, so the growth factor is $1 - 0.15 = 0.85$. In the exponential growth equation $y = b \cdot g^x$, b is 21,000, the new car's price, and g is 0.85. An equation that gives the value of the car is $y = 21,000 \cdot (0.85)^x$.

b. The year 2005 is 7 years after 1998, so $x = 7$.
$21,000(0.85)^7 \approx 6,732.12$. The predicted value was $6,732.12. The deviation between the actual price and the predicted price was $7,050 - $6,732.12 = $317.88, so the model gives a fairly accurate prediction.

c. The table below shows the value of the car each year. The initial value, $21,000, is the y-intercept of the graph. In the table, to move from one y value to the next, you can multiply by 0.85. The decrease is seen in the graph by the fact that the curve goes downward as you go to the right.

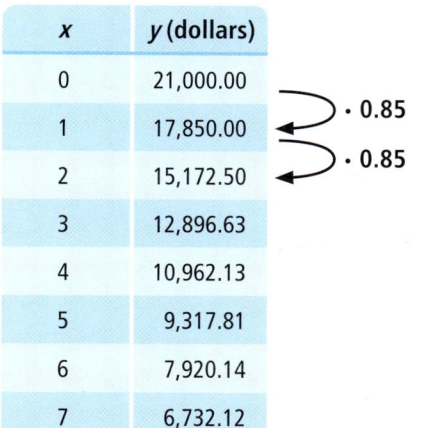
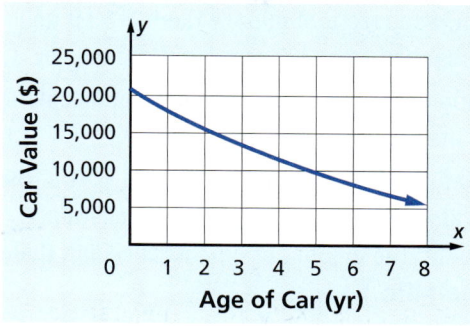

As with graphs of exponential growth, the points of an exponential decay relationship lie on a curve rather than a straight line. Another graph of the car value is shown below, with segments showing the rate of change between points. Notice that as you go from one year to the next, the amount of value is decreasing. For example, in going from new ($x = 0$) to 1 year old ($x = 1$), the value of the car dropped $3,150. But between $x = 7$ and $x = 8$, the value lost was only about $1,000.

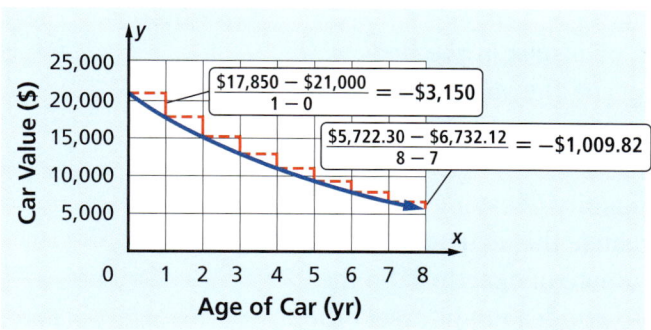

In each case, 15% of the value is lost. But as time passes, this is 15% of a smaller number.

 QY2

In Example 3 on the next page, the "population" is the amount of medication in a person's body.

> ▶ QY2
>
> A new boat costs $32,000. Its value depreciates by 8% each year. Give an equation for y, the value of the boat when it is x years old.

Exponential Decay

Chapter 7

> **GUIDED**
>
> ### Example 3
>
> A common medicine for people with diabetes is insulin. Insulin breaks down in the bloodstream quickly, with the rate varying for different types of the medication. Suppose that initially there are 10 units of insulin in a person's bloodstream and that the amount decreases by 3% each minute.
>
> a. Write an equation to describe y, the amount of insulin in the bloodstream, after x minutes have passed.
>
> b. Make a calculator table for the equation from Part a. Use the table to find when 5 units of insulin remain in the bloodstream. (This is half of the initial amount. The amount of time it takes half the quantity to decay is called the **half-life.**)
>
> c. How much insulin remains after 4 hours?
>
> d. According to the equation, when will the amount of insulin in the body be zero?
>
> **Solutions**
>
> a. The amount of insulin starts at __?__ units. Because 3% of the insulin is lost each minute, the growth factor is __?__. The exponential equation is $y =$ __?__ $\cdot$ __?__ ^{x}.
>
> b. The screen at the right shows the starting value of 10 units, with x increasing by 1 for each row. Scroll down the table to find where y is close to 5 units. This happens at $x =$ __?__. So 5 units of insulin remain after about __?__ minutes.
>
> c. First change 4 hours into __?__ minutes. When $x =$ __?__ minutes, then $y =$ __?__ $\cdot$ __?__ $^{?} \approx 0.007$ unit of insulin remain. So after 4 hours, there is almost no insulin left.
>
> d. Using this equation, the amount of insulin in the body will never be zero. No matter how great the value of x becomes, y will always be greater than zero.

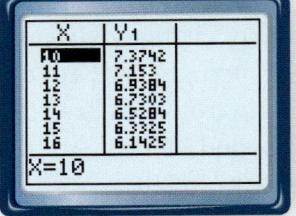

The above example illustrates that a quantity decaying exponentially will theoretically never reach zero. Because this is true, there will never be an ordered pair with a y-coordinate of exactly 0. So the graph of an exponential equation $y = b \cdot g^x$ does not intersect the x-axis.

Graphs and Growth Factors

Exponential growth and exponential decay are both described by an equation of the same form, $y = b \cdot g^x$.

The value of g determines whether the equation describes growth or decay. A third situation happens when there is no change. The three possibilities are graphed below.

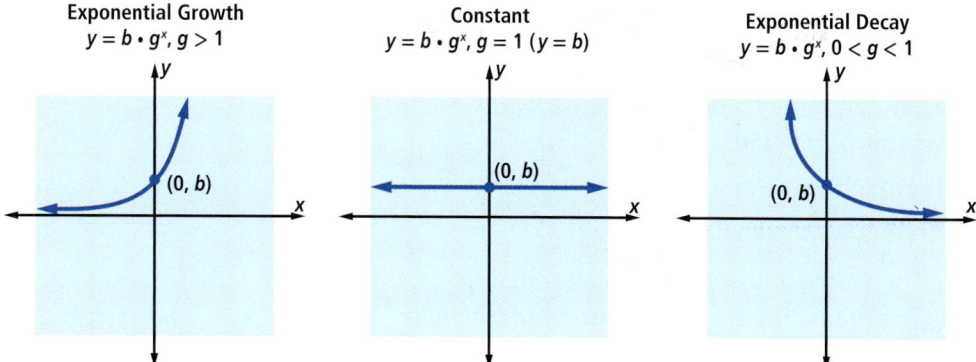

The growth factor can be given in words like "double," "triple," or "half," which indicate a factor of 2, 3, or $\frac{1}{2}$, respectively. But it is also common to describe the growth factor by using percent. If the quantity is growing by a percent r, the value of growth factor g is greater than 1 and the exponential equation is $y = b(1 + r)^x$. If the quantity is shrinking, g is less than 1 and the equation is $y = b(1 - r)^x$. Recall that the graph of equation $y = k$ is a horizontal line, like that shown in the middle graph above. In this situation, in which the original quantity remains constant, there is neither growth nor decay. It can be described by an exponential equation where g is 1. For example, the horizontal line $y = 5$ is also the graph of $y = 5 \cdot 1^x$, since $1^x = 1$.

Questions

COVERING THE IDEAS

In 1–3, give the growth factor for a quantity with the given characteristic.

1. decreases exponentially by 17%
2. increases exponentially by 2.5%
3. does not change
4. Many teachers have policies about late work that lower a student's grade for each day that it is late. Suppose a teacher lowers the grade of a 50-point assignment by 20% for every day late.
   a. Let x = the number of days late an assignment is and y = the number of points the assignment would earn. Make a table of values using x = 0, 1, 2, 3, 4, and 5.
   b. Write an equation to describe the relationship.

Chapter 7

5. Suppose a new car costs $32,000 in 2006. Find the value of the car in one year if the following is true.
   a. The car is worth 85% of its purchase price.
   b. The car depreciated 20% of its value.
   c. The value of the car depreciated d%.

6. A new piece of industrial machinery costs $2,470,000 and depreciates at a rate of 12% per year.
   a. Find the value of the machine after 15 years.
   b. Find the value of the machine after t years.

7. A person with diabetes requires a dose of 15 units of insulin. Assume that 3% of the insulin is lost from the bloodstream each minute. How much insulin remains in the bloodstream after 30 minutes? After x minutes?

8. a. Complete the table for the exponential decay situation at the right.
   b. Write an equation to describe the relationship.

x	y
0	160
1	?
2	?
3	?

(· 0.75 between each row)

In 9–11, classify the pattern in the table as exponential growth, exponential decay, or constant.

9.
x	y
0	35
1	?
2	?

(· 1 and · 1)

10.
x	y
0	0.26
1	?
2	?

(· 1.13 and · 1.13)

11.
x	y
0	458
1	?
2	?

(· 0.32 and · 0.32)

12. **Fill in the Blank** Fill in each blank with "decay" or "growth."
    a. $\frac{2}{3}$ can be the growth factor in an exponential ___?___ situation.
    b. $\frac{3}{2}$ can be the growth factor in an exponential ___?___ situation.

APPLYING THE MATHEMATICS

13. Suppose a school has 2,500 students and the number of students is decreasing by 2% each year.
    a. If this rate continues, write an equation for the number of students after x years.
    b. If this rate continues, how many students will the school have 10 years from now?

Using Algebra to Describe Patterns of Change

14. Imagine that you begin with a cutout of an equilateral triangle. If you fold on the dotted lines as shown below, each vertex will touch the midpoint of the opposite side. Four regions will be formed.

Before folding   **Fold each vertex like this.**   **After 1 set of folds**

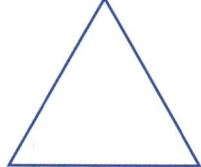 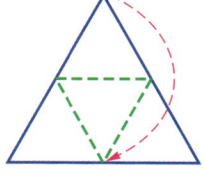 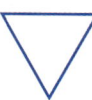

If you repeat this process, you will get a sequence of successively smaller equilateral triangles. Suppose the original triangle has an area of 100 square centimeters.

a. Complete the chart below.

Set of folds	0	1	2	3
Regions	1	4	?	?
Area of a Region (cm²)	100	?	?	?

b. Write an equation to describe the number of regions y after x sets of folds.

c. Write an equation to describe the area of each region y after x sets of folds.

15. Suppose one plate of tinted glass allows only 60% of light to pass through. The amount of light y that will pass through x panes of glass can then be described by the exponential decay equation $y = 0.6^x$.

    a. Plot and label 5 points on a graph of this equation.

    b. If enough panes of glass are put together, will the amount of light passing through the panes ever be zero according to this model?

16. The amount of a radioactive substance decreases over time. The *half-life* of a substance is the amount of time it takes half of the material to decay. Strontium-90 has a half-life of 29 years. This means that in each 29-year period, one half of the strontium-90 decays and one half remains. Suppose you have 2,000 grams of strontium-90.

    a. How much strontium-90 will remain after 5 half-life periods?

    b. How much strontium-90 will remain after 10 half-life periods?

    c. How many years equal 10 half-life periods of strontium-90?

Colored light patterns reflect on the wall of the Palais des Papes in Avignon, France.

17. Consider the equation $y = \left(\frac{1}{2}\right)^x$.
    a. Make a table of values giving y as both a fraction and a decimal when $x = 0, 1, 2, 3, 10,$ and 20.
    b. Find all solutions to the equation $\left(\frac{1}{2}\right)^x = 0$.

REVIEW

18. The rule of 72 is a simple finance method that can be used to estimate the number of periods that it will take an investment to double in value. The method is to divide 72 by the growth rate expressed as a percent, and the result will be the approximate number of periods. For example, if $50 is invested at 6% per year, then the rule of 72 says that after $\frac{72}{6} = 12$ years, a $50 investment will be worth $50 \cdot 2 = \$100$. Calculation shows that after 12 years, this investment is worth $100.61. (**Lessons 7-2, 7-1**)
    a. Use the rule of 72 to estimate the time it will take for $250 invested at 9% to double in value.
    b. Use the compound interest formula to give an exact value of a $250 investment after the number of periods found in Part a.

19. Refer to the table at the right that shows the average consumption of bottled water per person in the United States. (**Lesson 6-7, 6-4**)
    a. Draw a scatterplot with *year* on the *x*-axis and *gallons per person* on the *y*-axis.
    b. Find an equation of an eyeballed line to the data.
    c. Write the slope-intercept form of your equation for the line of fit from Part b.
    d. Use your equation to predict the consumption of bottled water in the United States in 2007.

Year	Bottled Water (gal/person)
2000	17.3
2001	18.8
2002	20.9
2003	22.4
2004	24.0
2005	25.7

Source: www.beveragemarketing.com

20. Solve $4.8q + 9.1 < 12.3q - 7.4$. (**Lessons 4-5, 3-8**)

EXPLORATION

21. Archaeologists use radioactivity to determine the age of ancient objects. Carbon-14 is a radioactive element that is often used to date fossils. Find the half-life of carbon-14 and describe how it is used to help date fossils.

QY ANSWERS

1. $100(0.80)^x$

2. $y = 32{,}000 \cdot (0.92)^x$

Lesson 7-4

Modeling Exponential Growth and Decay

Vocabulary

exponential regression

▶ **BIG IDEA** Situations of exponential growth and decay can be modeled by equations of the form $y = bg^x$.

As you saw with the population of California in Lesson 7-2, sometimes a scatterplot shows a data trend that can be approximated by an exponential equation. As with linear regression, your calculator can use a method called **exponential regression** to determine an equation of the form $y = b \cdot g^x$ to model a set of ordered pairs.

Mental Math

Each product is an integer. Write the integer.

a. $\frac{1}{4} \cdot 360$

b. $\frac{2}{3} \cdot 45$

c. $\frac{4}{5} \cdot 135$

d. $\frac{5}{12} \cdot 228$

Modeling Exponential Decay

In the Activity below, exponential regression models how high a ball bounces.

Activity

Each group of 5 or 6 students needs at least 3 different types of balls (kickball, softball, and so on), a ruler, markers or chalk, and large paper with at least 25 parallel lines that are 3 inches apart.

Step 1 Tape the paper to a wall or door so the horizontal lines can be used to measure height above the floor. To make measuring easier, number every fourth line (12 in., 24 in., 36 in., and so on).

Step 2 One student will drop each ball from the highest horizontal line and the other students will act as spotters to see how high the ball bounces. The 1st spotter will mark the height to which the ball rebounds after the 1st bounce. The 2nd spotter will mark the rebound height after the 2nd bounce, and so on, until the ball is too low to mark.

Step 3 Make a table similar to the one at the right and record the rebound heights after each bounce.

Step 4 For each ball, enter the data in a list and create a scatterplot on your calculator.

Step 5a. Use the linear regression capability of your calculator to find the line of best fit for the data. Graph this line on the same screen as your scatterplot. Sketch a copy of the graph.

(continued on next page)

Bounce	Ball Height (in.)
0 (drop height)	?
1	?
2	?
3	?
?	?

Modeling Exponential Growth and Decay

b. Find the deviation between the actual and predicted height of the ball after the 3rd bounce.

c. Use your linear regression model to predict the height of the ball after the 8th bounce.

Step 6a. Use the exponential regression capability of your calculator to find an exponential curve to fit the data. Graph this equation on the same screen as your scatterplot. Sketch a copy of the scatterplot and curve.

b. Find the deviation between the actual and predicted height of the ball after the 3rd bounce.

c. Use your exponential regression model to predict the height of the ball after the 8th bounce.

Step 7 Which seems to be the better model of the data, the linear equation or the exponential equation? Explain how you made your decision.

Step 8 Repeat Step 6 to find exponential regression equations that fit the bounces of the other balls.

Step 9 Write a paragraph comparing the "bounciness" of the balls you tested.

 QY

Modeling Exponential Growth

Advances in technology change rapidly. Some people say that if you purchase a computer today it will be out of date by tomorrow. When computers were first introduced to the public, they ran much more slowly. As computers have advanced over the years, the speed has increased greatly. On the next page is an example of data that a person collected to show the advancement in computer technology. The processing speed of a computer is measured in megahertz (MHz).

GUIDED

Example

The table and graph on the next page show the average speed of a computer and the year it was made.

a. Write an equation to model the data.

b. Find the deviation between the actual speed for the year 2000 and the predicted speed.

c. Use the model to predict the processing speed of a computer made in 2020.

▶ QY

A student dropped a ball from a height of 0.912 meter and used a motion detector to get the data below.

Bounce	Rebound Height (m)
0	0.912
1	0.759
2	0.603
3	0.496
4	0.411
5	0.328
6	0.271

a. Write an exponential equation to fit the data.

b. After the 8th bounce, how high will the ball rebound?

420 Using Algebra to Describe Patterns of Change

Lesson 7-4

Year	Years since 1976	Speed (MHz)
1976	0	2
1978	2	4
1980	4	5
1982	6	8
1984	8	13
1986	10	16
1988	12	20
1990	14	35
1992	16	48
1994	18	60
1996	20	85
1998	22	180
2000	24	420

Source: *Microprocessor Quick Reference Guide*

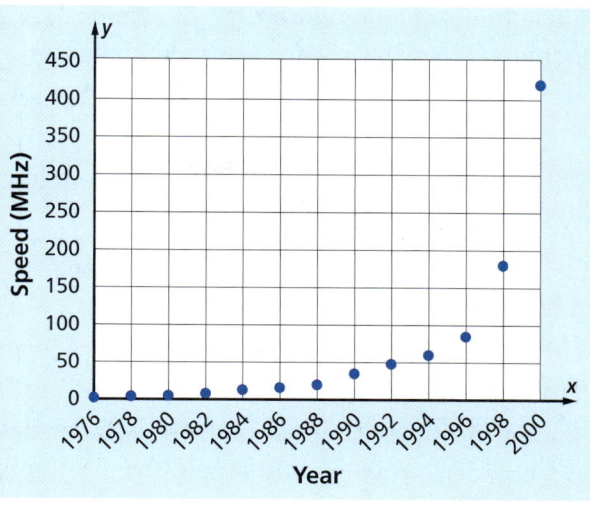

Solutions

a. First enter the data into your calculator lists. Instead of letting years be the x-values, let $x =$ the years since 1976. So for 1976 itself, $x = 0$ and for 1978, $x = 2$. Next, use exponential regression on your calculator to find an exponential equation to fit the data. For $y = b \cdot g^x$, the calculator gives $b \approx 2.241$ and $g \approx 1.218$. (Your calculator may call this equation $y = ab^x$.)
The exponential equation that best fits the data is
$y =$ ___?___.

b. For 2000, $x = 24$ and the actual speed was 420 MHz. Substitute 24 into the equation to find the predicted value. The predicted speed is $y =$ ___?___ MHz. The deviation is $420 -$ ___?___ $=$ ___?___. The actual processing speed in 2000 was ___?___ more than the predicted speed.

c. The year 2020 is $2020 - 1976$, or 44 years after 1976, so substitute 44 for x in your exponential equation. $y =$ ___?___ (___?___)^___?___, so the predicted processor speed for the year 2020 is ___?___ MHz.

On April 25, 1961, the patent office awarded the first patent for an integrated circuit to Robert Noyce while Jack Kilby's application was still being analyzed. Today, both men are acknowledged as having independently conceived of the idea.

Source: PBS

Questions

COVERING THE IDEAS

1. Suppose a ball is dropped and it rebounds to a height of y feet after bouncing x times, where $y = 6(0.55)^x$. Use the equation to
   a. give the height from which the ball was dropped, and
   b. give the percent the ball rebounds in relation to its previous height.

Modeling Exponential Growth and Decay

In 2–4, use the graph to answer the questions. The percent written above each bar represents the percent of the previous height to which each type of ball will rebound.

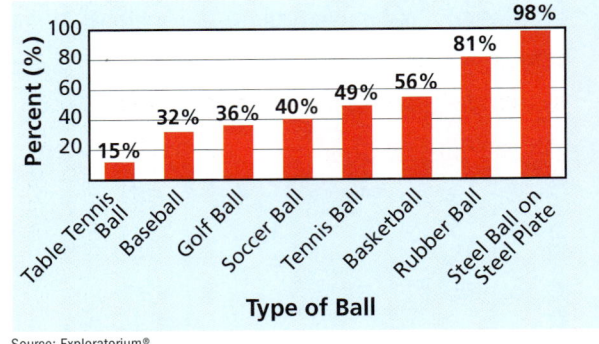

Source: Exploratorium®

2. Which ball's rebound height could be modeled by the equation $y = 10(0.49)^x$?

3. If a basketball is dropped from a height of 15 feet above the ground, how high will it rebound after the 1st bounce? After the 5th bounce?

4. Find and compare the rebound percentages in the Activity on page 419 to those in the graph. Are they similar or different?

5. A computer's memory is measured in terms of megabytes (MB). The table at the right shows how much memory an average computer had, based on the number of years it was made after 1977. Use exponential regression to predict the amount of memory for a computer made in 2020.

Years After 1977	Memory (MB)
0	0.0625
2	1.125
3	8
6	16
7	30
9	32
13	40
17	88
21	250
27	512

6. For each scatterplot, tell whether you would expect exponential regression to produce a good model for the data. Explain your reasoning.

a.
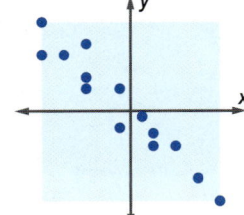

b.
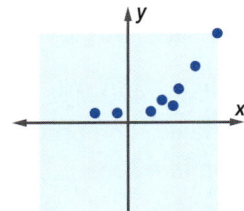

c.
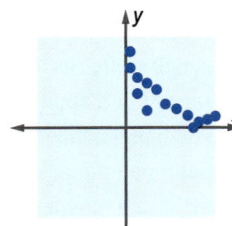

d.
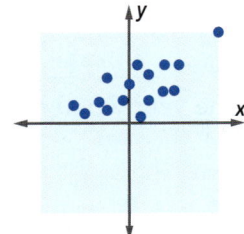

Lesson 7-4

APPLYING THE MATHEMATICS

Matching In 7–9, the graphs relate the bounce height of a ball to the number of times that it has bounced. Match a graph to the equation.

a. $y = 5.1(0.90)^x$ b. $y = 8.7(0.90)^x$ c. $y = 8.7(0.42)^x$

7.
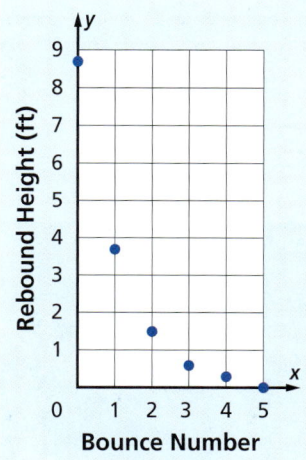

8.
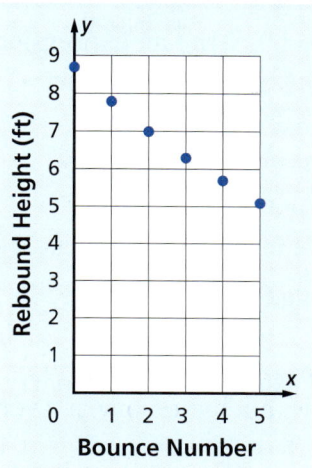

9.
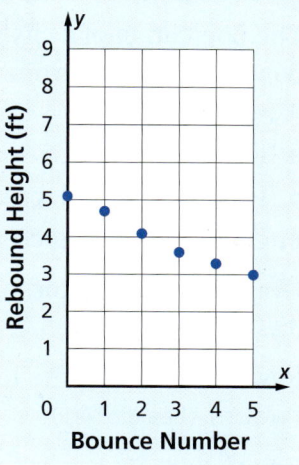

10. The table at the right shows the number of weeks a movie had played in theaters, how it ranked, and how much money it grossed each weekend. (Note that $x = 0$ is the weekend the movie opened.)
    a. Create a scatterplot with $y = $ gross sales after x weeks in theaters. Why is the exponential model a better model for these data than a linear model?
    b. Use exponential regression to find an equation to fit the data.
    c. What gross sales are predicted for the weekend of the 20th week?

Weeks in Theaters	Rank	Weekend Gross ($)
0	1	114,844,116
1	1	71,417,527
2	2	45,036,912
3	2	28,508,104
4	3	14,317,411
5	5	10,311,062
6	7	7,515,984
7	11	4,555,932
8	13	3,130,214
9	18	2,204,636
10	22	890,372
11	25	403,186

11. Lydia and Raul started with 2 pennies in a cup, shook them out onto the table, and added a penny for each coin that showed a head. They continued to repeat this process and their data are recorded in the table at the right.
    a. Create a scatterplot of their data.
    b. Use exponential regression to derive an equation relating the trial number to the number of pennies they will have on the table.

Trial Number	Number of Pennies
0	2
1	2
2	3
3	5
4	8
5	13
6	17
7	25
8	38
9	60

Modeling Exponential Growth and Decay   **423**

For 12 and 13, create a real-world problem that could be modeled by the given equation.

12. $y = 72(1.08)^x$

13. $y = 14(0.65)^x$

REVIEW

14. The population of a city is 1,250,000. Write an expression for the population y years from now under each assumption. (Lessons 7-3, 7-2, 6-1)
    a. The population grows 2.5% per year.
    b. The population decreases 3% per year.
    c. The population decreases by 1,500 people per year.

15. Graph $y = 125\left(\frac{2}{5}\right)^x$ for integer values of x from 0 to 5. (Lesson 7-3)

16. Rewrite $x^3 y^4$ using the Repeated Multiplication Property of Powers. (Lesson 7-1)

17. An art store buys a package of 40 bristle paintbrushes for $80.00 and a package of 30 sable paintbrushes for $150. If they plan to sell an art kit with 4 bristle paintbrushes and 3 sable paintbrushes, how much should they charge for the kit to break even on their costs? (Lesson 5-3)

18. **Skill Sequence** Divide and simplify each expression. (Lesson 5-2)
    a. $\frac{4}{x} \div \frac{5}{x}$
    b. $\frac{4}{x} \div \frac{5}{2x}$
    c. $\frac{4}{x} \div \frac{5}{x^2}$

19. Recall that if an item is discounted x%, you pay $(100 - x)$% of the original price. Calculate in your head the amount you pay for a camera that originally cost $300 and is discounted each indicated amount. (Lesson 4-1)
    a. 10%
    b. 25%
    c. $33\frac{1}{3}$%

20. Evaluate $(3a)^3 (4b)^2$ when $a = -2$ and $b = 6$. (Lesson 1-1)

a young artist at work

EXPLORATION

21. In the ball-drop activity on pages 419–420 and Questions 2–4 on page 422, you explored the rebound height of a ball as a percent of its previous height. Different types of balls have different percents. Does the height from which the ball is dropped affect the percent a ball will rebound? Explain your answer.

22. Do the activity described in Question 11 on page 423. How close is your exponential model to the one in that question?

QY ANSWERS

a. $y = 0.917(0.816)^x$

b. approximately 0.18 m

Lesson 7-5

The Language of Functions

Vocabulary

function
input
output
value of the function
squaring function
independent variable
dependent variable
domain of a function
range of a function
relation

▶ **BIG IDEA** A function is a relationship between two variables in which the value of the first variable is associated with, or determines, a unique value of the second variable.

In this course, you have seen many situations that involve two variables. In an investment situation, the length of time that money has been invested determines the value of the investment. In temperatures, the Fahrenheit temperature determines the Celsius temperature or vice versa. In a sequence of dot patterns, the term number determines the number of dots. When the value of a first variable determines the value of a second variable, we call the relationship between the variables a *function*.

Mental Math

To the nearest mile per gallon, estimate the mpg of a car that went

a. 300 miles on 11 gallons of gas.

b. 250 miles on 9 gallons of gas.

c. 400 miles on 14 gallons of gas.

A Squaring Function

Consider the squares of the integers from 1 to 10. There are two variables. The first is the integer. The second is its square. We can describe the relationship between these integers and their squares in many ways.

Table or List

Integer	Square
1	1
2	4
3	9
4	16
5	25
6	36
7	49
8	64
9	81
10	100

Graph

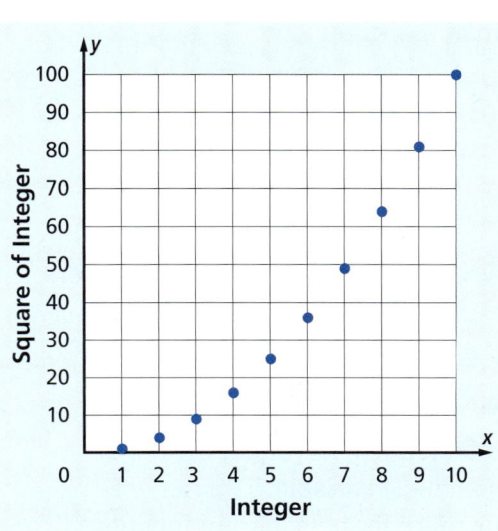

Equation $y = x^2$, where x is an integer from 1 to 10

Words The square of an integer from 1 to 10 is the result of multiplying the integer by itself.

The Language of Functions 425

In general, you can think of functions either as special kinds of correspondences or as special sets of ordered pairs. A **function** is a correspondence in which each value of the first variable (the **input**) corresponds to *exactly one* value of the second variable (the **output**), which is called a **value of the function.** We think of the first variable as determining the value of the second variable. The table, graph, equation, and words on page 425 describe a **squaring function.** The value of a number determines the value of its square. For example, when $x = 3$, the value of the squaring function is 9.

 QY1

The graph on page 425 shows the squaring function as a set of ordered pairs. A function is a set of ordered pairs in which each first coordinate appears with *exactly one* second coordinate. That is, once you know the value of the first variable (often called x), then there is only one value for the second variable (often called y). For this reason, the first variable is called the **independent variable** and the second variable is called the **dependent variable.**

> ▶ QY1
>
> What is the value of the squaring function when $x = 7$?

The Domain and Range of a Function

Suppose in the squaring function that x can be any real number from −10 to 10. You cannot list all the ordered pairs of the function, but the function still can be described by the equation $y = x^2$, where x is a real number from −10 to 10. The function can still be described in words: The square of any real number from −10 to 10 is the result of multiplying the number by itself. And the function can still be described by a graph, as shown at the right.

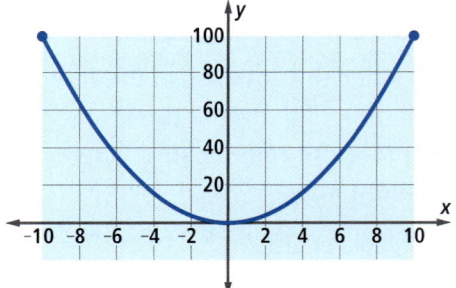

The difference between this squaring function and the one on page 425 is in the *domain of the function*. The **domain of a function** is the set of allowable inputs in the function, that is, the set of possible values of the first (independent) variable. In the squaring function on page 425, the domain is {1, 2, 3, 4, 5, 6, 7, 8, 9, 10}. In the squaring function on this page, the domain is the set of real numbers from −10 to 10.

If a function has a graph, you can read its domain from the graph. The domain is the set of x-coordinates of the points of the graph.

Corresponding to the domain of a function is its **range,** the set of possible values of the second (dependent) variable. The range is the set of possible values of the function. In the squaring function on page 425, the range is {1, 4, 9, 16, 25, 36, 49, 64, 81, 100}. In the squaring function on this page, the range is the set of real numbers from 0 to 100.

QY2

When a set of numbers is not specifically given for the domain of a function, you should assume that the domain is the set of all numbers possible in the situation.

▶ **QY2**

A third squaring function is graphed below. What is its domain? What is its range?

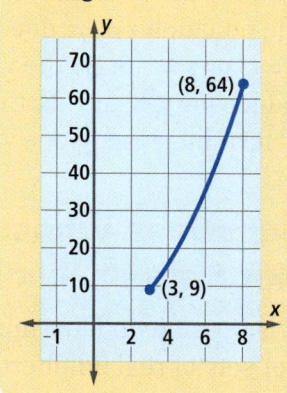

Example 1

Consider the reciprocal function $y = \frac{1}{x}$, which pairs real numbers with their reciprocals.

a. Give the domain.
b. Find the value of the function when $x = 4$.

Solution

a. The domain is the set of all values that can replace x, the independent variable. Any number except 0 can be used. (Because $\frac{1}{0}$ is undefined, 0 has no reciprocal.) So the domain is all real numbers except 0.

b. Substitute 4 for x. The value of the function is $\frac{1}{4}$.

GUIDED

Example 2

What are the domain and range of the function described by the equation $y = 4x - 3$?

Solution The domain is the set of allowable values of x. Because no situation is given for x, you should assume that its domain is ___?___.

The range is the set of possible values of y. The graph of $y = 4x - 3$ is an oblique line. So any value of y is possible, and the range is ___?___.

In many places in this book, you have seen one function modeling another.

Example 3

In Lesson 7-3, the equation $y = 21,000(0.85)^x$ describes a function that models a car's value y when it is x years old if it was purchased for $21,000 and depreciates 15% a year. What is the range of this function?

Solution You can think of the car's value as decreasing constantly even though you have not yet studied values of powers when the exponent is not an integer. Refer to the graph in Lesson 7-3 on page 413. According to the model, the value of the car keeps decreasing but never reaches 0. So the range of the function is the set of real numbers y with $0 < y \leq 21,000$, which is written in set-builder notation as $\{y: 0 < y \leq 21,000\}$.

 QY3

▶ **QY3**

Use a calculator to estimate the value of $21,000(0.85)^x$ to the nearest penny when $x = 25, 25.3,$ and 26.

Chapter 7

Relations That Are Not Functions

The word **relation** describes any set of ordered pairs. It is possible to have relations between variables that are not functions. This happens when the first variable x in a relation corresponds to more than one value of the second variable y. For example, the relation described by the equation $x = y^2$ does *not* describe a function. When $x = 4$, then $y = 2$ or $y = -2$. So the value $x = 4$ corresponds to two different values for y, 2 and -2. Because a value of x does not always determine exactly one value of y, the relation is not a function.

Functions Whose Domains or Ranges Are Not Sets of Numbers

It is possible to have functions whose domains and ranges are not sets of real numbers, or even that have little to do with mathematics. For example, every person on Earth has a unique blood type. So there is a function whose domain is the set of all living people on Earth and whose range is the set of all blood types. Each ordered pair of this function is of the form (a person, that person's blood type). This function has no equation and cannot be graphed, but it is still a function with a domain (the set of all people) and a range (the set of all blood types).

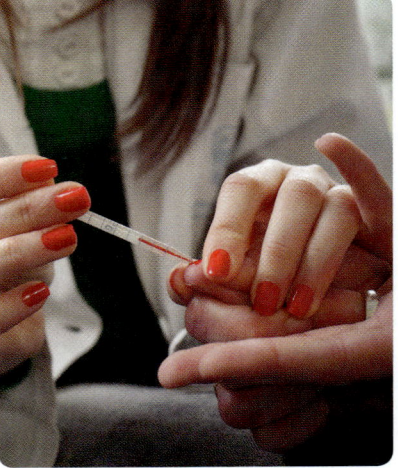

Everybody has a blood type. The most common blood-type classification system is the ABO system invented by Karl Landsteiner in the early 1900s.

Source: University of Utah

Questions

COVERING THE IDEAS

1. Consider the function described by $y = x^2$ with domain the set of integers from 1 to 5.
   a. What is the value of this function when $x = 3$?
   b. The point (4, 16) is on the graph of this function. Which of these coordinates is the input and which is the output?
   c. Which variable is the independent variable and which is the dependent variable?
   d. What is the range of this function?

2. Consider a cubing function described by $y = x^3$ with domain the set of real numbers from -50 to 50. Find the value of this function when
   a. $x = 36$.   b. $x = -36$.   c. $x = 0$.

3. a. What is the value of the reciprocal function when $x = -1$?
   b. What is the value of the reciprocal function when $x = 3.5$?

4. Give the two definitions of *function* stated in this lesson.

428 Using Algebra to Describe Patterns of Change

5. Explain why 0 is not in the range of the reciprocal function.

In 6 and 7, the graph of a function is given.
a. From the graph, determine the domain of the function.
b. From the graph, determine the range of the function.

6.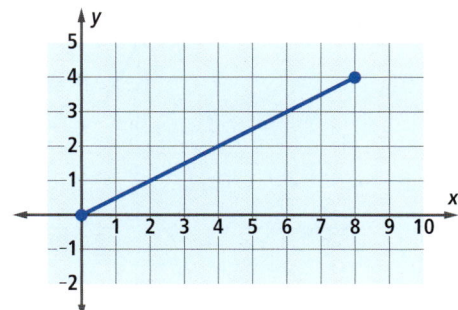

7.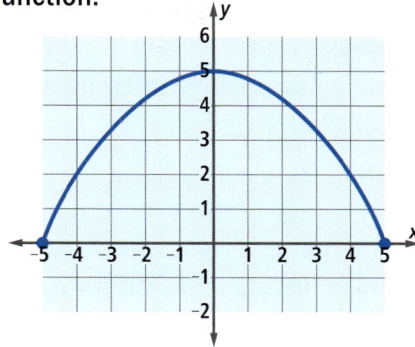

8. Determine if the following statement is *always*, *sometimes but not always*, or *never* true. The graph of a function may contain both points (6, 5) and (6, 7).

9. **Multiple Choice** Which table does *not* describe a function?

A			B			C			D		
x	y		x	y		x	y		x	y	
1	6		1	6		6	1		6	6	
2	53		2	6		6	2		6	6	
3	8		3	6		6	3		6	6	

APPLYING THE MATHEMATICS

10. The graph at the right is of a function showing the distance walked by a hiker over time.
    a. Find the value of the function when $x = 2$ P.M.
    b. Find the value of the function when $x = 3:30$ P.M.
    c. Find x for which the value of the function is 5.5 miles.
    d. Use inequalities to describe the domain and range of this function.

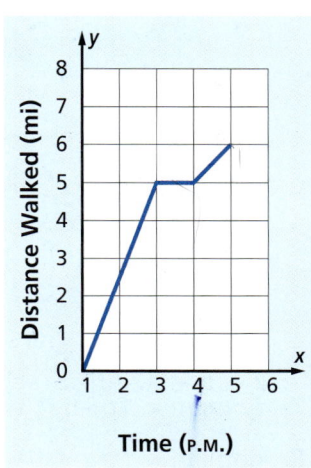

In 11 and 12, determine if the graph of ordered pairs (x, y) is that of a function. Justify your answer.

11.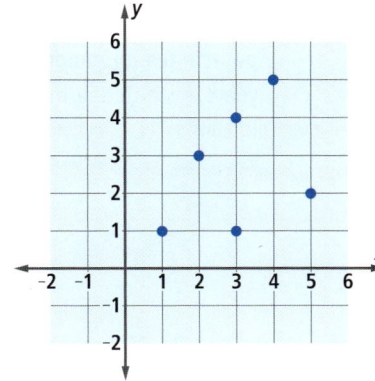

12.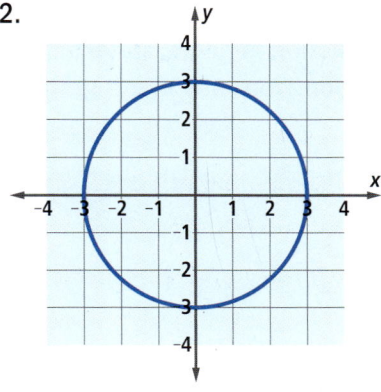

The Language of Functions 429

13. The function $y = 5 \cdot 0.8^x$ is graphed at the right.

    a. **True or False** Zero is in the range of this function.
    b. What is the value of this function when $x = 0$?
    c. What is the domain of the function that is graphed?
    d. What is the range of the function that is graphed?
    e. Suppose x can be any positive integer. What is the greatest possible value of y?

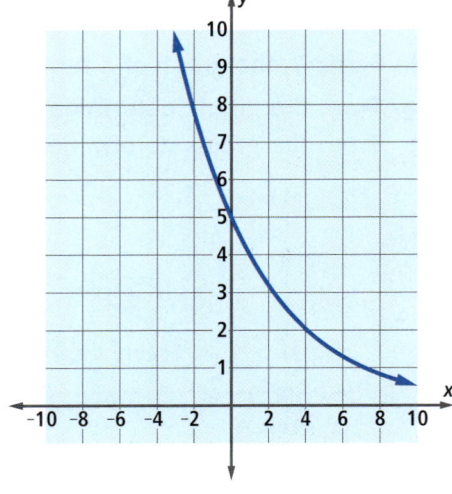

In **14** and **15**, a situation is described in which one quantity can be used to predict values of the second quantity. Tell which quantity you wish to be the input and which should be the output in order to have a function. Sketch a reasonable graph. Do not mark numbers on the axes. Think only about the basic shape of the graph.

14. the amount of time since a cup of hot coffee was poured and its temperature

15. the height of a skydiver who has jumped from an airplane

In **16–18**, use the graph of an absolute value function below. Find the range for the part of the function whose domain is given.

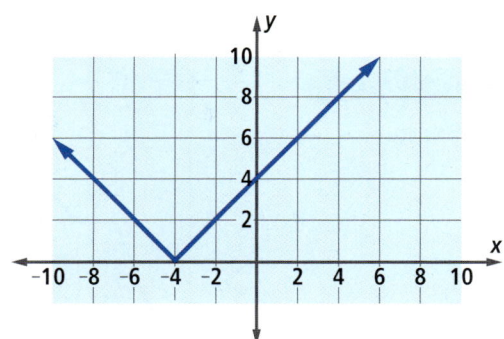

16. domain = $\{x: x \geq 1\}$
17. domain = $\{x: -6 \leq x \leq -2\}$
18. domain = $\{x: x \leq 0\}$

In **19** and **20**, an equation for a function is given. Determine the domain and the range of the function. You may use a graphing calculator to help you.

19. $y = \frac{1}{75}x - 3$
20. $y = 100 \cdot \left(\frac{1}{2}\right)^x$, when $x \geq 0$

People have been using parachutes for hundreds of years, even during the 1100s in China.

Source: United States Parachute Association

21. Since 1980, world records in the men's marathon have been set many times.

Date	Record-Setter	Country	Time	Location of Race
Dec. 6, 1981	Robert de Castella	Australia	2 hr, 8 min, 18 sec	Fukuoka, Japan
Oct. 21, 1984	Steve Jones	Britain	2 hr, 8 min, 5 sec	Chicago, USA
Apr. 20, 1985	Carlos Lopes	Portugal	2 hr, 7 min, 12 sec	Rotterdam, Netherlands
Apr. 17, 1988	Belayneh Dinsamo	Ethiopia	2 hr, 6 min, 50 sec	Rotterdam, Netherlands
Sept. 20, 1998	Ronaldo de Costa	Brazil	2 hr, 6 min, 5 sec	Berlin, Germany
Oct. 24, 1999	Khalid Khannouchi	Morocco	2 hr, 5 min, 42 sec	Chicago, USA
Apr. 14, 2002	Khalid Khannouchi	USA	2 hr, 5 min, 38 sec	London, England
Sept. 28, 2003	Paul Tergat	Kenya	2 hr, 4 min, 55 sec	Berlin, Germany

Source: www.marathonguide.com

  a. Consider the function using the pairs (date, time). This function is an example of a *decreasing function*. Why do you think it is called a decreasing function?

  b. Consider the eight ordered pairs (record-setter, time). By examining the definition of function, explain why these eight ordered pairs do *not* make up a function.

REVIEW

22. In China, most families are allowed to have only one child. This policy was implemented to reduce the population, with a goal of reaching 700 million citizens by 2050. Suppose the 2005 population of 1.3 billion decreases by 1% each year. (**Lesson 7-3**)

  a. Write an expression for the population of China x years after 2005.

  b. Will the goal of having 700 million citizens or less in the year 2050 be met?

23. Find the slope of the line given by the equation $\frac{7}{20}(x + 18) = \frac{8}{3}(y - 11)$. (**Lessons 6-2, 4-4, 3-8**)

24. Triangle 1 has an area of 12 cm² and is similar to Triangle 2, which has an area of 108 cm². What is the ratio of similitude of Triangle 1 to Triangle 2? (**Lesson 5-10**)

25. The maximum number p of people allowed on a certain elevator times the average weight w of an adult should not exceed 1,500 pounds. Write an inequality describing the rule and solve for p. (**Lesson 3-6**)

EXPLORATION

26. Find equations for two different functions with the same domain that contain both the ordered pairs (1, 1) and (2, 6).

QY ANSWERS

1. 49

2. domain: {x: 3 ≤ x ≤ 8}; range: {y: 9 ≤ y ≤ 64}

3. $361.15; $343.97; $306.98

Chapter 7

Lesson 7-6

Function Notation

Vocabulary

f(x) notation

function notation

> **BIG IDEA** When a function f contains the ordered pair (x, y), then y is the value of the function at x, and we may write y = f(x).

Refer to the graph on page 397. Recall that each equation represents a different prediction of the population of a town x years in the future. Each model describes a function. The functions are of three types.

Possibility 1 $P = 100{,}000$ describes a *constant function* where the population does not change.

Possibility 2 $P = 100{,}000 + 3{,}000x$ describes a *linear function* with 3,000 new people per year.

Possibility 3 $P = 100{,}000(1.02)^x$ describes an *exponential function* with a growth rate of 2% per year.

It is important to see these functions on the same axes because we want to compare them. But in talking about three functions, we might easily get confused. If we say the letter P, which P are we talking about? It would be nice to be able to name a function in a simple and useful manner.

Mental Math

Give the coordinates of a solution to the inequality.

a. $y \leq -22x + 6$

b. $5m - 4n > 3$

c. $-b + 3 < a - 4.5$

f(x) Notation

Conveniently, mathematics does have another way to name functions. With this method, Possibility 3 can be written $E(x) = 100{,}000(1.02)^x$.

We chose the letter E as a name for the function as a reminder of exponential growth. The symbol $E(x)$ shows that x is the input variable. It is read "E of x." What is the purpose of using this new symbol? It allows us to show the correspondence between specific pairs of values for the input (number of years x) and output (population predicted by the exponential growth model). For example, when $x = 3$, the output is $E(3) = 100{,}000(1.02)^3 = 106{,}120.8$. So $E(3) = 106{,}120.8 \approx 106{,}121$ people. When $x = 20$, the output is $E(20) = 100{,}000(1.02)^{20} \approx 148{,}595$ people.

The other population models can be written in function notation as well.

Possibility 2 could be written $L(x) = 100{,}000 + 3{,}000x$.
Possibility 1 could be written $C(x) = 100{,}000$.
Now each model has a different name.

It is important to know that $E(x)$ *does not* denote the multiplication of E and x. The parentheses indicate the input of a function.

Example 1

Given a function with equation $f(x) = 5x - 19$, find $f(2)$.

Solution $f(x) = 5x - 19$ is a general formula that tells how to find the output for any input. The symbol $f(2)$ stands for "the output of function f when the input is 2." So substitute 2 for x and evaluate the expression on the right side.

$f(2) = 5(2) - 19$
$\quad\quad = 10 - 19 = -9$

So $f(2) = -9$.

In Example 1, we say that -9 is the *value of the function* when $x = 2$.

GUIDED

Example 2

Use the three functions given earlier for population models to find $E(10)$, $L(10)$, and $C(10)$. Explain what the results mean in the context of the population situation.

Solution

$E(x) = 100{,}000(1.02)^x$
$E(10) = 100{,}000(1.02)^{\underline{\ ?\ }}$

After 10 years, the population based on the exponential model is predicted to be about __?__ people.

$L(x) = 100{,}000 + 3{,}000x$
$L(10) = \underline{\ ?\ } + \underline{\ ?\ }(\underline{\ ?\ }) = \underline{\ ?\ }$

After 10 years, the population based on the linear model is predicted to be __?__ people.

$C(x) = 100{,}000$
$C(10) = 100{,}000$

After __?__ years, the population based on the __?__ model is predicted to be __?__ people.

Chapter 7

In working with functions, questions arise in which you are given the value of one variable and are asked to find the value of the other variable. In Example 1, you were given the *input* value. You substituted to find the output value. In the next two examples, you are given the value of the *output*. This results in an equation to solve. When you have a formula for a function, symbolic methods may be used to solve the equation to find the input value.

Example 3

Possibility 2 used the linear function $L(x) = 100{,}000 + 3{,}000x$ to model the population of a town x years in the future. According to this model, in how many years will the population reach 150,000?

Solution In $L(x) = 100{,}000 + 3{,}000x$, replace $L(x)$ with 150,000. Then solve for x.

$150{,}000 = 100{,}000 + 3{,}000x$

$50{,}000 = 3{,}000x$

$16.67 \approx x$

The linear model predicts that in about 17 years the population will be 150,000. So $L(x) = 150{,}000$ when $x \approx 17$.

Because functions can also be described with tables and graphs, tables and graphs are useful in solving problems in which you are given the output and need to find the input.

Example 4

For $E(x) = 100{,}000(1.02)^x$, use the graph to find when the population reaches 150,000.

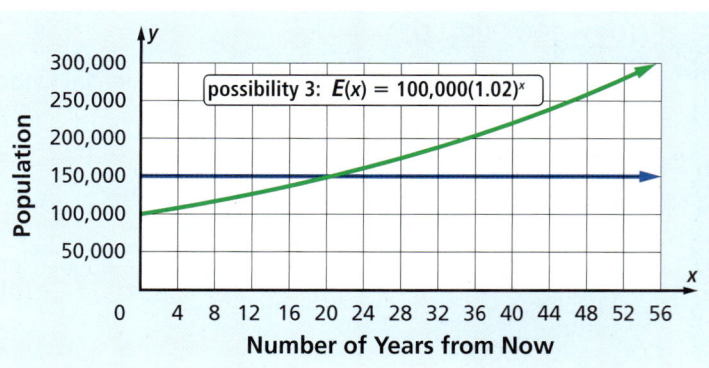

Solution Graph $y = 100{,}000(1.02)^x$. To help you see the point on this graph whose y-coordinate is 150,000, also graph the horizontal line $y = 150{,}000$. Trace on the graph to find where these two graphs intersect. When x is between 20 and 21 years, $E(x)$ is approximately 150,000.

Check Substitute 20 and 21 for x in the equation $E(x) = 100{,}000(1.02)^x$.

Using $x = 20$, $E(x) = E(20) = 100{,}000(1.02)^{20} \approx 148{,}595$.

Using $x = 21$, $E(x) = E(21) = 100{,}000(1.02)^{21} \approx 151{,}567$.

$148{,}595 < 150{,}000 < 151{,}567$, so the answer is reasonable.

Unless the situation suggests a better letter, the most common letter used to name a function is f. That is, instead of writing $y = 3x + 5$, we might write $f(x) = 3x + 5$. Then f is the linear function with slope 3 and y-intercept 5. It is read "f of x equals 3 times x plus 5." This way of writing a function is called **f(x) notation** or **function notation.** The symbol $f(x)$ is attributed to the great Swiss mathematician Leonhard Euler (1707–1783).

Activity

The CAS allows you to work with functions.

1. a. Use the DEFINE command to define $f(x) = 3x^2 + 2x + 10$ in your CAS.
   b. Find $f(2)$.
   c. Find $f(3) + f(-6)$.
   d. Find $f(2006) - f(2005)$.
   e. Find $f(a)$.
   f. Find $f(y)$.
   g. Find $f(math)$. (Do not type in multiplication symbols between the letters. CAS sees *math* as just one big variable called a string variable.)
   h. Find f(your name).
2. a. Define $g(x) = \dfrac{x^3 + 999}{8x - 1}$.
   b. Find $g(7)$, $g(\pi)$, $g(t)$, and $g(mom)$.
3. Explain the relationships between function notation and substitution.
4. Without using a CAS, find $h(4)$ and $h(algebra)$ if $h(x) = 5x + 2x^2 + 1$.

Questions

COVERING THE IDEAS

1. How is the symbol $f(x)$ read?

In 2 and 3, let $E(x) = 100{,}000(1.02)^x$, $L(x) = 100{,}000 + 3{,}000x$, and $C(x) = 100{,}000$.

2. a. Without a calculator, find the values of $E(1)$, $L(1)$, and $C(1)$.
   b. What do these values mean in the population projection situation?

3. a. With a calculator if necessary, find the values of $E(25)$, $L(25)$, and $C(25)$.
   b. What do these values mean in the population projection situation?

4. If $f(x) = 4 \cdot 0.12^x$, find each value.
   a. $f(1)$
   b. $f(3)$
   c. $f(5)$

Chapter 7

5. Let p be a function with $p(n) = 100 \cdot 2^n$. Find the value of $p(n)$ when $n = 7$.

6. Determine if the statement is *always true*, *sometimes but not always true*, or *never true*. If $g(x) = 4x^2$ and $g(x) > 0$, then $x > 0$.

APPLYING THE MATHEMATICS

7. The table below shows the amount of money earned by a person who worked x hours.

Hours Worked x	Total Amount Earned $f(x)$
5	$45
10	$90
15	$135
20	$180
25	$225
30	$270

   a. What is the value of $f(5)$?
   b. What is the value of $f(15)$?
   c. Find x if $f(x) = \$180$.

8. Let $h(n) = 0.5n$.
   a. Calculate $h(0)$, $h(1)$, and $h(2)$.
   b. Calculate $h(-1) + h(-2)$.
   c. Why do you think the letter h was chosen for this function?

9. A computer purchased for $1,600 is estimated to depreciate at a rate of 25% per year. The computer's value after t years is given by $W(t) = 1,600(0.75)^t$.
   a. Evaluate $W(4)$ and explain what the value means.
   b. Graph the function W for values of t with $0 \le t \le 7$.
   c. Use your graph to estimate the solutions to $W(t) < 1,000$ and explain what your answer means.

10. Suppose $L(x) = 12x - 18$.
    a. Calculate $L(5)$.
    b. Calculate $L(3)$.
    c. Evaluate $\dfrac{L(5) - L(3)}{5 - 3}$.
    d. What is the meaning of your calculation in Part c?

In 2005, approximately 89% of U.S. public middle and junior high schools had the use of computers in the classroom.

Source: Quality Education Data, Inc.

436 Using Algebra to Describe Patterns of Change

11. Oven temperature T varies with the length of time t the oven has been on. An oven, whose initial temperature was 80°, was set for 325°. The actual temperature was measured and then graphed over a 45-minute interval. Below is the graph of the function f where $T = f(t)$.

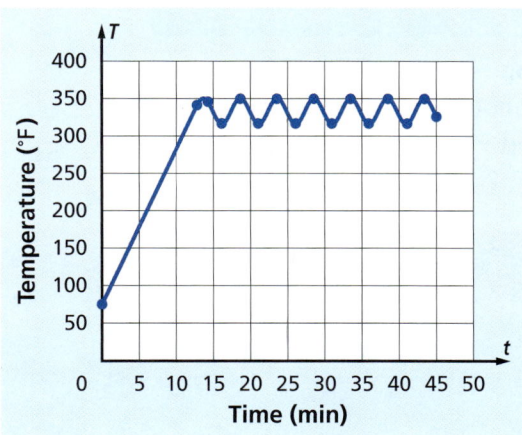

a. Estimate $f(25)$.
b. What is the meaning of $f(25)$?
c. Estimate the solution to the equation $f(t) = 200$.
d. What is the meaning of the solution in Part c?

REVIEW

In 12 and 13, could the table of values represent a function with x as the independent variable and y as the dependent variable? Why or why not? (Lesson 7-5)

12.

x	-3	-2	-1	0	1	2	3
y	0	1	5	1	6	0	2

13.

x	0	1	5	1	6	0	2
y	-3	-2	-1	0	1	2	3

14. Consider the relation described by the equation $x^2 + 3y^2 = 31$. (Lesson 7-5)
    a. Is (2, 3) a solution?
    b. Is (2, –3) a solution?
    c. Is $x^2 + 3y^2 = 31$ the equation of a function? Explain.

15. Rewrite $8y - 4x + 1 = 25 + 6x$ in each form.
    (Lessons 6-8, 6-4)
    a. standard form
    b. slope-intercept form

16. **Skill Sequence** (Lessons 4-4, 2-2, 2-1)
    a. Simplify $3(5 - 2m)$.
    b. Simplify $3(5 - 2m) - 2(7m + 1)$.
    c. Solve $3(5 - 2m) - 2(7m + 1) = 43$.

EXPLORATION

17. Some functions involve more than one input variable. The chart below shows wind chill as a function of the wind speed and the temperature. For wind speed V and temperature T, let $W(V, T)$ = wind chill and let $F(V, T)$ = frostbite time.

	Temperature (°F)																
Calm	40	35	30	25	20	15	10	5	0	-5	-10	-15	-20	-25	-30	-35	-40	-45
5	36	31	25	19	13	7	1	-5	-11	-16	-22	-28	-34	-40	-46	-52	-57	-63
10	34	27	21	15	9	3	-4	-10	-16	-22	-28	-35	-41	-47	-53	-59	-66	-72
15	32	25	19	13	6	0	-7	-13	-19	-26	-32	-39	-45	-51	-58	-64	-71	-77
20	30	24	17	11	4	-2	-9	-15	-22	-29	-35	-42	-48	-55	-61	-68	-74	-81
25	29	23	16	9	3	-4	-11	-17	-24	-31	-37	-44	-51	-58	-64	-71	-78	-84
30	28	22	15	8	1	-5	-12	-19	-26	-33	-39	-46	-53	-60	-67	-73	-80	-87
35	28	21	14	7	0	-7	-14	-21	-27	-34	-41	-48	-55	-62	-69	-76	-82	-89
40	27	20	13	6	-1	-8	-15	-22	-29	-36	-43	-50	-57	-64	-71	-78	-84	-91
45	26	19	12	5	-2	-9	-16	-23	-30	-37	-44	-51	-58	-65	-72	-79	-86	-93
50	26	19	12	4	-3	-10	-17	-24	-31	-38	-45	-52	-60	-67	-74	-81	-88	-95
55	25	18	11	4	-3	-11	-18	-25	-32	-39	-46	-54	-61	-68	-75	-82	-89	-97
60	25	17	10	3	-4	-11	-19	-26	-33	-40	-48	-55	-62	-69	-76	-84	-91	-98

    (Wind (mph) on vertical axis)

    Frostbite Times: 30 minutes, 10 minutes, 5 minutes

    Source: National Weather Service

    a. Pick three (V, T) pairs and find $W(V, T)$ and $F(V, T)$.
    b. Find two solutions for the equation $W(V, T) = -55$.
    c. Find two solutions for $W(V, T) = -39$ that have different values for $F(V, T)$.

18. Let $m(x)$ = the mother of person x and $f(x)$ = the father of person x. Using yourself for x, find $m(f(x))$ and $f(f(x))$. (*Hint:* Start with the inner-most parentheses.) What are simpler descriptions for each of these two functions?

Lesson 7-7

Comparing Linear Increase and Exponential Growth

> **BIG IDEA** In the long run, exponential growth always overtakes linear (constant) increase.

In the patterns that are constant increase/decrease situations, a number is repeatedly *added*. In exponential growth/decay situations, a number is repeatedly *multiplied*. In this lesson, we compare what happens as a result.

Mental Math

What is the date of the xth day of the year in a nonleap year when

a. $x = 100$?

b. $x = 200$?

c. $x = 300$?

GUIDED

Example

Suppose you have $10. For two weeks, your rich uncle agrees to do one of the following.
Option 1: Increase what you had the previous day by $50.
Option 2: Increase what you had the previous day by 50%.
Which option will give you more money?

Solution Make a table to compare the two options for the first week. Use the Now/Next method to fill in the table. The exponential growth factor is 1.50.

Start = $10
Next = Now + $50

Day	Option 1: Add $50.
0	$10
1	?
2	?

+ $50

Start = $10
Next = Now • 1.50

Day	Option 2: Multiply by 1.50.
0	$10
1	?
2	?

• 1.50

Continue the table until day 14. You should find that at first, you get more money from Option 1. But the table shows that starting on day __?__, Option 2 gives more money. **In the long run, Option 2, increasing by 50% each day, is the better choice.**

Above, the two options were described by telling how the amounts changed each day. In that situation, the Now/Next method works well. But to graph the situation on your calculator, you need equations for these functions.

Comparing Using a Graph

To find the equations for the functions in the Guided Example, you can make a table to compare the two options during the first week. The exponential growth factor is 1.50. Let $L(x) = $ the amount given to you under Option 1, and let $E(x) = $ the amount given to you under Option 2.

Day	Option 1: Add $50.	Option 2: Multiply by 1.50.
0	$L(0) = 10 = \$10.00$	$E(0) = 10 = \$10.00$
1	$L(1) = 10 + 50 \cdot 1 = \$60.00$	$E(1) = 10 \cdot 1.50^1 = \$15.00$
2	$L(2) = 10 + 50 \cdot 2 = \$110.00$	$E(2) = 10 \cdot 1.50^2 = \$22.50$
3	$L(3) = 10 + 50 \cdot 3 = \$160.00$	$E(3) = 10 \cdot 1.50^3 = \$33.75$
4	$L(4) = 10 + 50 \cdot 4 = \$210.00$	$E(4) = 10 \cdot 1.50^4 = \$50.63$
5	$L(5) = 10 + 50 \cdot 5 = \$260.00$	$E(5) = 10 \cdot 1.50^5 = \$75.94$
x	$L(x) = 10 + 50x$	$E(x) = 10 \cdot 1.50^x$

We see how the values compare by graphing the two functions E and L. The graphs of $L(x) = 10 + 50x$ and $E(x) = 10 \cdot 1.5^x$ are shown at the right.

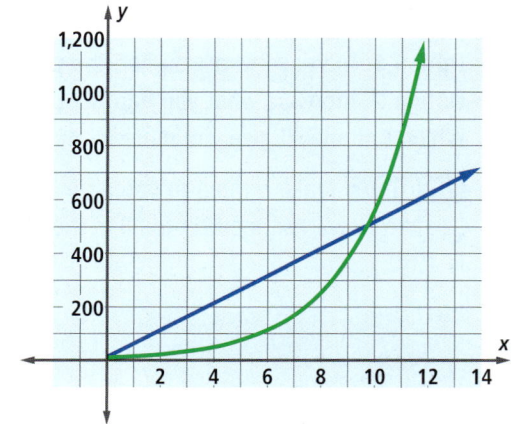

The line $L(x) = 10 + 50x$ has a constant rate of change. The graph of $y = 10 \cdot 1.5^x$ is a curve that gets steeper and steeper as you move to the right. Notice that at first the exponential curve is below the line. But toward the middle of the graph, it intersects the line and passes above it. On later days, the graph of the curve rises farther and farther above the line.

The longer your uncle gives you money, the better Option 2 is compared to Option 1.

Comparing Using Spreadsheets

Activity 1 shows how to use a spreadsheet to confirm the results of Guided Example 1.

Activity 1

Step 1 Create a spreadsheet similar to the one at the right. Be sure to have titles in row 1. In cells A2 through A16 enter the numbers 0 to 14.

	A	B	C
1	Day x	Option 1	Option 2
2	0	10	10
3	1	=B2+50	

440 Using Algebra to Describe Patterns of Change

Step 2 Type =B2+50 in cell B3. Press ENTER. What appears in cell B3?

Step 3 Type the formula for Option 2 into cell C3. (*Hint:* What is the Now/Next formula for Option 2?)

An advantage of spreadsheets is that you don't have to type a formula into each cell. When you type the formula =B2+50 into cell B3, the spreadsheet remembers this as: "Into this cell put 50 plus the number that is in cell B2 above." For example, if you copy cell B3 to cell D5, the formula copied will change to =D4+50 because one cell above D5 is D4. This way of copying in spreadsheets is called *replication*.

Step 4 Replicate the formula in cell B3 into cells B4 through B16.

Step 5 Replicate the formula in cell C3 into cells C4 through C16.

Step 6 Compare your spreadsheet to the table on page 440. Experiment by changing the starting amount of $10 to other values. Then go back to the original starting amount of $10 before doing the next step.

Step 7 Add two more columns to your spreadsheet.

◇	A	B	C	D	E
1	Day x	Option 1	Option 2	$L(x) = 10 + 50x$	$E(x) = 10 \cdot 1.5^x$
2	0	10	10		
3	1	=B2+50			

Step 8 Type =10+50*A2 in cell D2. This puts into D2 the value of the function L for the domain value in cell A2.

Step 9 Replicate the formula in cell D2 into cells D3 through D16.

Step 10 Compare the values in columns B and D. If they are the same, then you know that you have done the previous steps correctly.

Step 11 Type =10*1.5^A2 into cell E2. This puts into E2 the value of the function E for the domain value in cell A2.

Step 12 Replicate the formula in cell E2 into cells E3 through E16. What should happen? Have you done the previous steps correctly?

Chapter 7

Activity 2

In 1993, Florida introduced 19 mountain lions into its northern region. With animals in the wild, there are two scenarios. If there are no limiting factors, the population of animals tends to grow exponentially. However, if limiting factors are established, the population growth tends to be linear. Limiting factors can be things such as climate, availability of food, predators, and hunting.

Suppose the scientists who introduced the mountain lions into northern Florida used one of the following options to model the population growth.

Option 1: There are limiting factors so that 2 more mountain lions appear each year.

Option 2: There are no limiting factors so that the population grows by 6% each year.

Step 1 First, create a spreadsheet similar to the one below.

	A	B	C	D	E
1	Year	Option 1	Option 2	$L(x)$	$E(x)$
2	1993	19	19		
3	1994				

A typical male mountain lion inhabits 50 to 300 square miles, depending on how plentiful food is.

Source: *USAToday*

Step 2 Enter formulas into B3 and C3 to calculate the population using the Now/Next method.

Step 3 Copy and paste B3 into cells B4 and lower. Also copy and paste C3 into cells C4 and lower. Be sure to gather enough data and compare the populations in column B to those in column C.

Step 4 Enter formulas for $L(x)$ and $E(x)$ in columns D and E and copy these for as many rows as you used in Step 3.

Step 5 Graph each option's population equation on the same axes. Let the x-coordinates be the number of years since 1993. (Let $x = 0$ be 1993.) You can use the chart feature of the spreadsheet to create the graphs.

Step 6 Answer the following questions using the collected data and the graphs.

1. In 2010, which option would provide a larger population of mountain lions?

2. In which year (if ever), would Option 2 create a larger population of mountain lions?

A Summary of Constant Increase and Exponential Growth

In this lesson, you have seen that differences between linear and exponential models can be seen in the equations that describe them, the tables that list ordered pairs, and the graphs that picture them. You have seen that if the growth factor is greater than 1, exponential growth always overtakes constant increase. Here is a summary of their behavior.

Constant Increase	Exponential Growth
• Begin with an amount b.	• Begin with an amount b.
• Add m (the slope) in each of the x time periods.	• Multiply by g (the growth factor) in each of the x time periods.
• After x time periods, the amount is given by the function $L(x) = mx + b$.	• After x time periods, the amount is given by the function $E(x) = b \cdot g^x$.

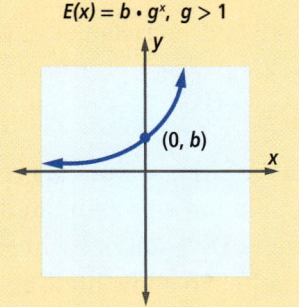

Questions

COVERING THE IDEAS

1. What is the difference between a constant increase situation and an exponential growth situation?

In 2–5, let $L(x) = 20 + 3x$ and $E(x) = 20(1.03)^x$.

2. Calculate $L(5)$ and $E(5)$.
3. Sketch a graph of both functions on the same axes.
4. Give an example of a value of x for which $E(x) > L(x)$.
5. What kind of situation could have led to these equations?

6. Two friends found $100 and split it equally between them. Alexis put her half in a piggy bank and added $7 to it each year. Lynn put her half in a bank with an annual yield of 7%.
   a. Make a spreadsheet to illustrate how much money each friend has at the end of each year for the next 25 years. Have one column represent Alexis and one column represent Lynn.
   b. Sketch a graph to represent the amount of money each friend has over the next 25 years.

7. Rochelle started to make the following spreadsheet. She replicated the formula in cell A2 into cells A3 and A4.

	A	B
1	28	6
2	=A1+13	=1.2*B1
3		
4		
5		

   a. Give the formulas that will occur in cells A3 and A4.
   b. What numbers result from the formulas in A3 and A4?
   c. How will the values in A3 and A4 change if Rochelle changes the start value in A1 to −5?
   d. Does column A illustrate constant increase or exponential growth? Explain.

8. Repeat Question 7 for column B.

9. The number of deer in the state of Massachusetts is a problem. In 1998, the deer population was estimated to be about 85,000. The Massachusetts Division of Fisheries and Wildlife had to decide whether to allow hunting (a limiting factor) or to ban hunting (no limiting factor). If hunting is allowed, they predict the deer population to increase at a constant rate of about 270 deer a year. If hunting is not allowed, the prediction is the deer population would grow exponentially by 15% each year.
   a. Write a Now/Next formula for the deer population if hunting is allowed.
   b. Write a Now/Next formula for the deer population if hunting is banned.
   c. Let $L(x)$ = the number of deer x years after 1998 if hunting is allowed. Find a formula for $L(x)$.
   d. Let $E(x)$ = the number of deer x years after 1998 if hunting is not allowed. Find a formula for $E(x)$.
   e. The state allowed hunting. The 2006 deer population was estimated between 85,000 and 95,000. Was the prediction correct?

In 1906, the U.S. deer population was a sparse 500,000. Today, experts estimate that at least 25 million deer roam the nation.

Source: AP

Lesson 7-7

APPLYING THE MATHEMATICS

10. Refer to the spreadsheet at the right.
    a. What Now/Next formula could be used to generate the numbers in column A?
    b. Let $f(x)$ be the value in column A at time x. What is a formula for $f(x)$?

11. Suppose you are reading a 900-page novel at the rate of 25 pages per hour. You are currently at page 67.
    a. Is the number of pages you read in the book an example of constant increase or exponential growth? Explain.
    b. Write an equation to describe the pages finished x hours from now.
    c. How many hours will it take you to finish the book?

12. The graph at the right shows the number of territories in which bald eagles nest around the five Great Lakes.
    a. Would you describe the graphs as constant increase or exponential growth? Explain your answer.
    b. The graph of which lake can be represented by $y = 14.5 \cdot 1.053^x$, where x is the years since 1962? Explain your answer.

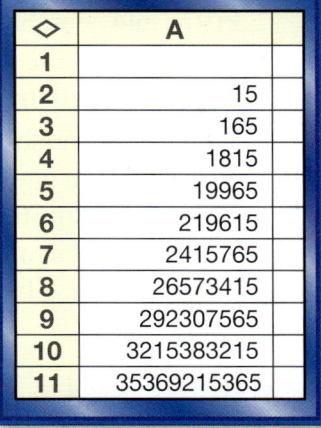

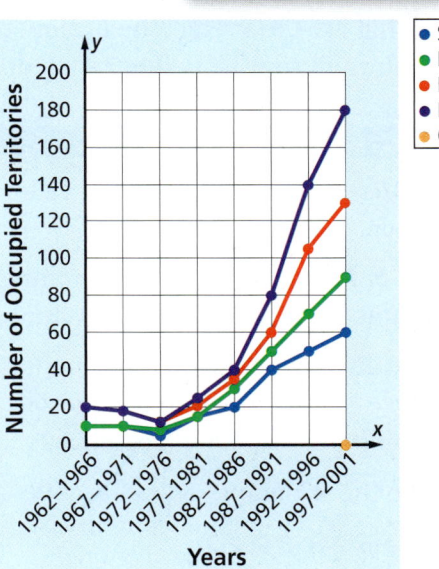

MATCHING In 13–16, each graph is drawn on the window $-2 \leq x \leq 15$, $0 \leq y \leq 2{,}000$. Match the graph with its equation.

    a. $f(x) = 100 \cdot 1.25^x$ b. $g(x) = 100 + 125x$
    c. $h(x) = 100 + 60x$ d. $j(x) = 100 \cdot 1.1^x$

13.
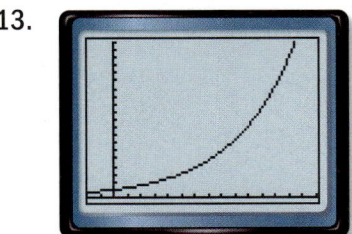

14.
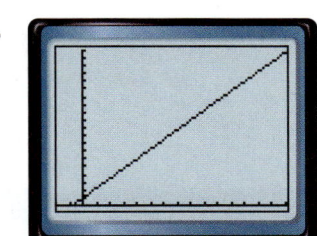

15.
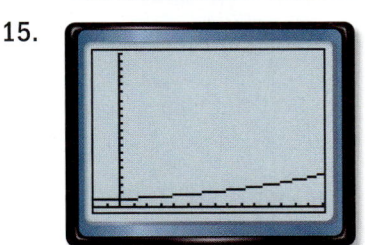

16.
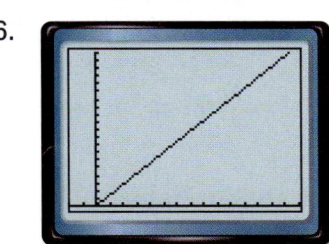

17. The principal of a high school is making long-range budget plans. The number of students dropped from 2,410 students to 2,270 in one year. Student enrollment is dropping, as shown in rows 2 and 3 of the spreadsheet. The situation may be modeled by a linear function or an exponential function.

a. Make a spreadsheet similar to the one at the right. Show the future enrollments in the two possible situations.

◇	A	B	C
1		Constant Decrease	Exponential Decrease
2	0	2410	2410
3	1	2270	2270

b. What are the predicted enrollments 5 years from the year shown in row 3? By how much do they differ?

c. What are the predicted enrollments 15 years from year shown in row 3? By how much do they differ?

The safest way to transport children to and from school and school-related activities is in a school bus.

Source: National Association of State Directors of Pupil Transportation Services

REVIEW

18. Let $M(x) = 13x - 18$. Find the value of x for which $M(x) = -1.2$. (Lesson 7-6)

19. **True or False** If a is any real number, then a is in the range of the function with equation $y = 3x$. (Lesson 7-5)

20. Write an exponential expression for the number, which, written in base 10, is 7 followed by n zeroes. (For example when $n = 3$, this number is 7,000.) (Lesson 7-2)

21. Write the equation $\frac{y - 3x + 2}{15} - \frac{1}{3} = \frac{7x + 2y}{5}$ in standard form. (Lessons 6-8, 3-8, 2-2)

22. Do the points (-1, 3), (5, 4), and (0, -8) lie on a line? How can you tell? (Lesson 6-6)

23. If you ask a random person the date of his or her birth, what is the probability that it will be one of the first ten days of the month? (Lesson 5-6)

24. Calculate $5(-5)^5$. (Lesson 2-4)

EXPLORATION

25. The statement, "If the growth factor g is greater than 1, exponential growth always overtakes constant increase," was made at the start of this lesson. Write a similar statement that could describe the relationship between exponential decay and constant decrease.

Chapter 7 Projects

1 Reciprocal Functions

For a function f, the function g with $g(x) = \frac{1}{f(x)}$ is called the *reciprocal function* of f.

a. On the same pair of axes, graph the function with equation $y = x$, and its reciprocal, the function with equation $y = \frac{1}{x}$. Do the same for $y = 2$, $y = x^2$, and $y = 2^x$.

b. For each function in Part a, for what values of x is the reciprocal not defined?

c. What is the reciprocal of the reciprocal of the function f?

d. Give an example of a function h that is equal to its reciprocal function. Give an example of a function that is always greater than its reciprocal function and a function that is always less than its reciprocal function.

e. Describe any general patterns you noticed when graphing reciprocal functions.

2 A Famous Snowflake

The drawing below illustrates the construction of a famous shape called the Koch snowflake. In the first stage, you begin with a triangle. At every other stage, you draw a small triangle in the middle of each of the sides of the previous stage.

Stage 1　　Stage 2　　Stage 3　　Stage 4

   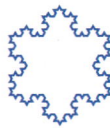

a. On a large piece of paper, draw the fifth stage of the Koch snowflake.

b. The first stage of the Koch snowflake has three sides. The next stage has 12 sides. The step after that has 48 sides. This number seems to grow exponentially. Explain why this is indeed the case, and find the growth factor and the initial value.

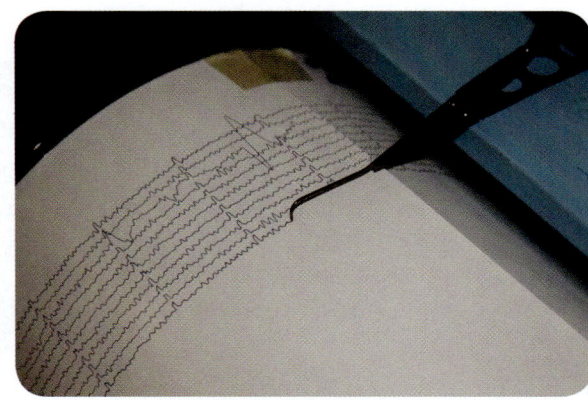

3 Richter Scale

The Richter scale is a scale used to measure the intensity of earthquakes. Look up the meaning of the Richter scale.

a. In 1992, an earthquake in Landers, California, measured 7.5 on the Richter scale. The largest earthquake ever recorded was the Great Chile earthquake of 1960, which measured 9.5 on the Richter scale. How much more powerful was the Chilean earthquake than the Landers earthquake?

b. Suppose a scientist plotted the magnitude of the most powerful annual earthquakes in a certain region over time and found that the measurements on the Richter scale increased linearly. How did the magnitude of the earthquakes increase?

c. Can an earthquake have a negative measure on the Richter scale? What does this mean?

4 Powers of Ten

In 1977, Charles and Ray Eames created a short documentary movie called *Powers of Ten*, which began with a picture of a picnic, and zoomed out by a factor of ten every ten seconds. (The movie is available on the Internet.) You can create a similar effect yourself. Download a free program that allows you to view satellite images of Earth at different levels of magnification.

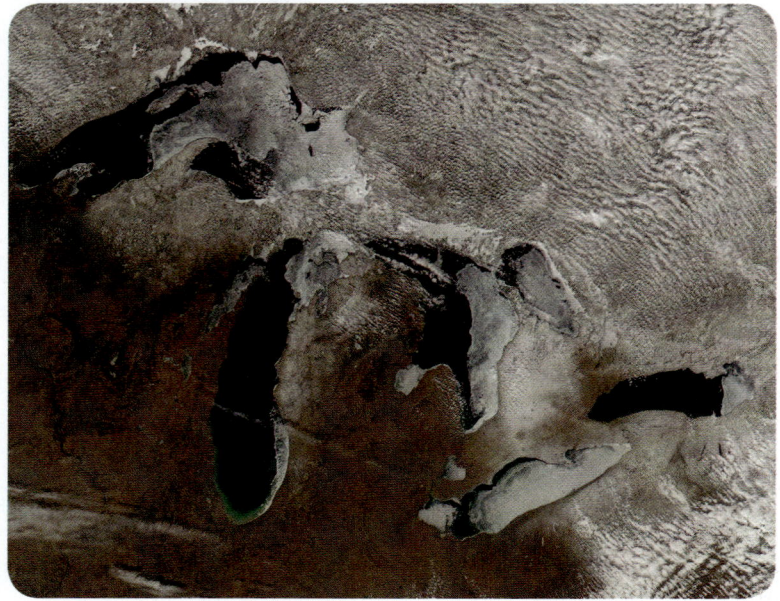

a. Find your school on the software, and zoom in as much as the software allows. Note the altitude from which you are viewing. Zoom out to an altitude ten times as high. Describe what you see. Do this as many times as the program allows.

b. Find out how many times you would have to zoom out before you could see the entire solar system, and how many times you would have to zoom out before you could see the entire galaxy.

c. Describe what you would see in the first three pictures if you zoomed in instead of out.

5 The Logistic Function and Growth

In this chapter you saw exponential growth used as a model for population growth. In nature, it is impossible for a population to continue to grow exponentially. (Otherwise it would overrun the entire Earth and still continue to grow!) When a population reaches a certain size, the resources available do not allow it to grow any larger. One way in which scientists model this kind of growth is called a *logistic function*. This is a function of the form $P(n) = a\frac{1 + b2^{-cn}}{1 + d2^{-cn}}$. This function gives the size of the population after n years.

a. Use a spreadsheet or a calculator to calculate the first 10 values of P, when $a = 200$, $b = 1$, $c = 0.5$, and $d = 20$. Plot the graph of the function and use exponential regression to find the exponential function that best fits the data. Do you think that in the first 10 years the population grows approximately exponentially?

b. For the values given in Part a, graph the first 100 values of the function. Does the population still appear to grow exponentially?

c. Explain why this function gives a better model of population growth than the exponential function.

Chapter 7 Summary and Vocabulary

- In Chapter 6, you saw many examples of constant-increase and constant-decrease patterns of change. They give rise to equations of the form $y = mx + b$. The change is called **linear** because the graph is a line. Now in Chapter 7, we turned our attention to patterns of change called **exponential growth** and **exponential decay**. They give rise to equations of the form $y = b \cdot g^x$.

- Graphs of exponential functions are curves. The change is called **exponential** because the independent variable is in the exponent. In exponential change, the number g is the growth factor. If $g > 1$, the situation is **exponential growth.** Among the common applications of exponential growth are compound interest and population growth. In the long run, exponential growth will always overtake a situation of linear increase. If $0 < g < 1$, the situation is **exponential decay.**

- These and other patterns can be described using the mathematical idea of a function. A **function** is a set of ordered pairs in which each first coordinate appears with exactly one second coordinate. Thus, functions exist whenever the value of one variable determines a unique value of another variable.

- A function may be described by a list of ordered pairs, a graph, an equation, or a written rule. If a function f contains the ordered pair (a, b), then we write $f(a) = b$. We say that b is the value of the function at a. If you know a formula for the function, you can obtain values and graphs of functions using calculators, spreadsheets, or paper and pencil.

- Constant-increase or constant-decrease situations are described by **linear functions.** Constant growth or decay situations are described by **exponential functions.** Repeatedly adding a quantity m to an initial value b gives rise to values of the linear function $f(x) = mx + b$. Repeatedly multiplying an initial value b by the growth factor g gives rise to values of the exponential function $f(x) = b \cdot g^x$. Spreadsheets are particularly useful for finding values of functions.

Vocabulary

7-1
power, nth power
base
exponent
principal
interest
annual yield
compound interest

7-2
exponential growth
growth factor
exponential growth equation

7-3
exponential decay
half-life

7-4
exponential regression

7-5
function
input, output
value of the function
squaring function
independent variable
dependent variable
domain of a function
range of a function
relation

7-6
$f(x)$ notation
function notation

Theorems and Properties

Repeated Multiplication Property of Powers (p. 398)
Compound Interest Formula (p. 400)
Growth Model for Powering (p. 405)
Zero Exponent Property (p. 405)

Chapter 7 Self-Test

Take this test as you would take a test in class. You will need a calculator. Then use the Selected Answers section in the back of the book to check your work.

1. Evaluate $x^2 + x^0$ when $x = \frac{1}{5}$.
2. Write $8 \cdot 8 \cdot 8 \cdot 8 \cdot d \cdot d \cdot d \cdot d \cdot d \cdot d$ using exponents.
3. If $f(x) = 3x^0$, find $f(1{,}729)$.
4. If $g(y) = 3y - y^2$, find $g(-2)$.

In 5–7, Tyrone deposits $400 into a savings account that pays 4.4% interest per year.

5. Write and evaluate an expression for the amount of money Tyrone will have after 7 years, assuming he doesn't deposit or withdraw money from the account.
6. At the same time that Tyrone makes his deposit, his sister Oleta deposits $400 in a highly unusual savings account. The account pays exactly $22 interest each year. Who will have more money after 10 years?
7. Who will have more money after 25 years?

In 8–10, use the following information. A particular new 2006 car costs $34,975. Suppose its value depreciates 16% each year.

8. What is the growth factor of the value of the car?
9. Write a function $m(x)$ that approximates the car's value in x years. Specify the domain of your function.
10. Find $m(5)$, the approximate value of the car in 5 years.

For 11–13, $f(x) = 5 \cdot 0.74^x$. Calculate the value.

11. $f(1)$
12. $f(5)$
13. $f(7)$

In 14–16, use the absolute value function $f(x) = |x - 3|$ graphed below. Consider the domain as the set of all real numbers.

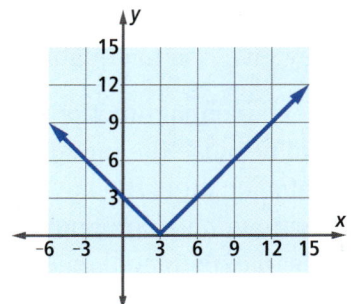

14. Determine $f(-12)$.
15. What is the range of the function f?
16. If the domain is restricted to $\{x : x \geq 5\}$, what is the range?

In 17–19, let $E(x) = 30(1.05)^x$ and $L(x) = 30 + 2x$.

17. Sketch a graph of these functions.
18. Which is greater, $L(9)$ or $E(9)$?
19. Give an example of a value of x when $L(x) < E(x)$.

In 20 and 21, write an equation describing the situation and graph the equation.

20. The population p of a country increases by 2.5% per year. In 1980, it had 76 million residents. Let k be the number of years since 1980.
21. The circulation c of a newspaper has decreased by 1% each month since January 2000, when it was 880,000. Let x be the number of months since January 2000.

In 22 and 23, graph the function on the given domain.

22. $h(k) = 1 - 3.5k$, $-10 \leq k \leq 8$

23. $c(x) = 10 \cdot 2^x$, $0 \leq x \leq 5$

24. **Matching** Decide which of the situations the function with the given equation describes.

   i. constant increase

   ii. constant decrease

   iii. exponential growth

   iv. exponential decay

   a. $f(x) = -4x + 18$

   b. $g(x) = 0.4(5)^x$

   c. $h(x) = 5(0.4)^x$

   d. $m(x) = \frac{2}{3}x - 7$

25. It is estimated that a house purchased in 1990 for $100,000 has increased in value about 4% a year since that time. Suppose you want to use a spreadsheet to display the estimated value of the house from 1990 to 2010.

	A	B
1	Year	Value of House
2	1990	$100,000
3		
4		
5		

   a. What formula could you enter in cell A3 to get the appropriate value using cell A2?

   b. Explain the process by which you would obtain appropriate amounts in cells B4 to B22.

Chapter 7 Chapter Review

SKILLS **P**ROPERTIES **U**SES **R**EPRESENTATIONS

SKILLS Procedures used to get answers

OBJECTIVE A Evaluate functions. (Lesson 7-6)

In 1–4, suppose $f(x) = 10 - 3x$. Evaluate the function.

1. $f(2)$
2. $f(-4)$
3. $f(1) + f(0)$
4. $f(3 + 6)$

5. If $g(x) = \left(\frac{11}{6}\right)^x$, give the value of $g(2)$.

6. If $h(x) = 2x^3$, calculate $h(4)$.

7. If $f(t) = -8t$ and $g(t) = 6t$, give the value of $f(-1) + g(-2)$.

8. If $E(m) = 6^m$ and $L(m) = m + 5$, find a value for m for which $E(m) < L(m)$.

OBJECTIVE B Calculate function values in spreadsheets. (Lesson 7-7)

In 9 and 10, use the spreadsheet below.

	A	B	C
1	x	1000(1.05)^x	1000+50x
2	0	1000	1000
3	1		
4	2		
5	3		

9. Rani wants to put values of the function with equation $y = 1,000(1.05)^x$ in column B of the spreadsheet.
   a. What formula can she enter in cell B3?
   b. What number will appear in cell B3?
   c. What should she do to get values of y in column B when $x = 2, 3, 4, 5, \ldots, 10$?

10. Olivia wants to put values of the function with equation $f(x) = 1,000 + 50x$ in column C of the spreadsheet.
    a. What formula can she enter in cell C3?
    b. What number will appear in cell C3?
    c. What should she do to get values of $f(x)$ in column C when $x = 2, 3, 4, 5, \ldots, 10$?

PROPERTIES The principles behind the mathematics

OBJECTIVE C Use the language of functions. (Lessons 7-5, 7-6)

11. Suppose $y = f(x)$.
    a. What letter names the independent variable?
    b. What letter names the function?

12. A linear function L is graphed below.

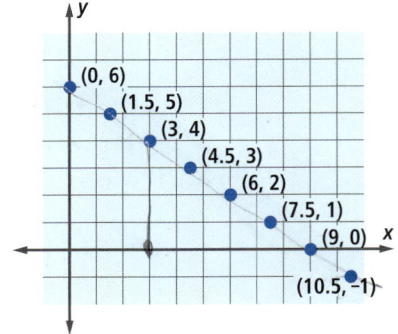

a. What is $L(3)$? = 4
b. What is the domain of L?
c. What is the range of L?
d. Find a formula for $L(x)$ in terms of x.

452 Using Algebra to Describe Patterns of Change

13. Suppose a function f consists of only the ordered pairs (2, 200), (4, 400), (5, 500), and (10, 1,000).

    a. What is the domain of f?

    b. What is the range of f?

    c. Give a formula for $f(x)$ in terms of x.

USES Applications of mathematics in real-world situations

OBJECTIVE D Calculate compound interest. (Lesson 7-1)

14. An advertisement indicated that a 3-year certificate of deposit would yield 4.53% per year. If $2,000 is invested in this certificate, what will it be worth at the end of 3 years?

15. When Brie was born, she received a gift of $500 from her grandparents. Her parents put it into an account at an annual yield of 5.2%. Brie is now 12 years old. How much is this gift worth now?

16. In 2004, the endowment of Harvard University (the value of the university's assets) was reported to be about $22.14 billion. Suppose the trustees of the university feel they can grow this endowment by 6% a year. What would be the value of the endowment in 2010?

17. Which investment yields more money: (a) x dollars for 4 years at an annual yield of 8% or (b) the same amount of money at an annual yield of 4% for 8 years? Explain your reasoning.

OBJECTIVE E Solve problems involving exponential growth and decay. (Lessons 7-2, 7-3, 7-4)

18. In 1990, there were 4.4 million cell phone subscribers in the United States; by 2006, there were 219.4 million subscribers. The table below shows the number of cell phone subscribers for each year from 1990 to 2006.

Year	Cell Phone Subscribers (in thousands)
1990	4,369
1991	3,380
1992	8,893
1993	13,067
1994	19,283
1995	28,154
1996	38,195
1997	48,706
1998	60,831
1999	76,285
2000	97,036
2001	118,398
2002	134,561
2003	148,066
2004	169,467
2005	194,479
2006	219,420

a. Create a scatterplot with y = cell phone subscribers and x = the year since 1990. Why is the exponential model a better model for these data than a linear model?

b. Use exponential regression to find an equation to fit the data.

c. Use your equation from Part b to predict the number of cell phone subscribers for the year 2013.

19. Twelve fish were introduced into a large lake. In 3 years, the population had multiplied by a factor of 20.
    a. In 15 years, at this growth rate, by how much would the population be multiplied?
    b. **Multiple Choice** If P is the number of fish t years after introduction, which formula relates P and t?
       A $P = 12(20)^t_t$ **B** $P = 12(20)^{3t}$
       C $P = 12(20)^{\frac{t}{3}}$

20. Suppose a car depreciates 20% in value each year, and its purchase price was $22,000.
    a. What is the growth factor in the situation?
    b. What is the car's value 1 year after purchase?
    c. What is its value n years after purchase?

OBJECTIVE F Determine whether a situation is constant increase, constant decrease, exponential growth, exponential decay, or a nonconstant change. (Lesson 7-4)

In 21–24, does the equation describe a situation of constant increase, constant decrease, exponential growth, or exponential decay?

21. $y = \frac{1}{5}x - 10$
22. $m = -3n + 4$
23. $p = \frac{2}{3}(3)^r$
24. $y = 3\left(\frac{2}{3}\right)^x$

25. A store is going out of business. It advertises that it is reducing prices 1% on day 1, then 2% more on day two, then 3% more on day 3, and so on for 100 days. Is this a situation of constant decrease, exponential decay, or neither of these?

26. Is the sequence: $\frac{1}{12}, \frac{1}{6}, \frac{1}{4}, \frac{1}{3}, \ldots$ one of constant increase, constant decrease, exponential growth, exponential decay, or a different kind of increase or decrease?

OBJECTIVE G Compare linear increase with exponential growth. (Lesson 7-7)

27. Country A has 10 million people and its population is growing by 2% each year. Country B has 20 million people and its population is growing by 1 million people per year.
    a. Give an equation for the population $A(n)$ of country A, n years from now.
    b. Give an equation for the population $B(n)$ of country B, n years from now.
    c. In 30 years, if these trends continue, which country would have the greater population? Explain why.
    d. In 100 years, if these trends continue, which country would have the greater population? Explain why.

28. Explain why exponential growth always overtakes linear increase if the time frame is long enough.

REPRESENTATIONS Pictures, graphs, or objects that illustrate concepts

OBJECTIVE H Graph exponential relationships. (Lessons 7-2, 7-3)

In 29–32, graph the equation and describe a situation that it might represent.

29. $A = 1,000(1.02)^x$
30. $C = 0.07 + 0.03t$
31. $y = 50 - 3x$
32. $V = 10,000(0.90)^t$

33. Graph the function of Question 20 for integer values of the domain from 1 to 10.

Chapter Wrap-Up

OBJECTIVE I Graph functions.
(Lessons 7-5, 7-6)

In 34–37, graph the function on the domain $-5 \leq x \leq 5$.

34. $f(x) = 30 - 2x$

35. $g(x) = x^2$

36. $h(x) = 3 \cdot 2^x$

37. $m(x) = x^3 - x$

38. **Multiple Choice** Which is the graph of the function A when $A(x) = \left(\frac{1}{3}\right)^x$?

A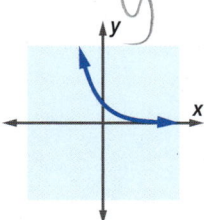

B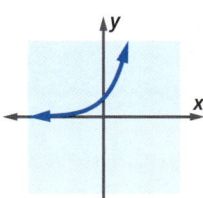

C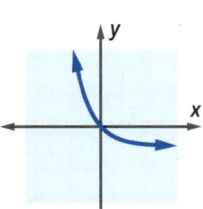

D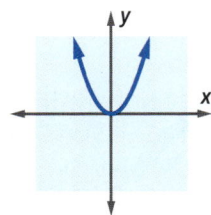

39. Refer to the graph of a linear function L below.

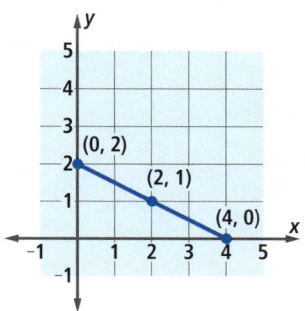

a. What is the value of $L(2)$?
b. What is the value of $L(0)$?
c. What is the domain of L?
d. What is the range of L?

40. On the grid below, each tick mark is one unit.

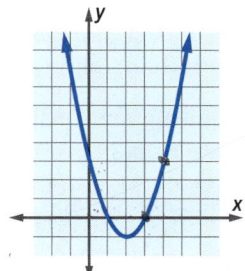

a. What is $f(0)$?
b. What is $f(3)$?
c. What is the domain of f?
d. What is the range of f?

Chapter 8
Powers and Roots

Contents

- 8-1 The Multiplication Counting Principle
- 8-2 Products and Powers of Powers
- 8-3 Quotients of Powers
- 8-4 Negative Exponents
- 8-5 Powers of Products and Quotients
- 8-6 Square Roots and Cube Roots
- 8-7 Multiplying and Dividing Square Roots
- 8-8 Distance in a Plane
- 8-9 Remembering Properties of Powers and Roots

Visible light, infrared and ultraviolet radiation, x-rays, microwaves, and radio waves are all parts of the electromagnetic spectrum. Waves differ only in their wavelengths. Radio waves can be as long as 1,000 meters or longer; x-rays can be as short as one billionth of a meter or less.

To describe these numbers, we use powers of 10. On the scale shown below, each tick mark is 10 *times* the length of the tick mark to its left. You have seen these powers of 10 used in scientific notation.

The Electromagnetic Spectrum

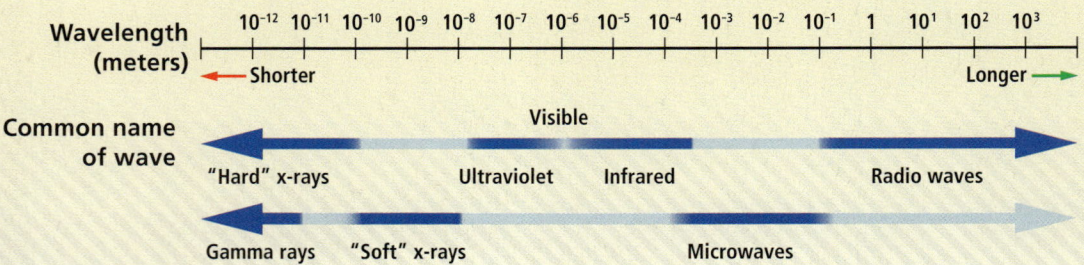

Here is a graph of $y = 10^x$ on the window $-5 \leq x \leq 5, -20 \leq y \leq 20$. You can see that as x increases by 1, y is multiplied by 10. The coordinates of some points on this graph are easily found by hand. When $x = 0$, $y = 10^0 = 1$. When $x = 1, y = 10^1 = 10$. When $x = 2, y = 10^2 = 100$. When $x = 3$, $y = 10^3 = 1{,}000$, too large to be on the graph. But the graph contains values of the function when x is negative. What is the meaning of 10^{-1}, 10^{-2}, 10^{-3}, and so on?

Also, the graph computes values for y when x is not an integer. Although we do not discuss all the powers of x in this chapter, the meanings of $\frac{1}{2}$ and $\frac{1}{3}$ are discussed and found to be related to square roots and cube roots. Additionally, there is the question of how all these powers and roots are related. For example, how is x^7 related to x^{-7}? Are the 7th powers of different numbers related in any way? The answers to these questions provide additional understanding of some of the applications of powers that you saw in the preceding chapter. They also shed light on some important formulas for lengths, area, and volume found in geometry.

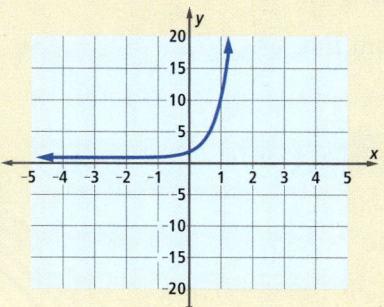

Chapter 8

Lesson 8-1
The Multiplication Counting Principle

Vocabulary

scientific notation

▶ **BIG IDEA** Expressions involving powers result from certain counting problems and are used in scientific notation.

Mental Math

Simplify $\dfrac{n \cdot 3n \cdot 6n \cdot 9n}{n \cdot 2n \cdot 4n \cdot 6n}$.

The Jaipur Friendship and Knitting Society decided that it would call itself by the 3-letter acronym JFK. (An *acronym*, like USA or NCAA, is a "word" made up of the first letter of each word in a phrase.) They were disappointed to learn that JFK was already a popular acronym. It is the initials of President John Fitzgerald Kennedy and identifies one of the airports in New York City as well as some highways throughout the country. So the society decided to use JFKS.

Members of the society realized that there is only a certain number of 3-letter acronyms. So they wondered, how many 3-letter acronyms are there in the English language?

To answer this question, we apply a very useful problem-solving strategy. We consider a simpler problem we may be able to solve and then apply its solution to the problem we want to solve.

Example 1
How many 2-letter acronyms are there?

Solution Count the acronyms in an organized manner, alphabetically. AA, AB, AC, ..., AZ gives 26. BA, BB, BC, ..., BZ gives another 26. There will be 26 groups of 26, so the total number is $26 \cdot 26 = 26^2 = 676$.

Knitting was first introduced to Europe in the 5th century c.e.

Source: Fine Living TV Network

Notice that we gave the answer to Example 1 in three forms: as a *product* $26 \cdot 26$, as a *power* 26^2, in our customary *base-10 decimal* system as 676. Each of these forms is useful, so you need to be able to move back and forth from one way of writing a number to another.

Example 1 applies multiplication in a manner that is so important that it has a special name, the *Multiplication Counting Principle*.

> **Multiplication Counting Principle**
>
> If one choice can be made in *m* ways and then a second choice can be made in *n* ways, then there are *mn* ways of making the first choice followed by the second choice.

The Multiplication Counting Principle was applied in Example 1. There were 26 choices for the first letter. After that choice was made, there were 26 choices for the second letter. So $m = 26$ and $n = 26$, and the number of 2-letter acronyms is $26 \cdot 26$, or 676.

Example 2
How many 3-letter acronyms are there?

Solution 1 Apply the Multiplication Counting Principle to the result of Example 1, which found that there are 676 different 2-letter acronyms. For each one, there are 26 possible third letters. So the total number is $676 \cdot 26 = 17{,}576$.

Solution 2 Keep the result from Example 1 in factored form. There are $26 \cdot 26$ different 2-letter acronyms. Now, with 26 possible third letters, the total number is $26 \cdot 26 \cdot 26 = 26^3$. This is also equal to 17,576.

The idea behind the solutions to Example 2 is very powerful and can be continued. To get the number of 4-letter acronyms, you can work from the number of 3-letter acronyms. Each 3-letter acronym is the beginning of 26 4-letter acronyms, so the number of 4-letter acronyms is $26^3 \cdot 26$, or 26^4, and so on. This thinking is much like the Now/Next thinking you used in the spreadsheets of Lesson 7-8.

 QY1

Choosing From *n* Objects Repeatedly

A sequence of two objects is said to have "length" 2; a sequence of three objects has length 3, and so on. In the previous examples you counted ways to make acronyms of length 2 and 3. Order matters with acronyms. President John Fitzgerald Kennedy had initials JFK, but not KFJ. Order also matters with acronyms such as NASA (National Aeronautics and Space Administration) and scuba (self-contained underwater breathing apparatus).

The process used to find the number of different acronyms of length 2 or 3 can be generalized, leading to the following result.

Arrangements Theorem
If there are *n* ways to select each object in a sequence of length *L*, then n^L different sequences are possible.

 QY2

▶ **QY1**

The Russian alphabet has 33 letters. How many 4-letter Russian acronyms are possible?

▶ **QY2**

A test has 20 multiple-choice questions with 5 choices each. How many different sets of answers are possible?

The Multiplication Counting Principle **459**

In Example 3, not all of the objects have the same number of choices possible.

> **GUIDED**
>
> ### Example 3
> Suppose a standardized test has 20 questions with 4 choices and 10 questions with 5 choices. What is the probability that a person could guess on every one of the 30 questions and answer them all correctly?
>
> **Solution** Think of the test as having 2 parts. Count the number of ways each part can be created.
>
> **Part 1** This part has 20 questions with 4 choices. We want to know how many sequences m of length 20 there are with these 4 letters.
>
> $$m = \underline{}$$
>
> **Part 2** This part has 10 questions with 5 choices. Let n be the number of sequences of length 10 with 5 letters.
>
> $$n = \underline{}$$
>
> Now apply the Multiplication Counting Principle and multiply the results from Part 1 and Part 2 to determine how many different sets of answers are possible.
>
> $$mn = \underline{}$$
>
> In base 10 this number is 10,737,418,240,000,000,000. If a person is guessing, then we assume that each one of these sets of answers is equally likely, and only one of them has all the correct answers. So the probability of having a perfect test is $\frac{1}{mn} = \underline{}$.

The national average mathematics score on the National Assessment of Educational Progress (NAEP) at grade 8 was 16 points higher in 2005 than in 1990.

Source: National Assessment of Educational Progress

Writing Large Numbers in Scientific Notation

Depending on your calculator and the mode it is in, if you enter mn to calculate the answer to Example 3, the result will be displayed either as the long base-10 number or in *scientific notation*. You should try this on your calculator. Presumably you have used scientific notation in other mathematics or science classes. Recall that in **scientific notation,** a number is represented as $x \cdot 10^n$, where n is an integer and $1 \leq x < 10$. In scientific notation, $10{,}737{,}418{,}240{,}000{,}000{,}000 \approx 1.0737 \cdot 10^{19}$.

A major advantage of scientific notation is that it quickly tells you the size of a number. The exponent is one less than the number of digits in the whole number. A whole number $x \cdot 10^n$ has $n + 1$ digits. Notice that the exponent is 19 in the scientific notation form of the number above, and the base-10 form has 20 digits.

Questions

COVERING THE IDEAS

1. a. Write all the 2-letter acronyms that can be made from the five vowels A, E, I, O, and U.
   b. **Fill in the Blanks** In Part a, you have found the number of sequences of length __?__ of __?__ objects.

2. The Greek alphabet is about 2,750 years old and is used by about 12 million people in Greece and other countries around the world. It contains 24 letters.
   a. How many 2-letter acronyms are there using Greek letters?
   b. How many 4-letter acronyms are there using Greek letters?

3. Write the number 5^6 in base-10 and in scientific notation.

4. a. Write an example of a 6-letter acronym made from the five letters A, B, C, D, and E.
   b. How many of these 6-letter acronyms are possible?

5. Suppose part of a spreadsheet has 6 columns and 15 rows.
   a. How many cells are in the spreadsheet?
   b. Explain how your answer applies the Multiplication Counting Principle by indicating how choices are involved in finding the number of cells.

6. a. Draw three horizontal lines and four vertical lines. In how many points do these lines intersect?
   b. If you drew 30 horizontal lines and 40 vertical lines, in how many points would they intersect?
   c. If you drew h horizontal lines and v vertical lines, in how many points would they intersect?

7. a. A quiz consists of 10 true-or-false questions. How many different sets of answers are possible? Write your answer in exponential form, in base-10, and in scientific notation.
   b. If you guess on all 10 questions, what is the probability of getting all the questions correct?
   c. Answer Parts a and b if there are Q true or false questions on the test.

8. A test has 5 true-or-false questions and 15 multiple-choice questions with 4 choices each.
   a. How many different sets of answers are possible? Write your answer in exponential form, in base-10, and in scientific notation.
   b. If you guessed on every question, what is the probability you would get all 20 questions correct?

The Multiplication Counting Principle

Chapter 8

9. Use the information on the electromagnetic spectrum on page 456. Write the number in base-10 notation.
   a. the longest wavelength marked on the spectrum
   b. the shortest wavelength marked on the spectrum

APPLYING THE MATHEMATICS

10. Radio station call letters in the United States must start with either W or K.
    a. How many choices are there for the first letter?
    b. How many choices are there for the second letter?
    c. How many different 4-letter station names are possible?

11. The Cayuga Indians played a game called *Dish* using 6 peach pits. The pits were blackened on one side and plain on the other. When pits were tossed, they landed on the blackened and plain sides with about the same frequency. When the six pits were tossed, a player scored if either all blackened or all plain sides landed up. What is the probability that a player would score points on one toss of the pits?

12. How many 6-digit whole numbers are there? Answer the question in two ways.
    a. by subtracting the least 6-digit number from the greatest 6-digit number and working from that
    b. by thinking of the problem as a series of choices: 9 choices for the left digit (because it cannot be zero) and 10 choices for every other

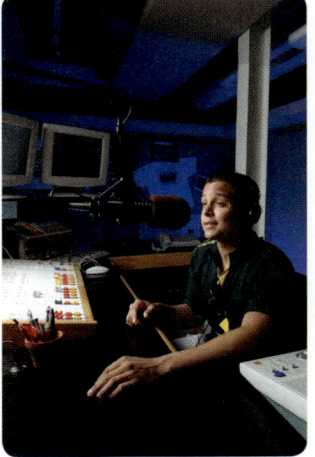

In June 2005, there were 2,019 commercial U.S. radio stations.

Source: Federal Communications Commission

13. Assume that everyone in the United States has a first name, a middle name, and a last name. Therefore, everyone has a 3-letter acronym of his or her initials. If your initials are typical, about how many of the 300 million people in the United States have your initials?

14. How many different sets of answers are possible for each of the following tests?
    a. a group of P true-or-false questions
    b. a group of Q questions that can be answered "sometimes," "always," or "never"
    c. a test made up of two parts: P true or false questions and Q *always*, *sometimes but not always*, or *never* questions

REVIEW

15. Do the ordered pairs (x, y) that satisfy $y < -4x + 8$ describe a function? Why or why not? (**Lesson 7-5**)

16. Consider these points (0, –2), (6, 4), and (–10, –12). (Lessons 6-6, 3-4)
    a. Write an equation for the line containing these three points.
    b. If the point (x, 20) lies on this line, find the value of x.

17. Refer to the similar triangles at the right. If the ratio of similitude of the smaller triangle to the larger triangle is $\frac{1}{3}$, find the area of the larger triangle. (Lesson 5-10)

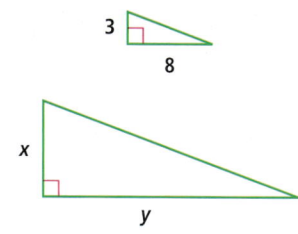

18. A biologist captured, tagged, and released 40 fish caught in a lake. Three weeks later, the biologist caught 28 fish. Of these, 8 had tags. Based on these findings, estimate the total number of fish in the lake. (Lesson 5-9)

19. Solve $\frac{1}{5}x - \frac{3}{10} = \frac{9}{10}$. (Lesson 3-8)

In 20 and 21, write the number in scientific notation. (Previous Course)

20. seven thousandths
21. 2.8 billion

EXPLORATION

22. Many of the cabinet-level departments in the United States government are identified by acronyms. Tell what department each of these acronyms stands for. As a hint, we have put the departments (not the acronyms) in alphabetical order.

    a. USDA   b. DOC   c. DOD   d. ED
    e. DOE   f. HHS   g. DHS   h. HUD
    i. DOJ   j. DOL   k. DOS   l. DOI
    m. DOT   n. VA

The Pentagon has three times the floor space of the Empire State Building in New York.
Source: Pentagon Tours

23. The Greek alphabet has 24 letters and the Russian alphabet has 33 letters.
    a. Are there more 4-letter Greek acronyms or 3-letter Russian acronyms?
    b. What is the least value of n for which there are *fewer* n-letter Greek acronyms compared to (n – 1)-letter Russian acronyms?

QY ANSWERS

1. $33^4 = 1,185,921$

2. $5^{20} =$ 95,367,431,640,625 ≈ $9.5 \cdot 10^{13}$

Lesson 8-2

Products and Powers of Powers

> **BIG IDEA** Because of the relationship between repeated multiplication and powers, products and powers of powers can themselves be written as powers.

Multiplying Powers with the Same Base

When n is a positive integer, $x^n = \underbrace{x \cdot x \cdot \ldots \cdot x}_{n \text{ factors}}$. From this, a number of important properties can be developed. They all involve multiplication in some way because of the relationship between exponents and multiplication. Addition is different. In general, there is no way to simplify the sum of two powers. For example, $3^2 + 3^4 = 9 + 81 = 90$, and 90 is not an integer power of 3. But notice what happens when we multiply powers with the same base.

$$3^2 \cdot 3^4 = \underbrace{(3 \cdot 3)}_{2 \text{ factors}} \cdot \underbrace{(3 \cdot 3 \cdot 3 \cdot 3)}_{4 \text{ factors}} = \underbrace{(3 \cdot 3 \cdot 3 \cdot 3 \cdot 3 \cdot 3)}_{6 \text{ factors}} = 3^6$$

$$1.06^0 \cdot 1.06^3 = 1 \cdot \underbrace{(1.06 \cdot 1.06 \cdot 1.06)}_{3 \text{ factors}} = 1.06^3$$

$$(-6)^5 \cdot (-6)^5 = \underbrace{(-6 \cdot -6 \cdot -6 \cdot -6 \cdot -6)}_{5 \text{ factors}} \cdot \underbrace{(-6 \cdot -6 \cdot -6 \cdot -6 \cdot -6)}_{5 \text{ factors}} = (-6)^{10}$$

These three expressions involved multiplying powers of the same base, where the base was a specific number (3, 1.06, or –6). The same process is used to multiply powers of a variable.

Mental Math

Use the circle graph. Give each value in the indicated form.

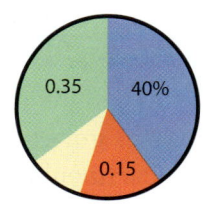

a. blue sector, decimal
b. green sector, fraction
c. red sector, percent
d. yellow sector, decimal
e. yellow sector, fraction

Activity

Step 1 Evaluate each expression.
  a. $z^7 \cdot z^4$ b. $y^6 \cdot y^8$ c. $y^2 \cdot y^4 \cdot y^3$ d. $x^3 \cdot x$
  e. $x^4 \cdot x^2 \cdot x$ f. $t \cdot t \cdot t$ g. $t^3 \cdot t^4 \cdot t \cdot t$ h. $z^0 \cdot z^5 \cdot z^2 \cdot z^2$

Step 2 Check your answers using a CAS.
  a. When multiplying powers with the same base, how is the exponent of the answer related to the exponents of the original factors?
  b. Some of the variables do not have visible exponents, like $x^3 \cdot x$. Does the relationship you described in Part a apply in this case?
  c. Refer to Part h. What does z^0 equal? How does this fit in with the answer to Part 2a?

464 Powers and Roots

The general pattern established in the activity leads us to the *Product of Powers Property*.

> **Product of Powers Property**
>
> For all m and n, and all nonzero b, $b^m \cdot b^n = b^{m+n}$.

The Product of Powers Property can be illustrated with a multiplication fact triangle. Notice that the powers are multiplied, but the exponents are added.

Activity 1 Part 1a:
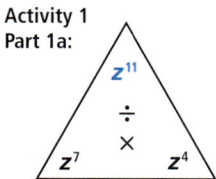

In general:
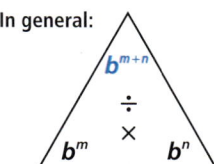

Here is a situation leading to multiplying powers with the same base.

Example 1

Suppose you fold an 8.5-inch by 11-inch piece of paper alternating the direction of the folds (fold down, fold to the left, fold down, fold to the left, and so on).

First Fold (Fold Down)
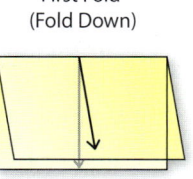

Second Fold (Fold Left)
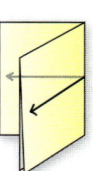

Third Fold (Fold Down)
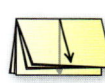

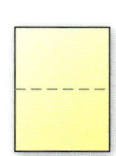
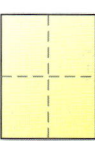

Imagine that you keep folding indefinitely. Write an expression for the number of regions created by first folding the paper n times, and then folding it *three times more*.

Solution You begin with 1 piece of paper, which is 1 region. Each time you fold the paper, you double the number of regions. After n folds, you have 2^n regions. Folding an additional 3 times doubles the number of regions 3 more times. The number of folds is $2^n \cdot 2 \cdot 2 \cdot 2$, or $2^n \cdot 2^3$. Applying the Product of Powers Property, $2^n \cdot 2^3 = 2^{n+3}$.

Chapter 8

Multiplying Powers with Different Bases

The Product of Powers Property tells how to simplify the product of two powers with the same base. A product with different bases, such as $a^3 \cdot b^4$, usually *cannot* be simplified.

Example 2
Simplify $r^9 \cdot s^5 \cdot r^7 \cdot s^2$.

Solution Use the properties of multiplication to group factors with the same base.

$r^9 \cdot s^5 \cdot r^7 \cdot s^2 = r^9 \cdot r^7 \cdot s^5 \cdot s^2$ Commutative Property of Multiplication

$\qquad\qquad\qquad = r^{9+7} \cdot s^{5+2}$ Product of Powers Property

$\qquad\qquad\qquad = r^{16} \cdot s^7$ Simplify.

$r^{16} \cdot s^7$ cannot be simplified further because the bases are different.

Check Perform the multiplication with a CAS.

A CAS indicates that $r^9 \cdot s^5 \cdot r^7 \cdot s^2 = r^{16} \cdot s^7$. It checks.

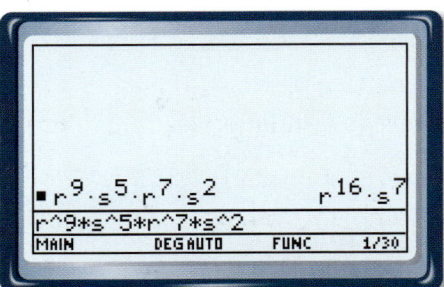

What Happens If We Take a Power of a Power?

When powers of powers are calculated, interesting patterns also emerge.

Example 3
Write $(5^2)^4$ as a single power.

Solution Think of 5^2 as a number that is raised to the 4th power.

$(5^2)^4 = 5^2 \cdot 5^2 \cdot 5^2 \cdot 5^2$ Repeated Multiplication Model for Powering

$\qquad\, = 5^{2+2+2+2}$ Product of Powers Property

$\qquad\, = 5^8$ Simplify.

The general pattern is called the *Power of a Power Property*.

> **Power of a Power Property**
>
> For all m and n, and all nonzero b, $(b^m)^n = b^{mn}$.

Some expressions involve both powers of powers and multiplication.

466   Powers and Roots

Lesson 8-2

Example 4
Simplify $3m(m^4)^2$.

Solution 1 First rewrite $(m^4)^2$ as repeated multiplication.
$$3m(m^4)^2 = 3m^1 \cdot m^4 \cdot m^4 = 3m^9$$

Solution 2 First use the Power of a Power Property with $(m^4)^2$.
$$3m(m^4)^2 = 3m^1 \cdot m^8 = 3m^9$$

Questions

COVERING THE IDEAS

In 1 and 2, write the product as a single power.

1. $18^5 \cdot 18^4$
2. $(-7)^3 \cdot (-7)^2$

3. Write $w^4 \cdot w^3$ as a single power and check your answer by substituting 2 for w.

In 4–6, suppose you fold an 8.5-inch by 11-inch piece of paper as in Example 1. Calculate the number of regions created by folding the paper in the way described.

4. two times, then three more times
5. three times, then two more times, then two more times
6. m times, then n more times

7. Find the expression that completes the fact triangle at the right.

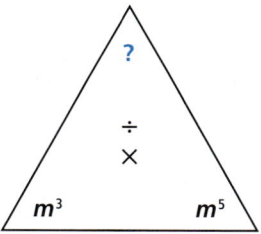

In 8–10, rewrite the expression as a single power.

8. $(2^3)^4$
9. $(m^5)^2$
10. $(y^2)^6$

In 11–16, simplify the power.

11. $3a^4 \cdot 5a^2$
12. $2(k^{10})^7$
13. $d(d^{13})$
14. $m^2 \cdot m^9 \cdot a^0 \cdot m^7 \cdot a^9$
15. $a^3(b^3a^5)$
16. $4k^2(k^3)^5$

APPLYING THE MATHEMATICS

17. A quiz has two parts. The first part has 5 multiple-choice questions. The second part has 3 multiple-choice questions. Each multiple-choice question has 4 choices. How many different sequences of answers are possible on the 8 questions?

In 18–20, solve the equation. Show all work.

18. $2^4 \cdot 2^n = 2^{12}$
19. $(5^6)^x = 5^6$
20. $(a^7 \cdot a^n)^2 = a^{24}$

Chapter 8

21. Suppose a population P of bacteria triples each day.
    a. Write an expression for the number of bacteria after 4 days.
    b. How many days after the 4th day will the bacteria population be $P \cdot 3^{20}$?

22. Does the Product of Powers Property work for fractions? Write each expression as a power and a simple fraction.
    a. $\frac{3}{5} \cdot \frac{3}{5} \cdot \frac{3}{5}$
    b. $\frac{3}{5} \cdot \frac{3}{5} \cdot \frac{3}{5} \cdot \frac{3}{5}$
    c. $\frac{3}{5} \cdot \frac{3}{5} \cdot \frac{3}{5} \cdot \frac{3}{5} \cdot \frac{3}{5} \cdot \frac{3}{5} \cdot \frac{3}{5}$

REVIEW

23. Abigail is going to buy a new car. She has to choose the body style (sedan, SUV, or convertible), transmission (automatic or standard), and color (white, black, red, blue, or green). **(Lesson 8-1)**
    a. How many different ways can Abigail make her choices?
    b. If another color choice of silver is given to her, how many more choices does she have?

24. If $f(x) = 3x + 2$ and $g(x) = 3x^2 - 2$, find each value. **(Lesson 7-6)**
    a. $f(3)$
    b. $g(-2)$
    c. $f(5) - g(5)$
    d. $g(-4) + f(-4)$

25. A band sold 1,252 tickets for a concert that were priced at $35. The band decided to lower the ticket price to their next concert to $30 in hopes of attracting a larger audience. After lowering the price, 1,510 tickets were sold. **(Lessons 6-6, 3-4)**
    a. Write a linear equation that relates the price of the ticket x and the number of tickets sold y.
    b. Use your answer to Part a to predict the number of tickets that will be sold if the price is lowered to $20.

26. Consider the line $y = 4x - 5$. Find **(Lessons 6-4, 6-2)**
    a. its slope.
    b. its y-intercept.
    c. its x-intercept.

27. Write 0.00324 in scientific notation. **(Previous Course)**

28. Write these numbers as decimals. **(Previous Course)**
    a. $9.8 \cdot 10^0$
    b. $9.8 \cdot 10^{-1}$
    c. $9.8 \cdot 10^{-2}$
    d. $9.8 \cdot 10^{-3}$

In 2004, there were 4,236,736 passenger cars produced in the United States.

Source: Automotive News Data Center

EXPLORATION

29. There are prefixes in the metric system for some of the powers of 10. For example, the prefix for 10^3 is kilo-, as in kilogram, kilometer, and kilobyte. Give the metric prefix for each power.
    a. 10^6
    b. 10^9
    c. 10^{12}
    d. 10^{15}
    e. 10^{18}

Lesson 8-3

Quotients of Powers

▶ **BIG IDEA** Because of the relationship between multiplication and division, quotients of powers can themselves be written as powers.

As you know, $\frac{24}{3} = 8$ because $8 \cdot 3 = 24$. Similarly, $\frac{24}{3} = \frac{8 \cdot \cancel{3}^1}{1\cancel{3}} = 8$. Both of these methods can be helpful in understanding quotients of powers.

For example, suppose $\frac{x^{10}}{x^2} = x^?$. By rewriting this statement to read $x^? \cdot x^2 = x^{10}$, we can apply the Product of Powers Property, $x^{?+2} = x^{10}$. You can see that the unknown exponent is 8 because $8 + 2 = 10$. So $\frac{x^{10}}{x^2} = x^8$. Another way of finding the unknown exponent in $\frac{x^{10}}{x^2} = x^?$ is to write both the numerator and denominator in expanded form and simplify the fraction.

$$\frac{x^{10}}{x^2} = \frac{{}^1\cancel{x} \cdot {}^1\cancel{x} \cdot x \cdot x \cdot x \cdot x \cdot x \cdot x \cdot x \cdot x}{{}_1\cancel{x} \cdot \cancel{x}_1}$$

$$= x \cdot x \cdot x \cdot x \cdot x \cdot x \cdot x \cdot x$$

$$= x^8$$

Mental Math

True or false?
a. $x^{100} + x^{101} = x^{201}$
b. $x^{100} \cdot x^{101} = x^{201}$
c. $x^{100} + x^{101} = 2x^{101}$
d. $x^{100} + x^{100} = 2x^{100}$

Activity

1. Simplify each expression.

   a. $\frac{a^7}{a^4}$
   b. $\frac{m^{14}}{m^5}$
   c. $\frac{y^{12}}{y}$
   d. $\frac{n^{13}}{n^{13}}$

2. When dividing powers of the same base, how is the exponent of the answer related to the exponents of the original division?

The general pattern established in the Activity is the *Quotient of Powers Property*.

Quotient of Powers Property

For all m and n, and all nonzero b, $\frac{b^m}{b^n} = b^{m-n}$.

▶ **QY**

Simplify $\frac{z^{50}}{z^{10}}$.

 QY

Quotients of Powers **469**

Fact triangles are another way of representing the Quotient of Powers Property.

A specific case:
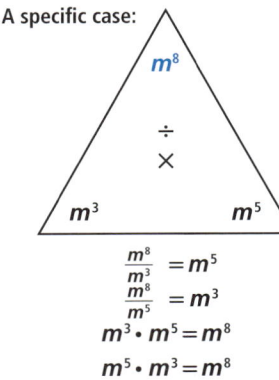

$$\frac{m^8}{m^3} = m^5$$
$$\frac{m^8}{m^5} = m^3$$
$$m^3 \cdot m^5 = m^8$$
$$m^5 \cdot m^3 = m^8$$

In general:
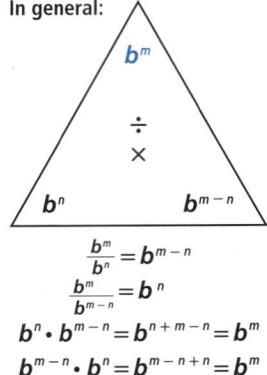

$$\frac{b^m}{b^n} = b^{m-n}$$
$$\frac{b^m}{b^{m-n}} = b^n$$
$$b^n \cdot b^{m-n} = b^{n+m-n} = b^m$$
$$b^{m-n} \cdot b^n = b^{m-n+n} = b^m$$

The Zero Power

In Question 1d of the Activity, you should have seen that $\frac{n^{13}}{n^{13}} = 1$. More generally, consider the fraction $\frac{b^m}{b^m}$. By the Quotient of Powers Property, $\frac{b^m}{b^m} = b^{m-m} = b^0$. But you also know that any nonzero number divided by itself is 1. So $1 = \frac{b^m}{b^m} = b^0$. This is another way of showing why $b^0 = 1$.

An Application of the Quotient of Powers Property

The Quotient of Powers Property is useful in dividing numbers written in scientific notation.

Example 1

The Gross Domestic Product (GDP) of a country is the total value of all the goods and services produced in the country. When the GDP is divided by the population of the country, the result is the GDP per person, often called the *GDP per capita*. In 2006, Denmark had a population of about 5.5 million and a GDP of $243.4 billion. What is Denmark's GDP per capita?

Solution Since GDP per capita is a rate unit, the answer is found by division.

$$\frac{\$243{,}400{,}000{,}000}{5.5 \text{ million people}} = \frac{2.434 \times 10^{11}}{5.5 \times 10^6} \quad \text{Write in scientific notation.}$$

$$= \frac{2.434}{5.5} \cdot \frac{10^{11}}{10^6} \quad \text{Multiplying Fractions Property}$$

$$\approx 0.44 \cdot 10^5 \quad \text{Quotient of Powers Property}$$

$$\approx \$44{,}000/\text{person} \quad \text{Write in base 10.}$$

Approximately 5.5 million people live in Denmark, making it one of the most densely populated nations in Northern Europe.

Source: Danish Tourist Board

Check Change the numbers to decimal notation and simplify the fraction.

$$\frac{243{,}400{,}000{,}000}{5{,}500{,}000} = \frac{2{,}434{,}000}{55} \approx 44{,}000$$

Dividing Powers with Different Bases

To use the Quotient of Powers Property, the bases must be the same. For example, $\frac{a^5}{b^2}$ cannot be simplified further. To divide two algebraic expressions that involve different bases, group powers of the same base together and use the Quotient of Powers Property to simplify each fraction.

Example 2

Simplify $\frac{30a^3n^6}{5a^2n}$.

Solution 1

$$\frac{30a^3n^6}{5a^2n} = \frac{30}{5} \cdot \frac{a^3}{a^2} \cdot \frac{n^6}{n} \quad \text{Multiplying Fractions Property}$$

$$= \frac{30}{5} \cdot a^{3-2} \cdot n^{6-1} \quad \text{Quotient of Powers Property}$$

$$= 6 \cdot a^1 \cdot n^5 = 6an^5 \quad \text{Arithmetic}$$

Solution 2

$$\frac{30a^3n^6}{5a^2n} = \frac{30 \cdot a \cdot a \cdot a \cdot n \cdot n \cdot n \cdot n \cdot n \cdot n}{5 \cdot a \cdot a \cdot n} \quad \text{Repeated Multiplication Property of Powers}$$

$$= \frac{6 \cdot a \cdot n \cdot n \cdot n \cdot n \cdot n}{1} \quad \text{Equal Fractions Property}$$

$$= 6an^5 \quad \text{Arithmetic}$$

Check Use a CAS to check your answer, as shown below.

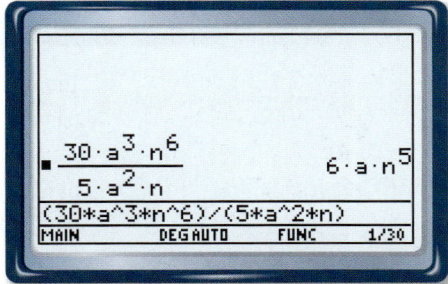

Questions

COVERING THE IDEAS

In 1–3, write the quotient as a single power.

1. $\frac{2^7}{2^4}$

2. $\frac{8^5}{8^m}$

3. $\frac{3^m}{3^n}$

In 4–9, use the Quotient of Powers Property to simplify the fraction.

4. $\dfrac{x^{12}}{x^2}$

5. $\dfrac{a^{20}}{a^{20}}$

6. $\dfrac{6.5 \times 10^{21}}{3.1 \times 10^{19}}$

7. $\dfrac{12a^2 b^{12}}{2ab^7}$

8. $\dfrac{2a^5 b^9}{8a^3 b}$

9. $\dfrac{24a^{10} b^5}{6a^4 b^5}$

10. In 2006, the African country of Burundi had a population of about 8.1 million and a GDP of about $5.7 billion. What is Burundi's GDP per capita?

11. Why can't $\dfrac{a^3}{b^4}$ be simplified?

12. If $\dfrac{b^n}{b^m} = 1$, how must m and n be related?

APPLYING THE MATHEMATICS

In 13–15, write the quotient as a single power.

13. $\dfrac{4^3}{2^6}$

14. $\dfrac{16 \cdot 2^m}{2^6}$

15. $\dfrac{4^3}{8 \cdot 2^6}$

In 16–19, rewrite the expression so that it has no fraction.

16. $\dfrac{(7m)^5}{(7m)^3}$

17. $\dfrac{(7+3m)^7}{(7+3m)^6}$

18. $\dfrac{x^{5a-10}}{x^{3-3a}}$

19. $\dfrac{2a^6 + 6a^6}{2a^5}$

20. In Norway in 2002, $4.7 \cdot 10^7$ kilograms of ground coffee were consumed, and the total population was approximately 4,525,000. To brew a typical cup of coffee, you need 10.6 grams of ground coffee. Determine the coffee consumption (in terms of cups per person) in Norway in 2002.

21. In 2005, the world's population was approximately $6.446 \cdot 10^9$ people. In the same year, global oil output was approximately $8.0 \cdot 10^7$ million barrels per day. A barrel is equivalent to 35 gallons of oil.

    a. How many barrels of oil was this per person per day?

    b. At this rate, how many gallons of oil were consumed per person during 2005?

22. Write an algebraic fraction that can be simplified to $12a^2 b$ using the Quotient of Powers Property.

The number of cups of coffee consumed per capita per year in the Nordic countries of Norway, Sweden, Denmark, and Finland is among the highest in the world.

Source: nationmaster.com

REVIEW

In 23–26, simplify the expression. (Lesson 8-2)

23. $3x \cdot x^2$

24. $n \cdot n^2 \cdot n^3$

25. $2h^3 \cdot 6h^4 + 3h \cdot 4h^6$

26. $a^x \cdot a^y \cdot a^z$

27. Suppose each question on a 5-question, multiple-choice quiz has four choices. (**Lesson 8-1**)

   a. Give the probability of guessing all the correct answers as the reciprocal of a power.

   b. Give the probability of guessing all wrong answers.

28. Distances after various times when traveling at 65 miles per hour are shown on the spreadsheet below. (**Lesson 6-1**)

	A	B
1	Time (hours)	Distance (miles)
2	1.0	?
3	1.5	97.5
4	2	130
5	2.5	162.5
6	3	?
7	3.5	?

   a. Complete the spreadsheet.

   b. Name two ways to get the value in cell B7.

29. Calculate the total cost in your head. (**Lesson 2-2**)

   a. 8 cans of tuna fish at $2.99 per can

   b. 5 tickets to a movie at $10.50 per ticket

EXPLORATION

30. The average 14-year-old has a volume of about 3 cubic feet.

   a. Consider all the students in your school. Would their total volume be more or less than the volume of one classroom? Assume the classroom is 10 feet high, 30 feet long, and 30 feet wide.

   b. Assume the population of the world to be 6.4 billion people and the average volume of a person to be 4 cubic feet. Is the volume of all the people more or less than 1 cubic mile? How much more or less? (There are $5{,}280^3$ cubic feet in a cubic mile.)

QY ANSWER

z^{40}

Chapter 8

Lesson 8-4

Negative Exponents

▶ **BIG IDEA** The numbers x^{-n} and x^n are reciprocals.

What Is the Value of a Power with a Negative Exponent?

You have used base 10 with a negative exponent to represent small numbers in scientific notation. For example, $10^{-1} = 0.1 = \frac{1}{10^1}$, $10^{-2} = 0.01 = \frac{1}{10^2}$, $10^{-3} = 0.001 = \frac{1}{10^3}$, and so on.

Now we consider other powers with negative exponents. That is, we want to know the meaning of b^n when n is negative. Consider this pattern of the powers of 2.

$$2^4 = 16$$
$$2^3 = 8$$
$$2^2 = 4$$
$$2^1 = 2$$
$$2^0 = 1$$

Each exponent is one less than the one above it. The value of each power is half that of the number above. Continuing the pattern suggests that the following are true.

$$2^{-1} = \tfrac{1}{2}$$
$$2^{-2} = \tfrac{1}{4} = \tfrac{1}{2^2}$$
$$2^{-3} = \tfrac{1}{8} = \tfrac{1}{2^3}$$
$$2^{-4} = \tfrac{1}{16} = \tfrac{1}{2^4}$$

A general description of the pattern is simple: $2^{-n} = \frac{1}{2^n}$. That is, 2^{-n} is the reciprocal of 2^n. We call the general property the *Negative Exponent Property*.

> ### Negative Exponent Property
> For any nonzero b and all n, $b^{-n} = \frac{1}{b^n}$, the reciprocal of b^n.

Mental Math
Give the area of
a. a square with side $\frac{s}{2}$.
b. a circle with radius $3r$.
c. a rectangle with $\frac{3}{4}x$ and $\frac{8}{3}y$ dimensions.

474   Powers and Roots

Notice that even though the exponent in 2^{-4} on the previous page is negative, the number 2^{-4} is still positive. All negative integer powers of positive numbers are positive.

 QY

▶ QY

Write 5^{-4} as a simple fraction without a negative exponent.

Example 1

Rewrite $a^7 \cdot b^{-4}$ without negative exponents.

Solution

$a^7 \cdot b^{-4} = a^7 \cdot \frac{1}{b^4}$ Substitute $\frac{1}{b^4}$ for b^{-4}.

$\phantom{a^7 \cdot b^{-4}} = \frac{a^7}{b^4}$

Because the Product of Powers Property applies to all exponents, it applies to negative exponents. Suppose you multiply b^n by b^{-n}.

$b^n \cdot b^{-n} = b^{n + -n}$ Product of Powers Property
$\phantom{b^n \cdot b^{-n}} = b^0$ Property of Opposites
$\phantom{b^n \cdot b^{-n}} = 1$ Zero Exponent Property

To multiply b^n by b^{-n}, you can also use the Negative Exponent Property.

$b^n \cdot b^{-n} = b^n \cdot \frac{1}{b^n}$ Negative Exponent Property
$\phantom{b^n \cdot b^{-n}} = 1$ definition of reciprocal

In this way, the Product of Powers Property verifies that b^{-n} must be the reciprocal of b^n. In particular, $b^{-1} = \frac{1}{b}$. That is, the –1 power (read "negative one" or "negative first" power) of a number is its reciprocal.

Suppose the base b is a fraction, $b = \frac{x}{y}$. Then the reciprocal of b is $\frac{y}{x}$. Consequently, this gives us a different form of the Negative Exponent Property that is more convenient when the base is a fraction. The simplest way to find the reciprocal of a fraction $\frac{a}{b}$ is to invert it, producing $\frac{b}{a}$.

Negative Exponent Property for Fractions

For any nonzero x and y and all n, $\left(\frac{x}{y}\right)^{-n} = \left(\frac{y}{x}\right)^n$.

Chapter 8

GUIDED

Example 2
Write each expression without negative exponents.

a. $\left(\dfrac{5}{4}\right)^{-2}$

b. $\left(\dfrac{1}{m^2}\right)^{-3}$

Solution

a. Use the Negative Exponent Property for Fractions.

$$\left(\dfrac{5}{4}\right)^{-2} = \left(\underline{\ ?\ }\right)$$

$$= \underline{\ ?\ }$$

b. Take the reciprocal to the opposite power.

$$\left(\dfrac{1}{m^2}\right)^{-3} = \left(\dfrac{\underline{\ ?\ }}{1}\right)^{3}$$

$$= (\underline{\ ?\ })^3$$

$$= \underline{\ ?\ }$$

Recall the compound interest formula $A = P(1 + r)^t$. In this formula, negative exponents stand for unit periods going back in time.

Example 3
Ten years ago, Den put money into a college savings account at an annual yield of 6%. If the money is now worth $9,491.49, what was the amount initially invested?

Solution

Here $P = 9{,}491.49$, $r = 0.06$, and $t = -10$ (for 10 years ago).

So, $A = 9{,}491.49(1.06)^{-10} \approx 5{,}300$.

So, Den originally started with approximately $5,300.

Check Use the Compound Interest Formula. If Den invested $5,300, he would have $5{,}300(1.06)^{10}$, which equals $9,491.49. It checks.

Quotient of Powers and Negative Exponents

The last lesson involved fractions in which two powers of the same base are divided. When the denominator contains the greater power, negative exponents can be used to simplify the expression. For example, $\dfrac{x^5}{x^9} = x^{5-9} = x^{-4}$.

476 Powers and Roots

This can be verified using repeated multiplication.

$$\frac{x^5}{x^9} = \frac{\overset{1}{\cancel{x}}\cdot\overset{1}{\cancel{x}}\cdot\overset{1}{\cancel{x}}\cdot\overset{1}{\cancel{x}}\cdot\overset{1}{\cancel{x}}}{\underset{1}{\cancel{x}}\cdot\underset{1}{\cancel{x}}\cdot\underset{1}{\cancel{x}}\cdot\underset{1}{\cancel{x}}\cdot\underset{1}{\cancel{x}}\cdot x \cdot x \cdot x \cdot x} = \frac{1}{x^4}$$

In this way, you can see again that $b^{-n} = \frac{1}{b^n}$.

> **GUIDED**
>
> ### Example 4
> Simplify $\frac{5a^4b^7c^2}{15a^{11}b^5c^3}$. Write the answer without negative exponents.
>
> **Solution**
>
> $\frac{5a^4b^7c^2}{15a^{11}b^5c^3} = \frac{5}{15} \cdot \frac{a^4}{a^{11}} \cdot \frac{b^7}{b^5} \cdot \frac{c^2}{c^3}$ Group factors with the same base together.
>
> $= \frac{1}{3} \cdot a^{\underline{\ ?\ }} \cdot b^{\underline{\ ?\ }} \cdot c^{\underline{\ ?\ }}$ Quotient of Powers Property
>
> $= \frac{1}{3} \cdot \frac{1}{a^{\underline{\ ?\ }}} \cdot \frac{b^{\underline{\ ?\ }}}{1} \cdot \frac{1}{c^{\underline{\ ?\ }}}$ Negative Exponent Property
>
> $= \underline{\ ?\ }$ Multiply the fractions.

Applying the Power of a Power Property with Negative Exponents

Consider $(x^3)^{-2}$, a power of a power. Wanda wondered if the Power of a Power Property would apply with negative exponents. She entered the expression into a CAS and the screen below appeared.

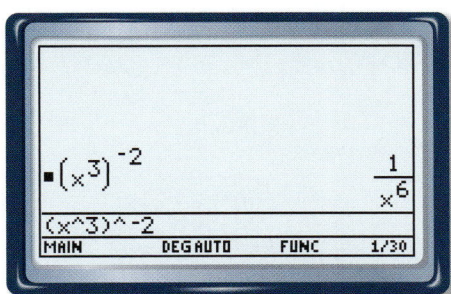

This is the answer that would result from applying the Power of a Power Property.

$$(x^3)^{-2} = x^{3 \cdot -2} = x^{-6}$$

Then you can rewrite the power using the Negative Exponent Property.

$$x^{-6} = \frac{1}{x^6}$$

All the properties of powers you have learned can be used with negative exponents. They can translate an expression with a negative exponent into one with only positive exponents.

Negative Exponents

Chapter 8

Example 5
Simplify $(y^{-4})^2$. Write without negative exponents.

Solution

$(y^{-4})^2 = y^{-8}$ Power of a Power Property

$= \dfrac{1}{y^8}$ Negative Exponent Property

Questions

COVERING THE IDEAS

1. **Fill in the Blanks** Complete the last four equations in the pattern below. Then write the next equation in the pattern.

$$3^4 = 81$$
$$3^3 = 27$$
$$3^2 = 9$$
$$3^1 = 3$$
$$3^0 = 1$$
$$3^{-1} = \underline{\ ?\ }$$
$$3^{-2} = \underline{\ ?\ }$$
$$3^{-3} = \underline{\ ?\ }$$
$$3^{-4} = \underline{\ ?\ }$$

In 2–5, write as a simple fraction.

2. 7^{-2} 3. 5^{-3} 4. $\left(\dfrac{2}{3}\right)^{-1}$ 5. $(y^6)^{-4}$

In 6–9, write as a negative power of an integer.

6. $\dfrac{1}{36}$ 7. $\dfrac{1}{81}$ 8. 0.1 9. 0.0001

10. Eight years ago, Abuna put money into a college savings account at an annual yield of 5%. If there is now $7,250 in the account, what amount was initially invested? Round your answer to the nearest penny.

11. Rewrite each expression without negative exponents.

    a. w^{-1} b. $w^{-1}x^{-2}$ c. $w^{-1}y^3$ d. $5w^{-1}x^{-2}y^3$

In 12–14, write each expression without negative exponents.

12. $9^2 \cdot 9^{-2}$ 13. $n^a \cdot n^{-a}$ 14. $(m^{-5})^3$

15. Simplify $\dfrac{32a^8bc^3}{8a^6b^4c}$. Write without negative exponents.

16. a. Graph $y = 2^x$ when the domain is $\{-4, -3, -2, -1, 0, 1, 2, 3, 4\}$.
    b. Describe what happens to the graph as x decreases.

17. Graph $y = 10^x$ as on page 457. Describe what happens as x goes from 0 to -12.

APPLYING THE MATHEMATICS

18. If the reciprocal of $a^{-12}b^5$ is $a^n b^m$, find m and n.

19. Use properties of algebra to justify the answer shown on the CAS screen below.

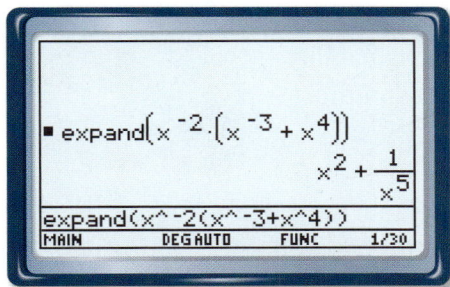

In 20 and 21, solve and check each equation.

20. $7^a \cdot 7^3 = 7^{-6}$

21. $5^m \cdot \frac{1}{25} = 5^{-3}$

22. Suppose you draw a square with area 25 square units and connect the midpoints of each side to create a smaller square inside the original. A sequence of successively smaller squares may be created by repeating the process with the most recently created square. The shaded regions show squares in the sequence.

Step 0	Step 1	Step 2	Step 3
	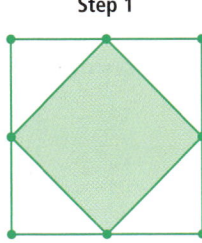	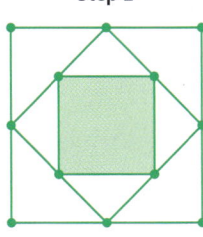	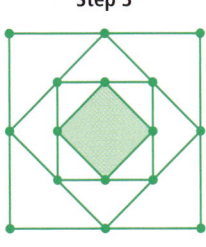
Area = 25 units²	Area = ?	Area = ?	Area = ?

Write the area of the shaded square for each step as 25 times a power of 2.

    a. Step 1   b. Step 2   c. Step 10   d. Step n

REVIEW

In 23–25 first simplify. Then evaluate when $a = 2$ and $b = 5$. (Lessons 8-3, 8-2)

23. $\dfrac{a^2 \cdot a^5 \cdot a^3}{a^4}$
24. $(b^2 a^{-2})^3$
25. $(2b^3)^a$

26. Some people use randomly generated passwords to protect their computer accounts. Suppose a Web site uses random passwords that are six characters long. They allow only lower-case letters and the digits 0 through 9 to be used. (Lessons 8-1, 5-6)

    a. What is the total number of possible passwords?
    b. Jacinta forgot her password. What is the probability that she will guess her password correctly on the first try?
    c. Myron says there would be more possibilities available if the site switched to passwords four characters long but allowed the use of upper-case letters as well. Is Myron correct? Why or why not?

Nearly 49 million laptop computers were sold worldwide in 2004, almost double the number sold in 2000.

Source: *USA Today*

27. Tyra is learning addition and multiplication. For practice, Tyra's teacher gives her a whole number less than 13. Tyra then multiplies the number by 8, adds 25, and states her answer. (Lessons 7-6, 7-5)

    a. Describe the situation with function notation, letting x be the number Tyra is given and $m(x)$ the number Tyra states.
    b. What is the domain of the function you wrote?
    c. What are the greatest and least values the function can have?

EXPLORATION

28. Objects in the universe can be quite small. Do research to find objects of the following sizes.

    a. 10^{-3} meter
    b. 10^{-6} meter
    c. 10^{-9} meter
    d. 10^{-12} meter

QY ANSWER

$\dfrac{1}{625}$

Lesson 8-5
Powers of Products and Quotients

> **BIG IDEA** Because of the relationship among multiplication, division, and powers, powers distribute over products and quotients.

The Power of a Product

The expression $(3x)^4$ is an example of a power of a product. It can be rewritten using repeated multiplication.

$$
\begin{aligned}
(3x)^4 &= (3x) \cdot (3x) \cdot (3x) \cdot (3x) && \text{Repeated Multiplication Model for Powering} \\
&= 3 \cdot 3 \cdot 3 \cdot 3 \cdot x \cdot x \cdot x \cdot x && \text{Associative and Commutative Properties} \\
&= 3^4 \cdot x^4 && \text{Repeated Multiplication Model for Powering} \\
&= 81x^4 && \text{Arithmetic}
\end{aligned}
$$

You can check this answer using a CAS.

Mental Math

Find the slope of the line through
a. (-4.5, 19) and (90, 19).
b. (4, -1.5), (-3.5, -1.5).
c. (0, 0) and $\left(\frac{3}{4}, \frac{7}{4}\right)$.

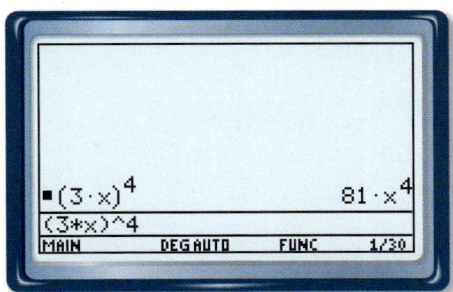

In general, any positive integer power of a product can be rewritten using repeated multiplication.

$$
\begin{aligned}
(ab)^n &= \underbrace{(ab) \cdot (ab) \cdot \ldots \cdot (ab)}_{n \text{ factors}} \\
&= \underbrace{a \cdot a \cdot \ldots \cdot a}_{n \text{ factors}} \cdot \underbrace{b \cdot b \cdot \ldots \cdot b}_{n \text{ factors}} \\
&= a^n \cdot b^n
\end{aligned}
$$

When a and b are nonzero, this result holds for all values of n.

> **Power of a Product Property**
>
> For all nonzero a and b, and for all n, $(ab)^n = a^n b^n$.

Powers of Products and Quotients **481**

Chapter 8

This property can be applied to simplify the expression $(3x)^4$ from page 481. The power is applied to each factor of $3x$, so $(3x)^4 = 3^4 x^4$, resulting in $81x^4$.

Example 1
Simplify $(-4x)^3$.

Solution Use the Power of a Product Property.

$(-4x)^3 = (-4)^3 \cdot x^3 = -64x^3$

Check Substitute a test value for x and follow order of operations.

Let $x = 1.5$. Does $(-4x)^3 = -64x^3$?

$(-4 \cdot 1.5)^3 = -64(1.5)^3$
$(-6)^3 = -64 \cdot (3.375)$
$-6 \cdot -6 \cdot -6 = -216$
$-216 = -216$ It checks.

Remember that in the order of operations, powers take precedence over opposites. In $-64x^3$, the power is done before the multiplication. In $(-4x)^3$, the multiplication is inside parentheses so it is done before the power.

 QY1

> ▶ QY1
>
> Simplify $(3xy)^4$.

GUIDED

Example 2
Simplify $(-5x^2 y^3 z)^3$.

Solution

$(-5x^2 y^3 z)^3$
$= (-5)\underline{}(x^2)\underline{}(y^3)\underline{}z\underline{}$ Apply the Power of a Product Property.
$= (-5)\underline{}x\underline{}y\underline{}z\underline{}$ Apply the Power of a Power Property.
$= \underline{}$ Evaluate the numerical power.

The Power of a Quotient

The expression $\left(\dfrac{a}{b}\right)^n$ is the power of a quotient. By using the properties of the previous lessons, you can write this without parentheses.

$\left(\dfrac{a}{b}\right)^n = \left(a \cdot \dfrac{1}{b}\right)^n = (a \cdot b^{-1})^n = a^n \cdot (b^{-1})^n = a^n \cdot b^{-n} = \dfrac{a^n}{b^n}$

Power of a Quotient Property

For all nonzero a and b, and for all n, $\left(\dfrac{a}{b}\right)^n = \dfrac{a^n}{b^n}$.

482   Powers and Roots

The Power of a Quotient Property enables you to find powers of fractions more quickly.

Example 3

Write $\left(\frac{3}{4}\right)^5$ as a simple fraction.

Solution 1 Use the Power of a Quotient Property.

$\left(\frac{3}{4}\right)^5 = \frac{3^5}{4^5} = \frac{243}{1,024}$

Solution 2 Use repeated multiplication.

$\left(\frac{3}{4}\right)^5 = \frac{3}{4} \cdot \frac{3}{4} \cdot \frac{3}{4} \cdot \frac{3}{4} \cdot \frac{3}{4} = \frac{3^5}{4^5} = \frac{243}{1,024}$

Check Change the fractions to decimals.

$\left(\frac{3}{4}\right)^5 = 0.75^5 = 0.2373046875$

$\frac{243}{1,024} = 0.2373046875$

They are equal.

🛑 **QY2**

Powers are found in many formulas for area and volume.

▶ **QY2**

Rewrite $11 \cdot \left(\frac{2}{m}\right)^6$ as a simple fraction.

Activity

You will need two pieces of 8.5-inch by 11-inch paper, tape, scissors, and a ruler.

Step 1
Begin with one sheet of paper, positioned so that it is taller than it is wide. Fold it into fourths lengthwise and tape the long edges together to form the sides of a tall box with a square base.

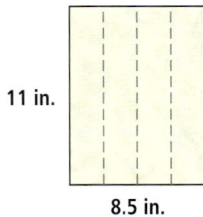

11 in.

8.5 in.

8.5 in.
perimeter of base

(continued on next page)

Powers of Products and Quotients 483

Chapter 8

Step 2 Cut the other piece of paper in half to create two 8.5-in. by 5.5-in. pieces. Fold each half, as shown by the dotted lines. Tape these pieces together to form a 17-in. by 5.5-in. piece of paper. Tape the short edges together to form the sides of a short box with a square base.

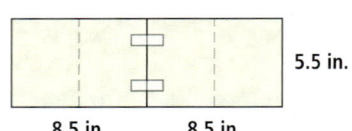
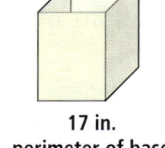

17 in.
perimeter of base

Step 3 **Multiple Choice** Which of the following do you think is true?
  A The tall, skinny box has more volume.
  B The short, wide box has more volume.
  C Both boxes have the same volume.

  In several sentences, justify your conjecture with a logical argument.

Step 4 Test your conjecture using the formula $V = s^2 h$ for the volume V of a box with height h and a square base whose sides have length s.
  a. Calculate the length of the sides of the base of the tall prism. $s = \underline{\ ?\ }$
  b. Calculate the volume of the tall prism. $V = \underline{\ ?\ }$
  c. Repeat Parts a and b for the short prism. $s = \underline{\ ?\ }$, $V = \underline{\ ?\ }$

Step 5 According to your calculations, which is the correct answer to Step 3? $\underline{\ ?\ }$

  Do you think you would get the same result if you started with a sheet of paper of a different size? Why or why not?

Using Powers of Quotients to Explain the Activity Results

Suppose you begin with a sheet of paper with height h and width p. The shorter box has half the height of the taller box, but the perimeter of its base is twice as long. Each side of the base of the tall box has length $\frac{p}{4}$. Each side of the base of the short box has length $\frac{2p}{4}$. So for the short prism, $2p$ = perimeter and $\frac{h}{2}$ = height.

Tall Box Short Box

h

$\frac{h}{2}$

$\frac{p}{4}$ $\frac{2p}{4}$

484 Powers and Roots

The volume of a box with a square base is given by the formula $V = s^2h$, where the height is h and the side of the base is s. So, the volume of the tall box $= \left(\frac{p}{4}\right)^2 \cdot h$, and the volume of the short box $= \left(\frac{2p}{4}\right)^2 \cdot \frac{h}{2}$.

To compare these volumes, we use properties of powers to simplify the expressions.

Example 4

The tall box has volume $\left(\frac{p}{4}\right)^2 \cdot h$ and the short box has volume $\left(\frac{2p}{4}\right)^2 \cdot \frac{h}{2}$.

a. Show that the volume of the tall box is always less than or equal to the volume of the short box.

b. The volume of the short box is how many times the volume of the tall one?

Solution

a. First apply the Power of a Quotient Property to simplify each volume.

Tall Box

$$V = \left(\frac{p}{4}\right)^2 \cdot h$$
$$= \frac{p^2}{4^2} \cdot h$$
$$= \frac{p^2}{16} \cdot h$$
$$= \frac{p^2 h}{16}$$

Short Box

$$V = \left(\frac{2p}{4}\right)^2 \cdot \frac{h}{2}$$
$$= \left(\frac{(2p)^2}{4^2}\right) \cdot \frac{h}{2}$$
$$= \left(\frac{4p^2}{16}\right) \cdot \frac{h}{2}$$
$$= \left(\frac{p^2}{4}\right) \cdot \frac{h}{2} = \frac{p^2 h}{8}$$

Volume of the tall box $= \frac{p^2 h}{16}$

$$= \frac{1}{2} \cdot \frac{p^2 h}{8}$$

$$= \frac{1}{2} \cdot \text{volume of short box}$$

Because the volume of the tall box is half the volume of the short one, the volume of the tall box is less than the volume of the short box.

b. The volume of the short box is 2 times the volume of the tall box.

Questions

COVERING THE IDEAS

1. a. Rewrite $(6x)^3$ without parentheses.
   b. Check your answer by letting $x = 2$.

In 2–5, rewrite the expression without parentheses.

2. $(5t^2)^3$
3. $8(-7xy)^3$
4. $2(x^2y)^4$
5. $(-t)^{93}$

6. Aisha made a common error when she wrote $(3x)^4 = 12x^4$. Show her this is incorrect by substituting 2 in for x. Then, write a note to Aisha explaining what she did wrong.

In 7–9, write as a simple fraction.

7. $\left(\frac{2}{3}\right)^4$ 8. $5\left(\frac{n^5}{10}\right)^3$ 9. $\left(\frac{19}{2y}\right)^3$

10. What is the area of a square with perimeter p?

APPLYING THE MATHEMATICS

11. The area A of an isosceles right triangle with leg L can be found using the formula $A = \frac{1}{2}L^2$. If L is multiplied by 6, what happens to the area of the triangle?

12. Suppose you tape a 3-inch by 5-inch notecard to a pencil widthwise (as shown in Figure 1). Assume that the radius of the round pencil is $\frac{3}{16}$ in.

    a. If you rotate the pencil, what shape is traced by point A? Find the area of the shape.

    b. If you rotate the pencil, the entire notecard in Figure 1 traces a cylinder. Cylinders with height h and a base of radius r have volume $V = \pi r^2 h$. Calculate the volume of this region.

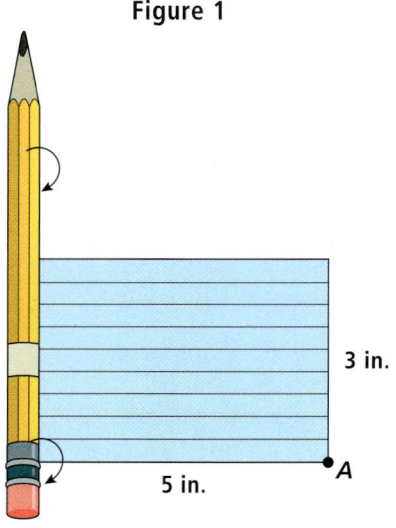

Figure 1

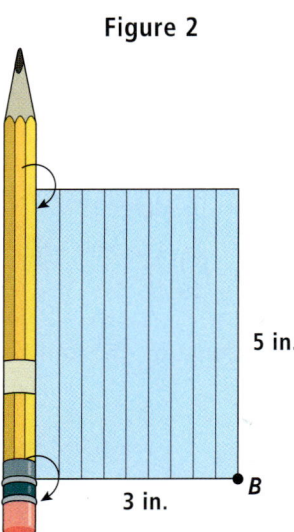

Figure 2

  c. Suppose you tape a 3-inch by 5-inch notecard to a pencil heightwise, as shown in Figure 2. If you rotate the pencil, what shape is traced by point B? Find the area of the shape.

  d. If you rotate the pencil, what shape is traced by the entire notecard? Calculate the volume of this region.

  e. **True or False** Changing the taping of the notecard does not change the volume of the shape that is traced by the notecard when the pencil is rotated.

Lesson 8-5

In 13–15, rewrite without parentheses and simplify.

13. $(xy)^2\left(\dfrac{x}{y}\right)^3$
14. $(abc)^0 \cdot \dfrac{(ab)^2}{abc}$
15. $(2w)^4(3w^3)^2$

In 16–18, fill in the blank with an exponent or an expression that makes the statement true for all values of the variables.

16. $(3x^2y)^{\underline{?}} = 27x^6y^3$ 17. $(2xy^2)^{\underline{?}} = 1$ 18. $(\underline{\ ?\ })^3 = 64x^6y^9$

19. If $x = 5$, what is the value of $\dfrac{(3x)^9}{(3x)^7}$?

REVIEW

In 20 and 21, simplify the expression so that your answer does not contain parentheses or negative exponents. Then evaluate when $r = 1.5$ and $s = 1$. (Lessons 8-4, 8-3)

20. $r^4s^9r^{-3}s^7$
21. $\dfrac{17s^{-2}}{5^5} \cdot r^{-2}$

22. On each day (Monday through Friday) this week, Antoine will do one of three activities after school: play tennis, walk his dog, or read. How many different orders of activities are possible? (Lesson 8-1)

23. Solve $9(p - 2) < 47p - 2(5 - p)$ for p. (Lesson 4-5)

The World Junior Tennis competition, the international team competition for players aged 14 and under, was started by the International Tennis Federation in 1991.

Source: International Tennis Federation

EXPLORATION

24. A list of some powers of 3 is shown below. Look carefully at the last digit of each number.

$3^0 = 1$ $3^4 = 81$
$3^1 = 3$ $3^5 = 243$
$3^2 = 9$ $3^6 = 729$
$3^3 = 27$

a. Predict the last digit of 3^{10}. Check your answer with a calculator.

b. Predict the last digit of 3^{20}. Check your answer with a calculator.

c. Describe how you can find the last digit of any positive integer power of 3.

d. Does a similar pattern happen for powers of 4? Why or why not?

QY ANSWERS

1. $81x^4y^4$
2. $\dfrac{704}{m^6}$

Powers of Products and Quotients   487

Chapter 8

Lesson 8-6
Square Roots and Cube Roots

Vocabulary

square
squared
square root
radical sign ($\sqrt{}$)
cube
cubed
cube root

▶ **BIG IDEA** If a first number is the square (or the cube) of a second number, then the second number is a square root (or cube root) of the first.

Mental Math

In each set, which does not equal the others?

a. $\frac{3x}{5}, \frac{3}{5}x, \frac{3}{5x}$

b. $\frac{y}{9}, y \cdot \frac{1}{9}, \frac{1}{9} \cdot \frac{1}{y}, y \div 9$

Areas of Squares and Powers as Squares

The second power x^2 of a number x is called the **square** of x, or x **squared**, because it is the area of a square with side length x. This is not a coincidence. The ancient Greek mathematicians pictured numbers as lengths, and they pictured the square of a number as the area of a square.

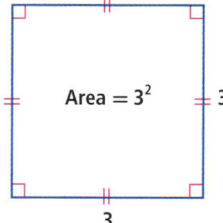 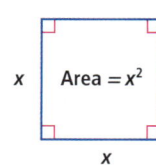

It is easy to calculate the area of a square on a grid if the square's sides are horizontal and vertical, but what if the square's sides are slanted?

Activity 1

Follow these steps to determine the area of the square *EFGH*, at the right.

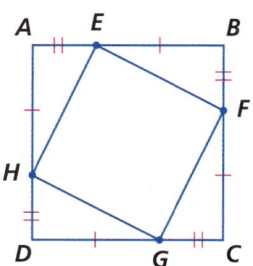

Step 1 Square *ABCD* is 3 units on a side. What is its area?

Step 2 Triangle *AEH* is a right triangle with legs of length 1 unit and 2 units. What is the area of △*AEH*?

Step 3 Subtract the areas of the four corner triangles from the area of *ABCD* to get the area of *EFGH*.

488 Powers and Roots

Lesson 8-6

Sides of Squares and Square Roots

You should have found that the area of *EFGH* is 5 square units. If the area of the square *EFGH* is 5 square units, what is the length of one of its sides? The Greek mathematicians could do the previous calculations easily. But now they were stumped. Can *GH* be $2\frac{1}{2}$? No, because $\left(2\frac{1}{2}\right)^2 = 2.5^2 = 6.25$, which is greater than 5. In fact, the Greeks were able to show that it is impossible to find any simple fraction whose square is exactly 5. So they simply called the length the *square root* of 5. We still do that today. The length of a side of a square whose area is *x* is called a square root of *x*. The length of *GH* is a square root of 5. Similarly, a square root of 9 is 3, because a square with area 9 has side 3.

> **Definition of Square Root**
>
> If $A = s^2$, then s is a **square root** of A.

If two numbers have the same absolute value, such as 3 and −3, then they have the same square, 9. Although −3 cannot be the length of a side of a square, every positive number but 0 has two square roots, one positive and one negative. We denote the square roots of *A* by the symbols $\sqrt{A}$ (the positive root) and $-\sqrt{A}$ (the negative root). So the square roots of 9 are $\sqrt{9} = 3$ and $-\sqrt{9} = -3$. The two square roots of 5 are $\sqrt{5}$ and $-\sqrt{5}$. In the figure on the previous page, $GH = \sqrt{5}$.

The Radical Sign $\sqrt{}$

The **radical sign** $\sqrt{}$ indicates that a square root is being found. The horizontal bar attached to it, called a *vinculum*, acts like parentheses. The order of operations applies, so work is done inside the radical sign before the square root is taken. For example, $\sqrt{16 - 9} = \sqrt{7}$. On the other hand, $\sqrt{16} - \sqrt{9} = 4 - 3 = 1$.

In dealing with square roots, it helps to know the squares of small positive integers: 1, 4, 9, 16, 25, 36, 49, 64, 81, 100, 121, 144, ….

> **Example 1**
>
> What are the square roots of each number?
>
> a. 64
>
> b. 17.3
>
> *(continued on next page)*

Square Roots and Cube Roots

Solutions

a. Because $8^2 = 64$ and $(-8)^2 = 64$, the square roots of 64 are 8 and −8. We can write $\sqrt{64} = 8$ and $-\sqrt{64} = -8$.

b. Because there is no decimal that multiplied by itself equals 17.3, just write $\sqrt{17.3}$ and $-\sqrt{17.3}$. A calculator shows $\sqrt{17.3} \approx 4.1593$ and so $-\sqrt{17.3} \approx -4.1593$.

Square Roots That Are Not Whole Numbers

The Greek mathematician Pythagoras and his followers, the Pythagoreans, were able to prove that numbers like $\sqrt{5}$ are not equal to simple fractions or ratios. Today we know that there is no finite or repeating decimal that equals $\sqrt{5}$. While $\sqrt{5}$ is approximately 2.23606797..., the decimal does not end nor repeat. You should check that the squares of truncated forms of 2.23606797..., are very close to 5. For example, $2.236 \cdot 2.236 = 4.999696$. But only $\sqrt{5}$ and $-\sqrt{5}$ square to be exactly 5, so $\sqrt{5} \cdot \sqrt{5} = 5 = -\sqrt{5} \cdot -\sqrt{5}$.

> **Square of the Square Root Property**
>
> For any nonnegative number x, $\sqrt{x} \cdot \sqrt{x} = \sqrt{x^2} = x$.

You can use this property to simplify or evaluate expressions that are exact, rather than use your calculator to deal with approximations.

🛑 **QY1**

▶ **QY1**
Explain why
$4\sqrt{10} \cdot 3\sqrt{10} = 120$.

A Positive Square Root of x Is a Power of x

Suppose $m = \frac{1}{2}$ and $n = \frac{1}{2}$ in the Product of Powers Property $x^m \cdot x^n = x^{m+n}$. Then, $x^{\frac{1}{2}} \cdot x^{\frac{1}{2}} = x^{\frac{1}{2} + \frac{1}{2}} = x^1 = x$. This means that $x^{\frac{1}{2}}$ is a number which, when multiplied by itself, equals x. Thus $x^{\frac{1}{2}}$ is a square root of x, and we identify $x^{\frac{1}{2}}$ as the positive square root of x. So, for any positive number x, $x^{\frac{1}{2}} = \sqrt{x}$. For example, $100^{\frac{1}{2}} = \sqrt{100} = 10$ and $64.289^{\frac{1}{2}} = \sqrt{64.289} \approx 8.02$.

Activity 2

Use a calculator to verify that $x^{\frac{1}{2}} = \sqrt{x}$.

Step 1a. Enter 16^(1/2). What number results?

    b. You have calculated the square root of what number?

Powers and Roots

Lesson 8-6

Step 2a. Enter 8^0.5. What number results?

   b. You have calculated the square root of what number?

Step 3 Enter (−4)^(1/2). What results, and why?

Activity 3

Use the idea of Activity 1 to determine the length of a side of square *IJKL* shown below. Show your work.

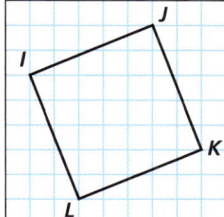

Activity 4

Three squares are drawn on a coordinate grid at the right.

1. Use the idea of Activities 1 and 3 to determine the area of square III. Explain your work.
2. What is the area of square I?
3. What is the area of square II?
4. How are the areas of the three squares related to each other?

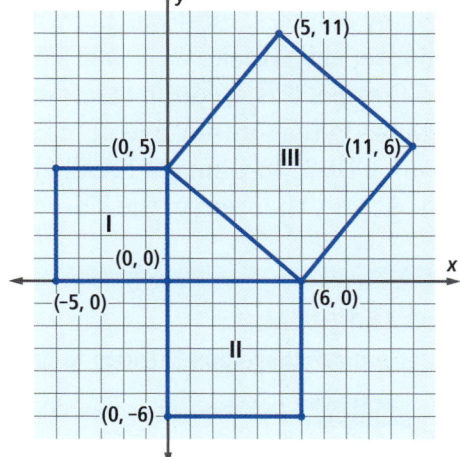

The Pythagorean Theorem

The result of Activity 4 is one example of the *Pythagorean Theorem*. We state this theorem in terms of area first, and then in terms of powers.

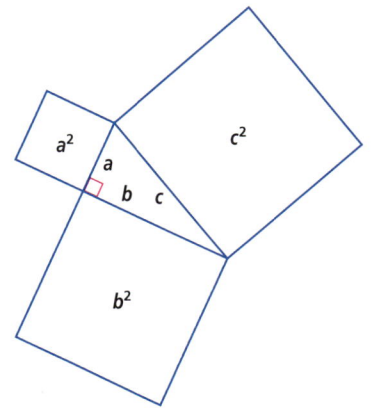 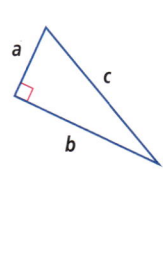

Square Roots and Cube Roots   491

Chapter 8

> **Pythagorean Theorem**
>
> (*In terms of area*) In any right triangle, the sum of the areas of the squares on its legs equals the area of the square on its hypotenuse.
>
> (*In terms of length*) In any right triangle with legs of lengths a and b and a hypotenuse of length c, $a^2 + b^2 = c^2$.

For example, in $\triangle GDH$ from Activity 1, $HD^2 + DG^2 = GH^2$.

$1^2 + 2^2 = GH^2$
$1 + 4 = GH^2$
$5 = GH^2$

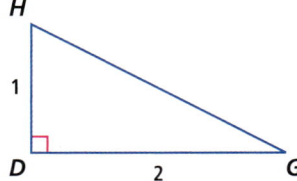

By the definition of square root, $GH = \sqrt{5}$.

The Pythagorean Theorem is perhaps the most famous theorem in all of mathematics. It seems to have been discovered independently in many cultures, for it was known to the Babylonians, Indians, Chinese, and Greeks well over 2,500 years ago. In the United States and Europe, this theorem is known as the Pythagorean Theorem because Pythagoras or one of his students proved it in the 6th century BCE. In China, it is called the Gougu Theorem. In Japan, it is called "The Theorem of the Three Squares."

Example 2

Use the Pythagorean Theorem to find the length of the missing side.

Solution Use the Pythagorean Theorem to write an equation involving the lengths of the three sides of the right triangle.

$m^2 + 12^2 = 14^2$
$m^2 + 144 = 196$
$m^2 + 144 - 144 = 196 - 144$
$m^2 = 52$
$m = \sqrt{52}$
$m \approx 7.21$

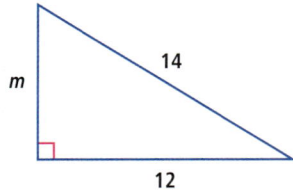

Check Substitute the solution into the original triangle and apply the Pythagorean Theorem.

Does $\left(\sqrt{52}\right)^2 + 12^2 = 14^2$?
$52 + 144 = 196$ Yes, it checks.

492   Powers and Roots

Cubes and Cube Roots

The third power x^3 of a number x is called the **cube** of x, or x **cubed,** because it is the volume of a cube with edge x. So, for example, the volume of a cube with edge of length 6 inches is 6 · 6 · 6, or 216 cubic inches. We write $6^3 = 216$, and we say "6 cubed equals 216." Like the square, this is not a coincidence. The ancient Greek mathematicians pictured the cube of a number s as the volume of a cube whose edge is s.

Also, in a manner like that of a square, if the volume of a cube is V, then an edge of the cube is called a **cube root** of V.

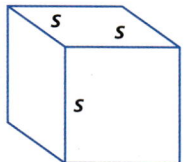

> **Definition of Cube Root**
>
> If $V = s^3$, then s is a cube root of V.

Since $6^3 = 216$, 6 is a cube root of 216. Unlike square roots, cube roots do not come in pairs. For example, –6 is not a cube root of 216, since $(-6)^3 = -216$. In the real numbers, all numbers have exactly one cube root.

 QY2

The cube root of V is written using a radical sign as $\sqrt[3]{V}$. For example, $\sqrt[3]{216} = 6$ and $\sqrt[3]{-216} = -6$. Many calculators have a $\sqrt[3]{}$ command, though it may be hidden in a menu. You should try to locate this command on your calculator. However, you will learn an alternate method for calculating cube roots in the next lesson.

> ▶ **QY2**
>
> **Fill in the Blanks**
> Since $4^3 = 64$, __?__ is the cube root of __?__.

> **Cube of the Cube Root Property**
>
> For any nonnegative number x, $\sqrt[3]{x} \cdot \sqrt[3]{x} \cdot \sqrt[3]{x} = \sqrt[3]{x^3} = x$.

For example, $1.2^3 = 1.2 \cdot 1.2 \cdot 1.2 = 1.728$. This means:

- 1.728 is the cube of 1.2.
- 1.2 is the cube root of 1.728.
- $1.2 = \sqrt[3]{1.728}$

When the value of a square root or cube root is not an integer, your teacher may expect two versions: (1) the exact answer written with a radical sign and (2) a decimal approximation rounded to a certain number of decimal places.

Chapter 8

Questions

COVERING THE IDEAS

1. a. A side of a square is 16 units. What is its area?
   b. The area of a square is 16 square units. What is the length of a side?

2. Rewrite the following sentences, substituting numbers for x and y to produce a true statement. *A square has a side of length x and an area y. Then y is the square of x, and x is the square root of y.*

In 3–6, write or approximate the number to two decimal places.

3. $\sqrt{36}$
4. $\sqrt{121}$
5. $50^{\frac{1}{2}}$
6. $10^{0.5}$

In 7–10, evaluate the expression to the nearest thousandth.

7. $\sqrt{1{,}000}$
8. $\sqrt{100 + 100}$
9. $\sqrt{5} \cdot \sqrt{5}$
10. $2 \cdot \left(\frac{3}{4}\right)^{\frac{1}{2}} \left(\frac{3}{4}\right)^{\frac{1}{2}}$

11. a. Approximate $\sqrt{11}$ to the nearest hundred-thousandth.
    b. Multiply your answer to Part a by itself.
    c. What property is validated by Parts a and b?

In 12–14, find the length of the missing side of the right triangle. If the answer is not an integer, give both its exact value and an approximation to the nearest hundredth.

12.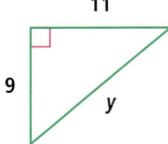
13.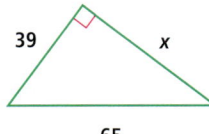
14.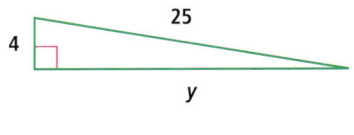

15. Write the cubes of the integers from 1 to 10.

16. 2 is a cube root of 8 because ___?___.

17. a. Write the exact cube root of 1,700.
    b. Estimate the cube root of 1,700 to the nearest thousandth.
    c. Check your answer to Part b by multiplying your estimate by itself three times.

In 18 and 19, evaluate the expression.

18. $\sqrt[3]{2.197}$
19. $\sqrt[3]{45} \cdot \sqrt[3]{45} \cdot \sqrt[3]{45}$

APPLYING THE MATHEMATICS

20. Suppose p is a positive number.

    a. What is the sum of the square roots of p?

    b. What is the product of the square roots of p?

21. In Chapter 7, the equation $P = 100,000(1.02)^x$ gave the population x years from now of a town of 100,000 today with a growth rate of 2% per year. Calculate P when $x = \frac{1}{2}$, and tell what the answer means.

22. A small park is shown below. If you want to go from one corner to the other corner, how many fewer feet will you walk if you go diagonally through the park rather than walk around it? Round your answer to the nearest foot.

    600 ft

    300 ft

23. In the movie *The Wizard of Oz*, the scarecrow recites the following after receiving his diploma. "The sum of the square roots of any two sides of an isosceles triangle is equal to the square root of the remaining side." The scarecrow was attempting to recite the Pythagorean Theorem.

    a. Write several sentences explaining how this statement differs from the Pythagorean Theorem.

    b. Is the scarecrow's statement accurate? If not, produce a counterexample.

24. A dog is on a leash that is 10 meters long and attached to a pole 2.5 meters above the the dog's collar. To the nearest tenth of a meter, how far from the pole can the dog roam?

REVIEW

25. As you know, $4 \cdot 9 = 36$. So the square of 2 times the square of 3 equals the square of 6. Determine the general pattern. (**Lesson 8-5**)

26. Simplify $a^6 \cdot \left(\frac{3}{a}\right)^3$. (**Lesson 8-5**)

In 27 and 28, solve. (**Lessons 8-4, 8-2**)

27. $3^4 \cdot 3^x = 3^{12}$

28. $\frac{1}{512} = 2^a$

29. After the sun, the nearest star to us is Proxima Centauri, about $4 \cdot 10^{13}$ km away. Earth's moon is about $3.8 \cdot 10^5$ km from us. If it took astronauts about 3 days to get to the moon in 1969, at that speed how long would it take them to get to Proxima Centauri? (**Lesson 8-3**)

30. **Skill Sequence** Solve each equation for *y*. Assume $a \neq 0$. (**Lesson 4-7**)

    a. $3x + 4y = 2$
    b. $6x + 8y = 4$
    c. $9x + 12y = 6$
    d. $3ax + 4ay = 2a$

You would have to circumnavigate Earth $9\frac{1}{2}$ times to equal the distance from Earth to the moon.

EXPLORATION

31. Make a table to evaluate $n^{\frac{1}{3}}$ on your calculator when *n* is 1, 2, 3, ..., up to 7. What do you think $n^{\frac{1}{3}}$ is equivalent to? Give a reason for your answer.

QY ANSWERS

1. $4\sqrt{10} \cdot 3\sqrt{10}$
   $= 4 \cdot 3 \cdot \sqrt{10} \cdot \sqrt{10}$
   $= 12 \cdot 10$
   $= 120$

2. 4; 64

Lesson 8-7

Multiplying and Dividing Square Roots

Vocabulary

radicand

▶ **BIG IDEA** Like powers, square roots distribute over products and quotients.

Activity 1

Step 1 Compute these square roots to the nearest thousandth either individually or using the list capability of a calculator.

$\sqrt{1} = 1.000$ $\sqrt{2} \approx 1.414$ $\sqrt{3} \approx 1.732$ $\sqrt{4} =$ __?__

$\sqrt{5} \approx$ __?__ $\sqrt{6} \approx 2.449$ $\sqrt{7} \approx$ __?__ $\sqrt{8} \approx$ __?__

$\sqrt{9} =$ __?__ $\sqrt{10} \approx$ __?__ $\sqrt{11} \approx$ __?__ $\sqrt{12} \approx$ __?__

$\sqrt{13} \approx$ __?__ $\sqrt{14} \approx$ __?__ $\sqrt{15} \approx$ __?__ $\sqrt{16} =$ __?__

$\sqrt{17} \approx$ __?__ $\sqrt{18} \approx$ __?__ $\sqrt{19} \approx$ __?__ $\sqrt{20} \approx$ __?__

Step 2 Consider the product $\sqrt{2} \cdot \sqrt{3}$. Find the product of the decimal approximations, rounded to 3 decimal places.

Decimal approximations: __?__ · __?__ ≈ 2.449

Is the decimal product found in the table above? __?__
If so, write the equation that relates the product of the square roots.
Square roots: __?__ · __?__ = __?__

Step 3 Repeat Step 2 but use a product of two different square roots from the list $\sqrt{2}, \sqrt{3}, \sqrt{4}, \sqrt{5}$.

Square roots: __?__ · __?__
Decimal approximations: __?__ · __?__ ≈ __?__

Is the decimal product found in the table above? __?__
If so, write the equation that relates the product of the square roots.
Square roots: __?__ · __?__ = __?__

Step 4 Multiply another pair of square roots in the table. __?__ · __?__
Predict what their product will be. __?__ Is your prediction correct?
__?__

Mental Math

Given $f(x) = 611x^2 + 492x - 1{,}000$, calculate the following.

a. $f(0)$

b. $f(1)$

Multiplying and Dividing Square Roots 497

Chapter 8

In Activity 1, you should have discovered that when the product of two numbers a and b is a third number c, it is also the case that the product of the square root of a and the square root of b is the square root of c. That is, if $ab = c$, then $\sqrt{a} \cdot \sqrt{b} = \sqrt{c} = \sqrt{ab}$. For example, because $5 \cdot 6 = 30$, $\sqrt{5} \cdot \sqrt{6} = \sqrt{30}$. You can check this by using decimal approximations to the square roots.

> **Product of Square Roots Property**
>
> For all nonnegative real numbers a and b, $\sqrt{a} \cdot \sqrt{b} = \sqrt{ab}$.

The Product of Square Roots Property may look unusual when the square roots are written in radical form. But when the square roots are written using the exponent $\frac{1}{2}$, the property takes on a familiar look.

$$a^{\frac{1}{2}} \cdot b^{\frac{1}{2}} = (ab)^{\frac{1}{2}}$$

It is just the Power of a Product Property, with $n = \frac{1}{2}$! This is further evidence of the appropriateness of thinking of the positive square root of a number as its $\frac{1}{2}$ power.

Activity 2

Step 1 Pick a square root from $\sqrt{6}$, $\sqrt{12}$, and $\sqrt{18}$.
Pick a square root from $\sqrt{2}$, $\sqrt{3}$, and $\sqrt{6}$.
Find the quotient of the decimal approximations.

Square roots __?__ ÷ __?__
Decimal approximations __?__ ÷ __?__ ≈ __?__
Is the quotient found in the table in Activity 1? __?__
If so, write the quotient as a square root. __?__
If not, is the quotient close to a number in the table? __?__
What square root is it closest to? __?__

Step 2 Repeat the process in Step 1 using a different square root from each group.

Step 3 Repeat the process again using a third pair of square roots.

In Activity 2, you should have discovered that when the quotient of two numbers c and a is a third number b, it is also the case that the quotient of the square root of c and the square root of a is the square root of b. That is, if $\frac{c}{a} = b$, then $\frac{\sqrt{c}}{\sqrt{a}} = \sqrt{\frac{c}{a}} = \sqrt{b}$. For example, since $\frac{24}{8} = 3$, $\frac{\sqrt{24}}{\sqrt{8}} = \sqrt{\frac{24}{8}} = \sqrt{3}$.

> **Quotient of Square Roots Property**
>
> For all positive real numbers a and c, $\dfrac{\sqrt{c}}{\sqrt{a}} = \sqrt{\dfrac{c}{a}}$.

Fact triangles can be used to visualize the Product of Square Roots Property and the Quotient of Square Roots Property. For all positive numbers a, b, and c:

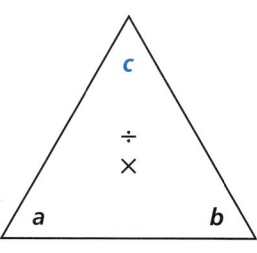

$a \cdot b = c$
$b \cdot a = c$
$\dfrac{c}{a} = b$
$\dfrac{c}{b} = a$

$\sqrt{a} \cdot \sqrt{b} = \sqrt{c}$
$\sqrt{b} \cdot \sqrt{a} = \sqrt{c}$
$\dfrac{\sqrt{c}}{\sqrt{a}} = \sqrt{b}$
$\dfrac{\sqrt{c}}{\sqrt{b}} = \sqrt{a}$

 QY1

> ▶ **QY1**
>
> Use either the Product or Quotient of Square Roots Property to evaluate each expression.
>
> a. $\sqrt{8} \cdot \sqrt{2}$
>
> b. $\dfrac{\sqrt{45}}{\sqrt{5}}$
>
> c. $\dfrac{\sqrt{80}}{\sqrt{40}}$

"Simplifying" Radicals

A radical expression is said to be simplified if the quantity under the radical sign, called the **radicand**, has no perfect square factors other than 1.

Just as you can multiply square roots by using the Product of Square Roots Property, $\sqrt{4} \cdot \sqrt{10} = \sqrt{4 \cdot 10} = \sqrt{40}$, you can rewrite a square root as a product by factoring the radicand.

$$\sqrt{40} = \sqrt{4 \cdot 10}$$
$$= \sqrt{4} \cdot \sqrt{10}$$
$$= 2 \cdot \sqrt{10}$$

Many people consider $2\sqrt{10}$ to be simpler than $\sqrt{40}$ because it has a smaller radicand. This process is called *simplifying a radical*. The key to the process is to find a perfect square factor of the radicand.

Example 1

Simplify $\sqrt{27}$.

Solution Perfect squares larger than 1 are 4, 9, 16, 25, 36, 49, …. Of these, 9 is a factor of 27.

(continued on next page)

Multiplying and Dividing Square Roots

Chapter 8

$\sqrt{27} = \sqrt{9 \cdot 3}$ Factor 27.
$= \sqrt{9} \cdot \sqrt{3}$ Product of Square Roots Property
$= 3\sqrt{3}$ $\sqrt{9} = 3$

Check Using a calculator we see $\sqrt{27} \approx 5.196152423$ and $3\sqrt{3} \approx 5.196152423$.

Is $3\sqrt{3}$ really simpler than $\sqrt{27}$? It depends. For estimating purposes, $\sqrt{27}$ is simpler since we can easily see it is slightly larger than $\sqrt{25}$ or 5. But for seeing patterns, $3\sqrt{3}$ may be simpler. In the next example, the answer $7\sqrt{2}$ is related to the given information in a useful way that is not served by leaving it in the unsimplified form $\sqrt{98}$.

GUIDED

Example 2
Each leg of the right triangle below is 7 cm long.

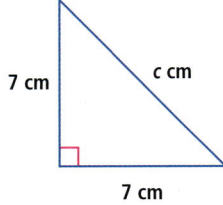

a. Find the exact length of the hypotenuse.
b. Put the exact length in simplified radical form.

Solutions

a. Use the Pythagorean Theorem.
$c^2 = \underline{}^2 + \underline{}^2$ Substitute the lengths of the legs.
$c^2 = 98$ Add.
$c = \underline{}$ Use a radical sign to write the exact answer.

b. Now use the Product of Square Roots Property to simplify the result. Note that the perfect square 49 is a factor of 98.

$c = \sqrt{\underline{} \cdot 2}$
$c = \sqrt{\underline{}} \cdot \sqrt{\underline{}}$
$c = \underline{} \sqrt{2}$

The exact length of the hypotenuse is $\sqrt{98}$, or $\underline{}$ cm.

The Product of Square Roots Property also applies to expressions containing variables.

Lesson 8-7

GUIDED

Example 3
Assume x and y are positive. Simplify $\sqrt{48x^2y^2}$.

Solution
$$\sqrt{48x^2y^2} = \sqrt{\underline{\ ?\ }} \cdot \sqrt{3} \cdot \sqrt{x^2} \cdot \sqrt{y^2}$$
$$= \underline{\ ?\ } \cdot \sqrt{3} \cdot x \cdot y$$
$$= \underline{\ ?\ } xy\sqrt{3}$$

Check Substitute values for x and y. We choose $x = 4$ and $y = 3$.

$\sqrt{48x^2y^2} = \sqrt{48 \cdot 16 \cdot 9}$ $\quad 4xy\sqrt{3} = 4 \cdot 4 \cdot 3\sqrt{3}$
$\qquad\quad = \sqrt{\underline{\ ?\ }}$ $\qquad\qquad\qquad = \underline{\ ?\ }\sqrt{3}$
$\qquad\quad \approx 83.14$ $\qquad\qquad\qquad \approx 83.14$

It checks.

🛑 **QY2**

▶ **QY2**

a. Assume x and y are positive.

Simplify $\sqrt{25x^2y}$.

b. Assume x is positive.

Simplify $\dfrac{\sqrt{24x^2}}{\sqrt{6x^2}}$.

Although square roots were first used in connection with geometry, they also have important applications in physical situations. One such application is with the pendulum clock.

In a pendulum clock, a clock hand moves each time the pendulum swings back and forth. The first idea for a pendulum clock came from the great Italian scientist Galileo Galilei in 1581. (At the time of Galileo, there was no accurate way to tell time; watches and clocks did not exist. People used sand timers but they were not very accurate.) Galileo died in 1642, before he could carry out his design. The brilliant Dutch scientist Christiaan Huygens applied Galileo's concept of tracking time with a pendulum swing in 1656.

A very important part of constructing the clock was calculating the time it takes a pendulum to complete one swing back and forth. This is called the *period* of the pendulum. The formula $p = 2\pi\sqrt{\dfrac{L}{32}}$ gives the time p in seconds for one period in terms of the length L (in feet) of the pendulum.

Dutch mathematician, Christiaan Huygens (1629–1695), patented the first pendulum clock, which greatly increased the accuracy of time measurement.

Source: University of St. Andrews

Example 4
A pendulum clock makes one "tick" for each complete swing of the pendulum. If a pendulum is 2 feet long, how many ticks would the clock make in one minute?

Solution First calculate p when $L = 2$.

(continued on next page)

Multiplying and Dividing Square Roots

Chapter 8

$$p = 2\pi\sqrt{\frac{2}{32}} = 2\pi\sqrt{\frac{1}{16}} = 2\pi\frac{\sqrt{1}}{\sqrt{16}} = 2\pi \cdot \frac{1}{4} = \frac{\pi}{2}$$

It takes $\frac{\pi}{2}$ seconds for the pendulum to go back and forth.

$$\frac{1 \text{ tick}}{\frac{\pi}{2} \text{ s}} \cdot \frac{60 \text{ s}}{1 \text{ min}} = \frac{60}{\frac{\pi}{2}} \frac{\text{tick}}{\text{min}} \approx 38.2 \text{ ticks/min}$$

So the clock makes about 38.2 ticks per minute.

Questions

COVERING THE IDEAS

In 1–4, use the Product or Quotient of Square Roots Property to evaluate the expression.

1. $\sqrt{8} \cdot \sqrt{2}$
2. $\sqrt{36 \cdot 81 \cdot 100}$
3. $\frac{\sqrt{40}}{\sqrt{10}}$
4. $\frac{\sqrt{6^3}}{\sqrt{6}}$

5. If $\sqrt{3} \cdot \sqrt{6} = \sqrt{x} = y\sqrt{z}$, what is x, what is y, and what is z?

6. If $\frac{\sqrt{63}}{\sqrt{7}} = \sqrt{x} = y$, what is x and what is y?

7. **Multiple Choice** Which is *not* equal to $\sqrt{50}$?
   A $\sqrt{5} \cdot \sqrt{10}$
   B $\sqrt{25} + \sqrt{25}$
   C $\sqrt{2} \cdot \sqrt{25}$
   D $5\sqrt{2}$

8. a. Use the formula $p = 2\pi\sqrt{\frac{L}{32}}$ to calculate the time p for one period of a pendulum of length $L = 8$ feet.
   b. If the clock makes one tick for each pendulum swing back and forth, how many ticks are there in one minute?

In 9–12, simplify the square root.

9. $\sqrt{18}$
10. $\sqrt{24}$
11. $\sqrt{50}$
12. $8\sqrt{90}$

13. Assume m and n are positive. Simplify each expression.
    a. $\sqrt{150m^2n}$
    b. $\frac{\sqrt{112m^7}}{\sqrt{7m^3}}$

14. The length of each leg of a right triangle is 8 cm. What is the exact length of the hypotenuse?

15. Let $m = \frac{1}{2}$ in the Power of a Quotient Property $\left(\frac{x}{y}\right)^m = \frac{x^m}{y^m}$. What property of this lesson is the result?

APPLYING THE MATHEMATICS

In 16–19, write the exact value of the unknown in simplified form. Then approximate the unknown to the nearest hundredth.

16.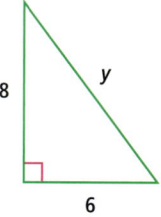

17.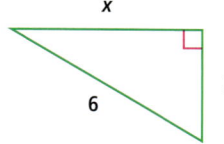

18.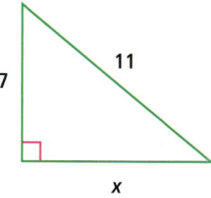

19.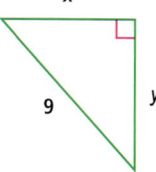

20. Find the area of a triangle with base $\sqrt{18}$ and height $6\sqrt{2}$.

21. The radical $\sqrt{50}$ is equivalent to $5\sqrt{2}$. Explain why it is easier to tell that $\sqrt{50}$ is slightly larger than 7 than it is to tell $5\sqrt{2}$ is slightly larger than 7.

In 22–25, explore adding square roots. You can add square roots using the Distributive Property if their radicands are alike. So, $3\sqrt{11} + 5\sqrt{11} = 8\sqrt{11}$, but $2\sqrt{11} + 4\sqrt{3}$ cannot be simplified. In each expression below, simplify terms if possible, then add or subtract.

22. $2\sqrt{25} + \sqrt{49}$

23. $\sqrt{12} - 10\sqrt{3}$

24. $\sqrt{45} - \sqrt{20}$

25. $4\sqrt{50} + 3\sqrt{18}$

REVIEW

In 26 and 27, consider the rectangular field pictured here. (Lesson 8-6, Previous Course)

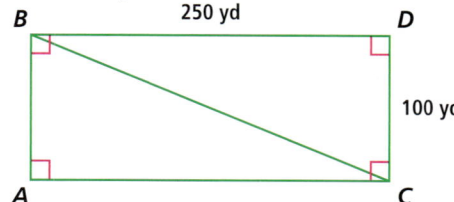

26. How much shorter would it be to walk diagonally across the field as opposed to walking along the sides to get from B to C?

27. Suppose A is the origin of the coordinate system with the y-axis on $\overleftrightarrow{AB}$ and the x-axis on $\overleftrightarrow{AC}$. Give the coordinates of point D.

In 28–30, write the expression as a power of a single number.
(Lessons 8-4, 8-3, 8-2)

28. $\dfrac{k^{15}}{k^9}$ 29. $x^4 \cdot x$ 30. $(w^2)^{-3}$

31. Which is greater, $(6^4)^2$, or $6^4 \cdot 6^2$? (**Lesson 8-2**)

32. In 1995, Ellis invested $5,000 for 10 years at an annual yield of 8%. In 2005, Mercedes invested $7,000 for 5 years at 6%. By the end of 2010, who would have more money? Justify your answer. (**Lesson 7-1**)

33. After x seconds, an elevator is on floor y, where $y = 46 - 1.5x$. Give the slope and y-intercept of $y = 46 - 1.5x$, and describe what they mean in this situation. (**Lesson 6-4**)

34. A box with dimensions 30 cm by 60 cm by 90 cm will hold how many times as much as one with dimensions 10 cm by 20 cm by 30 cm? (**Lesson 5-10**)

The Taipei 101 skyscraper houses the world's fastest elevator, with a maximum speed of 37 mph.

Source: ScientificAmerican.com

EXPLORATION

35. Is there a Product of Cube Roots Property like the Product of Square Roots Property? Explore this idea and reach a conclusion. Describe your exploration and defend your conclusion.

36. Use the formula $p = 2\pi\sqrt{\dfrac{L}{32}}$ to determine the length of a pendulum that will make 1 tick each second. Answer to the nearest hundredth of an inch.

QY ANSWERS

1a. 4

b. 3

c. $\sqrt{2}$

2a. $5x\sqrt{y}$

b. 2

Lesson 8-8

Distance in a Plane

▶ **BIG IDEA** Using the Pythagorean Theorem, the distance between any points in a plane can be found if you know their coordinates.

Competitions involving small robots (sometimes called Robot Wars™ or BotBashes™) began in the late 1990s as engineering school projects but quickly spread to competitions open to the public. Even television programs have featured the battling 'bots. The competitions take place in an enclosed arena that is laid out in a grid pattern like the one below.

Mental Math

What is 20% of each quantity?

a. $40x$

b. $5y$

c. $40x + 5y$

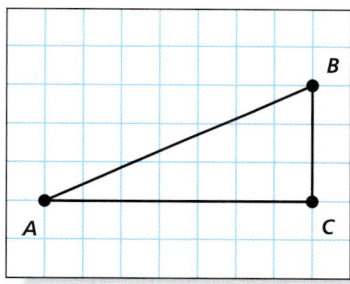

To get from point A to point B in the arena, robots can be maneuvered manually by their "driver," but because the shortest distance between A and B is the straight line segment, robot designers like to program direction and distance commands into their robots. The distance traveled is an application of the Pythagorean Theorem because side $\overline{AB}$ is the hypotenuse of a right triangle.

Australia's humanoid robot kicks a ball at the 2013 RoboCup in Eindhoven, Netherlands.

Distances along Vertical and Horizontal Lines

To find the distance between any two points in the coordinate plane, we begin by examining the situation where points are on the same vertical or horizontal line.

You can find the distance between two points on vertical or horizontal lines by thinking of them as being on a number line. The distance can be obtained by counting spaces or by subtracting appropriate coordinates. Consider the rectangle $DEFG$ graphed on the next page.

Distance in a Plane   **505**

Horizontal Distance The distance DE can be found by counting spaces on the number line or it can be calculated by subtracting the x-coordinates and then taking the absolute value.

$$DE = |-2 - 3| = 5$$

Vertical Distance Similarly, the distance EF can be found by counting spaces or it can be calculated by subtracting the y-coordinates and taking the absolute value.

$$EF = |3 - (-1)| = 4$$

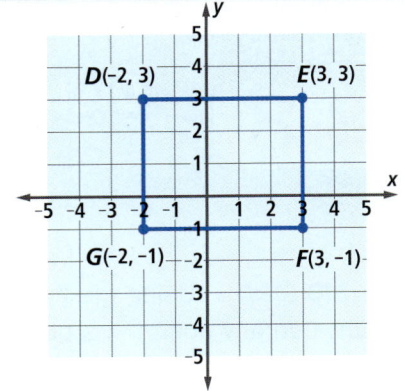

 QY1

The Distance between Any Two Points in a Plane

The Pythagorean Theorem enables you to find the distance between any two points in the plane.

> **QY1**
>
> Find the length of the segment whose endpoints are (50, 21) and (50, 46).

Example 1

Find AB in $\triangle ABC$ at the right.

Solution $\overline{AB}$ is the hypotenuse of $\triangle ABC$ whose legs, $\overline{AC}$ and $\overline{BC}$, are horizontal and vertical, respectively. The length of the legs can be calculated by subtracting appropriate coordinates.

$AC = |8 - 1| = 7 \qquad BC = |5 - 2| = 3$

Now apply the Pythagorean Theorem.

$(AB)^2 = (AC)^2 + (BC)^2$
$(AB)^2 = 7^2 + 3^2$
$(AB)^2 = 58$
$AB = \sqrt{58} \approx 7.62$

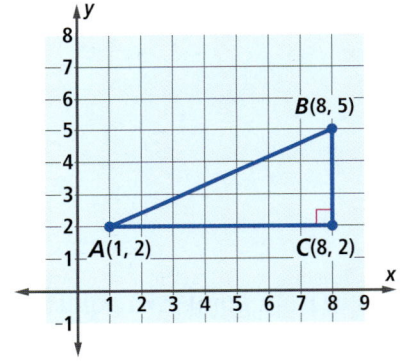

This method can be generalized to find the distance between any two points on a coordinate grid.

Let point $A = (x_1, y_1)$ and $B = (x_2, y_2)$, as shown at the right. Then a right triangle can be formed with a third vertex at $C = (x_2, y_1)$. Using these coordinates, $AC = |x_2 - x_1|$ and $BC = |y_2 - y_1|$. Now use the Pythagorean Theorem.

$AB^2 = AC^2 + BC^2 = |x_2 - x_1|^2 + |y_2 - y_1|^2$

Since a number and its absolute value have the same square, $|x_2 - x_1|^2 = (x_2 - x_1)^2$ and $|y_2 - y_1|^2 = (y_2 - y_1)^2$.

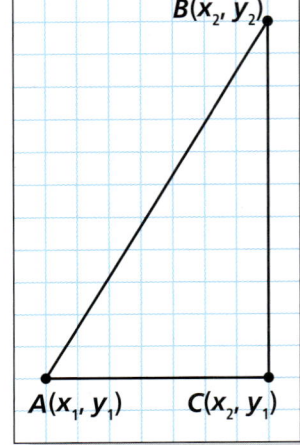

506 Powers and Roots

Thus $AB^2 = (x_2 - x_1)^2 + (y_2 - y_1)^2$.

Take the positive square root of each side. The result is a formula for the distance between two points in the coordinate plane.

> **Formula for the Distance between Two Points in a Coordinate Plane**
>
> The distance AB between the points $A = (x_1, y_1)$ and $B = (x_2, y_2)$ in a coordinate plane is $AB = \sqrt{(x_2 - x_1)^2 + (y_2 - y_1)^2}$.

GUIDED

Example 2

Find the distance between the points (23, 16) and (31, −11) to the nearest thousandth.

Solution Using the formula for the distance between two points in a plane, let $A = (x_1, y_1) = (\underline{\ ?\ }, \underline{\ ?\ })$ and $B = (x_2, y_2) = (\underline{\ ?\ }, \underline{\ ?\ })$.

$AB = \sqrt{(\underline{\ ?\ } - \underline{\ ?\ })^2 + (\underline{\ ?\ } - \underline{\ ?\ })^2}$

$ = \sqrt{(\underline{\ ?\ })^2 + (\underline{\ ?\ })^2}$

$ = \sqrt{\underline{\ ?\ } + \underline{\ ?\ }} = \sqrt{\underline{\ ?\ }} \approx \underline{\ ?\ }$

Example 3

Use the map at the right. It shows streets and the locations of three buildings in a city. The streets are 1 block apart.

a. Give the coordinates of all three buildings.
b. Find the distance from the train station to the zoo.

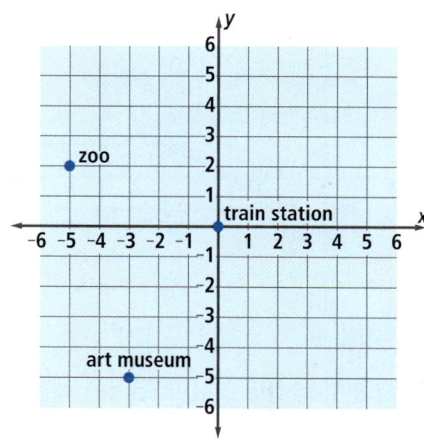

Solutions

a. The coordinates are as follows.

train station (0, 0)

zoo (−5, 2)

art museum (−3, −5)

b. We need to find the distance from (0, 0) to (−5, 2).

$\text{distance} = \sqrt{(0 - -5)^2 + (0 - 2)^2}$

$\phantom{\text{distance}} = \sqrt{(5)^2 + (-2)^2}$

$\phantom{\text{distance}} = \sqrt{25 + 4}$

$\phantom{\text{distance}} = \sqrt{29} \approx 5.39$ blocks

Distance in a Plane

Chapter 8

STOP QY2

> ▶ QY2
> In Example 3, find the distance from the zoo to the art museum.

Questions

COVERING THE IDEAS

1. Refer to △EFP below.

   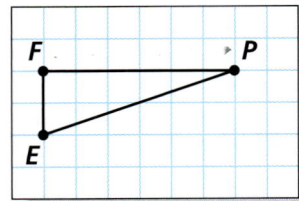

   a. Determine FE.
   b. Determine FP.
   c. Use the Pythagorean Theorem to calculate EP.

In 2–5, find PT.

2. $P = (3, 4)$ and $T = (3, 9)$
3. $P = (-3, 4)$ and $T = (4, 4)$
4. $P = \left(-\frac{2}{3}, \frac{1}{2}\right)$ and $T = \left(\frac{5}{3}, \frac{1}{2}\right)$
5. $P = (33, -4)$ and $T = (33, 18)$

In 6 and 7, find the length of $\overline{AB}$.

6.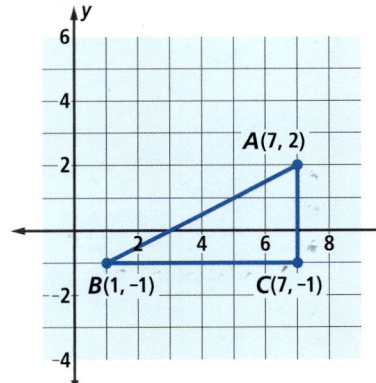

7.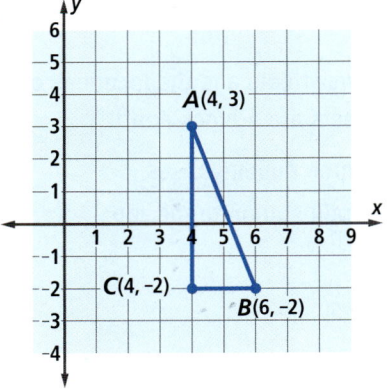

In 8–13, find the distance between the two points.

8. $E = (4, 9); D = (8, 6)$
9. $F = (15, 2); G = (20, -10)$
10. $H = (5, -1); I = (11, 2.2)$
11. $J = (-6, -7); K = (-2, 0)$
12. $L = (-1, 2); M = (-3, 4)$
13. $N = (-0.43, -0.91); P = (-0.36, -0.63)$

Powers and Roots

APPLYING THE MATHEMATICS

In 14 and 15, use the map at the right, which shows the locations of a house, school, and mall. Suppose each square of the grid is a half mile on a side.

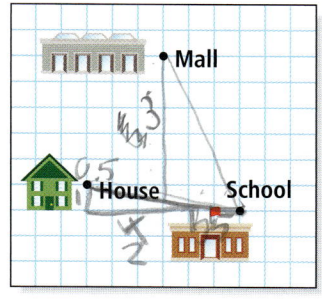

14. a. What is the distance from the house to the school?
    b. Which is closer to the house, the mall or the school?

15. How far is it from school to the mall?

16. Pirate Slopebeard has the treasure map below. How far is the treasure from the start if you travel along a straight path "as the crow flies?"

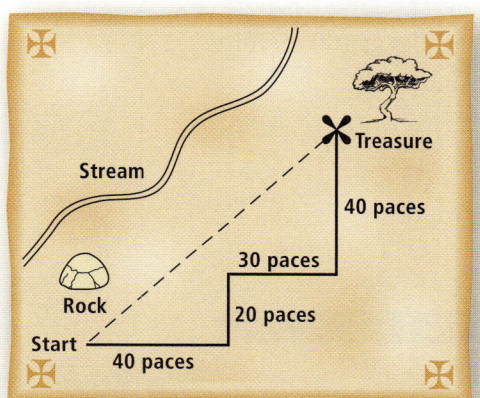

17. Write an expression for the distance between the points (0, 0) and (b, d).

18. The vertices of a triangle are (3, 4), (6, 9), and (9, 4). Is the triangle equilateral? How do you know?

19. Write the distance formula using a power instead of a radical sign.

20. Does it matter which ordered pair is first when using the distance formula? Choose two ordered pairs. Do the calculation both ways to verify your answer.

REVIEW

21. Evaluate $3^{\frac{1}{2}} \cdot 4^{\frac{1}{2}} \cdot 12^{\frac{1}{2}}$ in your head. (**Lesson 8-7**)

22. Consider the function f with $f(x) = 1.5^x$. Find the following values to the nearest hundredth. (**Lessons 8-6, 7-6**)
    a. $f(0)$
    b. $f\left(\frac{1}{2}\right)$
    c. $f(1)$
    d. $f(2)$
    e. $f(-1)$
    f. $f(-2)$

23. If $d = \sqrt{6}$, find the value of $\frac{d^5}{(3d)^2} \cdot 10d$. (**Lessons 8-6, 8-5**)

Distance in a Plane 509

24. Suppose Pythagoras Park is a rectangle 250 meters wide and 420 meters long. There are sidewalks around the edges of the park and a diagonal sidewalk connecting the southeast corner to the northwest corner. Esmeralda and Dory are standing at the southeast corner and want to get to the ice cream stand at the northwest corner. (**Lessons 8-6, 5-3**)

   a. How many meters would Esmeralda and Dory have to walk if they traveled along the diagonal sidewalk? Round your answer to the nearest meter.

   b. How many meters would Esmeralda and Dory have to walk if they traveled along the edge sidewalks?

   c. Esmeralda walks along the diagonal sidewalk at 60 meters per minute while Dory jogs along the edge sidewalks at 100 meters per minute. Who arrives at the ice cream stand first?

25. Suppose the graph of f is a line with slope $\frac{8}{5}$ and $f(7) = 1.2$. Write a formula for $f(x)$. (**Lessons 7-6, 6-2**)

26. Suppose $f(x) = 29x - 2$ and $g(x) = 2(34.5x + 17.75)$. For what value of x does $f(x) = g(x)$? (**Lessons 7-6, 4-4**)

EXPLORATION

27. You may have seen videos showing giant robotic arms maneuvering through space to perform a task. Did you ever wonder how the robot is controlled? One component is calculating how far to move the arm. To do this, designers extend the ideas in this lesson from 2 dimensions (x, y) to 3 dimensions (x, y, z) and calculate the distance between two *ordered triples* that describe locations in space. The distance between two points (x_1, y_1, z_1) and (x_2, y_2, z_2) in space is $\sqrt{(x_2 - x_1)^2 + (y_2 - y_1)^2 + (z_2 - z_1)^2}$. Calculate the distance between (3, 5, 2) and (−4, 3, −2). Draw a picture of the two points and the distance on the graph below.

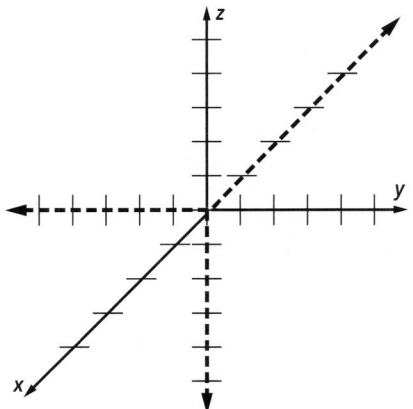

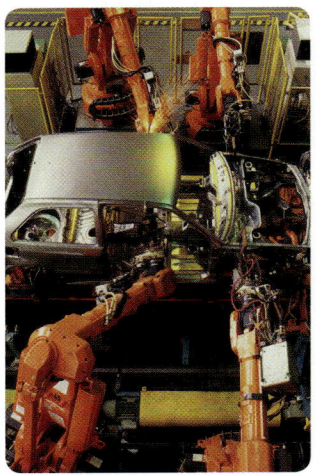

According to the U.N.'s 2004 World Robotics Survey, most industrial robots are used on assembly lines, chiefly in the auto industry.

QY ANSWERS

1. 25

2. $\sqrt{53} \approx 7.28$ blocks

Lesson 8-9

Remembering Properties of Powers and Roots

> **BIG IDEA** If you forget any of the properties of powers, you can recall them by testing special cases, following patterns, and knowing alternate ways of rewriting the same expression.

Here are six properties of powers that were presented in this chapter. They apply to all exponents m and n and nonzero bases a and b.

Properties of Powers

Product of Powers
$b^m \cdot b^n = b^{m+n}$

Negative Exponent
$b^{-n} = \frac{1}{b^n}$

Power of a Power
$(b^m)^n = b^{mn}$

Quotient of Powers
$\frac{b^m}{b^n} = b^{m-n}$

Power of a Product
$(ab)^n = a^n b^n$

Power of a Quotient
$\left(\frac{a}{b}\right)^n = \frac{a^n}{b^n}$

Mental Math

Evaluate.
a. $|\text{-}19 + \text{-}42|$
b. $|55.5 - 32| + |\text{-}9|$
c. $|\text{-}10 + 5| - |\text{-}30 + 15|$

In addition, two properties of square roots were studied. They apply to all nonnegative numbers a and b.

Properties of Square Roots

Product of Square Roots
$\sqrt{a} \cdot \sqrt{b} = \sqrt{ab}$

Quotient of Square Roots
$\frac{\sqrt{a}}{\sqrt{b}} = \sqrt{\frac{a}{b}}, b \neq 0$

With so many properties, some students confuse them. This lesson shows you how to use general problem-solving strategies to remember the properties and to test a special case to verify your reasoning.

Testing a Special Case

Because of calculators, a strategy called *testing a special case* is often possible. You can use this strategy to verify your reasoning.

Chapter 8

Example 1
Emily was not sure how to simplify $x^5 \cdot x^6$. She felt the answer could be x^{30}, $2x^{11}$, or x^{11}. Which is correct?

Solution 1 Use a special case. Let $x = 3$. Now calculate $x^5 \cdot x^6$ (with a calculator) and see if it equals the calculator result for x^{30} or $2x^{11}$ or x^{11}.

$x^5 \cdot x^6 = 3^5 \cdot 3^6 = 177{,}147$
$x^{30} = 3^{30} = 205{,}891{,}132{,}094{,}649 \approx 2.0589 \cdot 10^{14}$
$2x^{11} = 2 \cdot 3^{11} = 354{,}294$
$x^{11} = 3^{11} = 177{,}147$

So, the answer is x^{11}.

Solution 2 Use repeated multiplication to rewrite x^5 and x^6.

$x^5 \cdot x^6 = (x \cdot x \cdot x \cdot x \cdot x) \cdot (x \cdot x \cdot x \cdot x \cdot x \cdot x)$

Notice there are 11 factors of x in the product. So, $x^5 \cdot x^6 = x^{11}$.

Showing That a Pattern Is Not Always True

When testing a special case, you should be careful in choosing numbers. The numbers 0, 1, and 2 are not good for checking answers to problems involving powers, because a pattern may work for a few of these numbers but not for all numbers. Recall that a *counterexample* is a special case for which the answer is false. To show a pattern is not true, it is sufficient to find one counterexample.

Example 2
Sir Lancelot's assistant, Squire Root, noticed $2^3 + 2^3 = 2^4$ since $8 + 8 = 16$. He guessed that in general, there is a property $x^3 + x^3 = x^4$.

He tested a second case by letting $x = 0$ and found $0^3 + 0^3 = 0^4$. He concluded that the property is always true. Is he correct?

Solution Try a different value for x. Let $x = 5$.

Does $5^3 + 5^3 = 5^4$? Does $125 + 125 = 625$? No. $x = 5$ is a counterexample that shows that Squire's property is not always true.

If you have trouble remembering a property or are not certain that you have simplified an expression correctly, try using repeated multiplication or testing a special case.

In Example 3, using the properties of powers, a calculation that is complicated even with a calculator, is reduced to two calculations.

Example 3

The mean radius of Earth is about $3.96 \cdot 10^3$ miles. The mean radius of Jupiter is about $4.34 \cdot 10^4$ miles. Using the formula $V = \frac{4}{3}\pi r^3$ for the volume V of a sphere with radius r, how many times could Earth fit inside Jupiter?

Solution To answer the question, you need to divide the volume of Jupiter by the volume of Earth. So substitute for r in the formula for the volume of a sphere.

$$\frac{\text{volume of Jupiter}}{\text{volume of Earth}} \approx \frac{\frac{4}{3}\pi(4.34 \cdot 10^4)^3}{\frac{4}{3}\pi(3.96 \cdot 10^3)^3}$$

$$= \frac{(4.34 \cdot 10^4)^3}{(3.96 \cdot 10^3)^3} \qquad \text{Multiplication of Fractions}$$

$$= \left(\frac{4.34 \cdot 10^4}{3.96 \cdot 10^3}\right)^3 \qquad \text{Power of a Quotient Property}$$

$$= \left(\frac{43.4}{3.96}\right)^3 \qquad \text{Quotient of Powers Property}$$

$$\approx 1{,}316 \qquad \text{Arithmetic}$$

In Example 3, notice that by using the properties, we reduced a complicated calculation to two operations: division of 43.4 by 3.96 and then cubing of the quotient.

Earth could fit inside Jupiter about 1,300 times. (Jupiter is *very* big compared to Earth!)

It is important to realize that the properties of numbers and operations are consistent. If you apply them correctly, you can find many paths to a correct solution. The result you get using some properties will not disagree with the results another person gets by correctly using other properties.

GUIDED

Example 4

If $\left(\frac{9q^{-5}}{6q^{-3}}\right)^{-7} = aq^n$, what are the values of a and n?

Solution 1 This question requires that the expression on the left side be simplified into the form aq^n.

(continued on next page)

Chapter 8

$$\left(\frac{9q^{-5}}{6q^{-3}}\right)^{-7} = \frac{9^{\underline{\;?\;}} q^{\underline{\;?\;}}}{6^{\underline{\;?\;}} q^{\underline{\;?\;}}}$$ Apply the Power of a Power Property to eliminate the parentheses.

$$= \frac{9^{\underline{\;?\;}}}{6^{\underline{\;?\;}}} q^{\underline{\;?\;}}$$ Apply the Quotient of Powers Property.

$$= \frac{6^{\underline{\;?\;}}}{9^{\underline{\;?\;}}} q^{\underline{\;?\;}}$$ Apply the Negative Exponent Property $\left(b^{-n} = \frac{1}{b^n}\right)$.

$$= \left(\frac{6}{9}\right)^{\underline{\;?\;}} q^{\underline{\;?\;}}$$ Use the Power of a Quotient Property.

$$= \left(\frac{2}{3}\right)^{\underline{\;?\;}} q^{\underline{\;?\;}}$$ Rewrite the fraction in lowest terms.

Thus, $a = \left(\frac{2}{3}\right)^{\underline{\;?\;}}$ and $n = \underline{\;?\;}$.

Solution 2 Work with a partner and follow these steps.

1. Apply the Quotient of Powers Property inside the parentheses.
2. Write the fraction in lowest terms.
3. Apply the Power of a Power Property.

You should obtain the same answer as in Solution 1.

Questions

COVERING THE IDEAS

1. **Multiple Choice** For all nonzero values of n, $\frac{n^{40}}{n^{10}} =$

   A n^4.   B n^{30}.   C 1^{30}.   D 1^4.

2. **Multiple Choice** For all nonzero values of s, $\frac{s^4}{(2s)^2} =$

   A $4s^2$.   B $\frac{s^2}{2}$.   C $\frac{s^2}{4}$.   D 1.

3. **Multiple Choice** For all nonzero values of v, $\frac{v^8 \cdot v^{12}}{(v^8)^{12}} =$

   A 1.   B $-v^{76}$.   C v^{-76}.   D v^{76}.

4. **Multiple Choice** For all values of m and n, $(3m)^2 \cdot (2n)^3 =$

   A $6m^2n^3$.   B $48m^2n^3$.   C $72m^2n^3$.   D $3{,}125m^2n^3$.

5. What is a counterexample?

514 Powers and Roots

6. **True or False** If two special cases of a pattern are true, then the pattern is true.

7. **True or False** If one special case of a pattern is false, then the pattern is false.

8. Consider the equation $x^4 = 8x$.
   a. Is the equation true for the special case $x = 2$?
   b. Is the equation true for the special case $x = 0$?
   c. Is the equation true for the special case $x = 3$?
   d. Is the equation true for all values of x?

In 9–11, test special cases to decide whether the pattern is always true. Show all work.

9. $(r^3)^{-4} = r^{3-4}$
10. $(2n)^3 = 2n^3$
11. $5y^3 \cdot 4y^4 = 20y^7$

In 12–15, name the property or properties being used.

12. $\left(\frac{2}{q}\right)^{10} = \frac{2^{10}}{q^{10}}$
13. $(x+4)(x+4)^4 = (x+4)^5$
14. $(2x^2y)^3 = 2^3 x^6 y^3$
15. $\left(\frac{2}{41}\right)^{-3} = \left(\frac{1}{\frac{2}{41}}\right)^3$

In 16–18, write the expression without negative exponents.

16. $\left(\frac{3}{5}\right)^{-1}$
17. $\left(\frac{3x^2}{y^3}\right)^{-2}$
18. $\left(\frac{2m^{-2}}{12m}\right)^{-40}$

In 19 and 20, refer to Example 3 and use the fact that the mean radius of Mars is about $2.11 \cdot 10^3$ miles.

19. About how many times could Mars fit into Earth?

20. About how many times could Mars fit into Jupiter?

APPLYING THE MATHEMATICS

21. Describe two different ways to simplify $\left(\frac{x^6}{x^3}\right)^{-2}$.

22. Consider the pattern $\frac{2}{x} - \frac{1}{y} = \frac{2y-x}{xy}$.
    a. Is the pattern true when $x = 3$ and $y = 5$?
    b. Test the special case when $x = y$.
    c. Test another case. Let $x = 6$ and $y = 2$.
    d. Do you think the pattern is *always*, *sometimes but not always*, or *never true* for all nonzero x and y?

In 23 and 24, use the fact that the prime factorization of 24 is $2^3 \cdot 3$ and the prime factorization of 360 is $2^3 \cdot 3^2 \cdot 5$.

23. Give the prime factorization of 360^{25}.

24. Give the prime factorization of $\left(\frac{360}{24}\right)^8$.

Remembering Properties of Powers and Roots

REVIEW

25. Find the distance between the points (–3, –9) and (12, 56). Round to the nearest tenth, if necessary. (**Lesson 8-8**)

26. Let $y = \sqrt{3}$ and $z = \sqrt{5}$. Evaluate $\frac{y}{z} \cdot (2yz)^3$. (**Lesson 8-7**)

27. An isosceles right triangle is a right triangle in which both legs have the same length. Lenora says that an isosceles right triangle with legs of length s always has a hypotenuse of length $s \cdot \sqrt{2}$. Is Lenora correct? Why or why not? (**Lesson 8-6**)

28. Three days a week, Rollo rollerblades to work. He rollerblades eight and a half blocks. On each block, there are fourteen houses. Each house has three trees in its front yard. How many trees does Rollo rollerblade by on his way to work each week? (**Lesson 8-1**)

29. A line has an x-intercept of 12 and a y-intercept of 15. Write an equation of this line. (**Lesson 6-6**)

30. On a regular die, which is more likely: rolling an even number six times in a row, or rolling a number less than 3 four times in a row? (**Lesson 5-7**)

EXPLORATION

31. Six of the eight properties mentioned at the beginning of this lesson involve either multiplication or division. Show that *none* of these six properties is true if every multiplication is replaced by addition and every division is replaced by subtraction.

The first known roller skates were created in the 1760s and possessed a single line of wheels.

Source: National Museum of Roller Skating

Chapter 8 Projects

1. Estimating Square Roots

To estimate a square root without a calculator, you can use an algorithm called the *Babylonian method*. Suppose you want to estimate $\sqrt{r}$.

Step 1 Guess a number a_1 that is reasonably close to $\sqrt{r}$.

Step 2 Replace a_1 with $a_2 = \frac{1}{2}\left(a_1 + \frac{r}{a_1}\right)$.

Step 3 Replace a_2 by $a_3 = \frac{1}{2}\left(a_2 + \frac{r}{a_2}\right)$.

Doing this process again and again will make each a_n closer and closer to the square root. For example, suppose you wanted to estimate $\sqrt{5}$. A good starting guess for a_1, is 2. In Step 2, replace a_1 with $a_2 = \frac{1}{2}\left(2 + \frac{5}{2}\right) = 2.25$. In Step 3, replace 2.25 with $a_3 = \frac{1}{2}\left(2.25 + \frac{5}{2.25}\right) = 2.23611111\ldots$. Step 3 should already give you a very good approximation of $\sqrt{5}$.

a. Use the Babylonian method to estimate $\sqrt{10}$, $\sqrt{21}$, and $\sqrt{30.235}$. Then find these numbers using your calculator. In each step of the Babylonian method, find the absolute value of the difference between the number a_n and the square root. What do you conclude?

b. Calculate $\sqrt{5}$ again, but this time, in Step 1, choose the inappropriate guess $a_1 = 1{,}000{,}000$. Does the method still work?

c. Use the method to calculate $\sqrt{9}$, using the guess $a_1 = 3$. What do you notice?

2. Fraction Exponents

In this chapter you saw both $\frac{1}{2}$ and $-\frac{1}{2}$ used as exponents. In this project you will discover the meaning of exponents that are other fractions.

a. Recall the rule $b^m \cdot b^n = b^{m+n}$. According to this rule, what should the number $b^{\frac{1}{3}} \cdot b^{\frac{1}{3}} \cdot b^{\frac{1}{3}}$ equal? Check this with a calculator.

b. Remember the rule $b^{m \cdot n} = (b^m)^n$. Because $\frac{2}{3} = \frac{1}{3} \cdot 2$, what should $b^{\frac{2}{3}}$ be? Check your answer with a calculator.

c. What should the number $b^{\frac{m}{n}}$ be?

3. Interview with Pythagoras

Who was Pythagoras? Use a library or the Internet to find out about his life, philosophy, mathematics, and the people who followed him. Write an interview with him, as if it were to be printed in a local newspaper. Include responses you think Pythagoras might have given. You may want to perform your interview (with a partner) for your class.

Chapter 8

4 Moving in the Coordinate Plane

In a computer game, a spaceship is in the middle of the screen, which has coordinate (0, 0). The player can give it one of four instructions—up, down, left, or right—that cause the ship to move one unit in the appropriate direction. For example, the right command will cause the spaceship to move to (1, 0).

a. Suppose the player gave the spaceship 5 commands in a row (for example: up, down, right, right, up). How many different possibilities for this are there?

b. Draw all the possible points in the plane that the spaceship can be in at the end of the five commands. How many different points are there?

c. Complete Parts a and b for two commands, three commands, and four commands. What patterns do you notice? Check these patterns for six commands.

d. What is the minimal number of commands necessary for the ship to reach the point (m, n)?

5 Mathematics and Crossword Puzzles

When stumped on a definition in a crossword puzzle for which they are missing a few letters, many people resort to guessing.

a. Suppose you know three letters out of a five letter word. How many possible guesses are there? How many possible guesses are there if you know that the first missing letter is a vowel?

b. Find a dictionary on the Internet that is intended for use with crossword puzzles. These dictionaries often sort words by length, as well as alphabetically. For n from 2 to 10, what is the number of possible words of length n that can be spelled with the letters of the English alphabet? For each length, how many of these possibilities are actually words? What is the relative frequency of real words, for each word length?

6 Counting Braille Letters

Braille is a system of writing that can be read by touch. Each letter consists of six dots, some of which are raised and some are not. For example, the letter J is written ⠚ (the black dots represent raised dots), and the letter M is written ⠍.

a. Look up a table with a list of all Braille symbols and letters.

b. Why do you think Braille has six dots? Would four dots have been enough? Would five?

c. Modern versions of Braille have eight dots. How many different possible letters and symbols can be written with eight dots?

Chapter 8 Summary and Vocabulary

- In Chapter 7, you saw many applications of the operation of powering to situations of exponential growth and decay. Another important application of powering is to the counting of objects: If there are n ways to select each object in a sequence of length L, then n^L different sequences are possible. **Powers** are also convenient for writing very large and very small numbers and are essential in **scientific notation.**

- The operation of taking a number to a power has many properties that connect it with multiplication and division. If two powers of the same number are multiplied or divided, the result is another power of that number: $x^m \cdot x^n = x^{m+n}$ and $\frac{x^m}{x^n} = x^{m-n}$. From these results, we can verify again that when $x \neq 0$, $x^0 = 1$. We can also deduce that $x^{-n} = \frac{1}{x^n}$.

- The terms **square root** and **cube root** come from the historical origins of these ideas in geometry. If a square has side s, its area is s^2, "s squared." If its area is A, the length of its side is $\sqrt{A}$, or $A^{\frac{1}{2}}$. If a cube has edge e, then its volume is e^3 or, "e cubed." If a cube has volume V, then each of its edges has length $\sqrt[3]{V}$.

- The **nth power** of a product xy is the product of the nth powers: $(xy)^n = x^n \cdot y^n$. Closely related to squares are square roots, and so we have $\sqrt{xy} = \sqrt{x} \cdot \sqrt{y}$. Similarly, the nth power of a quotient is the quotient of the nth powers: $\left(\frac{x}{y}\right)^n = \frac{x^n}{y^n}$ and so $\sqrt{\frac{x}{y}} = \frac{\sqrt{x}}{\sqrt{y}}$. These properties help to simplify and rewrite expressions involving radicals.

- Two important applications of squares and square roots are the **Pythagorean Theorem** and the **distance formula** between two points in a coordinate plane, which is derived from that theorem.

Vocabulary

8-1
scientific notation

8-6
square
squared
square root
radical sign ($\sqrt{}$)
cube
cubed
cube root

8-7
radicand

Theorems and Properties

Multiplication Counting Principle (p. 458)
Arrangements Theorem (p. 459)
Product of Powers Property (p. 465)
Power of a Power Property (p. 466)
Quotient of Powers Property (p. 469)
Negative Exponent Property (p. 474)

Negative Exponent Property for Fractions (p. 475)
Power of a Product Property (p. 481)
Power of a Quotient Property (p. 482)
Square of the Square Root Property (p. 490)

Pythagorean Theorem (p. 492)
Cube of the Cube Root Property (p. 493)
Product of Square Roots Property (p. 498)
Quotient of Square Roots Property (p. 499)

Chapter 8 Self-Test

Take this test as you would take a test in class. You will need a calculator. Then use the Selected Answers section in the back of the book to check your work.

1. **Multiple Choice** $x^4 \cdot x^7 =$

   A x^{11} B x^{28} C $2x^{11}$ D $2x^{28}$

2. Rewrite 5^{-3} as a simple fraction.

3. Order from least to greatest: $(-4)(-3)$, $(-3)^4$, $(-4)^{-3}$

In 4–6, simplify the expression.

4. $\sqrt{600}$ 5. $\sqrt{25x}$ 6. $2^{\frac{1}{2}} \cdot 50^{\frac{1}{2}}$

In 7–12, simplify the expression.

7. $y^4 \cdot y^2$ 8. $(10m^2)^3$ 9. $\dfrac{a^{15}}{a^3}$

10. $\left(\dfrac{m}{6}\right)^3$ 11. $g^4 \cdot g \cdot g^0$ 12. $\dfrac{6n^2}{4n^3 \cdot 2n}$

13. Rewrite $4y^{-3}w^2$ without a negative exponent and justify your answer.

14. Rewrite $\dfrac{2}{x^2} \cdot \dfrac{5}{x^5}$ as a single fraction without negative exponents. Justify your steps.

15. The prime factorization of 288 is $2^5 \cdot 3^2$. Use this information to find the prime factorization of $10(288)^2$.

16. Rewrite $\left(\dfrac{3}{y^2}\right)^{-3}$ without parentheses or negative exponents.

17. Evaluate $\sqrt[3]{30}$ to the nearest thousandth.

18. If $f(x) = 1{,}000(1.06)^x$, estimate $f(-3)$ to the nearest integer.

19. State the general property that justifies each step.

    a. $\left(\dfrac{x^{-9}}{x^{-5}}\right)^3 = \dfrac{(x^{-9})^3}{(x^{-5})^3}$ b. $= \dfrac{x^{-27}}{x^{-15}}$

    c. $= x^{-12}$ d. $= \dfrac{1}{x^{12}}$

20. Find a value of m for which $m^2 \neq m^{-2}$.

21. What is the distance between the points $(9, 5)$ and $(1, -10)$?

22. Common notebook paper outside the United States is called A4, with dimensions 210 mm by 297 mm. A piece of A4 paper is placed on a grid as shown below. If the paper is cut along a diagonal, how long is the cut line?

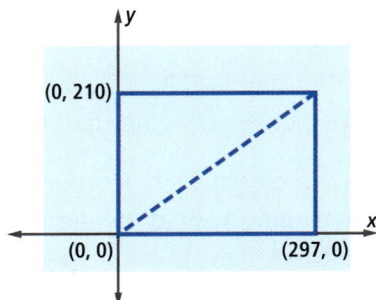

23. The volume of a cube is 30 cubic inches. What is the length of an edge of the cube?

24. A square has a diagonal with length 12 meters. What is the area of the square?

25. In some states, a license plate consists of 2 letters followed by 4 digits. How many different license plates are possible that fit this description? (Note: There are ten digits: 0, 1, 2, . . ., 9.)

Chapter 8 Chapter Review

SKILLS
PROPERTIES
USES
REPRESENTATIONS

SKILLS Procedures used to get answers

OBJECTIVE A Simplify products, quotients, and powers of powers. (Lessons 8-2, 8-3, 8-4, 8-5)

In 1–7, simplify the expression.

1. $2m^2 \cdot 3m^3 \cdot 4m^4$
2. $\dfrac{6y^8}{12y^4}$
3. $\dfrac{7xy^2}{3x^2y^{-2}}$
4. $b^{-2}(4b^5)$
5. $\left(\dfrac{v^4}{v^8}\right)^6$
6. $\dfrac{7.28 \cdot 10^{115}}{8.3 \cdot 10^{72}}$
7. $a^{10}(3a^3) + 5a^{13}(a^7 - 2)$

In 8 and 9, rewrite the expression a. without fractions and b. without negative exponents.

8. $\dfrac{a^{-1}b^4}{b^{-2}c^3}$
9. $\dfrac{60x^2}{45x^{-2}}$

10. Rewrite $-ab^{-1}$ without a negative exponent.

11. Rewrite $(7x^{-7})(6y^{-3})$ without a negative exponent.

OBJECTIVE B Evaluate negative integer powers of real numbers. (Lessons 8-4, 8-5)

In 12–15, rewrite the expression as a decimal or fraction without an exponent.

12. 4^{-3}
13. 6^{-2}
14. $\left(\dfrac{1}{10}\right)^{-2}$
15. $\left(\dfrac{2}{3}\right)^{-5}$

16. If $f(x) = 2x^{-4}$, what is $f(-3)$?

17. The value of a house today is estimated at $150,000 and has been growing at 3% a year. Its value x years from now will be $150,000(1.03)^x$. What was its value 5 years ago? Round your answer to the nearest thousand dollars.

18. Write $603.8 \cdot 10^{-4}$ in decimal form.

19. If $0.0051 = 5.1 \cdot 10^n$, what is n?

In 20 and 21, tell whether each expression names a positive number, a negative number, or zero.

20. a. $5^3 + 3^{-5}$
    b. $3^5 - 3^{-5}$
21. a. $\left(-\dfrac{1}{2}\right)^{-3}$
    b. $-\left(\dfrac{1}{2}\right)^{-3}$

OBJECTIVE C Rewrite powers of products and quotients. (Lessons 8-5, 8-9)

In 22–29, rewrite the expression without parentheses.

22. $(xy)^5$
23. $(60m^2n^3)^2$
24. $30\left(\dfrac{1}{3}u^4v\right)^3$
25. $\left(\dfrac{3}{4}\right)^{11} \cdot \left(\dfrac{6}{8}\right)^{-4}$
26. $\dfrac{1}{2}\left(\dfrac{1}{v}\right)^3 - (2v)^{-3}$
27. $\left(\dfrac{-8s}{t^2}\right)^{-4}$
28. $(m^3n^{-2})(m^2n^{-3})^3$
29. $\left(\dfrac{2y^2z^{-3}}{6y^{-3}z^4}\right)^{-2}$

OBJECTIVE D Simplify square roots. (Lessons 8-6, 8-7)

In 30–36, simplify the expression. Assume the variables stand for positive numbers.

30. $\sqrt{6} \cdot \sqrt{24}$
31. $(4^3 + 4^3)^{\frac{1}{2}}$
32. $\sqrt{3^2 + 4^2}$
33. $5\sqrt{7} \cdot 2\sqrt{3}$
34. $\sqrt{17m} \cdot \sqrt{17m}$
35. $\sqrt{\dfrac{4x^2}{y^2}}$
36. $\sqrt{36a^{36}b^4}$

OBJECTIVE E Evaluate cube roots. (Lesson 8-6)

In 37–40, give the exact cube root of the number or the cube root rounded to the nearest thousandth.

37. -8
38. 1
39. 200
40. $1{,}330$

In 41 and 42, estimate the number to the nearest thousandth and check your answer by an appropriate multiplication.

41. $\sqrt[3]{0.05}$

42. $\sqrt[3]{10}$

PROPERTIES The principles behind the mathematics

OBJECTIVE F Test a special case to determine whether a pattern is true. (Lesson 8-9)

43. Tell whether the pattern $x^4 = x^3$ is true for the given value of x.
    a. $x = 1$ b. $x = 0$ c. $x = -1$
    d. Based on your answers to Parts a–c, do you have evidence that the pattern is true, or are you sure it is not always true? Explain your reasoning.

44. Consider the pattern $(xy)^{-2} = \frac{1}{x^2y^2}$, where x and y are not zero.
    a. Is the pattern true when $x = 5$ and $y = 3$?
    b. Is the pattern true when $x = -4$ and $y = 0.5$?
    c. Do you have evidence that the pattern is true for all nonzero real number values of x and y? Explain your reasoning.

In 45 and 46, find a counterexample to the pattern.

45. $-4a = a^{-4}$

46. $(x + y)^2 = x^2 + y^2$

OBJECTIVE G Identify properties of powers that justify a simplification, from the following list: Zero Exponent Property (Chapter 7); Negative Exponent Property; Power of a Product Property; Power of a Quotient Property; Product of Powers Property; Quotient of Powers Property; Power of a Power Property. (Lessons 8-2, 8-3, 8-4, 8-5)

In 47–52, identify the property or properties that justify the simplification. Assume all variables represent positive numbers.

47. $\left(\frac{1}{2}\right)^3 = \frac{1}{8}$

48. $a^{10} = a^8 \cdot a^2$

49. $(b^4)^0 = 1$

50. $2^{-n} = \frac{1}{2^n}$

51. $4^{-2} \cdot 4^3 = 4$

52. $\frac{w^2 a^2}{w^2 h^{-1}} = aah$

53. Show and justify two different ways to simplify $\left(\frac{12}{13}\right)^{-4}$.

54. Show and justify two different ways to simplify $\frac{m^{-1}}{n^{-1}}$.

USES Applications of mathematics in real-world situations

OBJECTIVE H Use powers to count the number of sequences possible for repeated choices. (Lesson 8-1)

55. A test contains 5 questions where the choices are *always*, *sometimes but not always*, or *never*.
    a. How many different answer sheets are possible?
    b. If you guess, what is the probability that you will answer all 5 questions correctly?
    c. If you guess, what is the probability that you will answer all 5 questions incorrectly?

56. Excluding the five vowels A, E, I, O, and U in the English language, how many 4-letter acronyms are possible?

57. A restaurant serves two different types of pizza (thin crust and deep dish), three sizes, and nine toppings. How many different pizzas with one topping are possible?

58. Imagine that a fair coin is tossed 12 times. The result of each toss is recorded as H or T.
    a. How many different sequences of H and T are possible?
    b. What is the probability that all the tosses are heads?
    c. What is the probability of getting the sequence HTHHTTTHTTTH?

REPRESENTATIONS Pictures, graphs, or objects that illustrate concepts

OBJECTIVE I Represent squares, cubes, square roots, and cube roots geometrically. (Lesson 8-8)

59. The area of the tilted square below is 8 square units. What is the length of a side of the tilted square?

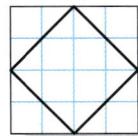

60. A square has area 1,000 square units.
    a. Give the exact length of a side.
    b. Estimate the length of a side to the nearest hundredth.

61. If a square has side xy, what is its area?

62. If a cube has an edge of length 2.3 cm, what is its volume?

63. A cube has a volume of 1 cubic meter. What is the length of an edge?

64. A cube has a volume of 2 cubic meters. What is the length of an edge?

OBJECTIVE J Calculate distances on the x-y coordinate plane. (Lesson 8-8)

In 65 and 66, use the graph below.

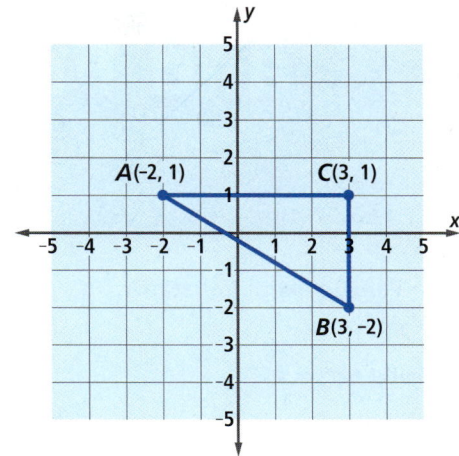

65. Find each length.
    a. AC b. BC c. AB

66. If O is the point $(0, 0)$, find OB and OC.

In 67–70, find the distance between the points.

67. $(2, 11)$ and $(11, 2)$ 68. $(6, 5)$ and $(3, -5)$

69. (a, b) and $(-1, 4)$ 70. (x, y) and (h, k)

71. Stanton, Nebraska is about 8 miles east and 3 miles south of the center of Norfolk, Nebraska.
    a. On a straight line distance, is it true that Stanton is less than 9 miles from the center of Norfolk?
    b. Draw a picture to justify your answer to Part a.

Chapter 9
Quadratic Equations and Functions

- 9-1 The Function with Equation $y = ax^2$
- 9-2 Solving $ax^2 = b$
- 9-3 Graphing $y = ax^2 + bx + c$
- 9-4 Quadratics and Projectiles
- 9-5 The Quadratic Formula
- 9-6 Analyzing Solutions to Quadratic Equations
- 9-7 More Applications of Quadratics: Why Quadratics Are Important

When an object is dropped from a high place, such as the roof of a building or an airplane, it does not fall at a constant speed. The longer it is in the air, the faster it falls. Furthermore, the distance d that a heavier-than-air object falls in a time t does not depend on its weight. In the early 1600s, the Italian scientist Galileo described the relationship between d and t mathematically. In our customary units of today, if d is measured in feet and t is in seconds, then $d = 16t^2$.

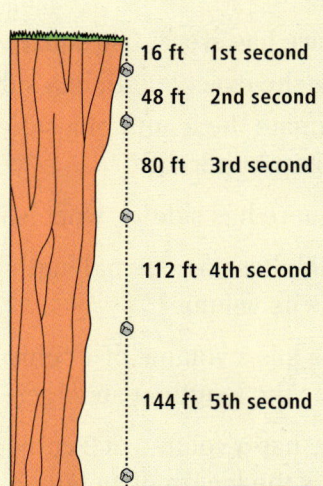

16 ft   1st second
48 ft   2nd second
80 ft   3rd second
112 ft 4th second
144 ft 5th second

A table of values and a graph of this equation are shown on the next page.

t (sec)	d (ft)
0	0
1	16
2	64
3	144
4	256
5	400

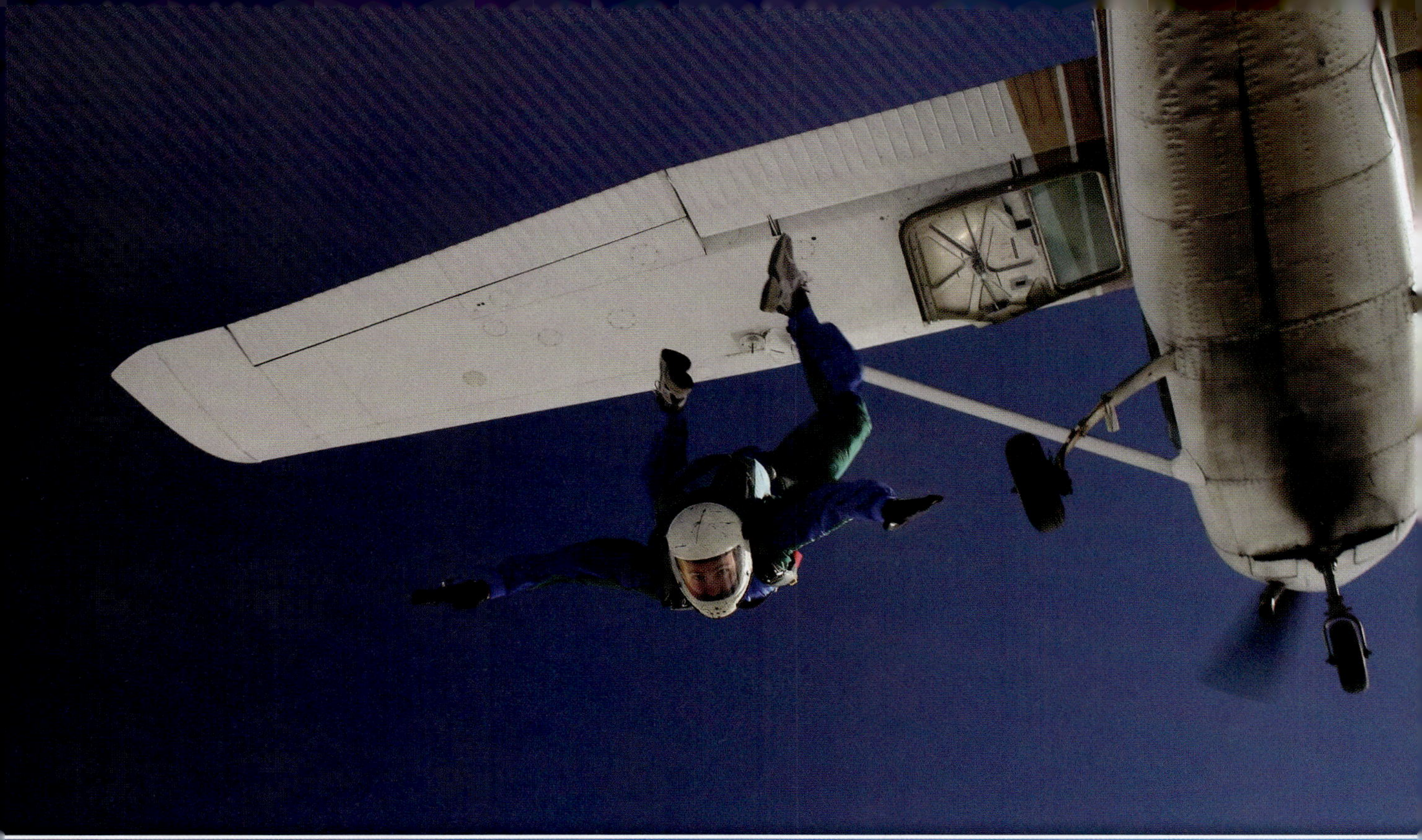

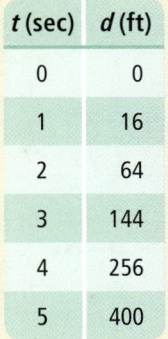

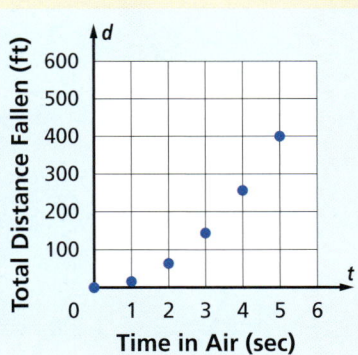

The expression $16t^2$ is a *quadratic expression*, the equation $d = 16t^2$ is an example of a *quadratic equation*, and the function whose independent variable is t and dependent variable is d is a *quadratic function*. The word "quadratic" comes from the Latin word *quadratum* for square. Think of the area x^2 of a square with side x.

From the time of the Ancient Greek mathematicians until about 1600, the only known physical applications of quadratic expressions were to the area of squares and other geometric figures. But, in the next hundred years, discoveries by Galileo, Kepler, Newton, Leibniz, and others found uses for quadratic expressions involving objects that were in motion. These discoveries explain everything from the path of a basketball shot to the orbits of planets around our sun. They enable us to talk to each other via cell phones and collect information about stars in distant space. They are important both for mathematics and for science. In this chapter you will learn about a wide variety of quadratic equations and functions, and their applications.

Lesson 9-1
The Function with Equation $y = ax^2$

Vocabulary

parabola
reflection-symmetric
axis of symmetry
vertex

▶ **BIG IDEA** The graph of any quadratic function with equation $y = ax^2$, with $a \neq 0$, is a parabola with vertex at the origin.

Graphing $y = x^2$

The simplest quadratic function has equation $y = x^2$. A table of values for $y = x^2$ is given below. Notice the symmetry in the second row of the table. Each x value and its opposite have the same square. For example, 3^2 and $(-3)^2$ are both equal to 9. The bottom row of the table shows that the output of the function is positive for a pair of opposite positive and negative input values.

Mental Math

If (a, b) is in the 2nd quadrant, in which quadrant is:

a. $(-a, -b)$?
b. $(-a, b)$?
c. $(a, -b)$?

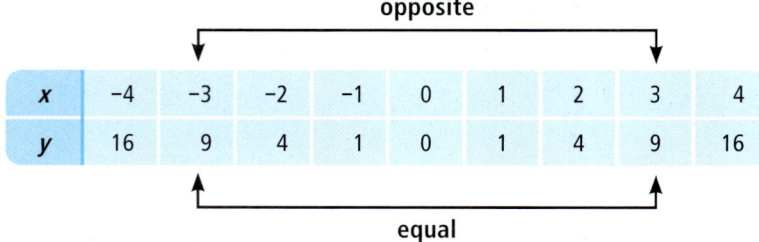

This symmetry can be seen in the graph of the equation $y = x^2$ at the right, which is a **parabola**. Every positive number is the y-coordinate of two points on the graph with opposite x-coordinates. For example, 25 is the y-coordinate of the points $(5, 25)$ and $(-5, 25)$. For this reason, the parabola is its own *reflection image* over the y-axis. For this reason we say the parabola is **reflection-symmetric** to the y-axis. The y-axis is called the **axis of symmetry** of the parabola.

The intersection point of a parabola with its axis of symmetry is called the **vertex** of the parabola. The vertex of the graph of $y = x^2$ is $(0, 0)$.

The function $y = x^2$ is of the form $y = ax^2$, with $a = 1$. You should be able to sketch the graph of any equation of this form.

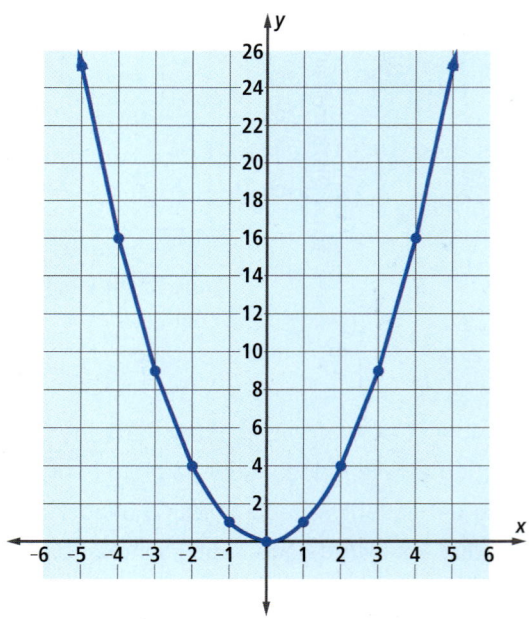

526 Quadratic Equations and Functions

Graphing $y = ax^2$

All equations of the form $y = ax^2$ have similar graphs.

> ### Activity
>
> **Step 1** Use the window $-20 \leq x \leq 20$, and $-20 \leq y \leq 20$ to graph all three equations on your calculator. Sketch the graphs on a single grid on a separate sheet of paper.
>
>   **a.** $f(x) = 3x^2$ **b.** $g(x) = -x^2$ **c.** $h(x) = -3x^2$
>
> **Step 2** Evaluate $f(2)$, $g(2)$, and $h(2)$.
>
> **Step 3** Is $(-2, -12)$ a point on the graph of $h(x)$? Explain how you know.
>
> **Step 4** Use a graphing calculator to make a sketch of the following functions. Use the same window you used for Step 1.
>
>   **a.** $j(x) = 0.2x^2$ **b.** $k(x) = -0.5x^2$
>
> **Step 5** Evaluate $j(3)$ and $k(3)$.
>
> **Step 6** Is $(-3, 2.9)$ a point on the graph of $j(x)$? Explain how you know.
>
> **Step 7** If a parabola is opening up, what must be true about the value of a in $y = ax^2$? If a parabola is opening down, what must be true about a in $y = ax^2$?

Properties of the Graph of $y = ax^2$

The graph of $y = ax^2$, where $a \neq 0$, has the following properties:

1. It is a parabola symmetric to the *y*-axis.
2. Its vertex is $(0, 0)$.
3. If $a > 0$, the parabola opens up. If $a < 0$, the parabola opens down.

Finding Points on the Graph of $y = ax^2$

If you know the *y*-coordinate of a point on the graph of a parabola and the equation, you can find the *x*-coordinate or coordinates. We illustrate this with a different parabola.

> ### GUIDED
>
> **Example 1**
>
> Consider the following situation: You know the area of a circle and want to find its radius.
>
> *(continued on next page)*

Chapter 9

Solution Sketch a graph of the familiar formula $A = \pi r^2$, where A is the area of a circle with radius r.

Step 1 Make a scale on each axis. In doing this, ask yourself: What are the possible values of r? What are the possible values of A?

Step 2 Make a table of values for A when $r = 1, 2, 3,$ and 4. Estimate each to the nearest hundredth. The first value has been done for you.

Step 3 Graph the points (r, A) from the table.

Step 4 Put an open circle at $(0, 0)$ because 0 is not in the domain of r. Connect $(0, 0)$ and the other points with a curve like a parabola.

Step 5 Use your graph to estimate the radius of a circle whose area is 12 square units.

r	A
1	3.14
2	?
3	?
4	?

Example 2

The graph of $f(x) = 1.5x^2$ is shown at the right. Estimate x if $f(x) = 10$.

Solution Draw the horizontal line $y = 10$. The graph intersects this line at two points. The x-coordinates of these points are approximately 2.5 and -2.5.

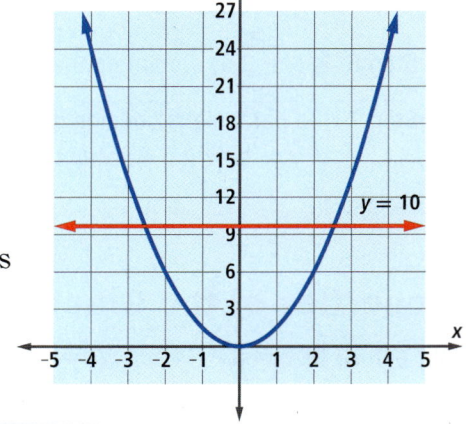

In the next lesson, you will see how to obtain the exact values of x with $f(x) = 10$.

Questions

COVERING THE IDEAS

In 1 and 2, an equation of a function is given.
  a. Make a table of x and y for integer values of x from -4 to 4.
  b. Graph the equation.
  c. Tell whether the graph opens up or down.

1. $g(x) = \frac{1}{2}x^2$
2. $f(x) = -\frac{1}{2}x^2$

3. Refer to the parabola at the right.
   a. Does the parabola open up or down?
   b. The parabola is the graph of a function. Which does the function have, a maximum value or a minimum value?
   c. Give the coordinates of the vertex.
   d. Give an equation of the axis of symmetry of the parabola.

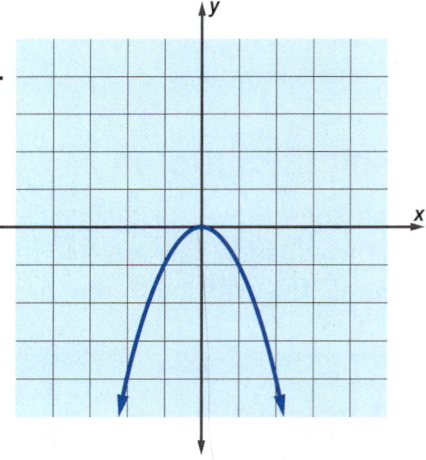

4. How are the graphs of $y = 7x^2$ and $y = -7x^2$ related to each other?

528 Quadratic Equations and Functions

5. Match each table with the graph it most accurately represents.

a.
x	y
-2	4
0	0
2	4

b.
x	y
-2	-4
0	0
2	-4

c.
x	y
-2	-10
0	0
2	-10

i.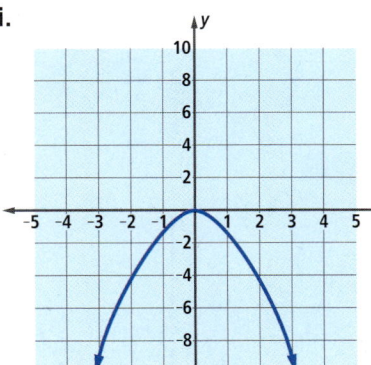

ii.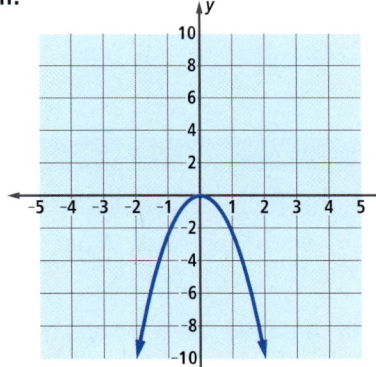

iii.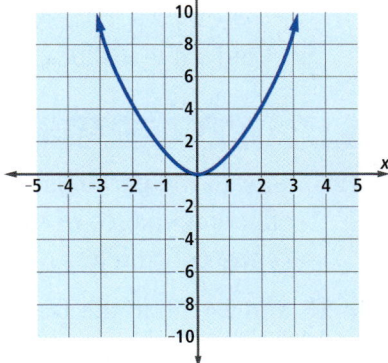

6. Match each graph at the right with one of the equations below.
   a. $y = x^2$
   b. $y = -0.25x^2$
   c. $y = -3x^2$

7. **Fill in the Blanks** Consider the graph of the function $f(x) = ax^2$.
   a. If a is positive, the graph is a parabola that opens ___?___.
   b. If a is negative, the graph is a parabola that opens ___?___.

8. Use the graph of $A = \pi r^2$ to estimate the radius of a circle whose area is 20 square units.

9. Consider the graph of the function defined by $y = 5x^2$.
   a. Without plotting any points, sketch what you think the graph of this function looks like.
   b. Make a table of values satisfying this function. Use $x = -2, -1.5, -1, -0.5, 0, 0.5, 1, 1.5,$ and 2.
   c. Draw a graph of this function from your table.
   d. From the graph, estimate the values of x for which $y = 14$.

The Function with Equation $y = ax^2$ 529

Chapter 9

10. Consider the formula $A = s^2$ for the area A of a square with a side of length s.
    a. Graph all possible values of s and A on a coordinate plane.
    b. Explain how the graph in Part a is like and unlike the graph of $y = x^2$ at the start of this lesson.

11. The parabola at the right has equation $y = -5x^2$.
    a. Find y if $x = 0$.
    b. Find x if $y = -5$.
    c. Find x if $y = -20$.

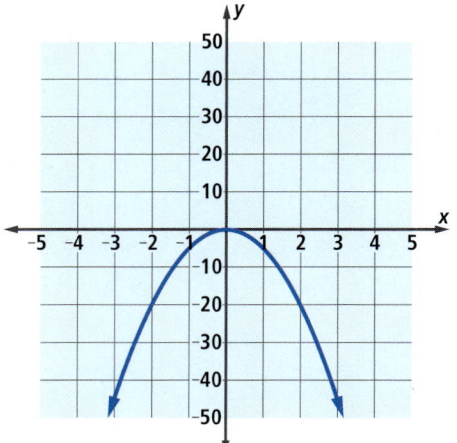

APPLYING THE MATHEMATICS

12. Refer to the parabola at the right. Points P and Q are reflection images of each other over the y-axis. What are the coordinates of Q?

In 13 and 14, fill in the blanks with *negative*, *zero*, or *positive*.

13. **Fill in the Blanks** Consider the expression $-1x^2$.
    a. If x is negative, $-1x^2$ is ___?___.
    b. If x is zero, $-1x^2$ is ___?___.
    c. If x is positive, $-1x^2$ is ___?___.
    d. What do Parts a–c tell you about the graph of $y = -1x^2$?

14. **Fill in the Blanks** Consider the expression $4x^2$.
    a. If x is negative, $4x^2$ is ___?___.
    b. If x is zero, $4x^2$ is ___?___.
    c. If x is positive, $4x^2$ is ___?___.
    d. What do Parts a–c tell you about the graph of $y = 4x^2$?

15. What is the only real number whose square is not a positive number?

In 16 and 17, a graph of a function f with $f(x) = ax^2$ is shown. Find the value of a.

 16.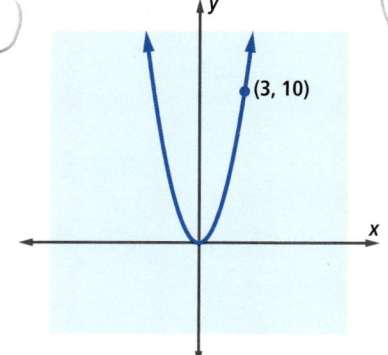

 17.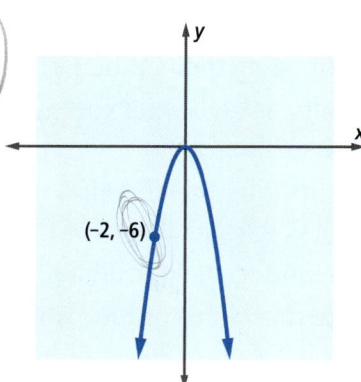

18. **Fill in the Blank** If $a = 0$, the graph of the function $y = ax^2$ is ___?___.

19. Consider the equation $d = 16t^2$, which gives the distance d in feet that an object dropped at time $t = 0$ will have fallen after t seconds.

    a. If an object falls 400 feet in t seconds, find t.

    b. Estimate how long it will take an object to fall 200 feet.

REVIEW

20. **Skill Sequence** Simplify each expression. (**Lesson 8-7**)

    a. $\sqrt{3} \cdot \sqrt{27}$ b. $\sqrt{3x} \cdot \sqrt{27x}$ c. $\sqrt{3x} + \sqrt{27x}$

21. Suppose a box has sides of length 6 inches, 8 inches, and 12 inches. Find the length of the longest thin pole, like the one shown at the right, which can fit inside the box. (*Hint:* First find the diagonal of the base.) (**Lesson 8-6**)

    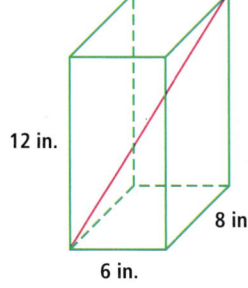

22. Suppose that t years ago, Kendra deposited P dollars into a savings account with an annual yield of 3%. If she has not deposited or withdrawn any additional money and the account now contains $500, write an equation involving t and P. (**Lesson 7-1**)

23. Derek is a tennis instructor at his local gym. He gives lessons to 3 people twice a week. If he charges $25 per person for a lesson, and he works for 15 straight weeks, how much money will Derek earn? (**Lesson 5-4**)

The United States Tennis Association is the largest tennis organization in the world, with more than 665,000 individual members and 7,000 organizational members.

Source: USTA

EXPLORATION

24. Draw a set of axes on graph paper. Aim a lit flashlight at the origin up the *y*-axis. What is the shape of the lit region? Keep the lit end of the flashlight over the origin but tilt the flashlight to raise its bottom. How does the shape of the lit region change?

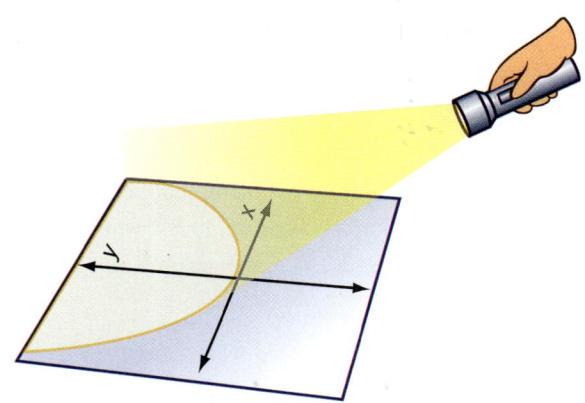

The Function with Equation $y = ax^2$

Chapter 9

Lesson 9-2
Solving $ax^2 = b$

▶ **BIG IDEA** When $\frac{b}{a}$ is not zero, the equation $ax^2 = b$ has two solutions, $x = \sqrt{\frac{b}{a}}$ and $x = -\sqrt{\frac{b}{a}}$.

Graphs and tables can be very helpful in seeing the behavior of equations of the form $ax^2 = b$.

Mental Math

Evaluate.
a. 5^0
b. $4^3 - 3^3$
c. $10^2 - 5^3 + 2^4$

Example 1
Solve $2x^2 = 32$.

Solution 1 Create and graph two functions from the equation. One is the parabola $y = 2x^2$. The other is the horizontal line $y = 32$. Notice that the graphs intersect at two points. This is because two different values of x make the equation true. Trace on the graph of the parabola to find the intersection points, $(-4, 32)$ and $(4, 32)$. Use the x-coordinates.

The solutions are $x = -4$ and $x = 4$.

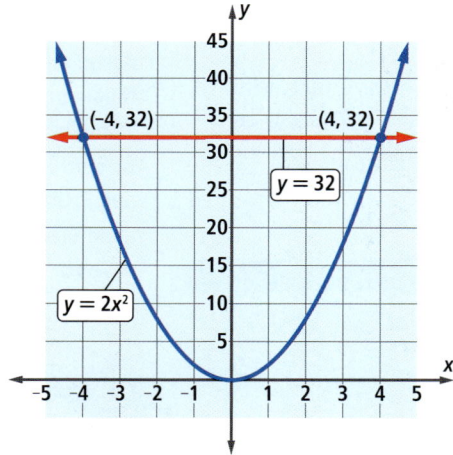

Solution 2 Create a table of values for the equation. Notice that there are two places where the expression $2x^2$ is equal to 32, when $x = -4$ and when $x = 4$. The solutions are $x = -4$ and $x = 4$.

Check

Does $2(4)^2 = 32$?   Does $2(-4)^2 = 32$?
$\quad 2(16) = 32$ $\quad\quad\quad\quad 2(16) = 32$
$\quad\quad 32 = 32$ $\quad\quad\quad\quad\quad 32 = 32$ Both -4 and 4 check.

x	$2x^2$
−5	50
−4	32
−3	18
−2	8
−1	2
0	0
1	2
2	8
3	18
4	32
5	50

532 Quadratic Equations and Functions

In Example 1, the graph and table helped to show that an equation like $2x^2 = 32$ has two solutions.

You can solve equations of the form $x^2 = b$ symbolically by recalling the meaning of *square root*. If $x^2 = b$, then $x = \sqrt{b}$ or $x = -\sqrt{b}$. Notice that it takes only one step to solve an equation of the form $x^2 = b$. With just one additional step, you can solve an equation of the form $ax^2 = b$.

Example 2

A quarter is dropped 60 feet from the roof of a school building. To determine how long the quarter will be in the air, use Galileo's equation $d = 16t^2$. In the equation, t is the time, in seconds, that it takes a heavier-than-air object to fall d feet.

Solution Here $d = 60$, so we need to solve $60 = 16t^2$.

$\frac{60}{16} = t^2$ Divide both sides by 16.

$t = \pm\sqrt{\frac{60}{16}}$ Take the square roots of both sides.

$t \approx \pm 1.936$ Approximate the square root.

$t \approx 1.936$ Only the positive solution makes sense in this situation.

The quarter will be in the air for approximately 1.9 seconds.

 QY

▶ **QY**

Find how long it will take an object to hit the ground if it falls from the top of Chicago's John Hancock Center, which is 1,127 feet tall.

You can combine your knowledge of solving linear equations with what was done in Example 2 to solve some equations that look quite complicated. In the next example, you should think of $2n + 11$ as a single number. Psychologists call this idea *chunking*. Chunking is what you do when you read an entire word without thinking of the individual letters.

Example 3

Solve $3(2n - 11)^2 = 75$.

Solution Think of $2n - 11$ as a single number, say x. Then this equation is $3x^2 = 75$. Some people like to write the x in place of $2n - 11$, but you do not have to do that.

$3(2n - 11)^2 = 75$

Divide both sides by 3.

$(2n - 11)^2 = 25$

(continued on next page)

Take the square roots of both sides of the equation.

$(2n - 11) = \pm\sqrt{25}$
$2n - 11 = \pm 5$

Thus either $2n - 11 = 5$ or $2n - 11 = -5$.

Now there are two linear equations to be solved. It is good to separate the two processes.

$2n - 11 = 5$	or	$2n - 11 = -5$
$2n = 16$		$2n = 6$
$n = 8$		$n = 3$

So there are two solutions, $n = 8$ or $n = 3$.

Check Substitute 8 for n in the original equation.

Does $3(2 \cdot 8 - 11)^2 = 75$? Yes, because $3(5)^2 = 75$.

Substitute 3 for n in the original equation.

Does $3(2 \cdot 3 - 11)^2 = 75$? Yes, because $3 \cdot (-5)^2 = 75$.

Questions

COVERING THE IDEAS

1. Solve $-4x^2 = -100$ using the graph at the right.
2. Solve $5x^2 + 7 = 7.8$ using the table below.

x	−0.6	−0.4	−0.2	0	0.2	0.4	0.6
$5x^2 + 7$	8.8	7.8	7.2	7	7.2	7.8	8.8

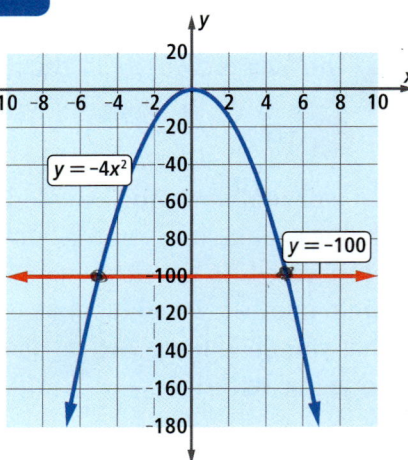

In 3–8, solve the equation.

3. $x^2 = 40$
4. $3,705 = y^2 + 436$
5. $12v^2 - 24 = 36$
6. $5w^2 = 400$
7. $3(a + 5)^2 = 12$
8. $(7v - 2)^2 = 81$

9. a. You drop a stone into a deep well and carefully time how long it takes until you hear the stone plop into the water. If it takes 2.4 seconds, about how far from the top is the water in the well?
   b. If the water was 64 feet from the top of the well, how long would it take until the stone hit the water?

10. A 30-foot ladder is placed against a wall so that its bottom is 9 feet away from the wall. How high up the wall is the top of the ladder?

APPLYING THE MATHEMATICS

11. If the area of a circle is 100 square units, then what is the radius of the circle, to the nearest hundredth of a unit?

In 12 and 13, solve the quadratic equation.

12. $\frac{1}{2}(2z + 3)^2 = 18$

13. $3v^2 + 10 = 7v^2 - 15$

14. Consider the figures drawn at the right. Suppose that the circle has the same area as the square. If the diameter of the circle is 4 feet, what is the length of the side of the square, to the nearest inch? (*Hint:* You will need to use the formulas for the area of a square and the area of a circle.)

15. You wish to make a bull's-eye target so that the area of the inner circle equals the area of the outer ring between the two circles. If the radius of the outer circle is 12 inches, what should the radius of the inner circle be?

REVIEW

16. **True or False** 0 is in the range of $y = ax^2$ for all values of a. (**Lesson 9-1**)

In 17–19, simplify the expression. (**Lessons 8-5, 8-3, 8-2**)

17. $n \cdot m^3 \cdot n^5 \cdot m^2$ 18. $\frac{12x^8}{8x^2}$ 19. $\left(\frac{3}{5a}\right)^2$

20. A company that manufactures combination locks wants each lock to have a unique 3-number combination. There are 36 numbers on each lock. (**Lesson 8-1**)

    a. How many locks can the company produce without having to use the same combination twice?

    b. Suppose a worker at the company forgets the combination of one of the locks, but he knows that it does not begin with 1. How many different combinations might he have to try to open the lock?

21. Let $f(x) = 2x^2$. (**Lessons 7-6, 7-5**)

    a. What is the domain of f?   b. What is the range of f?

22. a. Determine a real situation that can be answered by solving $\frac{18}{24} = \frac{x}{32}$.

    b. Answer your question from Part a. (**Lesson 5-9**)

Chapter 9

23. Tickets for a school play cost $6 for adults and $4 for children. The organizers of the play have determined that they must sell at least $975 worth of tickets in order to cover their expenses. (**Lesson 6-9**)

    a. Write an inequality that represents this situation.
    b. Graph the inequality.
    c. Suppose 86 adult tickets are sold. If exactly $1,000 were raised from ticket sales, how many children tickets must have been sold?

24. Jessica is driving between Cleveland and Chicago, a distance of about 350 miles. From Cleveland to Chicago, she averages 65 miles per hour. On the return trip, she averages x miles per hour. If the round trip takes Jessica 12 hours, find her average speed x on the way back. (**Lessons 5-3, 3-4**)

Since 1929, more than two million students have been honored for excellence in theater arts with invitations to join the International Thespian Society.

Source: Educational Theatre Association

25. **Multiple Choice** The graph below pictures solutions to which inequality? (**Lessons 3-7, 3-6**)

   A $h - 3 \leq -8$ B $h + 7 > 2$ C $h - 5 \geq -10$ D $h + 5 < 0$

EXPLORATION

26. On a piece of graph paper, carefully draw a circle with center at (0, 0) and radius 10 units. This circle will contain the point (10, 0) and three other points on the axes.

    a. What are the coordinates of those other three points?
    b. The circle will contain 8 other points whose coordinates are both integers. Identify those 8 points.
    c. Find the coordinates of three other points in the 1st quadrant that are on the circle. (*Hint:* You may need to describe the coordinates with square roots.)

QY ANSWER

about 8.39 sec

Lesson 9-3

Graphing $y = ax^2 + bx + c$

> **BIG IDEA** The graphs of any quadratic function with equation $y = ax^2 + bx + c$, $a \neq 0$, is a parabola whose vertex can be found from the values of a, b, and c.

If $a \neq 0$, the graph of $y = ax^2$ is a parabola with vertex $(0, 0)$. This lesson is about the graph of a more general function, $y = ax^2 + bx + c$. We begin with an important everyday use.

How Far Does a Car Travel after Brakes Are Applied?

When a driver decides to stop a car, it takes time to react and press the brake. Then it takes time for the car to slow down. The total distance traveled in this time is called the *stopping distance* of the car. The faster the car is traveling, the greater the distance it takes to stop the car. A formula that relates the speed x (in miles per hour) of a car and its stopping distance d (in feet) is $d = 0.05x^2 + x$.

This function is used by those who study automobile performance and safety. It is also important for determining the distance that should be maintained between a car and the car in front of it.

To find the distance needed to stop a car traveling 40 miles per hour, you can substitute 40 for x in the above equation.

$d = 0.05(40)^2 + 40$
$ = 0.05(1{,}600) + 40$
$ = 80 + 40$
$ = 120$

Thus, a car traveling 40 mph takes about 120 feet to come to a complete stop after the driver decides to apply the brakes.

A table of values and a graph for the stopping distance formula is shown on the next page. The situation makes no sense for negative values of x or d, so the graph has points in the first quadrant only. The graph is part of a parabola.

Mental Math

Simplify.

a. $\sqrt{18}$

b. $\sqrt{200}$

c. $\dfrac{\sqrt{18}}{\sqrt{200}}$

Traffic lights were used before the advent of the motorcar. In 1868, a lantern with red and green signals was used at a London intersection to control the flow of horse buggies and pedestrians.

Source: www.ideafinder.com

Chapter 9

Speed x (mph)	Distance $d = 0.05x^2 + x$ (ft)
10	15
20	40
30	75
40	120
50	175
60	240
70	315

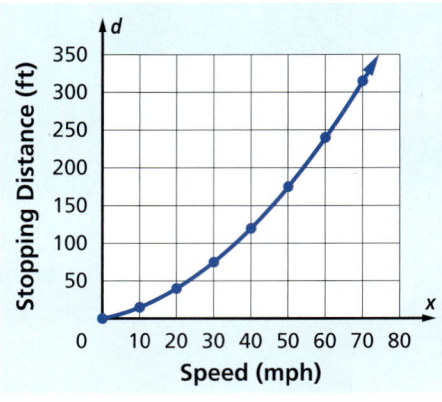

Properties of the Graph of $y = ax^2 + bx + c$

The equation $d = 0.05x^2 + x$ is of the form $y = ax^2 + bx + c$, with d taking the place of y, $a = 0.05$, $b = 1$, and $c = 0$. The graph of every equation of this form (provided $a \neq 0$) is a parabola. Moreover, every parabola with a vertical line of symmetry has an equation of this form. The values of a, b, and c determine where the parabola is positioned in the plane and whether it opens up or down. If $a > 0$ the parabola opens up, as in the situation above. If $a < 0$, the parabola opens down, as in Example 1.

Example 1

a. Graph $f(x) = -2x^2 - 3x + 8$. Use a window big enough to show the vertex of the parabola, the two x-intercepts, and the y-intercept.

b. Estimate its vertex, x-intercepts, and y-intercept.

Solutions

a. Here is a graph of $y = -2x^2 - 3x + 8$ on the window $-5 \leq x \leq 5$; $-10 \leq y \leq 10$.

b. From the window we have shown, we can only estimate the location of the vertex. **The vertex is near (-0.8, 9.1).** You do not have to estimate the y-intercept. It is $f(0)$, the value of y when $x = 0$, and is easily calculated. **The graph has y-intercept 8.** On the other hand, you can only estimate its x-intercepts. They are the values of x when the graph intersects the x-axis. **The x-intercepts are near -2.9 and 1.4.**

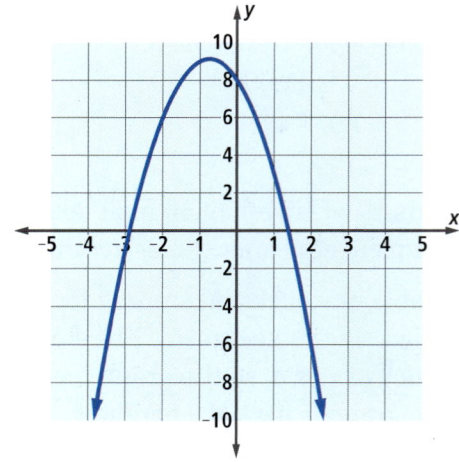

538   Quadratic Equations and Functions

Using Tables to Determine the Vertex of a Parabola

We could only estimate the vertex in Example 1 from the graph. Even if you use the trace function on a calculator or computer, you might not happen to find the vertex exactly. But in some cases, the symmetry of a parabola can be used to determine its vertex.

Example 2

Find the exact location of the vertex of the parabola that is the graph of $f(x) = -2x^2 - 3x + 8$ from Example 1.

Solution Because we know that the vertex is near the point $(-0.8, 9.1)$, we find y for values of x between -1 and 0. Look in the table at the right for a pair of points which have the same y-coordinates. Notice that there are three pairs of points whose y-coordinates are the same. These pairs occur on either side of -0.75. This indicates that the vertex is the point for which $x = -0.75$. Because $f(-0.75) = 9.125$, the vertex is $(-0.75, 9.125)$.

x	y
-1.0	9
-0.9	9.08
-0.8	9.12
-0.7	9.12
-0.6	9.08
-0.5	9
-0.4	8.88
-0.3	8.72
-0.2	8.52
-0.1	8.28
0	8

From a graph, it is often not as easy to locate the x-intercepts of a parabola as it is the vertex. The next two activities explore how the intercepts of the parabolas change as the values of b and c change in the equation $y = ax^2 + bx + c$.

Activity 1

Step 1 Graph $y = 2x^2$ and $y = 2x^2 - 10$ on the same axes.
   a. What is the y-intercept of $y = 2x^2$?
   b. What is the y-intercept of $y = 2x^2 - 10$?
   c. Describe how the two graphs are related to each other.

Step 2 Graph $y = -0.75x^2$ and $y = -0.75x^2 + 1$ on the same axes.
   a. What is the y-intercept of $y = -0.75x^2$?
   b. What is the y-intercept of $y = -0.75x^2 + 1$?
   c. Describe how the two graphs are related to each other.

Step 3 Graph $y = 2x^2 + c$ for three different values of c, with at least one of these values negative.
   a. Give the equations that you graphed.
   b. What is the effect of c on the graphs?

Graphing $y = ax^2 + bx + c$

Chapter 9

> ## Activity 2
>
> **Step 1** To determine the effect of b on the graph of a quadratic function, consider the function $y = x^2 + bx - 4$.
>
> a. Graph this function for the three different values of b given in the table below. Fill in the table.
>
> | Equation $y = ax^2 + bx + c$ | b | Vertex of Parabola | y-intercept of Parabola | x-intercepts (if any) |
> |---|---|---|---|---|
> | $y = x^2 + 2x - 4$ | ? | ? | ? | ? |
> | $y = x^2 + 4x - 4$ | ? | ? | ? | ? |
> | $y = x^2 - 3x - 4$ | ? | ? | ? | ? |
>
> b. What features of the graph of $y = x^2 + bx - 4$ does the value of b affect?
>
> c. What feature of the graph of $y = x^2 + bx - 4$ is not affected by the value of b?
>
> **Step 2** Find a window that shows a graph of $y = \frac{1}{2}x^2 - 6x + 4$, including its vertex, its y-intercept and its x-intercept(s).
>
> a. Describe your window.
>
> b. Use the trace feature of the calculator to estimate or determine the vertex, the y-intercept, and the x-intercepts.
>
> c. Tell how the x-coordinate of the vertex is related to the x-intercepts.

Questions

COVERING THE IDEAS

In 1–4, use the formula for automobile stopping distances given in this lesson, $d = 0.05x^2 + x$.

1. Define stopping distance.

2. Find the stopping distance for a car traveling 45 miles per hour.

3. Find the stopping distance for a car traveling 55 miles per hour.

4. **True or False** The stopping distance for a car traveling 50 mph is exactly double the stopping distance of a car traveling 25 mph.

5. The equation $d = 0.05x^2 + x$ is of the form $f(x) = ax^2 + bx + c$. What are the values of a, b, and c?

6. Explain how you can tell by looking at an equation of the form $y = ax^2 + bx + c$ whether its graph will open up or down.

In 7–9, an equation for a function is given.

a. Make a table of x and y values when x equals –3, –2, –1, 0, 1, 2, and 3.

b. Graph the equation.

c. Identify the y-intercept, x-intercept(s), and vertex.

d. Describe the range of the function.

7. $y = x^2 - 6$ 8. $y = x^2 - 2x + 5$ 9. $y = 3x^2 + 4$

10. The parabola at the right contains (–3, 0) and (–1, 0) and its vertex has integer coordinates.

    a. Find the coordinates of its vertex.

    b. Write an equation of its axis of symmetry.

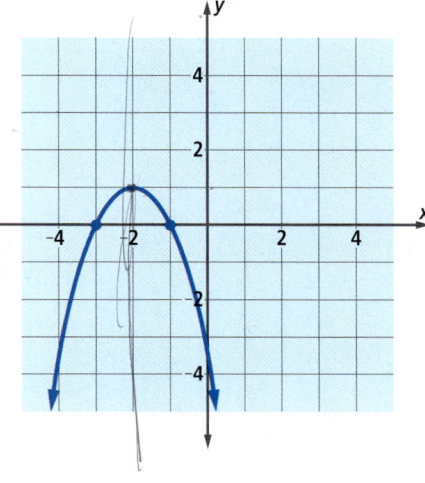

11. Consider this table of values for a parabola.

x	-8	-7	-6	-5	-4	-3	-2	-1	0
y	-28	0	20	32	36	32	20	?	?

a. What are the coordinates of the vertex of the parabola?

b. Use symmetry to find the coordinates of the points whose y values are missing in the table.

APPLYING THE MATHEMATICS

12. The parabola at the right contains points with integer coordinates as shown by the dots.

    a. Copy this graph on graph paper. Then use symmetry to graph more of the parabola.

    b. Give the coordinates of the vertex.

    c. Give an equation for the axis of symmetry.

    d. At what points does the parabola intersect the x-axis?

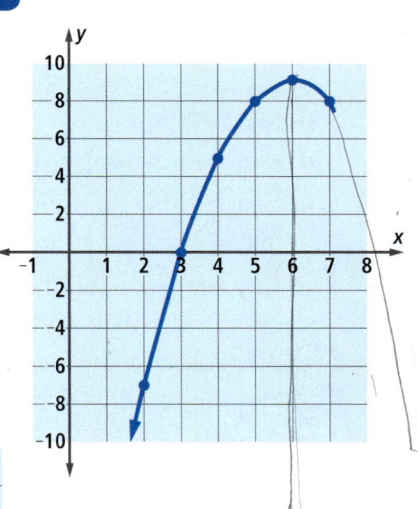

13. **Multiple Choice** Which of the two graphs is the graph of $y = x^2 - 6x + 8$? Justify your answer.

A 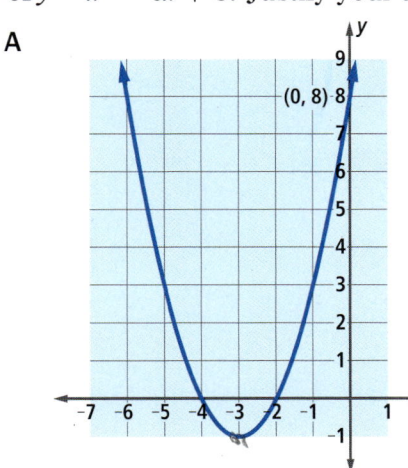   B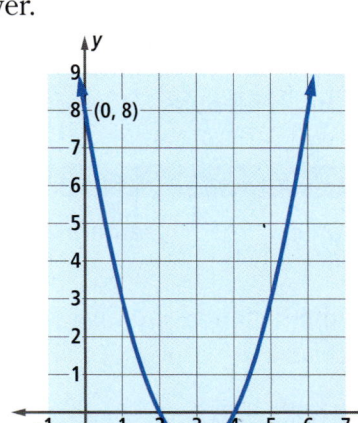

Matching In 14–17, match the graph with its equation.

a. $y = 2x^2$
b. $y = 2x^2 + 1$
c. $y = 2x^2 - 1$
d. $y = -2x^2 - 1$

14.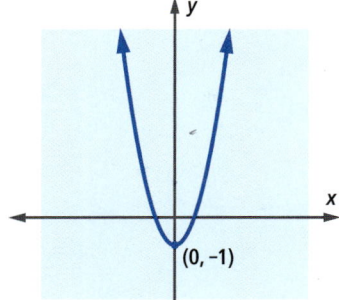

15.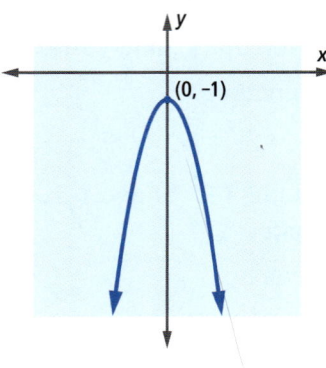

16.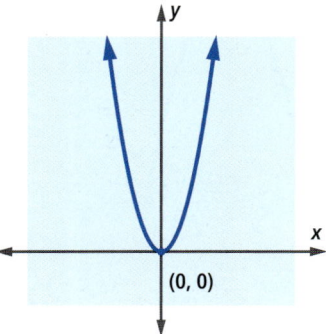

17.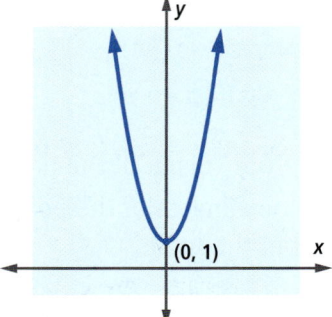

18. An insurance company reports that the equation $y = 0.4x^2 - 36x + 1{,}000$ relates the age of a driver x (in years) to the accident rate y (number of accidents per 50 million miles driven) when $16 \le x \le 74$.
    a. Graph this equation on your calculator. Give the window you used, the vertex, and the intercepts of the graph.
    b. Use the trace function of the graph to determine the age in which drivers have the fewest accidents per mile driven. About how many accidents do drivers of this age have per 50 million miles driven?
    c. According to this model, an 18-year-old driver is how many times as likely to have an accident as a 45-year-old driver?

REVIEW

19. Consider Galileo's equation $d = 16t^2$. **(Lesson 9-2)**
    a. Find t when $d = 350$.
    b. Write a question involving distance and time that can be answered using Part a.

20. Below are Charlotta's answers to questions about $y = \frac{x^2}{4}$. After she wrote this, she realized that she copied the equation incorrectly. It should be $y = -\frac{x^2}{4}$. What does Charlotta need to change to correct her work? (**Lesson 9-1**)

x	-4	-2	0	2	4
y	4	1	0	1	4

vertex = (0, 0)

axis of symmetry: $x = 0$

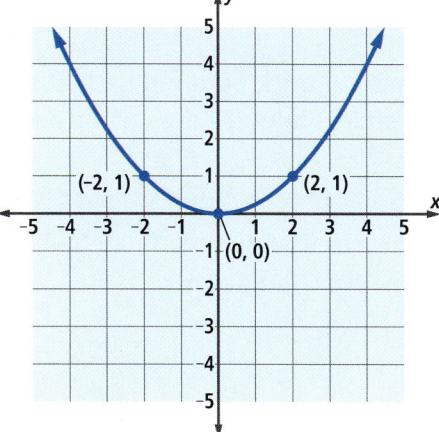

21. **Skill Sequence** If $a = -6$, $b = 8$, and $c = 12$, find the value of each expression. (**Lessons 8-6, 1-1**)

    a. $-4ac$ b. $b^2 - 4ac$ c. $\sqrt{b^2 - 4ac}$

22. A giant tortoise is walking at an average rate of $0.17 \frac{\text{mile}}{\text{hour}}$. Assume the tortoise continues walking at this rate.
    a. How long will it take the tortoise to travel 50 feet?
    b. How long will it take the tortoise to travel f feet?
    (**Lessons 5-4, 5-3**)

EXPLORATION

23. Begin with $y = -x^2$. With your calculator, experiment to find an equation for the parabola whose graph is used in Question 12.

In the past, giant species of *Geochelone* (tortoise) were found on all continents except Australasia, but today the giant forms are restricted to *G. elephantopus* in the Galapagos and *G. gigantea* on the island of Aldabara.

Source: Rochester Institute of Technology

Chapter 9

Lesson 9-4
Quadratics and Projectiles

Vocabulary

force of gravity
initial upward velocity
initial height

▶ **BIG IDEA** Assuming constant gravity, both the path of a projectile and the height of a projectile over time can be described by an equation of the form $y = ax^2 + bx + c$, $a \neq 0$.

A *projectile* is an object that is thrown, dropped, or launched, and then proceeds with no additional force on its own. A ball thrown up into the air is considered a projectile.

Mental Math

Estimate to the nearest 10.
a. 24% of 82
b. 5% of 206
c. 61% of 92

Equations for the Paths of Projectiles

When there is a constant force of gravity, the path of a projectile is a parabola. The parabola shows the height of the projectile as a function of the *horizontal distance* from the launch.

Activity

Step 1 A classroom board should be partitioned into rectangles by drawing evenly spaced lines, as shown. Work in a group and assign two students the tasks of tossing and catching a ball. Position the tosser at the left end of the board and the catcher at the right end of the board. During the experiment, the tosser will toss the ball to the catcher so that it does not go higher than the top of the board. For each vertical division line on the board, assign a student to act as a spotter. The diagram at the right would require 7 spotters. When the ball is tossed, these spotters will observe the ball's height when it crosses their vertical line.

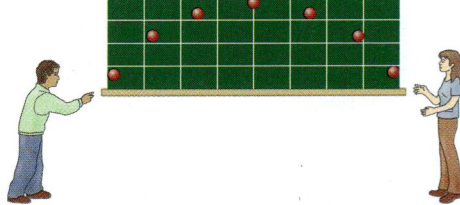

Step 2 Toss the ball while the spotters note its height as it passes each of the vertical lines. (It may take a few tosses to get everything right.) Following the toss, each spotter should mark the approximate point on the board at which the ball passed his or her vertical line. Using the board as a coordinate system, measure the horizontal and vertical distances to each point. Record the results.

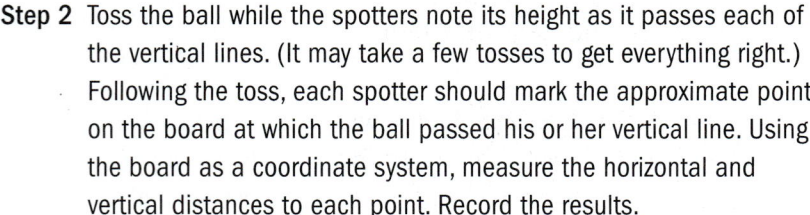

544 Quadratic Equations and Functions

Step 3 Enter the data into lists L1 and L2 on your calculator. Perform *quadratic regression* on the data. Quadratic regression fits a parabola "of best fit" to three or more points in the plane. Here is how to apply quadratic regression on one calculator.

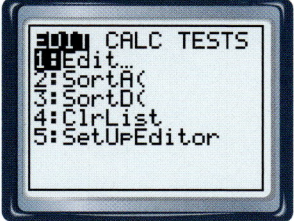 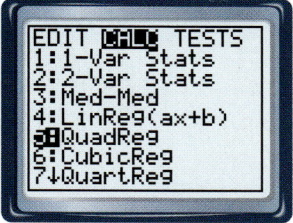 

The calculator returns coefficients *a*, *b*, and *c* of the quadratic equation that most closely fit the data entered into lists L1 and L2.

Step 4 Next, set up a reasonable plot window to fit your data. Enter the equation obtained in Step 3 into your Y= menu. Plot the scatterplot and function on the same grid. How close is the scatterplot to the parabola of best fit?

Equations for the Heights of Projectiles over Time

The graph of Galileo's formula $d = -16t^2$ is also a parabola. That parabola describes the height of the projectile as a function of the *length of time* since the projectile was launched.

When a projectile is launched, several factors determine its height above the ground at various times:

1. The **force of gravity** pulls the projectile back to Earth. By Galileo's formula, gravity pulls the projectile towards Earth $16t^2$ feet in t seconds.

2. The **initial upward velocity** with which the projectile is thrown or shot contributes to its height at time t. We use v to stand for the initial velocity. In t seconds, the projectile would go vt feet if there were no gravity.

3. The **initial height** of the projectile. We call this height s (for starting height).

Quadratics and Projectiles

Adding all these forces together yields a formula for the height of a projectile over time. Notice that the force of gravity is downward, so it has a negative effect on the height, while the launch velocity is considered to be upward and positive.

> **General Formula for the Height of a Projectile over Time**
>
> Let h be the height (in feet) of a projectile launched with an initial upward velocity v feet per second and an initial height of s feet. Then, after t seconds, $h = -16t^2 + vt + s$.

Since 16 feet ≈ 4.9 meters, if the units are in meters in the formula above, then $h = -4.9t^2 + vt + s$.

It is very easy to confuse the graph showing the height of a projectile as a function of time with a graph that represents the object's path, because they both are parabolas. However, even when a ball is tossed straight up and allowed to fall, its graph of height as a function of time is a parabola.

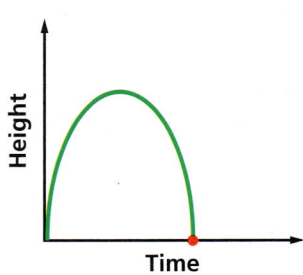

Ball's Path   **Ball's Height Graphed as a Function of Time**

Example 1

A ball is launched from an initial height of 6 feet with an initial upward velocity of 32 feet per second.
a. Write an equation describing the height h in feet of the ball after t seconds.
b. How high will the ball be 2 seconds after it is thrown?
c. What is the maximum height of the ball?

Solutions

a. Since units are provided in feet, use the general formula $h = -16t^2 + vt + s$. Substitute 6 for s and 32 for v.
$h = -16t^2 + 32t + 6$

b. Substitute 2 for t in $h = -16t^2 + 32t + 6$.

$$h = -16(2)^2 + 32(2) + 6$$
$$= -16(4) + 64 + 6$$
$$= -64 + 64 + 6$$
$$= 6$$

In 2 seconds, the ball will be 6 feet high.

c. Use a graphing calculator. Plot $y = -16x^2 + 32x + 6$. A graph of this function using the window $0 \leq x \leq 3$, and $0 \leq y \leq 25$ is shown at the right. The maximum height is the greatest value of h shown on the graph, the y-coordinate of the vertex. Trace along the graph, and read the y-coordinate as you go. When we did this using the window $0 \leq x \leq 3$, and $0 \leq y \leq 25$, our trace showed that (0.989, 21.998) and (1.021, 21.993) are on the graph. So try $t = 1$. This gives $h = 22$. The maximum height reached is 22 feet.

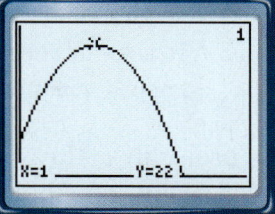

Check You can verify the maximum height by using the MAXIMUM or VERTEX command on your calculator. The screen at the right shows the vertex as (1, 22).

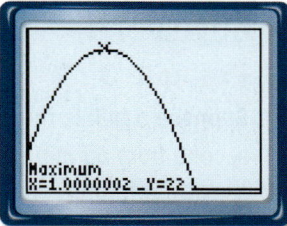

GUIDED

Example 2

An object is dropped from an initial height of 90 meters.
a. Write a formula describing the height of the object (in meters) after t seconds.
b. After how many seconds does the object hit the ground?
c. What is the maximum height of the object?

Solutions

a. Because units are provided in meters, use the general formula
$h = -4.9t^2 + v \cdot t + s$.

Substitute 90 for s and 0 for v.

$$h = -4.9t^2 + \underline{\ ?\ } \cdot t + \underline{\ ?\ }$$
$$= -4.9t^2 + \underline{\ ?\ }$$

b. The value(s) of t corresponding to $h = 0$ must be found when the object hits the ground.

$$0 = -4.9t^2 + \underline{\ ?\ }$$
$$4.9t^2 = \underline{\ ?\ }$$
$$t^2 \approx \underline{\ ?\ }$$

Because t is positive (it measures time after launch), ignore the negative square root.

$$t \approx \underline{\ ?\ }$$

(continued on next page)

You should find that the object hits the ground in about 4.3 seconds.

c. A graph of $y = -4.9x^2 + 90$ is shown at the right. The graph shows that the object is farthest from the ground when it is launched. That is when $t = 0$. Then $h = -4.9t^2 + 90 = -4.9(0)^2 + 90 = 90$. The maximum height is 90 meters. We knew this because the object was dropped from this height, and the equation and graph confirm it.

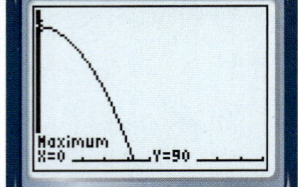

If a projectile is thrown upward and comes back down to Earth, then it will reach some heights twice.

GUIDED

Example 3
Suppose a ball is thrown upward with an initial velocity of 22 meters per second from an initial height of 2 meters.
a. Write a formula for the height in meters of the ball after t seconds.
b. Estimate when the ball is 20 meters high.

Solutions

a. Because units are provided in meters, use $h = -4.9t^2 + vt + s$. Substitute 2 for s and 22 for v.
$h =$ __?__

b. The values of t corresponding to $h = 20$ must be found. Graph the equation you found in Part a on the window $0 \le x \le 6$, $10 \le y \le 30$. Draw a horizontal line $y = 20$ to indicate $h = 20$ feet. Use the INTERSECT command on your calculator to find both intersections. Our calculator shows that when $x \approx$ __?__ and $x \approx$ __?__, $y \approx 20$.

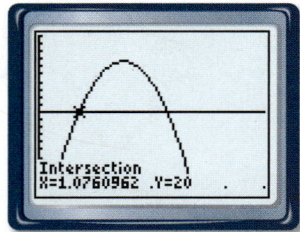 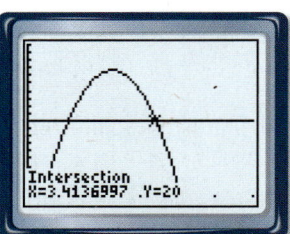

The ball is 20 meters off the ground at about __?__ seconds and __?__ seconds after being thrown.

A method for finding exact answers without finding the intersections of two graphs is discussed in Lesson 9-5.

Questions

COVERING THE IDEAS

1. In your own words, define the term projectile. Give several real-life examples of projectiles.

2. Use quadratic regression on your calculator to find an equation for the parabola passing through the points (0, 6), (5, 30), and (10, 6).

3. A ball is thrown from an initial height of 5 feet with an initial upward velocity of 30 feet per second.
   a. Write a function describing the height of the ball after t seconds.
   b. How high will the ball be 2 seconds after it is thrown?
   c. What is the maximum height of the ball?

4. Suppose a ball is batted with an initial upward velocity of 26 meters per second from an initial height of 1 meter.
   a. Write a function describing the height in meters of the ball after t seconds.
   b. Estimate when the ball is 5 meters high.

5. An object is dropped from an initial height of 40 feet.
   a. What is the object's initial velocity? What is its maximum height?
   b. Write a formula for the height h in feet of the object after t seconds.
   c. After how many seconds does the object hit the ground?

6. An object is dropped from an initial height of 150 meters.
   a. Write a formula for the height h in meters of the object after t seconds.
   b. After how many seconds does the object hit the ground?
   c. What is the maximum height of the object?

Los Angeles's Vladimir Guerrero

APPLYING THE MATHEMATICS

7. Use quadratic regression on your calculator to answer this question. A football kicker attempts to kick a 40-yard field goal. The kicker kicks the football from a height of 0 feet above the ground. The football is 26 feet above the ground at its peak (the vertex) at a distance of 22 yards from the kicker. The height of the crossbar (the bottom bar of the goal post) is 10 feet off the ground. Assuming the ball is kicked straight, will the kick clear the crossbar of the goal post? (*Hint:* Use the information given to determine a third point on the parabola and use quadratic regression.)

8. Refer to the graph at the right. It shows the height h in feet of a soccer ball t seconds after it is drop-kicked into the air.

   a. What is the approximate height of the soccer ball at 1 second? At 4 seconds?
   b. What is the greatest height the ball reaches?
   c. At what times is the ball 38.5 feet high?
   d. Approximately how long is the ball in the air?
   e. For how many seconds is the ball more than 38.5 feet above the ground?
   f. What does the h-intercept represent in this situation?
   g. What does the t-intercept represent in this situation?

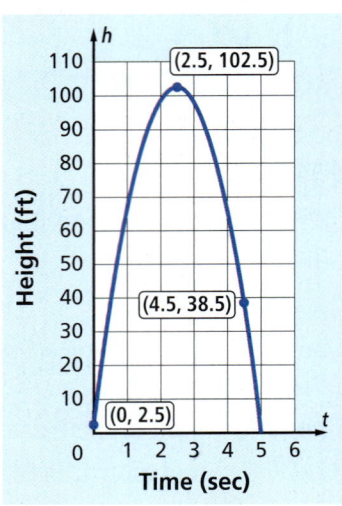

There were approximately 321,555 high school girl soccer players in the 2005–2006 school year.

Source: National Federation of High School Associations

9. A small rocket is shot from the edge of a cliff. Suppose that after x seconds, the rocket is y meters above the cliff, where $y = 25x - 5x^2$.

   a. Graph this equation using the window $0 \le x \le 8$, $-5 \le y \le 40$.
   b. What is the greatest height the rocket reaches?
   c. How far above the edge of the cliff is the rocket after 4 seconds?
   d. Between which times is the rocket more than 20 feet above the cliff's edge?
   e. What is the height of the rocket after 6 seconds?
   f. At approximately what time does the rocket fall below the height of the cliff's edge?

10. A ball is thrown from an initial height of 7 feet. After 5 seconds in the air, the ball reaches a maximum height of 18 feet above the ground.

    a. What third point on the graph can be deduced from this information?
    b. Use quadratic regression to find a formula for the height in feet of the ball after t seconds.
    c. Use the formula to find the initial velocity of the ball.
    d. Use the graph to approximate how long the ball is in the air.

REVIEW

In 11 and 12, an equation is given.
   a. Make a table of values for integer values of x from −3 to 3.
   b. Graph the equation. (Lessons 9-3, 9-2)

11. $y = \frac{1}{4}x^2$

12. $y = 8x - 3x^2$

In 13 and 14, use the graph of the parabola at the right. (Lessons 9-3, 9-1)

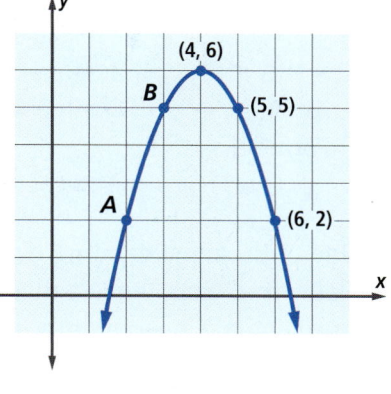

13. Use the symmetry of the parabola to find the coordinates of points A and B.

14. Write an equation for the axis of symmetry.

15. **Multiple Choice** Which of the following is *not* equal to uv^2? (Lessons 8-5, 8-2)

   A $u \cdot v \cdot v$ **B** $(u)v^2$ **C** $u(v)^2$ **D** $(uv)^2$

16. The table below estimates the number of calories people of various weights burn per minute while participating in various activities. (Lessons 6-8, 6-5, 3-4)

Activity	Weight (lb)			
	105–115	127–137	160–170	180–200
Full-court Basketball	9.8	11.2	13.2	14.5
Jogging (5 mph)	8.6	9.2	11.5	12.7
Running (8 mph)	10.4	11.9	14.2	17.3
Volleyball	7.8	8.9	10.5	11.6
Bicycling (10 mph)	5.5	6.3	7.8	14.5

Source: www.coolnurse.com

The state of California has produced the most high school All-American boy's basketball players (72) through the 2005–2006 season.

Source: mcdepk.com

   a. Suppose Vince, who weighs 168 pounds, works out by jogging and then playing basketball. Let x = the number of minutes he jogs, and let y = the number of minutes he plays basketball. If he burns a total of 445 calories, write an equation in standard form that describes x and y.
   b. Find the x- and y-intercepts of the line from Part a.
   c. If Vince plays basketball for 25 minutes, use your equation to calculate how long he must jog to burn 445 calories.

EXPLORATION

17. Infinitely many parabolas have x-intercepts at 0 and 6. Find equations for three such parabolas.

Chapter 9

Lesson 9-5

The Quadratic Formula

Vocabulary

quadratic equation

standard form of a quadratic equation

▶ **BIG IDEA** If an equation can be put into the form $ax^2 + bx + c = 0$, with $a \neq 0$, it can be solved using the Quadratic Formula.

One of the most exciting events in amateur sports is 10-meter platform diving. Once a diver leaves a platform, the diver becomes a projectile. Consequently, during a dive, a diver's height above the water at any given time can be determined using a quadratic equation. This is important because for a diver to practice spins and somersaults, he or she must know how much time will pass before entering the water.

Suppose a diver jumps upward at an initial velocity of 4.3 meters per second. Then the diver's height $h(t)$ in meters t seconds into the dive can be estimated using the equation $h(t) = -4.9t^2 + 4.3t + 10$ and graphed below. When the diver hits the water, $h(t)$ is zero. So solving the equation $0 = -4.9t^2 + 4.3t + 10$ gives the number of seconds from departing the platform to entering the water.

The equation $0 = -4.9t^2 + 4.3t + 10$ is an example of a *quadratic equation*. A **quadratic equation** is an equation that can be written in the form $ax^2 + bx + c = 0$ with $a \neq 0$. In this case, t is being used in place of x, $a = -4.9$, $b = 4.3$, and $c = 10$.

Mental Math

Write as a power of 10.

a. $65 \div 0.65$

b. $483 \div 48.3$

c. $7.2 \div 7,200$

The Quadratic Formula

You can find the solutions to *any* quadratic equation by using the *Quadratic Formula*. This formula gives the value(s) of x in terms of a (the coefficient of x^2), b (the coefficient of x), and c (the constant term). The formula states that there are at most two solutions to a quadratic equation.

If $ax^2 + bx + c = 0$ and $a \neq 0$, then
$$x = \frac{-b + \sqrt{b^2 - 4ac}}{2a} \text{ or } x = \frac{-b - \sqrt{b^2 - 4ac}}{2a}.$$
The calculations of the two solutions differ in only one way. $\sqrt{b^2 - 4ac}$ is added to $-b$ in the numerator of the first calculation, while $\sqrt{b^2 - 4ac}$ is subtracted from $-b$ in the second calculation.

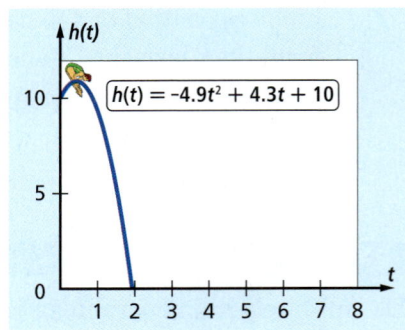

552 Quadratic Equations and Functions

The work in calculating the two solutions is almost the same. So, the two expressions can be written as one expression using the symbol ±, which means "plus or minus." This symbol means you do two calculations: one using the + sign to add and one using the − sign to subtract.

> **The Quadratic Formula**
>
> If $ax^2 + bx + c = 0$ and $a \neq 0$, then $x = \dfrac{-b \pm \sqrt{b^2 - 4ac}}{2a}$.

The quadratic equation $ax^2 + bx + c = 0$ and the Quadratic Formula $x = \dfrac{-b \pm \sqrt{b^2 - 4ac}}{2a}$ are equivalent equations. The quadratic equation was solved for x to generate the quadratic formula. In Chapter 13, you can see how this formula was found.

The quadratic formula is one of the most often used and most famous formulas in all mathematics. *You should memorize it today!*

Applying the Quadratic Formula

Example 1

Solve $x^2 + 8x + 7 = 0$.

Solution Recall that $x^2 = 1x^2$. So think of the given equation as $1x^2 + 8x + 7 = 0$ and apply the Quadratic Formula with $a = 1$, $b = 8$, and $c = 7$.

$$x = \frac{-b \pm \sqrt{b^2 - 4ac}}{2a}$$

$$= \frac{-8 \pm \sqrt{8^2 - 4 \cdot 1 \cdot 7}}{2 \cdot 1}$$

Follow the order of operations. Work under the radical sign (with its unwritten parentheses) first.

$$= \frac{-8 \pm \sqrt{64 - 28}}{2}$$

$$= \frac{-8 \pm \sqrt{36}}{2}$$

So, $x = \dfrac{-8 + 6}{2}$ or $x = \dfrac{-8 - 6}{2}$

$x = \dfrac{-2}{2} = -1$ or $x = \dfrac{-14}{2} = -7$.

Check Do -1 and -7 make the equation $x^2 + 8x + 7 = 0$ true? Substitute -1 for x.

Does $(-1)^2 + 8(-1) + 7 = 0$?

$1 + -8 + 7 = 0$ Yes, it checks.

(continued on next page)

> ▶ **READING MATH**
>
> On some calculators there is a key that is labeled ± or +/−. That key takes the opposite of a number. It does not perform the two operations + and − required in the Quadratic Formula.

Substitute −7 for x.

Does $(-7)^2 + 8(-7) + 7 = 0$?

$49 + -56 + 7 = 0$ Yes, it checks.

STOP QY

Sometimes you must decide whether or not a solution to a quadratic equation is reasonable, given the context of the problem. In Guided Example 2, we return to the diving situation described at the beginning of this lesson.

▶ **QY**

Solve $3y^2 − 10.5y + 9 = 0$.

The *x*-axis represents a height of 0. The graph crosses the *x*-axis a little to the left of 2, which is close to 1.93. The INTERSECT feature on a graph is also helpful for finding a solution.

GUIDED

Example 2

In 10-meter platform diving, the function $h(t) = -4.9t^2 + 4.3t + 10$ gives the approximate height $h(t)$ above the water in meters a diver is at t seconds after launching into the dive. How many seconds elapse from the time the diver leaves the 10-meter platform until the diver hits the water?

Solution The diver will hit the water when the diver's height above the water is zero, so solve the equation $0 = -4.9t^2 + 4.3t + 10$.

Apply the Quadratic Formula with $a = -4.9$, $b = 4.3$, and $c = 10$.

$$t = \frac{-4.3 \pm \sqrt{(\underline{})^2 \underline{} 4(\underline{})(\underline{})}}{2(\underline{})}$$

$$= \frac{-4.3 \pm \sqrt{\underline{}}}{-9.8}$$

So, $t = \dfrac{-4.3 + \sqrt{\underline{}}}{-9.8}$ or $t = \dfrac{-4.3 - \sqrt{\underline{}}}{-9.8}$.

These are exact solutions to the quadratic equation. However, since the given information is not exact, it is more reasonable to want an approximation.

$t \approx \dfrac{-4.3 + \underline{}}{-9.8}$ or $t \approx \dfrac{-4.3 - \underline{}}{-9.8}$

$t \approx -1.1$ or $t \approx 1.9$

The diver cannot reach the water in negative time, so the solution −1.1 seconds does not make sense in this situation. We therefore eliminate this as an answer. The diver will hit the water about 1.9 seconds after leaving the diving platform.

Check Substituting 1.9 in for t in the equation $0 = -4.9t^2 + 4.3t + 10$ is one method you can use to check a solution. Other methods are looking at a table or a graph.

Italian diver Tania Cagnotto on the platform during the final of the 2003 Swimming World Championship, in Barcelona, Spain

In Example 3, the equation has to be put in $ax^2 + bx + c = 0$ form before the Quadratic Formula can be applied. This form is called the **standard form of a quadratic equation.** When the number under the radical sign in the Quadratic Formula is not a perfect square, approximations are often used in the last step of the process.

Example 3
Solve $x^2 - 3x = 37$.

Solution Put $x^2 - 3x = 37$ into standard form.

$x^2 - 3x - 37 = 37 - 37$ Subtract 37 from both sides.

$x^2 - 3x - 37 = 0$

Apply the Quadratic Formula, with $a = 1$, $b = -3$, and $c = -37$.

$$x = \frac{-(-3) \pm \sqrt{(-3)^2 - 4(1)(-37)}}{2(1)}$$

$$x = \frac{3 \pm \sqrt{9 + 148}}{2}$$

$$x = \frac{3 \pm \sqrt{157}}{2}$$

So $x = \dfrac{3 + \sqrt{157}}{2}$ or $x = \dfrac{3 - \sqrt{157}}{2}$.

These are exact solutions. You can approximate the solutions using a calculator.

Because $\sqrt{157} \approx 12.5$, $x \approx \dfrac{3 + 12.5}{2}$ or $x \approx \dfrac{3 - 12.5}{2}$. So $x \approx 7.75$ or $x \approx -4.75$.

Check Do the two values found work in the equation $x^2 - 3x = 37$?
Substitute 7.75 for x.
$(7.75)^2 - 3(7.75) = 36.8$. This is close to 37.
Substitute -4.75 for x.
$(-4.75)^2 - 3(-4.75) = 36.8$. This is close to 37.

In both cases the checks are not exact, but the solutions are approximations, so the check is close enough.

Questions

COVERING THE IDEAS

1. State the Quadratic Formula.
2. Is it true that the Quadratic Formula can be used to solve *any* quadratic equation?
3. Find the two values of $\dfrac{-3 \pm 9}{2}$.

The Quadratic Formula

In 4–7, use the Quadratic Formula to solve the equation. Give the exact solutions and check both solutions.

4. $x^2 + 15x + 54 = 0$
5. $t^2 + 4t + 4 = 0$
6. $3m^2 + 2m = 4$
7. $3y^2 = 13y + 100$

In 8 and 9, use the Quadratic Formula to solve the equations. Round the solutions to the nearest hundredth and check both solutions.

8. $20n^2 - 6n - 2 = 0$
9. $3p^2 + 14 = -19p$

10. If a diver dives from a 20-foot platform with an initial upward velocity of 14 feet per second, then the diver's approximate height can be represented by the function $h(t) = -16t^2 + 14t + 20$, where $h(t)$ is the height and t is the time in seconds. (This formula is different from the one in this lesson because meters have been converted to feet.)

    a. Find $h(1)$. Write a sentence explaining what it means.
    b. Estimate to the nearest tenth of a second how much time the diver will be in the air before hitting the water.

APPLYING THE MATHEMATICS

11. The solutions to $ax^2 + bx + c = 0$ are the x-intercepts of the graph of $y = ax^2 + bx + c$.

    a. Use the Quadratic Formula to find the solutions to $3x^2 - 6x - 45 = 0$.
    b. Check your answers to Part a by graphing an appropriate function.

12. The graphs of $y = -0.5x^2 + 6$ and $y = 4$ intersect at two points.

    a. Find the x-coordinate of each of the intersecting points.
    b. Find both coordinates of the two points of intersection.
    c. Check your answers to Part b by graphing these equations.

13. In 1971, the astronaut Alan Shepard (who had been the first U.S. man in space 9 years earlier) snuck a collapsible golf club and a golf ball onto *Apollo 14*. Just before taking off from the moon to return to Earth, he hit two golf balls. In doing so, he vividly showed the difference between gravity on the moon and on Earth. On the moon the approximate height $h(t)$ of the ball (in feet) after t seconds is given by the function $h(t) = -0.8t^2 + 12t$.

    a. At what two times would the golf ball reach a height of 20 feet? (Round your answer to the nearest hundredth.)
    b. How long would it take for the ball to come back to the surface of the moon?

14. The area of a rectangle is 240 cm². The length is 14 cm more than the width. What are the length and width of the rectangle?

REVIEW

15. If the *x*-intercepts of a parabola are 8 and −4, what is the *x*-coordinate of its vertex? (**Lesson 9-3**)

16. **Multiple Choice** Which equation is graphed at the right? (**Lessons 9-3, 9-2**)

    A $y = 2x^2$
    B $y = -2x^2$
    C $y = 2x^2 + 2$
    D $y = -2x^2 - 2$

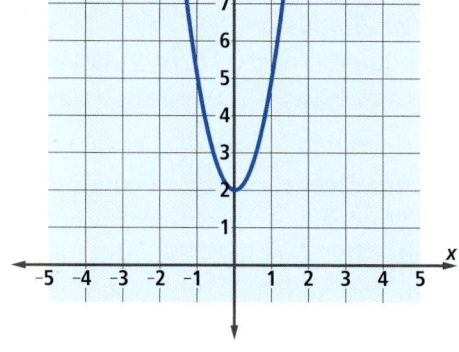

17. Find the radius of a circle if the center of the circle is at (−2, 6) and the point (1, 2) is on the circle. (**Lesson 8-8**)

18. Kristen is buying a new car. She can choose from 4 models, 2 transmission types, 9 exterior colors, and 6 interior colors. How many combinations of models, transmissions, exteriors, and interiors are possible? (**Lesson 8-1**)

19. **Skill Sequence** If $f(x) = \frac{5x - 6}{x}$, find each value. (**Lesson 7-6**)

    a. $f(2)$ b. $f(3)$ c. $f(x + 1)$

EXPLORATION

20. a. Solve the equation $ax^2 + bx + c = 0$ for x using a CAS. In what form does the CAS put the two solutions?
    b. Add the two solutions using the CAS. What is the sum?
    c. Multiply the two solutions using the CAS. What is the product?
    d. Use the results from Parts b and c to check the *exact* answers to Example 3 of this lesson.

QY ANSWER

$y = 1.5$ or $y = 2$

The Quadratic Formula

Chapter 9

Lesson 9-6

Analyzing Solutions to Quadratic Equations

Vocabulary

discriminant

▶ **BIG IDEA** The value of the discriminant $b^2 - 4ac$ of a quadratic equation $ax^2 + bx + c = 0$ can tell you whether the equation has 0, 1, or 2 real solutions.

In Acapulco, Mexico, cliff divers dive from a place called La Quebrada ("the break in the rocks") 27 meters above the water. As you have learned, a diver's path is part of a parabola that can be described using a quadratic equation. An equation that relates the distance x (in meters) away from the cliff and the distance y (in meters) above the water is $y = -x^2 + 2x + 27$. The graph at the right shows that when a diver pushes off the cliff, the diver arches upward and then descends.

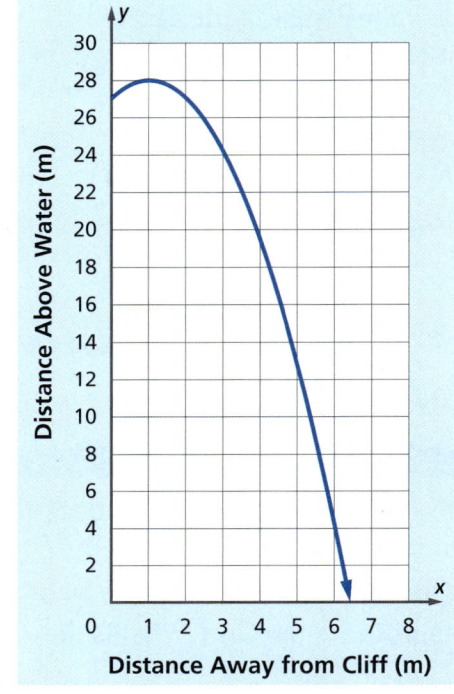

Mental Math

Solve the equation.

a. $a + 14 = 29$

b. $b^2 + 14 = 39$

c. $(c - 6)2 + 14 = 39$

Will the diver's height reach 27.75 meters? 28 meters? 30 meters? You can use the equation, the graph, or the table to determine whether or not a diver reaches a particular height.

Acapulco cliff diver

Using a Graph or Table to Determine Solutions to a Quadratic Equation

GUIDED

Example 1

Consider the situation of a La Quebrada diver. Graph and generate a table of the parabola with equation $y = -x^2 + 2x + 27$ to determine whether the diver will ever reach

a. 27.75 meters. b. 28 meters. c. 30 meters.

558 Quadratic Equations and Functions

Solutions

a. A graph of $y = -x^2 + 2x + 27$ is shown on the right using the window $0 \leq x \leq 4$ and $25 \leq y \leq 30$. Also graphed is the line $y = 27.75$. This line crosses the parabola twice.

The diver reaches 27.75 meters twice, once on the way up and once on the way down.

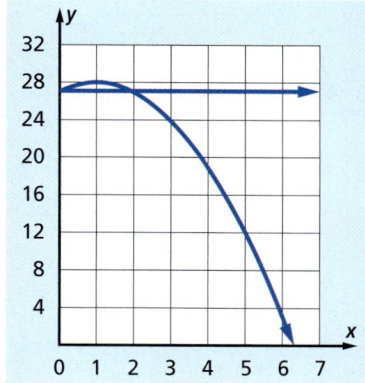

A table of $y = -x^2 + 2x + 27$ is shown below. From the table it is evident that there are two distances x when $y = 27.75$ meters: once 0.5 meter from the cliff and again 1.5 meters away.

b. Suppose you draw the line $y = 28$ on the graph. How many times does the line appear to intersect the graph? __?__

Now look at the table. It appears the diver reaches the height of 28 meters __?__ time(s). The diver reaches the height of 28 meters __?__ meter(s) from the cliff.

c. Suppose you draw the line $y = 30$ on the graph. How many times does the line appear to intersect the graph? __?__

Now look at the table below. It appears the diver reaches the height of 30 meters __?__ time(s).

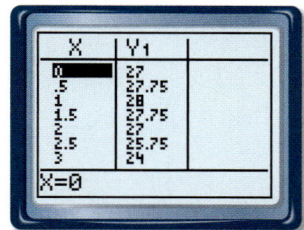

Using the Quadratic Formula to Find the Number of Real Solutions

We can answer the same questions about the height of the diver using the Quadratic Formula.

Example 2

Will the diver ever reach a height of

a. 27.75 meters?   b. 28 meters?   c. 30 meters?

Solutions

a. Let $y = 27.75$ in the equation $y = -x^2 + 2x + 27$.

$$27.75 = -x^2 + 2x + 27$$

Add -27.75 to both sides to put the equation in standard form.

$$0 = -x^2 + 2x - 0.75$$

(continued on next page)

Analyzing Solutions to Quadratic Equations

Let $a = -1$, $b = 2$, and $c = -0.75$ in the Quadratic Formula. Then

$$x = \frac{-2 \pm \sqrt{2^2 - 4(-1)(-0.75)}}{2(-1)}$$

$x = \frac{-2 + \sqrt{1}}{-2}$ or $x = \frac{-2 - \sqrt{1}}{-2}$

$x = 0.5$ or $x = 1.5$.

The diver reaches a height of 27.75 meters twice, first at 0.5 meter from the cliff and second at 1.5 meters from the cliff.

b. Let $y = 28$ in the equation $y = -x^2 + 2x + 27$.

$$28 = -x^2 + 2x + 27$$

Add -28 to both sides to place the equation in standard form.

$$0 = -x^2 + 2x - 1$$

Let $a = -1$, $b = 2$, and $c = -1$ in the Quadratic Formula.

$$x = \frac{-2 \pm \sqrt{2^2 - 4(-1)(-1)}}{2(-1)} = \frac{-2 \pm 0}{-2} = 1$$

So the diver reaches 28 meters just once, 1 meter up from the cliff. This agrees with the graph that shows the vertex to be (1, 28).

c. Let $y = 30$ in the equation $y = -x^2 + 2x + 27$.

$$30 = -x^2 + 2x + 27$$

Add -30 to both sides to place the equation in standard form.

$$0 = -x^2 + 2x - 3$$

Let $a = -1$, $b = 2$, and $c = -3$ in the Quadratic Formula. Then

$$x = \frac{-2 \pm \sqrt{2^2 - 4(-1)(-3)}}{2(-1)} = \frac{-2 \pm \sqrt{-8}}{-2}.$$

Because no real number multiplied by itself equals -8, there is no square root of -8 in the real number system. In fact, no negative number has a square root in the real number system. So $30 = -x^2 + 2x + 27$ does not have a real number solution. This means that the diver never reaches a height of 30 meters. This is consistent with the graph that shows there is no point on the parabola with a height of 30 meters.

The Discriminant of a Quadratic Equation

Look closely at the number under the square root in each of the solutions in Example 2. Notice that the number of solutions to a quadratic equation is related to this number, which is the number $b^2 - 4ac$ in the Quadratic Formula. In the solution to Part a, $b^2 - 4ac = 1$, which is positive. Adding $\sqrt{1}$ and subtracting $\sqrt{1}$ results in two solutions. In the solution to Part b, $b^2 - 4ac = 0$, and adding $\sqrt{0}$ and subtracting $\sqrt{0}$, yields the same result, 1. That quadratic equation has just one solution. There is no solution to the equation in Part c because $b^2 - 4ac = -8$, and -8 does not have a square root in the set of real numbers.

Because the value of $b^2 - 4ac$ *discriminates* among the various possible number of real number solutions to a specific quadratic equation, it is called the **discriminant** of the equation $ax^2 + bx + c = 0$. Stated below are the specific properties of the discriminant.

> **Discriminant Property**
>
> If $ax^2 + bx + c = 0$ and a, b, and c are real numbers ($a \neq 0$), then:
>
> When $b^2 - 4ac > 0$, the equation has exactly two real solutions.
>
> When $b^2 - 4ac = 0$, the equation has exactly one real solution.
>
> When $b^2 - 4ac < 0$, the equation has no real solutions.

An important use of the discriminant relates solutions of a quadratic equation to the *x*-intercepts of the related function. Specifically, the solutions to $ax^2 + bx + c = 0$ are the *x*-intercepts of $y = ax^2 + bx + c$. So the discriminant tells you how many times the function $f(x) = ax^2 + bx + c$ crosses the *x*-axis.

Quadratic Function	Value of $b^2 - 4ac$	Number of *x*-intercepts	Graph (All screens are shown in the standard viewing window.)
$y = x^2 + x - 6$	$1^2 - 4(1)(-6) = 25$ positive	two	
$y = x^2 - 6x + 9$	$(-6)^2 - 4(1)(9) = 0$ zero	one	
$y = x^2 + 2x + 7$	$2^2 - 4(1)(7) = -24$ negative	zero	

 QY

▶ QY

Determine the number of real solutions to $6x^2 + 3x = -7$.

Chapter 9

> **GUIDED**
>
> ### Example 3
> How many times does the graph of $y = 2x^2 + 16x + 32$ intersect the x-axis?
>
> **Solution** Find the value of the discriminant $b^2 - 4ac$. Here $a = \underline{}$, $b = \underline{}$, and $c = \underline{}$.
>
> So $b^2 - 4ac = \underline{}$.
>
> $ = \underline{} = 0$
>
> Because the discriminant is $\underline{}$, the graph of $y = 2x^2 + 16x + 32$ intersects the x-axis $\underline{}$ time(s).
>
> **Check** You should check your answer by graphing the equation with a calculator.

Questions

COVERING THE IDEAS

In 1 and 2, refer to the La Quebrada cliff diver equation $y = -x^2 + 2x + 27$ from Example 1.

1. a. What equation can be solved to determine how far away (horizontally) from the cliff the diver will be when the diver is 27 meters above the water?
   b. Will the diver reach a height of 27 meters above the water? If so, how many times?

2. How far from the cliff will the diver be at 10 meters above the water?

3. How many real solutions does a quadratic equation have when the discriminant is
   a. negative?   b. zero?   c. positive?

4. The discriminant of the equation $ax^2 + bx + c = 0$ is $-1{,}200$. What does this indicate about the graph of $y = ax^2 + bx + c$?

5. The equation $y = \frac{1}{2}x^2 - x - \frac{3}{2}$ is graphed at the right. Use the graph to determine the number of real solutions to each equation.

   a. $\frac{1}{2}x^2 - x - \frac{3}{2} = -2$ b. $\frac{1}{2}x^2 - x - \frac{3}{2} = -3$

   c. $\frac{1}{2}x^2 - x - \frac{3}{2} = 1$ d. $\frac{1}{2}x^2 - x - \frac{3}{2} = 5{,}000$

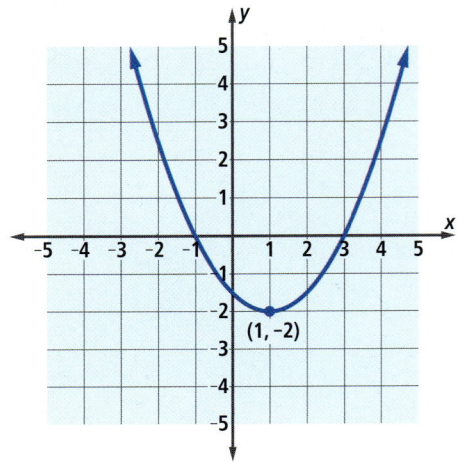

562 Quadratic Equations and Functions

In 6–9, a quadratic equation is given.
   a. Calculate the value of the discriminant.
   b. Give the number of real solutions.
   c. Find all the real solutions.

6. $2x^2 + x + 3 = 0$

7. $-4n^2 + 56n - 196 = 0$

8. $22q^2 = q + 3$

9. $x = \frac{x^2}{6} + \frac{1}{4}$

In 10 and 11, an equation of the form $y = ax^2 + bx + c$ is graphed. Tell whether the value of $b^2 - 4ac$ is *positive*, *negative*, or *zero*.

10.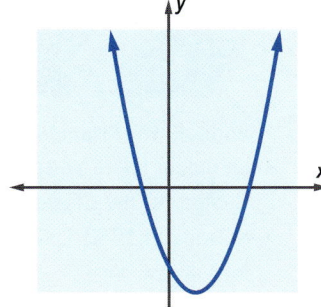

11.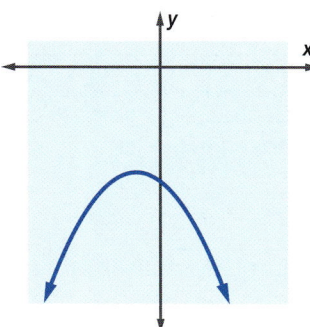

In 12 and 13, a quadratic function f is described.
   a. Calculate the value of the discriminant of the quadratic equation $f(x) = 0$.
   b. Give the number of x-intercepts of the graph of f.

12. $f(x) = 5x^2 + 20x + 20$

13. $f(x) = 2x^2 + x - 3$

APPLYING THE MATHEMATICS

14. For what value of h does $x^2 + 6x + h = 0$ have exactly one solution?

15. If the discriminant for the equation $2x^2 + 4x + c = 0$ is 8, what is the value of c?

16. In Lesson 9-5, a diver's height $h(t)$ above the water after t seconds was given by $h(t) = -4.9t^2 + 4.3t + 10$. Use the discriminant to find the time t when the diver reached the maximum height.

17. Can any parabolas with an equation of the form $y = ax^2 + bx + c$ *not* have a y-intercept? Why or why not?

18. By letting $x = m + 3$, solve $4(m + 3)^2 - 13(m + 3) - 35 = 0$ for m.

REVIEW

19. Solve $45x^2 - 100 = 0$. (**Lesson 9-5**)

In 20 and 21, use the following information. A softball pitcher tosses a ball to a catcher 50 feet away. The height h (in feet) of the ball when it is x feet from the pitcher is given by the equation $h = -0.016x^2 + 0.8x + 2$. (Lesson 9-4)

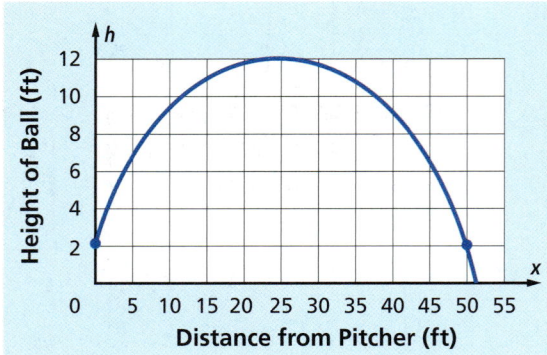

Softball was invented as an indoor sport by George Hamock of the Chicago Board of Trade in 1887.

Source: The History Channel

20. How high is the ball at its peak?

21. a. If the batter is 2 feet in front of the catcher, how far is the batter from the pitcher?
    b. How high is the ball when it reaches the batter?

In 22 and 23, state whether the parabola described by the equation opens up or down. (Lesson 9-3)

22. $y = -\frac{1}{3}x^2 - 6x + 1$
23. $y = 0.5x - 2x^2$

24. **Skill Sequence** In Parts a–d, simplify each statement. (Lesson 2-2)

a. $\dfrac{-4 + x}{2a} + \dfrac{-4 - x}{2a}$
b. $\dfrac{-b + y}{2a} + \dfrac{-b - y}{2a}$
c. $\dfrac{-b + \sqrt{z}}{2a} + \dfrac{-b - \sqrt{z}}{2a}$
d. $\dfrac{-b + \sqrt{b^2 - 4ac}}{2a} + \dfrac{-b - \sqrt{b^2 - 4ac}}{2a}$

e. What does Part d tell you about the solutions to a quadratic equation?

EXPLORATION

25. Create a parabola in the following way. Take a plain sheet of notebook paper and draw a dark dot in the middle of the paper. Draw a line anywhere on the paper that is parallel to the bottom of the paper. Make sure the line stretches to both edges of the paper. Now fold the paper so that the dot falls on the line. Unfold the paper, and then fold the paper so that the dot falls on another place on the line. Repeat this 20 times so that the dot has fallen in different places on the line each time. The folds should outline a parabola that can be seen by unfolding the paper. Take another sheet and see what happens if the dot is farther from the line or closer to the line than the dot you used the first time.

QY ANSWER

0

Lesson 9-7

More Applications of Quadratics: Why Quadratics Are Important

▶ **BIG IDEA** Quadratics have applications in engineering, geometry, and counting problems.

You may wonder why you are asked to memorize the Quadratic Formula or even why you need to know how to solve quadratic equations. There is a simple reason. Quadratic expressions, equations, and functions appear in a wide variety of situations. Furthermore, these situations are unlike those that lead to linear expressions. Many problems involving linear expressions can be solved by some people without algebra, using just intuition from arithmetic. Few people can solve problems that lead to quadratic expressions without using algebra.

Mental Math

Solve.
a. $E = mc^2$ for m.
b. $A = \pi r^2$ for π.

You have seen applications of quadratic expressions in areas of squares and of circles, and to describe paths of projectiles. There are too many other applications to describe them all here. In this lesson, we give just a few.

The total length of wires in the cables of the Brooklyn Bridge is approximately 3,600 miles.

Source: Endex Engineering, Inc.

Parabolas as Important Curves

In your future mathematics courses, you will study perhaps the most important property of a parabola: its reflective property. Because of its reflective property, a parabola is the shape of a cross-section of automobile headlights, satellite dishes, and radio telescopes.

Parabolas also appear on suspension bridges. When a chain is suspended between two fixed points, the curve it describes is a *catenary*. A catenary looks much like a parabola but is slightly deeper. But in a suspension bridge where the roadway is hung by support cables from the main cables, the shape of the main cable is a parabola.

Example 1

Suppose a team of engineers and construction workers are repairing a suspension bridge to strengthen it for use with increased traffic flow. The engineer uses scale models, such as the graph on the next page, to make decisions about repairs.

(continued on next page)

More Applications of Quadratics: Why Quadratics Are Important **565**

Let the roadway be along the *x*-axis. Place the *y*-axis at one end of the roadway. Then the ends of the bridge are at (0, 0) and (600, 0). The graph at the right shows that the parabola passes through points (0, 200), (300, 0), and (600, 200). By using quadratic regression, an equation for the parabola through these points can be found. In standard form, the equation is given by $y = \frac{1}{450}x^2 - \frac{4}{3}x + 200$, where *y* is the length of each vertical cable at the distance *x* from the end of the bridge. Suppose a support cable 82.5 feet long is delivered to the construction site. How far from the left end of the bridge should the cable be placed?

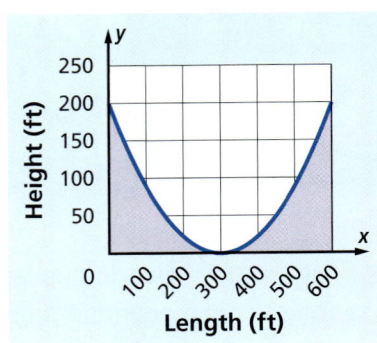

Solution 1 Substitute 82.5 for *y* in the equation of the parabola. Then solve the equation for *x*.

$82.5 = \frac{1}{450}x^2 - \frac{4}{3}x + 200$

$37{,}125 = x^2 - 600x + 90{,}000$ Multiply both sides by 450.

$0 = x^2 - 600x + 52{,}875$ Subtract 37,125 from both sides.

Now the equation $x^2 - 600x + 52{,}875 = 0$ can be solved using the Quadratic Formula, with $a = 1$, $b = -600$, and $c = 52{,}875$.

$x = \dfrac{-b \pm \sqrt{b^2 - 4ac}}{2a}$

$= \dfrac{-(-600) \pm \sqrt{(-600)^2 - 4(1)(52{,}875)}}{2(1)}$

$= \dfrac{600 \pm \sqrt{360{,}000 - 211{,}500}}{2}$

$= \dfrac{600 \pm \sqrt{148{,}500}}{2}$

$\approx \dfrac{600 \pm 385.4}{2}$

$x \approx 107.3$ ft or $x \approx 492.7$ ft

The cable can be placed either 107.3 ft or 492.7 ft from the left end of the bridge.

Solution 2 Enter the equation of the parabola and the desired *y* value into a calculator, as shown at the right.

Graph each equation over the domain $0 \leq x \leq 600$. The solutions to the problem are intersection points of the two functions.

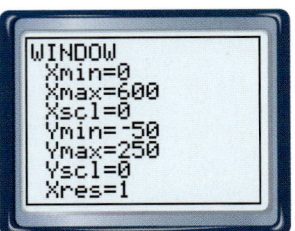
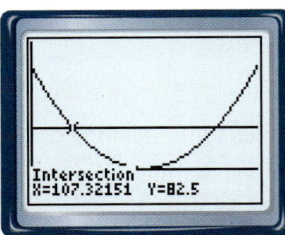
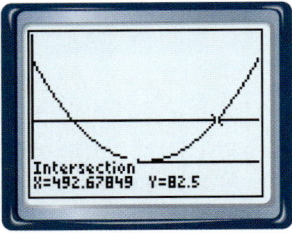

The x-coordinates of the intersections indicate that the cable is 82.5 feet long when x ≈ 107.3 ft and x ≈ 492.7 ft.

This makes sense because these distances are equally far from the center of the bridge at 300 ft.

A Geometry Problem Involving Counting

Many counting problems lead to quadratic equations. For example, the number d of diagonals of an n-sided convex polygon is given by the formula $d = \frac{n(n-3)}{2}$.

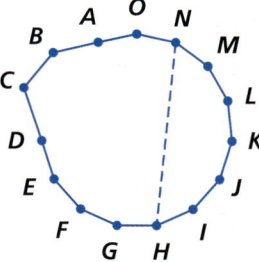

$\overline{NH}$ is one diagonal of this convex 15-sided polygon.

Example 2
a. How many diagonals does a convex polygon of 15 sides have?
b. Can a convex polygon have exactly 300 diagonals? If so, how many sides must that polygon have?

Solutions

a. When $n = 15$, $d = \frac{n(n-3)}{2} = \frac{15(15-3)}{2} = \frac{15 \cdot 12}{2} = 90$.
A 15-sided polygon has 90 diagonals.

b. Substitute 300 for d in the formula.

$300 = \frac{n(n-3)}{2}$

$600 = n(n-3)$ Multiply both sides by 2.

$600 = n^2 - 3n$ Distributive Property

To solve, write the equation in standard form.

$0 = n^2 - 3n - 600$ Subtract 600 from both sides.

This equation is in standard form with $a = 1$, $b = -3$, and $c = -600$. Use the Quadratic Formula.

$n = \frac{-b \pm \sqrt{b^2 - 4ac}}{2a}$

$= \frac{-(-3) \pm \sqrt{(-3)^2 - 4 \cdot 1 \cdot (-600)}}{2 \cdot 1}$

$= \frac{3 \pm \sqrt{9 + 2{,}400}}{2}$

$= \frac{3 \pm \sqrt{2{,}409}}{2}$

Because the discriminant 2,409 is not a perfect square, the values of n that we get are not integers. But n has to be an integer because it is the number of sides of a polygon. So the discriminant signals that there is no polygon with exactly 300 diagonals.

Chapter 9

Questions

COVERING THE IDEAS

1. A cable of length 100 feet is brought to the construction site in the situation of Example 1. How far from the left end of the bridge can that cable be placed?

2. Draw a convex decagon (10-sided polygon) and one of its diagonals. How many other diagonals does this polygon have?

3. Can a convex polygon have exactly 21 diagonals? If so, how many sides does that polygon have?

4. Can a convex polygon have exactly 2,015 diagonals? If so, how many sides does that polygon have?

5. The sum of the integers from 1 to n is $\frac{n(n+1)}{2}$. If this sum is 7,260, what is n?

6. The sum $5 + 6 + 7 + 8 + 9 + 10 = 45$ is an instance of the more general pattern that the sum of the integers from n to $2n$ is $1.5(n^2 + n)$.
    a. What is the sum of the integers from 100 to 200?
    b. If the sum of the integers from n to $2n$ is 759, what is n?

7. In the figure below, $\overline{PA}$ is *tangent* to the circle at point A. (It intersects the circle only at that point.) Another segment from P intersects the circle at points B and C. When you study geometry, you will learn that $PA^2 = PB \cdot PC$. Suppose $PA = 12$, $BC = 7$, and $PB = x$.

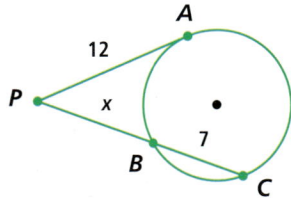

   a. Write an algebraic expression for PC.
   b. Substitute into $PA^2 = PB \cdot PC$ and solve the resulting quadratic equation to find PB.

APPLYING THE MATHEMATICS

8. In any circle O with diameter $\overline{PR}$, $SQ^2 = PQ \cdot RQ$, as shown at the right.
    a. If $PR = 10$ and $PQ = x$, write an algebraic expression for QR.
    b. Substitute the values from Part a into the formula $SQ^2 = PQ \cdot RQ$ to find SQ when $PR = 10$ and $PQ = 3$.

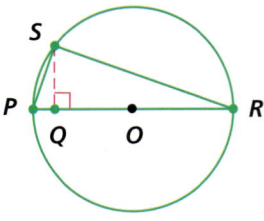

568   Quadratic Equations and Functions

9. Suppose an architect is designing a building with arched windows in the shape of a parabola. The area under the arch will be divided into windowpanes as shown in the diagram below. The architect needs to know the lengths of the four horizontal bars at heights 2, 4, 6, and 8 units. If the parabola has equation $y = -0.5x^2 + 8x - 22$, find the length, to the nearest tenth of a unit, of the bar at the 6-unit height.

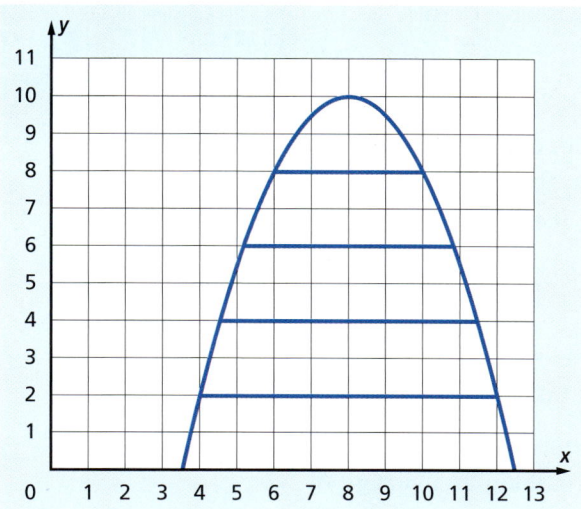

10. Secure telephone networks (ones in which each person is connected to every other person by a direct line) require $\frac{n(n-3)}{2} = \frac{1}{2}n^2 - \frac{1}{2}n$ cable lines for n employees. A company is interested in setting up a secure phone networking group. Each employee in the group is provided with a secure connection to all other employees in the group. Suppose enough money is allotted to provide cable for 500 such connections. How many employees can be enrolled in the group?

Radio and telecommunications equipment installers and repairers held about 222,000 jobs in 2004.

Source: U.S. Department of Labor

REVIEW

11. Suppose a quadratic equation has only one real solution. What can you conclude about its discriminant? **(Lesson 9-6)**

In 12–15, determine whether the equation has 0, 1, or 2 real number solutions. (You do not need to find the solutions.) **(Lesson 9-6)**

12. $-a^2 + 3a - 5 = 0$

13. $b^2 + 4b + 4 = 0$

14. $2c(c - 5) = 11$

15. $3(d - 6) = 5(d^2 + 11)$

16. Find an equation for a parabola with vertex (0, 0) that opens down. **(Lesson 9-1)**

17. **Skill Sequence** Solve the equation. **(Lessons 9-1, 8-6)**

   a. $\sqrt{a} = 6$
   b. $\sqrt{b + 8} = 6$
   c. $\sqrt{4c - 5} = 6$

In 18 and 19, use the figure at the right, which represents the front view of a building plan for a cottage. The cottage is to be 24 feet wide. The edges $\overline{AC}$ and $\overline{BC}$ of the roof are to be equal in length and to meet at a right angle. (Lessons 8-8, 8-7)

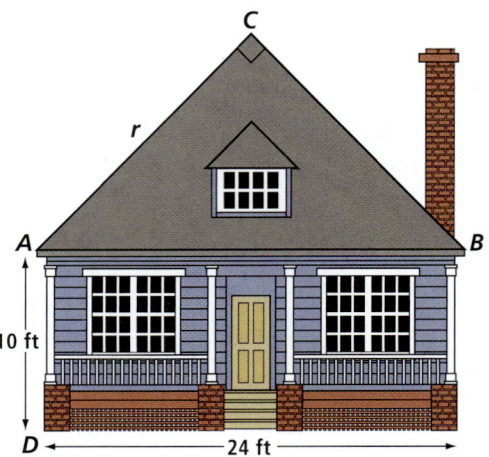

18. a. Find the length r of each edge as a simplified radical.
    b. Round the length of an edge to the nearest tenth of a foot.

19. Find BD.

20. **True or False** If the growth factor of an exponential growth situation is 2, then an equation that represents this situation is $y = 2 \cdot b^x$. (Lesson 7-2)

21. Consider the three points $(2, 1)$, $(-4, 31)$, and $(7, -24)$. (Lessons 6-8, 6-3)
    a. Show that these points lie on the same line.
    b. Write an equation for the line in standard form.

22. A school begins the year with 250 reams of paper. (A ream contains 500 sheets.) The teachers are using an average of 18 reams per week, and the school receives a shipment of 10 additional reams each week. (Lesson 2-2)
    a. How many reams will the school have after w weeks?
    b. Suppose the school year lasts 36 weeks. Assuming these rates continue, will the school run out of paper before the year ends?

EXPLORATION

23. Some telescopes use parabolic mirrors. Look on the Internet or in reference books to find out why it is useful to have mirrors shaped like parabolas and summarize your findings.

Chapter 9 Projects

1 Programming the Quadratic Formula

Use a computer or a graphing calculator to write a program that solves the equation $ax^2 + bx + c = 0$. The input of your program should be the numbers a, b, and c. Your program should state the number of real solutions and give them (if they exist). Your program should also work if $a = 0$.

2 Verifying Projectile Motion

For this project, you will need a device that can record video (such as a camcorder or a digital camera) and a television or computer on which to play the video you recorded. Have a friend throw a ball (or any other object), and use your recording device to film the path the ball takes. It is very important to keep the camera still the whole time, so be sure to stand far enough away from your friend. Tape a piece of tracing paper to a television. Play the video in slow motion and trace the path the ball takes. Does this path indeed seem to be a parabola? If possible, choose appropriate axes and find an equation for a parabola of best fit.

3 The Focus of a Parabola

Recall that a circle is the set of points at a fixed distance from a point called the center of the circle. Parabolas also have a definition involving distance from a special point called the *focus* of the parabola.

a. Look up this definition of a parabola, and write a paragraph that explains it.

b. Find out how the focus of a parabola is related to light reflecting from a mirror that is shaped like a parabola.

c. Parabolas are used in many manufactured items, such as satellite dishes and flashlights. Find three such uses that are related to the property you found in Part b, and explain why parabolas were used in these cases.

Chapter 9

4. Checking Whether Points Lie on a Parabola

a. Use a calculator or a spreadsheet to find the values of x^2, for $x = 1, 2, 3, \ldots, 100$. Then calculate the differences between the values for consecutive numbers, and plot them. For example, $2^2 - 1^2 = 3$, so you would plot the point $(2, 3)$, $3^2 - 2^2 = 5$, so your plot should include the point $(3, 5)$. There would be 99 points in all. What is an equation of a function that includes all 99 ordered pairs?

b. Choose three numbers: a, b, and c, and repeat Part a using $ax^2 + bx + c$ instead of x^2.

c. Repeat Part a using x^3 instead of x^2, and then using 2^x instead of x^2. Are the points on these plots still on a single line?

d. Parts a, b, and c suggest a method for checking if a collection of points are all on a parabola. Describe this method.

5. Formulas for Sums of Consecutive Integers

In this chapter, you saw the formula $S = \frac{n(n + 1)}{2}$ for the sum of the integers from 1 to n.

a. Graph this formula for values of n, from 1 to 10.

b. Find and graph a formula for the sum of the even integers from 2 to $2n$.

c. Find and graph a formula for the sum of the multiples of 3 from 3 to $3n$.

d. Generalize Parts a, b, and c to find a formula for the sum of the multiples of k, from k to kn.

6. Catenaries, Parabolas, and a Famous Landmark

The Gateway Arch in St. Louis, Missouri is one of the most famous landmarks in the United States. At first view, the shape of the Arch appears to be a parabola.

a. In Lesson 9-7, a shape called a catenary was mentioned. This shape looks very much like a parabola. Find out if the shape of the Gateway Arch is a parabola or a catenary, and why one is a better choice than the other.

b. Use the Internet to find a picture of a catenary. Print a large copy of this image on a piece of paper. Next, create a graph of a parabola that looks similar to the catenary, and print it. Make sure both printouts have the same size. Using these printouts as guidelines, use clay (or any other sculpting medium) to create a parabolic and a catenary model of the Gateway Arch. Which seems to be more stable?

Chapter 9

Summary and Vocabulary

- A **quadratic function** is a function f whose equation can be written in the form $f(x) = ax^2 + bx + c$ with $a \neq 0$. The simplest quadratic function is $f(x) = ax^2$. The **vertex** of the graph of $y = ax^2$ is at $(0, 0)$. The graph of $y = ax^2 + bx + c$ is a parabola symmetric to the vertical line through its vertex. If $a > 0$, the parabola opens up. If $a < 0$, the parabola opens down. To determine where this parabola crosses the horizontal line $y = k$, you can solve $ax^2 + bx + c = k$.

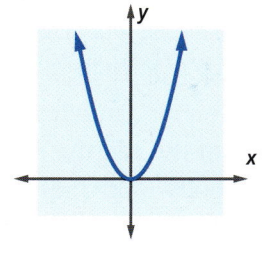
$y = ax^2$, if $a > 0$

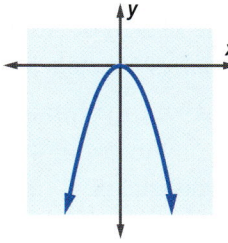
$y = ax^2$, if $a < 0$

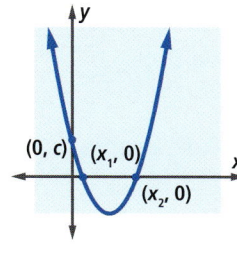
$y = ax^2 + bx + c$, if $a > 0$

- Quadratic expressions, equations, and functions appear in a variety of situations. The word "squaring" comes from applications of quadratic expressions in such formulas as $A = s^2$ and $A = \pi r^2$. The **path of a projectile** can be described by a quadratic equation. The function whose input is the time t (in seconds) since launch and whose output is the height h (in feet) of the projectile has the formula $h = -16t^2 + vt + s$, where s is the height at launch and v is the upward launch velocity (in feet per second).

- These situations, certain counting problems, and many geometric situations give rise to problems that can be solved by quadratic equations. The values of x that satisfy the equation $ax^2 + bx + c = 0$, where $a \neq 0$, can be found using the **Quadratic Formula**, $x = \dfrac{-b \pm \sqrt{b^2 - 4ac}}{2a}$.

- The **discriminant** of the quadratic equation $ax^2 + bx + c = 0$ is $b^2 - 4ac$. If the discriminant is positive, there are two real solutions to the equation; if it is zero, there is one solution; and if it is negative, there are no real solutions.

Vocabulary

9-1
parabola
reflection-symmetric
axis of symmetry
vertex

9-4
force of gravity
initial upward velocity
initial height

9-5
quadratic equation
standard form of a quadratic equation

9-6
discriminant

Theorems and Properties

General Formula for the Height of a Projectile over Time (p. 546)

The Quadratic Formula (p. 553)
Discriminant Property (p. 561)

Chapter 9 Self-Test

Take this test as you would take a test in class. You will need a calculator. Then use the Selected Answers section in the back of the book to check your work.

In 1–6, find all real solutions. Round your answers to the nearest hundredth. If there are no real solutions, write that.

1. $2x^2 = 162$
2. $n^2 - 8n - 10 = 0$
3. $5y^2 - 1 = 11y$
4. $24 = \frac{1}{6}z^2$
5. $v^2 = 16v - 64$
6. $3p^2 - 9p + 7 = 0$

7. If the discriminant of a quadratic equation is 6, how many solutions does the equation have?

8. **Multiple Choice** Which of these graphs is of the equation $y = 1.75x^2$?

A
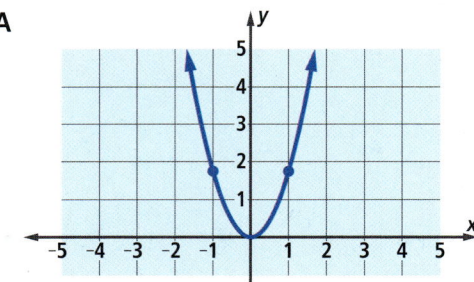

B
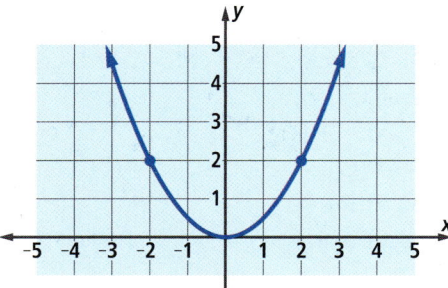

C
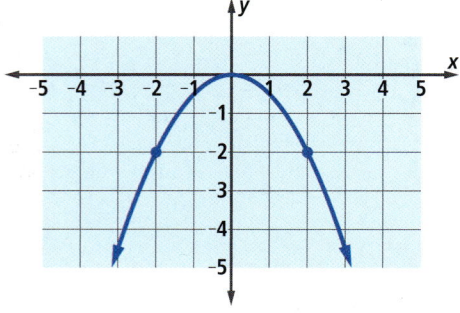

In 9 and 10, an equation is given.
   a. Make a table of values of x and y for integer values of x from $x = -3$ to $x = 3$.
   b. Graph the equation.

9. $y = 2x^2$
10. $y = -x^2 + 4x - 3$

In 11–13, consider the following. When a roller coaster goes down a hill, then $h = 0.049v^2$, where v is the velocity of the coaster (in meters per second) when it is h meters below the top of the hill.

11. Use the equation to determine at what distance below the top of the hill the roller coaster will reach a velocity of 20 meters per second.

12. Suppose the designer of the roller coaster builds the hill to be 44 meters high. At what velocity will the roller coaster be traveling when it reaches the bottom of the hill?

13. Currently the fastest wooden roller coaster is Son of Beast located at Kings Island in Cincinnati, Ohio. The maximum speed of the roller coaster is 35 meters per second. What is the height of the top of the hill?

14. The product of two consecutive integers, n and $n + 1$, is 1,722. If the integers are both negative, what are the numbers?

15. A circle and a rectangle have equal areas. One side of the rectangle is 6 inches longer than the other side, and the perimeter of the rectangle is 24. Calculate the radius of the circle, to the nearest hundredth.

In 16–18, use the parabola with vertex V below.

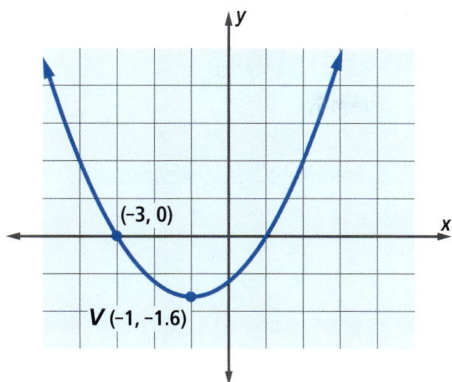

16. Find the minimum y value of the parabola.

17. Find the x-intercepts.

18. Find an equation for the axis of symmetry of the parabola.

In 19 and 20, a tennis ball is thrown from the top of a building. The graph below shows $h = -16t^2 + 40t + 50$, giving the height h of the ball in feet after t seconds.

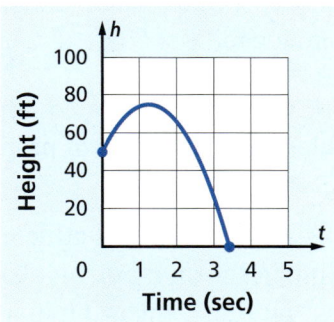

19. To the nearest hundredth of a second, how long does it take the ball to reach the ground?

20. At what times is the ball 70 feet above the ground? Give your answer to the nearest hundredth of a second.

21. **True or False** The parabola $y = \frac{1}{2}x^2 - 7x - 35$ opens down.

22. Suppose that the quadratic equation $ax^2 - 5x + 3 = 0$ has a discriminant of 1. Find the value of a.

In 23 and 24, use the discriminant to give the number of real solutions to the equation.

23. $-3x^2 + 12x - 7 = 0$

24. $x^2 - 4x = -4$

Chapter 9

Chapter Review

SKILLS
PROPERTIES
USES
REPRESENTATIONS

SKILLS Procedures used to get answers

OBJECTIVE A Solve quadratic equations of the form $ax^2 = b$. (Lesson 9-2)

In 1–8, solve without using the Quadratic Formula.

1. $4x^2 = 676$
2. $9 = \frac{1}{4}h^2$
3. $k^2 + 15 = 100$
4. $t^2 - 11 = 11$
5. $2(m + 3)^2 = 72$
6. $69 = 5 + 2y^2$
7. $\frac{63}{16} = 7(4 - v)^2$
8. $(6w - 1)^2 = \frac{25}{4}$

OBJECTIVE B Solve quadratic equations using the Quadratic Formula. (Lessons 9-5, 9-6)

In 9–18, solve the equation using the Quadratic Formula. Round your answers to the nearest hundredth.

9. $m^2 + 7m + 12 = 0$
10. $14x = x^2 + 49$
11. $y^2 - 6y = 3$
12. $r^2 - \frac{11}{7} = \frac{4}{5}r$
13. $0 = p^2 + 10(p + 2.5)$
14. $\frac{3}{4}x^2 - \frac{2}{3}x = 2$
15. $5n^2 + 9n = 2$
16. $2a^2 - 8a = -8$
17. $b^2 + 5.4b - 19.75 = 0$
18. $30 + 5(2z^2 - 10z) = 0$

PROPERTIES The principles behind the mathematics

OBJECTIVE C Identify and use the properties of solutions to quadratic equations. (Lesson 9-6)

19. Give the values of x that satisfy the equation $ax^2 + bx + c = 0$.

20. **True or False** If a quadratic equation has two solutions, then it has two x-intercepts.

In 21 and 22, calculate the discriminant.

21. $x^2 + 4x - 8 = 0$
22. $7y^2 - y = 1$

23. If the discriminant of the quadratic equation $x^2 + bx + 2 = 0$ is 8, find the possible value(s) of b.

In 24–27, find the number of real solutions to the equation by using the discriminant.

24. $g^2 - 3g - 8 = 0$
25. $3v = 2v^2 + 4$
26. $m^2 = 6m - 9$
27. $w(w - 2) = -8$

USES Applications of mathematics in real-world situations

OBJECTIVE D Use quadratic equations to solve problems about paths of projectiles. (Lessons 9-2, 9-4)

28. Regina is a track and field athlete competing in the shot put, an event that requires "putting" (throwing in a pushing motion) a heavy metal ball (the "shot") as far as possible. The height of the ball h when it is x feet from Regina can be described by the quadratic equation $h = -0.021x^2 + 0.6x + 6$.

   a. At what distances from Regina will the shot put be at a height of 8 feet? Round your answers to the nearest hundredth.

   b. Will Regina's shot put travel 38 feet, the distance needed to win the event? Justify your answer.

576 Quadratic Equations and Functions

In 29 and 30, when an object is dropped near the surface of a planet or moon, the distance d (in feet) it falls in t seconds is given by the formula $d = \frac{1}{2}gt^2$, where g is the acceleration due to gravity. Near Earth $g \approx 32$ ft/sec^2, and near Earth's moon $g \approx 5.3$ ft/sec^2.

29. A skydiver jumps from a plane at an altitude of 10,000 feet. She begins her descent in "free fall," that is, without opening the parachute.
   a. How far will she fall in 15 seconds?
   b. The diver plans to open the parachute after she has fallen 6,000 feet. How many seconds after jumping will this take place?

30. An astronaut on the moon drops a hammer from a height of 6 feet.
   a. How long will it take the hammer to hit the ground?
   b. Suppose the astronaut is back on Earth and drops a hammer from a height of 6 feet. How long will it take the hammer to hit the ground?

31. Refer to the graph below of $h = -4.9t^2 + 20t$, which shows the height (in meters) of a ball t seconds after it is thrown from ground level at an initial upward velocity of 20 meters per second.

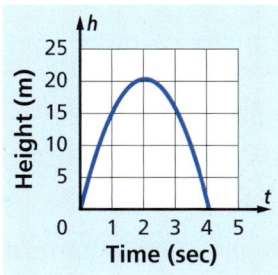

   a. Give the height of the ball after 1 second.
   b. Find when the ball will reach a height of 15 meters.
   c. Use the Quadratic Formula to calculate how long the ball will be in the air.

OBJECTIVE E Solve geometric problems involving quadratic equations. (Lessons 9-2, 9-7)

32. Suppose a rectangle has length $2x$ inches and width $x + 3$ inches. Find the length and width given that the area of the rectangle is 5.625 square inches.

33. Consider the rectangular region below. Luisa has 120 meters of fencing to build a fence around her yard.

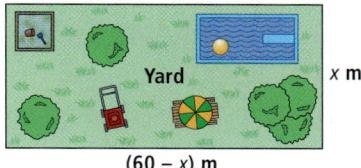

   a. Use the diagram to write an equation for the area A enclosed by the fencing.
   b. What value of x will result in the greatest possible area enclosed by the fencing?

34. Refer to the triangle below. If the area of the triangle is 18 square inches and the base of the triangle is 4 inches shorter than its height, find the length of the base of the triangle. Round your answer to the nearest hundredth of an inch.

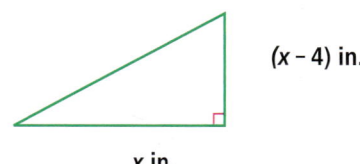

OBJECTIVE F Solve other real-world problems involving quadratic functions. (Lesson 9-7)

35. The relationship between elevation above sea level in kilometers, e, and the boiling point of water in degrees Celsius, t, can be approximated by the equation $e = t^2 - 200.58t + 10,058$. Water boils at lower temperatures at higher elevations. Find the boiling point of water at the top of Mt. Ararat in Turkey, which is 5,166 meters high.

36. Consider the formula $d = \frac{n(n-3)}{2}$, where d is the number of diagonals of an n-sided convex polygon.
    a. How many diagonals does a dodecagon (12-sided polygon) have?
    b. Is it possible for a polygon to have 27 diagonals? If so, how many sides does that polygon have?

37. A financial analyst working for an investment company projects the net profit P (in millions of dollars) of the company to be modeled by the equation $P = 4.23t^2 - 5.32t + 3.86$, where t is the number of years since 2005.
    a. Use the analyst's model to predict whether the net profit will reach 100 million dollars in 2011.
    b. Calculate, to the nearest tenth, when the company's net profit is projected to reach 75 million dollars.

REPRESENTATIONS Pictures, graphs, or objects that illustrate concepts

OBJECTIVE G Graph equations of the form $y = ax^2$ and interpret these graphs. (Lesson 9-1)

In 38 and 39, an equation is given.
    a. Make a table of values.
    b. Graph the equation.

38. $y = 7x^2$
39. $y = \frac{3}{5}x^2$

40. Consider the quadratic equation $y = -1.5x^2$.
    a. Graph the equation.
    b. Determine the coordinates of the vertex.
    c. Tell whether the vertex is a maximum or a minimum.

41. **Multiple Choice** Which equation is graphed below?

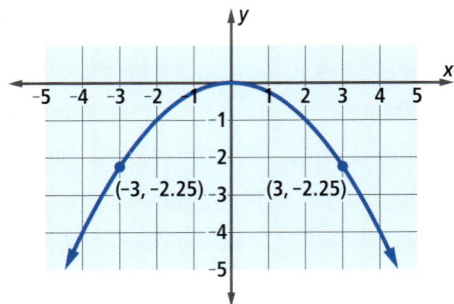

A $y = -4x^2$
B $y = -\frac{1}{4}x^2$
C $y = \frac{1}{4}x^2$
D $y = 4x^2$

42. **True or False** The axis of symmetry for the parabola with equation $y = 2x^2$ is the line $x = 0$.

OBJECTIVE H Graph equations of the form $y = ax^2 + bx + c$ and interpret these graphs. (Lesson 9-3)

43. Use the parabola with vertex V below.

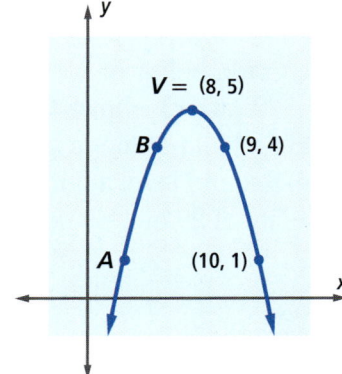

a. What is the maximum value of the function?
b. What is an equation for its axis of symmetry?
c. Find the coordinates of points A and B, the reflection images of the named points over the parabola's axis of symmetry.

578 Quadratic Equations and Functions

In 44 and 45, answer *true* or *false*.

44. Every parabola has a minimum value.

45. The parabola $y = -4x^2 + 2x - 13$ opens down.

46. What equation must you solve to find the x-intercepts of the parabola $y = ax^2 + bx + c$?

47. The parabola $y = \frac{1}{8}x^2 - 6x + 22$ has x-intercepts 4 and 44. Find the coordinates of its vertex without graphing.

48. A table of values for a parabola is given below.

x	0	2	4	6	8	10	12
y	?	?	−6	−8	−6	0	10

   a. Complete the table.
   b. Write an equation for the parabola's axis of symmetry.
   c. What are the coordinates of its vertex?

49. Consider the quadratic equation $y = -x^2 - 4x + 3$.

   a. Make a table of x and y values for integer values of $-5 \leq x \leq 1$.
   b. Graph the equation.
   c. Determine whether the vertex is a minimum or a maximum.

50. Which of these is the graph of $y = x^2 - 4x + 5$?

A
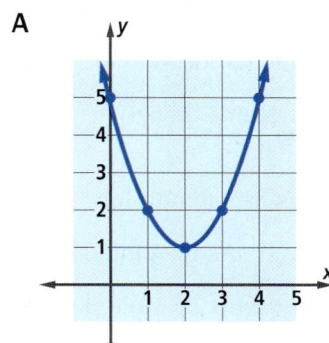

B
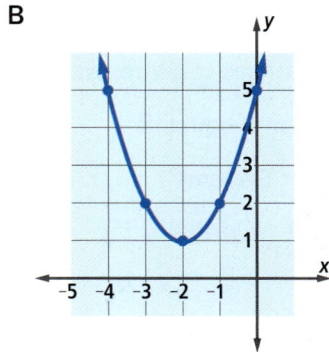

Chapter 10 Linear Systems

- **10-1** An Introduction to Systems
- **10-2** Solving Systems Using Substitution
- **10-3** More Uses of Substitution
- **10-4** Solving Systems by Addition
- **10-5** Solving Systems by Multiplication
- **10-6** Systems and Parallel Lines
- **10-7** Matrices and Matrix Multiplication
- **10-8** Using Matrices to Solve Systems
- **10-9** Systems of Inequalities
- **10-10** Nonlinear Systems

The table and graph below display the men's and women's winning times in the Olympic 100-meter freestyle swimming race for each Summer Olympic year from 1912 to 2004.

Year	Men's Time (sec)	Women's Time (sec)	Year	Men's Time (sec)	Women's Time (sec)
1912	63.4	82.2	1968	52.2	60.0
1920	60.4	73.6	1972	51.22	58.59
1924	59.0	72.4	1976	49.99	55.65
1928	58.6	71.0	1980	50.40	54.79
1932	58.2	66.8	1984	49.80	55.92
1936	57.6	65.9	1988	48.63	54.93
1948	57.3	66.3	1992	49.02	54.64
1952	57.4	66.8	1996	48.74	54.50
1956	55.4	62.0	2000	48.30	53.83
1960	55.2	61.2	2004	48.17	53.84
1964	53.4	59.5			

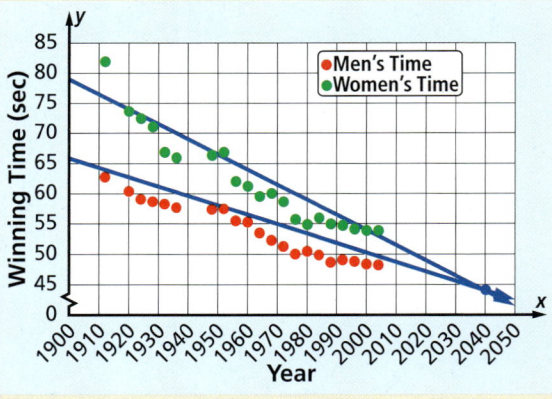

Source: *The World Almanac and Book of Facts*

The graph shows two trends. First, both men's and women's Olympic winning times have been decreasing rather steadily since 1912. Second, the women's winning time has been decreasing faster than the men's winning times. Regression lines have been fitted to the data. These lines have the following equations:
$y = -0.1627x + 372.99$ (men) and
$y = -0.269x + 589.83$ (women), where x is the year and y is the winning time in seconds. The lines intersect near (2040, 41).

This means that if the winning times were to continue to decrease at the rates they have been decreasing, the women's winning time will be about equal to the men's in the Olympic year 2040. The winning times will then each be about 41 seconds.

Finding points of intersection of lines or other curves by working with their equations is called *solving a system*. In this chapter you will learn various ways of solving systems.

Chapter 10

Lesson 10-1
An Introduction to Systems

Vocabulary

system
solution to a system
empty set, null set

> **BIG IDEA** Solving a system of equations means finding all the solutions that are common to the equations.

A **system** is a set of equations or inequalities joined by the word *and*, that together describe a single situation. The two equations at the beginning of this chapter describing the winning Olympic times of men and women in the 100-meter freestyle events are an example of a system of equations.

Systems are often signaled by using a single left-hand brace { in place of the word *and*. So we can write this system as $\begin{cases} y = -0.1627x + 372.99 \\ y = -0.269x + 589.83 \end{cases}$.
When you write a system in this way, it is helpful to align the equal signs under one another.

Mental Math

a. How many quarters make $10.50?

b. How many dimes make $10.50?

c. How many nickels make $10.50?

What Is a Solution to a System?

A **solution to a system** of equations with two variables is an ordered pair (x, y) that satisfies both equations in the system.

Hungary's Laszlo Cseh and U.S. swimmers Michael Phelps and Ryan Lochte stand on the victory podium for the 200-meter individual medley at the 2008 Summer Olympic Games in Beijing, China.

Example 1

Consider the system $\begin{cases} y = 3x - 7 \\ 2y - 2x = 10 \end{cases}$.

a. Verify that the ordered pair (6, 11) is a solution to the system.
b. Show that (1, −4) is *not* a solution to the system.

Solutions

a. In each equation, replace x with 6 and y with 11.

First equation: $y = 3x - 7$
Does $11 = 3 \cdot 6 - 7$?
$11 = 18 - 7$ Yes.

Second equation: $2y - 2x = 10$
Does $2 \cdot 11 - 2 \cdot 6 = 10$?
$22 - 12 = 10$ Yes.

(6, 11) is a solution because it satisfies both equations.

582 Linear Systems

b. Substitute 1 for x and −4 for y in both equations. The pair (1, −4) is a solution to the first equation because $-4 = 3 \cdot 1 - 7$. However, $2 \cdot -4 - 2 \cdot 1 = -10 \neq 10$. So (1, -4) is not a solution to the system.

 QY

> **QY**
>
> Is the ordered pair (2, −1) a solution to the system in Example 1?

Solving Systems by Graphing

You can find the solutions to a system of equations with two variables by graphing each equation and finding the coordinates of the point(s) of intersection of the graphs.

Example 2

A second-grade class has 23 students. There are 5 more boys than girls. Solve a system of equations to determine how many boys and how many girls are in the class.

Solution Translate the conditions into a system of two equations. Let x be the number of boys and y be the number of girls. Because there are 23 students, $x + y = 23$. Because there are 5 more boys than girls, $y + 5 = x$. The situation is described by the system $\begin{cases} x + y = 23 \\ y + 5 = x \end{cases}$.

Graph the equations and identify the point of intersection.

There are 14 boys and 9 girls in the class. The solution is (14, 9), as graphed on the next page.

Check To check that (14, 9) is a solution, $x = 14$ and $y = 9$ must be checked in both equations.

Is (14, 9) a solution to $x + y = 23$? Does $14 + 9 = 23$? Yes.

Is (14, 9) a solution to $y + 5 = x$? Does $9 + 5 = 14$? Yes.

A total of 49.6 million children attended public and private schools in the United States in 2003.

Source: U.S. Census Bureau

The two conditions about the numbers of boys and girls can be seen by looking at tables of solutions for each equation.

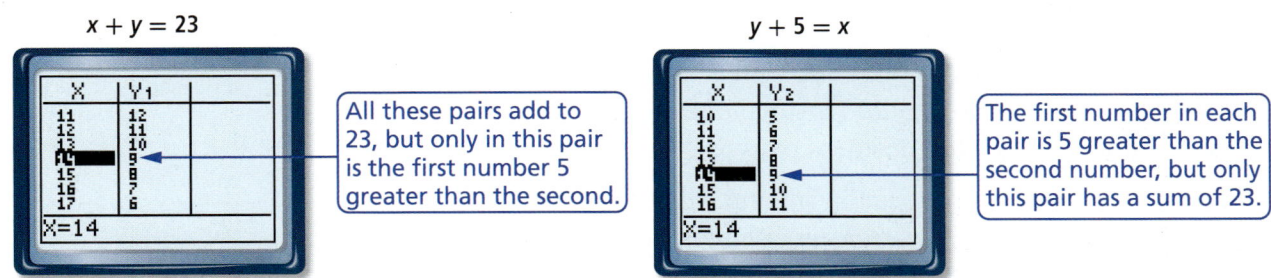

All these pairs add to 23, but only in this pair is the first number 5 greater than the second.

The first number in each pair is 5 greater than the second number, but only this pair has a sum of 23.

An Introduction to Systems

In general, there are four ways to indicate the solution to a system. They are shown below using the solution to the system in Example 2.

As an ordered pair	(14, 9)
As an ordered pair identifying the variables	$(x, y) = (14, 9)$
By naming the variables individually	$x = 14$ and $y = 9$
As a set of ordered pairs	$\{(14, 9)\}$

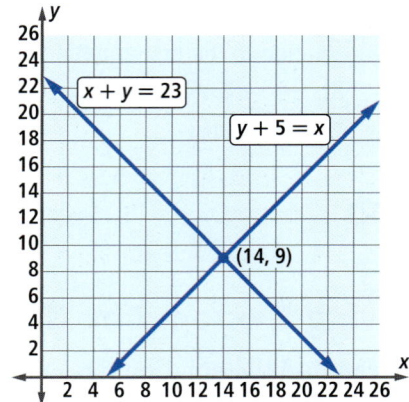

Systems with No Solutions

When the sentences in a system have no solutions in common, we say that there is no solution to the system. We cannot write the solution as an ordered pair or by listing the elements. The solution set is the set with no elements { }, written with the special symbol ∅. This set is called the **empty set** or **null set.**

GUIDED

Example 3

Find all solutions to the system $\begin{cases} y = -x + 4 \\ y = -x - 2 \end{cases}$.

Solution Graph both equations using a graphing calculator.

1. How many times do the lines appear to intersect? ___?___
2. What is the slope of each line? ___?___
3. If two lines in a plane do not intersect, they are called ___?___ lines.

Now look at the table of values on the graphing calculator.

4. Is there an ordered pair common to both lines? ___?___

This is an example of a system of equations for which there is no solution. **The solution set is ∅.**

Cost and Revenue Equations

In manufacturing, a *cost equation* describes the cost y of making x products. *Fixed costs* are things like rent and employee salaries, which must be paid regardless of the number of products made. *Variable costs* include materials and shipping, which depend upon how many products are made. The total of fixed and variable costs is the amount of money the business pays out each month.

A *revenue equation* describes the amount y that a business earns by selling x products. The *break-even point* is the point at which the revenue and total costs are the same. This point tells the manufacturer how many items must be sold in order to make a profit.

Example 4

A manufacturer of T-shirts has monthly fixed costs of $8,000, and the cost to produce each shirt is $3.40. Therefore, the cost y to produce x shirts is given by $y = 8{,}000 + 3.40x$. The business sells shirts to stores for $9 each. So the revenue equation is $y = 9x$. Find the break-even point for the shirt manufacturer.

A person is shown silk-screening T-shirts. Silk-screening is a process in which color is forced into material like fabric or paper through a silk screen.

Solution The break-even point can be found by solving the system $\begin{cases} y = 8{,}000 + 3.40x \\ y = 9x \end{cases}$.

Make a table to help you find reasonable x and y values to use in setting the window. The table below shows that the intersection point will be seen in the window $1{,}400 \le x \le 1{,}500$, $12{,}600 \le y \le 13{,}500$.

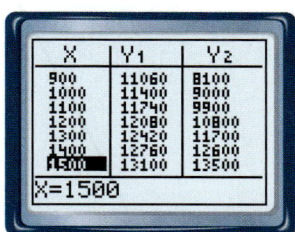

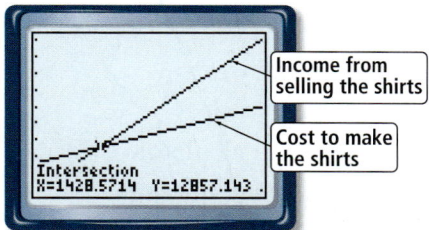

The INTERSECT command on a calculator shows that the point (1,429, 12,858) is an approximate solution. At the break-even point, 1,429 shirts are manufactured and sold. It costs about the same amount to make the shirts as the manufacturer earns from selling them. If more than 1,429 shirts are produced and sold, the business will earn a profit.

Check When $x = \$1{,}429$ in the cost equation, $y = \$8{,}000 + 3.40 \cdot \$1{,}429 = \$12{,}858.60$. When $x = 1{,}429$ in the revenue equation, $y = 9 \cdot \$1{,}429 = \$12{,}861$. These values are close enough to make 1,429 the first coordinate of the break-even point.

In Example 4, the solution is an approximation. When solutions do not have integer coordinates, it is likely that reading a graph will give you only an estimate. But graphs can be created quickly. In the next few lessons, you will learn algebraic techniques to find exact solutions to systems.

Chapter 10

Questions

COVERING THE IDEAS

1. **True or False** When a system has two variables, each solution to the system is an ordered pair.

2. What does the brace { represent in a system?

3. a. Verify that (8, −2) is a solution to the system $\begin{cases} -x = 4y \\ 2x + 3y = 10 \end{cases}$.
   b. Write this solution in two other ways.

4. Show that (11, 8) is *not* a solution to $\begin{cases} y = x - 3 \\ x + 5y = 50 \end{cases}$.

5. Refer to the graph at the right.
   a. What system is represented?
   b. What is the solution to the system?
   c. Verify your answer to Part b.

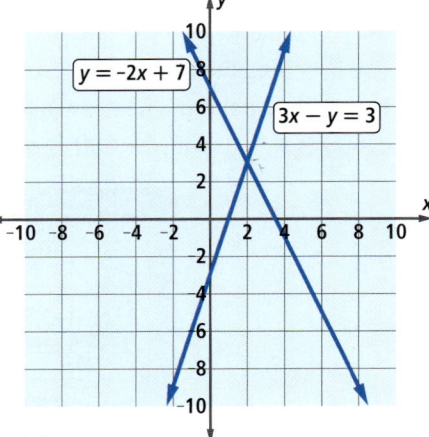

In 6 and 7, a system is given.
   a. Solve the system by graphing.
   b. Check your solution.

6. $\begin{cases} y = -\frac{1}{2}x + 3 \\ y = 2x - 7 \end{cases}$

7. $\begin{cases} y = x \\ 4x - 2y = 12 \end{cases}$

8. An elementary school has 518 students. There are 4 more girls than boys.
   a. If g is the number of girls and b is the number of boys, translate the given information into two equations.
   b. Letting $x = g$ and $y = b$, graph the equations on a calculator.
   c. Using the graph from Part b, use the INTERSECT command to find the number of boys and girls in the school.

9. A small business makes wooden toy trains. The business has fixed expenses of $3,800 each month. In addition to this, the production of each train costs $4.25. The business sells the trains to stores for $12.50 each.
   a. Write cost and revenue equations as a system.
   b. Use a calculator table to find a good window to display the graph. What window did you use?
   c. Find the break-even point.
   d. Last month the business made and sold 518 trains. Did the business earn a profit?

10. Find all solutions to the system $\begin{cases} y = 3x - 5 \\ y = 3x - 1 \end{cases}$.

Archaeologists have discovered jointed wooden dolls, carved horses, chariots, and even a crocodile with moveable jaws that date back to the year 1100 B.C.E.

Source: *TDmonthly*

586 Linear Systems

11. Consider the system $\begin{cases} y_1 = 2x + 5 \\ y_2 = 3x \end{cases}$.

    The screen at the right shows solutions to $y_1 = 2x + 5$. Make a column for y_2 and use it to find the ordered pair that also satisfies $y_2 = 3x$, and therefore is a solution to the system.

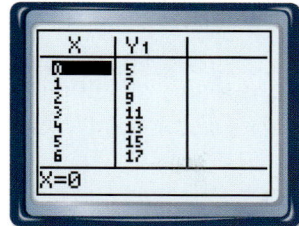

APPLYING THE MATHEMATICS

12. The sum of two numbers is –19 and their difference is –5. Write a system of equations and solve it using a graph.

13. Below are a table and graph of the winning times in seconds for the Olympic men's and women's 100-meter backstroke events.

Year	Men's Time	Women's Time	Year	Men's Time	Women's Time
1924	73.2	83.2	1972	56.58	65.78
1928	68.2	82.0	1976	55.49	61.83
1932	68.6	79.4	1980	56.33	60.86
1936	65.9	78.9	1984	55.79	62.55
1948	66.4	74.4	1988	55.05	60.89
1952	65.4	74.3	1992	53.98	60.68
1956	62.2	72.9	1996	54.10	61.19
1960	61.9	69.3	2000	53.72	61.21
1964	NA	67.7	2004	54.06	60.37
1968	58.7	66.2			

Source: International Olympic Committee

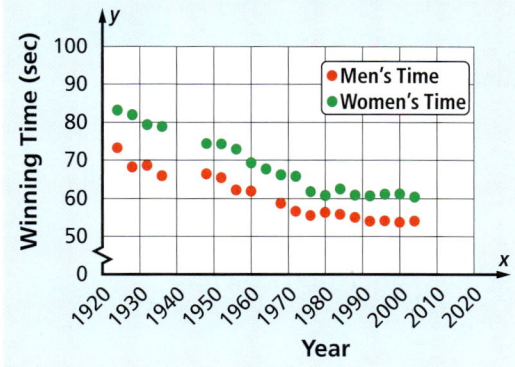

In Lesson 4-4, you were asked to estimate when the women's time might equal the men's time. Now repeat this question by finding equations for lines of best fit for the men's and women's times. Graph the two equations. According to these lines, will the women's winning time ever equal the men's winning time in the 100-meter backstroke? If yes, estimate the year when this will happen. If no, explain why not.

14. Buffy is hosting a meeting and plans to serve 4 dozen muffins. She wants to have twice as many blueberry muffins as plain muffins. The table at the right shows some of the possible ways to order 4 dozen muffins.

    a. Using x for the number of plain muffins and y for the number of blueberry muffins, write a system of equations to describe this situation.

    b. Graph your equations from Part a to find how many of each kind Buffy should order.

Number of Plain Muffins	Number of Blueberry Muffins
0	48
10	38
20	28
30	18
40	8
48	0

REVIEW

15. **Skill Sequence** Solve each equation. (Lessons 9-5, 4-4, 3-4)

    a. $5x + 6 = 3$ b. $5x + 6 = 2x + 3$
    c. $5x + 6 = 2x^2 + 3$ d. $5x + 6 = 2x(x + 1)$

In 16–18, simplify the expression. (Lessons 8-4, 8-3, 8-2)

16. $m^2 \cdot n^3 \cdot m \cdot n^4$ 17. $(-5x^7y^9)^4$ 18. $\dfrac{18r^2s^3}{6rs^4}$

19. Graph $\{(x, y): 4x - 8y < 2\}$. (Lesson 6-9)

20. Find the values of the variables so that the given point lies on the graph of $10x - 4y = 20$. (Lessons 6-8, 4-7)

    a. $(5, p)$ b. $(q, -2)$ c. $(r, 0)$

21. Two workers can dig a 20-foot well in 2 days. How long will it take 6 workers to dig a 90-foot well, assuming that each of these 6 workers dig at the same rate as each of the 2 workers? (Lessons 5-4, 5-3)

EXPLORATION

22. Some experts believe that even though the women's swim times are decreasing faster than the men's, it is the ratio of the times that is the key to predictions.

    a. Compute the ratio of the men's time to the women's time for the 100-meter freestyle for each Olympic year in Question 13.

    b. Graph your results.

    c. What do you think the ratio will be in the year 2020? Does this agree with the prediction in Question 13?

QY ANSWER

no

Lesson 10-2

Solving Systems Using Substitution

> **BIG IDEA** Substituting an expression that equals a single variable is an effective first step for solving some systems.

When equations for lines in a system are in $y = mx + b$ form, a method of solving called *substitution* can be very efficient. Example 1 illustrates this method.

Mental Math

Find the greatest common factor.

a. 15; 200
b. 1,500; 20,000
c. 14; 26; 53
d. 1,400; 2,600; 5,300

Example 1

Solve the system $\begin{cases} y = 7x + 25 \\ y = -5x - 11 \end{cases}$ using substitution.

Solution Because $7x + 25$ and $-5x - 11$ both equal y, they must equal each other. Substitute one of them for y in the other equation.

$7x + 25 = -5x - 11$ Substitution

$12x + 25 = -11$ Add $5x$ to both sides.

$12x = -36$ Subtract 25 from both sides.

$x = -3$ Divide both sides by 12.

Now you know $x = -3$. However, you must still solve for y. You can substitute -3 for x into either of the original equations. We choose the first equation.

$y = 7x + 25$

$y = 7(-3) + 25$

$y = -21 + 25$

$y = 4$

The solution is $x = -3$ and $y = 4$, or just $(-3, 4)$.

 QY

Check A graph shows that the lines with equations $y = 7x + 25$ and $y = -5x - 11$ intersect at $(-3, 4)$.

> **QY**
>
> Check that $(-3, 4)$ is a solution to $y = -5x - 11$.

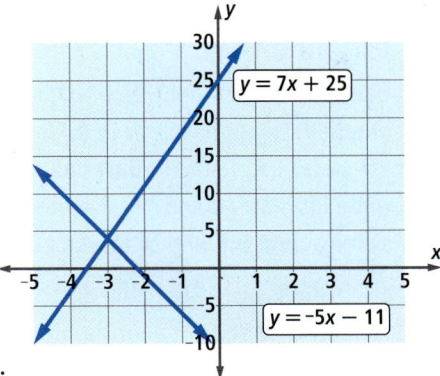

Suppose two quantities are increasing or decreasing at different constant rates. Then each quantity can be described by an equation of the form $y = mx + b$. To find out when the quantities are equal, you can solve a system using substitution. Example 2 illustrates this idea.

Solving Systems Using Substitution   589

Chapter 10

Example 2
The Rapid Taxi Company charges $2.15 for a taxi ride plus 20¢ for each $\frac{1}{10}$ mile traveled. A competitor, Carl's Cabs, charges $1.50 for a taxi ride plus 25¢ for each $\frac{1}{10}$ mile traveled. For what distance do the rides cost the same?

Solution Let d = the distance of a cab ride in tenths of a mile.

Let C = the cost of a cab ride of distance d.
Rapid Taxi: $C = 2.15 + 0.20d$
Carl's Cabs: $C = 1.50 + 0.25d$

The rides cost the same when the values of C and d for Rapid Taxi equal the values for Carl's Cabs, so we need to solve the system formed by these two equations. Substitute $2.15 + 0.20d$ for C in the second equation.

$2.15 + 0.20d = 1.50 + 0.25d$
Now solve.

$\quad\quad 0.65 = 0.05d \quad$ Add -1.50 and $-0.20d$ to both sides.
$\quad\quad\quad\ d = 13 \quad\quad$ Divide both sides by 0.05.

The two companies charge the same amount for a ride that is 13 tenths of a mile long, or 1.3 miles long.

Check Check to see if the cost will be the same for a ride of 13 tenths of a mile.
The cost for Rapid Taxi is $2.15 + 0.20 \cdot 13 = 2.15 + 2.60 = 4.75$.
The cost for Carl's Cabs is $1.50 + 0.25 \cdot 13 = 1.50 + 3.25 = 4.75$.
The cost is $4.75 from each company, so the answer checks.

The average taxi fare in New York City in 2013 was $13.60.

Source: 2014 Taxicab Fact Book, NYC.gov

In Example 2, Carl's Cabs is cheaper at first, but as the number of miles increases, the prices become closer. Eventually the price for Carl's Cabs catches up with Rapid Taxi's price, and then Carl's is more expensive than Rapid. The next example also involves "catching up."

GUIDED
Example 3
Bart was so confident that he could run faster than his little sister that he bragged, "I can beat you in a 50-meter race. I'm so sure that I'll give you a 10-meter head start!" Bart could run at a speed of 4 meters per second, while his sister could run 3 meters per second. Could Bart catch up to his sister before the end of the race?

Solution Let d be the distance that Bart and his sister have traveled after t seconds. Recall that distance = rate · time.

For Bart, $d = 4t$.

590   Linear Systems

Because Bart gives his sister a 10-meter head start, $d = 10 + 3t$. To know the time t when Bart will catch up to his sister, solve the system.

$$\begin{cases} d = \underline{\ ?\ } \\ d = \underline{\ ?\ } \end{cases}$$

Substitute $4t$ for d in the second equation.

$$\underline{\ ?\ } = \underline{\ ?\ } + \underline{\ ?\ }$$

Solve this equation as you would any other.

$$t = \underline{\ ?\ }$$

This means that after 10 seconds, Bart and his sister are at the same point. However, is the race finished at 10 seconds? Substitute 10 for t to find the distance. In 10 seconds Bart has run 40 meters. Because the race is 50 meters long, the race is not over when Bart catches up to his sister. Therefore, Bart wins.

Questions

COVERING THE IDEAS

In 1–5, a system is given.
  a. Use substitution to find the solution.
  b. Check your answer.

1. $\begin{cases} y = 3x - 4 \\ y = 5x - 10 \end{cases}$

2. $\begin{cases} b = 48 + a \\ b = 60 - a \end{cases}$

3. $\begin{cases} y = -\frac{1}{9}x + 6 \\ y = \frac{5}{3}x + 38 \end{cases}$

4. $\begin{cases} x = \frac{2}{3}y - 8 \\ x = -12.5y + 150 \end{cases}$

5. $\begin{cases} m = 8n + 33 \\ m = 3n - 78 \end{cases}$

6. Suppose that in Freeport, a taxi ride costs $2.50 plus 15¢ for each $\frac{1}{10}$ mile traveled. In Geneva, a taxi ride costs $1.70 plus 20¢ for each $\frac{1}{10}$ mile traveled. Write a system of equations and solve it to find the distance for which the costs are the same.

7. Recall from Example 3 that Bart ran 4 meters per second and his sister ran 3 meters per second. Bart's sister said to him, "You'll beat me if I have only a 10-meter head start. I'll race you if you give me a 15-meter head start." Solve a system of equations to find out if she would then beat him in a 50-meter race.

Solving Systems Using Substitution

Chapter 10

8. A tomato-canning company has fixed monthly costs of $4,200. There are additional costs of $2.35 to produce each case of canned tomatoes. The company sells tomatoes to grocery stores for $5.85 per case.
   a. Write a system of equations to describe this situation.
   b. How many cases must the company sell to break even?
   c. Check your solution.

In 2014, a total of 14.6 million tons of process tomatoes were grown on 277,000 acres in the U.S., with a total value of approximately $1.325 billion.

Source: Agricultural Marketing Resource Center

APPLYING THE MATHEMATICS

9. A car leaves a gas station traveling at 60 mph. The driver has accidentally left his credit card at the gas station. Six minutes later, his friend leaves the station with the credit card, traveling at 65 mph to catch up to him.
   a. Write two equations to indicate the distance d that each car is from the gas station t hours after the first car leaves.
   b. Solve the system to determine when the second car will catch up to the first car.
   c. How far will they have traveled from the gas station when they meet?

10. Cameron has $450 and saves $12 a week. Sean has only $290, but is saving $20 a week.
    a. After how many weeks will they each have the same amount of money?
    b. How much money will each person have then?

11. In 2000, the metropolitan area of Dallas had about 5,200,000 people and was growing at about 120,000 people a year. In 2000, the metropolitan area of Boston had about 4,400,000 people and was growing at about 25,000 people a year.
    a. If these trends had been this way for quite some time, in what year did Dallas and Boston have the same population?
    b. What was this population?

12. In July 2005, Philadelphia approved taxi fares with an initial charge of $2.30 and an additional charge of $0.30 for each $\frac{1}{7}$ mile. If $P(x)$ is the cost for taking a taxi x miles, then $P(x) = 2.30 + 0.30 \cdot 7x$. In October 2005, Atlanta established new taxi fares with an initial charge of $2.50 and an additional charge of $0.25 for each $\frac{1}{8}$ mile. If $A(x)$ is the cost of taking a taxi x miles in Atlanta, then $A(x) = 2.50 + 0.25 \cdot 8x$. Solve a system to approximate at what distance the fares for Philadelphia and Atlanta are the same.

13. One plumbing company charges $55 for the first half hour of work and $25 for each additional half hour. Another company charges $35 for the first half hour and then $30 for each additional half hour. For how many hours of work will the cost of each company be the same?

In 14 and 15, a system that involves a quadratic equation is given. Each system has two solutions.
   a. Solve the system by substitution.
   b. Check your answers.

14. $\begin{cases} y = \frac{1}{9}x^2 \\ y = 4x \end{cases}$

15. $\begin{cases} y = 2x^2 + 5x - 3 \\ y = x^2 - 2x + 5 \end{cases}$

REVIEW

16. Consider the system $\begin{cases} y = 20x + 8 \\ 24x - y = -6 \end{cases}$. Verify that $\left(\frac{1}{2}, 18\right)$ is a solution to the system, but that $(1, 20)$ is not. (Lesson 10-1)

In 17 and 18, solve the system of equations by graphing. (Lesson 10-1)

17. the system in Question 3

18. the system in Question 4

19. a. Simplify $y(y - 9) + 4y + 1$.
    b. Solve $y(y - 9) = 4y + 1$. (Lesson 9-5)

20. **Skill Sequence** Solve each equation. (Lessons 9-1, 8-6)
    a. $n^2 = 16$ b. $\sqrt{n} = 16$ c. $\sqrt{n^2} = 16$

21. What is the cost of x basketballs at $18 each and y footballs at $25 each? (Lessons 5-3, 1-2)

EXPLORATION

22. Find the taxi rates where you live or in a nearby community. Graph the rates to show how they compare to those in Question 12.

QY ANSWER

Does $4 = -5(-3) - 11$?
$4 = 15 - 11$
$4 = 4$ Yes, it checks.

Chapter 10

Lesson 10-3

More Uses of Substitution

▶ **BIG IDEA** Substitution is a reasonable method to solve systems whenever you can easily solve for one variable in an equation.

In the previous lesson, you saw how to use substitution as a technique to solve systems of equations when the same variable was alone on one side of each equation. Substitution may also be used in other situations. Here is a typical situation that lends itself to substitution.

Mental Math

Estimate between two consecutive integers.

a. $\sqrt{26}$
b. $\sqrt{171}$
c. $-\sqrt{171}$

Example 1

A grandfather likes to play guessing games with his grandchildren. One day he tells them, "I have only dimes and quarters in my pocket. They are worth $3.85. I have 14 fewer quarters than dimes. How many of each coin do I have?" Use a system of equations to answer his question.

Solution Translate each condition into an equation. Let D equal the number of dimes and Q equal the number of quarters. Dimes are worth $0.10 each, so D dimes are worth $0.10D$. Quarters are worth $0.25 each, so Q quarters are worth $0.25Q$. The total value of all the coins is $3.85. So $0.10D + 0.25Q = 3.85$.

There are 14 fewer quarters than dimes. That leads to a second equation $Q = D - 14$. Together the two equations form a system.

$$\begin{cases} 0.10D + 0.25Q = 3.85 \\ Q = D - 14 \end{cases}$$

Because $Q = D - 14$, substitute $D - 14$ for Q in the first equation.

$0.10D + 0.25(D - 14) = 3.85$	Substitution
$0.10D + 0.25D - 0.25(14) = 3.85$	Distributive Property
$0.10D + 0.25D - 3.5 = 3.85$	Arithmetic
$0.35D - 3.5 = 3.85$	Collect like terms.
$0.35D = 7.35$	Add 3.5 to both sides.
$D = 21$	Divide both sides by 0.35.

A grandfather posing with his two grandchildren

594 **Linear Systems**

To find Q, substitute 21 for D in either equation. We use the second equation because it is solved for Q. When $D = 21$, $Q = D - 14 = 21 - 14 = 7$. So $(D, Q) = (21, 7)$. The grandfather has 21 dimes and 7 quarters.

Check The 21 dimes are worth $2.10 and the 7 quarters are worth $1.75.
$2.10 + $1.75 = $3.85

In Chapter 6, you used the slope and *y*-intercept to find an equation of a line that passes through two given points. A different method for finding an equation through two points makes use of a system of equations.

Example 2

Find an equation of the line that passes through the points (3, 26) and (−2, 1).

Solution In slope-intercept form, the equation of the line is $y = mx + b$. If the values of *m* and *b* were known, each point on the line would make the equation true.

Substitute the coordinates of each given point for *x* and *y* to get two equations.

Using (3, 26), $26 = m \cdot 3 + b$.

Using (−2, 1), $1 = m \cdot -2 + b$.

This gives the system $\begin{cases} 26 = 3m + b \\ 1 = -2m + b \end{cases}$.

Either equation can be solved for *b*. From the second equation, $b = 2m + 1$. Now substitute $2m + 1$ for *b* in the first equation.

$26 = 3m + (2m + 1)$

$26 = 5m + 1$

$25 = 5m$

$5 = m$

This is the slope of the line. To find *b*, substitute 5 for *m* in either of the original equations. We use the second equation.

$1 = -2m + b$

$1 = -2 \cdot 5 + b$

$1 = -10 + b$

$11 = b$

This is the *y*-intercept. Thus, an equation of the line through (3, 26) and (−2, 1) is $y = 5x + 11$.

(continued on next page)

Chapter 10

Check Does each ordered pair satisfy the equation of the line?

Does $26 = 5 \cdot 3 + 11$? Yes, $26 = 15 + 11$.
Does $(3, 26)$ satisfy $y = 5x + 11$? Yes, $(3, 26)$ is on the line.
Does $1 = 5 \cdot -2 + 11$? Yes, $1 = -10 + 11$.
Does $(-2, 1)$ satisfy $y = 5x + 11$? Yes, $(-2, 1)$ is on the line.

Some situations have been around for generations. Example 3 is taken from an 1881 algebra text; the prices are out of date, but the situation is not.

GUIDED

Example 3

A farmer purchased 100 acres of land for $2,450. He paid $20 per acre for part of it and $30 per acre for the rest. How many acres were there in each part?

Solution You want to find two amounts, so use two variables.

Let x = the number of acres at $20/acre,
and y = the number of acres at $30/acre.

The farmer purchased a total of 100 acres, so $x + y = 100$.

The total cost is $2,450. So $20x + 30y = 2,450$.

Solve the system of these two equations.

$$\begin{cases} x + y = 100 \\ 20x + 30y = 2,450 \end{cases}$$

Although neither equation is solved for a variable, the first equation is equivalent to $y =$ __?__.

$20x + 30(\underline{}) = 2,450$ Substitute __?__ for y in the second equation.

$20x + \underline{} - \underline{} = 2,450$ Distributive Property

$3,000 - \underline{} = 2,450$ Collect like terms.

$\underline{} = -550$ Add $-3,000$ to both sides.

$x = 55$ Divide both sides by -10.

To find y, substitute 55 for x in either of the original equations. We use the first equation because it is simpler.

$$x + y = 100$$
$$55 + y = 100$$
$$y = 45$$

The farmer bought 55 acres at $20/acre, and 45 acres at $30/acre.

Linear Systems

Check Substitute 55 for x and 45 for y into the second equation.

Does $20x + 30y = 2{,}450$?
Yes, $20(55) + 30(45) = 1{,}100 + 1{,}350 = 2{,}450$.

A CAS can be used to solve systems of equations even when neither equation has an isolated variable.

Example 4

Use a CAS to solve $\begin{cases} y = \frac{2}{3}x + \frac{10}{3} \\ 5x + 2y = 32 \end{cases}$.

Solution Since the first equation is already solved for y, simply substitute that expression for y into the second equation and solve the second equation for x. To do that quickly on a CAS, you can use the SOLVE command.

Step 1 Find the SOLVE command on your calculator and place it on the entry line. On some calculators you may find it in the Algebra menu.

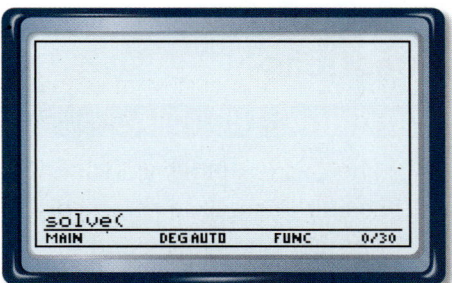

Step 2 Enter the second equation with $\left(\frac{2}{3}x + \frac{10}{3}\right)$ in place of y as you would if you were solving the equation by hand. That is, enter

$5x + 2\left(\frac{2}{3}x + \frac{10}{3}\right) = 32$.

Step 3 The SOLVE command on many CAS requires you to specify the variable for which to solve. In this case it is x. You may need to enter a comma followed by x before you close the parentheses and hit ENTER. You should get $x = 4$.

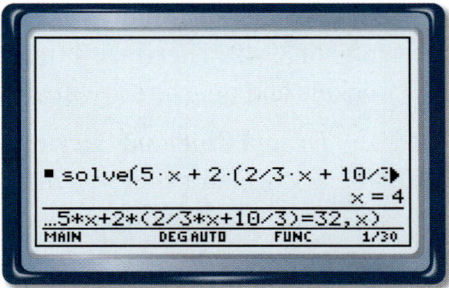

Step 4 Now substitute this x value into the first equation to get $y = 6$.

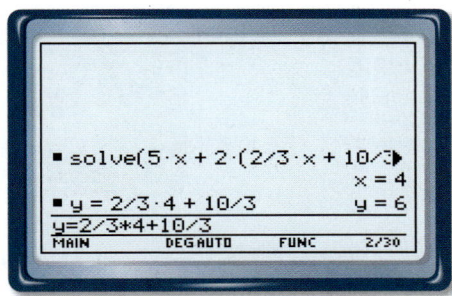

So, the solution to the system is $(4, 6)$.

(continued on next page)

More Uses of Substitution

Check Determining if a point is a solution to a system is easy to do on a CAS. Enter the two original equations and the specific values for each variable. The symbol " | " on one CAS means "with" or "such that" and indicates these values. The response "true" indicates that (4, 6) checks.

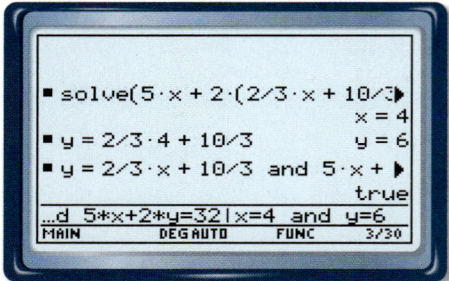

▶ **QY**

Below is the solution to a system found by using a CAS.
solve($4 \cdot x + 3 \cdot (2 - 5 \cdot x) = 17, x$)
$x = -1$
$y = 2 - 5 \cdot -1$
$y = 7$

a. Find the system of equations.
b. Check the solution by hand.

QY

Questions

COVERING THE IDEAS

1. The owners of a carnival have found that twice as many children as adults come to the carnival. Solve a system to estimate the number of children and the number of adults at the carnival when 3,570 people attend.

2. A jar of coins has only nickels and quarters, which are worth a total of $9.40. There are 4 more quarters than nickels. How many nickels and quarters are in the jar?

3. The Drama Club and Service Club held a charity car wash. There were four times as many Service Club members as Drama Club members working, so the Service Club earned four times as much money. The car wash raised $280 in all. How much did each club earn for their charity?

In 4–7, a system is given.
   a. Solve each system of equations by substitution.
   b. Check your answer.

4. $\begin{cases} y = 2x \\ 3x + 2y = 21 \end{cases}$

5. $\begin{cases} n + 5w = 6 \\ n = -8w \end{cases}$

6. $\begin{cases} a - b = 2 \\ a + 5b = 20 \end{cases}$

7. $\begin{cases} y = x - 1 \\ 4x - y = 19 \end{cases}$

8. Here is another problem from the 1881 algebra textbook. A farmer bought 100 acres of land, part at $37 per acre and part at $45 per acre, at a total cost of $4,220. How much land was there in each part?

Americans gave a total of $298.42 billion in contributions to charities in 2011.
Source: Giving USA Foundation

598 Linear Systems

APPLYING THE MATHEMATICS

In 9 and 10,
   a. solve each system by substitution.
   b. check your answer by graphing the system.

9. $\begin{cases} x + y = 8 \\ y = -3x \end{cases}$

10. $\begin{cases} x = 2y - 10 \\ 5x + 3 = 15 \end{cases}$

11. Solve the system $\begin{cases} b = -1.36a + 4.4 \\ 1.2a + 4.58b = -181 \end{cases}$ using a CAS.

12. Angles P and Q are complementary. If m$\angle P = 10x$ and m$\angle Q = 15x$, find x, m$\angle P$, and m$\angle Q$.

13. A business made $120,000 more this year than it did last year. This was an increase of 16% over last year's earnings. If T and L are the earnings (in dollars) for this year and last year, respectively, then $\begin{cases} T = L + 120{,}000 \\ T = 1.16L \end{cases}$. Find the profits for this year and last year.

14. Mrs. Rodriguez leaves money to her two favorite charities in her will. Charity A is to get 2.5 times as much money as Charity B. The total amount of money donated in the will is $28,000.
   a. Write a system of equations describing this situation.
   b. Solve to find the amount of money each charity will get.

15. Anica received her results for mathematics and verbal achievement tests. Her mathematics score is 40 points higher than her verbal score. Her total score for the two parts is 1,230.
   a. Let $v =$ Anica's verbal score, and $m =$ her mathematics score. Write a system of equations for this situation.
   b. Find Anica's two scores.

REVIEW

In 16 and 17 solve the system and check. (Lessons 10-2, 10-1)

16. $\begin{cases} y = x - 5 \\ y = -4x + 10 \end{cases}$

17. $\begin{cases} y = 6x + 6 \\ y = 6x - 2 \end{cases}$

18. One hot-air balloon takes off from Albuquerque, New Mexico and rises at a rate of 110 feet per minute. At the same time, another balloon takes off from Santa Fe, New Mexico and rises at a rate of 80 feet per minute. The altitude of Albuquerque is 4,958 feet and the altitude of Santa Fe is 6,950 feet. (Lesson 10-2)
   a. When are the two balloons at the same altitude?
   b. What is their altitude at that time?

19. If $a^3b^{-4}c$ is equal to the reciprocal of $\dfrac{a^{-5}b^2c^3}{a^{-2}b^xc^4}$, find x.
   (Lessons 8-4, 8-3)

Hot-air balloons hold from 19,000 to 211,000 cubic feet of air and are from 50 to 90 feet tall.

Source: hotairballoons.com

In 20 and 21, graph the solution set
  a. on a number line.
  b. in the coordinate plane. (Lessons 6-9, 3-6)

20. $x < 6$

21. $-4y + 2 < 6$

22. In 2002, India ended its Police Pigeon Service. This is a system in which trained pigeons transport messages. The service was used when traditional communication broke down during natural disasters. Suppose a trained pigeon flies 41.3 mph in still air. (Lesson 5-3)
   a. How far can it fly in m minutes in still air?
   b. How fast can it fly *with* the wind if the wind speed is s mph?
   c. How fast can it fly *against* the wind if the wind speed is s mph?
   d. If the pigeon is flying down a highway that has a speed limit of 65 mph and there is a 21.9 mph tailwind, would you give it a speeding ticket?

EXPLORATION

23. Here is a nursery rhyme whose earliest traceable publication date is around 1730 in *Folklore*, now in the library of the British Museum. (St. Ives is a village in England.)

   As I was going to St. Ives,
   I met a man with seven wives.
   Each wife had seven sacks,
   Each sack had seven cats,
   Each cat had seven kits:
   Kits, cats, sacks, and wives,
   How many were going to St. Ives?

   a. Let W = the number of wives, S = the number of sacks, C = the number of cats, and K = the number of kits. Write three equations that relate two of these variables to each other.
   b. Find the value of $K + C + S + W$.
   c. What is an answer to the riddle?

QY ANSWERS

a. $\begin{cases} y = 2 - 5x \\ 4x + 3y = 17 \end{cases}$

b. $4 \cdot -1 + 3 \cdot 7 = -4 + 21 = 17$ and $2 - 5 \cdot -1 = 2 + 5 = 7$.
So $(-1, 7)$ checks.

Lesson 10-4
Solving Systems by Addition

Vocabulary

addition method for solving a system

▶ **BIG IDEA** The sum of the left sides of two equations equals the sum of the right sides of those equations.

The numbers $\frac{1}{4}$ and 25% are equal even though they may not look equal; so are $\frac{17}{20}$ and 85%. If you add them, the sums are equal.

$$\frac{1}{4} = 25\% \qquad \frac{17}{20} = 85\%$$

So $\frac{1}{4} + \frac{17}{20} = 25\% + 85\%$.

Adding on each side, $\frac{22}{20} = 110\%$.

This is one example of the following generalization of the Addition Property of Equality.

> **Generalized Addition Property of Equality**
>
> For all numbers or expressions a, b, c, and d: If $a = b$ and $c = d$, then $a + c = b + d$.

Mental Math

Find the perimeter of

a. a square with sides of length $6.2x$.

b. a regular octagon with sides of length $21ab$.

c. a regular pentagon with sides of length $4.5m + 1.5n$.

The Generalized Addition Property of Equality can be used to solve some systems. Consider this situation: The sum of two numbers is 5,300. Their difference is 1,200. What are the numbers?

If x and y are the two numbers, with x the greater number, we can write the following system.

$$\begin{cases} x + y = 5{,}300 \\ x - y = 1{,}200 \end{cases}$$

Notice what happens when the left sides are added (combining like terms) and the right sides are added.

$$\begin{array}{r} x + y = 5{,}300 \\ + \, x - y = 1{,}200 \\ \hline 2x + 0 = 6{,}500 \end{array}$$

Because y and $-y$ sum to 0, the sum of the equations is an equation with only one variable. Solve $2x = 6{,}500$ as usual.

$$x = 3{,}250$$

Solving Systems by Addition

To find y, substitute 3,250 for x in one of the original equations. We choose $x + y = 5,300$.

$$x + y = 5,300$$
$$3,250 + y = 5,300$$
$$y = 2,050$$

The ordered pair (3,250, 2,050) checks in both equations:
$3,250 + 2,050 = 5,300$ and $3,250 - 2,050 = 1,200$.

So the solution to the system $\begin{cases} x + y = 5,300 \\ x - y = 1,200 \end{cases}$ is (3,250, 2,050).

Using the Generalized Addition Property of Equality to eliminate one variable from a system is sometimes called the **addition method for solving a system**. The addition method is an efficient way to solve systems when the coefficients of the same variable are opposites.

Example 1

A pilot flew a small plane 180 miles from North Platte, Nebraska to Scottsbluff, Nebraska in 1 hour against the wind. The pilot returned to North Platte in 48 minutes $\left(\frac{48}{60} = \frac{4}{5} \text{ hour}\right)$ with the wind at the plane's back. How fast was the plane flying (without wind)? What was the speed of the wind?

Solution Let A be the average speed of the airplane without wind and W be the speed of the wind, both in miles per hour. The total speed against the wind is then $A - W$, and the speed with the wind is $A + W$. There are two conditions given on these total speeds.

From North Platte to Scottsbluff the average speed of the plane was $\frac{180 \text{ miles}}{1 \text{ hour}} = 180 \frac{\text{miles}}{\text{hour}}$.

This was against the wind, so $A - W = 180$.

From Scottsbluff to North Platte the average speed of the plane was $\frac{180 \text{ miles}}{\frac{4}{5} \text{ hour}} = 225 \frac{\text{miles}}{\text{hour}}$.

This was with the wind, so $A + W = 225$.

We have the system $\begin{cases} A - W = 180 \\ A + W = 225 \end{cases}$.

Now solve the system. Since the coefficients of W are opposites (1 and −1), add the equations.

$$\begin{array}{r} A - W = 180 \\ \underline{A + W = 225} \\ 2A = 405 \quad \text{Add.} \\ A = 202.5 \quad \text{Divide by 2.} \end{array}$$

There are more than 8,100 airports in the United States used only by small planes. They have runways shorter than 3,000 feet.

Source: Aircraft Owners and Pilots Association

Substitute 202.5 for A in either of the original equations. We choose the second equation.

$$202.5 + W = 225$$
$$W = 22.5$$

The average speed of the airplane was about 202.5 mph and the speed of the wind was 22.5 mph.

Check Refer to the original question. Against the wind, the plane flew at $202.5 - 22.5$ or 180 mph, so it flew 180 miles in 1 hour. With the wind, the plane flew at $202.5 + 22.5$ or 225 mph. At that rate, in 48 minutes the pilot flew $\frac{48}{60}$ hr $\cdot$ 225 $\frac{mi}{hr} = 180$ miles, which checks with the given conditions.

Sometimes the coefficients of the same variable are equal. In this case, use the Multiplication Property of Equality to multiply both sides of one of the equations by –1. This changes all the numbers in that equation to their opposites. Then you can use the addition method to find solutions to the system.

Example 2

Solve $\begin{cases} 5x + 17y = 1 \\ 5x + 8y = -26 \end{cases}$.

Solution We rewrite the equations and number them to make it easy to refer to them later.

$$\begin{cases} 5x + 17y = 1 & \text{Equation \#1} \\ 5x + 8y = -26 & \text{Equation \#2} \end{cases}$$

Notice that the coefficients of x in the two equations are equal.

Multiply the second equation by –1. Call the resulting Equation #3.

$-5x - 8y = 26$ Equation #3

Now use the addition method with the first and third equations.

$$\begin{array}{rl} 5x + 17y = 1 & \text{Equation \#1} \\ +\ -5x - 8y = 26 & \text{Equation \#3} \\ \hline 9y = 27 & \text{Equation \#1 + Equation \#3} \\ y = 3 & \end{array}$$

To find x, substitute 3 for y in one of the original equations.

$5x + 17(3) = 1$ We use Equation #1.
$5x + 51 = 1$
$5x = -50$
$x = -10$

So $(x, y) = (-10, 3)$.

(continued on next page)

Check Substitute in both equations.

Equation #1 Does $5 \cdot -10 + 17 \cdot 3 = 1$? Yes.

Equation #2 Does $5 \cdot -10 + 8 \cdot 3 = -26$? Yes.

GUIDED

Example 3

A resort hotel offers two weekend specials.

Plan A: 3 nights with 6 meals for $564

Plan B: 3 nights with 2 meals for $488

At these rates, what is the cost of one night's lodging and what is the average cost per meal? (Assume there is no discount for 6 meals.)

Solution Let $N =$ price of one night's lodging.

Let $M =$ average price of one meal.

Write an equation to describe each weekend special.

From Plan A: $3N + 6M = 564$ Equation #1

From Plan B: ___?___   Equation #2

Notice the coefficients of N are the same, so multiply Equation #2 by -1.

___?___   Equation #3

___?___   Add Equations #1 and #3.

Does your last equation have only one variable? If so, solve this equation. If not, ask someone for help.

$M =$ ___?___

Substitute this value of M in either equation, and solve for N.

$(N, M) = ($ ___?___ , ___?___ $)$

What is the price of one night's lodging? ___?___

What is the average cost of a meal? ___?___

The average hotel room rate in the United States in 2006 was $96.42 per night.

Source: Smith Travel Research

Questions

COVERING THE IDEAS

1. a. When is adding equations an appropriate method for solving systems?

   b. What is the goal in adding equations to solve systems?

2. Which property allows you to add to both sides of two equations to get a new equation?

In 3 and 4, a system is given.
  a. Solve the system.
  b. Check your solution.

3. $\begin{cases} 3x + 9y = 75 \\ -3x - y = 15 \end{cases}$

4. $\begin{cases} a + b = -22 \\ a - b = 4 \end{cases}$

5. The sum of two numbers is 1,776 and their difference is 1,492. What are the numbers?

6. Find two numbers whose sum is 20 and whose difference is 20.

7. When is it useful to multiply an equation by –1 as a first step in solving a system?

8. Airlines schedule about 5.5 hours of flying time for an A320 Airbus to fly from Dulles International Airport near Washington, D.C. to Los Angeles International Airport. Airlines schedule about 4.5 hours of flying time for the reverse direction. The distance between these airports is about 2,300 miles. They allow about 0.4 hour for takeoff and landing.
  a. From this information, estimate (to the nearest 5 mph) the average wind speed the airlines assume in making their schedule.
  b. What average airplane speed (to the nearest 5 mph) do the airlines assume in making their schedule?

In 9 and 10, solve the system.

9. $\begin{cases} 14x - 5y = 9 \\ 17x - 5y = 27 \end{cases}$

10. $\begin{cases} 17m + 7n = 8 \\ 17m + 5n = 13 \end{cases}$

11. $(N, M) = (150, 19)$ is the solution to the system of equations $\begin{cases} 3N + 6M = 564 \\ 3N + 2M = 488 \end{cases}$ in Example 3. Check this solution.

12. A hotel offers the following specials. Plan A includes a two-night stay and one meal for $199. Plan B includes a 2-night stay and 4 meals for $247. What price is this per night and per meal?

APPLYING THE MATHEMATICS

In 13 and 14, solve the system.

13. $\begin{cases} 2x - 6y = 34 \\ x = 2 - 6y \end{cases}$

14. $\begin{cases} \frac{1}{4}z + \frac{3}{4}w = \frac{1}{2} \\ \frac{7}{4}w + \frac{1}{4}z = \frac{3}{8} \end{cases}$

Solving Systems by Addition

15. As you know, $\frac{3}{5} = 60\%$ and $\frac{3}{8} = 37.5\%$.

    a. Is it true that $\frac{3}{5} - \frac{3}{8} = 60\% - 37.5\%$? Justify your answer.

    b. Is it true that $\frac{3}{5} \cdot \frac{3}{8} = 60\% \cdot 37.5\%$? Justify your answer.

16. In 2008, the tallest person playing professional basketball in the Women's National Basketball Association (WNBA) was Margo Dydek. The shortest person was Shannon Bobbitt. Margo was 23 in. taller than Shannon. If one stood on the other's head, they would have stood 12 ft 5 in. tall. How tall is each player?

REVIEW

In 17 and 18, solve by using any method. (Lessons 10-3, 10-2, 10-1)

17. $\begin{cases} y = 2x - 3 \\ y = -8x + 6 \end{cases}$

18. $\begin{cases} A = -5n \\ B = 6n \\ 4A + B = 39 \end{cases}$

19. a. Solve $x^2 + 3x - 28 = 0$.

    b. Find the x-intercepts of the graph of $y = x^2 + 3x - 28$.
    (Lesson 9-5)

20. The formula $d = 0.04s^2 + 1.5s$ gives the approximate distance d in feet needed to stop a particular car traveling on dry pavement at a speed of s miles per hour. How much farther will this car travel before stopping if it is traveling at 65 mph instead of 50 mph? (Lesson 9-3)

21. Let $f(x) = \sqrt{2x - 9}$. (Lessons 8-6, 7-6, 7-5)

    a. What is the domain of f?

    b. What is the range of f?

22. Simplify $x^{-1} + x - \frac{1}{x}$. (Lessons 8-4, 8-3)

23. Find the slope of line ℓ pictured below. (Lesson 6-2)

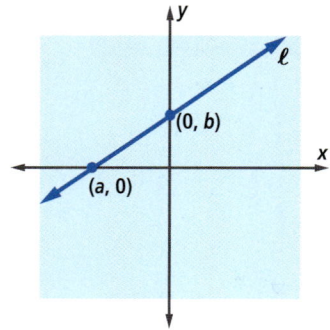

Margo Dydek

Shannon Bobbitt

24. In 2005, the total revenues of a cell phone company increased 5.5% from the previous year to $36.84 billion. What were the company's revenues in 2004? (**Lesson 4-1**)

25. Solve $38(212 - x) = 0$ in your head. (**Lesson 3-4**)

EXPLORATION

26. Subtracting equations is part of a process that can be used to find simple fractions for repeating decimals. For example, to find a fraction for $0.\overline{72} = 0.7272727272...$, first let $d = 0.\overline{72}$. Then multiply both sides of the equation by an appropriate power of 10. Here we multiply by 10^2 because $0.\overline{72}$ has a two-digit block that repeats.

$$100d = 72.\overline{72} \quad \text{Equation \#1}$$
$$d = 0.\overline{72} \quad \text{Equation \#2}$$

Subtract the second equation from the first.

$$99d = 72 \quad \text{Equation \#1} - \text{Equation \#2}$$

Solve for d and simplify the fraction.

$$d = \frac{72}{99} \text{ or } d = \frac{8}{11}$$

A calculator shows that $\frac{8}{11} = 0.7272727272...$.

a. Use the above process to find a simple fraction equal to $0.\overline{15}$.

b. Modify the process to find a simple fraction equal to $0.9\overline{02}$.

c. Find a simple fraction equal to $0.\overline{123456}$.

Chapter 10

Lesson 10-5
Solving Systems by Multiplication

Vocabulary

equivalent systems
multiplication method for solving a system

▶ **BIG IDEA** An effective first step in solving some systems is to multiply both sides of one of the equations by a carefully chosen number.

Recall that there are three common forms for equations of lines.

	Form	Example
Standard	$Ax + By = C$	$3x + 8y = 20$
Slope-Intercept	$y = mx + b$	$y = -2x + 1$
Point-Slope	$y - k = m(x - h)$	$y - 50 = \frac{3}{4}(x - 20)$

Mental Math

Classify the angle with the given measure as acute, right, or obtuse.

a. 134°

b. 84°

c. 0.23°

The substitution method described in Lessons 10-2 and 10-3 is convenient for solving systems in which one or both equations are in slope-intercept form. The addition method studied in Lesson 10-4 is convenient for solving systems in which both equations are in standard form and the coefficients of one variable are either equal or opposites. However, not all systems fall into one of these two categories.

Consider the following system.

$$\begin{cases} 3x - 4y = 7 \\ 6x - 5y = 20 \end{cases}$$

Adding or subtracting the two equations will not result in an equation with just one variable, because the x and the y terms are neither equal nor opposites. Substitution could be used, but it introduces fractions.

An easier method uses the Multiplication Property of Equality to create an *equivalent system* of equations. **Equivalent systems** are systems with exactly the same solutions. Notice that if you multiply both sides of the first equation by –2, the x terms of the resulting system have opposite coefficients.

608   Linear Systems

Lesson 10-5

Example 1

Solve the system $\begin{cases} 3x - 4y = 7 \\ 6x - 5y = 20 \end{cases}$.

Solution 1 Multiply both sides of the first equation by -2 and apply the Distributive Property.

$$\begin{cases} 3x - 4y = 7 \\ 6x - 5y = 20 \end{cases} \xrightarrow{\text{multiply by } -2} \begin{cases} -2(3x - 4y) = -2(7) \\ 6x - 5y = 20 \end{cases}$$

$$\begin{cases} -6x + 8y = -14 \\ 6x - 5y = 20 \end{cases}$$

$3y = 6$ Add the equations.

$y = 2$ Solve for y.

To find x, substitute 2 for y in one of the original equations.

$3x - 4y = 7$

$3x - 4 \cdot 2 = 7$

$3x - 8 = 7$

$3x = 15$

$x = 5$

So the solution is $(x, y) = (5, 2)$.

Solution 2 Multiply both sides of the second equation by $-\frac{1}{2}$. This also makes the coefficients of x opposites.

$$\begin{cases} 3x - 4y = 7 \\ 6x - 5y = 20 \end{cases} \xrightarrow{\text{multiply by } -\frac{1}{2}} \begin{cases} 3x - 4y = 7 \\ -\frac{1}{2}(6x - 5y) = -\frac{1}{2}(20) \end{cases}$$

$$\begin{cases} 3x - 4y = 7 \\ -3x + \frac{5}{2}y = -10 \end{cases}$$

$-\frac{3}{2}y = -3$ Add.

$y = 2$

Proceed as in Solution 1 to find x. Again $(x, y) = (5, 2)$.

Example 1 shows that the solution is the same no matter which equation is multiplied by a number. The goal is to obtain opposite coefficients for one of the variables in the two equations. Then the resulting equations can be added to eliminate that variable. This technique is sometimes called the **multiplication method for solving a system**.

 QY

▶ **QY**

$\begin{cases} 7x + 3y = 22.5 \\ 2x - 12y = 45 \end{cases}$

↓

$\begin{cases} 28x + 12y = 90 \\ 2x - 12y = 45 \end{cases}$

a. Explain what operation occurred to go from the first system to the second system.

b. Finish solving the system.

Solving Systems by Multiplication   **609**

Sometimes it is necessary to multiply *each* equation by a different number before adding.

Example 2
Solve the system $\begin{cases} -3m + 2n = 6 \\ 4m + 5n = -31 \end{cases}$.

Solution The idea is to multiply by a number so that one variable in the resulting system has a pair of opposite coefficients. To make the coefficients of m opposites, multiply the first equation by 4 and the second equation by 3.

$$\begin{cases} -3m + 2n = 6 \\ 4m + 5n = -31 \end{cases} \xrightarrow[\text{multiply by 3}]{\text{multiply by 4}} \begin{cases} -12m + 8n = 24 \\ 12m + 15n = -93 \end{cases}$$

Now add. $23n = -69$
$n = -3$

To find m, substitute -3 for n in either original equation. We use the first equation.

$-3m + 2 \cdot (-3) = 6$
$-3m - 6 = 6$
$-3m = 12$
$m = -4$

So $(m, n) = (-4, -3)$.

Check You should check your solution by substituting for m and n in each original equation.

Many situations naturally lead to linear equations in standard form. This results in a linear system that can be solved using the multiplication method.

Example 3
A marching band currently has 48 musicians and 18 people in the flag corps. The drum majors wish to form hexagons and squares like those diagrammed at the right. Are there enough members to create the formations with no people left over? If so, how many hexagons and how many squares can be made? If not, give a recommendation for the fewest people the drum majors would need to recruit and how many hexagons and how many squares could be made.

Hexagon
Flag bearer in center

Square
Two musicians in the center

Solution Consider the entire formation to include h hexagons and s squares. There are two conditions in the system: one for musicians and one for the flag corps.

There are 6 $\frac{\text{musicians}}{\text{hexagon}}$ and 2 $\frac{\text{musicians}}{\text{square}}$.

So $6h + 2s = 48$ musicians.

There are 1 $\frac{\text{flag bearer}}{\text{hexagon}}$ and 4 $\frac{\text{flag bearers}}{\text{square}}$.

So $h + 4s = 18$ flag bearers.

To find h, multiply the first equation by −2, and add the result to the second equation.

$$-12h + -4s = -96$$
$$\underline{h + 4s = 18}$$
$$-11h = -78$$
$$h = 7.\overline{09}$$

Marching bands perform in competitions, at sporting events, and in parades.

Because h is not a positive integer in this solution, these formations will not work with 48 musicians and 18 flag bearers.

−78 is not divisible by −11, but −77 is. By recruiting one additional member to the flag corps we get a number divisible by −11. This would create the following system.

$$-12h + -4s = -96$$
$$\underline{h + 4s = 19}$$
$$-11h = -77$$
$$h = 7$$

Now $h = 7$. Substituting for h in the second equation of this new system, you find that $s = 3$.

All 48 musicians and 19 flag bearers could be arranged into 7 hexagons and 3 squares, so the drum major needs to recruit 1 more flag bearer.

Check Making 7 hexagons would use 42 musicians and 7 flag bearers. Making 3 squares would use 6 musicians and 12 flag bearers. This setup uses exactly 48 musicians and 19 flag bearers.

When the equations in a system are not given in either standard or slope-intercept form, it is wise to rewrite the equations in one of these forms before proceeding. For example, to solve the system below, you could use one of three methods.

$$\begin{cases} n - 3 = \frac{3}{2}m \\ 4m + 5n = -31 \end{cases}$$

Solving Systems by Multiplication **611**

Method 1 Multiply the first equation by 2 to eliminate fractions.

$$\begin{cases} 2n - 6 = 3m \\ 4m + 5n = -31 \end{cases}$$

Add $-3m$ and 6 to both sides of the first equation. The result is the system of Example 2, which is in standard form.

Method 2 Add 3 to both sides of the first equation.

$$\begin{cases} n = \frac{3}{2}m + 3 \\ 4m + 5n = -31 \end{cases}$$

To finish solving this system you could use substitution by substituting n into the second equation.

Method 3 Use substitution on a CAS to solve the system.

Step 1 Use the SOLVE command to solve one of the equations for one of the variables. We choose the first equation and solve for n.

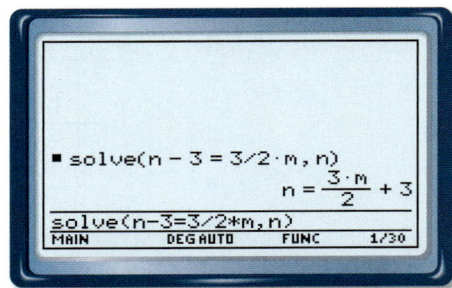

Step 2 Substitute this value for n into the second equation and solve for m. Most CAS will allow you to copy and paste so that you do not have to type expressions multiple times. The display shows $m = -4$.

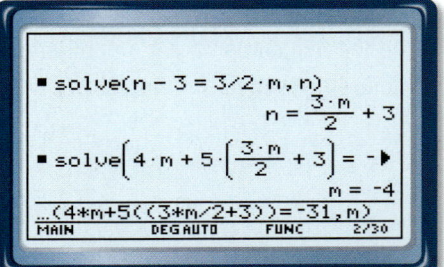

Step 3 Then substitute -4 for m into the first equation to get $n = -3$.

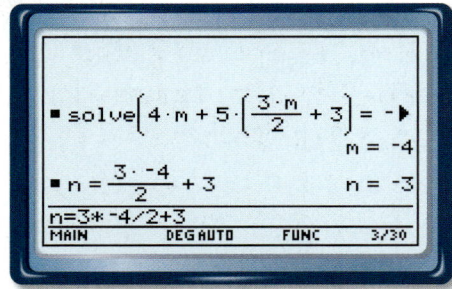

Questions

COVERING THE IDEAS

1. Consider the system $\begin{cases} 5x + 3d = 9 \\ 2x + d = 26 \end{cases}$.

    a. **Fill in the Blanks** If the ___?___ equation is multiplied by ___?___, then adding the equations will eliminate ___?___.

    b. Solve the system.

2. A problem on a test was to solve the system $\begin{cases} -8n + m = -19 \\ 4n - 3m = -8 \end{cases}$.

   Three students used three different methods to solve the system. Their first steps are shown.

   Annisha's Method
   $\begin{cases} -8n + m = -19 \\ 8n - 6m = -16 \end{cases}$

   Maxandra's Method
   $\begin{cases} m = -19 + 8n \\ 4n - 3m = -8 \end{cases}$

   Victor's Method
   $\begin{cases} -24n + 3m = -57 \\ 4n - 3m = -8 \end{cases}$

   a. Which student(s) used substitution to solve the system?
   b. Which variable will Annisha's method eliminate? Explain what she did to make an equivalent system.
   c. Which variable will Victor's method eliminate? Explain what he did to make an equivalent system.
   d. Pick one of the methods and finish solving the system.

3. Consider the system $\begin{cases} 7r - 3s = 9 \\ 2r + 5s = 26 \end{cases}$.

   a. By what two numbers can you multiply the equations so that, if you add the results, you will eliminate r?
   b. By what two numbers can you multiply the equations so that, if you add the results, you will eliminate s?
   c. Use one of these methods to solve the system.

4. Consider the system $\begin{cases} 10t + u = 85 \\ 2t + 3u = 31 \end{cases}$.

   a. Write an equivalent system that would eliminate t first.
   b. Write an equivalent system that would eliminate u first.
   c. Use one of the methods to solve the system.

5. A marching band has 60 musicians and 30 flag bearers. They wish to form pentagons and squares like those diagrammed at the right.

   a. If the formation has 3 pentagons and 4 squares, how many musicians and flag bearers will be involved?
   b. Is it possible to change the numbers of pentagons and squares so that every person will have a spot? If so, how many of each formation will be needed?

6. Solve the system $\begin{cases} n + 7 = \frac{1}{3}m \\ 7m - 3n = 57 \end{cases}$.

In 7–10, solve the system.

7. $\begin{cases} 24x + 15y = 20 \\ 4x + 3y = 5 \end{cases}$

8. $\begin{cases} 7a - 8b = 1 \\ 6a - 7b = 1 \end{cases}$

9. $\begin{cases} 9y + x = -8 \\ 2 = y - x \end{cases}$

10. $\begin{cases} 113.2 = 4x - 2y \\ 331.4 = 6x + 5y \end{cases}$

APPLYING THE MATHEMATICS

11. Solve the system by first rewriting each equation in standard form. $\begin{cases} 0.2x + 0.3(x + 4) = 0.16y \\ 0.04y - 0.07 = 0.08x \end{cases}$

12. Milo feels that the probability that he will be elected to the student council is $\frac{1}{10}$ of the probability that he will not be elected. What does Milo think is the probability that he will be elected? (Remember that the sum of the probabilities that he will be elected and not be elected is 1.)

13. A test has m multiple-choice (MC) questions and e extended-response (ER) questions. If the MC questions are worth 2 points each and the ER questions are worth 7 points each, the test will be worth a total of 95 points. If the MC questions are worth 3 points each and the ER questions are worth 8 points each, the test will be worth a total of 130 points. How many MC questions and how many ER questions are on the test?

Students are participating in student council elections.

14. A security guard counted 82 vehicles in a parking lot. The only vehicles in the lot were cars and motorcycles. To double-check his count, the security guard counted 300 wheels. How many motorcycles and how many cars are in the parking lot?

15. Delise and Triston's class went on a field trip to a local farm. The farm raised cows and chickens. Delise counted 27 heads and Triston counted 76 legs. How many cows and how many chickens are on the farm?
    a. Answer this question by solving a system.
    b. Write a few sentences explaining how to answer the question without using algebra.

REVIEW

In 16 and 17, solve the system using any method.
(Lessons 10-4, 10-3, 10-2, 10-1)

16. $\begin{cases} 6x + 2y = 26 \\ 4x + 2y = 8 \end{cases}$

17. $\begin{cases} y = \frac{2}{3}x - 4 \\ y = \frac{1}{4}x + 1 \end{cases}$

18. The two diagrams below illustrate a system of equations.
(**Lessons 10-4, 10-2**)

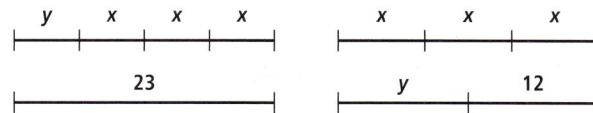

   a. Write an equation for the diagram at the left.
   b. Write an equation for the diagram at the right.
   c. Solve the system for x and y.
   d. Check your work.

19. Ashlyn has $600 and saves $30 each week. Janet has $1,500 and spends $30 each week. (**Lesson 10-2**)
   a. How many weeks from now will they each have the same amount of money?
   b. What will this amount be?

20. A 16-foot ladder leans against a house. If the base of the ladder is 6 feet from the base of the house, at what height does the top of the ladder touch the house? (**Lessons 9-1, 8-6**)

In 21 and 22, consider a garage with a roof pitch of $\frac{3}{12}$ at the right. The garage is to be 20 feet wide. (**Lessons 6-3, 5-9**)

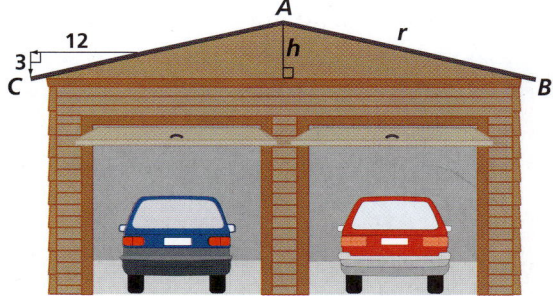

21. What is the slope of $\overline{AB}$?

22. a. What is the height h of the roof?
    b. Find the length r of one rafter.

23. **True or False** (**Lesson 5-6**)
    a. Probabilities are numbers from 0 to 1.
    b. A probability of 1 means that an event must occur.
    c. A relative frequency of −1 cannot occur.

EXPLORATION

24. Create formations of a college band consisting of 110 musicians and a flag corps of 36 with no members left over.

QY ANSWERS

a. The first equation was multiplied by 4.

b. $(x, y) = (4.5, -3)$

Chapter 10

Lesson 10-6
Systems and Parallel Lines

Vocabulary

coincident lines

▶ **BIG IDEA** Systems having 0, 1, or infinitely many solutions correspond to lines having 0 or 1 point of intersection, or being coincident.

The idea behind parallel lines is that they "go in the same direction." So we call two lines parallel if and only if they are in the same plane and either are the same line or do not intersect. All vertical lines are parallel to each other. So are all horizontal lines. But not all oblique lines are parallel. For oblique lines to be parallel, they must have the same slope.

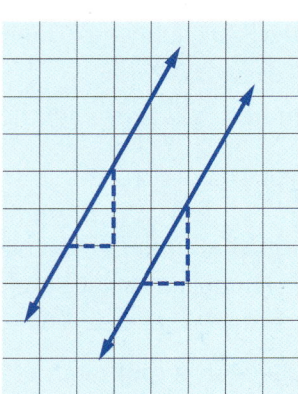

Mental Math

Evaluate.

a. $\frac{7 \cdot 6 \cdot 5}{3!}$

b. $\frac{9!}{9 \cdot 8 \cdot 7 \cdot 6}$

Slopes and Parallel Lines Property

If two lines have the same slope, then they are parallel.

Nonintersecting Parallel Lines

You have learned that when two lines intersect in exactly one point, the coordinates of the point of intersection can be found by solving a system. But what happens when the lines are parallel? Consider this linear system.

$$\begin{cases} 5x - 2y = 11 \\ 15x - 6y = -25 \end{cases}$$

You can solve the system by multiplying the first equation by –3, and adding the result to the second equation.

$$\begin{array}{r} -15x + 6y = -33 \\ + 15x - 6y = -25 \end{array}$$

Notice that when you add you get $0 = -58$.

This is impossible! When an equation with no solution (such as $0 = -58$) results from correct applications of the addition and multiplication methods on a system of linear equations, the original conditions must also be impossible. There are no pairs of numbers that work in *both* equations.

616 Linear Systems

Thus, the system has no solutions. The lines do not intersect. The graph of the system is two parallel nonintersecting lines, as shown at the right. As another check, rewrite the equations for the lines in slope-intercept form.

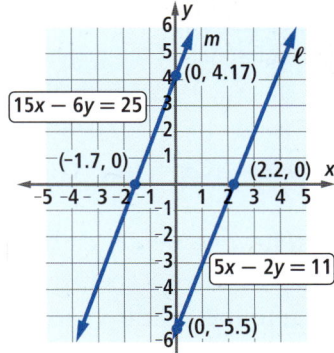

line ℓ: $5x - 2y = 11$
$-2y = -5x + 11$
$y = 2.5x - 5.5$

line m: $15x - 6y = -25$
$-6y = -15x - 25$
$y = 2.5x + 4.1\overline{6}$

Both lines ℓ and m have the same slope of 2.5, but different y-intercepts. Thus, they are parallel.

 QY1

Coincident Lines

Some systems have infinitely many solutions. They can be solved using any of the techniques you have studied in this chapter.

> ▶ **QY1**
>
> Show that the system
> $$\begin{cases} -4x + 2y = -16 \\ 6x - 3y = 18 \end{cases}$$
> has no solution.

Example 1

Solve the system $\begin{cases} 4x + 2y = 6 \\ y = -2x + 3 \end{cases}$.

Solution 1 Rewrite the first equation in slope-intercept form.

$4x + 2y = 6$
$2y = -4x + 6$ Add $-4x$ to each side.
$y = -2x + 3$ Divide each side by 2.

Notice that this equation is identical to the second equation in the system. So, any ordered pair that is a solution to one equation is also a solution to the other equation. The graphs of the two equations are the same line.

Solution 2 Use substitution. Substitute $-2x + 3$ for y in the first equation.

$4x + 2(-2x + 3) = 6$
$4x - 4x + 6 = 6$
$6 = 6$

The sentence $6 = 6$ is *always* true. So, any ordered pair that is a solution to one equation is also a solution to the other equation in the system. The graphs of the two equations are the same line.

The solution set consists of all ordered pairs on the line with equation $y = -2x + 3$, as shown at the right.

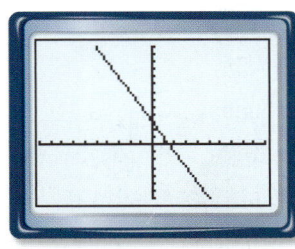

$-10 \leq x \leq 10, -7 \leq y \leq 13$

(continued on next page)

Check As a partial check, find an ordered pair that satisfies one of the equations of the original system. We use $y = -2x + 3$ to find the ordered pair (0, 3). Check that this ordered pair also satisfies the other equation. Does $4 \cdot 0 + 2 \cdot 3 = 6$? Yes, $6 = 6$.

 QY2

▶ QY2

Find a second ordered pair that satisfies one of the equations in Example 1 and show that it satisfies the other equation.

Whenever a sentence that is always true (such as $0 = 0$ or $6 = 6$) occurs from correct work with a system of linear equations, the system has infinitely many solutions. We say that the lines *coincide* and that the graph of the system is two **coincident lines.**

You have now studied all the ways that two lines in the plane can be related, and all the types of solutions a system of two linear equations might have. The table below summarizes these relationships.

Description of System	Graph	Number of Solutions to System	Slopes of Lines
Two intersecting lines		1 (the point of intersection)	Different
Two parallel and nonintersecting lines		0	Equal
One line (parallel and coincident lines)		Infinitely many	Equal

GUIDED

Example 2

Find all solutions to $\begin{cases} 12x - 10y = 2 \\ -18x + 15y = -3 \end{cases}$.

Solution

Step 1 Multiply both sides of the first equation by 3. ___?___

Step 2 Multiply both sides of the second equation by 2. ___?___

Step 3 Add the equations from Steps 1 and 2. ___?___

618 Linear Systems

The solution set consists of all ordered pairs on the line with equation $12x - 10y = 2$.

Questions

COVERING THE IDEAS

1. What is true about the slopes of parallel lines?
2. Which two lines among Parts a–d are parallel?
   a. $y = 8x + 500$
   b. $y = 2x + 500$
   c. $y = 8x + 600$
   d. $x = 2y + 500$
3. a. Graph the line with equation $y = \frac{1}{3}x + 5$.
   b. Draw the line parallel to it through the origin.
   c. What is an equation of the line you drew in Part b?
4. Give an example of a system with two nonintersecting lines.
5. Give an example of a system with two coincident lines.

In 6 and 7, a system is given.
   a. Determine whether the system includes *nonintersecting* or *coincident* lines.
   b. Check your answer to Part a by graphing.

6. $\begin{cases} 12a = 6b - 3 \\ 4a - 2b = -3 \end{cases}$

7. $\begin{cases} y - x = 5 \\ 3y - 3x = 15 \end{cases}$

8. **Matching** Match the description of the graph with the number of solutions to the system.
   a. lines intersect in one point   i. no solution
   b. lines do not intersect   ii. infinitely many solutions
   c. lines coincide   iii. one solution

APPLYING THE MATHEMATICS

9. a. How many pairs of numbers M and N satisfy both conditions i and ii below?
      i. The sum of the numbers is -2.
      ii. The average of the numbers is 1.
   b. Explain your answer to Part a.

10. Could the situation described here have happened? Justify your answer by using a system of equations. A pizza parlor sold 36 pizzas and 21 gallons of soda for $456. The next day, at the same prices, they sold 48 pizzas and 28 gallons of soda for $608.

Chapter 10

In 11–14, describe the graph of the system as two intersecting lines, two parallel nonintersecting lines, or coincident lines.

11. $\begin{cases} a = b \\ b - a = 0 \end{cases}$

12. $\begin{cases} y = 5 - 3x \\ 6x + 2y - 10 = 0 \end{cases}$

13. $\begin{cases} 10x + 20y = 30 \\ y + 2x = 3 \end{cases}$

14. $\begin{cases} \frac{4}{5}c - \frac{3}{5}d = 3.6 \\ 8c = 6d + 72 \end{cases}$

15. Melissa is the costume manager for a theater company and is supposed to receive a 15% professional discount from a fabric store. Last week, she bought 40 yards of a red material and 35 yards of a black fabric and paid $435.63. Two friends of hers went to the same store the following week. One bought 20 yards of red material and 10 yards of black fabric for $185, and the other bought 15 yards of red material and 40 yards of black fabric for $447.50. Did Melissa receive a discount? If so, was it the correct percentage?

Dancers are performing in a stage production.

REVIEW

16. A band has 59 musicians (M) with an additional 24 flag bearers (F). They plan to form pentagons and squares with one person in the middle, as shown below.

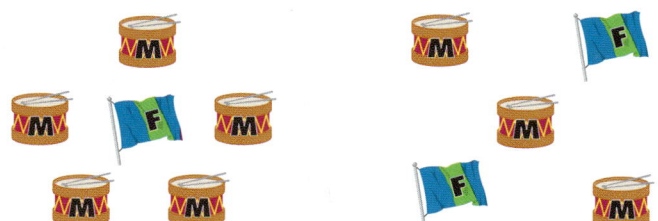

Let p be the number of pentagons formed and s be the number of squares formed. Can all musicians and flag bearers be accommodated into these formations? Why or why not? (Lesson 10-5)

17. Each diagram at the right represents an equation involving lengths t and u. (Lesson 10-4)

    a. Write a pair of equations describing these relationships.
    b. Use either your equations or the diagrams to find the lengths of t and u. Explain your reasoning.
    c. Check your work.

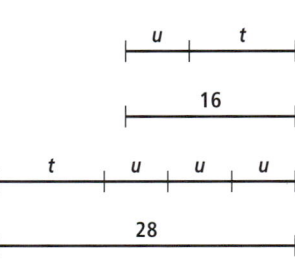

18. Find the x-intercepts of the graph of $y = x^2 + 9x - 5$ using each method. (**Lessons 9-5, 9-3**)
    a. Let $y = 0$ and use the Quadratic Formula.
    b. Use a graphing calculator and zoom in on the intercepts.

19. a. Simplify $\sqrt{d^8 + 3d^8}$.
    b. Check your answer from Part a by testing the special case where $d = 2$. (**Lesson 8-9**)

20. In December of 1986, Dick Rutan and Jeana Yeager flew the *Voyager* airplane nonstop around Earth without refueling, the first flight of its kind. The average rate for the 24,987-mile trip was 116 mph. How many days long was this flight?
    (**Lesson 5-3**)

21. a. Identify an equation of the vertical line through the point (–6, 18).
    b. Identify an equation of the horizontal line through the point (–6, 18). (**Lesson 4-2**)

A chase plane follows the *Voyager* as it flies over the clouds.

EXPLORATION

22. Consider the general system $\begin{cases} ax + by = e \\ cx + dy = f \end{cases}$ of two linear equations in two variables, x and y.

    a. Find y by multiplying the first equation by c and the second equation by $-a$ and then adding.

    b. Find x by multiplying the first equation by d and the second equation by $-b$ and then adding.

    c. Your answers to Parts a and b should be fractions. Write them with the denominator $ad - bc$ if they are not already in that form.

    d. Use a CAS to solve this system and compare the CAS solution with what you found by hand.

QY ANSWERS

1. Multiply the top equation by 3 and the bottom equation by 2, then add the equations. You should get $0 = -48 + 36 = -12$. This is never true, so the system has no solution.

2. Answers vary. Sample answer: (1, 1) satisfies $4x + 2y = 6$. Since $-2(1) + 3 = 1$, (1, 1) also satisfies $y = -2x + 3$.

Chapter 10

Lesson 10-7
Matrices and Matrix Multiplication

Vocabulary

matrix (matrices)
elements
dimensions
matrix form
2 × 2 identity matrix

▶ **BIG IDEA** Rectangular arrays called matrices can sometimes be multiplied and represent systems of linear equations.

When you use a method like multiplication or addition to solve systems of linear equations, you do the same steps over and over. Once a linear system to be solved for x and y is in the form $\begin{cases} ax + by = e \\ cx + dy = f \end{cases}$, the processes for solving it are the same. The different solutions are caused by the numbers a, b, c, and d that are the coefficients and the numbers e and f that are the constants.

A mathematical tool called a *matrix* allows you to separate those numbers from the overall structure of the problem. A **matrix** (the plural is **matrices**) is a rectangular array, such as $\begin{bmatrix} 3 & -4 \\ 15 & 0 \end{bmatrix}$.

The brackets [ ] identify the numbers that are in the matrix. The objects in the array are the **elements** of the matrix. The elements of the matrix $\begin{bmatrix} 3 & -4 \\ 15 & 0 \end{bmatrix}$ are 3, −4, 15, and 0. They are identified by the row and the column of the matrix they are in. The rows are counted from the top; the columns from the left. So −4 is the element in the 1st row and 2nd column.

The number of rows and the number of columns of a matrix are its **dimensions**. Because it has 2 rows and 2 columns, the matrix $\begin{bmatrix} 3 & -4 \\ 15 & 0 \end{bmatrix}$ is a 2 × 2 (read "2 by 2") matrix, while the matrix $\begin{bmatrix} x \\ y \end{bmatrix}$ is a 2 × 1 matrix because it has 2 rows and 1 column.

The linear system $\begin{cases} 2x + 6y = 2 \\ x + 4y = -5 \end{cases}$ is described by three matrices: the 2 × 2 *coefficient matrix* $\begin{bmatrix} 2 & 6 \\ 1 & 4 \end{bmatrix}$, the 2 × 1 *variable matrix* $\begin{bmatrix} x \\ y \end{bmatrix}$, and the 2 × 1 *constant matrix* $\begin{bmatrix} 2 \\ -5 \end{bmatrix}$.

Mental Math

What is the probability that

a. a randomly chosen one-digit number is odd?

b. a randomly selected day in the year 2015 is in June?

c. a fair, 6-sided die shows a 3 or 5?

622 Linear Systems

In *matrix form,* the system on the previous page is

$\begin{bmatrix} 2 & 6 \\ 1 & 4 \end{bmatrix} \cdot \begin{bmatrix} x \\ y \end{bmatrix} = \begin{bmatrix} 2 \\ -5 \end{bmatrix}$. In general, the **matrix form** of the system $\begin{cases} ax + by = e \\ cx + dy = f \end{cases}$ is $\begin{bmatrix} a & b \\ c & d \end{bmatrix} \cdot \begin{bmatrix} x \\ y \end{bmatrix} = \begin{bmatrix} e \\ f \end{bmatrix}$.

Example 1

Write $\begin{cases} 4x = 5y + 10 \\ 2x - 3y = 20 \end{cases}$ in matrix form.

Solution The first step is to rewrite the system with each equation in standard form: $\begin{cases} 4x - 5y = 10 \\ 2x - 3y = 20 \end{cases}$. Then form three matrices to describe the coefficients, variables, and constants in the standard-form system.

$$\begin{bmatrix} 4 & -5 \\ 2 & -3 \end{bmatrix} \cdot \begin{bmatrix} x \\ y \end{bmatrix} = \begin{bmatrix} 10 \\ 20 \end{bmatrix}$$

coefficient matrix • variable matrix = constant matrix

Matrix Multiplication

Above, we have put a dot between the coefficient and variable matrices. This is because these matrices are multiplied. Matrices can be added, subtracted, and multiplied, but in this lesson you will learn only about matrix multiplication.

The matrix form of the system in Example 1 shows that the matrix $\begin{bmatrix} 4 & -5 \\ 2 & -3 \end{bmatrix}$ is multiplied by $\begin{bmatrix} x \\ y \end{bmatrix}$. What does it mean to multiply these matrices? To multiply these matrices, we combine each row of the 2×2 matrix $\begin{bmatrix} 4 & -5 \\ 2 & -3 \end{bmatrix}$ with the 2×1 matrix $\begin{bmatrix} x \\ y \end{bmatrix}$ to form a product 2×1 matrix. Each element in the product matrix is the product of the first entries plus the product of the second entries. Multiplying the top row of $\begin{bmatrix} 4 & -5 \\ 2 & -3 \end{bmatrix}$ by $\begin{bmatrix} x \\ y \end{bmatrix}$ gives $4x + -5y$. Multiplying the bottom row of $\begin{bmatrix} 4 & -5 \\ 2 & -3 \end{bmatrix}$ by $\begin{bmatrix} x \\ y \end{bmatrix}$ gives $2x + -3y$.

Matrices and Matrix Multiplication

So $\begin{bmatrix} 4 & -5 \\ 2 & -3 \end{bmatrix} \cdot \begin{bmatrix} x \\ y \end{bmatrix} = \begin{bmatrix} 4x - 5y \\ 2x - 3y \end{bmatrix}$. This is why we say that the system $\begin{cases} 4x - 5y = 10 \\ 2x - 3y = 20 \end{cases}$ is equivalent to the matrix equation $\begin{bmatrix} 4 & -5 \\ 2 & -3 \end{bmatrix} \cdot \begin{bmatrix} x \\ y \end{bmatrix} = \begin{bmatrix} 10 \\ 20 \end{bmatrix}$. In the same way, two matrices can be multiplied when both matrices contain numbers.

Example 2
Perform the multiplication $\begin{bmatrix} 10 & 3 \\ -2 & 5 \end{bmatrix} \cdot \begin{bmatrix} 4 \\ 11 \end{bmatrix}$.

Solution The result will be a 2 × 1 matrix. So write down places for the elements of this matrix.

$$\begin{bmatrix} 10 & 3 \\ -2 & 5 \end{bmatrix} \cdot \begin{bmatrix} 4 \\ 11 \end{bmatrix} = \begin{bmatrix} \underline{?} \\ \underline{?} \end{bmatrix}$$

Multiply the top row by the column to obtain the top element:

$10 \cdot 4 + 3 \cdot 11 = 73$. Multiply the bottom row by the column to obtain the bottom element: $-2 \cdot 4 + 5 \cdot 11 = 47$.

$$\begin{bmatrix} 10 & 3 \\ -2 & 5 \end{bmatrix} \cdot \begin{bmatrix} 4 \\ 11 \end{bmatrix} = \begin{bmatrix} 10 \cdot 4 + 3 \cdot 11 \\ -2 \cdot 4 + 5 \cdot 11 \end{bmatrix} = \begin{bmatrix} 73 \\ 47 \end{bmatrix}$$

Multiplying 2 × 2 Matrices

Not all matrices can be multiplied. For a product AB of two matrices A and B to exist, each row of A must have the same number of elements as each column of B. This is so that row-by-column multiplication can be performed. The element in row i and column j of the product is the result of multiplying row i of A and the column j of B.

Example 3
Find the product $\begin{bmatrix} 1 & 2 \\ 5 & 3 \end{bmatrix} \cdot \begin{bmatrix} -4 & 6 \\ 30 & 5 \end{bmatrix}$.

Solution The product will be a 2 × 2 matrix. First write down the spaces for the elements of the product. The product will have the same number of rows as the first matrix and the same number of columns as the second matrix.

$$\begin{bmatrix} 1 & 2 \\ 5 & 3 \end{bmatrix} \cdot \begin{bmatrix} -4 & 6 \\ 30 & 5 \end{bmatrix} = \begin{bmatrix} \underline{?} & \underline{?} \\ \underline{?} & \underline{?} \end{bmatrix}$$

Pick an element of the product matrix.

For the element in the 1st row, 1st column of the product, multiply the 1st row of the left matrix by the 1st column of the right matrix.

$1 \cdot -4 + 2 \cdot 30 = 56$

$$\begin{bmatrix} 1 & 2 \\ 5 & 3 \end{bmatrix} \cdot \begin{bmatrix} -4 & 6 \\ 30 & 5 \end{bmatrix} = \begin{bmatrix} 56 & ? \\ ? & ? \end{bmatrix}$$

For the element in the 1st row, 2nd column of the product, multiply the 1st row of the left matrix by the 2nd column of the right matrix.

$1 \cdot 6 + 2 \cdot 5 = 16$

$$\begin{bmatrix} 1 & 2 \\ 5 & 3 \end{bmatrix} \cdot \begin{bmatrix} -4 & 6 \\ 30 & 5 \end{bmatrix} = \begin{bmatrix} 56 & 16 \\ ? & ? \end{bmatrix}$$

The other two elements are found in a similar manner.

$5 \cdot -4 + 3 \cdot 30 = 70$ and $5 \cdot 6 + 3 \cdot 5 = 45$

$$\begin{bmatrix} 1 & 2 \\ 5 & 3 \end{bmatrix} \cdot \begin{bmatrix} -4 & 6 \\ 30 & 5 \end{bmatrix} = \begin{bmatrix} 56 & 16 \\ 70 & 45 \end{bmatrix}$$

In matrix multiplication, the left and right matrices play different roles. So you should not expect that reversing the order of the matrices will give the same product. For the matrices of Example 3, $\begin{bmatrix} -4 & 6 \\ 30 & 5 \end{bmatrix} \cdot \begin{bmatrix} 1 & 2 \\ 5 & 3 \end{bmatrix} = \begin{bmatrix} 26 & 10 \\ 55 & 75 \end{bmatrix}$. Matrix multiplication is not commutative.

Questions

COVERING THE IDEAS

1. Consider the matrix $\begin{bmatrix} a & b & c \\ d & e & f \end{bmatrix}$.

   a. What are the dimensions of this matrix?
   b. Name the elements in the first row.
   c. Which element is in the 2nd row, 3rd column?

2. The matrix equation $\begin{bmatrix} -4 & 6 \\ 3 & -7 \end{bmatrix} \cdot \begin{bmatrix} d \\ g \end{bmatrix} = \begin{bmatrix} 18 \\ 54 \end{bmatrix}$ describes a system of equations. Write the system.

In 3 and 4, a system is given.
   a. Write the coefficient matrix.
   b. Write the constant matrix.

3. $\begin{cases} 5a - 2b = -4 \\ 3a + 4b = 34 \end{cases}$

4. $\begin{cases} 5x + 3(y + 1) = 85 \\ 2x = 7y \end{cases}$

Matrices and Matrix Multiplication

5. What is the result when the row [−4 6] is combined with the column $\begin{bmatrix} 0.25 \\ -0.50 \end{bmatrix}$ in a matrix multiplication?

In 6–8, multiply the two matrices.

6. $\begin{bmatrix} 3 & 5 \\ -2 & 4 \end{bmatrix} \cdot \begin{bmatrix} 6 \\ 1 \end{bmatrix}$

7. $\begin{bmatrix} 5 & -8 \\ 4 & 11 \end{bmatrix} \cdot \begin{bmatrix} 0.5 & 0 \\ -2 & 4 \end{bmatrix}$

8. $\begin{bmatrix} 0 & -1 \\ 1 & 2 \end{bmatrix} \cdot \begin{bmatrix} 3 & 4 \\ 5 & -6 \end{bmatrix}$

9. Give an example different from the one provided in this lesson to show that multiplication of 2 × 2 matrices is not commutative.

APPLYING THE MATHEMATICS

10. The matrix $\begin{bmatrix} 1 & 0 \\ 0 & 1 \end{bmatrix}$ is called the **2 × 2 identity matrix** for multiplication. To see why, calculate the products in Parts a and b.

    a. $\begin{bmatrix} 1 & 0 \\ 0 & 1 \end{bmatrix} \cdot \begin{bmatrix} a & b \\ c & d \end{bmatrix}$

    b. $\begin{bmatrix} a & b \\ c & d \end{bmatrix} \cdot \begin{bmatrix} 1 & 0 \\ 0 & 1 \end{bmatrix}$

    c. **True or False** Matrix multiplication of a 2 × 2 matrix with the 2 × 2 identity matrix is commutative.

11. Solve $\begin{bmatrix} -9 & 2 \\ 0 & 15 \end{bmatrix} \cdot \begin{bmatrix} x \\ 5 \end{bmatrix} = \begin{bmatrix} 100 \\ 75 \end{bmatrix}$ for x.

12. Solve $\begin{bmatrix} a & b \\ c & d \end{bmatrix} \cdot \begin{bmatrix} x \\ y \end{bmatrix} = \begin{bmatrix} 3x - 4y \\ 2x + y \end{bmatrix}$ for a, b, c, and d.

13. Create three different 2 × 2 matrices M, N, and P.
    a. Calculate MN.
    b. Calculate $(MN)P$.
    c. Calculate NP.
    d. Calculate $M(NP)$.
    e. Do your answers to Parts b and d tell you that matrix multiplication is definitely not associative, or do they tell you that matrix multiplication might be associative?

14. When a matrix M is multiplied by itself, the product $M \cdot M$ is called M^2 for short. $M^2 \cdot M = M^3$, $M^3 \cdot M = M^4$, and so on. Let $M = \begin{bmatrix} 0 & -1 \\ 1 & 0 \end{bmatrix}$. Show that M^4 is the identity matrix of Question 10.

REVIEW

15. Without drawing any graphs, explain how you can tell whether the graphs of $17x + 20y = 84$ and $16x + 20y = 85$ are two intersecting lines, one line, or two nonintersecting parallel lines. (**Lesson 10-6**)

In 16 and 17, solve by using any method. (**Lessons 10-6, 10-5, 10-4, 10-2, 10-1**)

16. $\begin{cases} 3x + 2y = 40 \\ 9x + 6y = 120 \end{cases}$

17. $\begin{cases} y = 10 - 4x \\ y = 4x - 10 \end{cases}$

18. A hardware store placed two orders with a manufacturer. The first order was for 18 hammers and 14 wrenches, and totaled $582. The second order was for 12 hammers and 10 wrenches, and totaled $396. What is the cost of one hammer and of one wrench? (**Lesson 10-5**)

19. Consider the equations $2x + y = 4$, $x = 5$, and $y = 3$. The graph of these equations forms a triangle. (**Lessons 10-1, 8-8, 8-6**)
    a. Find the vertices of the triangle.
    b. Find the length of each side of the triangle.
    c. Find the area of the triangle.

20. **Skill Sequence** Find an equivalent expression without a fraction. (**Lessons 8-4, 8-3**)
    a. $\dfrac{x^5}{x}$ b. $\dfrac{x^5}{x^3}$ c. $\dfrac{x^5}{y^3}$

21. An Aztec calendar is being placed on a rectangular mat that is 1.25 times as high and 2.25 times as wide as the calendar. What percent of the mat is taken up by the calendar? (**Lesson 5-7**)

Matrices and Matrix Multiplication

EXPLORATION

22. Rows and columns with 3 elements are multiplied as follows.

$$[a \ b \ c] \cdot \begin{bmatrix} d \\ e \\ f \end{bmatrix} = [ad + be + cf]$$

3×3 matrices can be multiplied using the same row-by-column idea as is used with 2×2 matrices. The product MN of two 3×3 matrices M and N is a 3×3 matrix. The element in row i and column j of MN is the result of multiplying row i of M and the column j of N.

a. Use this idea to find MN when $M = \begin{bmatrix} 2 & 1 & 0 \\ -1 & 5 & 2 \\ 0 & 3 & 10 \end{bmatrix}$ and

$N = \begin{bmatrix} 0 & 3 & -3 \\ 4 & 6 & 1 \\ -0.5 & 0 & 12 \end{bmatrix}$.

b. Show that $\begin{bmatrix} 1 & 0 & 0 \\ 0 & 1 & 0 \\ 0 & 0 & 1 \end{bmatrix}$ is the identity matrix for 3×3 matrix multiplication. (*Hint:* Calculate products as in Question 10.)

Lesson 10-8

Using Matrices to Solve Systems

Vocabulary

inverse (of a matrix)

▶ **BIG IDEA** By finding the inverse of a matrix, you can solve systems of linear equations.

The matrix method for solving systems follows a pattern like the one used to solve the equation $\frac{2}{7}x = 28$.

To solve this equation, you would multiply both sides of the equation by the number that makes the coefficient of x equal to 1. This is the multiplicative inverse of $\frac{2}{7}$, or $\frac{7}{2}$.

$$\frac{7}{2} \cdot \frac{2}{7}x = \frac{7}{2} \cdot 28$$
$$1 \cdot x = 98$$

When the coefficient is 1, the equation simplifies to become a statement of the solution, $x = 98$.

Mental Math

If $g(t) = -4t^2$, calculate

a. $g(10)$.
b. $g(5)$.
c. $\frac{g(10)}{g(5)}$.
d. $g(2)$.

The 2 × 2 Identity Matrix

Refer to the solution to the above equation. Working backwards, the key to the solution $x = 98$ is to have obtained $1 \cdot x = 98$ in the previous step. For a system of two linear equations, if (e, f) is the solution, then the solution can be written $\begin{cases} x = e \\ y = f \end{cases}$. What is the previous step?

Working backwards, this is the same as $\begin{cases} 1x + 0y = e \\ 0x + 1y = f \end{cases}$. The coefficient matrix of this system is $\begin{bmatrix} 1 & 0 \\ 0 & 1 \end{bmatrix}$. Recall from Question 10 in Lesson 10-7 that the matrix $\begin{bmatrix} 1 & 0 \\ 0 & 1 \end{bmatrix}$ is called the 2 × 2 identity matrix because when it multiplies a 2 × 2 or 2 × 1 matrix, it does not change that matrix.

$$\begin{bmatrix} 1 & 0 \\ 0 & 1 \end{bmatrix} \cdot \begin{bmatrix} h \\ k \end{bmatrix} = \begin{bmatrix} h \\ k \end{bmatrix}$$

Thus, you can solve a system with matrices if you can convert it into an equivalent system in which the coefficient matrix is the identity matrix. This is done by multiplying both sides of the original matrix equation by a new matrix, called the **inverse** of the coefficient matrix. You can use technology to help you find the inverse matrix.

Using Matrices to Solve Systems

Chapter 10

Example

Use matrices to solve the system $\begin{cases} 2x - 5y = 4 \\ -4x + 11y = -6 \end{cases}$.

Solution First, write the system in matrix form.

$$\begin{bmatrix} 2 & -5 \\ -4 & 11 \end{bmatrix} \cdot \begin{bmatrix} x \\ y \end{bmatrix} = \begin{bmatrix} 4 \\ -6 \end{bmatrix}$$

Use technology to find the inverse of the coefficient matrix as shown on one particular calculator below.

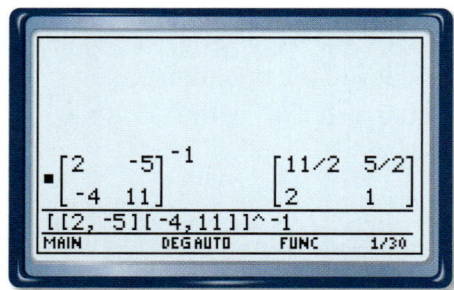

So the inverse of $\begin{bmatrix} 2 & -5 \\ -4 & 11 \end{bmatrix}$ is $\begin{bmatrix} 5.5 & 2.5 \\ 2 & 1 \end{bmatrix}$. Multiply each side of the matrix equation by $\begin{bmatrix} 5.5 & 2.5 \\ 2 & 1 \end{bmatrix}$ on the left.

$$\underbrace{\begin{bmatrix} 5.5 & 2.5 \\ 2 & 1 \end{bmatrix} \cdot \begin{bmatrix} 2 & -5 \\ -4 & 11 \end{bmatrix}} \cdot \begin{bmatrix} x \\ y \end{bmatrix} = \underbrace{\begin{bmatrix} 5.5 & 2.5 \\ 2 & 1 \end{bmatrix} \cdot \begin{bmatrix} 4 \\ -6 \end{bmatrix}}$$

This produces the identity matrix. Multiply these matrices.

$$\begin{bmatrix} 1 & 0 \\ 0 & 1 \end{bmatrix} \cdot \begin{bmatrix} x \\ y \end{bmatrix} = \begin{bmatrix} 5.5 \cdot 4 + 2.5 \cdot -6 \\ 2 \cdot 4 + 1 \cdot -6 \end{bmatrix}$$

$$\begin{bmatrix} x \\ y \end{bmatrix} = \begin{bmatrix} 7 \\ 2 \end{bmatrix}.$$

The solution is $(x, y) = (7, 2)$.

Check Substitute $x = 7$ and $y = 2$ into each of the original equations.

Does $2 \cdot 7 - 5 \cdot 2 = 4$? Yes, $14 - 10 = 4$.

Does $-4 \cdot 7 + 11 \cdot 2 = -6$? Yes, $-28 + 22 = -6$.

Linear Systems

Lesson 10-8

Inverse 2 × 2 Matrices

In the previous example, we asserted that the inverse of $\begin{bmatrix} 2 & -5 \\ -4 & 11 \end{bmatrix}$ is $\begin{bmatrix} 5.5 & 2.5 \\ 2 & 1 \end{bmatrix}$. While a calculator or computer may automatically give you the inverse, you still need to be able to check that what you are given is correct. This is done by doing the row-by-column multiplication,

$$\begin{bmatrix} 2 & -5 \\ -4 & 11 \end{bmatrix} \cdot \begin{bmatrix} 5.5 & 2.5 \\ 2 & 1 \end{bmatrix} = \begin{bmatrix} 2 \cdot 5.5 + -5 \cdot 2 & 2 \cdot 2.5 + -5 \cdot 1 \\ -4 \cdot 5.5 + 11 \cdot 2 & -4 \cdot 2.5 + 11 \cdot 1 \end{bmatrix} = \begin{bmatrix} 1 & 0 \\ 0 & 1 \end{bmatrix}.$$

When two matrices are inverses, you can multiply them in either order and you will still get the identity matrix.

 QY

The inverse of the matrix A is denoted by the symbol A^{-1}. We write $\begin{bmatrix} 2 & -5 \\ -4 & 11 \end{bmatrix}^{-1} = \begin{bmatrix} 5.5 & 2.5 \\ 2 & 1 \end{bmatrix}$. With powers of real numbers, $x \cdot x^{-1} = 1$, the multiplicative identity for real numbers. With matrices,

$A \cdot A^{-1} = \begin{bmatrix} 1 & 0 \\ 0 & 1 \end{bmatrix}$, the multiplicative identity for 2 × 2 matrices.

> **QY**
>
> Show that
> $\begin{bmatrix} 5.5 & 2.5 \\ 2 & 1 \end{bmatrix} \cdot \begin{bmatrix} 2 & -5 \\ -4 & 11 \end{bmatrix}$
> $= \begin{bmatrix} 1 & 0 \\ 0 & 1 \end{bmatrix}.$

Summary of the Matrix Method

The matrix method of solving a system of linear equations is useful because it can be applied to systems with more than two variables in exactly the same way as it is applied to systems with two variables. You will work with these larger systems in later courses. The process is always the same.

$A \cdot \begin{bmatrix} x \\ y \end{bmatrix} = B$ Write the system as the product of three matrices:
coefficients · variables = constants.

$A^{-1} \cdot A \cdot \begin{bmatrix} x \\ y \end{bmatrix} = A^{-1} \cdot B$ Multiply (on the left) each side by the inverse of the coefficient matrix.

$\begin{bmatrix} 1 & 0 \\ 0 & 1 \end{bmatrix} \cdot \begin{bmatrix} x \\ y \end{bmatrix} = A^{-1} \cdot B$ Because A and A^{-1} are inverses, their product is the identity matrix.

$\begin{bmatrix} x \\ y \end{bmatrix} = A^{-1} \cdot B$

Using Matrices to Solve Systems

The left side gives the variables. The right side gives the solutions.
Caution: Just as 0 has no multiplicative inverse, not all 2 × 2 matrices have multiplicative inverses. When a coefficient matrix does not have an inverse, then there is not a unique solution to the system. There may be infinitely many solutions, or there may be no solution.

Questions

COVERING THE IDEAS

In 1–3, consider the system $\begin{cases} 1x + 0y = 7 \\ 0x + 1y = -3 \end{cases}$.

1. Solve this system.
2. Write the coefficient matrix for the system.
3. Write the matrix form for the system.
4. Write the 2 × 2 identity matrix for multiplication.
5. Show that the inverse of $\begin{bmatrix} 1 & -2 \\ 5 & 4 \end{bmatrix}$ is $\begin{bmatrix} \frac{2}{7} & \frac{1}{7} \\ -\frac{5}{14} & \frac{1}{14} \end{bmatrix}$.
6. **Multiple Choice** Which of these matrices is the inverse of $\begin{bmatrix} 3 & 4 \\ 5 & 7 \end{bmatrix}$?

   A $\begin{bmatrix} -3 & -4 \\ -5 & -7 \end{bmatrix}$ B $\begin{bmatrix} -2 & -4 \\ -5 & -6 \end{bmatrix}$ C $\begin{bmatrix} 7 & -4 \\ -5 & 3 \end{bmatrix}$ D $\begin{bmatrix} \frac{1}{3} & 0 \\ 0 & \frac{1}{17} \end{bmatrix}$

In 7–9, use the given system.
   a. Write the system in matrix form.
   b. Use technology to find the inverse of the coefficient matrix.
   c. Solve the system.

7. $\begin{cases} 3x + 5y = 27 \\ 2x + 3y = 17 \end{cases}$
8. $\begin{cases} 2x + 3y = 18 \\ 3x + 4y = 21 \end{cases}$
9. $\begin{cases} 2m - 6t = -6 \\ 7.5m - 15t = -37.5 \end{cases}$

APPLYING THE MATHEMATICS

10. a. Show that $\begin{bmatrix} 1 & 0 \\ 0 & -1 \end{bmatrix}$ equals its multiplicative inverse.
    b. What real numbers equal their multiplicative inverses?

In 11 and 12, a system is given.
  a. Using inverse matrices, write the product of two matrices that will find the solution.
  b. Give the solution.
  c. Check the solution.

11. $\begin{cases} 2.3x + y = -5.5 \\ 3.1x + 2.4y = -1.1 \end{cases}$

12. $\begin{cases} 0.5m + 1.5t = 10 \\ t + 14 = 0.5m \end{cases}$

13. Consider the three systems below.

  i. $\begin{cases} x - 8y = -35 \\ 5x + 8y = 65 \end{cases}$
  ii. $\begin{cases} y = x - 3 \\ 2x + 3y = 16 \end{cases}$
  iii. $\begin{cases} 4x - 5y = 8 \\ 12x - 15y = 3 \end{cases}$

  a. Choose the system which describes two parallel lines, and write its coefficient matrix.
  b. What is your calculator's response when you try to find the inverse of the coefficient matrix of the system you chose in Part a?

14. Show that the matrix $\begin{bmatrix} 3 & 5 \\ 9 & 15 \end{bmatrix}$ has no multiplicative inverse by writing the matrix equation $\begin{bmatrix} 3 & 5 \\ 9 & 15 \end{bmatrix} \cdot \begin{bmatrix} a & b \\ c & d \end{bmatrix} = \begin{bmatrix} 1 & 0 \\ 0 & 1 \end{bmatrix}$ as two systems of equations and showing that those systems have no solution.

REVIEW

In 15 and 16, multiply the given matrices. (Lesson 10-7)

15. $\begin{bmatrix} 5 & 6 \\ 2 & -3 \end{bmatrix} \cdot \begin{bmatrix} -4 & 1 \\ 0 & 2 \end{bmatrix}$

16. $\begin{bmatrix} 1 & 2 \\ 4 & 3 \end{bmatrix} \cdot \begin{bmatrix} 8 \\ -2 \end{bmatrix}$

17. Solve $\begin{bmatrix} -4 & 1 \\ 3 & 6 \end{bmatrix} \cdot \begin{bmatrix} 5 \\ a \end{bmatrix} = \begin{bmatrix} -12 \\ 63 \end{bmatrix}$ for a. (Lesson 10-7)

In 18 and 19, determine whether the lines are parallel and nonintersecting, coincident, or intersecting in only one point. (Lesson 10-6)

18. $\begin{cases} y = 4x - 6 \\ 28x - 7y = -10 \end{cases}$

19. $\begin{cases} 8x + 6y = 10 \\ 4x - 3y = -5 \end{cases}$

20. A jar of change contains 64 coins consisting only of quarters and dimes. The total value of the coins in the jar is $13.60. Let q = the number of quarters, and d = the number of dimes. (Lesson 10-4)

  a. Write two equations that describe the information given above.
  b. How many of each type of coin is in the jar?

Each dime has 118 reeds and each quarter has 119 reeds along its outer edge.

Source: U.S. Mint

21. **Skill Sequence** Solve each equation. (Lessons 9-5, 9-2, 8-6)
    a. $x^2 = 144$
    b. $x^2 + 44 = 144$
    c. $4x^2 + 44 = 144$
    d. $4x^2 + 44x + 265 = 144$

22. Simplify the expression $\sqrt{x} \cdot \sqrt{x} \cdot \sqrt{x^3}$. (Lessons 8-7, 8-6)

23. Suppose a bank offers a 4.60% annual yield on a 4-year CD. What would be the amount paid at the end of the 4 years to an investor who invests $1,800 in this CD? (Lesson 7-1)

24. Ms. Brodeur wants to lease about 2,000 square meters of floor space for a business. She noticed an advertisement in the newspaper regarding a set of 3 vacant stores. Their widths are given in the floor plan at the right. How deep must these stores be to meet Ms. Brodeur's required area? (Lesson 2-1)

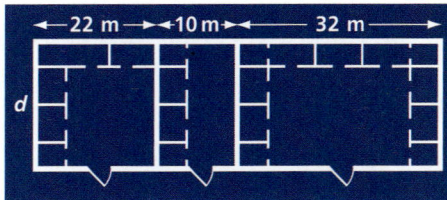

EXPLORATION

25. A formula for the inverse of a 2×2 matrix $\begin{bmatrix} a & b \\ c & d \end{bmatrix}$ can be found by following these steps.

    Step 1 Write $\begin{bmatrix} a & b \\ c & d \end{bmatrix} \cdot \begin{bmatrix} x & u \\ y & v \end{bmatrix} = \begin{bmatrix} 1 & 0 \\ 0 & 1 \end{bmatrix}$ as two systems of equations. Keep a, b, c, and d as the coefficients. One system will have the variables x and y and the other will have u and v.

    Step 2 Solve for x, y, u, and v in terms of a, b, c, and d. Find a formula in this way and check it with at least two different matrices. You may want to use a CAS to find the formula.

QY ANSWER

$\begin{bmatrix} 5.5 & 2.5 \\ 2 & 1 \end{bmatrix} \cdot \begin{bmatrix} 2 & -5 \\ -4 & 11 \end{bmatrix} =$

$\begin{bmatrix} 5.5 \cdot 2 + 2.5 \cdot -4 & 5.5 \cdot -5 + 2.5 \cdot 11 \\ 2 \cdot 2 + 1 \cdot -4 & 2 \cdot -5 + 1 \cdot 11 \end{bmatrix}$

$= \begin{bmatrix} 1 & 0 \\ 0 & 1 \end{bmatrix}$

Lesson 10-9

Systems of Inequalities

▶ **BIG IDEA** The graph of the solutions to a system of linear inequalities is a region bounded by lines.

In Lesson 6-9, you graphed linear inequalities like $y < 11$ and $y \geq 4x + 3$ on a coordinate plane. These inequalities describe half-planes. In this lesson, you will see how to graph regions described by a system of two or more inequalities. Solving a system of inequalities involves finding the common solutions of two or more inequalities.

Mental Math

Refer to the figure.
Find c if

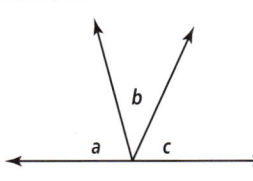

a. $a = 75°$ and $b = 40°$
b. $a = 75°$ and $b = 39°$
c. $a = b$

Example 1

Graph all solutions to the system $\begin{cases} x \geq 0 \\ y \geq 0 \end{cases}$.

Solution First graph the solution to $x \geq 0$. It is shown at the left below. Then graph the solution $y \geq 0$, shown at the right below.

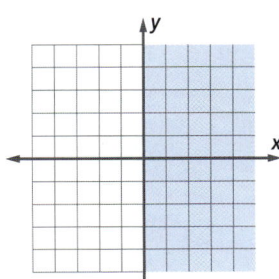 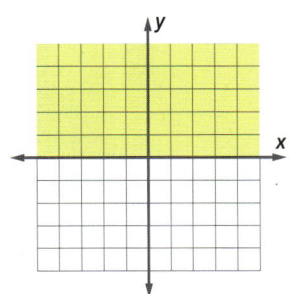

The solution to the system is the set of points common to both of the sets above. At the right we show the solutions to the two inequalities superimposed. At the far right is the solution to the system. The solution is the intersection of the two solution sets above shown in green. It consists of the first quadrant and the nonnegative parts of the x- and y-axes.

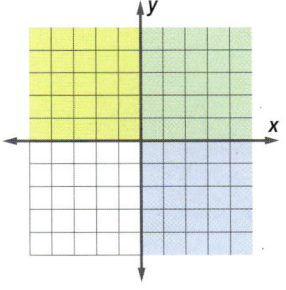 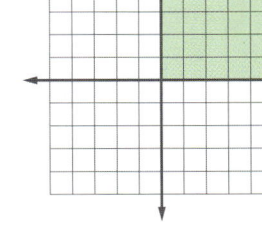

Recall that in general, the graph of $Ax + By < C$ is a half-plane, and that it lies on one side of the boundary line $Ax + By = C$.

Systems of Inequalities **635**

Chapter 10

Example 2

Graph all solutions to the system $\begin{cases} y \leq 3 \\ x > -2 \\ y > x - 1 \end{cases}$.

Solution The graph of the system is the set of points in common to all three half-planes. So graph all three inequalities.

Inequality 1 Graph $y \leq 3$. First graph the boundary line $y = 3$. Graphically, $y \leq 3$ consists of all points below (less than) or on (equal to) the solid boundary line. This is shaded in blue at the right.

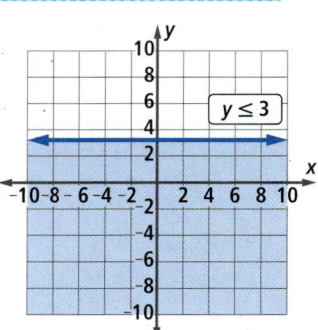

Inequality 2 Graph $x > -2$. Graph the boundary line $x = -2$. This line is dashed since we want only values greater than -2. $x > -2$ tells us to shade to the right of the line in red where values are greater than -2.

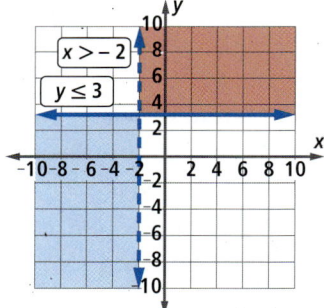

Inequality 3 Graph $y > x - 1$. Graph the dashed boundary line $y = x - 1$ and shade above the line as shown in yellow below.

The solution is the region shown below and includes one of the sides of the triangle. The graph at the far right shows what your final graph should look like.

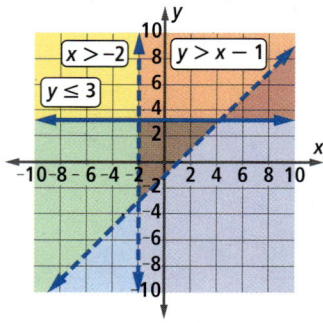

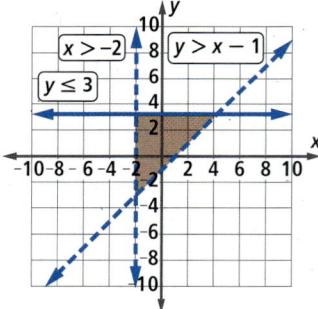

Check A partial check is to choose a point in the intersection region. We choose $(1, 2)$. Substitute the point into all the inequalities.

Is $2 \leq 3$? Yes.

Is $1 > -2$? Yes.

Is $2 > 1 - 1$? Yes.

When an intersection point cannot be found easily from a graph, you need to solve a system of equations to find it.

GUIDED

Example 3

Graph all solutions to the system $\begin{cases} y < 2x + 3 \\ y > 2x - 2 \end{cases}$.

Solution

1. The boundary lines $y = 2x + 3$ and $y = 2x - 2$ have the same slope. So they are ___?___ lines.

2. Graph the lines.

636 Linear Systems

3. The solutions to $y < 2x + 3$ comprise the half-plane ___?___ (above/below) the boundary line. The solutions to $y > 2x - 2$ make up the half-plane ___?___ (above/below) the line.

   You saw in Lesson 10-6 that a system whose graph is made up of nonintersecting and parallel lines has no solutions. Is this necessarily true for a system of inequalities whose boundary lines are nonintersecting and parallel? Explain. ___?___

4. Graph the intersection of the two half-planes in Step 3.

Questions

COVERING THE IDEAS

1. The graph of all solutions to $\begin{cases} x > 0 \\ y < 0 \end{cases}$ consists of all points in which quadrant?

2. Graph the solutions to the system $\begin{cases} x < 0 \\ y > 0 \end{cases}$.

3. Consider the system $\begin{cases} y \leq 4x + 1 \\ y > 2x + 1 \end{cases}$.

   a. How is the graph of all solutions to the system related to the graphs of $y \leq 4x + 1$ and $y > 2x + 1$?

   b. **Fill in the Blank** The graph of $y > 2x + 1$ is a ___?___.

   c. Why does the graph of $y \leq 4x + 1$ include its boundary line?

   d. Is (2, 5) a solution to this system? How can you tell?

In 4 and 5, graph the solution to the system.

4. $\begin{cases} x > 0 \\ y > 0 \\ 3x + y < 10 \end{cases}$

5. $\begin{cases} x \geq -3 \\ y \leq 4 \\ y \geq 2x + 1 \end{cases}$

6. Consider the system $\begin{cases} x > 0 \\ y > 0 \\ x + y < 30 \\ x + y > 20 \end{cases}$.

   a. The graph of all solutions to this system is the interior of a quadrilateral with what vertices?

   b. Name two points that are solutions to the system.

   c. Is (9, 10) a solution to the system? Why or why not?

7. The solution to a system of equations that involves different parallel lines is Ø (empty set). You saw in Example 3 this is not necessarily true of a system of inequalities that involves parallel lines. Is it possible to have a system of inequalities whose solution is Ø? Why or why not?

APPLYING THE MATHEMATICS

In 8 and 9, describe the shaded region with a system of inequalities.

8.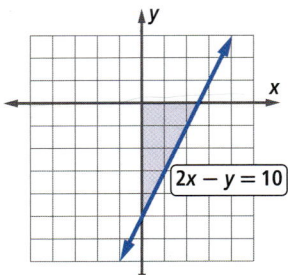

9.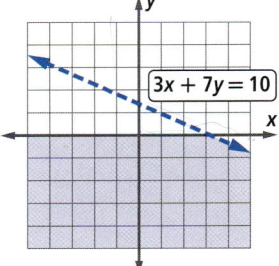

10. Write the system of inequalities whose solution is the interior of the rectangle with vertices $(-3, 10)$, $(5, 10)$, $(5, -1)$, and $(-3, -1)$.

11. It takes Tippie about 10 minutes to type a letter of moderate length and about 8 minutes to type a normal double-spaced page.
    a. Write a system of inequalities that describes the total number of letters L and pages P Tippie can type in an hour or less. Assume the number of letters typed is greater than or equal to zero, as is the number of pages.
    b. Accurately graph the set of points that satisfies the system.

12. An actress is paid $300 per day to understudy a part and $750 per day to perform the role before an audience. During one run, an actress earned between $4,000 and $7,000.
    a. At most, how many times might she have performed the role?
    b. What is the maximum number of times she might have been an understudy?
    c. Graph all possible ways she might have earned her salary.

13. A hockey team is scheduled to play 14 games during a season. Its coach estimates that it needs at least 20 points to make the playoffs. A win is worth 2 points and a tie is worth 1 point.
    a. Make a graph of all the combinations of wins w and ties t that will get the team into the playoffs.
    b. How many ways are there for the team to make the playoffs?

Actors' Equity Association, founded in 1913, is the labor union that represents more than 45,000 actors and stage managers in the United States.

Source: Actors' Equity Association

638 Linear Systems

REVIEW

In 14 and 15, a system of equations is given.
   a. Write the system in matrix form.
   b. Use technology to find the inverse of the coefficient matrix.
   c. Solve the system. (Lesson 10-8)

14. $\begin{cases} y = 3x - 8 \\ 4x + 4y = 12 \end{cases}$

15. $\begin{cases} 6x + 4y = 14 \\ -2x - 3y = -18 \end{cases}$

16. Let ℓ be the line with equation $y = -4x + 9$. Write an equation of a line that (Lesson 10-6)
   a. coincides with ℓ.
   b. does not intersect ℓ.

17. Solve the system $\begin{cases} 2.5x + y = 6 \\ 9x + 4y = 17.5 \end{cases}$ using any method.

(Lessons 10-5, 10-4, 10-2, 10-1)

In 18–20, use the formula $h = -16t^2 + 94t + 2$ for the height h in feet of a model rocket t seconds after being fired straight up from a stand 2 feet off the ground. (Lessons 9-5, 9-4, 9-3)

18. a. Graph the equation.
    b. Use the graph to find the maximum height of the rocket.

19. a. Find the height of the rocket after 6 seconds.
    b. Interpret your answer to Part a.

20. Use the Quadratic Formula to calculate at what time(s) the rocket is at a height of 100 feet.

EXPLORATION

21. Find a system of linear inequalities whose solution on a coordinate plane is a region like the one shown at the right.

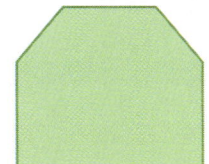

Chapter 10

Lesson 10-10

Nonlinear Systems

Vocabulary

nonlinear system

Mental Math

Refer to the rectangle.

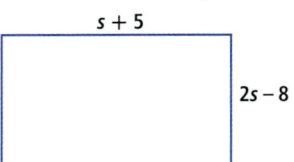

a. What is the area of the rectangle?

b. What is the perimeter of the rectangle?

▶ **BIG IDEA** The ideas used to solve systems of linear equations can be applied to solve some systems of nonlinear equations.

Previously in this chapter you have worked with systems of linear equations. In these systems, every graph involved is a line. In this lesson, we consider systems that involve curves. A **nonlinear system** is a system of equations or inequalities in which at least one of the equations or inequalities is nonlinear.

GUIDED

Example 1

Solve the system $\begin{cases} y = x^2 + 3 \\ y = x + 9 \end{cases}$.

Solution Look at the graphs at the right. The solutions occur at the points of intersection. These graphs intersect at (−2, _?_) and (3, _?_); therefore, the system has two solutions.

The solutions are (_?_) and (_?_).

🛑 QY

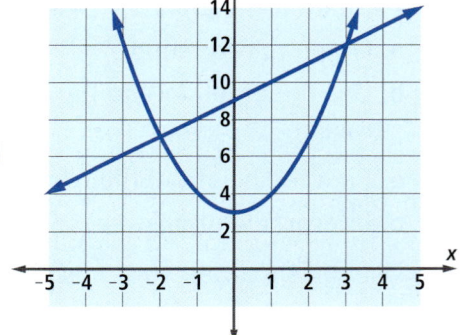

Solving algebraically using substitution is also an option. You may need to use the Quadratic Formula.

▶ **QY**

Check your answer to Guided Example 1 by substituting the coordinates of each point into both equations to verify that they are solutions to the system.

Example 2

Solve the system $\begin{cases} y = -2x + 8 \\ y = -x^2 + 4x - 1 \end{cases}$.

Solution Because both expressions in x are equal to y, they are equal to each other at the point of intersection.

$-2x + 8 = -x^2 + 4x - 1$ Substitution

$8 = -x^2 + 6x - 1$ Add 2x to both sides.

$0 = -x^2 + 6x - 9$ Subtract 8 from both sides.

$x = \dfrac{-6 \pm \sqrt{6^2 - 4(-1)(-9)}}{2(-1)}$ Quadratic Formula

$x = \dfrac{-6 \pm \sqrt{36 - 36}}{-2} = \dfrac{-6 \pm 0}{-2}$ or 3

640 Linear Systems

Now find the *y*-coordinate of the point of intersection.
$y = -2x + 8$
$y = -2(3) + 8 = -6 + 8 = 2$
The solution is (3, 2).

Check The graph reinforces that there is only one solution and that it is at (3, 2), as shown on the screen at the right.

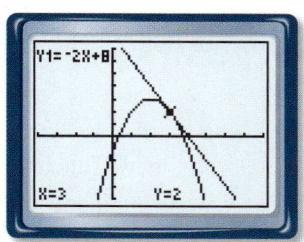

Some nonlinear systems have no solutions.

Example 3

Solve the system $\begin{cases} y = x^2 + 6x + 11 \\ y = -x^2 + 6x - 9 \end{cases}$.

Solution 1 Use substitution.

$x^2 + 6x + 11 = -x^2 + 6x - 9$
$2x^2 + 6x + 11 = 6x - 9$ Add x^2.
$2x^2 + 11 = -9$ Subtract 6x.
$2x^2 = -20$ Subtract 11.
$x^2 = -10$ Divide by 2.
$x = \pm\sqrt{-10}$ Take the square root.

The solutions $\sqrt{-10}$ and $-\sqrt{-10}$ are not real numbers, so there is no solution for this system in the set of real numbers.

Solution 2 Graph the equations.

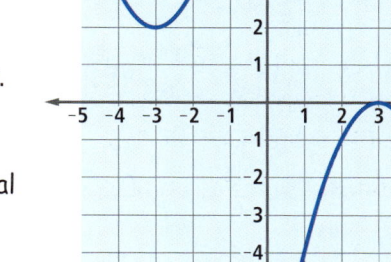

The parabolas never intersect, so there is no solution to this system.

When you don't know how to solve a system of equations algebraically, you can use a graphing approach to make very good approximations.

Example 4

Jonas is working on a science fair project. He wants to study the result of combining equal populations (in weight) of two types of bacteria. The first bacteria (in grams) grow according to the function $f(x) = 15(0.8)^x$ where *x* = the time from now (in hours). The other bacteria growth (in grams) is modeled by $g(x) = 4(1.5)^x$, where *x* = the time from now (in hours). If Jonas wants to combine them when the two populations are equal, how long will he have to wait? How many grams of each type of bacteria will there be at that time?

(continued on next page)

Nonlinear Systems

Solution This situation can be represented by the following system.

$$\begin{cases} f(x) = 15(0.8)^x \\ g(x) = 4(1.5)^x \end{cases}$$

Graph the two functions on your calculator. Be sure to include any intersections in the viewing window. Use the INTERSECT command to find an accurate approximation for the solution.

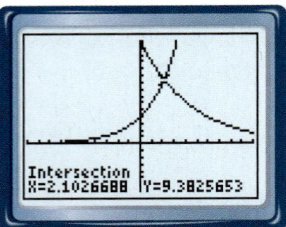

Jonas should combine the bacteria in approximately 2.10 hours (or 2 hours and 6 minutes). There will be about 9.4 grams of bacteria in each population at that time.

Questions

COVERING THE IDEAS

In 1 and 2, solve the system using substitution.

1. $\begin{cases} y = 2x^2 + 12x + 17 \\ y = -x^2 - 6x - 10 \end{cases}$

2. $\begin{cases} y = x^2 + 4x + 3 \\ y = 3x + 1 \end{cases}$

3. Solve the system $\begin{cases} y = 3x^2 \\ y = -4x^2 \end{cases}$ by picturing the graph in your head.

In 4–6, solve the system.

4. $\begin{cases} y = x^2 - 6x + 1 \\ y = -x^2 + 4x + 1 \end{cases}$

5. $\begin{cases} p = 9q + 25 \\ p = 2q^2 + 7q + 1 \end{cases}$

6. $\begin{cases} y = x^2 + 3 \\ y = -x + 1 \end{cases}$

In 7 and 8, solve the system by graphing. Round solutions to the nearest hundredth.

7. $\begin{cases} y = 5x^2 + 5 \\ y = 0.5x^2 \end{cases}$

8. $\begin{cases} y = 2.718^x \\ y = 3.14(0.25)^x \end{cases}$

APPLYING THE MATHEMATICS

In 9 and 10, solve the system by graphing.

9. $\begin{cases} y = |x - 3| \\ y = -2x^2 + 12x - 18 \end{cases}$

10. $\begin{cases} y = x^2 \\ y = 2^x \end{cases}$

11. Solve the system of Example 3 by addition.

12. A quarterback passes a football to a receiver downfield. The path of the ball is described by the equation $h = -0.025x^2 + x + 6$, where $x =$ the horizontal distance (in yards) of the ball from the quarterback and h is the height of the ball (in feet) above the ground. The receiver catches the ball 40 yards downfield. A bird flies over the quarterback in the same direction as the ball. Its flight is described by the equation $y = \frac{7}{60}x + 10$, where $x =$ the distance from the quarterback (in yards) and y is the bird's height (in feet).

   a. When the quarterback first threw the ball, how high was it off the ground?
   b. How high was the bird when it was right over the quarterback?
   c. How many times did the path of the bird and the football cross?
   d. Give these points of intersection.
   e. Notice that the time when the bird or the football is moving is not given. Explain why this means the bird and ball might not have hit each other.

Cleveland Browns quarterback Jason Campbell (#17) in a game against the New York Jets in 2013

13. During a spring training game in 2001, baseball pitching great Randy Johnson threw a fastball. A bird flew in the ball's path and was accidentally struck by the ball. This is the only time in Major League Baseball that a bird has been struck by a pitch. A possible equation for the height of Randy Johnson's pitch is $h = -5.56t^2 - 1.39t + 4.25$, where t is the time in seconds since he threw the ball and h is the ball's height in feet above the ground. A possible equation for the bird's height is $h = -24.42t^2 + 3.94t + 5.14$, where t is the time in seconds since Randy's pitch and h is the bird's height above the ground in feet.

   a. For the ball and the bird to collide, what must be true of the system $\begin{cases} h = -5.56t^2 - 1.39t + 4.25 \\ h = -24.42t^2 + 3.94t + 5.14 \end{cases}$?
   b. Find how long after the pitch was thrown that it struck the bird.
   c. What was the height of the bird when it got struck by the ball?
   d. Does the graph of the parabola $h = -5.56t^2 - 1.39t + 4.25$ represent the flight of the ball? Explain.

Chapter 10

14. Company One's stock values during the month of January can be represented by $V = 40(0.9)^x$, where x is time in days since the beginning of the month. Company Two's stock values can be represented by $V = |2x - 9|$ for the same time period.
    a. Approximate when Companies One and Two had the same stock value and give that value.
    b. Describe how well each company's stock values changed over the month (from $x = 0$ to $x = 31$).
    c. When would have been the best time to buy Company Two's stock? Explain your answer.

REVIEW

15. Solve $\begin{bmatrix} 1 & -2 \\ 3 & 1.5 \end{bmatrix} \cdot \begin{bmatrix} -3 \\ 2 \\ n \end{bmatrix} = \begin{bmatrix} -11.5 \\ 3 \end{bmatrix}$ for n. (**Lesson 10-8**)

16. Solve $\begin{cases} x - 1.5y = 11 \\ 5x + 9y = 11 \end{cases}$ by any method. (**Lessons 10-5, 10-4, 10-3**)

17. Suppose Rodney bought a car 32 years ago for $8,000. It lost 6% of its value each year for the first 15 years. Then its value stayed the same for 3 years. When it was 18 years old, the car became a collector's item and its value increased 21% each year. Find the value of the car now. (**Lessons 7-3, 7-2**)

18. Graph the solutions to the inequality $-4.5y - 18x + 3 < 12$ on a coordinate grid. (**Lesson 6-9**)

EXPLORATION

19. An equation for a circle with radius 2 centered at the origin is $x^2 + y^2 = 4$. Use your calculator to find a pair of values of m and b so that the number of solutions to the system $\begin{cases} x^2 + y^2 = 4 \\ y = mx + b \end{cases}$ is
    a. 2.
    b. 1.
    c. 0.

QY ANSWER

For $(-2, 7)$:
Does $7 = (-2)^2 + 3$? Yes.
Does $7 = -2 + 9$? Yes.

For $(3, 12)$:
Does $12 = 3^2 + 3$? Yes.
Does $12 = 3 + 9$? Yes.
Both solutions check.

Chapter 10 Projects

1. Cars and Computers

For the years 1985, 1990, 1995, 2000, and 2005, find the number of cars in the United States and the number of personal computers.

a. Find lines of best fit for these two sets of data.

b. Find the year in which these two lines meet.

c. Do you think your calculation in Part b is a good prediction of when the number of computers in the United States will be the same as the number of cars? Explain your answer.

2. When Do Systems Have Integer Solutions?

Suppose you have a system of the form
$$\begin{cases} ax + by = e \\ cx + dy = f \end{cases}$$
where a, b, c, d, e, and f are all integers.

a. Create three systems of this form that have a single solution whose coordinates are integers. In each case calculate $ad - bc$.

b. Create three systems of this form that have a single solution whose coordinates are not two integers. In each case calculate $ad - bc$.

c. When do you think that a system as described above has a solution with integer coordinates?

3. Systems with More Variables

In mathematics and in many of its applications, sometimes you need to solve systems with more than just two variables. Just as a system with two variables had two equations, a system with n variables has n equations. The methods for solving these systems are basically the same as the methods you studied in this chapter.

a. Explain how you could solve the system
$$\begin{cases} x + y + z = 3 \\ x + 2y = 5 \\ 3x - 2y = -1 \end{cases}$$
(*Hint:* You might want to solve a certain system with two variables in the process.)

b. Using methods similar to the methods in this chapter, what could you do to the system
$$\begin{cases} 3x - 2y - 2z = 0 \\ x + y + z = 0 \\ 7x + 10y + z = 9 \end{cases}$$
to make it look more like the system in Part a, in which only one equation had the variable z in it? Do this, then solve the system.

c. Write a short explanation on how to solve systems with three variables for someone who knows how to solve systems with only two variables.

4 Adding and Subtracting Equations

Write two linear equations and do the following.

a. Graph the two linear equations and find their intersection.

b. Graph the equation found by adding the two selected equations.

c. Graph the equation found by subtracting them.

d. Multiply each equation by a number, and add them again.

e. Do the graphs have any common features? If so, describe them.

f. Verify your calculations starting with a different pair of equations.

5 Finding the Counterfeit

A math-savvy pirate finds 12 gold coins, and he fears that several might be counterfeit. (Counterfeit coins are lighter.) He measures all 12 coins on a scale and finds that together they weigh 1.3 kilograms. He removes one coin, and finds that together they weigh 1.15 kilograms. He removes yet another coin, and finds that together they weigh 1 kilogram. At this point, his scale broke. Find out how many counterfeit coins (if any) the pirate found. Explain how systems of equations can be used to find the solution.

Chapter 10 Summary and Vocabulary

- A **system** is a set of sentences that together describe a single situation. Situations that lead to linear equations can lead to a **linear system**. All that is needed is that more than one condition must be satisfied. This chapter discusses ways of solving systems in which the sentences are equations or inequalities in two variables.

- The **solution set to a system** is the set of all solutions common to all of the sentences in the system. A solution to a system of two linear equations is an ordered pair (x, y) that satisfies each equation. Systems of two linear equations may have zero, one, or infinitely many solutions. Other systems may have other numbers of solutions.

- One way to solve a system is by graphing. There are as many solutions as intersection points. Graphing is also a way to describe solutions of systems that have infinitely many solutions, for example, systems of **coincident lines** and **systems of linear inequalities,** with overlapping half-planes.

- However, graphing does not always yield exact solutions. In this chapter, four strategies are presented for finding exact solutions to systems of linear equations. 1. Substitution is a good method to use if at least one equation is given in $y = mx + b$ form. 2. Addition is appropriate if the same term has opposite signs in the two equations in the system. 3. Multiplication is a good method when both equations are in $Ax + By = C$ form. Each of these methods changes the system into an equivalent system whose solutions are the same as those of the original system. 4. With **matrices,** the system $\begin{cases} ax + by = e \\ cx + dy = f \end{cases}$ becomes $\begin{bmatrix} a & b \\ c & d \end{bmatrix} \cdot \begin{bmatrix} x \\ y \end{bmatrix} = \begin{bmatrix} e \\ f \end{bmatrix}$. This matrix equation is of the form $AX = B$, where A is the coefficient matrix $\begin{bmatrix} a & b \\ c & d \end{bmatrix}$, X is the variable matrix $\begin{bmatrix} x \\ y \end{bmatrix}$, and B is the constant matrix $\begin{bmatrix} e \\ f \end{bmatrix}$. Multiplying both sides of $AX = B$ by the **multiplicative inverse** A^{-1} of the matrix A results in $X = A^{-1}B$.

Vocabulary

10-1
system
solution to a system
empty set, null set

10-4
addition method for solving a system

10-5
equivalent systems
multiplication method for solving a system

10-6
coincident lines

10-7
matrix (matrices)
elements
dimensions
matrix form
2×2 identity matrix

10-8
inverse (of a matrix)

10-10
nonlinear system

Theorems and Properties

Generalized Addition Property of Equality (p. 601)
Slopes and Parallel Lines Property (p. 616)

Chapter 10 Self-Test

Take this test as you would take a test in class. You will need a calculator. Then use the Selected Answers section in the back of the book to check your work.

In 1–4, solve the system by the indicated method.

1. $\begin{cases} y = x - 7 \\ y = 1.5x + 2 \end{cases}$ substitution

2. $\begin{cases} -4d + 9f = 3 \\ 4d - 5f = 9 \end{cases}$ addition

3. $\begin{cases} 7h + 3g = 4 \\ 2h - g = 2 \end{cases}$ multiplication

4. $\begin{cases} 3a - b = 6 \\ \frac{3}{5}b = 37 - 4a \end{cases}$ graphing

5. Solve $\begin{cases} y = x^2 + 3x - 5 \\ y = 6x - 7 \end{cases}$ by using any method.

6. Determine whether the system $\begin{cases} 3s = 2t - 5 \\ \frac{2}{3}t - \frac{1}{3}s = -5 \end{cases}$ has 0, 1, or infinitely many solutions.

In 7 and 8, multiply.

7. $\begin{bmatrix} 2 & 7 \\ 1 & 0 \end{bmatrix} \cdot \begin{bmatrix} 3 \\ 4 \end{bmatrix}$

8. $\begin{bmatrix} 3 & 5 \\ 4 & 6 \end{bmatrix} \cdot \begin{bmatrix} 2 & 8 \\ 1 & 7 \end{bmatrix}$

9. Solve the system $\begin{cases} 3p + 5q = 5 \\ p - q = 7 \end{cases}$ using matrices. Use a calculator to find the inverse of the matrix.

10. An electronics store receives two large orders. The first order is for 6 high-definition televisions and 3 DVD players, and totals $6,795. The second order is for 4 high-definition televisions and 4 DVD players, and totals $4,860. What is the cost of one high-definition television and what is the cost of one DVD player?

11. Give values for m, n, c, and d so that the system $\begin{cases} y = mx + c \\ y = nx + d \end{cases}$ has infinitely many solutions.

12. Rosie is buying posies and roses. She wants to spend no more than $40. Roses are $5 each, and posies are $2 each. Accurately graph all combinations of flowers that she can buy.

13. Solve the system $\begin{cases} y = 3x + 50 \\ y = -2x + 70 \end{cases}$ by graphing.

14. A passenger airplane took 2 hours to fly from St. Louis, Missouri to Orlando, Florida in the direction of the jet stream. On the return trip against the jet stream, the airplane took 2 hours and 30 minutes. If the distance between the two cities is about 1,000 miles, find the airplane's speed in still air and the speed of the jet stream.

15. Akando bought 80 feet of chicken wire to make a coop on his farm. He needs the coop to be at least 10 feet wide and 15 feet long.

    a. Draw a graph to show all possible dimensions (to the nearest foot) of the coop.

    b. At most, how wide can the coop be?

16. Graph all solutions to the system $\begin{cases} y \leq x + 3 \\ y \geq -2x + 4 \end{cases}$.

17. Kele paid for his lunch with 15 coins. He used only quarters and dimes. If his lunch cost $2.40, how many of each coin did he use?

18. Write a system of inequalities to describe the shaded region below.

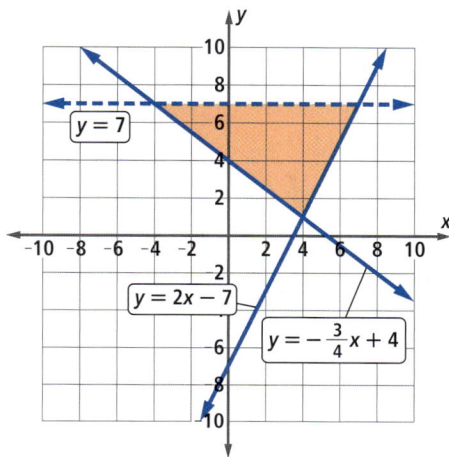

Chapter 10

Chapter Review

SKILLS
PROPERTIES
USES
REPRESENTATIONS

SKILLS Procedures used to get answers

OBJECTIVE A Solve systems using substitution. (Lessons 10-2, 10-3)

In 1 and 2, solve the system by using substitution.

1. $\begin{cases} m = n \\ 3m - 4 = n \end{cases}$
2. $\begin{cases} 2p = q + 5 \\ 2q = 4(p + 2) \end{cases}$

In 3 and 4, two lines have the given equations. Find the point of intersection, if any.

3. Line ℓ: $y = 2x + 5$;
   Line m: $y = -3x + 4$
4. Line p: $y = \frac{2}{3}x + \frac{1}{9}$;
   Line q: $y = \frac{1}{5}x - 4$

OBJECTIVE B Solve systems by addition and multiplication. (Lessons 10-4, 10-5)

In 5 and 6, solve the system by addition.

5. $\begin{cases} 3b + 4 = a \\ -3b - 5 = 2a \end{cases}$
6. $\begin{cases} 0.4x + 0.75y = 2.7 \\ 0.4x - 2y = 2.5 \end{cases}$

In 7–10, solve the system by multiplication.

7. $\begin{cases} 3f + g = 41 \\ f - 2g = 20 \end{cases}$
8. $\begin{cases} 3t - u = 5 \\ 6t + 3u = 7 \end{cases}$
9. $\begin{cases} 5w + 2v = 6 \\ 7 + 5v = w \end{cases}$
10. $\begin{cases} 3x - 5 = y \\ 2x - 7y = 9 \end{cases}$

OBJECTIVE C Multiply 2 × 2 matrices by 2 × 2 or 2 × 1 matrices. (Lesson 10-7)

In 11–14, multiply.

11. $\begin{bmatrix} 2 & -4 \\ 3 & 5 \end{bmatrix} \cdot \begin{bmatrix} 7 \\ 4 \end{bmatrix}$

12. $\begin{bmatrix} 6 & -0.5 \\ 0.5 & 2 \end{bmatrix} \cdot \begin{bmatrix} 6 \\ 12 \end{bmatrix}$

13. $\begin{bmatrix} -6 & 8 \\ -9 & 7 \end{bmatrix} \cdot \begin{bmatrix} 1 & 3 \\ 4 & 2 \end{bmatrix}$

14. $\begin{bmatrix} 2 & 11 \\ 0 & 7 \end{bmatrix} \cdot \begin{bmatrix} 0 & -4 \\ 5 & 9 \end{bmatrix}$

OBJECTIVE D Solve systems using matrices. (Lesson 10-8)

In 15–18, a system is given.
a. Write the system in matrix form.
b. Use technology to find the inverse.
c. Solve the system.

15. $\begin{cases} 3x + 2y = 7 \\ 5x + 7y = 9 \end{cases}$
16. $\begin{cases} 6m + 4d = 7 \\ 4m + 3d = 13 \end{cases}$
17. $\begin{cases} 5p - 7q = 20 \\ 4p - 8q = 14 \end{cases}$
18. $\begin{cases} w + 3z = 5 \\ 4z - 5w = 9 \end{cases}$

OBJECTIVE E Solve nonlinear systems. (Lesson 10-10)

In 19 and 20, solve by substitution.

19. $\begin{cases} y = -2x^2 \\ y + x^2 = -1 \end{cases}$

20. $\begin{cases} 2x^3 - 2y = x^2 - 5 \\ y = x^3 - 2x \end{cases}$

PROPERTIES Principles behind the mathematics

OBJECTIVE F Determine whether a system has 0, 1, or infinitely many solutions. (Lesson 10-6)

In 21–24, determine whether the given system has 0, 1, or infinitely many solutions.

21. $\begin{cases} 2y + 3x = 5 \\ 2y = 4 - 3x \end{cases}$

22. $\begin{cases} y + 3x = 7 \\ y = 3x + 7 \end{cases}$

23. $\begin{cases} 3p = 7q + 2 \\ 6p = 10q + 5 \end{cases}$

24. $\begin{cases} 4p + 5q = 7 \\ 2p = 3\frac{1}{2} - \frac{5}{2}q \end{cases}$

25. When will the system $\begin{cases} y = mx + a \\ y = mx + b \end{cases}$ have no solution? Explain your answer.

26. Can the given set of points be the intersection of two lines?
    a. exactly one point
    b. exactly two points
    c. infinitely many points
    d. no points

27. **Fill in the Blank** Two lines are parallel only if they have the same ___?___.

28. **True or False** Two lines can intersect in more than one point if they have different y-intercepts.

USES Applications of mathematics in real-world situations

OBJECTIVE G Use systems of linear equations to solve real-world problems. (Lessons 10-2, 10-3, 10-4, 10-5, 10-6)

29. Austin drove four times as far as Antonio. Together, they drove 350 miles. How far did they each drive?

30. Car A costs $2,800 down and $100 per month. Car B costs $3,100 down and $50 per month. After how many months is the amount paid for the cars equal?

31. Good Job offers $30,000 per year, plus a $1,000 raise each year. Nice Job offers $32,000 per year, plus a $500 raise each year. When will you make more money per year with Good Job?

32. Tickets to see an orchestra cost $30 for adults and $15 for students. One night, the total number of tickets sold was 633. If they sold $15,945 worth of tickets, how many adults and how many students attended?

33. A chemist wishes to mix a 15% acid solution with 30% acid solution to make a 25% acid solution. If the chemist wants to make 8 pints of the solution, how many pints of each solution should the chemist use?

34. From 1990 to 2000, the population of Seattle, Washington grew at a rate of about 4,700 people per year, to a population of about 565,000. Baltimore, Maryland decreased at a rate of about 8,400 people per year, to a population of about 650,000. If these rates continue, in about how many years will Seattle and Baltimore have the same population?

OBJECTIVE H Use systems of linear inequalities to solve real-world problems. (Lesson 10-9)

35. A chef has 10 ducks and wants to make Peking duck and duck salad. It takes 1 duck to make Peking duck, and 2 ducks to make duck salad. Make a graph to show all the combinations of duck dishes that the chef can make.

36. Suppose 40 students want to play 2 games. If each game must have at least 10 students and at most 25 students, make a graph of the number of ways the students could divide up to play the games.

37. Jackie is running around a rectangular track with a perimeter of at most 500 feet. The track is at least 80 feet wide and 100 feet long.
    a. Draw a graph to show all possible dimensions (to the nearest foot) of the track.
    b. What is the maximum length of the track?
    c. What is the maximum width of the track?

38. Nihad won $300 in a soccer all-stars contest. She wants to buy soccer balls for $25 and pairs of soccer shoes for $30. She wants to buy at least two balls and at least two pairs of shoes.
    a. Graph all the combinations she could buy.
    b. What is the maximum number of pairs of shoes she can buy?
    c. What is the maximum number of balls she can buy?

REPRESENTATIONS Pictures, graphs, or objects that illustrate concepts

OBJECTIVE I Find solutions to systems of equations by graphing. (Lessons 10-1, 10-6, 10-10)

In 39–43, solve the system by graphing. Round your answers to the nearest tenth.

39. $\begin{cases} 5x + 4y = 7 \\ 3x + 2y = 6 \end{cases}$

40. $\begin{cases} 16x - 16y = 16 \\ \frac{1}{2}x + \frac{1}{2}y = -2 \end{cases}$

41. $\begin{cases} 0.5x - 0.4y = 0.8 \\ x - 1.6 = 0.8y \end{cases}$

42. $\begin{cases} y = 2^x \\ y = 1.32x + 2.67 \end{cases}$

43. $\begin{cases} 2y - 3x = 7 \\ x^2 + y = 15 \end{cases}$

OBJECTIVE J Graphically represent solutions to systems of linear inequalities. (Lesson 10-9)

In 44–47, graph all solutions to the system.

44. $\begin{cases} y \leq 3x \\ y \geq 2x + 1 \end{cases}$

45. $\begin{cases} x + 4 > 7 + 2y \\ x - 4 < y + 2 \end{cases}$

46. $\begin{cases} y > -2 \\ x + y < 0 \\ x - y \geq 1 \end{cases}$

47. $\begin{cases} x \geq 0 \\ y \geq 0 \\ x + y < 9 \end{cases}$

In 48 and 49, accurately graph the set of points that satisfies the situation.

48. An elephant can eat 70 pounds of food in a meal. If the elephant eats P sacks of peanuts averaging 7 pounds each and L bunches of leaves weighing 5 pounds each, how many sacks and bunches can the elephant eat?

49. Conan wants to watch t television shows and m movies. Each show lasts 30 minutes, and each movie lasts 90 minutes. If he has 5 hours of viewing time available, how many full shows and movies can he watch?

OBJECTIVE K Write a system of inequalities given a graph. (Lesson 10-9)

In 50–52, write a system of inequalities to describe the shaded region.

50.
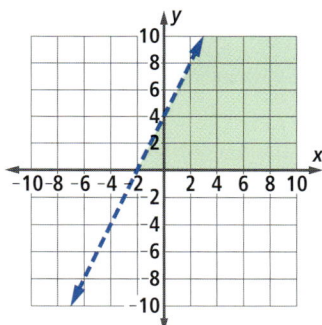

51.
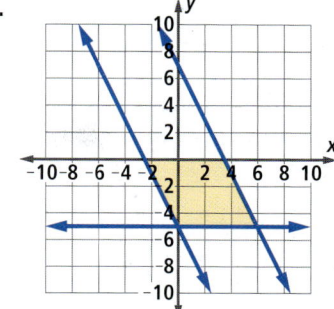

52.
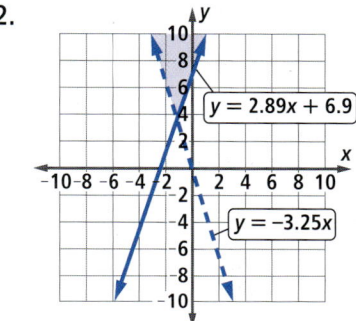

Chapter 11 Polynomials

- 11-1 Investments and Polynomials
- 11-2 Classifying Polynomials
- 11-3 Multiplying a Polynomial by a Monomial
- 11-4 Common Monomial Factoring
- 11-5 Multiplying Polynomials
- 11-6 Special Binomial Products
- 11-7 Permutations
- 11-8 The Chi-Square Statistic

Expressions such as those below are *polynomials*.

$1 \cdot 10^3 + 4 \cdot 10^2 + 9 \cdot 10^1 + 2$

s^3

$2\ell w + 2wh + 2\ell h$

$1{,}000x^4 + 500x^3 + 100x^2 + 200x$

Polynomials form the basic structure of our base 10 arithmetic. The expanded form of a number like 1,492, or
$1 \cdot 10^3 + 4 \cdot 10^2 + 9 \cdot 10^1 + 2 \cdot 10^0$,
is a polynomial in x with the base 10 substituted for x.

Polynomials are also found in geometry. For example, the monomial s^3 represents the volume of a cube with edge s, as shown below. The trinomial $2\ell w + 2wh + 2\ell h$ represents the surface area of a box of dimensions ℓ by w by h, as shown below.

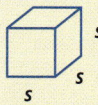 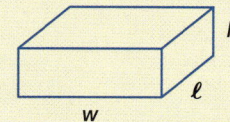

Algebra is filled with polynomials. The linear expression $ax + b$ is a polynomial, as is the quadratic expression $ax^2 + bx + c$. In Chapter 7, you calculated compound interest for a single deposit. When several deposits are made, the total amount of money accumulated can be expressed as a polynomial. For example, the polynomial $1{,}000x^4 + 800x^3 + 600x^2 + 250x$ represents the amount of money you would have if you had invested $1,000 four years ago, added $800 to it three years ago, added $600 to it two years ago, and added $250 to it one year ago, all earning at the same rate $x - 1$.

In this chapter you will study these and other situations that give rise to polynomials and how to add, subtract, multiply, and factor them.

Chapter 11

Lesson 11-1
Investments and Polynomials

Vocabulary

polynomial in x

standard form for a polynomial

▶ **BIG IDEA** When amounts are invested periodically and earn interest from the time of investment, the total value can be represented by a polynomial.

Among the most important money matters adults commonly deal with are salary or wages, savings, payments on loans for cars or trips or other items, and home mortgages or rent.

Each of these items involves paying or receiving money each month, every few months, or every year. But what is the total amount paid or received? The answer is not easy to calculate because interest starts at different times. Here is an example of this kind of situation.

Mental Math

Find the distance between

a. (x, y) and $(0, 0)$.

b. $(a, 5)$ and $(a, -11)$.

c. $(m, m - n)$ and (m, m).

Example 1

Each birthday from age 12 on, Jessica has received $500 from her grandparents. She saves the money in an account that pays an annual yield of 6%. How much money will she have by the time she is 18?

Solution Write down how much Jessica has on each birthday. On her 12th birthday she has $500. She then receives interest on that $500. She receives an additional $500 on her 13th birthday. So on her 13th birthday she has $500(1.06) + 500 = \$1,030.00$.

Each year interest is paid on all the money previously saved and each year another $500 gift is added. The totals for her 12th through 15th birthdays are given below.

Birthday	Expression	Total
12th	500	= $500
13th	500(1.06) + 500	= $1,030.00
14th	$500(1.06)^2 + 500(1.06) + 500$	= $1,591.80
15th	$500(1.06)^3 + 500(1.06)^2 + 500(1.06) + 500$	= $2,187.31

from 12th birthday ↑ from 13th birthday ↑ from 14th birthday ↑ from 15th birthday ↑

Lesson 11-1

You can see the pattern. By her 18th birthday, Jessica will have three more gifts of $500 and earn interest on this money for three more years. The total will be $500(1.06)^6 + 500(1.06)^5 + 500(1.06)^4 + 500(1.06)^3 + 500(1.06)^2 + 500(1.06) + 500 = \$4{,}196.91$.

This total of $4,196.91 that she has by her 18th birthday is $696.91 more than the total $3,500 she received as gifts because of the interest earned.

Letting $x = 1.06$, the amount of money Jessica has (in dollars) after her 18th birthday is given by the polynomial
$500x^6 + 500x^5 + 500x^4 + 500x^3 + 500x^2 + 500x + 500$.

This expression is called a *polynomial in x*. A **polynomial in x** is a sum of multiples of powers of x. In this situation the polynomial is useful because if the interest rate is different, you only have to substitute a different value for x. We call x in this situation a *scale factor*. For example, had Jessica invested her money at an annual yield of 4%, the scale factor would be 104% = 1.04. At the end of 6 years, Jessica's investment (in dollars) would be $500(1.04)^6 + 500(1.04)^5 + 500(1.04)^4 + 500(1.04)^3 + 500(1.04)^2 + 500(1.04) + 500$.

You should verify with a calculator that this sum equals $3,949.14.

> **GUIDED**
>
> ### Example 2
> Suppose Rajib's parents gave him $100 on his 12th birthday, $120 on his 13th, $140 on his 14th, and $160 on his 15th. If he invests all the money in an account with a yearly scale factor x, how much money will he have on his 15th birthday?
>
> **Solution** By his 15th birthday, the __?__ from Rajib's 12th birthday will earn 3 years' worth of interest. It will have grown to __?__ $\cdot x^?$.
>
> The __?__ from his 13th birthday will have grown to __?__.
>
> The __?__ from his 14th birthday will have grown to __?__.
>
> On his 15th birthday he receives __?__.
>
> Through his 15th birthday, the total dollar amount Rajib will have from his birthday gifts is $100x^3 + 120x^2 + 140x + 160$.

Rajib's aunt gave him $50 on each of these 4 birthdays. If he puts this money into the same account, the amount available from the aunt's gifts would be $50x^3 + 50x^2 + 50x + 50$.

Investments and Polynomials

The total amount he would have from all these gifts is found by adding these two polynomials.

$(100x^3 + 120x^2 + 140x + 160) + (50x^3 + 50x^2 + 50x + 50)$

Recall that this sum can be simplified. First, use the Associative and Commutative Properties of Addition to rearrange the polynomials so that like terms are together.

$= (100x^3 + 50x^3) + (120x^2 + 50x^2) + (140x + 50x) + (160 + 50)$

Then use the Distributive Property to add like terms.

$= (100 + 50)x^3 + (120 + 50)x^2 + (140 + 50)x + (160 + 50)$
$= 150x^3 + 170x^2 + 190x + 210$

Notice what the answer means in relation to Rajib's birthday presents. The first year he got $150 ($100 from his parents, $50 from his aunt). The $150 has 3 years to earn interest. The $170 from his next birthday earns interest for 2 years. And so on. Also notice that in these examples we have written the polynomials in the form of decreasing powers of x. This is called **standard form for a polynomial.** Polynomials are often written in standard form.

When comparing investments, it is often useful to make a table or construct a spreadsheet.

Example 3

Kelsey and Chip plan to save money for a round-the-world trip when they retire 10 years from now. Kelsey plans to save $2,000 per year for the first 5 years, and then will stop making deposits. Chip plans to wait 5 years to begin saving, but then hopes to save $2,500 per year for 5 years. They will each deposit their savings at the beginning of the year into a special account earning 6% interest compounded annually. How much will each have after 10 years?

Solution Make a spreadsheet showing the amount of money each person will have at the end of each year. At the end of the first year Kelsey will have $1.06(2,000) = 2,120$. At the end of the second year, she will have 106% of the sum of the previous balance and the new deposit of $2,000. In all, she will have $1.06(2,120 + 2,000) = \$4,367.20$. This pattern continues. But after 5 years, she deposits no more money. So her money only accumulates interest. Kelsey's end-of-year balance in the spreadsheet on the next page was computed by entering the formula =1.06*B2 into cell C2 and the formula =1.06*(C2+B3) into cell C3. The formula in cell C3 was then replicated down column C to C11. A similar set of formulas generated Chip's end-of-year balance.

Miami was the top cruise ship departure port of the United States in 2004 with 641 departures.

Source: Bureau of Transportation Statistics

Lesson 11-1

	A	B	C	D	E
1	Year	Kelsey's Deposits ($)	Kelsey's End-of-Year Balance ($)	Chip's Deposits ($)	Chip's End-of-Year Balance ($)
2	1	2,000	2,120.00	0	0
3	2	2,000	4,367.20	0	0
4	3	2,000	6,749.23	0	0
5	4	2,000	9,274.19	0	0
6	5	2,000	11,950.64	0	0
7	6	0	12,667.68	2,500	2,650.00
8	7	0	13,427.74	2,500	5,459.00
9	8	0	14,233.40	2,500	8,436.54
10	9	0	15,087.40	2,500	11,592.73
11	10	0	15,992.65	2,500	14,938.30

Ten years from now Kelsey will have about $16,000 and Chip will have about $15,000.

In Example 3, notice that even though Kelsey deposits $10,000 and Chip deposits $12,500 at the same rate of interest, compounding interest over a longer period of time gives Kelsey about $1,055 more than Chip. Here is what has happened. After 10 years:

Kelsey has $2,000x^{10} + 2,000x^9 + 2,000x^8 + 2,000x^7 + 2,000x^6$.

Chip has $2,500x^5 + 2,500x^4 + 2,500x^3 + 2,500x^2 + 2,500x$.

When $x = 1.06$, Kelsey has more than Chip.

Questions

COVERING THE IDEAS

1. Refer to Example 1. Suppose Jessica is able to get an annual yield of 5% on her investment.
   a. How much money will she have in her account by her 18th birthday?
   b. How much less is this than what she would have earned with a 6% annual yield?

2. Mary's grandfather will receive a $3,000 bonus from his employer on each of his birthdays from age 61 to age 65 if he indicates he will retire at 65.
   a. If he saves the money in an account paying an annual yield of 6.5%, how much will he have by the time he retires at age 65?
   b. How much will he have accumulated by the time he retires at age 65 if his investment grows by a scale factor of x each year?

Investments and Polynomials 659

3. Refer to Example 2. Suppose Rajib also gets $75, $85, $95, and $105 from cousin Lilly on his four birthdays. He puts this money into his account also.

   a. By his 15th birthday, how much money will Rajib have from just his cousin?

   b. What is the total Rajib will have saved by his 15th birthday from all of his birthday presents?

In 4–6, refer to Example 3.

4. Explain why Chip had less money saved than Kelsey at the end of the 10-year period even though he put more money into his account than Kelsey.

5. Chip said to Kelsey, "I might have less money now, but I am catching up to you, and even if we put no more money into our accounts, the amount in my account will be greater after a few more years." Kelsey said, "You have less now and you will always have less." Who is right?

6. Suppose Kelsey and Chip were able to earn 3% on their investments. Recalculate the balances in the spreadsheet and describe the end result.

7. Refer to page 654. Write the number 84,267 as a polynomial in base 10.

APPLYING THE MATHEMATICS

In 8–11, Clara, Mona, and Odella are friends who have the same birthday. They received the following cash presents on their birthdays. Each put all her money into a bank account that paid a 6% annual yield.

	Clara	Mona	Odella
In 2003	$200	$250	$100
In 2004	$300	$250	$500
In 2005	$250	$250	nothing

8. How much money did Clara have on her birthday in 2003?

9. How much did Mona have on her birthday in 2004?

10. How much did Odella have on her birthday in 2005?

11. In 2006, Clara received $300 on her birthday. If all the money from 2003 to 2006 had been and remains in an account with scale factor x, how much would she have had by her birthday in each of the following years?

   a. 2005
   b. 2006
   c. 2007
   d. 2008

Out of a class of 24 students, the probability of any 3 sharing a birthday is about 16.6%.

12. Suppose in 1999 Tanya received $100 on her birthday. From 2000 to 2003 she received $150 on her birthday. She put the money in a shoe box. The money is still there.

    a. How much money did Tanya have after her 2003 birthday?

    b. How much more would she have had if she had invested her money at an annual yield of 4% each year?

13. **Multiple Choice** Which is the sum of $x^4 + x^3 + x^2$?

    A x^9
    B $3x^9$
    C x^{24}
    D None of these

In 14–17, simplify the expression.

14. $(2y^2 + 13y - 14) + (4y^2 - 3y - 24)$

15. $6(11n + 8n^2 - 2) + (6n^2 - n - 9)$

16. $(7w^2 - 2w + 16) - 4(7w^2 + 15)$

17. $(x^3 + 2x^2 + 8) - (2x - 5x^3 + 6)$

18. Solve the equation $(3x^2 + 2x + 4) + (3x^2 + 11x + 2) = 0$.

19. Solve the equation $(3x^2 + 2x + 4) - (3x^2 + 11x + 2) = 0$.

In 20 and 21, find the missing polynomial.

20. $(91x^2 + 4x - 15) + (\underline{\ ?\ }) = 110x^2 + 62$

21. $(3y^2 - 2y - 1) - (\underline{\ ?\ }) = -4y^2 - 6y + 21$

22. A *cord* of wood is an amount of wood equal to about 128 cubic feet. A wood harvester has planted trees in a forest each spring for four years, as shown in the table at the right.

Year	Number of Trees Planted
1	10,000
2	15,000
3	20,000
4	18,000

    Suppose each tree contains 0.01 cord of wood when planted, and the cordage grows with a scale factor x each year. How many cords of wood are in the forest after planting the fourth spring?

Forests cover 747 million acres in the United States.
Source: U.S. Department of Agriculture

REVIEW

23. If 8 pencils and 5 erasers cost $4.69 and 3 pencils and 4 erasers cost $2.80, find the cost of 2 pencils. (**Lesson 10-2**)

24. Write an equation for the line which passes through the points (4, –8) and (–10, 6). (**Lesson 6-6**)

25. Consider the equation $\frac{30(4h-2)}{6(2h-1)} = 10$. (**Lesson 5-5**)

    a. For what value of h is $\frac{30(4h-2)}{6(2h-1)}$ undefined?

    b. Solve for h.

26. In one of his studies, discussed in the book *The Effects of Cross- and Self-Fertilization in the Vegetable Kingdom*, Charles Darwin compared the heights of two groups of plants. One group was cross-pollinated, meaning they were fertilized by pollen from other plants. The other group was self-fertilized. For the cross-fertilized plants, the mean height was 20.2, and for the self-fertilized plants the mean height was 17.6. Darwin's data are shown below. (*Note:* The values were rounded before plotting in Question 27.) (**Lessons 1-7, 1-6**)

Cross-Fertilized (in.)	23.5	12.0	21.0	22.0	19.1	21.5	22.1	20.4	18.3	21.6	23.3	21.0	22.1	23.0	12.0
Self-Fertilized (in.)	17.4	20.4	20.0	20.0	18.4	18.6	18.6	15.3	16.5	18.0	16.3	18.0	12.8	15.5	18.0

   a. Find the range of the data for each type of plant.

   b. Find the mean for each type of plant.

   c. Find the mean absolute deviation for each type of plant.

27. Use the two dot plots below. (**Lesson 1-7**)

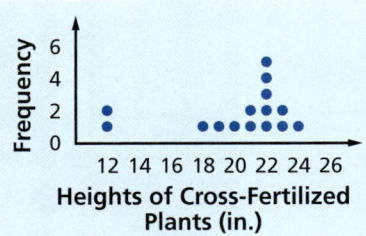 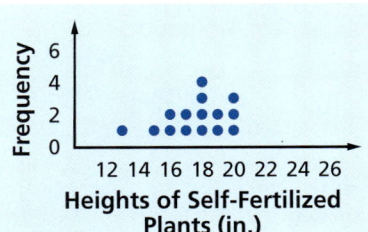

   a. Classify each dot plot as skewed right, skewed left, symmetric, or uniform.

   b. Based on Darwin's data, what advice would you give to a farmer who wants to grow tall plants?

EXPLORATION

28. Refer to Kelsey and Chip in Example 3 and Questions 4–6.

    a. For what interest rates does Kelsey end up with more money after 10 years than Chip?

    b. For what interest rates does Chip end up with more money than Kelsey?

Lesson 11-2
Classifying Polynomials

▶ **BIG IDEA** Polynomials are classified by their number of terms and by their degree.

Vocabulary
monomial
polynomial
binomial
trinomial
degree of a monomial
degree of a polynomial
linear polynomial
quadratic polynomial

Mental Math

Refer to the graph of a function.

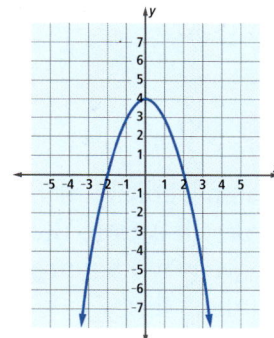

a. State the domain of the function.

b. State the range of the function.

c. State the x-intercepts.

d. State the y-intercept.

Classifying Polynomials by Numbers of Terms

Recall that a *term* can be a single number, variable, or product of numbers and variables. In an expression, addition (or subtraction, which is "adding the opposite") separates terms.

Polynomials are identified by their number of terms. A **monomial** is a single term in which the exponent for every variable is a positive integer. A **polynomial** is an expression that is either a monomial or sum of monomials. Polynomials with two or three terms are used so often they have special names. A **binomial** is a polynomial that has two terms. A **trinomial** is a polynomial that has three terms. Here are some examples.

Monomials	Not Monomials	
$6x$	$6x + y$	(a binomial)
$-16t^2$	$-16t^{-2}$	(negative exponent on a variable)
x^2y^4	$\dfrac{x^2}{y^4}$	(variables divided)

Binomials	Not Binomials	
$x + 26\sqrt{2}$	$26x\sqrt{2}$	(monomial)
$\dfrac{x}{3} - y^3$	$-\dfrac{xy^3}{3}$	(monomial)
$0.44 - 2^{-10}pq^4$	$0.44 - 2^{-10}p + q^4$	(trinomial)

Trinomials	Not Trinomials	
$18x^2 + 5x + 9$	$(15x^2)(5x)(9)$	(monomial)
$a^2 + 2ab - b^{20}$	$a^{-2} + 2ab - b^{-20}$	(negative exponent on variables)
$pq + qr + rp$	$\dfrac{1}{pq + qr + rp}$	(variables divided)

There are no special names for polynomials with more than three terms.

Classifying Polynomials **663**

Chapter 11

Classifying Polynomials by Degree

Every nonconstant term of a polynomial has one or more exponents. For example, $3x^2$ has 2 as its exponent. $10t$ has an unwritten exponent of 1, since $t^1 = t$. $15a^2b^3c^4$ has 2, 3, and 4 as its exponents.

The **degree of a monomial** is the sum of the exponents of the variables in the expression.

$3x^2$ has degree 2.

$10t$ has degree 1.

$15a^2b^3c^4$ has degree $2 + 3 + 4$, or 9.

The degree of a single number, such as 15, is considered to be 0 because $15 = 15x^0$. However, the number 0 is said not to have any degree, because $0 = 0 \cdot x^n$, where n could be any number. The **degree of a polynomial** is the highest degree of any of its monomial terms after the polynomial has been simplified. For example, $6x - 17x^4 + 8 + x^2$ has degree 4. $p + q^2 + pq^2 + p^2q^3$ has degree 5 (because $2 + 3 = 5$).

 QY1

> ▶ QY1
>
> Classify each polynomial by the number of its terms and its degree.
>
> a. $x^6 + x^7 + x^5$
> b. $8y^3z^2 - 40yz^6$
> c. $\frac{4}{3}\pi r^3$

When a polynomial has only one variable, writing it in standard form makes it easy to determine its degree. When the polynomial in x above is written in standard form, the degree is the exponent of the leftmost term.

$-17x^4 + x^2 + 6x + 8$ has degree 4.

Function notation can be used to represent a polynomial in a variable. For example, let $p(x) = -17x^4 + x^2 + 6x + 8$. Then values of the polynomial are easily described. For example, $p(2) = -17 \cdot 2^4 + 2^2 + 6 \cdot 2 + 8 = -248$.

 QY2

> ▶ QY2
>
> If $p(x) = -x + 3 + 4x^4$, what is $p(-2)$?

The polynomial $p + q^2 + pq^2 + p^2q^3$ is a polynomial in p and q. There is no standard form for writing polynomials that have more than one variable, like this one. However, sometimes one variable is picked and the polynomial is written in decreasing powers of that variable. For example, written in decreasing powers of q, this polynomial is $p^2q^3 + pq^2 + q^2 + p$, or, to emphasize the powers of q, $p^2q^3 + (p + 1)q^2 + p$.

A polynomial of degree 1, such as $13t - 6$, is called a **linear polynomial**. A polynomial of degree 2, such as $2x^2 + 3x + 1$ or ℓw, is called a **quadratic polynomial**. Linear and quadratic polynomials whose coefficients are positive integers can be represented by tiles.

664 Polynomials

The tiles below represent the polynomial $2x^2 + 3x + 1$ because this polynomial is the area of the figure.

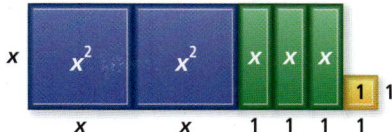

Using the Degree of a Polynomial to Check Operations with Polynomials

The simplest polynomials in one variable are the monomials x, x^2, x^3, x^4, and so on. You know how to add, subtract, multiply, and divide these monomials. For example, $x^2 \cdot x^3 = x^5$, so in this case a polynomial of degree 2 multiplied by a polynomial of degree 3 gives a polynomial of degree 5. In general, the degree of an answer to a polynomial computation is as easy to determine with complicated polynomials as it is with the simplest ones. Consider these examples of polynomial addition and subtraction.

GUIDED

Example
Collect like terms and determine the degree of the resulting polynomial.

1. $(17w + 14) - (6 - 5w) = $ __?__
   degree 1   degree 1   degree 1

2. $(6ab - 22) + (2a + 8b) = 6ab + 2a + 8b - 22$
   degree 2   degree 1   degree __?__

3. $(4x + x^3 - 7) - (x^2 + 4x + 5) = $ __?__
   degree 3   degree 2   degree __?__

4. $(x^7 + 4x - 5) + (3 - 2x - x^7) = $ __?__
   degree 7   degree 7   degree __?__

Notice that the degree of the sum or difference of two polynomials is never greater than the highest degree of the polynomial addends. Can you see why this is so?

Questions

COVERING THE IDEAS

1. Explain why $3x^2 + 4$ is a polynomial but $\frac{3}{x^2} + 4$ is not.

2. **Fill in the Blank** A binomial is a polynomial with __?__ term(s).

In 3–6, an expression is given.

    a. Tell whether the expression is a monomial.

    b. If it is a monomial, state its degree.

3. $17x^{11}$

4. $2w^{-4}$

5. $\frac{1}{2}bh$

6. $2a^4b^5$

7. Is *xyz* a trinomial? Explain your reasoning.

8. Classify each polynomial by its degree and number of terms.

    a. $x^2 + 10$
    b. $x^2 + 10x + 21$
    c. $x^2 + 10xy + y^2$
    d. $x^3 + 10x^2 + 21x$

9. Write the polynomial $12 - 4x - 3x^5 + 8x^2$ in standard form.

10. a. Write the polynomial $a^3 - 3ab^2 - b^3 - 3a^2b$ in standard form as a polynomial in *a*.

    b. Write the polynomial $a^3 - 3ab^2 - b^3 - 3a^2b$ in standard form as a polynomial in *b*.

In 11 and 12, what polynomial is represented by the tiles?

11.

12.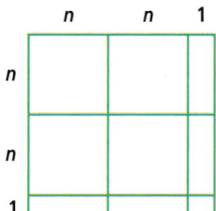

13. Fill in the blank with *always, sometimes but not always,* or *never*. Explain your answer. The degree of the sum of two polynomials is ___?___ greater than the degree of either polynomial addend.

APPLYING THE MATHEMATICS

In 14–18, an expression is given.

    a. Show that the expression can be simplified into a monomial.

    b. Give the degree of the monomial.

14. $10x - 14x$

15. $10x(-14x)$

16. $(5n^3)(6n)^2$

17. $xy + yx$

18. $12x^4 - (3x^4 + 2x^4 + x^4)$

19. Let $p(x) = 50x^3 + 50x^2 + 50x + 50$ and $q(x) = 100x^3 + 120x^2 + 140x + 160$. Give the degree of each polynomial.

   a. $p(x)$ b. $q(x)$ c. $p(x) + q(x)$ d. $p(x) - q(x)$

20. Repeat Question 19 if $p(x) = x^{200} - x^{100} + 1$ and $q(x) = x^{100} - x^{200} + 1$.

In 21–24, give the degree of these polynomials used to find length, area, and volume of geometric figures.

21. perimeter of a triangle $= a + b + c$

22. volume of a circular cone $= \frac{1}{3}\pi r^2 h$

23. area of a trapezoid $= \frac{1}{2}hb_1 + \frac{1}{2}hb_2$

24. surface area of a cylinder $= 2\pi r^2 + 2\pi rh$

25. a. Write a monomial with one variable whose degree is 70.
    b. Write a monomial with two variables whose degree is 70.

26. Complete the fact triangle below and write the related polynomial addition and subtraction facts.

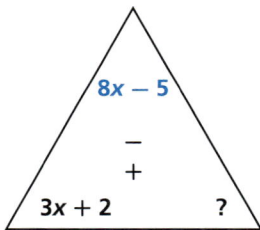

27. a. Give an example of two trinomials in x of degree 5 whose sum is of degree 5.
    b. Give an example of two trinomials in x of degree 5 whose sum is not of degree 5.

28. a. Write 318 and 4,670 as polynomials with 10 substituted for the variable.
    b. Add your polynomials from Part a. Is your sum equal to the sum of 318 and 4,670?

REVIEW

29. a. If you received $1,000 as a present on the day you were born, and the money was put into an account at an annual scale factor of x, how much would be in your account on your 18th birthday?
    b. Evaluate the amount in Part a if $x = 1.05$. **(Lesson 11-1)**

30. Consider the system $\begin{cases} a - 2b = 50 \\ b = -4c \end{cases}$. To solve this system, one student substituted $-4c$ for b in the first equation. The student then wrote $a - 8c = 50$. (Lessons 10-3, 10-2)

   a. Is the student's work correct?

   b. If it is correct, finish solving the system. If not, describe the error the student made.

31. A parking lot with length 70 meters and width 40 meters is to have a pedestrian sidewalk surrounding it, increasing its total area to 3,256 square meters. What will be the width of the sidewalk? (Lesson 9-7)

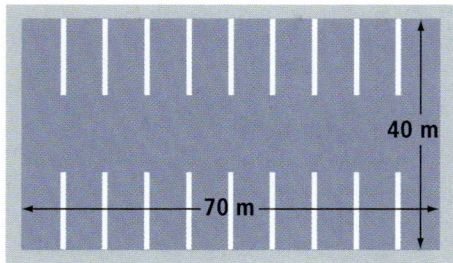

In 32–34, use the Distributive Property to expand the expression. (Lesson 2-1)

32. $4x(x - 9)$ 33. $n(n + 52)$ 34. $(3m + 19.2)80$

EXPLORATION

35. Suppose three polynomials of the same degree n are added.
   a. What is the highest possible degree of their sum? Explain your answer.
   b. What is the lowest possible degree of their sum?

QY ANSWERS

1a. trinomial, degree 7

b. binomial, degree 7

c. monomial, degree 3

2. 69

Lesson 11-3

Multiplying a Polynomial by a Monomial

▶ **BIG IDEA** To multiply a polynomial by a monomial, multiply each term of the polynomial by the monomial and add the products.

In earlier chapters, you saw several kinds of problems involving multiplication by a monomial. To multiply a monomial by a monomial, you can use properties of powers.

$$(9a^4b^5)(8a^3b) = 9 \cdot 8 \cdot a^{4+3}b^{5+1}$$
$$= 72a^7b^6$$

To multiply a monomial by a binomial, you can use the Distributive Property $a(b + c) = ab + ac$.

$$2x(5x + 3) = 2x \cdot 5x + 2x \cdot 3$$
$$= 10x^2 + 6x$$

Some products of monomials and binomials can be pictured using the Area Model for Multiplication. For example, the product $2x(5x + 3)$ is the area of a rectangle with dimensions $2x$ and $5x + 3$. Such a rectangle is shown below at the left.

Mental Math

Find the multiplicative inverse of each number.

a. $-5x$

b. $\frac{4q}{-9}$

c. 0

d. $a + b$

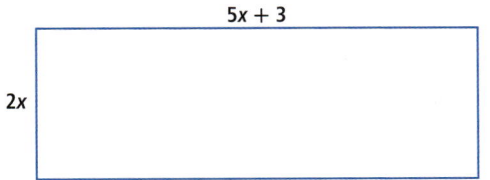

This rectangle can be split into tiles as shown above at the right. The total area of the rectangle is $10x^2 + 6x$, which agrees with the result obtained using the Distributive Property.

Example 1
Find two equivalent expressions for the total area of the rectangle pictured at the right.

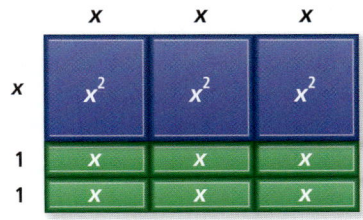

(continued on next page)

Multiplying a Polynomial by a Monomial **669**

Solution The total area is the same as the sum of the areas of the individual tiles, or $3x^2 + 6x$. Also, the total area is length times width, or $3x(x + 2)$.

So this drawing shows $3x(x + 2) = 3x^2 + 6x$.

The area representation of a polynomial shows how to multiply a monomial by any other polynomial. The picture shows a view of some storefronts at a shopping mall.

The displays in the windows are used to attract shoppers, so store owners and mall managers are interested in the areas of storefronts. Note that the height h of each storefront is a monomial, and the sum of the lengths of the storefronts $(L_1 + L_2 + L_3 + L_4)$ is a polynomial.

The total area of the four windows can be computed in two ways. One way is to consider all the windows together. They form one big rectangle with length $(L_1 + L_2 + L_3 + L_4)$ and height h. Thus, the total area equals $h \cdot (L_1 + L_2 + L_3 + L_4)$.

A second way is to compute the area of each storefront and add the results. Thus, the total area also equals $hL_1 + hL_2 + hL_3 + hL_4$.

These areas are equal, so
$h \cdot (L_1 + L_2 + L_3 + L_4) = hL_1 + hL_2 + hL_3 + hL_4$.

In general, to multiply a monomial by a polynomial, extend the Distributive Property: multiply the monomial by each term in the polynomial and add the results.

Example 2

Expand $6r(x^2 - \sqrt{3}x + 7rx)$.

Solution Multiply each term in the trinomial by the monomial $6r$.

$$6r(x^2 - \sqrt{3}x + 7rx) = 6r \cdot x^2 - 6r \cdot \sqrt{3}x + 6r \cdot 7rx$$
$$= 6rx^2 - 6\sqrt{3}rx + 42r^2x$$

Check 1 Test a special case by substituting for both r and x. We let $r = 5$ and $x = 3$.

Does
$6 \cdot 5(3^2 - \sqrt{3} \cdot 3 + 7 \cdot 5 \cdot 3) = 6 \cdot 5 \cdot 3^2 - 6\sqrt{3} \cdot 5 \cdot 3 + 42 \cdot 5^2 \cdot 3?$
Remember to follow order of operations on each side.

Does $30(114 - 3\sqrt{3}) = 270 - 90\sqrt{3} + 3{,}150?$

A calculator shows that each side has the value 3,264.115….

Check 2 Use a CAS. Enter
EXPAND(6*r*(x^2−[√](3)*x+7*r*x)).
You should get an expression equivalent to the answer.

As always, you must be careful with the signs in polynomials.

Activity

A student was given the original expressions below and asked to expand them. The student's answers are shown below.

Original Expressions	Student's Expanded Expressions
1. $2x(3x^2y^3z^7)$	1. $6x^3 + 2xy^3 + 2xz^7$
2. $-3a^2(4a^2b + 7ab - 5a^3b^2)$	2. $-12a^4b - 21a^3b + 15a^5b^2$
3. $7m^3n(4mn^4)$	3. $28m^4n^5$
4. $7xy(2x^3y - 5xy^5 + x^2y)$	4. $14x^4y^2 - 5xy^5 + x^2y$
5. $\frac{1}{4}a^5b(8ab^2 + 2a^2 - 20a^3b)$	5. $2a^6b^3 + \frac{1}{2}a^7b - 5a^8b^2$

Step 1 Identify the expressions you believe the student expanded correctly.

Step 2 Expand the original expressions. Did you accurately find the expanded expressions with mistakes?

Step 3 For each expression the student did not expand correctly, write a sentence explaining what the student did incorrectly.

Explaining a Rule from Arithmetic

Recall the rule for multiplying a decimal by a power of 10: To multiply by 10^n, move the decimal point n places to the right. Multiplication of a monomial by a polynomial can show why this rule works. For example, suppose 81,026 is multiplied by 1,000. Write 81,026 as a polynomial in base 10, and 1,000 as the monomial 10^3.

$$1{,}000 \cdot 81{,}026 = 10^3 \cdot (8 \cdot 10^4 + 1 \cdot 10^3 + 2 \cdot 10 + 6)$$

Multiplying a Polynomial by a Monomial

Chapter 11

Now use the Distributive Property.

$= 10^3 \cdot 8 \cdot 10^4 + 10^3 \cdot 1 \cdot 10^3 + 10^3 \cdot 2 \cdot 10 + 10^3 \cdot 6$

The products can be simplified using the Product of Powers Property and the Commutative and Associative Properties of Multiplication.

$= 8 \cdot 10^7 + 1 \cdot 10^6 + 2 \cdot 10^4 + 6 \cdot 10^3$

Now simplify the polynomial.

$= 81{,}026{,}000$

This same procedure can be repeated to explain the product of any decimal and any integer power of 10.

Questions

COVERING THE IDEAS

In 1 and 2, find the product.

1. $(5x)(11x)$
2. $(200xy^3)(3x^2y)$

In 3 and 4,
  a. find the product, and
  b. draw a rectangle to represent the product.

3. $3h(h + 5)$
4. $4n(n + 3)$

In 5 and 6, a large rectangle is shown.
  a. Express its area as the sum of areas of smaller rectangles.
  b. Express its area as length times width.
  c. Write an equality from Parts a and b.

5.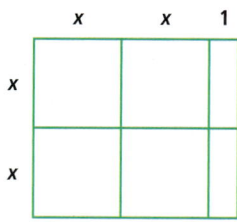
6.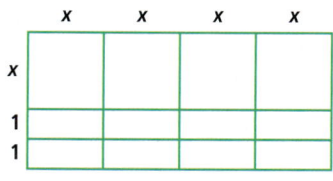

7. **Fill in the Blank** Using the Distributive Property, $a(b - c + d) = $ __?__ .

In 8–11, expand the expression.

8. $3x^2(x^2 + 2x - 8)$
9. $5x(-5x^2 - x + 6.2)$
10. $p(2 + p^2 + p^3 + 5p^4)$
11. $-0.5ab(4b - 2a + 10)$

12. Use multiplication of a monomial by a polynomial to explain why the product of 7,531 and 100,000 is 753,100,000.

Lesson 11-3

APPLYING THE MATHEMATICS

13. Suppose the building below had to increase its height by 2 feet.

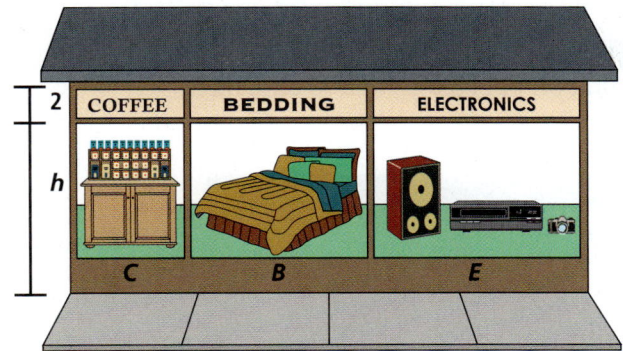

   a. Express the entire building's new storefront area as the sum of the three individual stores' storefront areas.
   b. Express the entire building's new storefront area as new height times length.
   c. Express the new storefront area as the sum of its old area and the additional area.

14. The arrangement of rectangles at the right is used by children in many countries for playing hopscotch. What is the total area?

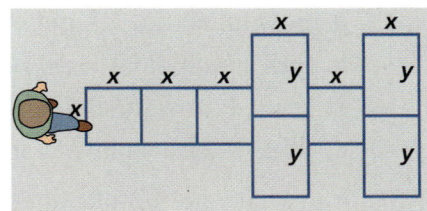

In 15–17, write the expression as a polynomial in standard form.

15. $(6x)(3x) - (5x)(2x) - (4x)(x)$

16. $2(x^2 + 3x) + 3x^2$

17. $m^3(m^2 - 3m + 2) - m^2(m^3 - 5m^2 - 6)$

In 18 and 19, simplify the expression.

18. $a(2b - c) + b(2c - a) + c(2a - b)$

19. $(x^2 + 2xy + y^2) - (x^2 - 2xy + y^2)$

20. At the right is a circle in a square.
   a. What is the area of the square?
   b. What is the area of the circle?
   c. What is the area of the shaded region?
   d. If a person had 3 copies of the shaded region, how much area would be shaded?
   e. If a person had c copies of the shaded region, how much area would be shaded?

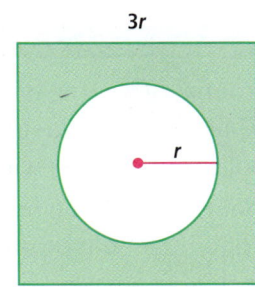

Multiplying a Polynomial by a Monomial

21. **Fill in the Blank** Find the missing polynomial in the given equation.

$$3n^2 \cdot (\ ?\ ) = 60n^4 + 27n^3 - 30n^2$$

22. a. What is the rule for dividing a decimal by 1,000?
    b. Make up an example like the one on pages 671–672 to explain why the rule works.

REVIEW

In 23 and 24, an expression is given.
   a. Tell whether the expression is a polynomial.
   b. If it is a polynomial, give its degree. If it is not a polynomial, explain why not. (Lesson 11-2)

23. $4a^4 + 2a^{-2}$

24. $n^3m^5 + n^2m^4 + nm^3$

25. Write a trinomial with degree 4. (Lesson 11-2)

26. After five years of birthdays, T.J. has received and saved $50x^4 + 70x^3 + 45x^2 + 100x + 80$ dollars. He put the money in a savings account at a yearly scale factor x. (Lessons 11-1, 7-1)
    a. How much did T.J. get on his last birthday?
    b. How much did T.J. get on the first of these birthdays?
    c. If $x = 1$, how much has T.J. saved?
    d. What does a value of 1 for x mean?

27. For what value(s) of c does the quadratic equation $3x^2 - 4x + c = 0$ have no solutions? (Lesson 9-6)

In 28 and 29, simplify the expression. (Lessons 8-4, 5-2)

28. $\dfrac{18xy^3}{6xy}$

29. $\dfrac{4m}{9} \div \dfrac{6m^3}{15m}$

In 30 and 31, write an equation for the line with the given characteristics.

30. contains the points (8, –2) and (3, 13) (Lesson 6-6)

31. slope $\dfrac{1}{2}$, x-intercept 1 (Lesson 6-4)

EXPLORATION

32. a. Suppose a monomial of degree 3 is multiplied by a monomial of degree 4. What must be true about the degree of the product? Support your answer with an example.
    b. Suppose a monomial of degree m is multiplied by a monomial of degree n. What must be true about the degree of the product? Support your answer with an example.

Lesson 11-4
Common Monomial Factoring

Vocabulary

factoring
trivial factors
greatest common factor
factorization
prime polynomials
complete factorization

▶ **BIG IDEA** Common monomial factoring is the process of writing a polynomial as a product of two polynomials, one of which is a monomial that factors each term of the polynomial.

When two or more numbers are multiplied, the result is a single number. *Factoring* is the reverse process. In **factoring,** we begin with a single number and express it as the product of two or more numbers. For example, the product of 7 and 4 is 28. So, factoring 28, we get $28 = 7 \cdot 4$. In Lesson 11-3, you multiplied monomials by polynomials to obtain polynomials. In this lesson you will learn how to reverse the process.

If factors are not integers, then every number has infinitely many factors. For example, 8 is not only $4 \cdot 2$ and $8 \cdot 1$, but also $24 \cdot \frac{1}{3}$ and $2.5 \cdot 3.2$. For this reason, in this book all factoring is *over the set of integers*.

Mental Math

Consider the parabola $y = 3(x - 6)^2 + 4$.

a. What is its vertex?

b. Does the parabola open up or down?

c. True or False The parabola is congruent to $y = -3x^2$.

d. What is an equation for its line of symmetry?

Factoring Monomials

Every expression has itself and the number 1 as a factor. These are called the **trivial factors.** If a monomial is the product of two or more variables or numbers, then it will have factors other than itself and 1.

> **Example 1**
> What are the factors of $49x^3$?
>
> **Solution** The factors of 49 are 1, 7, and 49. The monomial factors of x^3 are 1, x, x^2, and x^3. The factors of $49x^3$ are the 12 products of a factor of 49 with a factor of x:
>
> 1, 7, 49, x, $7x$, $49x$, x^2, $7x^2$, $49x^2$, x^3, $7x^3$, $49x^3$

 QY

The **greatest common factor** (GCF) of two or more monomials is the product of the greatest common factor of the coefficients and the greatest common factors of the variables.

▶ **QY**

Which of the factors of $49x^3$ are trivial factors?

Common Monomial Factoring **675**

Chapter 11

Example 2
Find the greatest common factor of $6xy^2$ and $18y$.

Solution The GCF of 6 and 18 is 6. The GCF of xy^2 and y is y. Because the factor x does not appear in all terms, it does not appear in the GCF.

So the GCF of $6xy^2$ and $18y$ is $6 \cdot y$, which is $6y$.

Notice that the GCF of the monomials includes the GCF of the coefficients of the monomials. It also includes any common variables raised to the *least* exponent of that variable found in the terms.

As with integers, the result of factoring a polynomial is called a **factorization.** Here is a factorization of $6x^2 + 12x$.

$$6x^2 + 12x = 2x(3x + 6)$$

Again, as with integers, a factorization with two factors means that a rectangle can be formed with the factors as its dimensions. Here is a picture of the factorization.

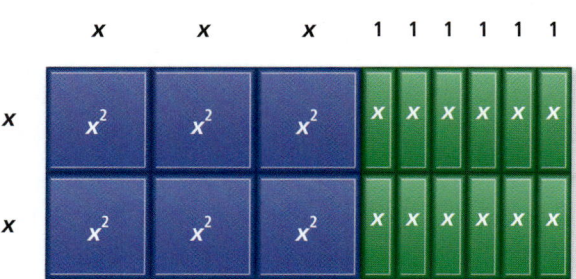

Activity

Step 1 Build or draw two other rectangles with an area of $6x^2 + 12x$.

Step 2 Write the factorization that is shown by each rectangle.

Step 3 Do any of the rectangles have the greatest common factor of $6x^2$ and $12x$ as a side length? If so, which rectangle?

The Activity points out that there is more than one way to factor $6x^2 + 12x$. When factoring a polynomial, the goal is that the GCF of all the terms is one factor. In $6x^2 + 12x$, $6x$ is the greatest common factor, so $6x^2 + 12x = 6x(x + 2)$.

Monomials such as $6x$, and polynomials such as $x + 2$ that cannot be factored into polynomials of a lower degree, are called **prime polynomials.** To factor a polynomial completely means to factor it into prime polynomials. When there are no common numerical factors in the terms of any of the prime polynomials, the result is called a **complete factorization.** The complete factorization of $6x + 12$ is $6(x + 2)$.

676 Polynomials

Lesson 11-4

There are many techniques for factoring polynomials, and you will study several in this and the next chapter. No matter which technique you use, you will obtain the same answer. As with integers, the prime factorization of a polynomial is unique except for order.

Unique Factorization Theorem for Polynomials

Every polynomial can be represented as a product of prime polynomials in exactly one way, disregarding order and integer multiples.

GUIDED

Example 3
Factor $20a^3b + 8a - 12a^5b^2$ completely.

Solution The greatest common factor of 20, 8, and −12 is __?__. The greatest common factor of a^3, a, and a^5 is __?__. Because the variable b does not appear in all terms, b does not appear in the greatest common factor.

The greatest common factor of $20a^3b$, $8a$, and $-12a^5b^2$ is __?__.

Divide each term by the GCF to find the terms in parentheses. With practice you'll be able to do these steps in your head.

$$\frac{20a^3b}{4a} = \underline{\ ?\ } \quad \frac{8a}{4a} = \underline{\ ?\ } \quad \frac{-12a^5b^2}{4a} = \underline{\ ?\ }$$

So $20a^3b + 8a - 12a^5b^2 = 4a(\underline{\ ?\ } + \underline{\ ?\ } - \underline{\ ?\ })$.

Factoring provides a way of simplifying some fractions.

Example 4
Simplify $\frac{22m + 4m^2}{m}$. ($m \neq 0$)

Solution 1 Factor the numerator, simplify the fraction, and multiply.

$$\frac{22m + 4m^2}{m} = \frac{2m(11 + 2m)}{m} = 2(11 + 2m) = 22 + 4m$$

Solution 2 Separate the given expression into the sum of two fractions and then simplify each fraction.

$$\frac{22m + 4m^2}{m} = \frac{22m}{m} + \frac{4m^2}{m} = 22 + 4m$$

Check The solutions give the same answer, so they check.

Common Monomial Factoring

Chapter 11

Questions

COVERING THE IDEAS

1. List all the factors of $33x^4$.

In 2 and 3, find the GCF.

2. $25y^5$ and $40y^2$
3. $17a^2b^2$ and $24ba^2$

4. Represent the factorization $12x^2 + 8x = 4x(3x + 2)$ with rectangles.

5. a. Factor $15c^2 + 5c$ by finding the greatest common factor of the terms.
   b. Illustrate the factorization by drawing a rectangle whose sides are the factors.

6. Showing tiles, draw two different rectangles each with area equal to $16x^2 + 4x$.

7. Explain why $x^2 + xy$ is not a prime polynomial.

8. In Parts a–c, complete the products.
   a. $36x^3 + 18x^2 = 6(\underline{\ ?\ } + \underline{\ ?\ })$
   b. $36x^3 + 18x^2 = 18(\underline{\ ?\ } + \underline{\ ?\ })$
   c. $36x^3 + 18x^2 = 18x^2(\underline{\ ?\ } + \underline{\ ?\ })$
   d. Which of the products in Parts a–c is a complete factorization of $36x^3 + 10x^2$? Explain your answer. $2x^2(18x + 5)$

9. Simplify $\dfrac{24n^6 + 20n^4}{4n^2}$.

10. Find the greatest common factor of $28x^5y^2$, $-14x^4y^3$, and $49x^3y^4$.

In 11–14, factor the polynomial completely.

11. $33a - 33b + 33ab$
12. $x^{2,100} - x^{2,049}$
13. $12v^9 + 16v^{10}$
14. $46cd^3 - 69cd^2 + 18c^2d^2$

APPLYING THE MATHEMATICS

15. The area of a rectangle is $14r^2h$. One dimension is $2r$. What is the other dimension?

16. The top vertex in the fact triangle at the right has the expression $27abc - 45a^2b^2c^2$. What expression belongs in the position of the question mark?

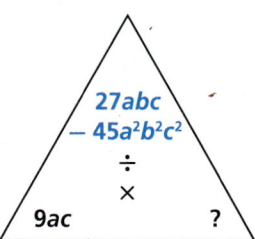

17. a. Graph $y = 2x^2 - 8x$.
    b. Graph $y = 2x(x - 4)$.
    c. What do you notice about the graphs of the equations? Explain why this occurs.

In 18 and 19, a circular cylinder with height *h* and radius *r* is pictured at the right. Factor the expression giving its surface area.

18. $\pi r^2 + 2\pi rh$, the surface area with an open top

19. $2\pi r^2 + 2\pi rh$, the surface area with a closed top

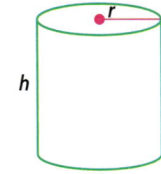

In 20 and 21, simplify the expression.

20. $\dfrac{9x^2y + 54xy - 9xy^2}{9xy}$

21. $\dfrac{-100n^{100} - 80n^{80} + 60n^{60}}{2n^2}$

REVIEW

In 22 and 23, simplify the expression. (Lesson 11-3)

22. $-4x^3(3 - 5x^2 + 7x^4)$

23. $k(k + 2k^2 + n) - 2n(k - 2n) - k^2$

24. Which investment plan is worth more at the end of 10 years if the annual yield is 6%? Justify your answer. (**Lesson 11-1**)

    Plan A: Deposit $50 each year on January 2, beginning in 2008.

    Plan B: Deposit $100 every other year on January 2, beginning in 2008.

25. **Multiple Choice** Which system of inequalities describes the shaded region in the graph at the right? (**Lesson 10-9**)

    A $\begin{cases} y - x < 6 \\ x \leq 0 \\ y \leq 6 \end{cases}$ B $\begin{cases} y + 2x \geq 6 \\ x \leq 6 \\ y \leq 0 \end{cases}$ C $\begin{cases} y + 2x \leq 6 \\ x \leq 0 \\ y \leq 0 \end{cases}$ D $\begin{cases} y \leq 6 + 4x \\ x \geq 3 \\ y \geq 1 \end{cases}$

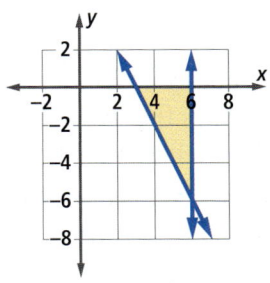

26. Simplify $\sqrt{50x^3y^4}$. (**Lesson 8-6**)

27. There are 4 boys, 7 girls, 6 men, and 5 women on a community youth board. How many different leadership teams consisting of one adult and one child could be formed from these people? (**Lesson 8-1**)

EXPLORATION

28. The number 6 has four factors: 1, 2, 3, and 6. The number 30 has eight factors: 1, 2, 3, 5, 6, 10, 15, and 30.

    a. Find five numbers that each have an odd number of factors.
    b. Give an algebraic expression that describes all numbers with an odd number of factors. Explain why you think these numbers have an odd number of factors.

QY ANSWER

1 and $49x^3$

Chapter 11

Lesson 11-5
Multiplying Polynomials

▶ **BIG IDEA** To multiply a polynomial by a polynomial, multiply each term of one polynomial by each term of the other polynomial and add the products.

Picturing the Multiplication of Polynomials with Area

The Area Model for Multiplication shows how to multiply two polynomials with many terms. For example, to multiply $a + b + c + d$ by $x + y + z$, draw a rectangle with length $a + b + c + d$ and width $x + y + z$.

Mental Math

Evaluate.
a. $27^{\frac{1}{3}}$
b. $27^{\frac{2}{3}}$
c. $27^{-\frac{2}{3}}$

	a	b	c	d
x	ax	bx	cx	dx
y	ay	by	cy	dy
z	az	bz	cz	dz

The area of the largest rectangle equals the sum of the areas of the twelve smaller rectangles.

total area = $ax + ay + az + bx + by + bz + cx + cy + cz + dx + dy + dz$

But the area of the biggest rectangle also equals the product of its length and width.

total area = $(a + b + c + d) \cdot (x + y + z)$

The Distributive Property can be used to justify why the two expressions must be equal. Distribute $(x + y + z)$ over $(a + b + c + d)$ to get $(a + b + c + d) \cdot (x + y + z) = a(x + y + z) + b(x + y + z) + c(x + y + z) + d(x + y + z) = ax + ay + az + bx + by + bz + cx + cy + cz + dx + dy + dz$.

Because of the multiple use of the Distributive Property, we call this general property the *Extended Distributive Property*.

Extended Distributive Property

To multiply two sums, multiply each term in the first sum by each term in the second sum and then add the products.

680   Polynomials

If one polynomial has m terms and the second has n terms, there will be mn terms in their product. This is due to the Multiplication Counting Principle. If some of these are like terms, you can simplify the product by combining like terms.

GUIDED

Example 1

Cassandra used the EXPAND feature on a CAS to multiply the polynomials $x^2 - 4x + 8$ and $5x - 3$. Her result is shown at the right.

The CAS does not display steps that most people would show in order to find the answer. Using the Extended Distributive Property, show the steps that the CAS does not display to expand $(x^2 - 4x + 8)(5x - 3)$.

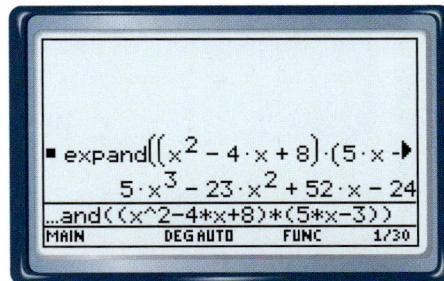

Solution $(x^2 - 4x + 8)(5x - 3)$

$= x^2 \cdot \underline{} + x^2 \cdot \underline{} + (-4x) \cdot \underline{} + (-4x) \cdot \underline{} + 8 \cdot \underline{} + 8 \cdot \underline{}$

$= \underline{} + \underline{} + \underline{} + \underline{} + \underline{} + \underline{}$

Now combine like terms.

$= \underline{} - \underline{} + \underline{} - \underline{}$

Example 2

Expand $(4x + 3)(x - 6)$.

Solution Think of $x - 6$ as $x + -6$. Multiply each term in the first polynomial by each in the second. There will be four terms in the product.

$(4x + 3)(x - 6) = 4x \cdot x + 4x \cdot (-6) + 3 \cdot x + 3 \cdot (-6)$

$= 4x^2 + (-24)x + 3x + (-18)$

Now simplify by adding or subtracting like terms.

$= 4x^2 - 21x - 18$

Check 1 Let $x = 10$. (Ten is a nice value to use in checks because powers of 10 are so easily calculated.) Then $(4x + 3)(x - 6) = (4 \cdot 10 + 3)(10 - 6) = 43 \cdot 4 = 172$.
When $x = 10$, $4x^2 - 21x - 18 = 4 \cdot 10^2 - 21 \cdot 10 - 18 = 400 - 210 - 18 = 172$; so it checks.

(continued on next page)

Chapter 11

Check 2 It is possible that when $x = 10$, the expression $(4x + 3)(x - 6)$ just happened to have the same value as $4x^2 - 21x - 18$. A better check is to set each expression equal to y and graph the resulting equation.

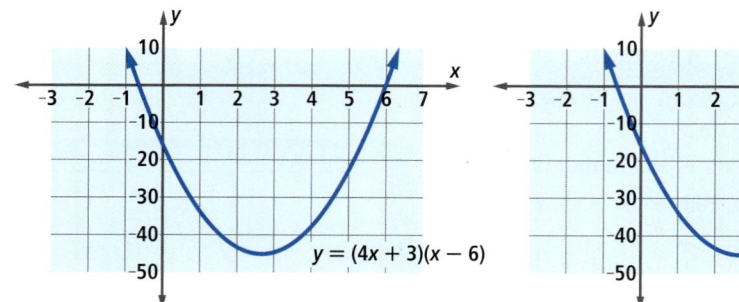

The two graphs are identical, so it checks.

STOP QY

Because multiplication is associative and commutative, to multiply three polynomials you can start by multiplying any two of them, and then multiply their product by the third polynomial.

Example 3
Expand $n(n - 1)(n - 2)$.

Solution 1 Multiply n by $n - 1$ first.

$$n(n - 1)(n - 2) = n(n - 1) \cdot (n - 2)$$
$$= (n^2 - n)(n - 2)$$
$$= n^2 \cdot n + n^2(-2) - n \cdot n - n(-2)$$
$$= n^3 - 2n^2 - n^2 + 2n$$
$$= n^3 - 3n^2 + 2n$$

Solution 2 Multiply $n - 1$ by $n - 2$ first.

$$n(n - 1)(n - 2) = n \cdot (n - 1)(n - 2)$$
$$= n \cdot (n \cdot n + n \cdot (-2) - 1 \cdot n - 1 \cdot (-2))$$
$$= n \cdot (n^2 - 2n - n + 2)$$
$$= n \cdot (n^2 - 3n + 2)$$
$$= n^3 - 3n^2 + 2n$$

▶ **QY**

Austin used a CAS to multiply $a^2 + 5n - 14$ by $4n + 1$. The screen shows $4a^2n + a^2 + 20n^2 - 51n - 14$.

a. How many terms did the CAS combine to get the product shown?

b. Which like terms produced a^2?

c. Which like terms produced $20n^2$?

d. Which like terms produced $-51n$?

682   Polynomials

Lesson 11-5

Questions

COVERING THE IDEAS

1. a. What multiplication is shown at the right?
   b. Do the multiplication.

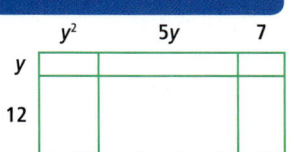

2. Simone used a CAS to multiply $2n - 5$ by $n^2 + 3n + 6$. The screen at the right shows her result. Explain which multiplications were done and combined to get each term in the product.

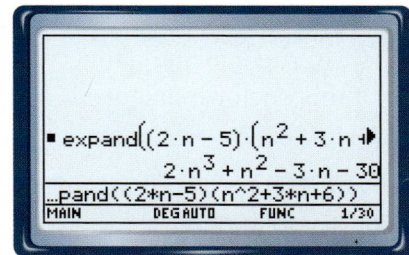

3. a. Multiply $2x - 5$ by $6x + 4$.
   b. Check your answer by letting $x = 10$.
   c. Check your answer by graphing $y = (2x - 5)(6x + 4)$ and graphing your answer.

In 4–7, expand and simplify the expression.

4. $(3x^2 + 7x + 4)(x + 6)$
5. $(n + 1)(2n^2 + 3n - 1)$
6. $(m^2 - 10m - 3)(2m^2 - 5m - 4)$
7. $4(x^2 + 2x + 2)(x^2 - 2x + 2)$

8. Find the area of the rectangle at the right, and simplify the result.

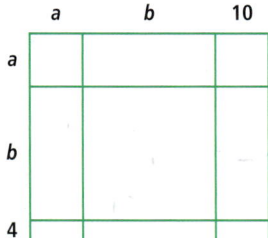

9. Expand $(5c - 4d + 1)(c - 7d)$.

10. a. Expand $(n - 3)(n + 4)(2n + 5)$ by first multiplying $n - 3$ by $n + 4$, then multiplying that product by $2n + 5$.
    b. Do the same expansion starting with a different multiplication.

APPLYING THE MATHEMATICS

11. a. Expand $(3x + 5)(4x + 2)$.
    b. Expand $(3x - 5)(4x - 2)$.
    c. Make a generalization from the pattern of answers in Parts a and b.

12. a. Expand $(4p - 1)(2p + 3)$.
    b. Expand $(4p + 1)(2p - 3)$.
    c. Make a generalization from the pattern of answers in Parts a and b.

13. Expand $\left(\frac{1}{5}x - 2.7\right)^2$ by writing the power as $\left(\frac{1}{5}x - 2.7\right)\left(\frac{1}{5}x - 2.7\right)$.

Multiplying Polynomials 683

14. How much larger is the volume of a cube with edges of length $n + 1$ than the volume of a cube with edges of length n?

15. Solve the equation $(2n - 7)(n + 14) = 57$ by multiplying the binomials and then using the Quadratic Formula.

REVIEW

16. The sector at the right is one-third of a circle with radius r. Write a formula for the perimeter p of this sector in factored form. (**Lesson 11-4**)

    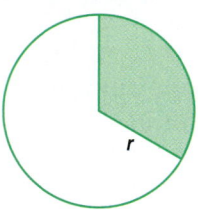

17. If $p(x) = 2x^2 + 5$ and $q(x) = 3x^2 + 6$, simplify $2 \cdot p(x) - 3 \cdot q(x)$. (**Lesson 11-3**)

18. Twice the larger of two numbers is six more than four times the smaller. If the sum of eight times the smaller and three times the larger is 93, what are the two numbers? (**Lessons 10-5, 10-4, 10-2**)

19. Given $f(x) = 4^{-x} + 2$, find each value. (**Lessons 8-4, 7-6**)

    a. $f(2)$ b. $f(-1)$ c. $f(0) + f(-3)$

20. Find two algebraic fractions whose product is $\frac{14m}{39p}$. (**Lesson 5-1**)

EXPLORATION

21. Multiply each of the polynomials in Parts a–d by $x + 1$.

    a. $x - 1$
    b. $x^2 - x + 1$
    c. $x^3 - x^2 + x - 1$
    d. $x^4 - x^3 + x^2 - x + 1$

    e. Look for a pattern and use it to multiply
    $(x + 1)(x^8 - x^7 + x^6 - x^5 + x^4 - x^3 + x^2 - x + 1)$.

    f. Predict what you think the product of $(x + 1)$ and $(x^{100} - x^{99} + x^{98} - x^{97} + ... + x^2 - x + 1)$ is when simplified. Can you explain why your answer is correct?

22. A multidigit number in base 10 is shorthand for a polynomial in x. When $x = 10$, $436 = 4x^2 + 3x + 6$, and $2{,}187 = 2x^3 + 1x^2 + 8x + 7$.

    When you multiply 2,187 by 436, you are essentially multiplying two polynomials.

    $$(4x^2 + 3x + 6)(2x^3 + 1x^2 + 8x + 7)$$

    Multiply these polynomials and show how their product equals the product of 436 and 2,187 when $x = 10$.

QY ANSWERS

a. two like terms were combined

b. a^2 times 1

c. $5n$ times $4n$

d. $5n$ times 1 and -14 times $4n$

Lesson 11-6

Special Binomial Products

Vocabulary

perfect square trinomials
difference of squares

▶ **BIG IDEA** The square of a binomial $a + b$ is the expression $(a + b)^2$ and can be found by multiplying $a + b$ by $a + b$ as you would multiply any polynomials.

Can you compute $46 \cdot 54$ in your head? How about 103^2? Studying products of special binomials can help you find the answers quickly without a calculator. Two such products are used so frequently that they are given their own names: Perfect Squares and the Difference of Two Squares.

Mental Math

A circle has diameter 10 centimeters. Estimate

a. its circumference.

b. its area.

Perfect Squares: The Square of a Sum

Just as numbers and variables can be squared, so can algebraic expressions. Given any two numbers a and b, you can expand $(a + b)^2$ or $(a - b)^2$. These are read "a plus b, quantity squared" and "a minus b, quantity squared."

How can you expand $(a + b)^2$? One way is to write the power as repeated multiplication.

$$(a + b)^2 = (a + b)(a + b)$$

Next, use the Distributive Property.

$$= a(a + b) + b(a + b)$$

Then apply the Distributive Property again to the first and second products.

$$= (a^2 + ab) + (ba + b^2)$$

And finally combine like terms (because $ab = ba$).

$$= a^2 + 2ab + b^2$$

The square of a sum of two terms is the sum of the squares of the terms plus twice their product.

Geometrically, $(a + b)^2$ can be thought of as the area of a square with sides of length $a + b$. As the figure shows, its area is $a^2 + 2ab + b^2$.

🛑 **QY**

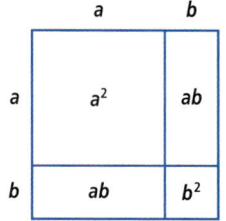

▶ **QY**

Expand $(x + 8)^2$.

Special Binomial Products **685**

Chapter 11

Example 1
Calculate 103^2.

Solution 1 Write 103 as the sum of two numbers whose squares you can calculate in your head. $103 = 100 + 3$, so $103^2 = (100 + 3)^2$. Then use the special binomial product rule for the square of a sum.
$(100 + 3)^2 = 100^2 + 2 \cdot 3 \cdot 100 + 3^2 = 10{,}000 + 600 + 9$
$= 10{,}609$

Solution 2 Write the square as a multiplication and expand.
$103^2 = (100 + 3)(100 + 3)$
$= 100 \cdot 100 + 100 \cdot 3 + 3 \cdot 100 + 3 \cdot 3$
$= 10{,}000 + 300 + 300 + 9 = 10{,}609$

With practice, either of the solutions to Example 1 can be done in your head.

GUIDED

Example 2
The area of a square with side $7c + 5$ is $(7c + 5)^2$. Expand this binomial.

Solution 1 Use the rule for the square of a binomial.

$(7c + 5)^2 = (7c)^2 + \underline{\ ?\ } + \underline{\ ?\ }$
$= \underline{\ ?\ } c^2 + \underline{\ ?\ } c + \underline{\ ?\ }$

Solution 2 Rewrite the square as a multiplication and expand using the Distributive Property.

$(7c + 5)^2 = (7c + 5)(7c + 5)$
$= \underline{\ ?\ }(7c + 5) + \underline{\ ?\ }(7c + 5)$
$= \underline{\ ?\ } \cdot 7c + \underline{\ ?\ } \cdot 5 + \underline{\ ?\ } \cdot 7c + \underline{\ ?\ } \cdot 5$
$= \underline{\ ?\ } c^2 + \underline{\ ?\ } c + \underline{\ ?\ }$

Solution 3 Draw a square with side $7c + 5$. Subdivide it into smaller rectangles and find the sum of their areas.

Check Test a special case. Let $c = 3$. Then $7c + 5 = \underline{\ ?\ }$ and $(7c + 5)^2 = \underline{\ ?\ }$.

Also $\underline{\ ?\ } c^2 + \underline{\ ?\ } c + \underline{\ ?\ } = \underline{\ ?\ } \cdot 9 + \underline{\ ?\ } \cdot 3 + \underline{\ ?\ } = \underline{\ ?\ }$. It checks.

686 Polynomials

Perfect Squares: The Square of a Difference

To square the difference $(a - b)$, think of $a - b$ as $a + -b$. Then apply the rule for the perfect square of a sum.

$$(a - b)^2 = (a + -b)^2$$
$$= a^2 + 2a(-b) + (-b)^2$$
$$= a^2 - 2ab + b^2$$

The square of a difference of two terms is the sum of the squares of the terms minus twice their product.

Squaring a binomial always results in a trinomial. Trinomials of the form $a^2 + 2ab + b^2$ or $a^2 - 2ab + b^2$ are called **perfect square trinomials** because each is the result of squaring a binomial.

> **Perfect Squares of Binomials**
>
> For all real numbers a and b, $(a + b)^2 = a^2 + 2ab + b^2$ and $(a - b)^2 = a^2 - 2ab + b^2$.

Activity 1

Complete the table.

$(a + b)^2$	$a^2 + 2ab + b^2$	$(a - b)^2$	$a^2 - 2ab + b^2$
$(x + 1)^2$	?	$(x - 1)^2$	?
$(x + 2)^2$	?	$(x - 2)^2$	?
$(x + 3)^2$	?	$(x - 3)^2$	?
$(x + 4)^2$	?	$(x - 4)^2$	?
$(x + 15)^2$	?	$(x - 15)^2$	?
$(x + n)^2$	?	$(x - n)^2$	?

The Difference of Two Squares

Another special binomial product is the sum of two numbers times their difference. Let x and y be any two numbers. What is $(x + y)(x - y)$?

$(x + y)(x - y) = x(x - y) + y(x - y)$ Distributive Property
$ = x^2 - xy + yx - y^2$ $-xy$ and yx are opposites.
$ = x^2 - y^2$

The product of the sum and difference of two numbers is the **difference of squares** of the two numbers.

Chapter 11

> **Difference of Two Squares**
>
> For all real numbers x and y, $(x + y)(x - y) = x^2 - y^2$.

Activity 2

Complete the table at the right.

The difference of two squares can be used to multiply two numbers that are equidistant from a number whose square you know.

$(a + b)(a - b)$	$a^2 - b^2$
$(x + 1)(x - 1)$	?
$(x + 2)(x - 2)$	?
$(x + 3)(x - 3)$	?
$(x + 4)(x - 4)$	?
$(x + 15)(x - 15)$	?
$(x + n)(x - n)$	?

Example 3

Compute 46 · 54 in your head.

Solution 46 and 54 are the same distance from 50. So think of 46 · 54 as $(50 - 4)(50 + 4)$. This is the product of the sum and difference of the same numbers, so the product is the difference of the squares of the numbers.

$$(x - y)(x + y) = x^2 - y^2$$
$$(50 - 4)(50 + 4) = 50^2 - 4^2 = 2{,}500 - 16 = 2{,}484$$

Example 4

Expand $(8x^5 + 3)(8x^5 - 3)$.

Solution This is the sum of and difference of the same numbers, so the product is the difference of squares of the numbers.

$$(8x^5 + 3)(8x^5 - 3) = (8x^5)^2 - 3^2 = 64x^{10} - 9$$

Check Let $x = 2$.
$$(8x^5 + 3)(8x^5 - 3) = (8 \cdot 2^5 + 3)(8 \cdot 2^5 - 3)$$
$$= (8 \cdot 32 + 3)(8 \cdot 32 - 3)$$
$$= 259 \cdot 253 = 65{,}527$$
$64x^{10} - 9 = 64 \cdot 2^{10} - 9 = 64 \cdot 1{,}024 - 9 = 65{,}527$, so it checks.

Questions

COVERING THE IDEAS

In 1–3, expand and simplify the expression.

1. $(g + h)^2$
2. $(g - h)^2$
3. $(g + h)(g - h)$

4. What is a *perfect square trinomial*?

5. Give an example of a perfect square trinomial.

In 6 and 7, a square is described.
   a. Draw a picture to describe the situation.
   b. Write the area of the square as the square of a binomial.
   c. Write the area as a perfect square trinomial.

6. A square with sides of length $2n + 1$.

7. A square with sides of length $5p + 11$.

8. Verify that $(a - b)^2 = a^2 - 2ab + b^2$ by substituting numbers for a and b.

In 9–16, expand and simplify the expression.

9. $(x - 5)^2$
10. $(3 + n)(3 - n)$
11. $(n^2 + 4)(n^2 - 4)$
12. $(13s + 11)^2$
13. $(9 - 2x)^2$
14. $\left(10 + \frac{1}{2}t\right)^2$
15. $(3x + yz)(3x - yz)$
16. $(2a + 5b)(-5b + 2a)$

17. Compute in your head. Then write down how you did each computation.
   a. 30^2
   b. $29 \cdot 31$
   c. $28 \cdot 32$
   d. $27 \cdot 33$

In 18–20, compute in your head. Then write down how you did each computation.

18. $16 \cdot 24$
19. 201^2
20. $75 \cdot 65$

APPLYING THE MATHEMATICS

In 21–25, tell whether the expression is a perfect square trinomial, difference of squares, or neither of these.

21. $u^2 - 2uj + j^2$
22. $9 - v^2$
23. $2sd + s^2 + d^2$
24. $xy - 16$
25. $-i^2 + p^2$

26. Solve $\frac{x - 4}{7} = \frac{6}{x + 4}$.

27. The numbers being multiplied in each part of Question 17 add to 60. Use the pattern found there to explain why, of all the pairs of numbers that add to 100, the largest product occurs when both numbers are 50.

In 28 and 29, expand and simplify the expression.

28. $(\sqrt{11} + \sqrt{13})(\sqrt{11} - \sqrt{13})$
29. $(3x + y)^2 + (3x - y)^2$

REVIEW

30. a. Expand $(x - 12)(x + 10)$.
    b. Solve $(x - 12)(x + 10) = 85$. (**Lessons 11-6, 9-5**)

31. After 7 years of putting money into a retirement account at a scale factor x, Lenny has saved $800x^6 + 1{,}000x^5 + 1{,}500x^4 + 1{,}200x^3 + 1{,}400x^2 + 1{,}800x + 2{,}000$ dollars. (**Lesson 11-1**)
    a. How much did Lenny put in during the most recent year?
    b. How much did Lenny put in during the first year?
    c. Give an example of a reasonable value for x, and evaluate the polynomial for that value of x.

32. Richard wants to construct a rectangular prism with height and width of x inches and length of 5 inches. He wants his prism to have the same volume as surface area. Construct a system with equations for the volume and surface area. Then solve for x. (**Lesson 10-10**)

In 33–35, describe a situation that might yield the given polynomial. (**Lesson 8-2**)

33. e^3
34. $6x^2$
35. $\pi r^2 - \pi s^2$

36. In 1965, Gordon Moore stated that computing speed in computers doubles every 24 months (Moore's Law). Computing speed is measured by transistors per circuit. (**Lesson 7-2**)
    a. In 1971, engineers could fit 4,004 transistors per circuit. Use Moore's Law to write an expression for the number of transistors per circuit that were possible in 1979.
    b. Many experts believe that Moore's Law will hold until 2020. Estimate the number of transistors per circuit possible in 2020, given that processors developed in 2000 had about 100 million transistors per circuit.

EXPLORATION

37. A CAS will be helpful in this question. After collecting terms, the expansion of $(a + b)^2$ has 3 unlike terms. Expand $(a + b + c)^2$. You should find that the expansion of $(a + b + c)^2$ has 6 unlike terms. How many unlike terms does the expansion of $(a + b + c + d)^2$ have? Try to generalize the result.

QY ANSWER

$x^2 + 16x + 64$

Lesson 11-7

Permutations

Vocabulary

permutation

n!, n factorial

circular permutation

> **BIG IDEA** From a set of n symbols, the number of permutations of length r without replacement is given by the product of the polynomials $\underbrace{n(n-1)(n-2)\ldots}_{r \text{ factors}}$.

In Lesson 8-1, you saw the following problem: How many 3-letter acronyms (like JFK, IBM, BMI, or TNT) are there in English? The answer is 26^3 because the first letter can be any one of the 26 letters of the alphabet, and so can the second letter, and so can the third letter. Think of the spaces to be filled and use the Multiplication Counting Principle: $\underline{26} \cdot \underline{26} \cdot \underline{26} = 17{,}576$.

Now suppose that the letters in the acronym have to be different. Then TNT and other acronyms with duplicate letters are not allowed. The first letter can still be any one of the 26 letters of the alphabet, but the second letter can only be one of the 25 letters remaining, and the third letter can only be one of the 24 letters remaining. So the total number of 3-letter acronyms in English with different letters is $\underline{26} \cdot \underline{25} \cdot \underline{24} = 15{,}600$.

The first situation is called an *arrangement with replacement* because a letter can be used more than once. The second situation, where a letter cannot be used more than once, is called a **permutation.** So, a permutation is an arrangement without replacement. The above situation shows that, with 26 letters, there are 17,576 arrangements with replacement of length 3, and 15,600 permutations of length 3.

Suppose there were only three letters, A, B, and C. Notice the difference between the two types of arrangements.

Length 2 with replacement	AA, AB, AC, BA, BB, BC, CA, CB, CC
Length 2 without replacement	AB, AC, BA, BC, CA, CB
Length 3 with replacement	AAA, AAB, AAC, ABA, ABB, ABC, ACA, ACB, ACC, 9 starting with B, and 9 more starting with C, a total of 3^3 or 27 arrangements
Length 3 without replacement	ABC, ACB, BAC, BCA, CAB, CBA
Length 4 with replacement	AAAA, AAAB, AAAC, ..., AABA, AABB, AABC, ..., and so on, a total of 3^4 or 81 arrangements
Length 4 without replacement	None! (Do you see why?)

Mental Math

Suppose n is an integer. Determine if the following statements are *always, sometimes but not always,* or *never* true.

a. If $x > 0$, $x^n > 0$.

b. If $x < 0$, $x^n > 0$.

c. If $x < 0$, $2x^n < 0$.

Permutations 691

Chapter 11

Example 1

Dori saw a license plate with the numbers 15973. She noticed that all the digits were different odd digits. She wondered if this was unusual.

a. How many 5-digit numbers are there with only odd digits?
b. How many 5-digit numbers are there with only odd digits, all of them different?

Solution

a. This is a situation of arrangements with replacement. Each digit could be any of 5 numbers. Think of spaces to be filled.
$$\underline{5} \cdot \underline{5} \cdot \underline{5} \cdot \underline{5} \cdot \underline{5} = 3{,}125$$

b. This is a situation of permutations. The first digit can be any one of the 5 odd numbers. But then the second digit must be different, so it can only be one of the 4 odd numbers that remain. The third digit can only be one of the 3 odd numbers that remain after the first two have been chosen. The 4th digit can only be one of the 2 that remain. The fifth digit can only be the remaining digit. Filling spaces, you can picture this as
$$\underline{5} \cdot \underline{4} \cdot \underline{3} \cdot \underline{2} \cdot \underline{1} = 120.$$

There is quite a difference between the two kinds of arrangements! So it is not so unusual that Dori saw a license plate with all odd digits, but rather unusual that all the digits would be different.

GUIDED

Example 2

Three of the 11 members of a jazz band will each perform a solo at a concert. In how many orders can three people be picked from the band to perform a solo?

Solution This is an arrangement __?__ replacement situation.
(with/without)

Any of __?__ people could perform the first solo.

After that soloist, any of __?__ people could perform the second solo.

Then any of __?__ people could perform the third solo.

So there are __?__ · __?__ · __?__ or __?__ possible orders.

A Connection with Polynomials

Suppose you begin with n letters. *Without* replacement, there are $\underline{n} \cdot \underline{(n-1)}$ acronyms of length 2. Using the Distributive Property, you can see that $n(n-1) = n^2 - n$, which is a polynomial of degree 2. There are $\underline{n} \cdot \underline{(n-1)} \cdot \underline{(n-2)}$ acronyms of length 3.

Jazz music developed in the last part of the 19th century in New Orleans.

Source: *Columbia Encyclopedia*

Using the Extended Distributive Property, this product equals $n^3 - 3n^2 + 2n$, a polynomial of degree 3. Example 1b asked for permutations of length 5. Since $n = 5$ in Example 1b, the multiplication was $\underline{5} \cdot \underline{(5-1)} \cdot \underline{(5-2)} \cdot \underline{(5-3)} \cdot \underline{(5-4)}$. In general, there are $\underline{n} \cdot \underline{(n-1)} \cdot \underline{(n-2)} \cdot \underline{(n-3)} \cdot \underline{(n-4)}$ such permutations.

This product equals $n^5 - 10n^4 + 35n^3 - 50n^2 + 24n$, a 5th degree polynomial. In general, the number of permutations of length n can be calculated by evaluating a polynomial of degree n.

Permutations Using All the Items

With five different items, there cannot be permutations of length 6 because there are only 5 different items. However, the situation of permutations using all the items is quite common, as Example 3 shows.

Example 3
You have 13 books to put on a shelf. In how many ways can they be arranged?

Solution This is a permutation problem. Any one of the 13 books can be farthest left. Once it has been chosen, there are 12 choices for the book to its right. Then there are 11 choices for the book to the right of the first two; and so on. The total number of permutations of the books is $\underline{13} \cdot \underline{12} \cdot \underline{11} \cdot \underline{10} \cdot \underline{9} \cdot \underline{8} \cdot \underline{7} \cdot \underline{6} \cdot \underline{5} \cdot \underline{4} \cdot \underline{3} \cdot \underline{2} \cdot \underline{1}$, or 6,227,020,800.

The answers to Examples 1b and 3 each are the product of the integers from 1 to n. This product is denoted as **$n!$** and called **n factorial**. Using factorial notation, the answer to Example 1b is 5! and the answer to Example 3 is 13!.

$$n! = n(n-1)(n-2) \cdot \ldots \cdot 3 \cdot 2 \cdot 1$$

Specifically, $1! = 1$.

$2! = 2 \cdot 1 = 2$
$3! = 3 \cdot 2 \cdot 1 = 6$
$4! = 4 \cdot 3 \cdot 2 \cdot 1 = 24$
$5! = 5 \cdot 4 \cdot 3 \cdot 2 \cdot 1 = 120$
$6! = 6 \cdot 5 \cdot 4 \cdot 3 \cdot 2 \cdot 1 = 720$, and so on.

Notice how $n!$ gets large quite quickly as n grows.

 QY

▶ **QY**
a. A baseball manager is setting the batting order for the 9 starting players. How many different batting orders are possible?
b. What is 8!?

Factorial notation is very convenient for describing numbers of permutations. You saw at the beginning of this lesson the product of three numbers 26 · 25 · 24. This product can be written as the quotient of two factorials.

$$26 \cdot 25 \cdot 24 = 26 \cdot 25 \cdot 24 \cdot \frac{23 \cdot 22 \cdot 21 \cdot \ldots \cdot 3 \cdot 2 \cdot 1}{23 \cdot 22 \cdot 21 \cdot \ldots \cdot 3 \cdot 2 \cdot 1}$$

$$= \frac{26 \cdot 25 \cdot 24 \cdot 23 \cdot 22 \cdot 21 \cdot \ldots \cdot 3 \cdot 2 \cdot 1}{23 \cdot 22 \cdot 21 \cdot \ldots \cdot 3 \cdot 2 \cdot 1}$$

$$= \frac{26!}{23!}$$

So, with a factorial key on a calculator, you can calculate any number of permutations just by dividing two factorials.

Questions

COVERING THE IDEAS

1. a. Identify two of the permutations of length 4 from the six letters A, B, C, D, E, and F.
   b. How many permutations are there of length 4 from the six letters A, B, C, D, E, and F?
2. You have 8 of your favorite pictures that you would like to hang on a wall, but there is room for only 3 of the pictures. How many different permutations of 3 pictures are possible from your 8 favorites, assuming you arrange the pictures in a straight line so that their order matters?
3. How many different permutations of length 2 are there from n objects when $n \geq 2$?
4. How many different permutations of length 7 are there from n objects when $n \geq 7$?
5. a. Identify two of the permutations of length 5 from the five letters V, W, X, Y, and Z.
   b. How many permutations of length 5 are there from these five letters?
6. A volleyball coach is deciding the serving order for the six starting players. How many different starting orders are possible?
7. In how many different orders can n objects be arranged on a shelf?
8. Give the values of 7!, 8!, 9!, and 10!.

In 9–12, write as a single number in base 10 or as a simple fraction in lowest terms.

9. $\frac{8!}{6!}$ 10. $\frac{100!}{99!}$ 11. $\frac{15!}{17!}$ 12. $\frac{2!}{6!}$

The forearm pass, or "dig," is a type of shot used when receiving a serve or playing a hard, low hit ball.

Lesson 11-7

APPLYING THE MATHEMATICS

13. Consider the following pattern.

    $2! = 2 \cdot 1!$ $\qquad$ $3! = 3 \cdot 2!$ $\qquad$ $4! = 4 \cdot 3!$

    Are all three instances above true? If so, describe the general pattern using one variable. If not, correct any instances that are false.

14. Three hundred people enter a raffle. The first prize is a computer and the second prize is a cell phone.
    a. How many different ordered pairs of people are eligible to win these prizes?
    b. Suppose you and your best friend are among the 300. If the winning tickets are chosen at random, what is the probability that you will win the computer and your best friend will win the cell phone?

15. a. How many license plate numbers with 4 digits have all different odd digits, with the first digit being 3?
    b. Write down all these license plate numbers.

16. How many license plate numbers with 6 digits have all different odd digits?

17. A **circular permutation** is an ordering of objects around a circle. Some permutations that are different along a line are considered the same around a circle. For example, the two arrangements of A, B, C, and D pictured here are considered to be the same.

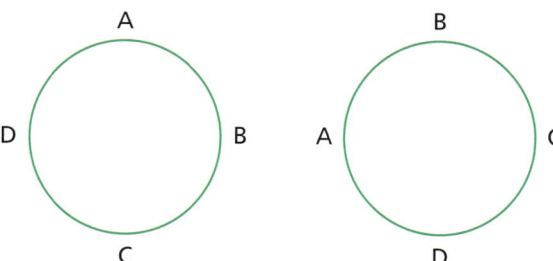

A raffle is used in many organizations looking to raise funds.

Write all the circular permutations of 4 objects around a circle.

REVIEW

18. Expand and simplify $(-a + 6b)(-a - 6b)$. **(Lesson 11-6)**

19. How much smaller is the area of a circle with radius $(r - 4)$ than that of a circle with radius r (assume $r > 4$)? **(Lesson 11-5)**

In 20 and 21, simplify the expression. **(Lessons 11-4, 8-3)**

20. $\dfrac{25n^2 - 21n}{n}$

21. $\dfrac{8x^3 + 16x^2 + 24x^6}{4x^2}$

Chapter 11

22. A rectangle with dimensions $4w$ and $2w + 1$ is contained in a rectangle with dimensions $10w$ and $6w + 2$, as shown in the diagram at the right. (**Lesson 11-3**)

    a. Write an expression for the area of the larger rectangle.
    b. Write an expression for the area of the smaller rectangle.
    c. Write a polynomial in standard form for the area of the shaded region.

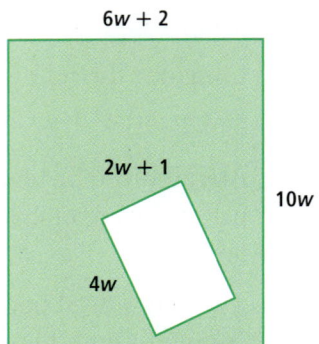

23. Consider the polynomial $3x^3 - 4x^2 + x + 6$. Give an example of a polynomial of degree 3 that when added to this will give a polynomial of degree 1. (**Lesson 11-2**)

24. The spreadsheet below shows an investment of $1,200 increasing in two different ways. (**Lesson 7-7**)

	A	B	C
1	Years From Now	Exponential Growth	Constant Increase
2	0	1200	1200
3	1	1260	1275
4	2	1323	1350
5	3		
6	4		
7	5		
8	6		
9	7		

    a. What formula should be entered in cell B5? C5?
    b. Describe the difference in investments after 7 years.
    c. Which investment will be worth more in 10 years? Justify your answer.

EXPLORATION

25. Here are polynomial expressions for the number of permutations of n symbols with various lengths:

    Length 2: $n(n - 1) = n^2 - n$
    Length 3: $n(n - 1)(n - 2) = n^3 - 3n^2 + 2n$
    Length 4: $n(n - 1)(n - 2)(n - 3) = n^4 - 6n^3 + 11n^2 - 6n$
    Length 5: $n(n - 1)(n - 2)(n - 3)(n - 4) = n^5 - 10n^4 + 35n^3 - 50n^2 + 24n$

    a. Use a CAS to find a polynomial expression for the number of permutations of n symbols with length 6.
    b. Identify some patterns in these polynomials that enable you to predict some features of the polynomial for the number of permutations of n symbols with length 7.

QY ANSWERS

1. 362,880
2. 40,320

Lesson 11-8

The Chi-Square Statistic

Vocabulary
expected number
deviation
chi-square statistic

▶ **BIG IDEA** The chi-square statistic is found by adding squares of binomials and provides evidence for whether data found in certain tables represent events that are occurring randomly.

The *chi-square* statistic is different from any statistic you have yet seen. This statistic compares actual frequencies with the frequencies that would be expected by calculating probabilities, as shown below.

The following table shows the average daily numbers of live births in California between 1995 and 1997. Are birthdays randomly distributed among the days of the week?

Day of the Week	Mon.	Tues.	Wed.	Thurs.	Fri.	Sat.	Sun.
Actual Numbers of Births	1,473	1,629	1,602	1,588	1,593	1,272	1,159

Source: *Journal of the American Medical Association*

Mental Math
Solve.
a. $|n| = 14$
b. $|n| = -14$
c. $|-n| = 14$
d. $-|n| = 14$

The **expected number** of births is the mean number of births for a given day that is predicted by a probability. If the births occurred randomly, then the expected number for each day would be the same. There were 10,316 births each week on average. So the expected number of births for each day is $\frac{10,316}{7}$ which, rounded to the nearest integer, is 1,474.

Day of the Week	Mon.	Tues.	Wed.	Thurs.	Fri.	Sat.	Sun.
Expected Numbers of Births	1,474	1,474	1,474	1,474	1,474	1,474	1,474

As you know, even if events occur randomly, it is not common for all events to occur with the same frequency. When you toss a coin ten times, you would not usually get 5 heads even if the coin were fair. Similarly, if there were 1,460 births on 4 days and 1,425 on the other three days, that would not seem to be much of a difference from the expected numbers. So the question is: Do the actual numbers deviate enough from the expected numbers that we should think that the births happen more often on certain days of the week?

If we let an expected number be e and an actual observed number be a, then $|e - a|$. The absolute value of the difference between these numbers is called the **deviation** of a from e. For example, the deviation on Saturday is $|1,474 - 1,272|$, or 202.

The Chi-Square Statistic **697**

Chapter 11

In 1900, the English statistician Karl Pearson introduced the **chi-square statistic** as a way of determining whether the difference in two frequency distributions is greater than that expected by chance. ("Chi" is pronounced *ky* as in *sky*.) The algorithm for calculating this statistic uses the squares of deviations, which is why we study it in this lesson.

Calculating the Chi-Square Statistic

Step 1 Count the number of events. Call this number n. In the above situation, there are 7 events, one each for Mon., Tues., Wed., Thurs., Fri., Sat., and Sun.

Step 2 Let $a_1, a_2, a_3, a_4, a_5, a_6$, and a_7 be the actual frequencies. In this example, $a_1 = 1{,}473$, $a_2 = 1{,}629$, $a_3 = 1{,}602$, $a_4 = 1{,}588$, $a_5 = 1{,}593$, $a_6 = 1{,}272$, and $a_7 = 1{,}159$.

Step 3 Let $e_1, e_2, e_3, e_4, e_5, e_6$, and e_7 be the expected frequencies. In this example, $e_1 = e_2 = e_3 = e_4 = e_5 = e_6 = e_7 = 1{,}474$.

Step 4 Calculate $\frac{(a_1 - e_1)^2}{e_1}, \frac{(a_2 - e_2)^2}{e_2}, \ldots, \frac{(a_n - e_n)^2}{e_n}$. Each number is the square of the deviation, divided by the expected frequency.

$\frac{(a_1 - e_1)^2}{e_1} = \frac{1}{1{,}474}, \frac{(a_2 - e_2)^2}{e_2} = \frac{24{,}025}{1{,}474}, \frac{(a_3 - e_3)^2}{e_3} = \frac{16{,}384}{1{,}474}, \frac{(a_4 - e_4)^2}{e_4} = \frac{12{,}996}{1{,}474}, \frac{(a_5 - e_5)^2}{e_5} = \frac{14{,}161}{1{,}474}, \frac{(a_6 - e_6)^2}{e_6} = \frac{40{,}804}{1{,}474}, \frac{(a_7 - e_7)^2}{e_7} = \frac{99{,}225}{1{,}474}$

Step 5 Add the n numbers found in Step 4. This sum is the chi-square statistic.

$\frac{1}{1{,}474} + \frac{24{,}025}{1{,}474} + \frac{16{,}384}{1{,}474} + \frac{12{,}996}{1{,}474} + \frac{14{,}161}{1{,}474} + \frac{40{,}804}{1{,}474} + \frac{99{,}225}{1{,}474} = \frac{207{,}596}{1{,}474} \approx 140.8$

The chi-square statistic measures how different a set of actual observed numbers is from a set of expected numbers. The larger the chi-square statistic is, the greater the difference. But is 140.8 unusually large? You can find that out by looking in chi-square tables. These tables give the values for certain values of n and certain probabilities. On the next page is such a table. In this table, n is the number of events. The other columns of the table correspond to probabilities of 0.10 (an event expected to happen $\frac{1}{10}$ of the time), 0.05 (or $\frac{1}{20}$ of the time), 0.01 (or $\frac{1}{100}$ of the time), and 0.001 (or $\frac{1}{1{,}000}$ of the time). You are not expected to know how the values in the table were calculated. The mathematics needed to calculate them is normally studied in college.

Critical Chi-Square Values

n − 1	0.10	0.05	0.01	0.001
1	2.71	3.84	6.63	10.8
2	4.61	5.99	9.21	13.8
3	6.25	7.81	11.34	16.3
4	7.78	9.49	13.28	18.5
5	9.24	11.07	15.09	20.5
6	10.6	12.6	16.8	22.5
7	12.0	14.1	18.5	24.3
8	13.4	15.5	20.1	26.1
9	14.7	16.9	21.7	27.9
10	16.0	18.3	23.2	29.6
15	22.3	25.0	30.6	37.7
20	28.4	31.4	37.6	45.3
25	34.4	37.7	44.3	52.6
30	40.3	43.8	50.9	59.7
50	63.2	67.5	76.2	86.7

How to Read a Chi-Square Table

Examine the number 14.1, which appears in column 0.05, row 7. This means that, with 8 events, a chi-square value greater than 14.1 occurs with probability 0.05 or less.

On page 698, we obtained a chi-square value of 140.8 with $n = 7$ events. So we look in row $n - 1$, which is row 6. A value as large as 140.8 would occur with probability less than 0.001, that is, less than 1 in 1,000 times. So we have evidence that the births in California are not evenly distributed among the days of the week. The data should be examined to determine why Saturdays and Sundays have fewer births.

Suppose the frequencies of the births had led to a chi-square value of 10.9. Then, looking across row 6, we would see that this value is between the listed values 10.6 and 12.6. So 10.9 has a probability between 0.10 and 0.05. That means that a chi-square value as high as 10.9 would occur between $\frac{1}{10}$ and $\frac{1}{20}$ of the time just by chance. Statisticians normally do not consider this probability to be low enough to think there is reason to question the expected values.

When a chi-square value is found that occurs with probability less than 0.05, statisticians question whether the assumptions that led to the expected values are correct. With this criterion, the above distribution of births is highly unusual. So we would question whether the births are occurring randomly.

 QY

▶ QY

a. Why do you think there are fewer births on Saturday than on Monday through Friday?
b. Why do you think there are even fewer births on Sunday?

Chapter 11

Example
Suppose 90 students were asked to name the United States President in 1950 from the names listed below. Suppose: 24 picked Dwight Eisenhower, 31 picked John Kennedy, and 35 picked Harry Truman (the correct answer). Is there evidence to believe the people were just guessing?

Solution Calculate the chi-square statistic following the steps given above.

Step 1 Find the number of events. $n = 3$.

Step 2 Identify the actual observed values. $a_1 = 24$; $a_2 = 31$; $a_3 = 35$.

Step 3 Calculate the expected values. If people were just guessing, we would expect each of the three names to be picked by the same number of people. Since there were 90 people in all, each name would be picked by 30. So, $e_1 = 30$; $e_2 = 30$; $e_3 = 30$.

Step 4 Calculate $\frac{(a_1 - e_1)^2}{e_1}$, $\frac{(a_2 - e_2)^2}{e_2}$, and $\frac{(a_3 - e_3)^2}{e_3}$.

$\frac{(a_1 - e_1)^2}{e_1} = \frac{(24 - 30)^2}{30} = \frac{36}{30}$; $\frac{(a_2 - e_2)^2}{e_2} = \frac{(31 - 30)^2}{30} = \frac{1}{30}$;

$\frac{(a_3 - e_3)^2}{e_3} = \frac{(35 - 30)^2}{30} = \frac{25}{30}$

Step 5 The sum of the numbers in Step 4 is $\frac{36 + 1 + 25}{30} = \frac{62}{30} \approx 2.07$.

Now examine the table. When $n = 3$, $n - 1 = 2$. So, look at the second row. The number 2.07 is less than the value 4.61 that would occur with probability 0.10. The numbers 24, 31, and 35 are like those that could randomly appear more than 10% of the time. It is quite possible that the people were guessing.

Harry S. Truman was the 33rd President of the United States.

The chi-square statistic can be used whenever there are actual frequencies and you have some way of calculating expected frequencies. However, the chi-square value is not a good measure of the deviation from the expected frequencies when there is an expected frequency that is less than 5.

Questions

COVERING THE IDEAS

1. What does the chi-square statistic measure?
2. When was the chi-square statistic developed, and by whom?
3. For what expected frequencies should the chi-square statistic not be used?

In 4 and 5, average number of traffic deaths per day of the week in the United States are given for a particular year.

a. Calculate the chi-square statistic assuming that traffic deaths occur randomly on days of the week.

b. Is there evidence to believe that the deaths are not occurring randomly on the days of the week?

		Mon.	Tues.	Wed.	Thurs.	Fri.	Sat.	Sun.
4.	Year 1985	100	105	105	110	145	170	140
5.	Year 1995	100	100	100	105	140	150	130

6. Suppose in the Example of this lesson that 40 students had picked Harry Truman, 30 had picked Dwight Eisenhower, and 20 had picked John Kennedy. Would there still be evidence that students were guessing randomly?

APPLYING THE MATHEMATICS

7. You build a spinner as shown at the right and spin it 50 times with the following outcomes. Use the chi-square statistic to determine whether or not the spinner seems to be fair.

Outcome	1	2	3	4	5
Frequency	13	13	9	8	7

8. A coin is tossed 1,000 times and lands heads up 537 times. Compare the numbers of heads and tails with what would be expected if the coin were fair. Use the chi-square statistic to test whether the coin is fair.

9. *The World Almanac and Book of Facts 2006* lists 64 notable tornadoes in the United States since 1925. The table at the right shows their frequencies by season of the year. Use the chi-square statistic to determine whether these figures support a view that more tornadoes occur at certain times of the year than at other times of the year.

Season	Number of Tornadoes
Autumn	8
Winter	15
Spring	38
Summer	3

10. Here are the total points scored in each quarter from the 16 National Football League games played December 17–19, 2005.

Quarter	1	2	3	4	Total
Points	122	196	133	150	601

Source: National Football League

Use the chi-square statistic to answer this question. Do football teams tend to score more points in one quarter than in any other?

Chapter 11

REVIEW

11. a. How many different permutations can be made using the letters of HORSE?

    b. How many different permutations can be made using the letters of MONKEY? (Lesson 11-7)

12. a. Which holds more, a cube with edges of length 6, or a rectangular box with dimensions 5 by 6 by 7? (Lessons 11-6, 11-5)

    b. Which holds more, a cube with edges of length x, or a box with dimensions $x - 1$ by x by $x + 1$? Justify your answer.

In 13–15, expand and simplify the expression. (Lessons 11-6, 11-5)

13. a. $(2x + y)^2$ b. $(2x - y)^2$ c. $(2x - y)(2x + y)$

14. $(8 - a)(a - 8)$ 15. $(3k^2 - 6km + 3m^2)^2$

16. Draw a picture of the following multiplication using rectangles. $4x(x + 6) = 4x^2 + 24x$ (Lesson 11-3)

17. On the fifth day after planting, a Moso bamboo tree was 38.5 cm. On the 14th day, the tree was 70.9 cm. (Lessons 6-1, 5-4)

    a. What was the average rate of change in height per day between the 5th and 14th days?

    b. Express the rate in Part a in cm/week.

EXPLORATION

18. A new high school for mathematics and science was opened in Cityville. Four hundred girls and 600 boys applied for 150 slots. A committee considered the applications and accepted 65 girls and 85 boys. Some people complained that there was discrimination.

    a. One complaint was that too few girls were accepted. This person's opinion was that there should have been equal numbers of boys and girls accepted and that the numbers accepted deviated too much from equality. Use the chi-square statistic to test whether this deviation could have occurred easily by chance.

    b. A second complaint was that too many girls were accepted. This person's position was that the numbers of boys and girls accepted should have been proportional to the number of applicants who were boys and girls. Use the chi-square statistic to test whether this deviation could have occurred easily by chance.

    c. If you were on the school board, would you agree with either complaint? Explain your answer.

Moso bamboo was introduced into the United States in about 1890.

Source: BAMBOO The Magazine of The American Bamboo Society

QY ANSWERS

a. Answers vary. Sample answer: With the ability to schedule deliveries, many physicians opt to schedule them during the week.

b. Answers vary. Sample answer: Many physicians choose Sunday as their day off and do not schedule deliveries for that day.

Chapter 11 Projects

1. Dividing Polynomials

If you wanted to check if 23 is a factor of 10,373, you could divide 10,373 by 23, and check if there is a remainder in the division. One way to do this would be to use long division. There is an algorithm for long division with polynomials, which is similar to an algorithm for long division with numbers. Here are two examples: one in which $x^3 - 4x^2 + 7x - 4$ is divided by $(x - 1)$, and the other in which $x^2 - 9$ is divided by $(x + 2)$.

$$
\begin{array}{r}
x^2 - 3x + 4 \\
x - 1 \overline{) x^3 - 4x^2 + 7x - 4} \\
\underline{-(x^3 - x^2)} \\
0x^3 - 3x^2 + 7x - 4 \\
\underline{-(-3x^2 + 3x)} \\
0x^2 + 4x - 4 \\
\underline{-(4x - 4)} \\
0
\end{array}
$$

$$
\begin{array}{r}
x - 2 \\
x + 2 \overline{) x^2 - 9} \\
\underline{-(x^2 + 2x)} \\
0x^2 - 2x - 9 \\
\underline{-(-2x - 4)} \\
-5
\end{array}
$$

In the first example, the result is $x^2 - 3x + 4$ with no remainder. In the second example the result is $(x - 2)$ with remainder −5.

a. Is $(x - 1)$ a factor of $x^3 - 4x^2 + 7x - 4$? Is $(x - 2)$ a factor of $x^2 - 9$?

b. Divide $x^3 + 3x^2 + 9x + 27$ by $(x + 3)$.

c. Write a description of the long division algorithm and explain it to a friend.

2. The Right Order

When performing complex actions, the order in which you do them is often important. For example, when dressing yourself in the winter, you might choose to put on pants, then a shirt, then a sweater, then socks, then shoes, and finally a jacket. The order in which you perform these actions is important.

a. How many different permutations of these actions are there?

b. Give four permutations of this order that would result in you not being dressed properly. Describe what happens in each case.

c. Give four permutations of this order that would result in you being dressed properly.

d. Give another sequence of events in which order matters. Answer Parts a, b, and c for this sequence.

3. Representing Positive Integers Using Powers

a. Every positive integer can be written as a sum of different powers of 2, where each power is added at most once. For example, $50 = 2^5 + 2^4 + 2^1$. You may be surprised to learn that for each number there is only one way to write it as a sum of powers of 2. Find the powers of 2 representation for the integers from 1 to 16. Explain why there is only one such representation and how this relates to writing numbers in base 10.

(continued on next page)

b. Every positive integer can be represented as a sum of powers of 3, where each power is added at most twice. For example, $50 = 3^3 + 3^2 + 3^2 + 3^1 + 3^0 + 3^0$. Find the powers of 3 representation for all the integers from 1 to 16. Explain why there is only one such representation for each number.

4 Differences of Higher Powers

In this chapter you saw why $x^2 - y^2 = (x - y)(x + y)$. This is only one instance of a general pattern.

a. Use the Extended Distributive Property to show that $x^3 - y^3 = (x - y)(x^2 + xy + y^2)$ and that $x^4 - y^4 = (x - y)(x^3 + x^2y + yx^2 + y^3)$.

b. Find similar factors for $x^5 - y^5$ and then generalize them to find similar factors for $x^n - y^n$.

c. Suppose $y = 1$. What does your answer from Part b say that $\frac{x^n - 1}{x - 1}$ is equal to?

d. Use the formula from Part c to calculate $1 + 2 + 2^2 + 2^3 + \ldots + 2^{10}$. (You can use the fact that $2^{10} = 1{,}024$.)

5 Switches and Sorting

One special type of permutation, in which the positions of two elements are changed, is called a *switch*. For example, a switch could change the acronym BADC into the acronym ABDC. Another switch could change ABDC to ABCD. Suppose you are given an acronym and you want to put it into alphabetical order. (This sort of problem comes up very often in computer programming.)

a. Find and describe an algorithm to put an acronym into alphabetical order using a sequence of switches.

b. Use your algorithm to sort the acronym UNESCO. How many switches were required to do this?

6 Testing Astrology

From a book (like *Who's Who*) or online source, find at least 100 famous people in a field and note their birthdays. Identify the astrological sign of each person. Then tabulate the number of people with each sign. Do these data lead you to believe that certain birth signs are more likely to produce famous people? Use a chi-square statistic assuming a random distribution of the birthdays among the 12 astrological signs is expected. Although $n - 1 = 11$ here, refer to row 10 from the chi-square table on page 699.

Chapter 11 Summary and Vocabulary

- A **monomial** is a product of terms. The **degree of a monomial** is the sum of the exponents of its variables. A **polynomial** is an expression that is either a monomial or a sum of monomials. The **degree of a polynomial** is taken to be the largest degree of its monomial terms. **Linear polynomials** are polynomials of degree 1. **Quadratic polynomials** are polynomials of degree 2.

- Polynomials emerge from a variety of situations. Our customary way of writing whole numbers in base 10 can be considered as a polynomial with 10 substituted for the variable. If different amounts of money are invested each year at a **scale factor** x, the total amount after several years is a **polynomial in x.** The number of permutations of n objects can be represented by a polynomial in n.

- Addition and subtraction of polynomials are based on adding like terms, one of the forms of the Distributive Property that you studied earlier in this book. Multiplication of polynomials is also justified by the Distributive Property. To multiply one polynomial by a second, multiply each term in the first polynomial by each term in the second polynomial, then add the products. For example:

    monomial by a polynomial: $a(x + y + z) = ax + ay + az$
    two polynomials: $(a + b + c)(x + y + z) =$
    $\qquad ax + ay + az + bx + by + bz + cx + cy + cz$
    two binomials: $(a + b)(c + d) = ac + ad + bc + bd$
    perfect square: $(a + b)^2 = (a + b)(a + b) = a^2 + 2ab + b^2$
    difference of two squares: $(a + b)(a - b) = a^2 - b^2$

- If each of the terms of a polynomial has a common factor, then so does their sum. It can be factored out using the Distributive Property.

- The square of the difference of actual and expected values in an experiment, $(a - e)^2$, appears in the calculation of the **chi-square statistic.** This statistic can help you decide whether the assumptions that led to the expected values are correct.

Vocabulary

11-1
polynomial in x
standard form for a polynomial

11-2
monomial
polynomial
binomial
trinomial
degree of a monomial
degree of a polynomial
linear polynomial
quadratic polynomial

11-4
factoring
trivial factors
greatest common factor
factorization
prime polynomials
complete factorization

11-6
perfect square trinomials
difference of squares

11-7
permutation
$n!$, n factorial
circular permutation

11-8
expected number
deviation
chi-square statistic

Theorems and Properties

Unique Factorization Theorem for Polynomials (p. 677)
Extended Distributive Property (p. 680)
Perfect Squares of Binomials (p. 687)
Difference of Two Squares (p. 688)

Chapter 11 Self-Test

Take this test as you would take a test in class. You will need a calculator. Then use the Selected Answers section in the back of the book to check your work.

In 1–6, expand and simplify the expression.
1. $3x(10 - 4x + x^3)$
2. $(2b - 5)^2$
3. $(8z + 3)(8z - 3)$
4. $6a(2a^2 + 9a - 1)$
5. $(5a^2 - a)(5a^2 - a)$
6. $(2 - 6c)(4 + 3c)$

In 7 and 8, consider the polynomial $8x^3 - 5 + 2x + 11x^3 - 9x^2$.

7. What is the degree of this polynomial?
8. Is the polynomial a *monomial*, *binomial*, *trinomial*, or *none of these*?
9. Factor completely: $12x^3y^2 - 24x^2y^3 + 30x^2y^4$.

In 10 and 11, write as a single polynomial.
10. $(20n^2 - 8n - 12) + (16n^3 - 7n^2 + 5)$
11. $9p^4 + p^2 - 5 - p(3p^3 + p - 2)$

12. **True or False** The expression $3v^2 + 3v - 1$ is a trinomial of degree 3.

13. Simplify the fraction $\frac{28w^3 - 18w}{2w}$, assuming that $w \neq 0$.

14. Write the area of the shaded portion of the rectangular region as a polynomial in standard form.

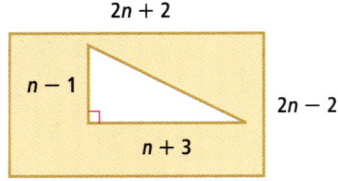

15. A swim team has 7 swimmers available to race a 4-person relay. How many different orders of swimmers are possible?

16. Twenty-five art students entered a competition in which 1st, 2nd, and 3rd place prizes were to be awarded. How many permutations of students could receive these prizes?

In 17 and 18, consider the following. On Ashley's 16th birthday, she received $200. She received $150 on her 17th birthday, and $300 on each of her 18th and 19th birthdays.

17. If she invested all of this money each year in a savings account with a yearly scale factor x, how much money would she have on her 21st birthday?

18. Evaluate your answer to Question 17 if the savings account had an interest rate of 3%.

19. a. What is the area of the largest rectangle below?

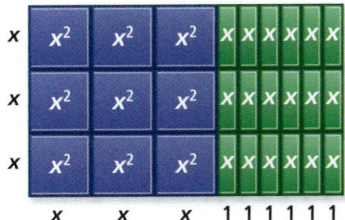

b. What factorization of the rectangle's area is shown?

c. Is this a complete factorization? Explain your answer.

20. Represent the product $(x + 2y)(5 + z)$ using areas of rectangles.

21. Jerry believes a die is weighted to favor certain numbers. The table below shows the outcome of 198 random tosses of the die.

Number	Outcomes
1	24
2	44
3	31
4	30
5	43
6	26

Use the chi-square statistic to provide evidence as to whether Jerry's view is correct. Justify your answer.

22. Represent the product $(x + 4)^2$ using areas of rectangles. Write your answer in standard form.

Chapter 11 Chapter Review

SKILLS
PROPERTIES
USES
REPRESENTATIONS

SKILLS Procedures used to get answers

OBJECTIVE A Add and subtract polynomials. (Lessons 11-1, 11-2)

In 1–4, simplify the expression and write the answer in standard form.

1. $(-k^3 + 2k^2 - 17k + 8) + (5k^3 - 2k^2 - 8)$
2. $\frac{4}{3}h^3 + 4 - \frac{2}{3}h^3 + \frac{1}{7}h^2 - 5$
3. $(5.4s^4 + 9.8s^2 - 8) - (-3.7s^3 - 4 + 5.2s)$
4. $(12w^3 - 3w^2 + -8w - 9) - (80w + 8)$

OBJECTIVE B Multiply polynomials. (Lessons 11-3, 11-5, 11-6)

5. **Fill in the Blank** $(-6x + 9) \cdot \left(\frac{5}{3}x + 6\right) = -10x^2 - \underline{\ ?\ }x + 54$

In 6–16, write as a single polynomial in standard form.

6. $m(7m^2 - 13m + 12)$
7. $9p(p^3 + p^2 - 6p + 5)$
8. $-3x^2\left(12x^2 + \frac{2}{3}x\right)$
9. $-\frac{1}{7}q^8(-q^3 + 14q^2 - 112q + 5)$
10. $2(g + 3g^2 + 19g^3 - g^6) + g(-3g^5 + 2g + 6g^3)$
11. $(b - 3)(b + 3)$
12. $(8x - 2)(2x - 1)$
13. $2(1 + 6w)(1 - 6w)$
14. $3(a^2 + a - 1)(a + 1)$
15. $(n - 2)(n - 3)(n - 4)$
16. $(c^2 + 10c - 4)(2c^2 - 8c + 1)$

17. The length of one leg of a right triangle is $(3x - 1)$, and the length of the other leg is $(17x + 2)$. Express the area of the triangle as a polynomial in standard form.

OBJECTIVE C Find common monomial factors of polynomials. (Lesson 11-4)

In 18–21, factor the polynomial completely.

18. $9k^3 + 6k^2$
19. $u^2v - uv^2$
20. $-84y^3 - 18y^2 + 93y$
21. $45a^9b^5 + 60a^6b^4 - 15a^5b^3 + 420a^3b^2$

22. **Multiple Choice** Which is a complete factorization of $24y^7 + 18y^5 - 90y^3$?

A $y^3(24y^4 + 18y^2 - 90)$
B $3y^3(8y^7 + 6y^5 - 30y^3)$
C $6y^7(4 + 3y^{-2} + 15y^{-4})$
D $6y^3(4y^4 + 3y^2 - 15)$

OBJECTIVE D Expand squares of binomials. (Lesson 11-6)

In 23–26, expand and simplify the expression.

23. $(p + 6)^2$
24. $(-u - 5)^2$
25. $(11 - 45z)^2$
26. $(-w + 2)^2$

27. **Multiple Choice** The square of which binomial below is $64c^4 + -80c^3 + 25c^2$?

A $8c - 5$
B $8c^2 - 5c$
C $8c^2 + 5c$
D $8c^3 + 5c$

PROPERTIES The principles behind the mathematics

OBJECTIVE E Classify polynomials by their degrees or number of terms. (Lesson 11-2)

708 Polynomials

28. Give an example of a monomial of degree 5.

29. Give an example of a trinomial of degree 5.

In 30–33, consider the polynomials below.

a. $n^2 - 5$
b. $2m^2 + 4m - 7$
c. $12a^4 - 16a^2 + 3$
d. $8w^2y + 9wy$

30. Which are binomials?

31. Which are trinomials?

32. Which have degree 2?

33. Which have degree 4?

USES Applications of mathematics in real-world situations

OBJECTIVE F Translate investment situations into polynomials. (Lesson 11-1)

34. Flora decides to open a retirement account with an annual interest rate y. In the first year, Flora invests $5,000. The second year, she invests $3,000, and then in the third year she invests $2,000. She keeps the money in the account at the same scale factor for five years after her last deposit.

    a. Write a polynomial that describes how much money she has in this account at the end of that time.

    b. If $y = 1.07$, how much money does Flora have in her retirement account after the five years?

35. Jeffrey is saving money during his high school years to go on a trip to Egypt after he graduates. The trip costs $3,000. At the end of his freshman year, he deposits $1,200 in a savings account with an annual scale factor of x. At the end of his sophomore year, he deposits $700 in the same account. At the end of his junior year, he deposits $500. It is now the end of his senior year.

    a. Write an expression that shows how much money Jeffrey has in his account.

    b. If the savings account pays 4% annual interest, how much more money does he need to afford the trip?

OBJECTIVE G Determine numbers of permutations. (Lesson 11-7)

In 36–38, Beth has 7 different blouses in her closet.

36. How many different ways can she select a blouse for each of the 7 days of the week?

37. How many different ways can she select a blouse for each of the 5 school days of the week?

38. Suppose Beth wants to wear her favorite blouse on Monday. How many different ways can she select a blouse for each of the remaining 4 school days of the week?

In 39–41, Kyle has recently opened a bank account and is asked to set the 4-digit PIN (Personal Identification Number) for his ATM card, in which each digit can be any number from 0 to 9.

39. How many different ways can Kyle select a PIN?

40. If the digits of the PIN must all be different, how many ways can Kyle select the PIN?

41. If the first digit of the PIN is *not* allowed to be zero, how many different ways can Kyle a select PIN, assuming the digits cannot be repeated?

42. Six guests arrive for a dinner party and are to be seated around a large circular table. How many different ways can the host of the party seat the guests around the table?

43. During a track meet, 9 runners are racing in the 100-yard dash, but only the top 3 runners receive an award. How many different ways are there for the top 3 runners to place?

Chapter 11

OBJECTIVE H Use a chi-square statistic to determine whether or not statistics support a conclusion. (Lesson 11-8)

44. Sixty people were surveyed in a taste test of two types of chocolate. 26 people preferred chocolate A and 34 people preferred chocolate B.
    a. If the chocolates were equally tasty, what would be the expected numbers of preference for each chocolate?
    b. Calculate the chi-square statistic for this situation using the actual numbers and the expected numbers from Part a.
    c. Use the chi-square table on page 699. Does the evidence support the fact that chocolate B is preferred to chocolate A? Explain why or why not.

45. A company was open only Monday through Friday. Because it was not open Saturday or Sunday, it expected that it would get about the same amount of mail each day Tuesday through Friday, but three times this amount on Monday. However, some people thought there was too much mail coming on Monday. When the numbers of pieces of mail for each day for a few weeks were totaled, here were the numbers on each day.

Day	Mon.	Tues.	Wed.	Thurs.	Fri.
Pieces of Mail	122	30	41	35	27

    a. How many pieces of mail did the company expect each day?
    b. Calculate the chi-square statistic for this situation using the actual numbers and the expected numbers from Part a.
    c. Use the chi-square table on page 699. Does the evidence support the company's expectations on how much mail to expect? Justify your answer.

46. A large factory believes that its floor manager is becoming careless towards the end of his shift each day. Below is a table that shows the hour of his shift and the number of accidents that occurred during that hour.

Hour	1	2	3	4	5	6	7	8
Accidents	5	4	4	7	9	8	11	10

    a. If the factory were to have an equal number of accidents each hour, find the expected number.
    b. Calculate the chi-square statistic for this situation using the actual numbers and the expected number from Part a.
    c. Use the chi-square table on page 699. Does the evidence support the factory's belief about the floor manager? Justify your answer.

REPRESENTATIONS Pictures, graphs, or objects that illustrate concepts

OBJECTIVE I Represent polynomials by areas. (Lessons 11-3, 11-5, 11-6)

47. a. Write the area of the largest rectangle below as the sum of 4 terms.

    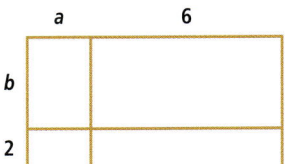

    b. Write the area of the largest rectangle as the product of 2 binomials.
    c. Are the answers to Parts a and b equal?

48. Represent $(n + m)(p + q)$ using areas of rectangles.

49. a. What polynomial multiplication is represented by the area of the largest rectangle below?

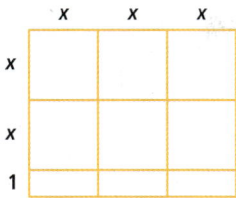

b. What is the product in standard form?

50. Show the product $(3x + 1)^2$ using areas of rectangles. Write your answer in standard form.

51. Show a factorization of $3x^2 + 9x$ by rearranging these tiles into a different rectangle.

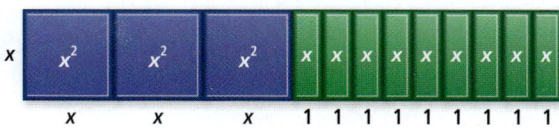

52. Show $4x^2 + 4x + 1 = (2x + 1)^2$ by rearranging these tiles into a square with sides $2x + 1$.

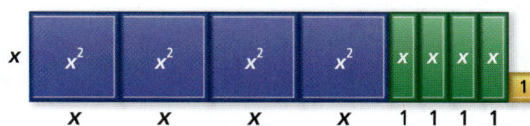

Chapter 12

More Work with Quadratics

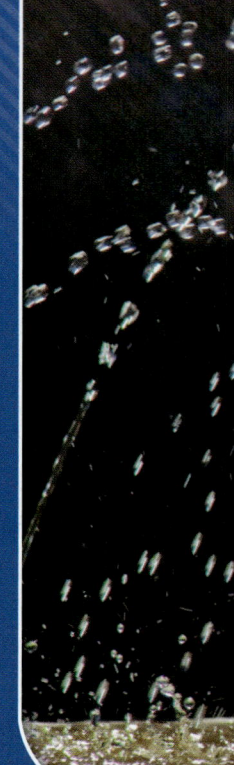

Contents

- 12-1 Graphing $y - k = a(x - h)^2$
- 12-2 Completing the Square
- 12-3 The Factored Form of a Quadratic Function
- 12-4 Factoring $x^2 + bx + c$
- 12-5 Factoring $ax^2 + bx + c$
- 12-6 Which Quadratic Expressions Are Factorable?
- 12-7 Graphs of Polynomial Functions of Higher Degree
- 12-8 Factoring and Rational Expressions

In writing it is important to know and use synonyms for the same idea. A picture may be beautiful or pretty, or it may be dazzling or brilliant, or it may be dull or drab or gray. Synonyms help convey ideas more clearly and one word may fit a situation just a little better than another.

Equivalent expressions in mathematics are like synonyms in writing. Equivalent expressions are often classified by their form.

Here are four ways to write the same number:

2,048	base 10
$2 \cdot 10^3 + 4 \cdot 10^1 + 8 \cdot 1$	expanded form
2^{11}	exponential form
$2.048 \cdot 10^3$	scientific notation

There are occasions when each of these forms is most appropriate or most helpful to understanding a situation. Base 10 is our normal compact way of writing numbers. Expanded form shows the meaning of base 10. Exponential form arises in many counting situations and situations of growth. Scientific notation is useful when comparing numbers that differ greatly in size.

Equivalent equations are also like synonyms. Consider the line that contains the two points (2, 9) and (8, 11). Three equivalent equations for this line are in forms that you saw in Chapter 6.

$y = \frac{1}{3}x + \frac{25}{3}$ slope-intercept form

$x - 3y = -25$ standard form

$y - 9 = \frac{1}{3}(x - 2)$ point-slope form

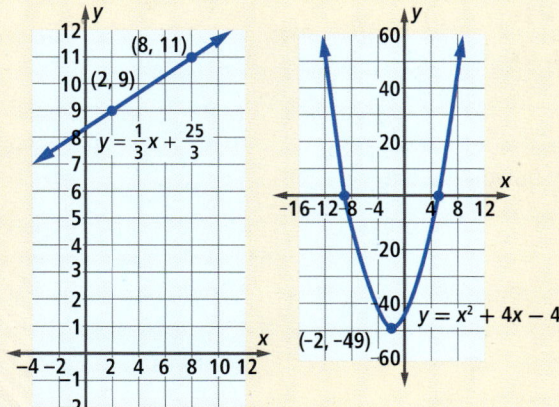

In this chapter, we return to quadratic equations and parabolas. Here are three equivalent equations for the parabola that is graphed at the left.

$y = x^2 + 4x - 45$ standard form

$y + 49 = (x + 2)^2$ vertex form

$y = (x - 5)(x + 9)$ factored form

In Chapter 9, you saw that standard form enables you to use the Quadratic Formula to find the x-intercepts of the graph. Standard form also is very useful for solving equations. The vertex form, which opens the chapter, lets you quickly find the vertex of the parabola. This is useful for graphing and for obtaining the minimum (or maximum) value of y. The factored form, which you will study later in this chapter, shows the x-intercepts of the graph.

713

Chapter 12

Lesson 12-1

Graphing $y - k = a(x - h)^2$

Vocabulary

vertex form of an equation for a parabola

▶ **BIG IDEA** The graph of the equation $y - k = a(x - h)^2$ is a parabola whose vertex can be easily found.

In Chapter 9, you graphed many parabolas with equations in the *standard form* $y = ax^2 + bx + c$. When an equation is in this form, the vertex of the parabola is not obvious. In Activity 1, you are asked to examine some parabolas with equations in the form $y - k = a(x - h)^2$. When an equation is in this form, its graph is a parabola whose vertex can be found rather easily.

Mental Math

Suppose $b(x) = 2|x| - 4$. Evaluate

a. $b(5)$.
b. $b(-5)$.
c. $b(5) - b(-5)$.
d. $\frac{b(5)}{b(-5)}$.

Activity 1

In 1–6, set a graphing calculator for the window $-15 \leq x \leq 15$, $-10 \leq y \leq 10$.

a. Graph the equation on your calculator. (You will have to solve the equation for y if it is not already solved for y.) Copy the graph by hand onto your paper.
b. Label the vertex with its coordinates.
c. Draw the axis of symmetry as a dotted line. Label the axis of symmetry with its equation.

1. $y - 4 = (x - 3)^2$
2. $y + 3 = (x - 5)^2$
3. $y - 1 = (x + 4)^2$
4. $y + 8 = -(x + 6)^2$
5. $y - 12 = -(x - 4)^2$
6. $y = (x - 0.35)^2$

7. Look back at your graphs for Questions 1–6 and the equations that produced them. Explain how to look at an equation like $y - k = (x - h)^2$ to help determine the vertex of its graph.

In 8 and 9, each graph is of an equation of the form $y - k = (x - h)^2$. The vertex and axis of symmetry of the parabola are given. Use what you learned in Questions 1–7.

a. Write an equation for the graph.
b. Check your equation by graphing it on your calculator. Do you get the graph you expected?

714 More Work with Quadratics

8.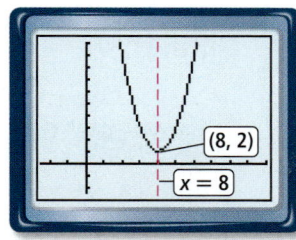
$-5 \leq x \leq 20, -5 \leq y \leq 20$

9.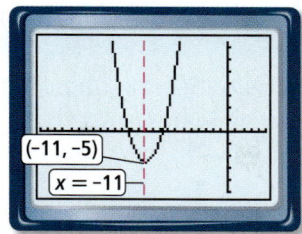
$-25 \leq x \leq 5, -10 \leq y \leq 14$

Example 1

Consider the graph of $y - 7 = 6(x - 15)^2$.

a. Explain why the graph is a parabola.
b. Find the vertex of this parabola without graphing.

Solution

a. If the equation can be written in the form $y = ax^2 + bx + c$, its graph is a parabola. Work with the given equation to get it into that form.

$y - 7 = 6(x - 15)^2$
$y - 7 = 6(x^2 - 30x + 225)$ Square of a Binomial
$y - 7 = 6x^2 - 180x + 1{,}350$ Distributive Property
$y = 6x^2 - 180x + 1{,}357$ Add 7 to both sides.

This equation is of the form $y = ax^2 + bx + c$, with $a = 6$, $b = -180$, and $c = 1{,}357$, so its graph is a parabola.

b. To find its vertex, examine the original equation $y - 7 = 6(x - 15)^2$. Add 7 to both sides.

$$y = 6(x - 15)^2 + 7$$

Since $(x - 15)^2$ is the square of a real number, $(x - 15)^2$ cannot be negative. The least value $(x - 15)^2$ can have is 0, and that occurs when $x = 15$. Thus the least value that $6(x - 15)^2$ can have is 0, and the least value that $6(x - 15)^2 + 7$ can have is 7. All of these least values occur when $x = 15$. As a consequence, the vertex of the parabola is at the point on the parabola with x-coordinate 15. When $x = 15$, substitution shows that $y = 7$. So the vertex is (15, 7). Since 7 is the least value of y, the parabola must open up.

Check You should graph the equation $y - 7 = 6(x - 15)^2$ with a graphing calculator, making sure that the window contains the point (15, 7). (You will likely have to solve the equation for y before graphing.)

The argument in Example 1 can be repeated in general. It demonstrates the theorem on the next page.

Chapter 12

> **Parabola Vertex Theorem**
>
> The graph of all ordered pairs (x, y) satisfying an equation of the form $y - k = a(x - h)^2$ is a parabola with vertex (h, k).

▶ **QY**

Give the vertex of the parabola with equation $y - 8 = 3(x + 15)^2$.

 QY

The form $y - k = a(x - h)^2$ is called the **vertex form of an equation for a parabola** because you can easily find the vertex from the equation.

Example 2

Multiple Choice Which of the four curves at the right is the graph of $y + 3 = (x - 4)^2$?

Solution Rewrite $y + 3 = (x - 4)^2$ so it corresponds to the general equation $y - k = a(x - h)^2$.

$$y - {-3} = (x - 4)^2$$

The vertex is given by (h, k). So the vertex of the graph is $(4, -3)$. The parabola with this vertex is choice D.

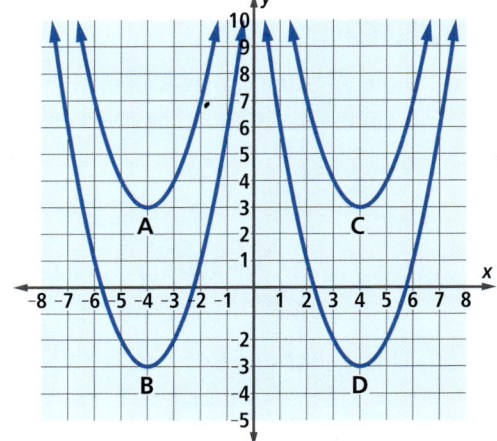

Activity 2

Use a dynamic graphing system to do the following.

Step 1 Create three sliders that include positive and negative numbers.

Name one slider *a*. (Or create a parameter *a* that is based on the slider.)

Name one slider *h*. (Or create a parameter *h* that is based on the slider.)

Name one slider *k*. (Or create a parameter *k* that is based on the slider.)

Step 2 Move the sliders so $a = 1$, $h = 0$, and $k = 0$.

Step 3 Create the equation $y - k = a(x - h)^2$, using the parameters for *a*, *h*, and *k*. Plot the equation. The graph should show
$y - 0 = 1(x - 0)^2$ or $y = x^2$.

    a. Make a table of values for $-3 \leq x \leq 3$.

    b. Find the differences between the *y*-coordinates as the *x*-coordinates increase by 1. Does this pattern look familiar?

716   More Work with Quadratics

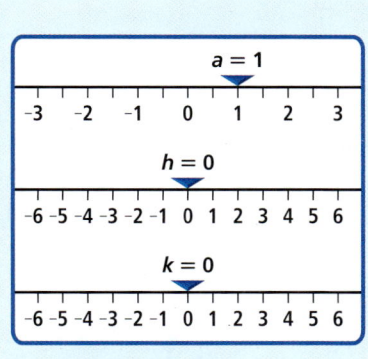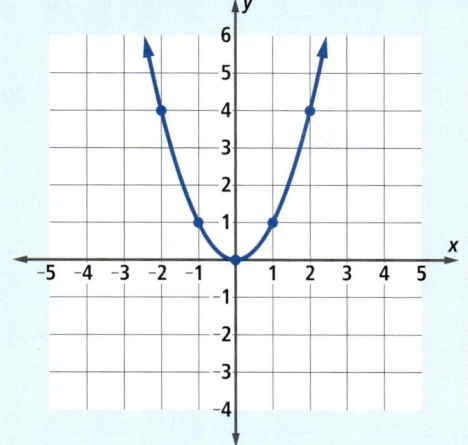

$y - k = a(x - h)^2$
$y - 0 = 1(x - 0)^2$

Step 4 Slowly move the *a* slider from 1 to 0.
   a. What happens to the graph for $0 < a < 1$?
   b. Move the *a* slider to 0.5. Make a table of values for $-3 \leq x \leq 3$. How do the *y*-coordinates compare to those in the table in Step 3?
   c. Find the differences between the *y*-coordinates as the *x*-coordinates increase by 1. How do the differences compare to those in Step 3?

Step 5 Now move the *a* slider to the right of 1.
   a. What happens to the graph when $a > 1$?
   b. Move the *a* slider to 2. Make a table of values for $-3 \leq x \leq 3$. How do the *y*-coordinates compare to those in the table in Step 3?
   c. Find the differences between the *y*-coordinates as the *x*-coordinates increase by 1. How do the differences compare to those in Step 3?

Step 6 Is the effect *a* has on the differences in the *y*-coordinates of $y = ax^2 + bx + c$ the same as or different from the effect *a* has on the differences in the *y*-coordinates of $y - k = a(x - h)^2$? Explain.

Step 7 Write a prediction of what you think happens to the graph when *a* is between -1 and 0.

Write a prediction of what you think happens to the graph when *a* is less than -1.

Step 8 Move the *a* slider to test your predictions. Were you correct? Explain how your predictions in Step 7 are similar to and different from what occurred in Steps 4 and 5.

Step 9 Sketch what you think the graph of each function will look like.

$y - 5 = 2(x - 1)^2$ $y + 3 = 0.25(x - 2)^2$ $y = -3(x + 4)^2$

(continued on next page)

Step 10 Check your predictions from Step 9 by moving sliders *a*, *h*, and *k* to match the values in the function.

Step 11 Write a possible equation in vertex form for the graph of the parabola at the right.

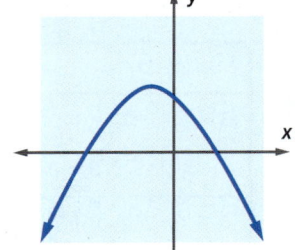

Example 3

Write an equation in vertex form for the graph of the parabola at the right.

Solution 1 First locate the vertex and substitute it into the vertex form of an equation for a parabola, $y - k = a(x - h)^2$.

The vertex is at (4, -1), so $y - -1 = a(x - 4)^2$.

Next find the value of *a*. Compare the pattern in the change of the *y*-coordinates as the *x*-coordinates change by 1 in the graph to that of $y = x^2$.

Pattern change in *y*-coordinates for $y = x^2$: 1, 3, 5, ...

Pattern change in *y*-coordinates for graph: 2, 6, 10, ...

The pattern change for the graph is double the pattern change of $y = x^2$, so $a = 2$.

The equation for the parabola is $y + 1 = 2(x - 4)^2$.

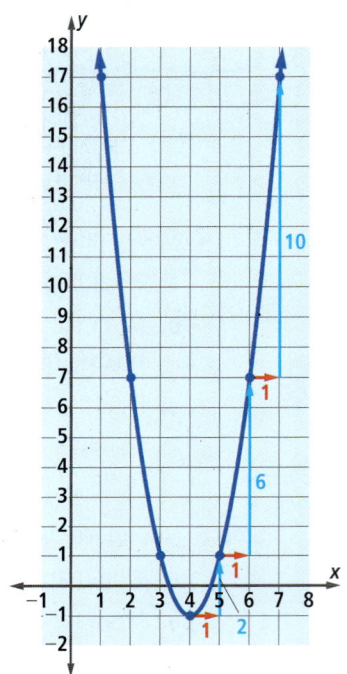

Solution 2 As in Solution 1, first locate the vertex and substitute it into the vertex form of an equation for a parabola, $y - k = a(x - h)^2$.

The vertex is at (4, -1), so $y - -1 = a(x - 4)^2$.

Next find the value of *a*. Pick a point on the graph that is not the vertex and substitute it into the equation. Then solve for *a*. We pick point (6, 7).

$$7 + 1 = a(6 - 4)^2$$
$$8 = a(2)^2$$
$$8 = 4a$$
$$2 = a$$

Substitute *a* and the vertex into the equation.

The equation for the parabola is $y + 1 = 2(x - 4)^2$.

Check Pick a point on the graph and substitute it into the equation. We'll use the point (3, 1).

$$1 + 1 = 2(3 - 4)^2$$
$$2 = 2(-1)^2$$
$$2 = 2 \cdot 1$$
$$2 = 2 \text{ So it checks.}$$

Problems involving area can lead to parabolas. Consider the following problem.

> **GUIDED**
>
> **Example 4**
> There are many possible rectangles with a perimeter of 24 units. Suppose the length of one side of such a rectangle is L.
> a. Find the area A of the rectangle in terms of L.
> b. Graph the equation from Part a.
> c. Use the graph to determine the maximum area of a rectangle with perimeter of 24 units.
>
> **Solutions**
> a. You know that the perimeter P of a rectangle is given by the formula $P = 2L + 2W$. So in this case, $24 = 2L + 2W$.
> Now solve this equation for W.
>
> $\underline{?} = 2W$
>
> Divide both sides by 2.
>
> $\underline{?} = W$
>
> Since the area $A = LW$, substituting $\underline{?}$ for W gives the following formula for A.
>
> $A = L\underline{?}$
>
> b. Letting $x = L$ and $y = A$, a graph of $y = x(12 - x)$ is shown here. The graph is a parabola because $y = 12x - x^2$ is of the form $y = ax^2 + bx + c$. The only part of the parabola that makes sense in this problem is for values of x between 0 and 12, so we use that window.
>
> c. Each value of y in $y = x(12 - x)$ is the area of a particular rectangle. If $x = L = 2$, then $W = 12 - 2 = 10$ units. The area is $y = 2 \cdot 10 = 20$ units^2. That is, the point (2, 20) on the parabola means that when one side of the rectangle is 2 units, the area of the rectangle is 20 units^2.
>
> If $x = L = 3$, then $W = \underline{?}$. The area $y = \underline{?} = \underline{?}$.
>
> If $x = L = 10$, then $W = \underline{?}$. The area $y = \underline{?} = \underline{?}$.
>
> The maximum value of y is at the vertex of the parabola. From the graph the vertex is (6, 36). So the maximum area of the rectangle is $\underline{?}$, occurring when $L = \underline{?}$ and $W = \underline{?}$, that is, when the rectangle is a square.

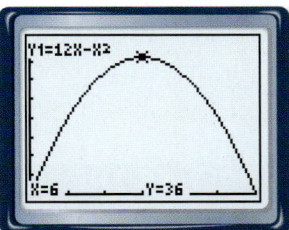

The equation graphed in Example 4 is not in vertex form. That makes it difficult to know the vertex. In Lesson 12-2, you will see how to convert an equation into vertex form.

Graphing $y - k = a(x - h)^2$ **719**

Chapter 12

Questions

COVERING THE IDEAS

1. The graph at the right shows a parabola.
   a. What are the coordinates of its vertex?
   b. Give an equation for its axis of symmetry.

2. Give the minimum value of each expression.
   a. x^2 b. $(x+6)^2$ c. $3(x+6)^2$ d. $3(x+6)^2 - 8$

3. Explain why 7 is the maximum value of the expression $-4(x-5)^2 + 7$.

In 4–7, an equation of a parabola is given.
   a. Find the coordinates of its vertex.
   b. Write an equation for its axis of symmetry.
   c. Tell whether the parabola opens up or down.

4. $y + 8 = -5(x - 9)^2$
5. $y - 21 = 0.2(x - 15)^2$
6. $y - 43 = 8x^2$
7. $y + \frac{1}{2} = -(x + 6)^2$

8. The equation $y = x(18 - x)$ gives the area of a rectangle with perimeter of 36 where x is the length of one of its sides. This equation is graphed at the right.
   a. Give the areas of the three rectangles for which $x = 4$, 5, and 11.
   b. Point P on the graph represents a rectangle. Give the rectangle's side lengths and area.
   c. What is the maximum area of a rectangle with perimeter 36?

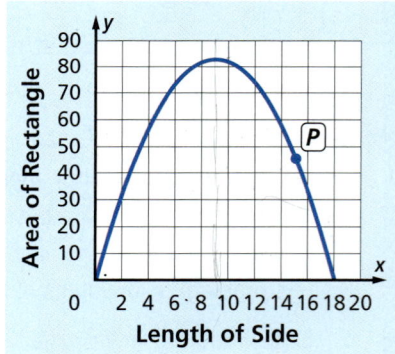

9. Explain the similarities and differences in the graphs of the functions.
   a. $y + 8 = (x - 4)^2$
   b. $y + 8 = 1.25(x - 4)^2$
   c. $y + 8 = 0.75(x - 4)^2$

APPLYING THE MATHEMATICS

10. A rectangle has perimeter 60.
    a. Find a formula for its area A in terms of the length L of one side.
    b. Graph the formula you found in Part a.
    c. What is the maximum area of this rectangle?

11. A parabola has vertex $(-12, 9)$ and contains the point $(-10, 5)$. Give an equation for the parabola.

More Work with Quadratics

In 12 and 13, an equation of the form $y = ax^2$ and its graph are given. A translation image of the parabola is graphed with a dashed curve. Write an equation for the image.

12.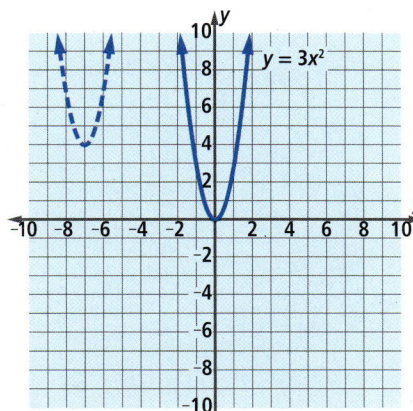

13.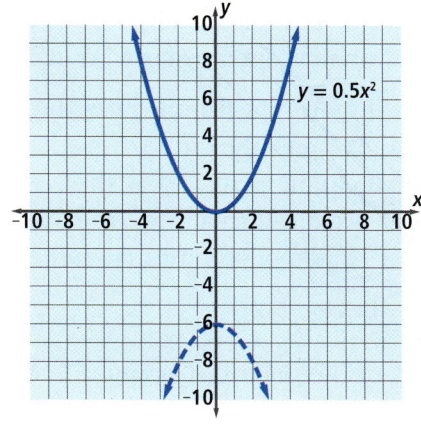

14. Write an equation for the graph of the parabola at the right.

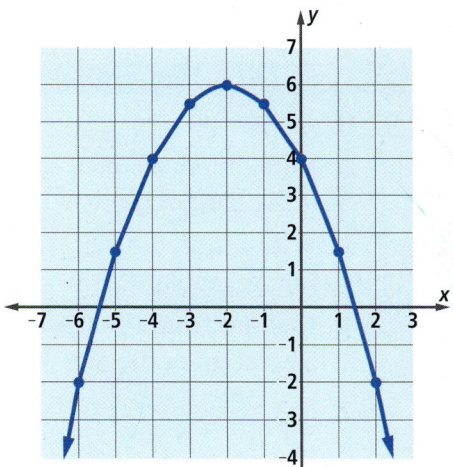

In 15 and 16, find equations for two different parabolas that fit the description.

15. The vertex is (5, −18) and the parabola opens down.

16. The axis of symmetry is $x = 2$ and the parabola opens up.

In 17 and 18, a graph of a parabola and a point on it are given. Find the coordinates of a second point on the parabola that has the same y-coordinate as the given point.

17.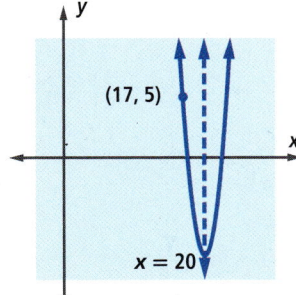

18.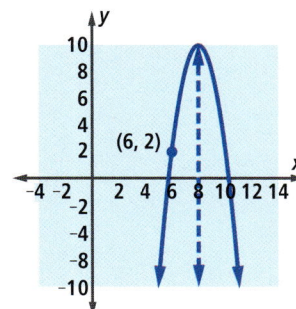

REVIEW

19. Draw rectangles picturing $3a(a + 8) = 3a^2 + 24a$. (**Lesson 11-3**)

20. The sum of the legs of a right triangle is 34 cm. If the hypotenuse is 26 cm, calculate the length of each of the legs. (**Lessons 10-2, 8-6**)

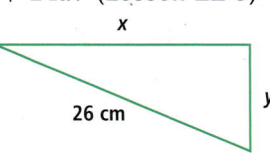

21. Find all values of m that satisfy $(m^2)^2 - 15m^2 + 36 = 0$. (**Lesson 9-5**)

22. Suppose a basketball team wins 9 of its first 11 games during a season. At this rate, how many games would you expect the team to win in a 28-game season? (**Lesson 5-9**)

23. A climber is ascending Mount Kilimanjaro, the highest mountain in Africa. At 9 A.M. the climber is at an elevation of 15,416 feet and at 10:15 A.M. the climber is at an elevation of 16,004 feet. At this rate, when would the climber reach the 19,336-foot summit? (**Lesson 5-5**)

24. Simplify $5\pi \div \frac{4\pi}{3}$. (**Lesson 5-2**)

25. Solve the equation or inequality. (**Lessons 4-5, 4-4**)
    a. $2x = 3x$
    b. $2x > 3x$

Mount Kilimanjaro is not only the highest peak on the African continent, it is also the tallest freestanding mountain in the world at 19,336 feet.

Source: Mount Kilimanjaro National Park

EXPLORATION

26. Tiger Woods drives golf balls 300 yards before they hit the ground. Suppose one of his drives is 80 feet high at its peak, and that the path of the ball is a parabola.
    a. With a suitable placement of coordinates, find an equation for this parabola.
    b. How far from the tee (where the drive begins) is the ball 50 feet up in the air?

QY ANSWER

$(-15, 8)$

Lesson 12-2
Completing the Square

▶ **BIG IDEA** Completing the square is a process that converts an equation for a parabola from standard form into vertex form.

Vocabulary

complete the square

Mental Math

A rocket's height h in feet t seconds after it is launched is shown below.

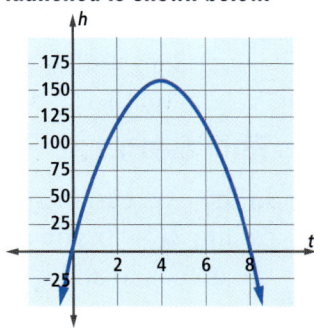

a. How long is the rocket in the air?

b. Estimate the greatest height it reaches.

c. When does it reach this greatest height?

You have now seen two forms of equations whose graphs are parabolas.

$$\text{Standard form } y = ax^2 + bx + c$$

$$\text{Vertex form } y - k = a(x - h)^2$$

From the vertex form you can read the vertex of the parabola and also the maximum or minimum possible value of y. For example, from this form, you could tell the highest point that a baseball or a rocket reaches if you have an equation for its path.

But equations for paths are usually found in standard form $y = ax^2 + bx + c$. So the goal of this lesson is for you to learn how to convert an equation in standard form to one in vertex form.

The Problem, Visually Stated

Consider the equation $y = x^2 + 6x + 14$. Visually, you can picture this quadratic expression as 1 square, 6 lengths, and 14 units as shown at the right.

We want to convert it into vertex form. The idea is to move half of the lengths to try to create a bigger square as shown at the right.

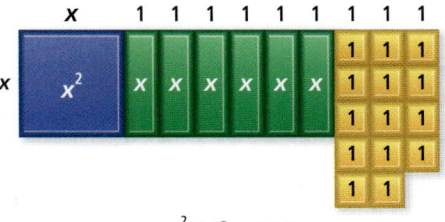

$x^2 + 6x + 14$

It will take 9 of the units to fill in the bottom right corner to complete the square. The new bigger square, pictured below, has an area $x^2 + 3x + 3x + 9$. But its length and width are each $x + 3$. So it has area $(x + 3)^2$.

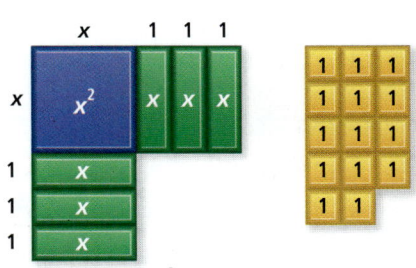

$x^2 + 3x + 3x + 14$

And because 5 units are left over, we have shown that $x^2 + 6x + 14 = (x + 3)^2 + 5$.

Now, if $y = x^2 + 6x + 14$, then $y = (x + 3)^2 + 5$, which means that $y - 5 = (x + 3)^2$.

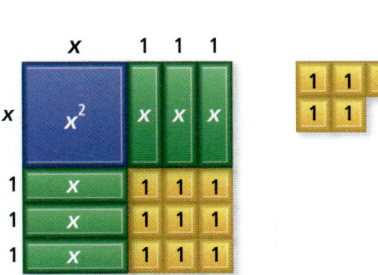

$(x + 3)^2 + 5$

Completing the Square 723

Chapter 12

> ## Activity
>
> In 1–4, an equation for a parabola is given.
>
> a. Using algebra tiles, build the given quadratic expression with a square, lengths, and units.
>
> b. Rearrange the square, lengths, and units to convert the equation to vertex form.
>
> 1. $y = x^2 + 4x + 18$
> 2. $y = x^2 + 10x + 30$
> 3. $y = x^2 + 10x + 25$
> 4. $y = x^2 + 14x + 52$

The General Process

In the expression $x^2 + 6x + 14$ on the previous page, we separated $6x$ into $3x + 3x$ and added 9 units to get the square. In general, the goal is to add a number to $x^2 + bx$ so that the right side of the equation contains a perfect square. We know, from the square of a binomial, that $(x + h)^2 = x^2 + 2hx + h^2$. Our goal is to find a number h^2 so that $x^2 + bx + h^2$ is a perfect square.

Comparing $x^2 + 2hx + h^2$ with $x^2 + bx$, we see that $b = 2h$. So $h = \frac{1}{2}b$. This means that $h^2 = \left(\frac{1}{2}b\right)^2$. And so $\left(x + \frac{1}{2}b\right)^2 = x^2 + bx + \left(\frac{1}{2}b\right)^2$.

Thus, to **complete the square** on $x^2 + bx$, add $\left(\frac{1}{2}b\right)^2$.

For example, in $x^2 + 6x$, $b = 6$ and $h = 3$. Then $h^2 = 9$.

Converting from Standard Form to Vertex Form

By completing the square, you can convert an equation in standard form to one in vertex form.

> ### Example 1
>
> a. Convert the equation $y = x^2 + 9x$ for a parabola into vertex form.
>
> b. Find the vertex of this parabola.
>
> **Solutions**
>
> a. Think of $x^2 + 9x$ as $x^2 + bx$. Then $b = 9$. So $\left(\frac{1}{2}b\right)^2 = (4.5)^2$. Thus, using the above argument, if you add 4.5^2 to $x^2 + 9x$, you will have the square of a binomial. But in an equation, you cannot add something to one side without adding it to the other.

$$y = x^2 + 9x$$
$$y + 4.5^2 = x^2 + 9x + 4.5^2 \quad \text{Add } (4.5)^2 \text{ to both sides.}$$
$$y + 4.5^2 = (x + 4.5)^2 \quad \text{square of a binomial}$$

b. From Part a, we see that the vertex is $(-4.5, -4.5^2)$, that is, $(-4.5, -20.25)$.

Check Graph the two parabolas with equations $y = x^2 + 9x$ and $y + 4.5^2 = (x + 4.5)^2$ on the same grid. The graphs are identical to the one shown at the right, and the vertex is $(-4.5, -20.25)$.

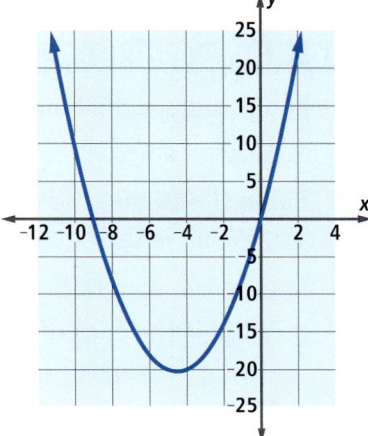

Completing the Square on $y = x^2 - bx$

Recall from Lesson 11-6 that $(x - h)^2 = x^2 - 2hx + h^2$.

Consequently, to complete the square on $x^2 - 2hx$ you add h^2. That is, to complete the square on $x^2 - bx$ you add the same amount as you do to complete the square on $x^2 + bx$.

Example 2
Without graphing, find the minimum value for y when $y = x^2 - 6x - 13$.

Solution You can find the minimum value of y if you know the vertex of the parabola that is the graph of $y = x^2 - 6x - 13$. First add 13 to both sides to isolate $x^2 - 6x$ on the right side.

$$y = x^2 - 6x - 13$$
$$y + 13 = x^2 - 6x$$

Now complete the square on $x^2 - 6x$. Here $b = -6$, so add $\left(\frac{-6}{2}\right)^2$, or 9, to both sides.

$$y + 13 + 9 = x^2 - 6x + 9$$
$$y + 22 = (x - 3)^2$$

So the vertex is $(3, -22)$.

Consequently, **the minimum value of y is -22.**

Check 1 Try values of x near the vertex and see what values of y result.

When $x = 4$, $y = 4^2 - 6 \cdot 4 - 13 = -21$.
When $x = 2$, $y = 2^2 - 6 \cdot 2 - 13 = -21$ also.

The symmetry confirms that -22 is a minimum value for y when $x = 3$, because it is less than -21.

Check 2 Graph the equation $y = x^2 - 6x - 13$. We leave that to you.

Completing the Square

Questions

COVERING THE IDEAS

In 1 and 2, square the binomial.

1. $x + 7$
2. $n - 6.5$

3. **Fill in the Blanks** To complete the square for $x^2 + 20x$, add __?__. The result is the square of the binomial __?__.

4. a. Give the sum of the areas of the three rectangles below.
   b. What is the area of the undrawn rectangle needed to complete the large square?
   c. What algebraic expression will the completed large square below picture?

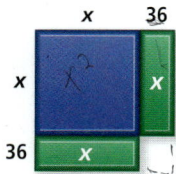

In 5–9, a quadratic expression is given.
   a. What number must be added to the expression to complete the square?
   b. After adding that number, the expression is the square of what binomial?

5. $x^2 + 2x$
6. $t^2 + 30t$
7. $r^2 - 7r$
8. $v^2 + bv$
9. $w^2 - bw$

10. a. Convert the equation $y = x^2 + 14x$ into vertex form.
    b. Find the vertex of this parabola.

11. a. Convert the equation $y = x^2 - 3x + 1$ into vertex form.
    b. Find the minimum value of y.

APPLYING THE MATHEMATICS

12. In this lesson, all the parabolas are graphs of equations of the form $y = x^2 + bx + c$. To deal with an equation of the form $y = -x^2 + bx + c$, first multiply both sides of the equation by -1, then complete the square, and finally multiply by -1 again so that y will be on the left side. Try this method to find the vertex of the parabola with equation $y = -x^2 + 5x + 2$.

13. In Lesson 9-4, the equation $h = -16t^2 + 32t + 6$ described the height h of a ball t seconds after being thrown from a height of 6 feet with an initial upward velocity of 32 feet per second. Put this equation into vertex form using the following steps.

    Step 1 Substitute y for h and x for t.

    Step 2 Divide both sides of the equation by −16 so that the coefficient of x^2 is 1.

    Step 3 Complete the square on the right side of the equation and add the appropriate amount to the left side.

    Step 4 Multiply both sides of the equation by −16 so that the coefficient of y on the left side of the equation is 1.

    a. What is the vertex of the parabola?

    b. Is this a minimum or a maximum?

Kevin Durant of the Oklahoma City Thunder shoots a free throw.

14. The equation $h = -0.12x^2 + 2x + 6$ describes the path of a basketball free throw, where h is the height of the ball in feet when the ball is x feet forward of the free-throw line.

    a. Use the steps in Question 13 to put this equation into vertex form.

    b. What is the greatest height the ball reaches?

15. If $y = x^2 - x + 1$, can y ever be negative? Explain your answer.

16. The process of completing the square can be used to solve quadratic equations. Consider the equation $y^2 - 10y + 24 = 0$.

    a. Add −24 to both sides.

    b. Complete the square on $y^2 - 10y$ and add the constant term to both sides.

    c. You now have an equation of the form $(y - 5)^2 = k$. What is k?

    d. Solve the equation in Part c by taking the square roots of both sides.

17. Use the process described in Question 16 to solve $x^2 + 24x + 7 = 0$.

REVIEW

18. Consider the parabola with quadratic equation $y + 8 = 3(x + 2)^2$. (Lesson 12-1)

    a. Find the vertex of the parabola.

    b. Graph the parabola.

19. Two parents of blood type AB will produce children of three possible blood types: A, B, and AB. One inheritance hypothesis argues that when parents of blood type AB produce children, 25% will have blood type A, 25% will have blood type B, and 50% will have blood type AB. Consider the table below that gives the blood types of 248 children born of 100 couples with both parents of blood type AB. Use a chi-square test to determine whether the data support the hypothesis. Justify your reasoning. (**Lesson 11-8**)

Blood Type	Number of Children
A	58
B	51
AB	139

20. a. How many solutions does the system $\begin{cases} y = |x| \\ y = 2 \end{cases}$ have?

    b. Find the solutions. (**Lesson 10-1**)

In 21 and 22, solve. (**Lessons 9-2, 5-2**)

21. $\dfrac{4}{x} = \dfrac{8}{15}$

22. $\dfrac{m}{7} = \dfrac{20}{m}$

23. A watch company increases the price of its watches by 8%. If their watch now sells for $130.50, what did it sell for before the increase? (**Lesson 4-1**)

24. Solve $6(3x^2 - 3x) - 9(2x^2 + 1) = 12$. (**Lessons 3-4, 2-1**)

EXPLORATION

25. Explain how the drawing below can be used to show $(x - b)^2 = x^2 - 2bx + b^2$.

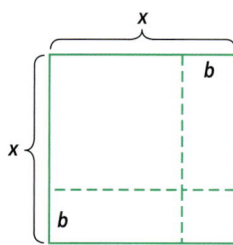

Lesson 12-3

The Factored Form of a Quadratic Function

Vocabulary

factored form (of a quadratic function)

▶ **BIG IDEA** The graph of the equation $y = a(x - r_1)(x - r_2)$ is a parabola that intersects the x-axis at $(r_1, 0)$ and $(r_2, 0)$.

You have seen two forms of equations for a quadratic function: standard form and vertex form. In this lesson, you will see some advantages of a third form called *factored form*. Below are graphs of three equations: $y + 4 = (x - 3)^2$, $y = (x - 1)(x - 5)$, and $y = x^2 - 6x + 5$.

Mental Math

Using one fair, 6-sided die, what is the probability of rolling

a. a 3?

b. an even number?

c. a number less than 3?

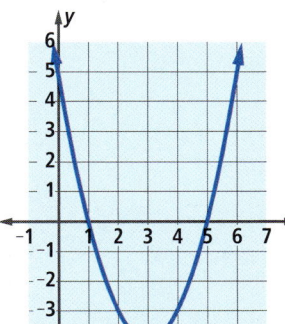
$y + 4 = (x - 3)^2$
Vertex form

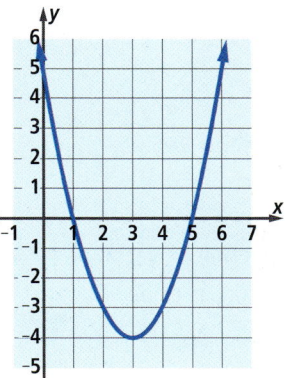
$y = (x - 1)(x - 5)$
Factored form

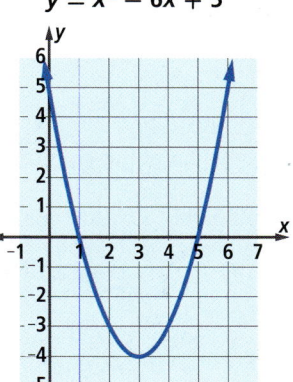
$y = x^2 - 6x + 5$
Standard form

They are in fact the same parabola described in three different ways. You can check this by converting the first two equations into standard form.

$y + 4 = (x - 3)^2$
$y + 4 = (x - 3)(x - 3)$
$y + 4 = x^2 - 6x + 9$
$y = x^2 - 6x + 5$

$y = (x - 1)(x - 5)$
$y = x^2 - 1x - 5x + 5$
$y = x^2 - 6x + 5$

Different key aspects of the graph are revealed by each form. From the vertex form, you can easily determine the vertex, (3, −4). From the factored form, you can easily determine the *x*-intercepts, 1 and 5. In standard form, the *y*-intercept is clearly 5.

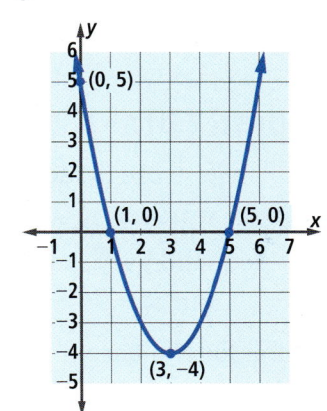

The Factored Form of a Quadratic Function 729

Chapter 12

Activity 1

Use a dynamic graphing system.

Step 1 Create two sliders with values between −6 and 6. Label one r_1 and the other r_2.

Step 2 Slide bars so $r_1 = 1$ and $r_2 = 4$.

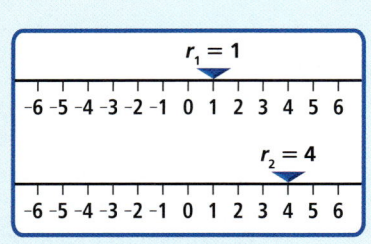

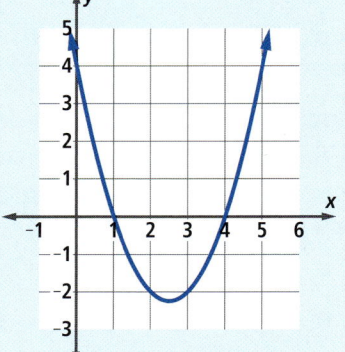

$f(x) = (x - r_1)(x - r_2)$
$f(x) = (x - 1)(x - 4)$

Step 3 Graph the function $f(x) = (x - r_1)(x - r_2)$.

Step 4 Give the points of intersection of the graph of f and the x-axis.

Step 5 Move the sliders to complete the table below.

r_1	r_2	$f(x) = (x - r_1)(x - r_2)$	Points of intersection of graph and x-axis
5	2	$f(x) = (x - 5)(x - 2)$	(?, 0) and (?, 0)
−4	−3	?	?
0	−1	?	?
3	3	?	?
?	?	$f(x) = (x + 2)(x - 4)$	?
?	?	$f(x) = (x + 5)(x + 5)$	?

Step 6 Explain how the factored form of a quadratic in the third column reveals the x-intercepts of the graph of that quadratic.

Step 7 Give the x-intercepts of the following functions using their graphs.

a. $f(x) = (x - 3)(x + 1)$ **b.** $g(x) = x(x + 6)$ **c.** $h(x) = (x - 2)(x - 2)$

How the Factored Form Displays the x-Intercepts

The equation $y = ax^2 + bx + c$ is in **factored form** when it is written as $y = a(x - r_1)(x - r_2)$.

For the function with equation $y = (x - 1)(x - 4)$ graphed in Activity 1, $a = 1$, $r_1 = 1$, and $r_2 = 4$.

The x-intercepts of the function are the values of x for which $y = 0$. So they are the values of x that satisfy the equation $0 = (x - 1)(x - 4)$.

730   More Work With Quadratics

Recall the Zero Product Property from Lesson 2-8: When the product of two numbers is zero, at least one of the numbers must be 0. In symbols, if $ab = 0$, then $a = 0$ or $b = 0$. Consequently, $y = 0$ when either $x - 1 = 0$ or $x - 4 = 0$. So $y = 0$ when either $x = 1$ or $x = 4$.

In general, the x-intercepts of a parabola can be determined from factored form in the same way that the vertex can be determined from vertex form.

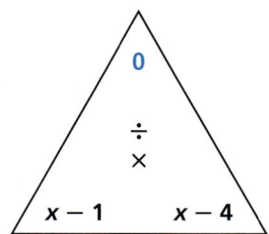

Factor Theorem for Quadratic Functions
The x-intercepts of the graph of $y = a(x - r_1)(x - r_2)$ are r_1 and r_2.

Example 1
Consider the equation $y = (x + 4)(x - 2)$.
a. Find the x-intercepts of its graph.
b. Graph the equation.

Solutions

a. The x-intercepts occur when $y = 0$. So solve $(x + 4)(x - 2) = 0$. By the Zero Product Property, either $x + 4 = 0$ or $x - 2 = 0$, so either $x = -4$ or $x = 2$. So the x-intercepts are -4 and 2.

b. Recall that the x-coordinate of the vertex is the mean of the x-intercepts -4 and 2. So the vertex has x-coordinate -1. When $x = -1$, $y = (-1 + 4)(-1 - 2) = -9$. So the vertex is $(-1, -9)$. With this information, you can sketch a graph.

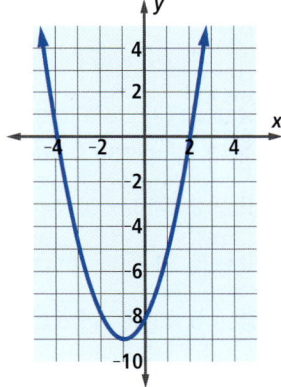

In the equation $y = (x + 4)(x - 2)$, the value of a, the coefficient of x^2, is 1. If the factors $x + 4$ and $x - 2$ remain the same but the value of a is changed, notice the similarities and changes in the graphs.

$y = (x + 4)(x - 2)$

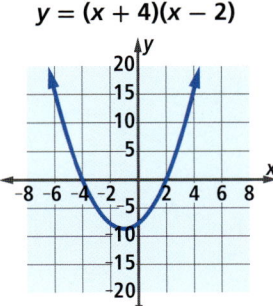

x-intercepts: −4 and 2
vertex: (−1, −9)

$y = 2(x + 4)(x - 2)$

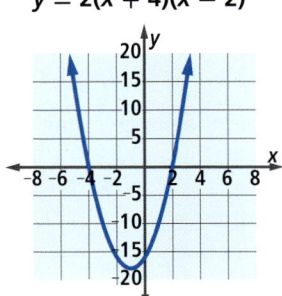

x-intercepts: −4 and 2
vertex: (−1, −18)

$y = -(x + 4)(x - 2)$

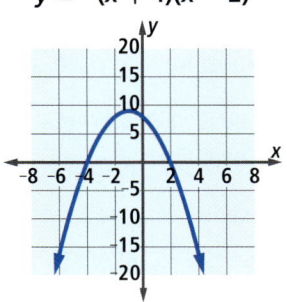

x-intercepts: −4 and 2
vertex: (−1, 9)

$y = -3(x + 4)(x - 2)$

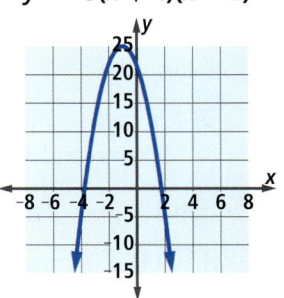

x-intercepts: −4 and 2
vertex: (−1, 27)

The Factored Form of a Quadratic Function

To see them better, all four equations can be placed on the same set of axes.

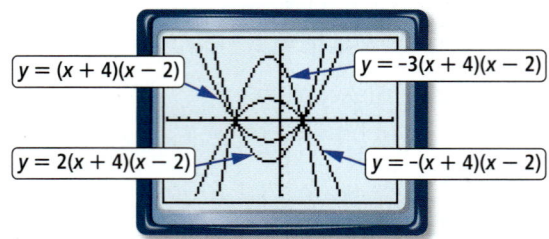

All four graphs have the same pair of x-intercepts, –4 and 2, so each goes through the points (–4, 0) and (2, 0).

Activity 2

Use a dynamic graphing system. You can use the previous Activity's set-up for this Activity.

Step 1 Create two sliders with values between –6 and 6. Label one r_1 and the other r_2.

Step 2 Create a third slider with values between –6 and 6 and label it a.

Step 3 Slide bars so $r_1 = 1$, $r_2 = 4$, and $a = 1$.

Step 4 Plot the function $f(x) = a \cdot (x - r_1)(x - r_2)$.

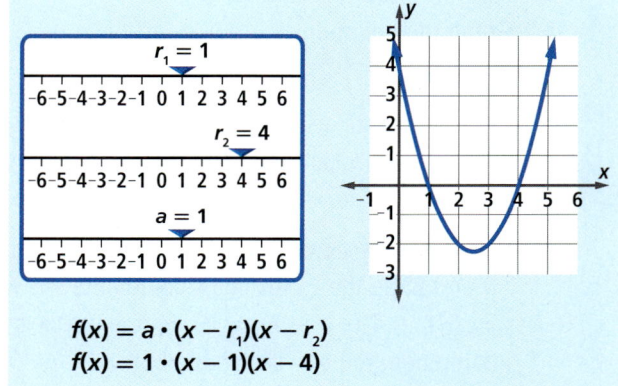

Step 5 Slide a. Do the x-intercepts change?

Step 6 Make $r_1 = -3$ and $r_2 = -3$. Slide a. Do the x-intercepts change?

Step 7 Move r_1 and r_2 to other values and then slide a. Explain your observations about the relationship between the value of a and the x-intercepts.

Step 8 Slide a into the positive values. What is true about the shape of the parabola when a is positive?

Step 9 Slide a into the negative values. What is true about the shape of the parabola when a is negative?

Step 10 Slide a to zero. Describe what happens to the graph.

Step 11 Complete the table. Verify your results by graphing.

a	r_1	r_2	$f(x) = a(x - r_1)(x - r_2)$	Does the parabola open up or down?
2	2	–3	$f(x) = 2(x - 2)(x + 3)$	?
–1	–4.1	5	?	?
5	–6	–6	?	up
–3	–5	0	?	?

732   More Work With Quadratics

Lesson 12-3

When a quadratic expression is in factored form and equal to 0, you can solve equations and find x-intercepts quite easily. You can also determine vertices and maximum and minimum values of the expression.

Example 2
a. Find the x-intercepts of the graph of $y = (3x - 5)(2x + 1)$.
b. Find the vertex of the parabola.

Solutions

a. Solve $0 = (3x - 5)(2x + 1)$. Use the Zero Product Property.

Either $3x - 5 = 0$ or $2x + 1 = 0$.

$$3x = 5 \quad \text{or} \quad 2x = -1$$
$$x = \frac{5}{3} \quad \text{or} \quad x = -\frac{1}{2}$$

Thus the x-intercepts are $\frac{5}{3}$ and $-\frac{1}{2}$.

b. The x-coordinate of the vertex is the mean of the x-intercepts.

$$\frac{\frac{5}{3} + \frac{-1}{2}}{2} = \frac{\frac{10}{6} - \frac{3}{6}}{2} = \frac{\frac{7}{6}}{2} = \frac{7}{12}$$

When $x = \frac{7}{12}$, $y = \left(3 \cdot \frac{7}{12} - 5\right)\left(2 \cdot \frac{7}{12} + 1\right) = \left(\frac{21}{12} - \frac{60}{12}\right)\left(\frac{7}{6} + \frac{6}{6}\right)$

$$= -\frac{39}{12} \cdot \frac{13}{6} = -\frac{169}{24}.$$

So the vertex of the parabola is $\left(\frac{7}{12}, -\frac{169}{24}\right) = \left(\frac{7}{12}, -7\frac{1}{24}\right)$.

Questions

COVERING THE IDEAS

1. Give the x-intercepts of the graph of $y = 3(x - 8)(x + 4)$.

2. If the product of two numbers is zero, what must be true of at least one of those numbers?

In 3–5, solve the equation.

3. $0 = -5(x - 32)(x + 89.326)$
4. $777(n + 198)(2n - 10) = 0$
5. $p(p + 19) = 0$

6. Consider the equations $y = 3(x - 20)(x + 80)$ and $y = -2(x - 20)(x + 80)$.
   a. What two points do the graphs of these equations have in common?
   b. What is the x-coordinate of the vertex of both graphs?
   c. What is the y-coordinate of the vertex for each graph?

The Factored Form of a Quadratic Function

In 7–10, an equation for a function is given.
   a. Find the x-intercepts of the graph of the function.
   b. Find the vertex of the graph of the function.
   c. Sketch a graph of the function.
   d. Check your work by writing the equation in standard form and graphing that equation.

7. $y = (x + 15)(x + 7)$

8. $y = -3(x + 8)(2x - 9)$

9. $f(x) = -x(4x + 11)$

10. $g(x) = (x - 3)^2$

11. A quadratic function is graphed at the right.
   a. Give an equation for the axis of symmetry of the parabola.
   b. Give 3 possible equations in factored form for the graph.

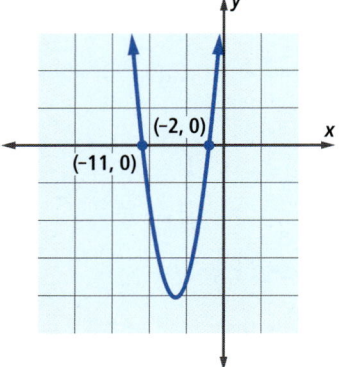

APPLYING THE MATHEMATICS

12. Down in a canyon there is a cannon that shoots cantaloupes straight up. Candice is standing on a cliff above the cannon. The cliff is at ground level, or a height of 0 feet, so that the cantaloupes are fired from a negative starting height. The cantaloupe is shot up into the air higher than the cliff (on which Candice is standing) and then comes back down past the cliff back into the canyon. The cantaloupe passes by Candice 1 second after it is fired on the way up and 2 seconds after it was fired on the way down.
   a. What part of the situation represents the x-intercepts (or where the cantaloupe has a height of 0 feet)?
   b. In projectile problems where the units are in feet and seconds, $a = -16$. Write an equation for the situation in factored form.
   c. Give the axis of symmetry for this graph.
   d. Give the coordinates of the vertex of the graph.
   e. What does the vertex represent in the scenario about the cantaloupe?

13. The vertex of a parabola is (−2, −18) and one of the x-intercepts is 1.
   a. Give the other x-intercept.
   b. Write an equation for the parabola in factored, vertex, and standard forms.

14. A formula that describes how many diagonals d that can be drawn in a polygon with n sides is $d = \frac{1}{2}n(n-3)$.

Number of sides	3	4	5	6	...
Number of diagonals	0	2	5	9	...

   a. What are the n-intercepts of the formula's graph?
   b. Why does the point $(2, -1)$ not make sense in this situation?
   c. The graph of the formula is part of a parabola. Find its vertex.

REVIEW

15. Consider the equation $y = -x^2 + 10x - 20$. (Lessons 12-2, 12-1)
    a. Rewrite the equation in vertex form.
    b. Give the vertex of the parabola.
    c. Graph the parabola.

In 16 and 17, multiply the expression. (Lessons 11-6, 11-5)

16. $(4a - 1)(3a + 6)$ 17. $(5n + 8)(5n - 8)$

18. If the cost of 15 pads of paper is $12.30, how many pads can be purchased with $3.75? (Lesson 5-5)

19. Give the coordinates of the point of intersection of the two lines. (Lesson 4-2)
    a. $x = 2, y = -4$ b. $x = a, y = 0$ c. $x = r, y = s$

20. A class of 34 students contains 2.5% of all the students in the school. How many students are in the school? (Lesson 4-1)

EXPLORATION

21. Consider the equation $y_1 = (x - 5)(x - 2)(x + 1)$.
    a. Graph this equation using a graphing calculator.
    b. Identify the x-intercepts of the graph.
    c. Use the results of Parts a and b to graph $y_2 = -(x - 5)(x - 2)(x + 1)$ without a graphing calculator.
    d. Use the results of Parts a and b to graph $y_3 = 3(x - 5)(x - 2)(x + 1)$ without a graphing calculator.
    e. Write a few sentences generalizing Parts a through d.

Chapter 12

Lesson 12-4
Factoring $x^2 + bx + c$

Vocabulary

square term
linear term
constant term
prime polynomial over the integers

> **BIG IDEA** Some quadratic trinomials of the form $x^2 + bx + c$ can be factored into two linear factors.

In Lesson 12-3 you saw the advantage of the factored form $y = a(x - r_1)(x - r_2)$ in finding the x-intercepts r_1 and r_2 of the graph of a quadratic function. In this lesson you will see how to convert quadratic expressions from the form $x^2 + bx + c$ into factored form.

Notice the pattern that results from the multiplication of the binomials of the forms $(x + p)$ and $(x + q)$. After combining like terms, the product is a trinomial.

Mental Math

Use the discriminant to determine the number of real solutions to

a. $-5y^2 + 6y + 7 = 0$.
b. $3h^2 + 10 - h = 0$.
c. $-9x^2 - 12x - 4 = 0$.

	square term	linear term	constant term
$(x + 4)(x + 3) = x^2 + 3x + 4x + 12 =$	$x^2 +$	$7x$	$+ 12$
$(x - 6)(x + 8) = x^2 + 8x - 6x - 48 =$	$x^2 +$	$2x$	$- 48$
$(x + p)(x + q) = x^2 + qx + px + pq =$	$x^2 +$	$(p + q)x$	$+ pq$

Examine the trinomials above. Their constant term pq is the product of the constant terms of the binomials. The coefficient $p + q$ of the linear term is the sum of the constant terms of the binomials. This pattern suggests a way to factor trinomials in which the coefficient of the square term is 1.

Example 1

Factor $x^2 + 11x + 18$.

Solution To factor, you need to identify two binomials, $(x + p)$ and $(x + q)$, whose product equals $x^2 + 11x + 18$. You must find p and q, two numbers whose product is 18 and whose sum is 11. Because the product is positive and the sum is positive, both p and q are positive. List the positive pairs of numbers whose product is 18. Then calculate their sums.

Product is 18	Sum of Factors
1, 18	19
2, 9	11
3, 6	9

The sum of the numbers 2 and 9 is 11. So $p = 2$ and $q = 9$.

Thus, $x^2 + 11x + 18 = (x + 2)(x + 9)$.

Check Factoring can always be checked by multiplication.
$(x + 2)(x + 9) = x^2 + 9x + 2x + 18 = x^2 + 11x + 18$; it checks.

736 More Work with Quadratics

Lesson 12-4

GUIDED

Example 2
Factor $x^2 - x - 30$.

Solution Think of this trinomial as $x^2 + -1x + -30$. You need two numbers whose product is –30 and whose sum is –1. Since the product is negative, one of the factors is negative. List the possibilities.

Product is -30	Sum of Factors
-1, 30	29
-2, ?	?
-3, ?	?
-5, ?	?
-6, ?	?
-10, ?	?
?, ?	?
?, ?	?

The only pair of factors of –30 whose sum is –1 is __?__ and __?__.

So $x^2 - x - 30 = (x - \underline{\;?\;})(x + \underline{\;?\;})$.

Check 1 Multiply $(x - \underline{\;?\;})(x + \underline{\;?\;}) = x^2 + \underline{\;?\;}x - \underline{\;?\;}x - 30 = x^2 - x - 30$. It checks.

Check 2 Graph $y = x^2 - x - 30$ and $y = (x - \underline{\;?\;})(x + \underline{\;?\;})$. The graphs should be identical. Below we show the output from a graphing calculator, with the window $-10 \leq x \leq 10, -35 \leq y \leq 35$.

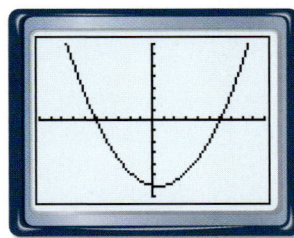

$y = x^2 - x - 30$

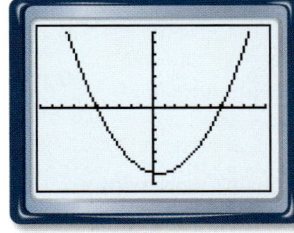

$y = (x - 6)(x + 5)$

The graphs appear to be identical.

As you know, some trinomials are perfect squares. You can use the method from Examples 1 and 2 to solve perfect square trinomials.

Factoring $x^2 + bx + c$

Chapter 12

Example 3

Factor $t^2 - 8t + 16$.

Solution Find factors of 16 whose sum is –8. Because the product is positive and the sum is negative, both numbers are negative. You need only to consider negative factors of 16.

Product is 16	Sum of Factors
-1, -16	-17
-2, -8	-10
-4, -4	-8

So $t^2 - 8t + 16 = (t - 4)(t - 4) = (t - 4)^2$.

Check Use a CAS to factor $t^2 - 8t + 16$.

Not all trinomials of the form $x^2 + bx + c$ can be factored into polynomials with integer coefficients. For example, to factor $t^2 - 12t + 16$ as two binomials $(t + p)(t + q)$, where p and q are integers, the product of p and q would have to be 16 and their sum would have to be –12. The table in Example 3 shows that there are no such pairs of numbers. We say that $t^2 - 12t + 16$ is *prime over the integers*. A **prime polynomial over the integers** is one that cannot be factored into factors of lower degree with integer coefficients.

GUIDED

Example 4

Factor $m^2 + 5m - 24$.

Solution

1. Think of factors of __?__ whose sum is __?__.
2. Because the product is negative, how many of the factors are negative?
3. $m^2 + 5m - 24 = ($__?__$)($__?__$)$

Check Check your solution by graphing $y = x^2 - 5x - 24$ and $y = ($__?__$)($__?__$)$.

Activity

Use a CAS and the FACTOR command to complete this Activity. Work with a partner, a team, or your entire class.

The entries in the table on the next page are quadratic expressions of the form $x^2 + bx + c$. We want to factor them into polynomials with integer coefficients.

738   More Work with Quadratics

Lesson 12-4

b	c = 1	c = 2	c = 3	c = 4	c = 5	c = 6	c = 7	c = 8	c = 9	c = 10
1	$x^2 + x + 1$	$x^2 + x + 2$	$x^2 + x + 3$	$x^2 + x + 4$	$x^2 + x + 5$	$x^2 + x + 6$	$x^2 + x + 7$	$x^2 + x + 8$	$x^2 + x + 9$	$x^2 + x + 10$
2	$x^2 + 2x + 1$	$x^2 + 2x + 2$	$x^2 + 2x + 3$	$x^2 + 2x + 4$	$x^2 + 2x + 5$	$x^2 + 2x + 6$	$x^2 + 2x + 7$	$x^2 + 2x + 8$	$x^2 + 2x + 9$	$x^2 + 2x + 10$
3	$x^2 + 3x + 1$	$x^2 + 3x + 2$	$x^2 + 3x + 3$	$x^2 + 3x + 4$	$x^2 + 3x + 5$	$x^2 + 3x + 6$	$x^2 + 3x + 7$	$x^2 + 3x + 8$	$x^2 + 3x + 9$	$x^2 + 3x + 10$
4	$x^2 + 4x + 1$	$x^2 + 4x + 2$	$x^2 + 4x + 3$	$x^2 + 4x + 4$	$x^2 + 4x + 5$	$x^2 + 4x + 6$	$x^2 + 4x + 7$	$x^2 + 4x + 8$	$x^2 + 4x + 9$	$x^2 + 4x + 10$
5	$x^2 + 5x + 1$	$x^2 + 5x + 2$	$x^2 + 5x + 3$	$x^2 + 5x + 4$	$x^2 + 5x + 5$	$x^2 + 5x + 6$	$x^2 + 5x + 7$	$x^2 + 5x + 8$	$x^2 + 5x + 9$	$x^2 + 5x + 10$
6	$x^2 + 6x + 1$	$x^2 + 6x + 2$	$x^2 + 6x + 3$	$x^2 + 6x + 4$	$x^2 + 6x + 5$	$x^2 + 6x + 6$	$x^2 + 6x + 7$	$x^2 + 6x + 8$	$x^2 + 6x + 9$	$x^2 + 6x + 10$
7	$x^2 + 7x + 1$	$x^2 + 7x + 2$	$x^2 + 7x + 3$	$x^2 + 7x + 4$	$x^2 + 7x + 5$	$x^2 + 7x + 6$	$x^2 + 7x + 7$	$x^2 + 7x + 8$	$x^2 + 7x + 9$	$x^2 + 7x + 10$
8	$x^2 + 8x + 1$	$x^2 + 8x + 2$	$x^2 + 8x + 3$	$x^2 + 8x + 4$	$x^2 + 8x + 5$	$x^2 + 8x + 6$	$x^2 + 8x + 7$	$x^2 + 8x + 8$	$x^2 + 8x + 9$	$x^2 + 8x + 10$
9	$x^2 + 9x + 1$	$x^2 + 9x + 2$	$x^2 + 9x + 3$	$x^2 + 9x + 4$	$x^2 + 9x + 5$	$x^2 + 9x + 6$	$x^2 + 9x + 7$	$x^2 + 9x + 8$	$x^2 + 9x + 9$	$x^2 + 9x + 10$
10	$x^2 + 10x + 1$	$x^2 + 10x + 2$	$x^2 + 10x + 3$	$x^2 + 10x + 4$	$x^2 + 10x + 5$	$x^2 + 10x + 6$	$x^2 + 10x + 7$	$x^2 + 10x + 8$	$x^2 + 10x + 9$	$x^2 + 10x + 10$

Step 1 Make a table like the one above with the same row and column headings, but keep the other cells blank.

Step 2 Factor each of the 100 entries in the table. Put the factored form in your table. If the quadratic cannot be factored over the integers, write "P," for prime, in the box. The expressions $x^2 + x + 1$ and $x^2 + 2x + 1$ have been done for you below.

b	c = 1
1	P
2	$(x + 1)(x + 1)$

Step 3 In your table, circle the factored expressions.

Step 4 For which values of c is there only one factorization of $x^2 + bx + c$ in that column? What type of numbers are these c values?

Step 5 When $c = 6$, there are two factorizations of $x^2 + bx + 6$. The factorizations occur when $b = 5$ and $b = 7$.

$$x^2 + 5x + 6 = (x + 2)(x + 3)$$

How are the 2 and 3 related to the 6?
How are the 2 and 3 related to the 5?

$$x^2 + 7x + 6 = (x + 1)(x + 6)$$

How are the 1 and 6 related to the 6?
How are the 1 and 6 related to the 7?

Step 6 If $x^2 + bx + c$ factors into $(x + p)(x + q)$ then $p + q =$ __?__ and $pq =$ __?__.

Step 7 For how many integer values of b is the expression $x^2 + bx + 20$ factorable? Explain. Give the values of b that allow $x^2 + bx + 20$ to be factored.

For how many integer values of b is the expression $x^2 + bx + 37$ factorable? Explain. Give the values of b that allow $x^2 + bx + 37$ to be factored.

Chapter 12

Questions

COVERING THE IDEAS

1. a. In order to factor $x^2 + 10x + 24$, list the possible integer factors of the last term and their sums.
   b. Factor $x^2 + 10x + 24$.
   c. Check your work.

2. Suppose $(x + p)(x + q) = x^2 + bx + c$.
   b. What must pq equal?
   c. What must $p + q$ equal?

3. Sandra, Steve, and Simona each attempted to factor $n^2 + 2n + 48$. Which student's factorization is correct? Explain what mistake the other two students made.

Sandra's	Steve's	Simona's
$(n + 6)(n - 8)$	$(n - 6)(n - 8)$	$(n - 6)(n + 8)$

In 4–9, write the trinomial as the product of two binomials.

4. $x^2 + 22x + 40$
5. $q^2 + 20q + 19$
6. $z^2 - z - 56$
7. $v^2 - 102v + 101$
8. $r^2 + 10r + 16$
9. $m^2 + 17m - 38$

10. Explain why the trinomial $x^2 + 6x + 4$ cannot be factored over the integers.

11. a. Factor $x^2 - 8x + 15$.
    b. What are the y-intercepts of the graph of $y = x^2 - 8x + 15$?

APPLYING THE MATHEMATICS

12. The diagram at the right uses tiles to the factorization of $x^2 + 4x + 3 = (x + 1)(x + 3)$.

    Make a drawing to show the factorization of $x^2 + 7x + 10$.

    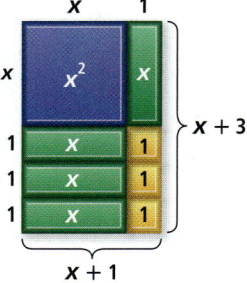

13. a. Find an equation of the form $y = (x + p)(x + q)$ whose graph is identical to the graph of $y = x^2 - 18x + 32$.
    b. Check your work by graphing both equations on the same set of axes.

14. a. Find the vertex of the parabola with equation $y = x^2 + 2x - 35$ by factoring to find the x-intercepts.
    b. Check Part a by completing the square to put the equation in vertex form.

15. Factor $-40 + 13x - x^2$.

16. If $m^2 + 13mn + 22n^2 = (m + pn)(m + qn)$, what are p and q?

740 More Work with Quadratics

REVIEW

17. Solve and check $(n - 10)\left(\frac{1}{2}n + 6\right) = 0$. (**Lesson 12-3**)

In **18** and **19**, find the value of c that makes each trinomial a perfect square. (**Lesson 12-2**)

18. $x^2 - 14x + c$

19. $x^2 + 9x + c$

In **20** and **21**, expand the expression. (**Lesson 11-6**)

20. $(4 - x)(4 + x)$

21. $(5a - 3)(5a + 3)$

22. Explain how $37^2 - 35^2$ can be calculated in your head. (**Lesson 11-6**)

In **23** and **24**, simplify the expression. (**Lessons 11-5, 11-4**)

23. $\dfrac{9z^3 + 10z}{z}$

24. $(n^2 + m^2) - (n - m)^2$

25. Factor $28b^4 + 8b^2 + 40$ completely. (**Lesson 11-4**)

26. A certain type of glass allows 85% of the light hitting it to pass through 1 centimeter of glass. The fraction y of light passing through x centimeters of glass is then $y = (0.85)^x$. (**Lesson 7-3**)
    a. Draw a graph of this equation for $0 \leq x \leq 10$.
    b. Use the graph to estimate the thickness of this glass you would need to allow only a quarter of the light hitting it to pass through.

In **27** and **28**, give the slope and y-intercept for each line. (**Lessons 6-8, 6-4**)

27. $y = \frac{1}{4}x$

28. $12x - 3y = 30$

Light is shining through tinted glass.

EXPLORATION

29. Using a CAS, make a table like that in the Activity but with integer values of c from -1 to -10. (Row 1 of the table is $x^2 + x - 1$, $x^2 + x - 2$, and so on.)
    a. Repeat Steps 1–3 from the Activity.
    b. How many of these 100 quadratic expressions can be factored over the integers?
    c. Describe a pattern in the table that could enable you to extend the table to more factors without doing any calculations.

Chapter 12

Lesson 12-5

Factoring $ax^2 + bx + c$

▶ **BIG IDEA** Some quadratic trinomials of the form $ax^2 + bx + c$ can be factored into two linear factors.

You have seen that some trinomials of the form $x^2 + bx + c$ can be factored into a product of two binomials.

$$x^2 + 0x - 100 = (x + 10)(x - 10)$$
$$x^2 + 12x + 36 = (x + 6)(x + 6)$$
$$x^2 - 9x + 14 = (x - 7)(x - 2)$$
$$x^2 + 7x - 8 = (x - 1)(x + 8)$$

In this lesson, we consider quadratic trinomials in which the coefficient of the square term is not 1. Again, we seek to factor the trinomial into binomials with integer coefficients.

In factoring such a trinomial, first check for a common factor of the three terms.

Mental Math

Find the greatest common factor of

a. $16t$ and 32.

b. $9a$, $6b$, and $10ab$.

c. x^2 and $4x^3$.

Example 1

Factor $50x^5 + 200x^4 + 200x^3$.

Solution $50x^3$ is a common factor of the three terms.

$$50x^5 + 200x^4 + 200x^3 = 50x^3(x^2 + 4x + 4)$$

To factor $x^2 + 4x + 4$, we need a binomial whose constant terms have a product of 4 and a sum of 4.

$$x^2 + 4x + 4 = (x + 2)(x + 2)$$

So, $50x^5 + 200x^4 + 200x^3 = 50x^3(x + 2)(x + 2)$.

The original polynomial is said to be factored completely.

Check

$$50x^3(x + 2)(x + 2) = (50x^4 + 100x^3)(x + 2)$$
$$= 50x^5 + 100x^4 + 100x^4 + 200x^3$$
$$= 50x^5 + 200x^4 + 200x^3$$

The factorization checks.

When the coefficient of the square term is not 1 and there is no common factor of the three terms, a different process is applied. Suppose $ax^2 + bx + c = (dx + e)(fx + g)$.

The product of *d* and *f*, from the first terms of the binomials, is *a*. The product of *e* and *g*, the constant terms of the binomials, is *c*. The task is to find *d*, *e*, *f*, and *g* so that the rest of the multiplication and addition gives *b*.

Example 2
Factor $5x^2 + 32x + 12$.

Solution

Step 1 Rewrite the expression as a product of two binomials.
$$5x^2 + 32x + 12 = (dx + e)(fx + g)$$
You need to find integers *d*, *e*, *f*, and *g*.

Step 2 The coefficient of $5x^2$ is 5, so $df = 5$. Assume either *d* or *f* is 5, and the other is 1. Now write the following.
$$5x^2 + 32x + 12 = (5x + e)(x + g)$$
The product of *e* and *g* is 12, so $eg = 12$. Because the middle term $32x$ is positive, *e* and *g* must be positive. Thus, *e* and *g* might equal 1 and 12, or 2 and 6, or 3 and 4, in either order. Try all six possibilities.

Can *e* and *g* be 1 and 12?
$$(5x + 1)(x + 12) = 5x^2 + 61x + 12$$
$$(5x + 12)(x + 1) = 5x^2 + 17x + 12$$
No, we want $b = 32$, not 61 or 17.

Can *e* and *g* be 3 and 4?
$$(5x + 3)(x + 4) = 5x^2 + 23x + 12$$
$$(5x + 4)(x + 3) = 5x^2 + 19x + 12$$
No. Again the middle term is not what we want.

Can *e* and *g* be 2 and 6?
$$(5x + 2)(x + 6) = 5x^2 + 32x + 12$$
Yes; here $b = 32$. This is what we want.
$$5x^2 + 32x + 12 = (5x + 2)(x + 6)$$

Check 1 The multiplication $(5x + 2)(x + 6) = 5x^2 + 32x + 12$ is a check.

Check 2 Graph $y = 5x^2 + 32x + 12$ and $y = (5x + 2)(x + 6)$ on the same set of axes. The graphs should be identical. It checks.

(continued on next page)

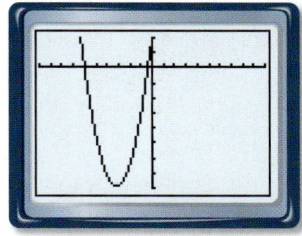

$y = 5x^2 + 32x + 12$
and
$y = (5x + 2)(x + 6)$

Check 3 Another check is to substitute a value for x, say 4.

Does $5x^2 + 32x + 12 = (5x + 2)(x + 6)$?
$5 \cdot 4^2 + 32 \cdot 4 + 12 = (5 \cdot 4 + 2)(4 + 6)$
$80 + 128 + 12 = 22 \cdot 10$

Yes. Each side equals 220.

In Example 2, there are not many possible factors because the coefficient of x^2 is 5 and all numbers are positive. Example 3 has more possibilities, but the idea is still the same. Try factors until you find the correct ones.

GUIDED

Example 3
Factor $15y^2 - 16y - 7$.

Solution First write down the form. $(ay + b)(cy + d)$. So $ac =$ __?__.
Thus either a and c are 3 and 5 or they are __?__ and __?__. The product $bd = -7$. So b and d are either __?__ and __?__, or __?__ and __?__.

List all the possible factors with $a = 3$ and $c = 5$, and multiply.

$(3y + 7)(5y - 1) =$ __?__

$(3y - 7)(5y + 1) =$ __?__

$(3y - 1)(5y + 7) =$ __?__

$(3y + 1)(5y - 7) =$ __?__

List all the possible factors with $a = 1$ and $c = 15$.

$(\underline{\ ?\ }y + 7)(\underline{\ ?\ }y - 1) =$ __?__

$(\underline{\ ?\ }y - 7)(\underline{\ ?\ }y + 1) =$ __?__

$(\underline{\ ?\ }y - 1)(\underline{\ ?\ }y + 7) =$ __?__

$(\underline{\ ?\ }y + 1)(\underline{\ ?\ }y - 7) =$ __?__

At most, you need to do these eight multiplications. If one of them gives $15y^2 - 16y - 7$, then that is the correct factoring.

So $15y^2 - 16y - 7 =$ __?__.

In Example 3, notice that each choice of factors gives a product that differs only in the coefficient of y (the middle term). If the original problem were to factor $15y^2 - 40y - 7$, this process shows that no factors with integer coefficients will work. The quadratic $15y^2 - 40y - 7$ is a prime polynomial over the set of integers.

Once a quadratic trinomial $f(x)$ has been factored, then solving $f(x) = 0$ is easy using the Zero Product Property.

Lesson 12-5

GUIDED

Example 4
Solve $15y^2 - 16y - 7 = 0$.

Solution Use the factorization of $15y^2 - 16y - 7$ in Example 3.

$15y^2 - 16y - 7 = 0$

$(\underline{\ ?\ })(\underline{\ ?\ }) = 0$

$\underline{\ ?\ } = 0 \quad \text{or} \quad \underline{\ ?\ } = 0$

$y = \underline{\ ?\ } \quad \text{or} \quad y = \underline{\ ?\ }$

Questions

COVERING THE IDEAS

1. Perform the multiplications in Parts a–d.
   a. $(2x + 3)(4x + 5)$
   b. $(2x + 5)(4x + 3)$
   c. $(2x + 1)(4x + 15)$
   d. $(2x + 15)(4x + 1)$
   e. Explain how these multiplications are related to factoring $8x^2 + 26x + 15$.

2. Suppose $ax^2 + bx + c = (dx + e)(fx + g)$ for all values of x.
   a. The product of d and f is $\underline{\ ?\ }$.
   b. The product of $\underline{\ ?\ }$ and $\underline{\ ?\ }$ is c.

3. Factor the trinomials completely.
   a. $2x^2 + 14x + 2$
   b. $5n^2 + 35n - 50$

In 4–9, factor the trinomial, if possible.

4. $5A^2 + 7A + 2$

5. $-3x^2 + 11x - 6$

6. $y^2 - 10y + 16$

7. $14w^2 - 9w - 1$

8. $-4x^2 - 11x + 3$

9. $17k^2 - 36k + 19$

10. Check the solutions to Example 1 by substitution.

11. Solve $20x^2 + 11x - 3 = 0$ by factoring.

Chapter 12

APPLYING THE MATHEMATICS

12. Jules solved the equation $2x^2 - 5x + 3 = 7$ in the following way.
    Step 1 He factored $2x^2 - 5x + 3$ into $(2x - 3)(x - 1)$.
    Step 2 He substituted the factored expression back into the equation $(2x - 3)(x - 1) = 7$.
    Step 3 He considered all the possibilities: $2x - 3 = 7$ and $x - 1 = 1$, in this case $x = 5$ or $x = 2$; or $2x - 3 = 1$ and $x - 1 = 7$, in this case $x = 2$ or $x = 8$.
    Step 4 He checked his work and found that none of these values of x check in the original equation.
    What did Jules do wrong?

13. Find the vertex of the parabola with equation $y = 8x^2 - 6x + 1$ by first factoring to obtain the x-intercepts.

14. Consider the equation $6t^2 + 7t - 24 = 0$.
    a. Solve the equation by using the Quadratic Formula.
    b. Solve the equation by factoring.
    c. Which method do you prefer to solve this problem? Why?

15. a. Solve the equation $3n^2 = 2 - 5n$ using the Quadratic Formula.
    b. Check your solution to Part a by solving the same equation using factoring.

16. a. Factor $14x^3 - 21x^2 - 98x$ into the product of a monomial and a trinomial.
    b. Give the complete factorization of $14x^3 - 21x^2 - 98x$.

In **17** and **18**, find the complete factorization.

17. $9p^2 + 30p^3 + 25p^4$
18. $-2x^2 + 23xy - 30y^2$

REVIEW

19. Rewrite $x^8 - 16$ as the product of
    a. two binomials.
    b. three binomials. (**Lessons 12-4, 11-6**)

20. Find two consecutive positive even integers whose product is 360. (**Lessons 12-4, 12-2**)

21. Find the x-intercepts of $y = x^2 - 8x + 3$ by completing the square. (**Lessons 12-3, 12-2**)

Lesson 12-5

22. **a.** Here are three instances of a pattern. Describe the general pattern using two variables, x and y. (**Lessons 11-6, 1-2**)

$$(48 + 32)(24 - 16) = 2(24^2 - 16^2)$$
$$(10 + 20)(5 - 10) = 2(5^2 - 10^2)$$
$$(4 + 1)(2 - 0.5) = 2(2^2 - 0.5^2)$$

   b. Does your general pattern hold for all real values of x and y? Justify your answer.

23. **Multiple Choice** Which equation is graphed below? (**Lesson 9-1**)

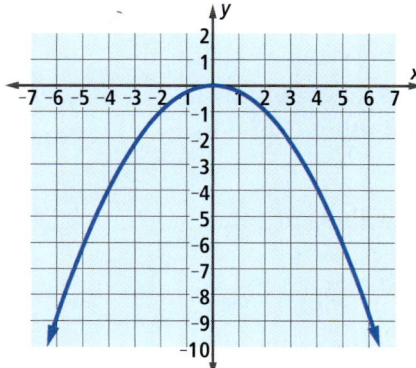

   A $y = -\frac{1}{4}x^2$
   B $y = 4x^2$
   C $y = -4x^2$
   D $y = \frac{1}{4}x^2$

24. When a fair, six-sided die is tossed, the probability of getting a 1 is $\frac{1}{6}$. If the die is tossed twice, the probability of getting a 1 both times is $\frac{1}{6} \cdot \frac{1}{6} = \left(\frac{1}{6}\right)^2$. (**Lessons 8-2, 8-1**)

   a. Write an expression to represent the probability of rolling a die m times and getting a 1 each time.
   b. Write your answer to Part a as a power of 6.

EXPLORATION

25. The polynomial $6x^3 + 47x^2 + 97x + 60$ can be factored over the integers into $(3x + a)(2x + b)(x + c)$. Find a, b, and c. (*Hint:* What is $a \cdot b \cdot c$?)

Chapter 12

Lesson 12-6

Which Quadratic Expressions Are Factorable?

> **BIG IDEA** A quadratic expression with integer coefficients is factorable over the integers if and only if its discriminant is a perfect square.

This lesson connects two topics you have seen in this chapter: factoring and solutions to quadratic equations. These topics seem quite different. Their relationship to each other is an example of how what you learn in one part of mathematics is often useful in another part.

You have seen four ways to find the real-number values of x that satisfy $ax^2 + bx + c = 0$.

1. You can graph $y = ax^2 + bx + c$ and look for its x-intercepts.
2. You can use the Quadratic Formula.
3. You can factor $ax^2 + bx + c$ and use the Zero Product Property.
4. You can set $f(x) = ax^2 + bx + c$ to 0 and look for values of x such that $f(x) = 0$.

The first two ways can always be done. But you know that it is not always possible to factor $ax^2 + bx + c$ over the integers, so it useful to know when it is possible.

Mental Math

Match each function graphed below with its type.

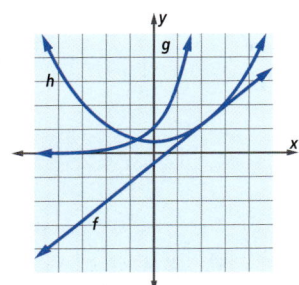

a. exponential growth

b. linear

c. quadratic

A Quadratic Equation with Rational Solutions

Consider the equation $9x^2 + 14x - 8 = 0$. Use the Quadratic Formula.

Step 1 $\quad x = \dfrac{-14 \pm \sqrt{14^2 - 4 \cdot 9 \cdot (-8)}}{2 \cdot 9}$

Step 2 $\quad = \dfrac{-14 \pm \sqrt{484}}{18}$

Step 3 $\quad = \dfrac{-14 \pm 22}{18}$

Step 4 So $x = \dfrac{-14 + 22}{18} = \dfrac{4}{9}$ or $x = \dfrac{-14 - 22}{18} = -2$

Notice that the solutions $\frac{4}{9}$ and -2 have no visible radical sign. This is because the number 484 under the square root sign is a perfect square (Step 2). So, after calculating the square root (Step 3), one integer is divided by another, and the solutions are rational numbers.

748   More Work with Quadratics

In general, for quadratic equations with integer coefficients, if the *discriminant* $b^2 - 4ac$ in the quadratic equation $ax^2 + bx + c = 0$ is a perfect square, then the solutions are rational. The square root of an integer that is not a perfect square is irrational. So the square root will remain in the solutions, and the solutions will be irrational. These results can be summarized in one sentence: *When a, b, and c are integers, the solutions to $ax^2 + bx + c = 0$ are rational numbers if and only if $b^2 - 4ac$ is a perfect square.*

Relating the Solving of $ax^2 + bx + c = 0$ to the Factoring of $ax^2 + bx + c$

Now we connect this with factoring. Consider the same equation as before: $9x^2 + 14x - 8 = 0$.

Factor the left side.

$(9x - 4)(x + 2) = 0$

Use the Zero Product Property.

$9x - 4 = 0$ or $x + 2 = 0$

$x = \frac{4}{9}$ or $x = -2$

You can see from the equation above that when a quadratic equation in standard form is factorable, the solutions are rational numbers. Combining this observation with the facts on the previous page leads us to the following conclusion, which we call the *Discriminant Theorem*. A formal proof of this theorem is given in Chapter 13.

Discriminant Theorem

When a, b, and c are integers, with $a \neq 0$, either all three of the following conditions hold, or none of these hold.

1. $b^2 - 4ac$ is a perfect square.
2. $ax^2 + bx + c$ is factorable over the set of polynomials with integer coefficients.
3. The solutions to $ax^2 + bx + c = 0$ are rational numbers.

Example 1

Is $8x - 5x^2 + 21$ factorable into polynomials with integer coefficients?

Solution First rewrite this expression in the standard form of a polynomial.

$$-5x^2 + 8x + 21$$

Thus $a = -5$, $b = 8$, and $c = 21$.

(continued on next page)

Then $b^2 - 4ac = (8)^2 - 4 \cdot (-5) \cdot 21 = 64 + 420 = 484$.

Since $484 = 22^2$, 484 is a perfect square. So the expression is factorable.

🛑 QY

▶ **QY**

Verify Example 1 by finding the factorization of $8x - 5x^2 + 21$.

GUIDED

Example 2
Is the polynomial $2x^2 - 10 + 5x$ factorable?

Solution

Step 1 Write the polynomial in standard form. __?__

Step 2 Identify a, b, and c. $a = $ __?__, $b = $ __?__, and $c = $ __?__

Step 3 Calculate $b^2 - 4ac$. __?__

Step 4 Is $b^2 - 4ac$ a perfect square? __?__

Step 5 What is your conclusion?

The phrase "with integer coefficients" is necessary in Example 1 because every quadratic expression is then factorable if noninteger coefficients are allowed.

Example 3
What can be learned by applying the Discriminant Theorem to the quadratic equation $x^2 - 29 = 0$?

Solution In this case, $a = 1$, $b = 0$, and $c = -29$, so $b^2 - 4ac = 0^2 - 4 \cdot 1 \cdot (-29) = 116$. Since 116 is not a perfect square, the solutions to $x^2 - 29 = 0$ are irrational and the polynomial $x^2 - 29$ cannot be factored into linear factors with integer coefficients.

Yet the polynomial $x^2 - 29$ in Example 3 can be factored as the difference of two squares.

$$x^2 - 29 = x^2 - (\sqrt{29})^2 = (x - \sqrt{29})(x + \sqrt{29})$$

The factors do not have integer coefficients, so we say that $x^2 - 29$ is prime over the set of polynomials with integer coefficients, but not over the set of all polynomials. It is just like factoring 7 into $3 \cdot \frac{7}{3}$. The integer 7 is prime over the integers but can be factored into rational numbers.

Lesson 12-6

Applying the Discriminant Theorem

Knowing whether an expression is factorable can help determine what methods are available to solve an equation.

> **Example 4**
> Solve $m^2 - 9m + 24 = 0$ by any method.
>
> **Solution** Because the coefficient of m^2 is 1, it is reasonable to try to factor the left side. But first evaluate $b^2 - 4ac$ to see whether this is possible.
>
> $a = 1$, $b = -9$, and $c = 24$. So, $b^2 - 4ac = (-9)^2 - 4 \cdot 1 \cdot (24) = -15$. This is not a perfect square, so the equation does not factor over the integers.
>
> In fact, because $b^2 - 4ac$ is negative, there are no real solutions to this equation.
>
> **Check** Graph $y = x^2 - 9x + 24$. You will see that the graph does not intersect the x-axis. There are no x-intercepts.

What percent of quadratic expressions are factorable? Try the following activity.

> **Activity**
>
> This activity can be done with a partner if a CAS is available, or as a whole-class activity otherwise.
>
> There are infinitely many quadratic expressions of the form $ax^2 + bx + c$, but there are only 8,000 of these in which a, b, and c are nonzero integers from –10 to 10. What percent of these are factorable? There are too many to try to factor by hand, even with a CAS. It is possible to determine this number by programming a computer to factor all of them. But it is also possible to estimate the percent by sampling.
>
> **Step 1** Set a calculator to generate random integers from −10 to 10.
>
> **Step 2** Generate three such nonzero integers. Call them a, b, and c. Record them and the expression $ax^2 + bx + c$ in a table like the one shown below.
>
> | Trial | a | b | c | $ax^2 + bx + c$ | Factorable? |
> |---|---|---|---|---|---|
> | 1 | 2 | –3 | 7 | $2x^2 - 3x + 7$ | Prime |
> | 2 | ? | ? | ? | ? | ? |
> | 3 | ? | ? | ? | ? | ? |
>
> *(continued on next page)*

Which Quadratic Expressions Are Factorable? **751**

Step 3 If you have a CAS, try to factor $ax^2 + bx + c$ over the integers. If you are working by hand, calculate $b^2 - 4ac$. If the expression is factorable, record the factors. If not, record the word *Prime*.

Step 4 Repeat Steps 2 and 3 at least 20 times. What percent of your quadratic expressions are factorable?

Step 5 Combine your results with those of others in the class. What is your class's estimate of the percent of these quadratic expressions that are factorable?

Questions

COVERING THE IDEAS

In 1–5, a quadratic expression is given. Calculate $b^2 - 4ac$ to determine if the quadratic is factorable or prime over the integers. If possible, factor the expression.

1. $x^2 - 9x - 22$
2. $36 - y^2$
3. $4n^2 - 12n + 9$
4. $-3 + m^2 + 2m$
5. $7x^2 - 13x - 6$

6. Consider the equation $ax^2 + bx + c = 0$ where a, b, and c are integers. If $b^2 - 4ac$ is a perfect square, explain why x is rational.

7. Suppose a, b, and c are integers. When will the x-intercepts of $y = ax^2 + bx + c$ be rational numbers?

8. Give an example of a quadratic expression that can be factored only if noninteger coefficients are allowed.

APPLYING THE MATHEMATICS

9. The sum of the integers from 1 to n is $\frac{1}{2}n(n + 1)$. Find n if the sum of the integers from 1 to n is 499,500.

10. What in this lesson tells you that the solutions to the quadratic equation $x^2 - 3 = 0$ are irrational? (This provides a way of showing that $\sqrt{3}$ is irrational.)

11. Find a value of k such that $4x^2 + kx - 5$ is factorable over the integers.

12. a. By multiplying, verify that $(x + 5 + \sqrt{2})(x + 5 - \sqrt{2}) = x^2 + 10x + 23$.
    b. Verify that the discriminant of the expression $x^2 + 10x + 23$ is not a perfect square.
    c. Parts a and b indicate that $x^2 + 10x + 23$ is factorable, yet its discriminant is not a perfect square. Why doesn't this situation violate the Discriminant Theorem?

REVIEW

In 13 and 14, factor the polynomial completely. (Lessons 12-5, 12-4)

13. $r^2 - 5r + 4$

14. $-2x^2 + 5x + 12$

15. Consider $w^2 + 9w + c$. (Lesson 12-2)
    a. Complete the square to find the value of c.
    b. Express the perfect square trinomial in factored form.

16. The surface area S of a cylinder with radius r and height h is given by the formula $S = 2\pi r^2 + 2\pi rh$. (Lesson 11-4)
    a. Factor the right hand side of this formula into prime factors.
    b. Calculate the exact surface area of a cylinder with a diameter of 12 cm and a height of 9 cm using either the given formula or its factored form. Which form do you think is easier for this purpose?

In 17–19, rewrite the expression with no negative exponents and each variable mentioned no more than once. (Lessons 8-5, 8-4, 8-3)

17. $\dfrac{8n^{-3}m^2}{6m^{-5}}$

18. $\left(\dfrac{a^2b}{4a^5}\right)^2$

19. $\left(\dfrac{3x^4y^{-1}}{15x^3y^0}\right)^2$

EXPLORATION

20. Look back at Question 11. Find *all* integer values of k such that $4x^2 + kx - 5$ is factorable in the set of polynomials over the integers. Explain how you know that you have found all values.

21. Use a CAS to factor the general expression $ax^2 + bx + c$. (You will likely have to indicate that x is the variable.)
    a. What factorization does the CAS give?
    b. Explain how you know that this factorization is correct.
    c. Why doesn't the existence of factors for any quadratic expression violate the Discriminant Theorem?

QY ANSWER

$-(5x + 7)(x - 3)$

Chapter 12

Lesson 12-7
Graphs of Polynomial Functions of Higher Degree

Vocabulary

cubic polynomial

▶ **BIG IDEA** The factored form of a polynomial is useful in graphing and solving equations.

Some of the ideas that you have seen in earlier lessons of this chapter extend to the graphs of polynomial functions of degrees 3 and higher. In particular, factoring a polynomial can be a powerful tool to uncover interesting features of graphs of polynomial functions.

Mental Math

A baseball player has a batting average of .245. State whether his batting average goes up or down if he plays a game where he goes

a. 1 for 4.
b. 1 for 5.
c. 2 for 6.

How Are Factors and x-Intercepts Related?

Activity 1

In 1–6, a polynomial function is given in factored form.
a. Graph the function with a graphing calculator.
b. Identify the x-intercepts of the graph.
c. How are the x-intercepts of the graph related to the factors?

1. $f(x) = 5(x-1)$
2. $g(x) = (x-1)(x+2)$
3. $h(x) = (x-1)(x+2)(x-4)$
4. $k(x) = 3(x-1)(x+2)(x-4)$
5. $m(x) = (x-1)^2(x+2)^2$
6. $q(x) = (x-1)(x+2)(x-4)(2x+15)$

7. Look back at your work. Make a conjecture about the factors of a polynomial and the x-intercepts of its graph. Test your conjecture with a different polynomial function than those above.

A graph of the function P with $P(x) = x(x-8)(x+6)$ is shown at the right, along with a table of values. Notice how the factors and the x-intercepts are related.

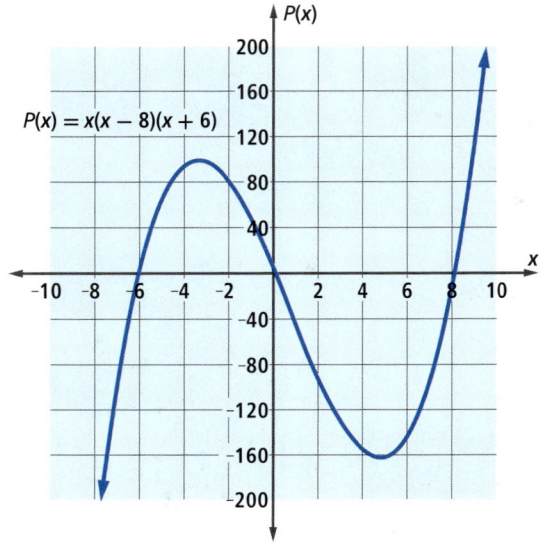

x	P(x)	x	P(x)	x	P(x)
−9	−459	−1	45	7	−13
−8	−256	0	0	8	0
−7	−105	1	−49	9	135
−6	0	2	−96	10	320
−5	65	3	−135	11	561
−4	96	4	−160	12	864
−3	99	5	−165		
−2	80	6	−144		

754 More Work with Quadratics

The graph of the function P has three x-intercepts: 0, 8, and −6. Would your conjecture from Activity 1 have predicted this result? Whether or not you predicted this, there is a simple but elegant relationship between factors and x-intercepts that holds for any polynomial function.

> **Factor Theorem**
>
> Let r be a real number and $P(x)$ be a polynomial in x.
>
> 1. If $x - r$ is a factor of $P(x)$, then $P(r) = 0$; that is, r is an x-intercept of the graph of P.
>
> 2. If $P(r) = 0$, then $x - r$ is a factor of $P(x)$.

The Factor Theorem is true because the x-intercepts of the graph of the function P are the values of x such that $P(x) = 0$. For the polynomial $P(x) = x(x - 8)(x + 6)$ graphed on the previous page, the equation $P(x) = 0$ means $x(x - 8)(x + 6) = 0$.

By the Zero Product Property, this is a true statement when
$x = 0$ or $x - 8 = 0$ or $x + 6 = 0$.
$\qquad\qquad\ \ x = 8$ or $\quad x = -6$

So, because x, $x - 8$, and $x + 6$ are factors of $P(x)$, 0, 8, and −6 are the x-intercepts of the graph of P.

GUIDED

Example 1

A polynomial function P has x-intercepts 5, 7.8, −46, and −200. What is a possible equation for the function?

Solution The polynomial must have at least four factors.

Because 5 is an x-intercept, $x - 5$ is a factor of the polynomial.

Because 7.8 is an x-intercept, ___?___ is a factor of the polynomial.

Because −46 is an x-intercept, ___?___ is a factor of the polynomial.

Because −200 is an x-intercept, ___?___ is a factor of the polynomial.

Possibly, $P(x) = (x - 5)(\_\_\_?\_\_\_)(\_\_\_?\_\_\_)(\_\_\_?\_\_\_)$.

Converting from Factored Form to Standard Form

The polynomial $P(x) = x(x - 8)(x + 6)$ is the product of three factors. To convert this polynomial to standard form, multiply any two of its factors. Then multiply the product of those factors by the third factor. Because multiplication is associative, it does not make any difference which two factors you multiply first.

$$P(x) = x(x-8)(x+6)$$
$$= (x^2 - 8x)(x+6)$$
$$= (x^2 - 8x) \cdot x + (x^2 - 8x) \cdot 6$$
$$= x^3 - 8x^2 + 6x^2 - 48x$$
$$= x^3 - 2x^2 - 48x$$

The standard form shows clearly that $P(x)$ is a polynomial of degree 3. It is a **cubic polynomial.** Just as a quadratic function has at most two x-intercepts, a cubic function has at most three x-intercepts.

The following example involves the function q from Activity 1. It is a polynomial function of 4th degree.

Example 2

Rewrite the polynomial $q(x) = (x-1)(x+2)(x-4)(2x+15)$ in standard form.

Solution There are four factors in the polynomial. Any two can be multiplied first. We multiply the first two and the last two.

$$q(x) = (x-1)(x+2)(x-4)(2x+15)$$
$$= (x^2 + x - 2)(2x^2 + 7x - 60)$$

Now use the Extended Distributive Property. Each term of the left factor must be multiplied by each term of the right factor. There are nine products.

$$= x^2(2x^2 + 7x - 60) + x(2x^2 + 7x - 60) - 2(2x^2 + 7x - 60)$$
$$= 2x^4 + 7x^3 - 60x^2 + 2x^3 + 7x^2 - 60x - 4x^2 - 14x + 120$$
$$= 2x^4 + 9x^3 - 57x^2 - 74x + 120$$

Check 1 In factored form it is easy to see that $q(1) = 0$. So substitute 1 for x in the standard form to see if the value of the polynomial is 0. The value is $2 + 9 - 57 - 74 + 120 = 0$. It checks.

Check 2 Use a CAS to factor the answer. See if the original factored form appears. Our CAS gives $q(x) = (x-1)(x+2)(x-4)(2x+15)$.

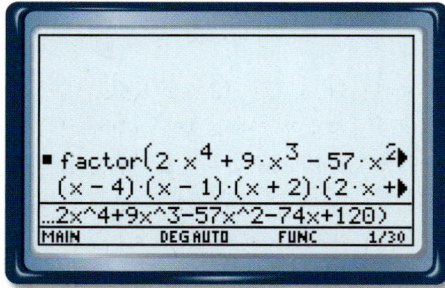

In Activity 2, you are asked to explore what happens when the same factor appears twice in a polynomial.

Lesson 12-7

Activity 2

Step 1 a. Graph the following polynomial functions on the window $-10 \leq x \leq 10$, $-500 \leq y \leq 500$. Make a sketch of each graph.

$f(x) = x(x - 3)(x + 7)$ $g(x) = x(x - 3)(x + 7)^2$
$h(x) = x(x - 3)^2(x + 7)^2$ $j(x) = x^2(x - 3)^2(x + 7)^2$

b. What are the x-intercepts of these graphs?

c. What is different about the graphs around the point $(-7, 0)$ when $(x + 7)^2$ is a factor rather than just $(x + 7)$?

Step 2 a. Multiply each polynomial by -1 and graph the resulting functions.

b. Describe what happens to the x-intercepts.

Step 3 Multiplying $P(x)$ by -1 means graphing $-P(x)$. The graph wiggles in the middle. But in all these functions, as you go farther to the right, the graph heads either up or down. As you go farther to the left, the graph also heads either up or down.

a. Copy and complete the table.

Polynomial	Far Right: Up or Down?	Far Left: Up or Down?	Polynomial	Far Right: Up or Down?	Far Left: Up or Down?
$f(x) = x(x - 3)(x + 7)$	?	?	$-f(x)$	?	?
$g(x) = x(x - 3)(x + 7)^2$	?	?	$-g(x)$	?	?
$h(x) = x(x - 3)^2(x + 7)^2$	?	?	$-h(x)$	?	?
$j(x) = x^2(x - 3)^2(x + 7)^2$	?	?	$-i(x)$	?	?

b. Describe the general pattern.

Step 4 a. Experiment to find what happens to the graph of $P(x) = ax(x - 3)(x + 7)$ as values of a change from positive to negative.

b. Does the same situation hold for the graph of $j(x) = ax^2(x - 3)^2(x + 7)^2$?

The Importance of Polynomial Functions

In this course, you have studied polynomial functions of degrees 1 and 2 in detail. A polynomial function of degree 1 has an equation of the form $y = mx + b$. Its graph is a line. A polynomial function of degree 2 has an equation of the form $y = ax^2 + bx + c$. Its graph is a parabola. The graphs of polynomial functions of degrees 3 and 4 have more varied shapes and graphs of higher degrees have still more varied shapes. This makes polynomial functions very useful in approximating data of many kinds and very useful in approximating other functions. Polynomials also appear as formulas in a number of situations, a few of which are mentioned in the Questions for this lesson.

Chapter 12

Questions

COVERING THE IDEAS

In 1–4, an equation for a function is given.
   a. Identify the x-intercepts of the graph of the function.
   b. Check your answer to Part a by graphing the function. Draw a rough sketch of the graph.
   c. Write the equation in standard form.

1. $f(x) = (x + 5)(2x - 3)$
2. $y = (x + 5.8)(2x - 3.4)$
3. $y = -3(x - 1)(2x + 1)^2$
4. $g(x) = x(8x + 5)(10x + 2)(x - 1)$

5. Give an equation for a polynomial function whose graph intersects the x-axis at $(-9, 0)$, $(4, 0)$, and nowhere else.

6. Suppose the graph of a polynomial function has three x-intercepts: 1, –4, and 5.
   a. Give an equation in factored form for the polynomial function.
   b. Rewrite your equation from Part a in standard form.

7. a. Give an equation in factored form for a polynomial function with four x-intercepts: 2, –2, 5, and –5.
   b. Rewrite your equation from Part a in standard form.

8. Match the following equations with their possible graphs.
   a. $a(x) = (x + 2)(x - 4)^2(x - 6)$
   b. $b(x) = -1(x + 2)(x - 4)(x - 6)$
   c. $c(x) = -1(x + 2)^2(x - 4)(x - 6)$
   d. $d(x) = -(x + 2)(x - 4)^2(x - 6)$

   i.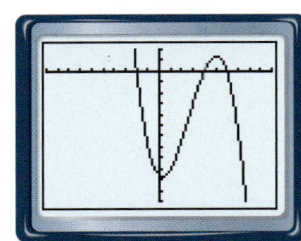
   ii.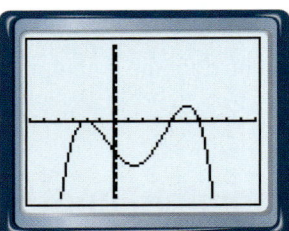
   iii.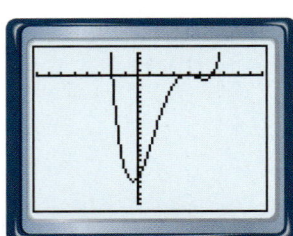
   iv.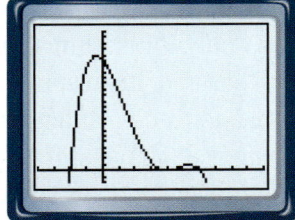

758 More Work with Quadratics

In 9–12, give a possible equation in factored form for each graph.

9.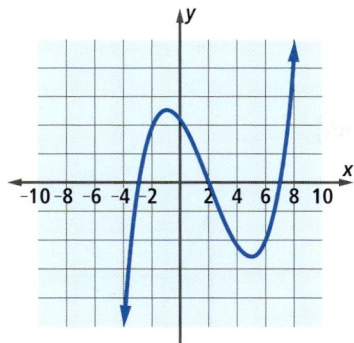

10.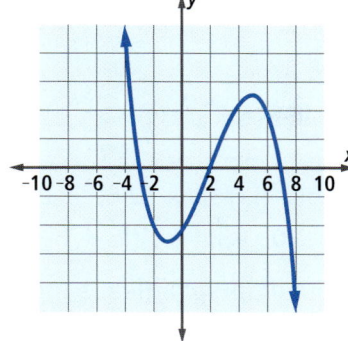

11.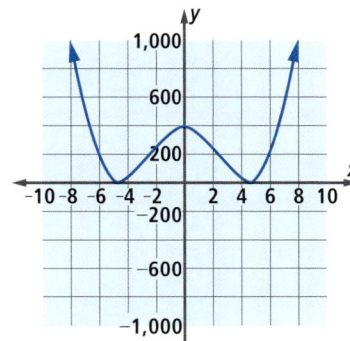

12.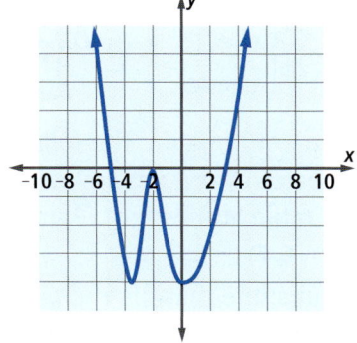

13. Let $S(n)$ = the sum of the squares of the integers from 1 to n. It is known that $S(n) = \frac{1}{6}n(n + 1)(2n + 1)$.

   a. Find $S(5)$ using the formula and verify your answer by using the definition of $S(n)$.

   b. What is the value of $S(n + 1) - S(n)$?

   c. Write the formula for $S(n)$ in standard form.

14. Explain what you know about each of the following aspects of the graph of the function $h(x) = -8(x - 11)^2(x + 5)(x + 10)^2$.

   a. x-intercepts

   b. when the graph changes direction at an x-intercept

   c. the direction of the far right side of the graph

APPLYING THE MATHEMATICS

15. a. Graph $y_1 = (x - 2)(x + 4)(x - 5)$ and $y_2 = (x - 2)^3(x + 4)(x - 5)$.

   b. Use the results of Part a to predict a characteristic of the graph of $y_3 = (x - 2)(x + 4)(x - 5)^3$.

   c. Use the results of Parts a and b to predict a characteristic of the graph of $y_4 = (x - 2)^3(x + 4)(x - 5)^3$.

   d. Generalize the results of Parts a, b, and c.

Graphs of Polynomial Functions of Higher Degree

Chapter 12

REVIEW

In 16–18, factor the polynomial. (Lessons 12-5, 11-4)

16. $9x^2y^2 - 27x^3y + 19xy$
17. $4a^2 - 16b^2$
18. $15n^2 + 1 - 8n$

In 19–22, solve by using any method. (Lessons 12-5, 12-2, 9-5)

19. $a^2 + 6a = 55$
20. $-3w^2 - 7w + 11 = 0$

21. **Multiple Choice** Which system of inequalities is graphed at the right? (Lesson 10-9)

A $\begin{cases} y \geq -2x - 8 \\ y \geq 0 \\ x \leq -9 \end{cases}$
B $\begin{cases} y \leq -2x - 8 \\ y \geq 0 \\ x \geq -9 \end{cases}$
C $\begin{cases} y \geq -2x - 8 \\ y \leq 0 \\ x \leq -9 \end{cases}$
D $\begin{cases} y \leq -2x - 8 \\ y \geq -9 \\ x \geq 0 \end{cases}$

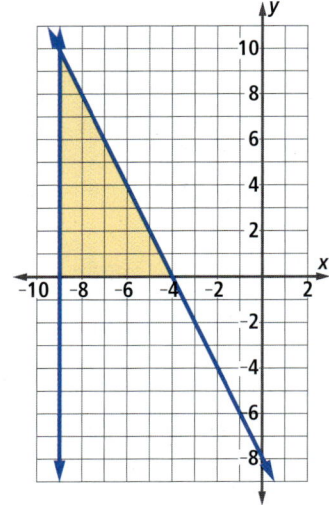

22. According to the U.S. Department of Labor, the following percentages of the adult population were employed in the particular years shown in the table at the right. Assume these trends continue. (Lesson 6-7)

   a. Use linear regression to find a line that fits the pairs (year, male). Write the equation of a line of best fit.
   b. Repeat Part a for the pairs (year, female). Write the equation of a line of best fit.
   c. Using your equation from Part b, predict what percentage of adult women will be employed in the year 2010.
   d. Using your equations from Parts a and b, predict when, if ever, the same percentage of adult men and adult women will be employed.

Year	Male	Female
1974	74.9	42.6
1979	73.8	47.5
1984	70.7	49.5
1989	72.5	54.3
1995	70.8	55.6
2001	70.9	57.0
2002	69.7	56.3

EXPLORATION

23. a. Explore the graphs of $y_1 = x^3$, $y_2 = x^4$, $y_3 = x^5$, and $y_4 = x^6$ and make a generalization about the graph of $f(x) = x^n$, when n is a positive integer.
    b. If x is replaced by $x - 5$ in each of the graphs of Part a, what happens to the graphs?
    c. Generalize the result in Part b.

760 More Work with Quadratics

Lesson 12-8

Factoring and Rational Expressions

Vocabulary

rational expression
lowest terms

> **BIG IDEA** The factored form of a polynomial is useful in writing rational expressions in lowest terms and in performing operations on rational expressions.

In Lesson 12-7, you saw how factoring can help find the solutions to polynomial equations. Factoring can also help in working with fractions that have polynomials in their numerators and denominators.

Mental Math

Estimate to the nearest dollar the interest earned in one year in a savings account with 2.02% annual interest rate containing

a. $501.

b. $4,012.

c. $9,998.

Activity

Step 1 Before reading Step 2, perform the addition of these fractions and write your answer in lowest terms. Do not use a calculator.

$$\frac{75}{100} + \frac{66}{99}$$

Step 2 Describe how you did the addition. Did you find a common denominator? If so, then you probably worked with fractions with denominator 9,900. Or, did you put each fraction in lowest terms first? Then you could do the addition with a common denominator of 12. Compare the way you added these fractions with the ways that others in your class added them.

Writing Rational Expressions in Lowest Terms

In the addition of fractions above, it is useful to write each fraction in lowest terms. As you know, this is done by dividing both the numerator and denominator of the fraction by one of their factors. Because 25 is a common factor of the numerator and denominator of the first fraction, $\frac{75}{100} = \frac{25 \cdot 3}{25 \cdot 4}$. Similarly, $\frac{66}{99} = \frac{33 \cdot 2}{33 \cdot 3} = \frac{2}{3}$.

The same idea can be used with *rational expressions*. A **rational expression** is the written quotient of two polynomials. Here are five examples of rational expressions.

$\frac{0.7819}{x + y}$

$\frac{3x^3y}{x^2y^4}$

$\frac{a + \pi}{b - \pi}$

$\frac{4n^2 + 4n + 1}{4n^2 - 1}$

$\frac{6k^3 - 12k^2 + 42k - 210}{3k^3 - 6k^2 + 21k - 105}$

Factoring and Rational Expressions **761**

A rational expression is in **lowest terms** when there is no polynomial that is a factor of its numerator and denominator. The second rational expression on the previous page is not in lowest terms because x^2y is a common factor of the numerator and denominator.

$$\frac{3x^3y}{x^2y^4} = \frac{x^2y \cdot 3x}{x^2y \cdot y^3} = \frac{3x}{y^3}$$

Technically, $\frac{3x}{y^3}$ is not exactly equivalent to $\frac{3x^3y}{x^2y^4}$ because the x in $\frac{3x}{y^3}$ can equal 0 while the x in $\frac{3x^3y}{x^2y^4}$ cannot equal 0. Often this is taken for granted, but sometimes people will write $x \neq 0$ and $y \neq 0$ because fractions do not allow 0 as the denominator.

The rational expressions $\frac{4n^2 + 4n + 1}{4n^2 - 1}$ and $\frac{6k^3 - 12k^2 + 42k - 210}{3k^3 - 6k^2 + 21k - 105}$ look more complicated than the others on page 761. But each expression can be put in lowest terms by factoring out the common factors.

GUIDED

Example 1

Write $\frac{4n^2 + 4n + 1}{4n^2 - 1}$ in lowest terms. Assume the denominator does not equal 0.

Solution

Step 1 Factor the numerator $4n^2 + 4n + 1$.

Step 2 The denominator $4n^2 - 1$ is the difference of two squares. Use this information to factor it.

Step 3 From Steps 1 and 2, fill in the blanks.

$$\frac{4n^2 + 4n + 1}{4n^2 - 1} = \frac{(\underline{\ ?\ }n + \underline{\ ?\ })(\underline{\ ?\ }n + \underline{\ ?\ })}{(\underline{\ ?\ }n + \underline{\ ?\ })(\underline{\ ?\ }n - \underline{\ ?\ })}$$

Step 4 The numerator and denominator in Step 3 have a factor in common. Divide them by that factor.

$$\frac{4n^2 + 4n + 1}{4n^2 - 1} = \frac{(\underline{\ ?\ }n + \underline{\ ?\ })}{(\underline{\ ?\ }n - \underline{\ ?\ })}$$

Because the numerator and denominator on the right have no common factor, the fraction is in lowest terms.

Check Check your answer by substituting 3 for n in the original rational expression and in the expression of Step 4.

 QY

A graphing calculator and CAS technology can be of great assistance when writing a fraction in lowest terms.

▶ QY

Explain why $\frac{6k^3 - 12k^2 + 42k - 210}{3k^3 - 6k^2 + 21k - 105}$ is not in lowest terms.

Example 2

Simplify the expression $\dfrac{x^3 - 2x^2 - 23x + 60}{x^3 + 8x^2 - 3x - 90}$.

Solution 1 To factor the numerator, graph $f(x) = x^3 - 2x^2 - 23x + 60$ with a graphing calculator. The graph has three x-intercepts: -5, 3, and 4, indicating that $(x - 4)$, $(x - 3)$, and $(x - -5)$ are factors of the numerator. To factor the denominator, graph $g(x) = x^3 + 8x^2 - 3x - 90$. This graph has x-intercepts 3, -5, and -6, indicating that $(x - 3)$, $(x - -5)$, and $(x - -6)$ are factors of the denominator. Thus,

$$\dfrac{x^3 - 2x^2 - 23x + 60}{x^3 + 8x^2 - 3x - 90} = \dfrac{(x + 5)(x - 4)(x - 3)}{(x - 3)(x + 5)(x + 6)} = \dfrac{x - 4}{x + 6}.$$

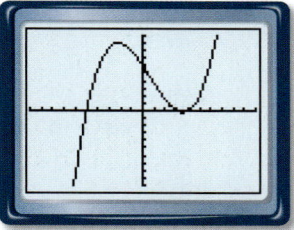

$-10 \leq x \leq 10, -100 \leq y \leq 100$

Solution 2 Use a CAS. If you enter the given rational expression into a CAS, you may immediately see the expression in lowest terms.

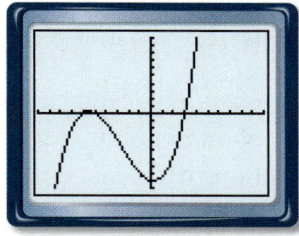

$-10 \leq x \leq 10, -100 \leq y \leq 100$

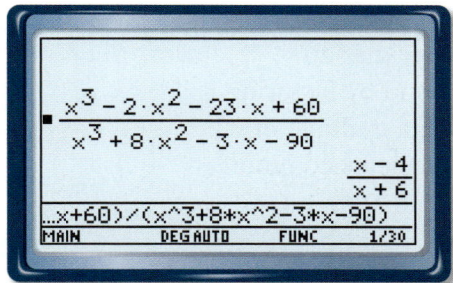

Solution 3 Use a CAS to show the steps of Solution 1. Factor the numerator and the denominator.

Thus $\dfrac{x^3 - 2x^2 - 23x + 60}{x^3 + 8x^2 - 3x - 90} = \dfrac{(x - 4)(x - 3)(x + 5)}{(x - 3)(x + 5)(x + 6)}.$

Divide numerator and denominator by the common factors.

The solution is $\dfrac{x - 4}{x + 6}$.

Caution: Notice that you *cannot* simplify $\dfrac{x - 4}{x + 6}$ by dividing by x because the x's are terms, not factors.

(continued on next page)

Check Graph $Y1 = \frac{x^3 - 2x^2 - 23x + 60}{x^3 + 8x^2 - 3x - 90}$ and $Y2 = \frac{x-4}{x+6}$ to see whether they appear to be equivalent expressions. Use the [WINDOW] $-25 \leq x \leq 25$, $-6 \leq y \leq 6$.

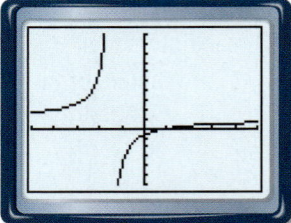

The two appear to form the same graph, so the result checks.

Adding and Subtracting Rational Expressions

To add $\frac{1}{9} + \frac{7}{15}$, you can either look for a calculator or look for a common denominator. If you do not have a calculator that does fractions, the least common denominator can be found by factoring each denominator and finding a product that will be a multiple of each denominator. Since $9 = 3 \cdot 3$ and $15 = 3 \cdot 5$, the least common denominator is $3 \cdot 3 \cdot 5$, or 45.

$$\frac{1}{9} + \frac{7}{15} = \frac{1}{3 \cdot 3} + \frac{7}{3 \cdot 5}$$
$$= \frac{1 \cdot 5}{3 \cdot 3 \cdot 5} + \frac{7 \cdot 3}{3 \cdot 5 \cdot 3}$$
$$= \frac{5}{45} + \frac{21}{45}$$
$$= \frac{26}{45}$$

You may have done most of this process in your head. We show the steps because you can add or subtract rational expressions in the same way.

GUIDED

Example 3

Write $\frac{a}{b} + \frac{c}{d}$ as a single rational expression.

Solution Since b and d have no common factors, their least common multiple is bd. This is the common denominator.

$\frac{a}{b} + \frac{c}{d} = \frac{a \cdot ?}{b \cdot d} + \frac{? \cdot c}{b \cdot d}$ Fraction Multiplication Property

$= \frac{?}{b \cdot d}$ Distributive Property

Check You can check by substituting numbers for a, b, c, and d. Try $a = 1$, $b = 9$, $c = 7$, and $d = 15$, since you already know the sum is $\frac{26}{45}$. The rest is left as a question in the Questions section.

Example 4

Write $\dfrac{6}{k+1} - \dfrac{3k+7}{k^2-1}$ as a single rational expression.

Solution 1 Use a CAS. You must be careful to include parentheses to identify numerators and denominators of the fractions. In one calculator, we entered `6/(k+1)-(3*k+7)/(k^2-1)`. The calculator returned the expression $\dfrac{3k-13}{(k-1)(k+1)}$.

Solution 2 Work by hand. Find the least common denominator. The denominator $k+1$ is prime. The denominator k^2-1 equals $(k+1)(k-1)$. So the least common denominator is $(k+1)(k-1)$. Rewrite the first fraction with that denominator.

$$\dfrac{6}{k+1} - \dfrac{3k+7}{k^2-1} = \dfrac{6k-1}{(k+1)(k-1)} - \dfrac{3k+7}{(k+1)(k-1)}$$

Subtract the fractions as you would any fractions with the same denominator. But notice that the subtracted numerator must be treated as a quantity.

$$= \dfrac{6(k-1)-(3k+7)}{(k+1)(k-1)}$$

$$= \dfrac{6k-6-3k-7}{(k+1)(k-1)}$$

$$= \dfrac{3k-13}{(k-1)(k+1)}$$

Check You can check by substitution or by graphing. We leave these checks to you in the Questions section.

Questions

COVERING THE IDEAS

1. What is the definition of *rational expression*?

2. **Multiple Choice** Which of the following are *not* rational expressions?

    A $\dfrac{2x}{3}$ B $\dfrac{4y^0}{5y^{15}}$ C $\dfrac{z\sqrt{6}}{7}$ D $\dfrac{8\sqrt{w}}{9w}$

In 3 and 4, simplify the rational expression and indicate all restrictions on values of the variables.

3. $\dfrac{30a^4b^2c^3}{12a^4bc^5}$

4. $\dfrac{28x-21x}{7xy}$

5. a. By factoring the numerator and denominator, write $\dfrac{2x^2-7x+6}{x^2-4}$ in lowest terms.
   b. What values can x not have in this expression?
   c. Check your answer by letting $x = 10$.

6. a. Write $\dfrac{7x^2 + 161x + 910}{14x^2 + 182x + 420}$ in lowest terms.

   b. Check your answer by graphing, as was done in Example 2.

7. Complete the check of Guided Example 3.

8. Check the answer to Example 4 by graphing.

9. Check the answer to Example 4 by substitution.

In 10–13, write as a single rational expression. Check your answer.

10. $\dfrac{a}{b} - \dfrac{c}{d}$

11. $\dfrac{3}{2n} + \dfrac{3}{8n}$

12. $\dfrac{z+1}{4z-3} + \dfrac{3z}{8z^2 - 18z + 9}$

13. $\dfrac{8x^2 + 16x + 8}{5x + 5} - \dfrac{x+1}{3x^2 - 3}$

APPLYING THE MATHEMATICS

In 14 and 15, use a CAS or graphing calculator to write the expression in lowest terms.

14. $\dfrac{-3x^3 - 15x^2 - 24x - 12}{x^2 + 3x + 2}$

15. $\dfrac{x^5 - 4x^4 - 37x^3 + 124x^2 + 276x - 720}{x^5 + 20x^4 + 155x^3 + 580x^2 + 1{,}044x + 720}$

In 16 and 17, the sum F of the integers from 1 to n is given by the formula $F = \dfrac{n^2 + n}{2}$. The sum S of the squares of the integers from 1 to n is given by the formula $S = \dfrac{2n^3 + 3n^2 + n}{6}$. The sum C of the cubes of the integers from 1 to n is given by the formula $C = \dfrac{n^4 + 2n^3 + n^2}{4}$.

16. a. Find the values of C, F, and $\dfrac{C}{F}$ when $n = 13$.

    b. Show that $\dfrac{C}{F} = F$ for all values of n.

17. a. Find the values of S, F, and $\dfrac{S}{F}$ when $n = 13$.

    b. Find a rational expression for $\dfrac{S}{F}$ in lowest terms.

    c. Explain why there are values of n for which $\dfrac{S}{F}$ is not an integer.

18. Generalize the following pattern and use addition of rational expressions to show why your generalization is true.

    $$\dfrac{1}{2} - \dfrac{1}{3} = \dfrac{1}{6}, \quad \dfrac{1}{3} - \dfrac{1}{4} = \dfrac{1}{12}, \quad \dfrac{1}{4} - \dfrac{1}{5} = \dfrac{1}{20}$$

More Work with Quadratics

REVIEW

19. If an object is thrown upward from a height of 0 meters at a speed of $v \frac{\text{meters}}{\text{second}}$, then its height after t seconds is $vt - 4.9t^2$ meters. **(Lesson 12-6)**
    a. If an object is thrown upward with speed v, how long will it take it to hit the ground?
    b. What must be true about v if the time it takes the object to hit the ground in seconds is an integer? Give an example of a v for which this happens and a v for which it does not happen.

20. Is $3x^2 - 2x + 5$ a prime polynomial? How do you know? **(Lesson 12-6)**

21. Gerardo is cutting a shape out of paper. He begins with a square piece of paper, and cuts out the upper right corner, as in the figure at the right. What is the area of the piece he cut out? **(Lesson 12-2)**

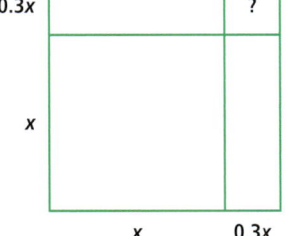

22. Nikki wants to call her friend Adelaide. She remembers that the last three digits of her number are 3, 4, and 7, but she doesn't remember the correct order. If she guesses an order at random, what is the probability that she will get the right number? **(Lesson 11-7)**

23. Marcus took a test in which there were 20 questions. In this test, every correct answer gave 6 points, and every incorrect answer subtracted 3 points from the grade. If Marcus got a 93 on the test, how many questions did he get right? **(Lesson 10-5)**

24. Is there a convex polygon with exactly 43 diagonals? Explain how you know. **(Lesson 9-7)**

EXPLORATION

25. A teacher, wanting to show students that their ideas could be used with very complicated rational expressions, used the following expression.
$$\frac{x^9 + 11x^8 - 84x^7 - 1{,}660x^6 - 4{,}874x^5 + 44{,}082x^4 + 400{,}140x^3 + 1{,}347{,}300x^2 + 2{,}156{,}625x + 1{,}366{,}875}{x^8 - 22x^7 + 90x^6 + 882x^5 - 5{,}508x^4 - 10{,}530x^3 + 74{,}358x^2 + 39{,}366x - 295{,}245}$$
However, the teacher forgot the operation sign between x^3 and 1,347,300.
    a. If the numerator was meant to be factored, what is the operation?
    b. Use the answer to Part a to write the expression in lowest terms.

QY ANSWER

The numerator and denominator have common factors.

$\dfrac{6k^3 - 12k^2 + 42k - 210}{3k^3 - 6k^2 + 21k - 105} =$

$\dfrac{6(k^3 - 2k^2 + 7k - 35)}{3(k^3 - 2k^2 + 7k - 35)} = 2$

Chapter 12 Projects

1. Combining Solutions

a. Choose an equation of the form $x^2 + bx + c = 0$ and call its two solutions: r and s. Calculate $r + s$ and $r \cdot s$. How are these related to your original equation? Do the same thing for three more equations, and write a short description of what you found.

b. Repeat Part a, but this time with equations of the form $ax^2 + bx + c = 0$.

c. The solutions to $x^3 - 6x^2 + 11x - 6 = 0$ are 1, 2, and 3. The solutions to $x^3 - 10x^2 + 24x = 0$ are 0, 4, and 6. How are these solutions related to the results you found in Part a? Can you find how the coefficient of x is related to the solutions?

d. Repeat Part c, but this time use equations of the form $ax^3 + bx^2 + cx + d = 0$.

2. Infinite Repeating Continued Fractions

Consider this sequence of complex fractions: $\dfrac{1}{5}, \dfrac{1}{5 + \frac{1}{5}}, \dfrac{1}{5 + \frac{1}{5 + \frac{1}{5}}}, \dfrac{1}{5 + \frac{1}{5 + \frac{1}{5 + \frac{1}{5}}}}, \ldots$

a. Calculate the values of the first five terms of this sequence.

b. As you calculate more and more terms of this sequence, the sequence approaches the value of x, where

$$x = \cfrac{1}{5 + \cfrac{1}{5 + \cfrac{1}{5 + \cfrac{1}{5 + \cfrac{1}{5 + \ldots}}}}}.$$

This x satisfies $x = \dfrac{1}{5 + x}$. Find the value of x.

c. Complete Parts a and b, this time with a 4 replacing each 5.

3. Prime Numbers, Prime Polynomials

One of the most famous discoveries of the Greek mathematician and astronomer Eratosthenes (276 B.C.E.–194 B.C.E.) is an algorithm to check if a number is a prime, called the sieve of Eratosthenes.

a. Look up the sieve of Eratosthenes. Write a description of how it works. If you can, write a computer (or calculator) program that checks if an integer is a prime.

b. Do you think a similar algorithm could be invented to check if a polynomial is prime over the integers? Why, or why not?

Projects

4 Public-Key Cryptography

The use of codes based on prime numbers to protect information is called *public-key cryptography*. Do research in your library or on the Internet and write an essay describing how public-key cryptography works.

5 Using Polynomials to Approximate

The graphs of polynomials of higher degree can have interesting shapes. Using a graphing calculator, to make graphs of polynomials that look like the letters of the English alphabet (both lower case and upper case).

a. For each letter you successfully create, write down the polynomial and the appropriate window.

b. Write some letters which you think cannot be drawn well this way. Why do you think this is so?

c. Which are harder to draw, lower case letters or upper case letters? Why?

6 The Bisection Method

For higher degree polynomials, there are no simple formulas that allow you to find the x-intercepts of the graphs. Nevertheless, in real life, it is often important to find those x-intercepts. One method in which approximate values can be found is called the *bisection method*.

a. Look up the bisection method in other books or on the Internet. Write a description of how it works. Include pictures to illustrate your explanation.

b. Suppose you know that there is an x-intercept of a graph between 0 and 1. How many times would you need to apply the bisection method to get an approximation with an accuracy of at least 0.01?

c. Use the bisection method to find the value of one of the x-intercepts of $y = x^5 - 2x + 5$ with an accuracy of at least 0.01.

Chapter 12 Summary and Vocabulary

- From the **standard form of a quadratic equation,** $y = ax^2 + bx + c$, the vertex of the parabola is not visible, but if the form is converted to $y - k = a(x - h)^2$, then the vertex is (h, k). The process of converting is called **completing the square.**

- The x-intercepts of this parabola are the solutions to the equation $ax^2 + bx + c = 0$. These can be found using the Quadratic Formula: $x = \frac{-b \pm \sqrt{b^2 - 4ac}}{2a}$. Letting

  $r_1 = \frac{-b + \sqrt{b^2 - 4ac}}{2a}$ and

  $r_2 = \frac{-b - \sqrt{b^2 - 4ac}}{2a}$,

  then $y = a(x - r_1)(x - r_2)$. This **factored form** is a special case of a more general theorem about polynomials: If $P(x)$ is a polynomial and $P(r) = 0$, then $x - r$ is a factor of $P(x)$.

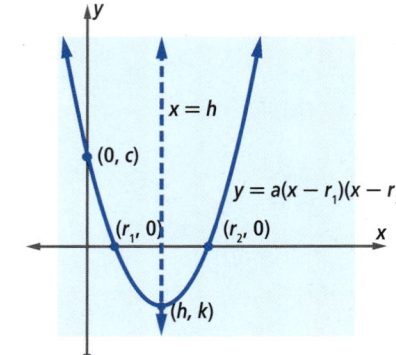

- Because the factored form of a polynomial enables you to see the x-intercepts, it is useful to be able to factor quadratic expressions $ax^2 + bx + c$ into two linear factors $dx + e$ and $fx + g$. In this chapter, we considered only factoring situations where $a, b, c, d, e, f,$ and g are integers. Most quadratic expressions with integer coefficients do *not* factor into linear factors with integer coefficients. The key is the value of the **discriminant** $b^2 - 4ac$. If $a, b,$ and c are real numbers and $b^2 - 4ac > 0$, then there are two real solutions to the equation $ax^2 + bx + c = 0$. If $b^2 - 4ac = 0$, then $r_1 = r_2$; there is exactly one solution to the equation, and the vertex of the parabola is on the x-axis. If $b^2 - 4ac < 0$, then there are no real solutions; the parabola does not intersect the x-axis.

- The factoring of polynomials also helps in work with **rational expressions.** By dividing out common factors from the numerator and denominator, you can write rational expressions in **lowest terms** and can be added or subtracted.

Vocabulary

12-1
vertex form of an equation for a parabola

12-2
complete the square

12-3
factored form (of a quadratic function)

12-4
square term
linear term
constant term
prime polynomial over the integers

12-7
cubic polynomial

12-8
rational expression
lowest terms

Theorems and Properties

Parabola Vertex Theorem (p. 716)
Factor Theorem for Quadratic Functions (p. 731)
Discriminant Theorem (p. 749)
Factor Theorem (p. 755)

Chapter 12 Self-Test

Take this test as you would take a test in class. You will need a calculator. Then use the Selected Answers section in the back of the book to check your work.

In 1–3, factor completely.

1. $x^2 + 3x - 40$
2. $m^2 - 17m + 72$
3. $-9h^2 + 9h - 2$

4. **Multiple Choice** Which of the following can be factored over the integers?
   A $x^2 + 22x - 9$
   B $x^2 + 9x - 22$
   C $x^2 - 9x + 22$
   D $x^2 - 22x + 9$

5. Determine what number must be added to $z^2 - 12z$ to complete the square.

6. Simplify the expression $\dfrac{3x^2 - 75}{2x^2 - 7x - 15}$ and indicate all restrictions on x.

7. **Multiple Choice** Which equation is graphed below?

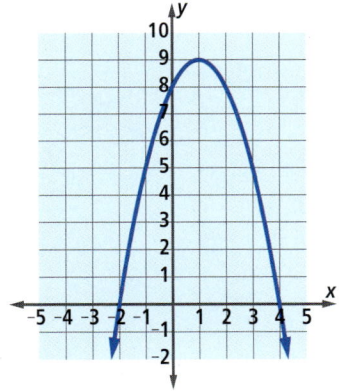

   A $y = (x + 2)(x - 4)$
   B $y = (x - 2)(x + 4)$
   C $y = (-x - 2)(x + 4)$
   D $y = -(x + 2)(x - 4)$

8. a. Give an equation in factored form for a polynomial function with 3 x-intercepts: 0, –3, and 9.
   b. Rewrite your equation from Part a in standard form.

9. A square frame is x feet on each side. The painting it holds is 1 foot shorter and 2 feet narrower than the frame. If the area of the painting is 12 square feet, what are the dimensions of the frame?

In 10 and 11, determine the vertex of the parabola with the given equation.

10. $y = x^2 - 2x - 3$
11. $y = 7 - 4x - 2x^2$

In 12 and 13, sketch the graph of the equation without a calculator.

12. $y - 4 = 2(x - 3)^2$
13. $y = (x - 5)(x + 4)$

14. Factor $3n^4 - 15n^3 + 18n^2$ completely over the integers.

15. Suppose a rectangle has an area of 486 cm², and one side is 9 cm longer than the other side. Find the dimensions of the rectangle.

16. **True or False** A quadratic expression with a positive discriminant is always factorable over the integers. Justify your answer.

17. Write a possible equation for the polynomial graphed below.

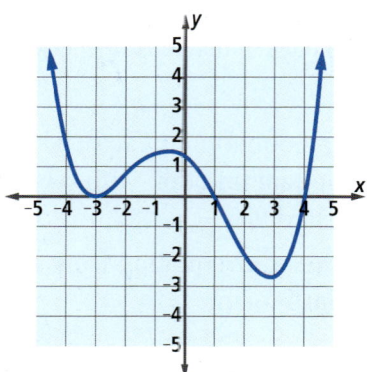

18. For the graph of the parabola, Marie wrote $y = (x + 3)(x + 1)$. Anthony wrote $y = (x + 2)^2 - 1$.

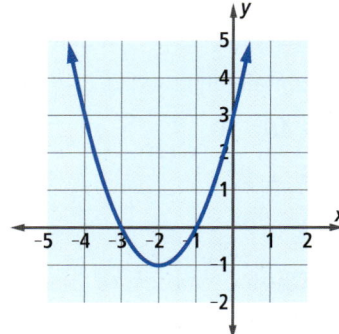

a. Are Marie and Anthony's equations equivalent? Explain.

b. What is Anthony's equation useful for finding on the graph?

Chapter 12 Chapter Review

SKILLS
PROPERTIES
USES
REPRESENTATIONS

SKILLS Procedures used to get answers

OBJECTIVE A Complete the square on a quadratic expression. (Lesson 12-2)

In 1–4, a quadratic expression is given.
a. Determine what number must be added to the expression to complete the square.
b. Complete the square.

1. $x^2 + 4x$
2. $t^2 - 5t$
3. $z^2 + bz$
4. $w^2 - \frac{3}{4}w$

OBJECTIVE B Factor quadratic expressions of the form $x^2 + bx + c$ and $ax^2 + bx + c$. (Lessons 12-4, 12-5)

In 5–8, factor the quadratic expression.

5. $x^2 - x - 6$
6. $y^2 + 60y + 800$
7. $m^2 - 2m - 24$
8. $n^2 + 4n - 12$

In 9–12, factor the trinomial.

9. $3x^2 - 2x - 8$
10. $5x^2 + 16x + 3$
11. $6d^2 - 8d - 8$
12. $8n^2 + 21 + 34n$

OBJECTIVE C Find the product of three or more binomials. (Lesson 12-7)

In 13–16, an equation of a function is given.
a. Identify the x-intercepts of its graph.
b. Put the equation into standard form.

13. $y = 6(x + 1)(x - 3)(2x - 11)$
14. $y = x(x - 5)(x + 2)$
15. $f(x) = (x - 4)(2x - 7)(x + 2)$
16. $g(x) = (3 - x)(4x - 1)(x + 3)(x + 1)$

OBJECTIVE D Use factoring to write rational expressions in lowest terms. (Lesson 12-8)

In 17 and 18, simplify the rational expression and indicate all restrictions on values of the variables.

17. $\frac{4n^2m - 8nm^2}{nm}$
18. $\frac{-7rst^2q^4}{(s+1)tq}$

In 19–21, write as a single rational expression. You may need a calculator or a CAS.

19. $\frac{-4n^2 + 23n - 15}{n^2 - 8n + 15}$
20. $\frac{3}{x+4} - \frac{3x-7}{x^2 + 3x - 4}$
21. $\frac{-m^3 + 5m^2 - 2m - 8}{2m^3 - 9m^2 + 3m + 14}$

PROPERTIES Principles behind the mathematics

OBJECTIVE E Determine whether a quadratic polynomial can be factored over the integers. (Lessons 12-4, 12-5, 12-6)

22. **True or False** Every quadratic polynomial whose discriminant is an integer is factorable over the integers.

23. For what values of the discriminant is a quadratic polynomial factorable?

In 24–27, determine whether the quadratic polynomial is factorable over the integers. If it is not, label it *prime*. If it is, factor the polynomial.

24. $5x^2 + 33x + 12$
25. $15x^2 + 69x + 72$
26. $4c^2 - 14c + 8$
27. $8x^2 + 34x + 182$

Chapter 12

OBJECTIVE F Apply the Factor Theorem. (Lesson 12-7)

28. $f(x) = 3x^3 - 15x^2 - 42x$.
   a. Write $f(x)$ in factored form.
   b. What does the Factor Theorem tell you about the graph of $f(x)$?

In 29 and 30, x-intercepts are given.
a. Give an equation in factored form for a polynomial function with the given x-intercepts.
b. Rewrite your equation in standard form.

29. 2 and –2   30. 0, 7, and –10

In 31 and 32, give a possible equation, in factored form, for the graph.

31.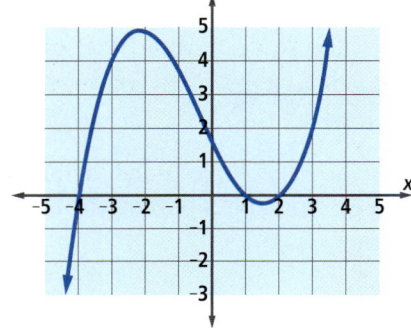

32.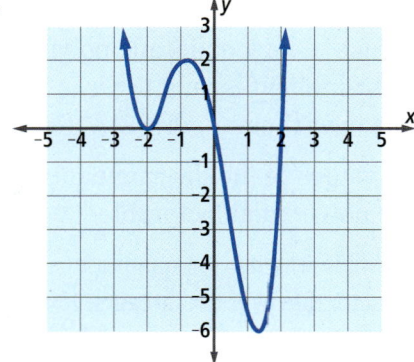

USES Applications of mathematics in real-world situations

OBJECTIVE G Solve problems involving areas and perimeters of rectangles that lead to quadratic functions or equations. (Lesson 12-1)

In 33 and 34, a rectangle is given.
a. Find its area A and perimeter P.
b. Determine the dimensions of the rectangle so that the perimeter and area are numerically equal.

33.

34.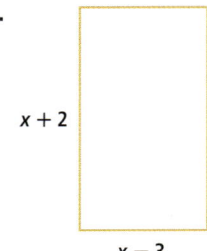

35. A rectangular soccer field has a perimeter of 390 yards.
   a. Write a formula for the area of the field in terms of its length x.
   b. If the width of the field must be at least 50 yards and at most 75 yards, what is the maximum area of the field?

36. A dairy cow is walking around the outside of her rectangular pasture at an average speed of 2 feet per second.
   a. If it takes the cow 20 minutes to walk around the pasture, what is the perimeter of the pasture?
   b. Write a formula for the area A of the pasture in terms of the length of one side L, and determine the dimensions of the rectangle that will maximize the cow's pasture.

774 More Work with Quadratics

37. The Irish sport of hurling can be played on a rectangular field with an area of 12,600 m². If the longer side of a hurling field is 50 m longer than the shorter side, what are the dimensions of the field?

38. The perimeter of a rectangle is 24 cm.
   a. Write a formula for the area A of the rectangle in vertex form in terms of the length x of one of its sides.
   b. What is the meaning of the vertex in this scenario?

REPRESENTATIONS Pictures, graphs, or objects that illustrate concepts

OBJECTIVE H Graph quadratic functions whose equations are given in vertex form. (Lesson 12-1)

In 39–42, sketch the graph of the equation without a calculator.

39. $y - 3 = (x - 5)^2$

40. $y - 5 = (x + 2)^2$

41. $y + 4 = 2(x + 8)^2$

42. $y + 2 = -3(x - 7)^2$

OBJECTIVE I Find the vertex of a parabola whose equation is given in standard form. (Lesson 12-2)

In 43–46, match the graph with one of the equations i. through iv. below.

i. $y = -x^2 + 6x - 14$ ii. $y = x^2 - 6x + 14$
iii. $y = -x^2 - 6x - 4$ iv. $y = x^2 + 6x + 9$

43.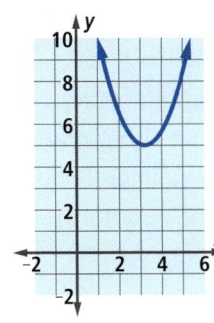

44.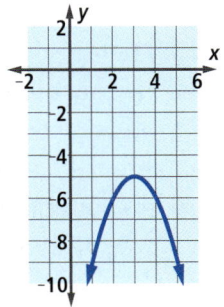

45.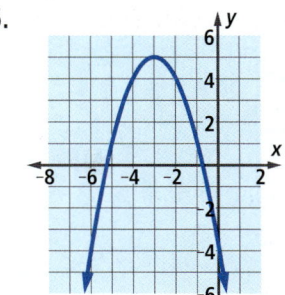

46.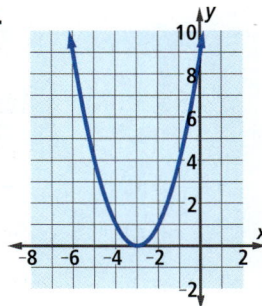

In 47–50, find the vertex of the parabola.

47. $y = 2x^2$

48. $y = x^2 + 3x - 2$

49. $y = -3x^2 + 2x - 5$

50. $y = -x^2 + 6x + 4$

OBJECTIVE J Graph quadratic functions whose equations are given in factored form. (Lesson 12-3)

In 51–54, sketch the graph of the equation without using a calculator.

51. $y = (x - 2)(x + 3)$

52. $y = (x + 4)(x - 6)$

53. $y = (2x - 1)(x + 4)$

54. $y = (3x + 7)(x + 8)$

Chapter 13
Using Algebra to Prove

- **13-1** If-Then Statements
- **13-2** The Converse of an If-Then Statement
- **13-3** Solving Equations as Proofs
- **13-4** A History and Proof of the Quadratic Formula
- **13-5** Proofs of Divisibility Properties
- **13-6** From Number Puzzles to Properties of Integers
- **13-7** Rational Numbers and Irrational Numbers
- **13-8** Proofs of the Pythagorean Theorem

In previous chapters, you looked at instances and made generalizations about patterns. For example, the instances $3 \cdot 5 + 3 = 3 \cdot 6$, $7.4 \cdot 5 + 7.4 = 7.4 \cdot 6$, and $\left(-\frac{8}{99}\right) \cdot 5 + \left(-\frac{8}{99}\right) = \left(-\frac{8}{99}\right) \cdot 6$ can be described by the general pattern $n \cdot 5 + n = n \cdot 6$. In making this generalization, you have used *inductive reasoning*. **Inductive reasoning** is the process of arriving at a general conclusion (not necessarily true) from specific instances. You use inductive reasoning quite often in everyday situations. For example, if every single family house you see on a block is yellow, you may want to conclude that every house in a city is also yellow. That conclusion is wrong. There are houses in a neighborhood or city that are not yellow. But the general

pattern $n \cdot 5 + n = n \cdot 6$ does happen to be true for all real numbers. We know this because we can *prove* it mathematically.

To prove a generalization, you must use *deductive reasoning*. Deductive reasoning starts from properties that are assumed to be true. For example, we assumed the Distributive Property of Multiplication over Addition to be true: If a, b, and c, are any real numbers, then $ab + ac = a(b + c)$.

If this is true for all real numbers, then it is true when $a = n$, $b = 5$, and $c = 1$. Substituting these values for a, b, and c, $n \cdot 5 + n \cdot 1 = n(5 + 1)$.

Using another assumed property, that if n is any real number, then $n \cdot 1 = n$. Adding 5 and 1 then gives $n \cdot 5 + n = n \cdot 6$.

This string of justified if-then statements has *proved* that for all real numbers n, $n \cdot 5 + n = n \cdot 6$.

You may not have written the words *if* and *then*, but in this course you have often strung if-then statements to follow each other.

A string of justified statements that follow from each other like these is a *proof*. In this chapter, you will see many examples of proofs that use the algebra that you have studied. These proofs involve the solving of equations, divisibility properties of arithmetic, and geometric figures. They comprise one of the most important uses of algebra: showing that a statement is true when there are infinitely many cases to consider.

Chapter 13

Lesson 13-1

If-Then Statements

Vocabulary

if-then statement
antecedent
consequent
generalization

▶ **BIG IDEA** Statements that are of the form *if . . . then* are the basis of mathematical logic, so it is important to know how to determine whether they are true or false.

If is one of the most important words in mathematics. (The words *given*, *when*, *whenever*, and *suppose* often have the same meaning.) The word *if* is often followed by the word *then*, which may or may not be written. The result is an **if-then statement**. Here are some examples.

1. If a bug is an insect, it has six legs.
2. Suppose a person likes outdoor football. Then the person will like arena football.
3. Every animal on land grows leaves.

In an if-then statement, the clause following *if* is called the **antecedent**. The clause following *then* is the **consequent**. Below we have underlined the antecedent once and the consequent twice.

$$\text{If a } \underline{\text{bug is an insect}}, \text{ then } \underline{\underline{\text{it has six legs}}}.$$
$$\text{antecedent} \qquad\qquad \text{consequent}$$

Some if-then statements are *generalizations*. A **generalization** is an if-then statement in which there is a variable in the antecedent and consequent. In Statements 1–3 above, the variable is not seen, but each statement can be thought of as an if-then statement with a variable.

If B is an insect, (then) B has six legs.

If P likes outdoor football, then P will like arena football.

 QY

Mental Math

A 28-page newspaper consists of a News section, a Classifieds section, and a Sports section. The News section is twice as long as the Sports section, and the Classifieds section is half as long as the Sports section. How long is

a. the News section?
b. the Sports section?
c. the Classifieds section?

▶ **QY**

Write Statement 3 above using a variable.

When Is an If-Then Statement True?

An if-then statement is true if its consequent is true for *every* value in the domain of the variables in its antecedent.

778 Using Algebra to Prove

Lesson 13-1

Example 1

True or False If *B* is an insect, (then) *B* has six legs.

Solution The statement is true because every insect has six legs. (Having six legs is one of the defining characteristics of insects.)

An if-then statement with a variable describes a pattern. As in any pattern, a value of the variable for which both the antecedent and consequent of an if-then statement is true is an instance of the statement. A beetle is an insect and a beetle has six legs. So a beetle is an instance of Statement 1.

Beetles are the largest group in the animal kingdom, representing one-fourth of all animals.
Source: San Diego Zoo

Situations in which the antecedent is false do not affect whether an if-then statement is true. A dining room table might have six legs, but a table is not an insect. So a dining room table is not an instance of Statement 1.

A true if-then statement can be represented with a Venn diagram. The set of insects is placed inside the set of things with six legs.

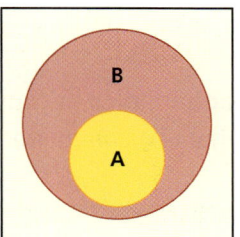

If A, then B.

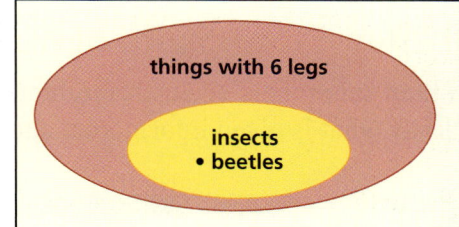

Example 2

True or False If *x* is a month of the year, then *x* has 31 days.

Solution This statement is not true because there are some months that have fewer than 31 days.

If there is a value of the variable for which the antecedent of an if-then statement is true and the consequent is not true, then the if-then statement is *false*. This value is a *counterexample* to the statement. The month of November has only 30 days, so it is a counterexample to the if-then statement. One counterexample is enough to cause an if-then statement to be false.

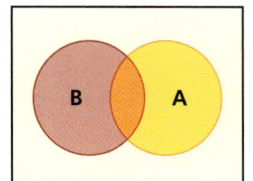

If A, then sometimes (but not always) B.

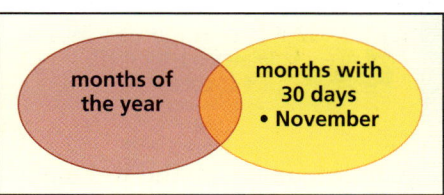

Even though there are values of the variable for which the statement of Example 2 is true, because there is a counterexample, the generalization is false.

If-Then Statements

Chapter 13

Example 3

Draw a Venn diagram for the statement: If G is a land animal, G grows leaves.

Solution A tiger is a land animal and a tiger does not grow leaves, so the statement is false. In fact, no land animals grow leaves. So if G is a land animal, G never grows leaves. The Venn diagram has two circles that do not overlap.

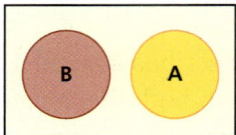

If A, then not B.

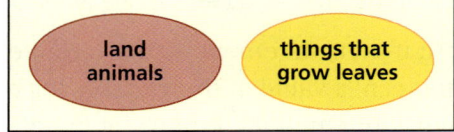

If-Then Statements in Mathematics

If-then statements occur throughout mathematics because they clarify what is given information and what are conclusions.

GUIDED

Example 4

Fill in the blanks.

If-Then Statement	True or False	Explanation
a. If $3n + 5 = 65$, then $n = 20$.	true	If $3n + 5 = 65$, then $3n = \underline{\ ?\ }$. If $3n = \underline{\ ?\ }$, then $n = 20$.
b. If a figure is a square, then it is a rectangle.	?	Definition of square: a square is $\underline{\ ?\ }$.
c. If x is a real number, then $x^2 > 0$.	?	0 is a real number and 0^2 is not greater than 0.
d. If a quadrilateral has 3 right angles, then it is a square.	false	?

Counterexamples to the two false statements in Guided Example 4 can be pictured. For Statement c, graph $y = x^2$. Notice that the graph is not always above the x-axis, so it is not the case that $x^2 > 0$ for every value of x. For Statement d, show a quadrilateral that has 3 right angles but is not a square.

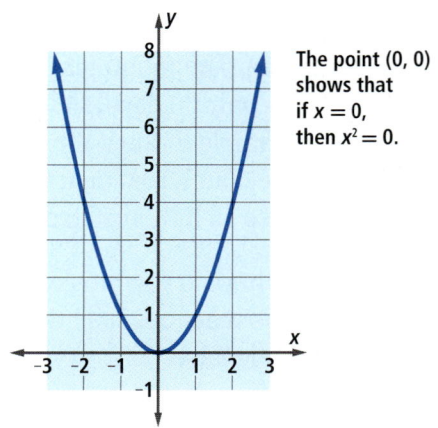

The point (0, 0) shows that if $x = 0$, then $x^2 = 0$.

This quadrilateral has 3 right angles and is not a square.

780   Using Algebra to Prove

Lesson 13-1

Putting Statements into If-Then Form

Statements with the words "all," "every," and "no" can be rewritten in if-then form without changing their meaning.

GUIDED

Example 5
Rewrite each statement in if-then form.

Statement	If-then Form with Variables
a. Every whole number is a real number.	If x is a whole number, then ___?___.
b. All people born in the United States are U.S. citizens.	If ___?___, then P is a U.S. citizen.
c. No power of a positive number is negative.	If p is a positive number and g is any real number, then ___?___.

Questions

COVERING THE IDEAS

In 1 and 2, identify the antecedent and the consequent of the if-then statement.

1. If the sun shines this afternoon, I will be happy.
2. The world would be a better place if people did not litter.

In 3 and 4, rewrite the statement as an if-then statement with a variable in it. Underline the antecedent once and the consequent twice.

3. Every integer greater than 1 is either a prime number or a product of prime numbers.
4. When a person drives over 35 miles per hour on that street, that person is speeding.
5. Explain why this if-then statement is true.
   If $3x + 16 > 10$, then $x > -2$.
6. Explain why this if-then statement is false.
   If $3x + 16 > 10$, then $x > 0$.

In 7–9, an antecedent is given. Complete the statement with two consequents that make it true.

7. If L and W are the length and width of a rectangle, then ___?___.
8. If $x^m \cdot x^n = x^{m+n}$, then ___?___.
9. If n is divisible by 10, then ___?___.

Interstate highways nationwide usually have posted speed limits between 55 and 75 mi/h.
Source: Federal Highway Administration

If-Then Statements   **781**

Chapter 13

In 10–13, an if-then situation is given.
a. Tell whether the statement is true or false.
b. Draw a Venn diagram picturing the situation.

10. If a figure is a rectangle, then the figure is a square.
11. If you are in the United States and in Chattanooga, then you are in Tennessee.
12. If a number is an even number, then it is a prime number.
13. If $x > 0$, then $-x > 0$.

APPLYING THE MATHEMATICS

In 14–17, a false statement is given.
a. Find a counterexample that shows it is false.
b. Find the largest domain of the variable for which the statement is true.

14. If w is a real number, then w^3 is positive.
15. If t is a real number, then t^4 is positive.
16. If r is a real number, then $2r \geq r$.
17. If x is a real number and $y = x^2 + x$, then $y \geq 0$.

In 18 and 19, draw a counterexample to the statement.
18. For all real numbers a, the graph of $y = ax^2$ is a parabola.
19. A triangle cannot have two angles each with measure over 75°.
20. Draw the following four true statements in one Venn diagram.
    a. Every rhombus is a parallelogram.
    b. Every rectangle is a parallelogram.
    c. If a figure is both a rhombus and a rectangle, then it is a square.
    d. Every square is both a rhombus and a rectangle.

In 21 and 22, rewrite the statement as an if-then statement and indicate why each statement is false.
21. All sentences have a subject, verb, and object.
22. Every president of the United States has served fewer than three terms.

REVIEW

23. **Skill Sequence** Factor each expression. (Lesson 12-4)
    a. $a^2 - 36$
    b. $n^2 - 5n - 36$
    c. $x^2 - 5xy - 36y^2$

Lesson 13-1

24. **Fill in the Blank** Do this problem in your head. Because one thousand times one thousand equals one million, then $1{,}005 \cdot 995 = $ __?__. **(Lesson 11-6)**

25. Tickets to a hockey game cost $22 for adults and $16 for children. The total attendance at one game was 3,150 and the total revenue from ticket sales for the game was $66,258. How many of each kind of ticket were sold for the game? **(Lesson 10-5)**

26. Write $2^{-3} + 4^{-3}$ as a simple fraction. **(Lesson 8-4)**

27. Let $g(x) = \frac{3}{4}x - 7$. **(Lesson 7-6)**
    a. Calculate $g(80)$.
    b. Calculate $g(-80)$.
    c. Describe the graph of g.

28. **True or False** The slope of the line through (x_1, y_1) and (x_2, y_2) is the opposite of the slope of the line through (x_2, y_2) and (x_1, y_1). **(Lesson 6-6)**

The Stanley Cup is the championship trophy awarded annually to the National Hockey League playoff winner.

EXPLORATION

29. Consider this statement: No four points in a plane can all be the same distance from each other.
    a. Write the statement in if-then form.
    b. Is the statement true or false?
    c. If "in a plane" is deleted from the statement, show that the resulting statement is false.

QY ANSWER

If A is a land animal, then A grows leaves.

Chapter 13

Lesson 13-2

The Converse of an If-Then Statement

Vocabulary

converse
equivalent statements
if and only if

▶ **BIG IDEA** The converse of the statement, "If A, then B." is the statement, "If B, then A."

You might have heard of the question, "Which came first, the chicken or the egg?" This question is difficult because: If there is a chicken, then there must have been an egg. If there is an egg, then there must have been a chicken.

These two if-then statements are *converses*. The *converse* of the statement, "If p, then q" is the statement, "If q, then p." Another way of stating this is: the **converse** of an if-then statement is found by switching the antecedent and the consequent of the statement.

🛑 **QY**

Mental Math

Evaluate when $a = 2$ and $b = -2$.
a. $(a + b)^2$
b. $a^2 + b^2$
c. $a^2 + 2ab + b^2$

▶ **QY**

Give the converse of the statement: If $\sqrt{2x} = 10$, then $x = 50$.

Converses of True Statements

Often the converse of a true statement is not true. Here is an example.

Statement If $x = 8$, then $x^2 = 64$. (true)

Converse If $x^2 = 64$, then $x = 8$. (False; x could also be –8.)

GUIDED

Example 1
The given statement is true. Write its converse and explain why the converse is not true.
a. If a quadrilateral has 4 sides of the same length and its diagonals have the same length, then the quadrilateral is a rectangle.
b. If you live in the state of North Dakota, then you live in the United States.

Solutions

a. Converse: If a quadrilateral is ___?___, then ___?___. This statement is false because ___?___.

b. Converse: If ___?___, then ___?___. The statement is false because ___?___.

784 Using Algebra to Prove

Sometimes the converse of a true statement is true. When a statement and its converse are both true, then the antecedent and consequent are **equivalent statements.** Equivalent statements can be connected by the phrase *if and only if*. Here is an example.

Statement If $x = 6$, then $2^x = 64$. (true)

Converse If $2^x = 64$, then $x = 6$. (true)

Because the statement and its converse are both true, you can write: $x = 6$ **if and only if** $2^x = 64$.

GUIDED

Example 2

The given statement is true and so is its converse. Write the converse and then combine the statement and its converse into one if-and-only-if statement.

a. If a quadrilateral has 4 sides of the same length and its diagonals have the same length, then the quadrilateral is a square.

b. If you live in the largest country in South America, then you live in Brazil.

Solution

a. Converse: If a quadrilateral ___?___, then ___?___. You can write: A quadrilateral has 4 sides of the same length and diagonals of the same length *if and only if* ___?___.

b. Converse: If you live in Brazil, then you live in the largest country in South America. You can write: ___?___ *if and only if* ___?___.

Brazil is the fifth most populous country in the world, with about 190 million people.

Source: infoplease.com

When you see the phrase "if and only if," then you can separate the sentence into two if-then statements.

> In Hardnox High School, a student is on the honor roll if and only if his or her grade point average is at least 3.75.

means

> In Hardnox High School, if a student's grade point average is at least 3.75, then the student is on the honor roll.

and

> If a student in Hardnox High School is on the honor roll, then his or her grade point average is at least 3.75.

Questions

COVERING THE IDEAS

1. State the converse of the statement: If there is smoke, then there is fire.

2. **Multiple Choice** If a statement is true, then its converse
   A must be true.   B may be true.   C must be false.

In 3–5, a statement is given.
   a. Is it true?
   b. State its converse.
   c. Is its converse true?
   d. If either the statement or its converse is not true, correct them so that they are both true.

3. If an integer is divisible by 3 and by 4, then it is divisible by 24.

4. If both the units and tens digits of an integer written in base 10 equal 0, then the integer is divisible by 100.

5. If $7u < 56$, then $u > 8$.

In 6–8, write the two if-then statements that are meant by the if-and-only-if statement.

6. You will receive full credit for this question if and only if you get both parts correct.

7. A quadrilateral is a rectangle if and only if it has four right angles.

8. $5x + 4y = 20$ if and only if $y = -1.25x + 5$.

In 9 and 10, rewrite the definition as an if-and-only-if statement.

9. Every linear function has an equation of the form $f(x) = ax + b$.

10. The reciprocal of a nonzero number x is the number y such that $xy = 1$.

In 11–14, are statements (1) and (2) equivalent? If not, why not?

11. (1) $x = 3$
    (2) $2x = 6$

12. (1) $y = 15$
    (2) $y^4 = 50{,}625$

13. (1) $z = 8$
    (2) $z^2 + 48 = 14z$

14. (1) $a + b = c$
    (2) $c - b = a$

APPLYING THE MATHEMATICS

In 15–18, a statement is given.
   a. Is it true?
   b. State its converse.
   c. Is its converse true?
   d. If either the statement or its converse is not true, correct them so that they are both true.

15. If $(x - 5)(2x + 3) = 45$, then $2x^2 - 7x + 30 = 90$.

16. If a polygon has 7 sides, then it is a hexagon.
17. If a parabola has an equation of the form $y = (x - 5)(x + 3)$, then its x-intercepts are –5 and 3.
18. If $3x + 4y = 6$ and $2x - 5y = 7$, then $x = 6$ and $y = -3$.

In 19–21, a statement is given.
   a. Tell whether the statement is true.
   b. If the statement is true, give an example. If it is not true, modify it so that it is true.

19. A line is a vertical line if and only if its slope is undefined.
20. $x^a \cdot x^b = x^{a+b}$ if and only if a and b are positive integers.
21. A person can be a U.S. citizen if and only if the person was born in the United States.

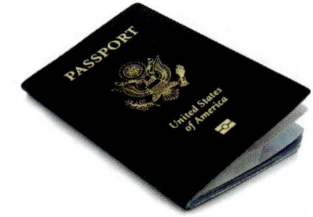

A valid U.S. passport is required for U.S. citizens to enter and leave most foreign countries.
Source: U.S. Department of State

REVIEW

In 22 and 23, tell whether the given if-then statement is true or false. If it is false, provide a counterexample to the statement. (Lessons 13-1, 9-1)

22. If a dining room table seats 8 people, then it seats 6 people.
23. If n is even, the graph of $y = ax^n$ crosses the x-axis twice.
24. **Multiple Choice** Rectangles of width x are cut off from two adjacent sides of a 12 in.-by-12 in. sheet of wrapping paper, as shown at the right. What is the area, in square inches, of the square region that remains? (Lesson 11-6)
    A $12 - x^2$ B $144 + x^2$
    C $144 - x^2$ D $144 - 24x + x^2$

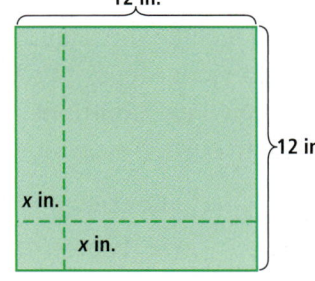

25. If 8 pencils and 5 erasers cost $4.69 and 3 pencils and 4 erasers cost $2.80, find the cost of 2 pencils. (Lesson 10-5)

In 26–28, simplify the expression. (Lessons 8-6, 8-5)

26. $5(3\sqrt{x})^2$ 27. $(6 \cdot 2^{-2})^3$ 28. $\dfrac{9 \pm \sqrt{18}}{3}$

EXPLORATION

29. Write two statements from outside of mathematics that are true but whose converses are false.

QY ANSWER

If $x = 50$, then $\sqrt{2x} = 10$.

Chapter 13

Lesson 13-3
Solving Equations as Proofs

Vocabulary

justifications
proof argument
deduction

▶ **BIG IDEA** Writing the steps in solving an equation and then checking the solutions together prove that you have found the only solutions to the equation.

An Example of a Proof Argument

Every time you solve an equation, you are proving or deducing something true. Consider solving the equation $8x + 19 = 403$. The solution is 48. You may write the steps as we show here.

i. $8x + 19 = 403$
ii. $8x + 19 + -19 = 403 + -19$
iii. $8x + 0 = 384$
iv. $8x = 384$
v. $\frac{1}{8} \cdot 8x = \frac{1}{8} \cdot 384$
vi. $1 \cdot x = 48$
vii. $x = 48$

Mental Math

Solve for *n*.
a. $5^n \cdot 5^4 = 5^{20}$
b. $(5^n)^4 = 5^{20}$

If you include the statements that show why each step follows from the previous steps, then you have written a *proof*. These statements are called **justifications.** They indicate why you can do what you have done to solve the equation.

	Conclusion	What Was Done	Justification
i.	$8x + 19 = 403$	Given	Given
ii.	$8x + 19 - 19 = 403 + -19$	Add -19 to both sides.	Addition Property of Equality
iii.	$8x + 0 = 384$	$19 + -19 = 0$	Additive Inverse Property
iv.	$8x = 384$	$8x + 0 = 8x$	Additive Identity Property
v.	$\frac{1}{8} \cdot 8x = \frac{1}{8} \cdot 384$	Multiply both sides by $\frac{1}{8}$.	Multiplication Property of Equality
vi.	$1x = 48$	$\frac{1}{8} \cdot 8 = 1$, $\frac{1}{8} \cdot 384 = 48$	Multiplicative Inverse Property
vii.	$x = 48$	$1 \cdot x = x$	Multiplicative Identity Property

Using Algebra to Prove

With this argument, you have proved: *If $8x + 19 = 403$, then $x = 48$.*
In this if-then statement, $8x + 19 = 403$ is the antecedent and $x = 48$
is the consequent. A **proof argument** in mathematics is a sequence
of justified conclusions, starting with the antecedent and ending
with the consequent. The use of a proof to show that one statement
follows from another is called **deduction**.

A Justification Is Different from What Was Done

What you *do* to solve an equation is different from the justification.
What was done applies to the specific equation being solved. The
justification is the general property. In proofs, some people prefer to
see what was done. Other people prefer the justification.

GUIDED

Example 1
Prove if $-6x - 14 = 118$, then $x = -22$.

Solution

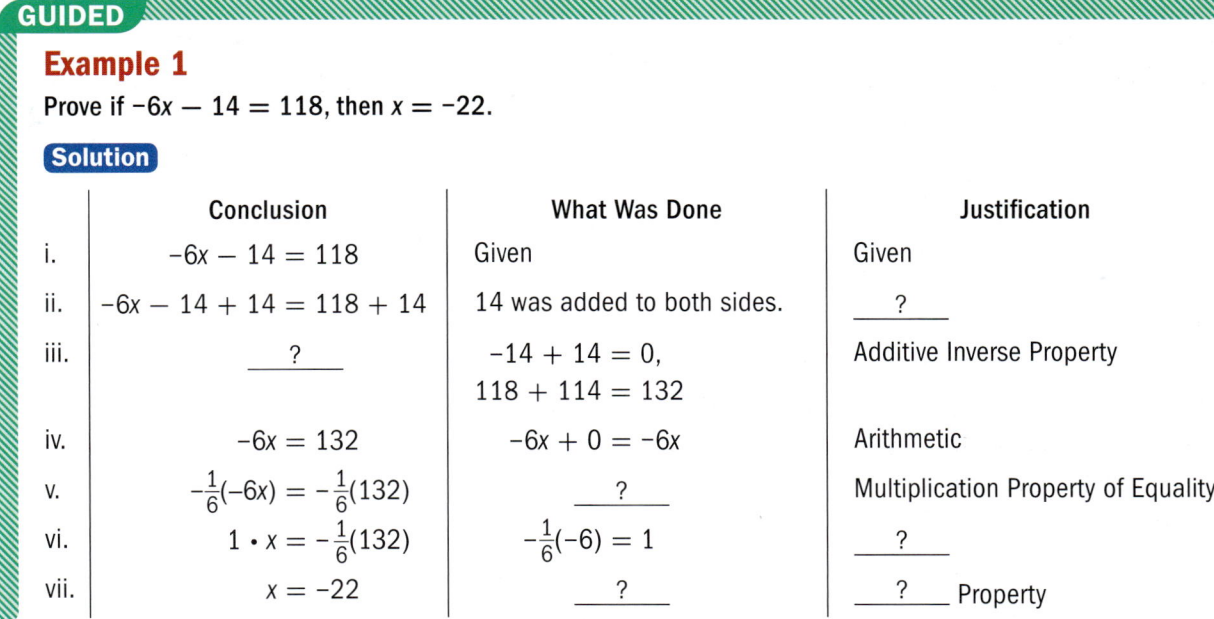

	Conclusion	What Was Done	Justification
i.	$-6x - 14 = 118$	Given	Given
ii.	$-6x - 14 + 14 = 118 + 14$	14 was added to both sides.	___?___
iii.	___?___	$-14 + 14 = 0$, $118 + 114 = 132$	Additive Inverse Property
iv.	$-6x = 132$	$-6x + 0 = -6x$	Arithmetic
v.	$-\frac{1}{6}(-6x) = -\frac{1}{6}(132)$	___?___	Multiplication Property of Equality
vi.	$1 \cdot x = -\frac{1}{6}(132)$	$-\frac{1}{6}(-6) = 1$	___?___
vii.	$x = -22$	___?___	___?___ Property

An Abbreviated Proof

Because work with additive and multiplicative inverses and identities
is so automatic, some people prefer abbreviated proofs that do not
show these steps. Here is an abbreviated proof of the statement:
If $8x + 19 = 403$, then $x = 48$.

	Conclusion	Justification
i.	$8x + 19 = 403$	Given
ii.	$8x + 19 + -19 = 403 + -19$	Addition Property of Equality
iii.	$8x = 384$	Arithmetic
iv.	$\frac{1}{8} \cdot 8x = \frac{1}{8} \cdot 384$	Multiplication Property of Equality
v.	$x = 48$	Multiplicative Inverse Property

Chapter 13

> **GUIDED**
>
> ### Example 2
> Prove that if $2x - 9 = 37$, then $x = 23$.
>
> **Solution** Supply the justifications for the following conclusions.
>
> | | Conclusion | Justification |
> |---|---|---|
> | i. | $2x - 9 = 37$ | ___?___ |
> | ii. | $2x - 9 + 9 = 37 + 9$ | ___?___ |
> | iii. | $2x = 46$ | ___?___ |
> | iv. | $\frac{1}{2}(2x) = \frac{1}{2}(46)$ | ___?___ |
> | v. | $x = 23$ | ___?___ |

Justifications and Properties

In solving an equation or inequality, every justification is one of the following:

1. *Given* information (the given equation or inequality to be solved)
2. A *property*, of which there are three types:
   a. a *defined property*, such as the definition of slope, absolute value, or square root
   b. an *assumed property* of numbers, such as the Product of Powers Property or the Distributive Property
   c. a *previously-proved property*, such as the Means-Extremes Property or the Power of a Product Property, that were not assumed but proved using definitions or other known properties
3. *Arithmetic* (a catch-all term for all the properties you have learned that help you compute results of operations)

The Check Is a Converse

On the previous page you saw the proof of the statement: *If $8x + 19 = 403$, then $x = 48$.* When you check your work by substitution, you are proving: *If $x = 48$, then $8x + 19 = 403$.* The check is the converse of the solution. Together, the solution and the check mean $8x + 19 = 403$ if and only if $x = 48$.

Another way of saying this is: $8x + 19 = 403$ exactly when $x = 48$.

This means that 48 is a solution and no other numbers are solutions. Solving an equation means proving both a statement (to find the possible solutions) and its converse (to check that the possible solutions do work). Example 3 illustrates the importance of the check.

Lesson 13-3

Example 3
Find all solutions to $\sqrt{x} = x - 6, x > 0$.

Solution

$\sqrt{x} = x - 6$	Given
$\sqrt{x} \cdot \sqrt{x} = (x - 6)(x - 6)$	Multiplication Property of Equality
$x = (x - 6)(x - 6)$	definition of square root
$x = x^2 - 12x + 36$	Extended Distributive Property
$0 = x^2 - 13x + 36$	Addition Property of Equality
$x = \dfrac{13 \pm \sqrt{(-13)^2 - 4 \cdot 1 \cdot 36}}{2}$	Quadratic Formula
$x = \dfrac{13 \pm 5}{2}$	Simplify.
$x = 9$ or $x = 4$	Simplify.

This argument proves: If $\sqrt{x} = x - 6$, then $x = 9$ or $x = 4$. So, 9 and 4 are possible values of x. To see if they are solutions, a check is necessary.

Check When $x = 9$: Does $\sqrt{9} = 9 - 6$? Yes, $\sqrt{9} = 3$ and $9 - 6 = 3$. It checks.

When $x = 4$: Does $\sqrt{4} = 4 - 6$? No, $\sqrt{4} = 2$ and $4 - 6 = -2$. It does not check.

Consequently, $x = 9$ is the only solution to $\sqrt{x} = x - 6$. Putting it another way, $\sqrt{x} = x - 6$ if and only if $x = 9$.

In Example 3, the solution has proved: *If $\sqrt{x} = x - 6$, then $x = 9$ or $x = 4$.* The check has shown *If $x = 4$, then $\sqrt{x} \neq x - 6$ and if $x = 9$, then $\sqrt{x} = x - 6$.* So, $\sqrt{x} = x - 6$ *if and only if $x = 9$.*

Questions

COVERING THE IDEAS

1. Here is part of a proof argument. Explain what was done to get to Steps a–e and supply the missing justification.

   $$40x + 12 = 3(6 + 13x)$$
   a. $\quad 40x + 12 = 18 + 39x$
   b. $40x + 12 + -12 = 18 + 39x + -12$
   c. $\quad 40x + 0 = 18 + 39x + -12$
   d. $\quad 40x = 18 + 39x + -12$
   e. $\quad 40x = 6 + 39x$
   f. The argument in Steps a–e proves what if-then statement?

Solving Equations as Proofs **791**

2. Steps a–c are the conclusions in an abbreviated proof argument. Write what was done to get to each step and provide the justifications.

$$29 - 3y \leq 44$$

a. $-29 + 29 - 3y \leq -29 + 44$

b. $\qquad -3y \leq 15$

c. $\qquad y \geq -5$

d. The argument in Steps a–c proves what if-then statement?

3. Explain the difference between inductive reasoning and deduction.

4. Give another example of inductive reasoning that is not written in this book.

5. After a month of timing her walks to school, Sula told her friend Lana, "It takes me 5 minutes less to get to school if I walk at a constant pace diagonally through the rectangular park than if I walk around two edges of its perimeter!" Lana replied, "That's a result of the Pythagorean Theorem!"
   a. Who used inductive reasoning?
   b. Who used deduction?

6. What five kinds of justifications are allowed in solving an equation or inequality?

7. a. Provide conclusions and justifications to prove that if $12m = 3m + 5$, then $m = \frac{5}{9}$.
   b. What else do you need to do in order to prove that $12m = 3m + 5$ if and only if $m = \frac{5}{9}$.

8. Find all solutions to the equation $\sqrt{2n + 1} = n - 7$. You do not have to give justifications.

APPLYING THE MATHEMATICS

In 9–14, two consecutive steps of a proof are shown as an if-then statement. State what was done and state the justification.

9. If $d = rt$, then $r = \frac{d}{t}$.

10. If $3x + 5x = 80$, then $8x = 80$.

11. If $gn^2 + hn + k = 0$, then $n = \frac{-h \pm \sqrt{h^2 - 4gk}}{2g}$.

12. If $\sqrt{t} = 400$, then $t = 160{,}000$.

13. If $3x + 4y = 6$ and $3x - 4y = -18$, then $6x = -12$.

14. If $\frac{8}{3}b = 12$, then $8b = 36$.

15. Prove: If $ax + b = c$ and $a \neq 0$, then $x = \frac{c - b}{a}$. (Hint: Follow the steps of the solution to the first equation in the lesson.)

16. Show that if $x = \frac{c - b}{a}$, then $ax + b = c$.

Using Algebra to Prove

17. Together, what do the statements in Questions 15 and 16 prove?

18. Use the definition of absolute value to find all values of x satisfying $|500x - 200| = 800$.

19. In Parts a and b, give abbreviated proofs.
    a. Prove: If a and k are both positive and $a(x - h)^2 = k$, then $x = h \pm \sqrt{\frac{k}{a}}$.
    b. Prove: If a and k are both positive and $x = h \pm \sqrt{\frac{k}{a}}$, then $a(x - h)^2 = k$.
    c. What has been proved in Parts a and b?

20. Prove: The slope of the line with equation $Ax + By = C$ is the reciprocal of the slope of the line with equation $Bx + Ay = D$.

REVIEW

21. Consider the following statement to be true: Every person under 8 years of age receives a reduced fare on the metro city bus. (**Lessons 13-2, 13-1**)
    a. Write this as an if-then statement.
    b. Write the converse.
    c. Is the converse true? Why or why not?

22. What value(s) can z not have in the expression $\frac{(2 - z)(1 + z)}{(4 + z)(3 - z)}$? (**Lessons 12-8, 5-2**)

23. a. Factor $3x^2 + 9x - 12$.
    b. Find a value for x for which $3x^2 + 9x - 12$ is a prime number. (**Lesson 12-5**)

24. a. Solve the system $\begin{cases} 2y + 3x = 7 \\ y = 6x - 1 \end{cases}$.
    b. Are the lines in Part a coincident, parallel, or intersecting? (**Lessons 10-6, 10-2**)

In 25–30, a property is stated. Describe the property using variables.

25. Product of Square Roots Property (**Lesson 8-7**)

26. Quotient of Square Roots Property (**Lesson 8-7**)

27. Power of a Power Property (**Lesson 8-2**)

28. Zero Product Property (**Lesson 2-8**)

29. Multiplication Property of –1 (**Lesson 2-4**)

30. Distributive Property of Multiplication over Subtraction (**Lesson 2-1**)

In 31 and 32, △ACB is similar to △XZY. (Lesson 5-10, Previous Course)

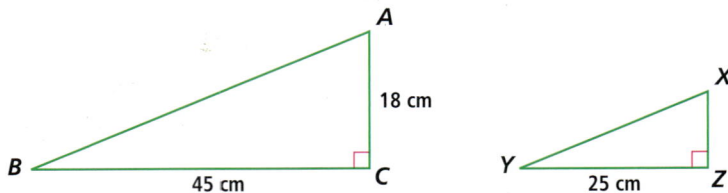

31. Find the missing lengths.

32. **Fill in the Blank** If m∠Y ≈ 22°, then m∠B ≈ __?__.

33. Two cards are drawn at random from a standard 52-card deck, without replacement. What is the probability of drawing an ace and a jack, in that order? (Lesson 5-8)

EXPLORATION

34. Use the Extended Distributive Property and other properties you have learned in this course to prove:
$(a + b + c)(a + b - c)(b + c - a)(c + a - b) = 2(a^2b^2 + b^2c^2 + c^2a^2) - (a^4 + b^4 + c^4)$.

The earliest playing cards are believed to have originated in Central Asia.

Source: The International Playing-Card Society

Lesson 13-4
A History and Proof of the Quadratic Formula

> **BIG IDEA** The Quadratic Formula can be proved using the properties of numbers and operations.

The Quadratic Formula $x = \frac{-b \pm \sqrt{b^2 - 4ac}}{2a}$ is quite complicated. You may wonder how people used to solve quadratic equations before they had this formula, and how they discovered the Quadratic Formula in the first place. Here is some of the history.

What Problem First Led to Quadratics?

Our knowledge of ancient civilizations is based only on what survives today. The earliest known problems that led to quadratic equations are on Babylonian tablets dating from 1700 B.C.E. In these problems, the Babylonians were trying to find two numbers x and y that satisfy the system $\begin{cases} x + y = b \\ xy = c \end{cases}$.

This suggests that some Babylonians were interested in finding the dimensions x and y of a rectangle with a given area c and a given perimeter $2b$. The historian Victor Katz suggests that maybe there were some people who believed that if you knew the area of a rectangle, then you knew its perimeter. In solving these problems, these Babylonians may have been trying to show that many rectangles with different dimensions have the same area.

Mental Math

Use the graph below to find each length.

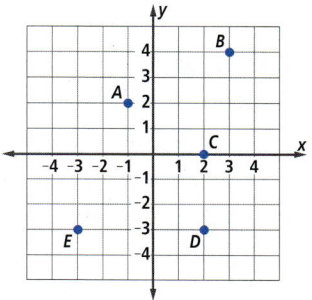

a. ED
b. CD
c. BC
d. AD

GUIDED
Example
Find the dimensions of a rectangular field whose perimeter is 300 meters and whose area is 4,400 square meters.

Solution Let L and W be the length and width of this rectangle.

Then $\begin{cases} \underline{} + \underline{} = 300 \\ \underline{} \cdot \underline{} = 4{,}400 \end{cases}$.

This system can be solved by substitution. First solve the top equation for W.

$\underline{} = 300 - \underline{}$

$W = 150 - \underline{}$

(continued on next page)

A History and Proof of the Quadratic Formula

Now, substitute __?__ for W in the second equation.

L(__?__) = 4,400

This is a quadratic equation and so it can be solved by using either the Quadratic Formula or factoring to get L = __?__ or L = __?__. Now substitute these values for L in either of the original equations to get W = __?__ or W = __?__. So, the dimensions of the field are __?__ m by __?__ m.

How the Babylonians Solved Quadratics

The Babylonians, like the Greeks who came after them, used a geometric approach to solve problems like these. Using today's algebraic language and notation, here is what they did. It is a sneaky way to solve this sort of problem. Look back at the Example.

Because $L + W = 150$, the average of L and W is 75. This means that L is as much greater than 75 as W is less than 75. So let $L = 75 + x$ and $W = 75 - x$. Substitute these values into the second equation.

$$L \cdot W = 4,400$$
$$(75 + x)(75 - x) = 4,400$$
$$5,625 - x^2 = 4,400$$
$$x^2 = 1,225$$

Taking the square root, $x = 35$ or $x = -35$.

If $x = 35$:
$L = 75 + x$, so $L = 75 + 35 = 110$
$W = 75 - x$, so $W = 75 - 35 = 40$

If $x = -35$:
$L = 75 + -35 = 40$
$W = 75 - -35 = 110$

Either solution tells us that the field is 40 meters by 110 meters.

 QY

Notice what the Babylonians did. They took a complicated quadratic equation and, with a clever substitution, reduced it to an equation of the form $x^2 = k$. That equation is easy to solve. Then they substituted the solution back into the original equation.

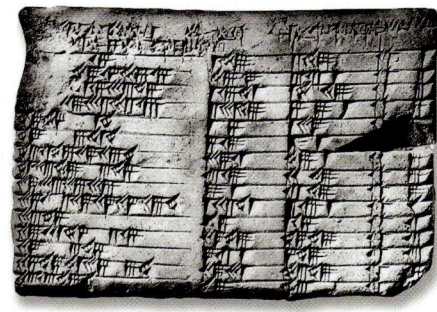

This tablet contains 14 lines of a mathematical text in cuneiform script.
Source: Iraq Museum

▶ QY

Use the Babylonian method to find two numbers whose sum is 72 and whose product is 1,007. (*Hint:* Let one of the numbers be $36 + x$, the other $36 - x$.)

The Work of Al-Khwarizmi

The work of the Babylonians was lost for many years. In 825 C.E., about 2,500 years after the Babylonian tablets were created, a general method that is similar to today's Quadratic Formula was authored by the Arab mathematician Muhammad bin Musa al-Khwarizmi in a book titled *Hisab al-jabr w'al-muqabala*. Al-Khwarizmi's techniques were more general than those of the Babylonians. He gave a method to solve any equation of the form $ax^2 + bx = c$, where a, b, and c are positive numbers. His book was very influential. The word "al-jabr" in the title of his book led to our modern word "algebra." Our word "algorithm" comes from al-Khwarizmi's name.

Muhammad bin Musa al-Khwarizmi

A Proof of the Quadratic Formula

Neither the Babylonians nor al-Khwarizmi worked with an equation of the form $ax^2 + bx + c = 0$, because they considered only positive numbers, and if a, b, and c are positive, this equation has no positive solutions.

In 1545, a Renaissance scientist, Girolamo Cardano, blended al-Khwarizmi's solution with geometry to solve quadratic equations. He allowed negative solutions and even square roots of negative numbers that gave rise to complex numbers, a topic you will study in Advanced Algebra. In 1637, René Descartes published *La Géometrie* that contained the Quadratic Formula in the form we use today.

Now we prove why the formula works. Examine the argument in the following steps closely. See how each equation follows from the preceding equation. The idea is quite similar to the one used by the Babylonians, but a little more general. We work with the equation $ax^2 + bx + c = 0$ until the left side is a perfect square. Then the equation has the form $t^2 = k$, which you know how to solve for t.

Given the quadratic equation: $ax^2 + bx + c = 0$ with $a \neq 0$. We know $a \neq 0$ because otherwise the equation is not a quadratic equation.

Step 1 Multiply both sides of the equation by $\frac{1}{a}$. This makes the left term equal to $x^2 + \frac{b}{a}x + \frac{c}{a}$. The right side remains 0 because $0 \cdot \frac{1}{a} = 0$.

$x^2 + \frac{b}{a}x + \frac{c}{a} = 0$

Step 2 Add $-\frac{c}{a}$ to both sides in preparation for completing the square on the left side.

$x^2 + \frac{b}{a}x = -\frac{c}{a}$

A History and Proof of the Quadratic Formula

Step 3 To complete the square add the square of half the coefficient of x to both sides. (See Lesson 12-2.)

$$x^2 + \frac{b}{a}x + \left(\frac{b}{2a}\right)^2 = -\frac{c}{a} + \left(\frac{b}{2a}\right)^2$$

Step 4 The left side is now the square of a binomial.

$$\left(x + \frac{b}{2a}\right)^2 = -\frac{c}{a} + \left(\frac{b}{2a}\right)^2$$

Step 5 Take the power of the fraction to eliminate parentheses on the right side.

$$\left(x + \frac{b}{2a}\right)^2 = -\frac{c}{a} + \frac{b^2}{4a^2}$$

Step 6 To add the fractions on the right side, find a common denominator.

$$\left(x + \frac{b}{2a}\right)^2 = -\frac{4ac}{4a^2} + \frac{b^2}{4a^2}$$

Step 7 Add the fractions.

$$\left(x + \frac{b}{2a}\right)^2 = \frac{b^2 - 4ac}{4a^2}$$

Step 8 Now the equation has the form $t^2 = k$, with $t = x + \frac{b}{2a}$ and $k = \frac{b^2 - 4ac}{4a^2}$. This is where the discriminant $b^2 - 4ac$ becomes important. If $b^2 - 4ac \geq 0$, then there are real solutions. They are found by taking the square roots of both sides.

$$x + \frac{b}{2a} = \pm\sqrt{\frac{b^2 - 4ac}{4a^2}}$$

Step 9 The square root of a quotient is the quotient of the square roots.

$$x + \frac{b}{2a} = \pm\frac{\sqrt{b^2 - 4ac}}{2a}$$

Step 10 This is beginning to look like the formula. Add $-\frac{b}{2a}$ to each side.

$$x = -\frac{b}{2a} \pm \frac{\sqrt{b^2 - 4ac}}{2a}$$

Step 11 Adding the fractions results in the Quadratic Formula.

$$x = \frac{-b \pm \sqrt{b^2 - 4ac}}{2a}$$

What if $b^2 - 4ac < 0$? Then the quadratic equation has no real number solutions. The formula still works, but you have to take square roots of negative numbers to get solutions. You will study these nonreal solutions in a later course.

Questions

COVERING THE IDEAS

1. **Multiple Choice** The earliest known problems that led to the solving of quadratic equations were studied about how many years ago?
   A 1,175
   B 1,700
   C 2,500
   D 3,700

2. In what civilization do quadratic equations first seem to have been considered and solved?

3. What is the significance of the work of al-Khwarizmi in the history of the Quadratic Formula?

In 4 and 5, suppose two numbers sum to 53 and have a product of 612. Show your work in finding the numbers.

4. Use the Quadratic Formula.

5. Use the Babylonian Method.

6. Suppose a rectangular room has a floor area of 54 square yards. Find two different lengths and widths that this floor might have.

In 7 and 8, suppose a rectangular room has a floor area of 144 square yards and that the perimeter of its floor is 50 yards.

7. Find its length and width by solving a quadratic equation using the Quadratic Formula or factoring.

8. Find its length and width using a more ancient method.

9. Find two numbers whose sum is 15 and whose product is 10.

10. In the proof of the Quadratic Formula, each of Steps 1–11 tells what was done but does not name the property of real numbers. For each step, name the property (or properties) from the following list.
    i. Addition Property of Equality
    ii. Multiplication Property of Equality
    iii. Distributive Property of Multiplication over Addition
    iv. Equal Fractions Property
    v. Power of a Quotient Property
    vi. Quotient of Square Roots Property
    vii. definition of square root

APPLYING THE MATHEMATICS

11. Solve the equation $7x^2 - 6x - 1 = 0$ by following the steps in the derivation of the Quadratic Formula.

12. Explain why there are no real numbers x and y whose sum is 10 and whose product is 60.

13. In a Chinese text that is thousands of years old, the following problem is given: The height of a door is 6.8 more than its width. The distance between its corners is 10. Find the height and width of the door.

14. Here is an alternate proof of the Quadratic Formula. Tell what was done to get each step.
$$ax^2 + bx + c = 0$$
a. $\quad 4a^2x^2 + 4abx + 4ac = 0$
b. $4a^2x^2 + 4abx + 4ac + b^2 = b^2$
c. $\quad 4a^2x^2 + 4abx + b^2 = b^2 - 4ac$
d. $\quad (2ax + b)^2 = b^2 - 4ac$
e. $\quad 2ax + b = \pm\sqrt{b^2 - 4ac}$
f. $\quad 2ax = -b \pm \sqrt{b^2 - 4ac}$
g. $\quad x = \dfrac{-b \pm \sqrt{b^2 - 4ac}}{2a}$

REVIEW

15. Consider the following statement. (**Lessons 13-2, 13-1**)

    A number that is divisible by 8 is also divisible by 4.
    a. Write the statement in if-then form.
    b. Decide whether the statement you wrote in Part a is true or false. If it is false, find a counterexample.
    c. Write the converse of the statement you wrote in Part a.
    d. Decide whether the statement you wrote in Part c is true or false. If it is false, find a counterexample.

16. Solve $x^2 + 5x = 30$. (**Lesson 12-6**)

In 17–19, an open soup can has volume $V = \pi r^2 h$ and surface area $S = \pi r^2 + 2\pi rh$, where r is the radius and h is the height of the can.

17. Use common monomial factoring to rewrite the formula for S. (**Lesson 11-4**)

18. Find each of the following. (**Lesson 11-2**)
    a. the degree of V
    b. the degree of S

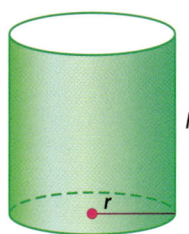

19. If the can has a diameter of 8 cm and a height of 12 cm, about how many milliliters of soup can it hold? ($1\text{ L} = 1{,}000\text{ cm}^3$) (**Lesson 5-4**)

20. Solve this system by graphing. $\begin{cases} y = |x| \\ y = \frac{1}{4}x^2 \end{cases}$ (**Lessons 10-1, 4-9**)

21. **Skill Sequence** Simplify each expression. (**Lessons 8-7, 8-6**)
    a. $\sqrt{8} + \sqrt{5}$
    b. $\sqrt{8} \cdot \sqrt{5}$
    c. $\dfrac{\sqrt{8} \cdot \sqrt{5}}{\sqrt{2}}$

EXPLORATION

22. In a book or on the Internet, research al-Khwarizmi and find another contribution he made to mathematics or other sciences. Write a paragraph about your findings.

QY ANSWER

19 and 53

Chapter 13

Lesson 13-5

Proofs of Divisibility Properties

Vocabulary

closed under an operation
even integer, even number
odd integer, odd number
semiperimeter

> **BIG IDEA** Using algebra, you can prove that even and odd numbers have certain general properties.

Mental Math

A function contains {(−16, 4.5), (−7, 4.5), (0, 2), (1, −5), (5.5, 10)}.

a. State the domain.

b. State the range.

In this lesson, the following statements are assumed to be true.

The sum of two integers is an integer.

The difference of two integers is an integer.

The product of two integers is an integer.

The three properties above are examples of *closure properties*. A set is **closed under an operation** if the results of that operation always lies in that set. So another way of saying the above statements is:

The set of integers is closed under addition.

The set of integers is closed under subtraction.

The set of integers is closed under multiplication.

But the set of integers is *not* closed under division. For example, 8 ÷ 3 is not an integer. When the quotient $a \div b$ is an integer, then we say that a is divisible by b, or that a is a multiple of b.

From the closure properties and other properties of real numbers that you know, it is possible to prove criteria that describe when one number is divisible by another.

Divisibility by 2

An **even integer** (or **even number**) E is an integer that is twice another integer; that is, it is an integer that can be written as $2n$, where n is an integer. As you know, the integers are the numbers 0, 1, −1, 2, −2, 3, −3, …. So, if you multiply these numbers by 2, the results are the *even integers* 0, 2, −2, 4, −4, 6, −6, 8, −8, ….

Geometrically, a positive even number of dots can be arranged in two rows of the same length. For example, 14 dots can be split into two rows of 7.

802 Using Algebra to Prove

To tell whether a large number is even, you cannot draw a pattern of dots. You must be able to show that it is twice another integer. For example, you can show that 5,734 is an even integer because 5,734 = 2 · 2,867, and 2,867 is an integer. How did we find 2,867? We divided 5,734 by 2. The number 0 is an even integer because 0 = 2 · 0. The negative number −88 is an even integer because −88 = 2 · −44.

Odd Integers

An integer that is not even is called *odd*. Geometrically, an odd number of dots cannot be arranged in two rows of the same length. An example of this is shown with 15 dots. Notice below that there are two rows of 7 dots plus an additional dot.

You can see that an odd integer is one more than an even integer. So we define an **odd integer** (or **odd number**) as an integer that can be written as $2n + 1$, where n is an integer. For example, $2 \cdot (-54) + 1 = -107$, so -107 is odd.

 QY

If two numbers m and n are positive, you know that their sum $m + n$ and their product mn are positive. But the difference $m - n$ might be positive or negative. What happens if you know whether m and n are even or odd?

By trying some numbers, fill in the following table with one of the words "odd" or "even."

> ▶ QY
>
> Let $m = 87{,}654$ and $n = 3{,}210$. Tell whether these numbers are even or odd.
>
> a. $m + n$
> b. $m - n$
> c. mn

Activity 1

m	n	m + n	m − n	mn
even	even	?	?	?
even	odd	?	?	?
odd	even	?	?	?
odd	odd	?	?	?

Testing pairs of numbers is not enough to show that a statement is true for *all* odd or even integers. Proofs are needed.

Chapter 13

Example 1
Prove that the sum of two even numbers is an even number.

Solution To prove this statement, we think of it as an if-then statement: If two even numbers are added, then their sum is an even number.

Let m and n be the even numbers. Then, by the definition of even number, there are integers p and q with $m = 2p$ and $n = 2q$. So $m + n = 2p + 2q$.

By the Distributive Property of Multiplication over Addition, $m + n = 2(p + q)$.

Since the sum of two integers is an integer, $p + q$ is an integer.

So $m + n$, equal to 2 times an integer, is even.

Example 1 shows that the set of even numbers is closed under addition.

Using the idea of Example 1, you can prove that the difference of two even numbers is an even number. That is, the set of even numbers is closed under subtraction. (You are asked to write a proof of this in one of the questions at the end of the lesson.) Also, the product of two even numbers is an even number.

GUIDED

Example 2
Prove that the difference of two odd numbers is an even number.

Solution Let m and n be odd numbers. Then, by the definition of odd number, there are integers p and q with $m = 2p + 1$ and $n = 2q + 1$.

So $m - n = (\underline{\ ?\ }) - (\underline{\ ?\ }) = 2p - 2q$.

Thus $m - n = 2(\underline{\ ?\ })$.

Since the difference of integers is an integer, $\underline{\ ?\ }$ is an integer.

Consequently, $m - n$ is $\underline{\ ?\ }$ times an integer, so $m - n$ is even.

You can similarly prove that the sum of two odd numbers is an even number. The set of odd integers is *not* closed under addition or subtraction.

Example 3 deals with products.

Example 3
Prove that the product of two odd numbers is an odd number.

804 Using Algebra to Prove

Solution Let m and n be odd numbers. Then, by the definition of odd number, there are integers p and q with $m = 2p + 1$ and $n = 2q + 1$.

Now multiply these numbers.

$$\begin{aligned} mn &= (2p + 1)(2q + 1) \\ &= 4pq + 2p + 2q + 1 \quad \text{Extended Distributive Property} \\ &= 2(2pq + p + q) + 1 \quad \text{Distributive Property} \\ &\qquad\qquad\qquad\qquad\quad \text{(Common monomial factoring)} \end{aligned}$$

Since the product of two integers is an integer, $2pq$ is an integer. Since p and q are integers, the sum $2pq + p + q$ is an integer. This means that mn is 1 more than twice an integer, so mn is odd.

Divisibility by Other Numbers

A number is *divisible by 3* if and only if it can be written as $3n$, where n is an integer. Similarly, a number is *divisible by 4* if and only if it can be written as $4n$, where n is an integer. In general, a number is divisible by m if and only if it can be written as mn, where n is an integer.

Activity 2

Step 1 Let n be an odd positive integer. Fill in the table.

n	1	3	5	7	9	11	13	15	17
n^2	?	?	?	?	?	?	?	?	?
$n^2 - 1$	?	?	?	?	?	?	?	?	?

Step 2 a. What is the greatest common factor of the integers in the bottom row?

b. Is this number a factor of $n^2 - 1$ for all odd integers?

You cannot answer the last question in Activity 2 by just writing down more odd integers, squaring them, and subtracting 1. You can never show by examples that the statement is true for all odd integers. A proof is needed.

Example 4

Prove that the square of an odd integer is always 1 more than a multiple of 4.

Solution First find the square of an odd integer.

(continued on next page)

Let n be an odd integer. By definition of an odd integer, there is an integer k such that $n = 2k + 1$.

$$\begin{align} \text{So } n^2 &= (2k + 1)^2 \\ &= (2k + 1)(2k + 1) \\ &= 4k^2 + 4k + 1 \\ n &= 4k(k + 1) + 1. \end{align}$$

This shows that n^2 is 1 more than a multiple of 4.

Questions

COVERING THE IDEAS

1. State the definition of *even integer*.
2. State the definition of *odd integer*.
3. Find a counterexample to show that this statement is not always true: *If two numbers are each divisible by 2, then their sum is divisible by 4.*

In 4 and 5, use Example 1 or Guided Example 2 as a guide to write a proof.

4. Prove: The difference of two even integers is an even integer.
5. a. Prove: The sum of two odd integers is an even integer.
   b. Is the set of odd integers closed under addition? Why or why not?
6. Use rows of dots to explain why the statement of Question 4 is true.
7. Use rows of dots to explain why the statement of Question 5a is true.
8. Prove that the set of even numbers is closed under multiplication. (*Hint:* Use Example 3 as a guide.)

APPLYING THE MATHEMATICS

In 9 and 10, complete the fact triangle. Then state the related facts.

9. 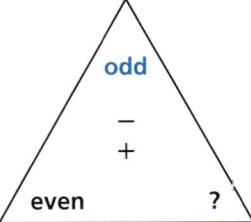   10.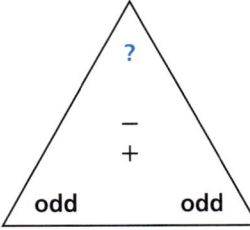

11. Prove: If a number is even, then its square is a multiple of 4.
12. Prove: If a number is divisible by 3, then its square is divisible by 9.

13. Prove: If the sum of two numbers is divisible by 35 and one of the two numbers is divisible by 70, then the other number is divisible by 35.

14. Prove or find a counterexample: If one number is divisible by 20 and a second number is divisible by 30, then their sum is divisible by 50.

15. Prove or find a counterexample: If one number is divisible by 4 and a second number is divisible by 6, then their product is divisible by 24.

REVIEW

16. Find two numbers whose sum is 562 and whose product is 74,865. **(Lesson 13-4)**

17. Find a value of b so that the quadratic expression $2x^2 - bx + 20$ is factorable over the integers. **(Lesson 12-6)**

In 18–23, solve the sentence. **(Lessons 12-5, 9-5, 8-6, 4-5, 4-4, 3-4)**

18. $100x^2 + 100x - 100 = 0$

19. $x^2 - 11x + 28 = 0$

20. $\frac{26}{N} = \frac{N}{0.5}$

21. $a \cdot 11^{\frac{1}{2}} = 99^{\frac{1}{2}}$

22. $4p - 12 \leq 60 - 5p$

23. $9.5 = 6x + 23.3$

24. Consider the system of equations $\begin{cases} 2x - 2y = 10 \\ -3x + 8y = -6 \end{cases}$. **(Lesson 10-8)**
    a. Write the system in matrix form.
    b. Use technology to find the inverse of the coefficient matrix.
    c. Solve the system.

25. The **semiperimeter** of a triangle is half the perimeter of the triangle. Heron's formula (also called Hero's formula) shown below can be used to calculate the area A of a triangle given the lengths of the three sides a, b, and c. **(Lesson 8-6)**

    $A = \sqrt{s(s-a)(s-b)(s-c)}$, where $s = \frac{1}{2}(a + b + c)$

    a. If the side lengths of a triangle are 15, 9, and 12 inches, calculate the semiperimeter s of the triangle.
    b. Find the area of the triangle in Part a.

26. If Emily reads 20 pages of a 418-page novel in 42 minutes, about how many hours will it take her to read the entire novel? **(Lesson 5-9)**

Hero of Alexandria

Chapter 13

EXPLORATION

27. The numbers 1, 4, 7, 10, ..., which increase by 3, can be pictured as 3 equal rows of dots with 1 left over. These numbers are of the form $3n + 1$. The numbers 2, 5, 8, 11, ..., which increase by 3, can be pictured as 3 equal rows of dots with 2 left over. These numbers are of the form $3n + 2$.

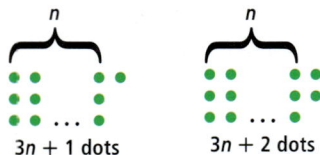

What happens if you add, subtract, and multiply numbers of these forms? Are the answers all of the same form? Try to prove any results you find.

28. In Activity 2, you found that, for the first 9 odd positive integers, the square of the odd number is 1 more than a multiple of 8. Prove that the result is true for all odd integers.

QY ANSWERS

a. even

b. even

c. even

Lesson 13-6

From Number Puzzles to Properties of Integers

▶ **BIG IDEA** Using algebra, you can show why divisibility tests and tricks relating to divisibility work.

In Lesson 2-3, you saw some number puzzles. In this lesson, you will see some unusual properties of divisibility that are like puzzles. Algebra shows why they work.

Activity 1

Step 1 Write down a 3-digit whole number, such as 175 or 220.

Step 2 Reverse the digits and subtract the new number from your original number.

$$\begin{array}{r} 175 \\ -\ 571 \end{array} \qquad \begin{array}{r} 220 \\ -\ 022 \end{array}$$

Step 3 Repeat Steps 1 and 2 with a few different numbers. You should find that the differences you get are always divisible by a large 2-digit number. What is that number?

Activity 2

1. Repeat Activity 1 with a few 4-digit numbers. Does the result you got in Activity 1 work for 4-digit numbers?
2. Does the result you got in Activity 1 work for 5-digit numbers?

Activity 3

Step 1 Write down a 3-digit whole number.

Step 2 Next to your number from Step 1 write the same 3 digits, creating a 6-digit number.

Step 3 The 6-digit number you get is always divisible by 3 different small prime numbers. What are those numbers?

Mental Math

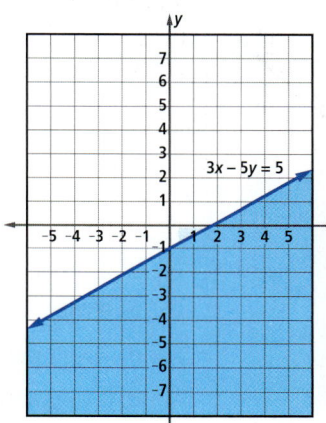

a. Describe the shaded region of the graph with an inequality.

b. Describe the unshaded region of the graph with an inequality.

Chapter 13

Activity 4

Step 1 Write down spaces for the digits of an 8-digit number.

__?__ __?__ , __?__ __?__ __?__ , __?__ __?__ __?__

Step 2 Choose numbers for these digits so that the sum of the 1st, 3rd, 5th, and 7th digits equals the sum of the 2nd, 4th, 6th, and 8th digits.

Step 3 Try this with a few numbers. Find a 2-digit number less than 25 that divides the 8-digit number. You may want to use the FACTOR feature of a CAS.

Activity 5

Step 1 Create a 10-digit number using each of the digits 0, 1, 2, 3, 4, 5, 6, 7, 8, and 9 once. For example, one such number is 8,627,053,914.

Step 2 Tell whether the statement is true or false.
  a. Every number created in Step 1 will be divisible by 3.
  b. Every number created in Step 1 will be divisible by 6.
  c. Every number created in Step 1 will be divisible by 9.
  d. Every number created in Step 1 will be divisible by 18.
  e. Every number created in Step 1 will be divisible by 27.

Divisibility Properties Depending on the Rightmost Digits of Numbers

You have known for a long time that in the base-10 number system the 4-digit number 5,902 is a shorthand for $5 \cdot 1{,}000 + 9 \cdot 100 + 0 \cdot 10 + 2 \cdot 1$ or, using exponents, it is a shorthand for $5 \cdot 10^3 + 9 \cdot 10^2 + 0 \cdot 10^1 + 2 \cdot 10^0$.

We say that 2 is the *units digit*, or the digit in the units place, 0 is the *tens digit*, 9 is the *hundreds digit*, and 5 is the *thousands digit* for the number 5,902. In general, if u is the units digit, t is the tens digit, h is the hundreds digit, and T is the thousands digit, then the value of the 4-digit number is

$$1{,}000T + 100h + 10t + u.$$

You can extend this idea using more variables to give the value of any integer written in base 10 in terms of its digits.

By representing the value of a number in terms of its digits, you can prove some divisibility tests that you have known for a long time. The proofs are quite similar to those used in Lesson 13-5.

Example 1
Prove that if the units digit of a number in base 10 is even, then the number is even.

Solution The proof here is for a 4-digit number N. The proof for numbers with fewer or more digits is very similar. A 4-digit number in base 10 with digits as named above has the value

$$N = 1{,}000T + 100h + 10t + u.$$

If the units digit u is even, then $u = 2k$, where k is an integer. Substituting $2k$ for u, $N = 1{,}000T + 100h + 10t + 2k$.

Notice that 2 is a common monomial factor of the polynomial on the right side. Factor out the 2.

$$N = 2(500T + 50h + 5t + k)$$

Since $500T + 50h + 5t + k$ is an integer, N is twice an integer, so it must be even.

In a similar way, you can prove divisibility tests for 4, 5, 8, and 10.

Divisibility Tests Based on the Sum of the Digits

There is a different type of divisibility test for 9: just add the digits of the number. The number is divisible by 9 if and only if the sum of its digits is divisible by 9. Proving this involves a variation of the approach taken in Example 1.

Example 2
Prove that if the sum of the digits of a 4-digit integer written in base 10 is divisible by 9, then the number is divisible by 9.

Solution Call the number N. Suppose N has digits T, h, t, and u as named above. (The same idea holds for any number of digits.)

$$N = 1{,}000T + 100h + 10t + u$$

Now separate the sum of the digits from the value of the number.

$$N = (T + h + t + u) + (999T + 99h + 9t)$$

If the sum of the digits is divisible by 9, then there is an integer k with $T + h + t + u = 9k$. Substitute $9k$ for $T + h + t + u$.

$$N = 9k + (999T + 99h + 9t)$$
$$N = 9(k + 111T + 11h + t)$$

Since $k + 111T + 11h + t$ is an integer, N is divisible by 9.

From Number Puzzles to Properties of Integers

Chapter 13

> **GUIDED**
>
> ## Example 3
> Prove that if the sum of the digits of a 4-digit integer written in base 10 is divisible by 3, then the number is divisible by 3.
>
> **Solution** Use Example 2 as a model for your solution.
>
> 1. Call the number N. Suppose N has digits T, h, t, and u.
> $N = 1{,}000T + \underline{} + \underline{} + \underline{}$
>
> 2. Now separate the sum of the digits from the value of the number.
> $N = (T + h + t + u) + (\underline{} + \underline{} + \underline{})$
>
> 3. If the sum of the digits is divisible by 3 then there is an integer k with
> $T + h + t + u = 3k$.
>
> 4. Substitute into Step 2.
> $N = 3k + (\underline{} + \underline{} + \underline{})$
> $N = 3(\underline{})$
>
> Because $\underline{}$ is an integer, N is divisible by 3.

Reversing Digits of a Number

Consider the 3-digit number $581 = 5 \cdot 100 + 8 \cdot 10 + 1$.

Reversing the digits of this number results in the number $185 = 1 \cdot 100 + 8 \cdot 10 + 5$.

So if a number has hundreds digit h, tens digit t, and units digit u, the number with the digits reversed has hundreds digit u, tens digit t, and units digit h. Whereas the first number has value $100h + 10t + u$, the number with its digits reversed has value $100u + 10t + h$.

Working with these numbers yields some surprising properties.

> ## Example 4
> Prove that if a 3-digit number is subtracted from the number formed by reversing its digits, then the difference is divisible by 99.
>
> **Solution** Suppose the original number has the value $100h + 10t + u$.
>
> Then the number with its digits reversed has value $100u + 10t + h$.
>
> Subtracting the reversed number from the original yields the difference D.

812 Using Algebra to Prove

$$D = 100h + 10t + u - (100u + 10t + h)$$
$$D = 100h + 10t + u - 100u - 10t - h$$
$$D = 99h - 99u$$
$$D = 99(h - u)$$

Since $h - u$ is an integer, the difference D is divisible by 99.

Questions

COVERING THE IDEAS

In 1–4, what is the value of the number?

1. The units digit of this 2-digit number is 7 and the tens digit is 5.

2. The units digit of this 2-digit number is u and the tens digit is t.

3. The thousands digit of the 4-digit number is A, the hundreds digit is B, the tens digit is C, and the units digit is D.

4. The millions digit of this 7-digit number is M, the thousands digit of this number is T, the units digit is 3, and all other digits are 0.

5. A 4-digit number has thousands digit T, hundreds digit h, tens digit t, and units digit u.
   a. What is the value of the number?
   b. What is the value of the number with its digits reversed?

6. Use Example 1 as a guide to prove: If the units digit of a 4-digit number in base 10 is 5, then the number is divisible by 5.

7. The proof in Example 2 is given for a 4-digit number. Adapt this proof for a 5-digit number, letting D be the ten-thousands digit.

8. **Fill in the Blanks** A number is divisible by 3 if and only if it can be written as __?__, where __?__ is an integer.

In 9–12, an integer is given.
   a. Tell whether the integer is divisible by 2 and state a reason why.
   b. Tell whether the integer is divisible by 5 and state a reason why.
   c. Tell whether the integer is divisible by 9 and state a reason why.

9. 259,259,259

10. 225

11. 522

12. 522 − 225

APPLYING THE MATHEMATICS

13. a. Find a counterexample: If a 4-digit number is subtracted from the number formed by reversing its digits, then the difference is divisible by 99.
    b. Prove: If a 4-digit number is subtracted from the number formed by reversing its digits, then the difference is divisible by 9.

14. a. Give an example of this statement and then prove it: If the units digit of a 5-digit number is 5 and the tens digit is 2, then the number is divisible by 25.
    b. Is the converse of the statement in Part a true?

15. In a certain 6-digit number, the hundred-thousands and hundreds digits are equal, the ten-thousands and tens digits are equal, and the thousands and units digits are equal. Prove that this number is divisible by 13.

16. The number $46x3$, written in base 10, is divisible by 9. What is the value of x?

17. The tens digit of a 3-digit number is 4 times the hundreds digit and the number is divisible by 19. Find the number.

REVIEW

In 18 and 19, a statement is given. Prove the statement to show that it is true or provide a counterexample to show that it is false. (Lesson 13-5)

18. If a number is divisible by 5, then its square is divisible by 25.

19. If one number is divisible by 3, and a second number is divisible by 4, then the product of the two numbers is divisible by 7.

20. Give an example of an if-then statement that is false but whose converse is true. (Lessons 13-2, 13-1)

21. The triangle below has an area of 45 square inches. Find the height h of the triangle if the base is $2h + 8$ inches. (Lesson 12-4)

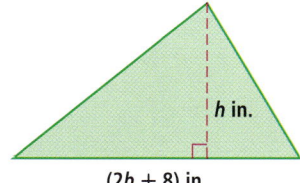

814 Using Algebra to Prove

22. A rectangular box has dimensions a, $a + 3$, and $2a + 1$.
    (**Lessons 11-5, 11-2**)
    a. Find a polynomial expression in standard form for the volume of the box.
    b. What is the degree of the polynomial in Part a?

23. Consider the quadratic equation $4m^2 - 20m + 25 = 0$. (**Lesson 9-6**)
    a. Find the value of the discriminant.
    b. Use your answer to Part a to determine the number of real solutions to the equation.

In 24–26, solve the sentence. (**Lessons 8-6, 5-9, 4-5**)

24. $\sqrt{m - 10} = 3$ 25. $5y - 2 > y$ 26. $\dfrac{w + 27}{9} = \dfrac{w}{3}$

27. What is the value of x in the equation $\dfrac{(h^5)^{10} \cdot h^{15}}{h^{20}} = h^x$?
    (**Lessons 8-4, 8-3, 8-2**)

EXPLORATION

28. Let h, t, and u be the hundreds, tens, and units digits of a 3-digit number in base 10.
    a. Find values of h, t, and u so that $hx^2 + tx + u$ is factorable over the integers.
    b. For your values of h, t, and u, is it true that $100h + 10t + u$ is factorable over the integers?
    c. **True or False** If h, t, and u are digits, and $hx^2 + tx + u$ is factorable over the set of polynomial with integer coefficients, then $100h + 10t + u$ is factorable over the integers.
    d. Explore this statement to decide whether it is true or false: If h, t, and u are digits, and $100h + 10t + u$ is a prime number, then $hx^2 + tx + u$ is a prime polynomial over the integers.

Chapter 13

Lesson 13-7
Rational Numbers and Irrational Numbers

Vocabulary

irrational number

> **BIG IDEA** The Distributive Property enables you to prove that repeating decimals represent rational numbers, and divisibility properties enable you to prove square roots of certain integers are irrational.

A number that can be represented by a decimal is a real number. All the real numbers are either rational or irrational. In this lesson, you will see how we know that some numbers are not rational numbers.

What Are Rational Numbers?

Recall that a *simple fraction* is a fraction with integers in its numerator and denominator. For example, $\frac{2}{3}$, $\frac{5,488}{212}$, $\frac{10}{5}$, $\frac{-7}{-2}$, and $\frac{-43}{1}$ are simple fractions.

Some numbers are not simple fractions, but are *equal* to simple fractions. Any mixed number equals a simple fraction. For example, $3\frac{2}{7} = \frac{23}{7}$. Also, any integer equals a simple fraction. For example, $-10 = \frac{-10}{1}$. And any finite decimal equals a simple fraction. For example, $3.078 = 3\frac{78}{1,000} = \frac{3,078}{1,000}$. All these numbers are *rational numbers*. A *rational number* is a number that can be expressed as a simple fraction.

All repeating decimals are also rational numbers. The Example below shows how to find a simple fraction that equals a given repeating decimal.

> **Mental Math**
>
> A school enrolled 120 freshmen, 110 sophomores, 125 juniors, and 100 seniors. What is the probability that
>
> **a.** a student at the school is a sophomore?
>
> **b.** a student at the school is a junior or senior?
>
> **c.** a student is not a junior?

Example

Show that $18.4\overline{23}$ is a rational number.

Solution Let $x = 18.4\overline{23}$. Multiply both sides by 10^n, where n is the number of digits in the repetend $\overline{23}$. Here there are two digits in the repetend, so we multiply by 10^2, or 100.

$$x = 18.4\overline{23}$$
$$100x = 1,842.3\overline{23}$$

816 Using Algebra to Prove

Subtract the top equation from the bottom equation. The key idea here is that the result is no longer an infinite repeating decimal; in this case, after the first decimal place the repeating parts subtract to zero.

$$100x = 1{,}842.3\overline{23}$$
$$-\ x = 18.4\overline{23}$$
$$99x = 1{,}823.900$$

Divide both sides by 99.

$$x = \frac{1{,}823.9}{99} = \frac{18{,}239}{990}$$

Since $x = \frac{18{,}239}{990}$, x is a rational number.

 QY1

> ▶ QY1
>
> a. Divide 18,239 by 990 to check the result of the Example.
>
> b. Write $4.\overline{123}$ as a simple fraction.

Rational numbers have interesting properties. They can be added, subtracted, multiplied, and divided; and they give answers that are also rational numbers.

What Are Irrational Numbers?

The ancient Greeks seem to have been the first to discover that there are numbers that are not rational numbers. They called them *irrational*. An **irrational number** is a real number that is not a rational number. Some of the most commonly found irrational numbers in mathematics are the square roots of integers that are not perfect squares. That is, numbers like $\sqrt{2}, \sqrt{3}, \sqrt{5}, \sqrt{6}, \sqrt{7}, \sqrt{8}, \sqrt{10}$, and so on, are irrational. But notice that $\sqrt{4}$ is rational, not irrational, because $\sqrt{4} = 2 = \frac{2}{1}$. All these numbers can arise from situations involving right triangles. Examine the array of right triangles shown below.

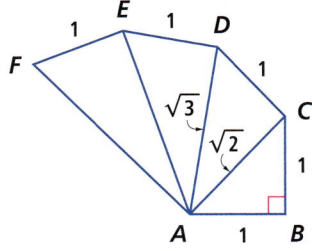

$\triangle ABC$ is a right triangle with legs of 1 and 1. Use the Pythagorean Theorem to find the side lengths AC and AD.

$$AC^2 = AB^2 + BC^2$$
$$AC^2 = 1 + 1$$
$$AC^2 = 2$$
$$AC = \sqrt{2}$$

Rational Numbers and Irrational Numbers

Chapter 13

△ACD is drawn with leg $\overline{AC}$, and another leg $CD = 1$. Use the Pythagorean Theorem.

$$AD^2 = AC^2 + CD^2$$
$$AD^2 = (\sqrt{2})^2 + 1^2$$
$$AD^2 = 2 + 1$$
$$AD^2 = 3$$
$$AD = \sqrt{3}$$

 QY2

> ▶ QY2
>
> Find *AE* and *AF* in the figure on the previous page.

How Do We Know That Certain Numbers Are Irrational?

If you evaluate $\sqrt{2}$ on a calculator, you will see a decimal approximation. One calculator shows 1.414213562. Another shows 1.41421356237. No matter how many decimal places the calculator shows, it is not enough to show the entire decimal because the decimal for $\sqrt{2}$ is infinite and does not repeat.

Is it possible the decimal could repeat after 1,000 decimal places, or after 1 million or 1 billion decimal places? How do we know that the decimal does not repeat? The answer is that we can *prove* the decimal does not repeat, because we can prove that $\sqrt{2}$ is not a rational number. The proof uses some of the ideas of divisibility you have seen in Lessons 13-5 and 13-6. In particular, we use the fact that if a number is even, then its square is divisible by 4. The idea of the proof is to show that there is no simple fraction in lowest terms equal to $\sqrt{2}$.

Here is the proof: Suppose $\sqrt{2}$ is rational. Then there would be two whole numbers *a* and *b* with $\sqrt{2} = \frac{a}{b}$ (with the fraction in lowest terms). Then, multiply each side of this equality by itself.

$\sqrt{2} \cdot \sqrt{2} = \frac{a}{b} \cdot \frac{a}{b}$ Multiplication Property of Equality

$2 = \frac{a^2}{b^2}$ definition of square root; Multiplication of Fractions

$2b^2 = a^2$ Multiply both sides by b^2.

So if you could find two numbers *a* and *b* with twice the square of *b* equal to the square of *a*, then $\sqrt{2}$ would be a rational number. (You can come close. 7^2 or 49 is one less than twice 5^2 or 25.)

Notice that since a^2 would be twice an integer, a^2 would be even. This means that *a* would be even (because the square of an odd number is odd). Because *a* would be even, there would be an integer *m* with $a = 2m$. This means that $a^2 = (2m)^2 = 4m^2$. Substitute in the bottom equation.

818   Using Algebra to Prove

$2b^2 = 4m^2$ Substitute $4m^2$ for a^2.

$b^2 = 2m^2$ Divide both sides by 2.

Now we repeat the argument used above. Because b^2 would be twice an integer, b^2 would be even. This means that b would have to be even. And because a and b would both be even, the fraction $\frac{a}{b}$ could not be in lowest terms. This shows that what we supposed at the beginning of this proof is not true.

For this reason, it is impossible to find two whole numbers a and b with $\sqrt{2} = \frac{a}{b}$ and with the fraction in lowest terms. Since any simple fraction can be put in lowest terms, it is impossible to find any two whole numbers a and b with $\sqrt{2} = \frac{a}{b}$.

Arguments like this one can be used to prove the following theorem.

> **Irrationality of $\sqrt{n}$ Theorem**
>
> If n is an integer that is not a perfect square, then $\sqrt{n}$ is irrational.

Johann Lambert

Today, we now know that there are many irrational numbers. For example, every number that has a decimal expansion that does not end or repeat is irrational. Among the irrational numbers is the famous number π. But the argument to show that π is irrational is far more difficult than the argument used above for some square roots of integers. It requires advanced mathematics, and was first done by the German mathematician Johann Lambert in 1767, more than 2,000 years after the Greeks had first discovered that some numbers were irrational.

There is a practical reason for knowing whether a number is rational or irrational. When a number is rational, arithmetic can be done with it rather easily because it can be represented as a simple fraction. Just work as you do with fractions. But if a number is irrational, then it is generally more difficult to do arithmetic with it. Rather than use its infinite decimal, we often leave it alone and just write π or $\sqrt{3}$, for example.

Questions

COVERING THE IDEAS

In 1–3, find an example of each.

1. a simple fraction
2. a fraction that is not a simple fraction
3. a rational number

Chapter 13

In 4–6, write the number as a simple fraction.

4. 98.6
5. $0.\overline{84}$
6. $14.0\overline{327}$

7. **Multiple Choice** Which *cannot* stand for a rational number?
   A a terminating decimal
   B a simple fraction
   C a repeating decimal
   D an infinite nonrepeating decimal

8. Refer to the proof that $\sqrt{2}$ is irrational.
   a. If $\sqrt{2}$ were rational, what would $\sqrt{2}$ have to equal?
   b. **True or False** If the square of an integer is even, then the integer is even.
   c. **True or False** If an integer is divisible by 2, then its square is divisible by 4.
   d. In the proof, what characteristic of both a and b shows that the fraction $\frac{a}{b}$ is not in lowest terms?

In 9–11, tell whether the number is a rational or an irrational number.

9. π
10. -220
11. $\sqrt{121}$

12. Draw a segment whose length is $\sqrt{5}$ units.

13. Draw a square whose diagonal has length $\sqrt{338}$ cm.

APPLYING THE MATHEMATICS

14. Is 0 a rational number? Why or why not?

15. Is it possible for two irrational numbers to have a sum that is a rational number? Explain why or why not.

16. Using the proof in this lesson as a guide, prove that $\sqrt{3}$ is irrational.

17. a. Draw a segment whose length is $1 + \sqrt{3}$ units.
    b. Is $1 + \sqrt{3}$ rational or irrational?

18. If a circular table has a diameter of 4 cm, is its circumference rational or irrational?

19. A diagonal of a square has a length of 42 cm. Find the perimeter of the square. Is the perimeter rational or irrational?

20. Determine whether the solutions to the equation $x^2 - 8x - 1 = 0$ are rational or irrational.

21. Refer to the right triangle at the right.
    a. Find the exact value of a.
    b. Is a rational or irrational?

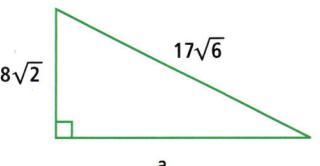

REVIEW

In 22–25, consider the spreadsheet below, which was used to compute the value of $f(x) = 3x^3 + 5x^2 - 2x$ for integer values of x from −5 to 5. (Lessons 12-7, 12-6, 11-4)

22. Complete the spreadsheet.

23. Graph the function f for $-5 \leq x \leq 5$.

24. Identify all x-intercepts.

25. Rewrite the equation in factored form.

	A	B
1	x	f(x)
2	-5	
3	-4	-104
4	-3	
5	-2	
6	-1	4
7	0	0
8	1	
9	2	40
10	3	120
11	4	
12	5	490

26. Solve $x^3 - 10x^2 + 16x = 0$. (Lessons 12-6, 12-5, 11-4)

27. Suppose $20x^2 + 9xy - 20y^2 = (ax + b)(cx + d)$. (Lesson 12-5)
    a. Find the value of $ad + bc$.
    b. Find b, c, and d if $a = 5$.

28. Find two numbers whose sum is 30 and whose product is 176. (Lessons 12-4, 11-6, 10-2)

29. Expand the expression $(\sqrt{25} - \sqrt{x^2})(\sqrt{25} + \sqrt{x^2})$. (Lessons 11-6, 8-6)

30. Calculate the area of the shaded region. (Lesson 11-3)

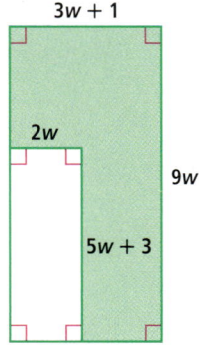

31. With a stopwatch and a stone, you can estimate the depth of a well. If the stone takes 2.1 seconds to reach the bottom, how deep is the well? Use Galileo's equation, $d = 16t^2$. (Lesson 9-1)

Dug wells typically used for drinking water are 10 to 30 feet deep.

Source: U.S. Environmental Protection Agency

EXPLORATION

32. Because $2 \cdot 5^2$ is one away from 7^2, 2 is close to $\frac{7^2}{5^2}$. That means that $\sqrt{2}$ is close to $\frac{7}{5}$, or 1.4. Find two other numbers c and d such that $2 \cdot c^2$ is one away from d^2. (*Hint:* There is a pair of such numbers with both of them greater than 2 less than 20.) What rational number estimate does that pair give for $\sqrt{2}$?

33. Shown here is a different way to draw a segment with length $\sqrt{n}$ from the one given in the lesson.

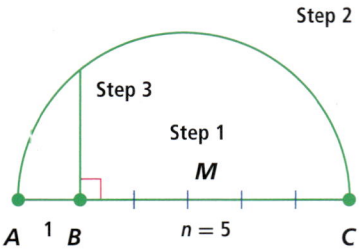

Step 1 Draw a segment $\overline{AB}$ with length 1, and then next to it, a segment $\overline{BC}$ with length n. (In the drawing here, $n = 5$.)

Step 2 Find the midpoint M of segment $\overline{AC}$. Draw the circle with center M that contains A and C. ($\overline{AC}$ will be a diameter of this circle.)

Step 3 Draw a segment perpendicular to $\overline{AC}$ from B to the circle. This segment has length $\sqrt{n}$. (In our drawing it should have length $\sqrt{5}$.)

  a. Try this algorithm to draw a segment with length $\sqrt{7}$.
  b. Measure the segment you constructed.
  c. How close is its length to $\sqrt{7}$?

QY ANSWERS

1a. $18{,}239 \div 990 = 18.4\overline{23}$

b. $\frac{1{,}373}{333}$

2. $AE = 2$, $AF = \sqrt{5}$

Lesson 13-8

Proofs of the Pythagorean Theorem

▶ **BIG IDEA** There are many ways to deduce the Pythagorean Theorem using algebra.

In this book you have seen how areas of rectangles can picture various forms of the Distributive Property. The idea is to calculate the area of a figure in two different ways. Here is a picture of $(a + b)(c + d + e) = ac + ad + ae + bc + bd + be$.

Mental Math

Tell whether the three numbers can be lengths of sides in a triangle.

a. 5, 13, 5

b. 2, 14, 15

c. 1, 2, 3

You could also say that this reasoning uses area to prove that $(a + b)(c + d + e) = ac + ad + ae + bc + bd + be$.

We close this book by showing how areas of figures provide proofs of the most famous theorem in geometry, the Pythagorean Theorem. If a and b are the lengths of the legs of a right triangle, and c is the length of its hypotenuse, then $a^2 + b^2 = c^2$.

For these proofs, you need to think of a^2, b^2, and c^2 as the areas of squares whose sides are a, b, and c. This is the form in which the theorem was discovered over 2,500 years ago in many different parts of the world.

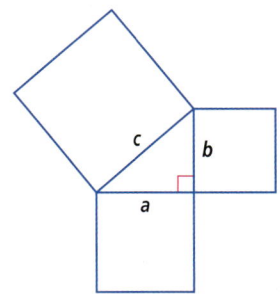

These proofs assume that you are familiar with the definitions and area formulas for some common figures. They are:

square: $A = s^2$ rectangle: $A = \ell w$

right triangle: $A = \frac{1}{2}ab$ triangle: $A = \frac{1}{2}bh$

trapezoid: $A = \frac{1}{2}h(b_1 + b_2)$

The proofs also use the properties of real numbers that you have seen in this course.

Proofs of the Pythagorean Theorem

Chapter 13

Bhaskara's Proof

Bhaskara's proof is a generalization of the idea that you saw in Lesson 8-6. Begin with right triangle DHK with side lengths a, b, and c. Make three copies of the triangle and place them so that quadrilateral DEFG is a square, as shown at the right. In $\triangle DHK$, $\angle DHK$ and $\angle DKH$ are complementary. Since corresponding parts of congruent triangles are congruent, m$\angle GKJ$ = m$\angle DHK$. So m$\angle DKH$ + m$\angle GKJ$ = 90°. Thus, m$\angle JKH$ = 180° − 90° = 90°. Likewise the other three angles of HIJK are right angles. So, the inside quadrilateral HIJK formed by the four hypotenuses has four right angles and four sides of length c, so it is also a square.

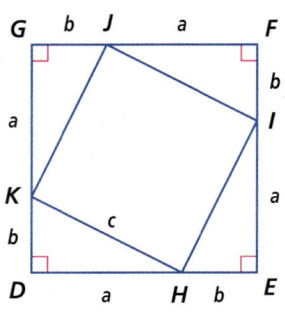

Let A be the area of quadrilateral DEFG. Each side of quadrilateral DEFG has length $a + b$. So $A = (a + b)^2$. But the area of DEFG can also be found by adding up the areas of the four right triangles $\left(4 \cdot \frac{1}{2}ab\right)$ and the square in the middle (c^2). So $A = 4 \cdot \frac{1}{2}ab + c^2$.

The two values of A must be equal.

$$(a + b)^2 = 4 \cdot \frac{1}{2}ab + c^2$$

Now use the formula for the square of a binomial on the left side and simplify the right side.

$$a^2 + 2ab + b^2 = 2ab + c^2$$

Add −2ab to each side of the equation.

$$a^2 + b^2 = c^2$$

This is the Pythagorean Theorem.

President Garfield's Proof

This proof of the Pythagorean Theorem was discovered by James Garfield in 1876 while he was a member of the U.S. House of Representatives. Five years later he became the 20th President of the United States.

Major General James Garfield won the 1880 Presidential election by only 10,000 popular votes, defeating Gen. Winfield Scott Hancock.

Source: The White House

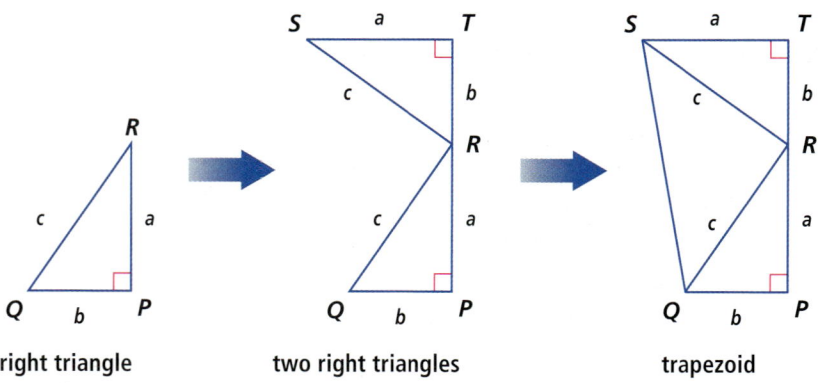

one right triangle　　　two right triangles　　　trapezoid

President Garfield's proof uses half the figure of the preceding proof. Begin with right triangle PQR as shown on the previous page. With one copy of $\triangle PQR$, create a trapezoid $PQST$ with bases a and b and height $a + b$. The area of any trapezoid is $\frac{1}{2}h(b_1 + b_2)$. Here the height $h = a + b$.

$$\text{Area of } PQST = \tfrac{1}{2}(a + b)(a + b)$$

But the area of $PQST$ is also the sum of the areas of three right triangles: PQR, RST, and QRS. Look at $\triangle QRS$. Because the sum of the measures of the angles of a triangle is $180°$, $m\angle QRP + m\angle RQP = 90°$. Consequently, $m\angle QRP + m\angle SRT = 90°$. This means that $\angle QRS$ is a right angle and so $\triangle QRS$ is a right triangle. Now add the areas of the three right triangles.

$$\text{Area of } PQST = \tfrac{1}{2}ab + \tfrac{1}{2}ab + \tfrac{1}{2}c^2$$

The area of the entire trapezoid must be the same regardless of how it is calculated.

$$\tfrac{1}{2}(a + b)(a + b) = \tfrac{1}{2}ab + \tfrac{1}{2}ab + \tfrac{1}{2}c^2$$

Now multiply both sides of the equation by 2.

$$(a + b)(a + b) = ab + ab + c^2$$

Multiply the binomials on the left side and collect terms on the right side.

$$a^2 + 2ab + b^2 = 2ab + c^2$$

Subtract $2ab$ from each side of the equation and the result is the Pythagorean Theorem.

$$a^2 + b^2 = c^2$$

Other Proofs

It takes only one valid proof of a theorem to make it true. Yet in mathematics you will often see more than one proof of a statement, just as you often see more than one way to solve a problem. Alternate methods can help you to understand better how the various parts of mathematics are related. In this lesson, you have seen how areas of triangles, trapezoids, and squares are put together with binomials to prove a statement about the lengths of the three sides of any right triangle. In your next course, likely to be more concerned with geometry than this one, you will see how this theorem is related to similar triangles. Later you will learn how important this theorem is in the study of trigonometry. The algebra you have learned this year is fundamental in these and every other area of mathematics.

Chapter 13

Questions

COVERING THE IDEAS

1. Picture the property that for all positive numbers a, b, and c, $a(b + c) = ab + ac$.

2. Picture the property that for all positive numbers a and b, $a(a + b) = a^2 + ab$.

In 3 and 4, refer to Bhaskara's proof of the Pythagorean Theorem.

3. a. Draw the figure of Bhaskara's proof when $a = 6$ and $b = 2$.
   b. What is the area of *DEFG*?
   c. Explain how to get the area of *HIJK*.
   d. What is the value of c?

4. $DH = a$ and $DK = b$ in the figure of Bhaskara's proof.
   a. What is the length of *EF*?
   b. What is the area of *EFGD*?
   c. What is the area of triangle *IJF*?
   d. What is the area of *HIJK*?
   e. What is the length of *HK* in terms of a and b?

5. a. Draw a trapezoid whose bases have lengths 1 in. and 2 in., and whose height is 1 in.
   b. What is the area of this trapezoid?

6. Draw a trapezoid with bases b_1 and b_2 and height h. Explain why the area of this trapezoid is $\frac{1}{2}hb_1 + \frac{1}{2}hb_2$.

7. Refer to President Garfield's proof of the Pythagorean Theorem. Let $a = 28$ and $b = 45$.
   a. Find the area of trapezoid *PQST*.
   b. Explain how to get the area of $\triangle RQS$.
   c. What is the value of c?
   d. Does the value of c agree with what you would get using the Pythagorean Theorem?

APPLYING THE MATHEMATICS

8. a. Find two expressions for the shaded region in the figure below.
   b. What property is illustrated by the answer to Part a?

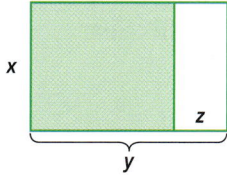

9. The square below has been split into two smaller squares and two rectangles. What property is pictured?

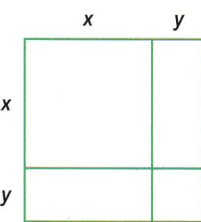

10. Quadrilateral *MNQP* at the right has perpendicular diagonals. Add the areas of the four triangles to show that the area of *MNQP* is one-half the product of the lengths of its diagonals.

11. A proof of the Pythagorean Theorem published by W.J. Dobbs in 1916 uses the figure at the right. △*ACB* and △*DAE* are right triangles, and $AC = b$, $BC = a$, and $AB = c$. Complete each step to show the proof.

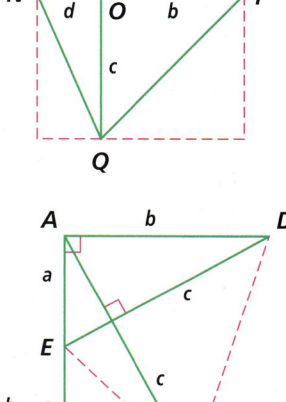

   a. What is the length of $\overline{EB}$ in terms of a and b?
   b. Find the area of △*EBC*.
   c. Find the area of *AEBD*, a quadrilateral with perpendicular diagonals, using the formula from Question 10.
   d. Add the areas in Parts b and c to find an expression for the area of *ACBD*.
   e. Use the formula for the area of a trapezoid to express the area of *ACBD*.
   f. Set the formulas from Parts d and e equal to each other to show that $c^2 = a^2 + b^2$.

REVIEW

12. Prove that if the last three digits of a 4-digit number form a number divisible by 8, then the entire number is divisible by 8. (**Lesson 13-6**)

13. **Multiple Choice** Consider the following statement. If the cost of 5 pounds of ice is $2.15, then at the same rate, the cost of 32 ounces of ice is 86 cents. (**Lessons 13-2, 13-1, 5-9**)
    A The statement and its converse are both true.
    B The statement and its converse are both false.
    C The statement is true but its converse is false.
    D The statement is false but its converse is true.

14. a. Find a value of c to complete the square for $4x^2 - 12x + c$.
    b. Use your answer to Part a to solve the equation $4x^2 - 2x = -9 + 10x$. (**Lessons 12-3, 12-2**)

15. Leonardo and Miranda are at an amusement park and are trying to decide in which order they want to ride the 9 roller coasters in the park. (**Lesson 11-7**)

    a. How many different orders can they ride all 9 roller coasters if they ride each coaster one time?

    b. If they only have time to ride six of the roller coasters, how many ways can they do this?

16. Consider the following number puzzle. (**Lessons 8-6, 2-3**)

    Step 1 Choose any whole number.
    Step 2 Square that number.
    Step 3 Add 4 times your original number.
    Step 4 Add 4 to the result of Step 3.
    Step 5 Take the square root of the result of Step 4.
    Step 6 Subtract your original number.

    a. Follow the number puzzle with any whole number. What is your result?

    b. Let x represent the number chosen. Write a simplified expression to represent each step of the puzzle and to show why your result will always be what you found in Part a.

17. A piece of landscaping machinery is valued at $15,000. If the machinery depreciates at a constant rate of 8% per year, what will be its value in 6 years? (**Lesson 7-3**)

Out of the 710 roller coasters in North America, 628 are in the United States.

Source: Roller Coaster Database

EXPLORATION

18. A different kind of proof of the Pythagorean Theorem is called a *dissection proof*. Dissection means cutting the squares on the legs of the right triangle shown on page 823 into pieces and then rearranging these pieces together to fill up the square on the hypotenuse. Find such a proof in a book or on the Internet and explain why it works.

Chapter 13 Projects

1 Squares Surrounding Triangles

Step 1 On a piece of grid paper, draw ten squares of different sizes. (*Note:* They don't *all* have to be different from one another.) Then, carefully cut them out.

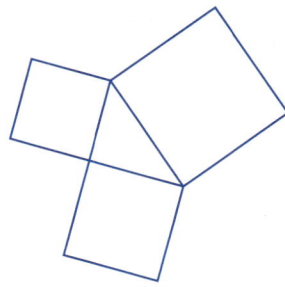

Step 2

a. Lay three of the squares on the table to form a triangle, as shown above. Refer to the longest side as c and the shorter sides as a and b.

b. Add the areas of the two smaller squares, and write down the sum. Is the sum $(a^2 + b^2)$ equal to the area of the largest square (c^2)? If not, is the sum greater or less?

c. Use a protractor to measure the angle across from the longest side of the triangle. Does it have the largest measure? Is it a right angle? If not, is it acute or obtuse?

Step 3 Repeat this procedure for at least ten different triangles. Record your information in a table, and look for any patterns in the data.

Step 4 Write a brief report about what you have learned about triangles, their largest angles, and the lengths of their sides.

2 Can Everything Be Proved?

At the beginning of the 20th century, a widespread belief among mathematicians was that any mathematical statement could either be proved, if it were true, or disproved, if it were false. But nobody knew for certain if this were true or not. In 1931, the mathematician Kurt Gödel settled the issue with his Incompleteness Theorems. Research to find out what Gödel discovered and how it impacted the world of mathematics.

3 Divisibility Tests

In this chapter, you saw several divisibility rules. These were just the tip of the iceberg. There are a great many known divisibility rules.

a. Find properties to check if a number is divisible by 4, by 8, by 25, and by 125. Explain how these rules work.

b. Look up a rule that can check if a number is divisible by 7, and one that can be used to check if a number is divisible by 11. Show why these methods work.

Chapter 13

4 Conjectures

A mathematical statement may be easy to write down and understand yet still be very difficult to prove. A statement that a mathematician believes to be true but is not yet proved is called a *conjecture* or a *hypothesis*. Sometimes conjectures remain unproved for many years. One of the most famous problems in mathematics is Fermat's Last Theorem. Research to learn about Fermat and his famous theorem, and about Andrew Wiles, the man who finally proved it. Finally, find at least two conjectures in mathematics that are still unproved today.

5 If-Then Statements in Games

When solving problems in life, you use many if-then statements.

a. Start with a single Sudoku puzzle. Write the first 5 if-then statements you can use to solve it. Estimate how many if-then statements it would take for you to solve the whole puzzle.

b. In 1997, the IBM supercomputer Deep Blue won 1 out of 3 matches against world Chess champion Garry Kasparov (above at the right). Find out how many different possible chess moves Deep Blue could consider each second. Given this number, how do you think that Kasparov was able to win 2 out of the 3 matches?

6 The Euclidean Algorithm

Given two positive integers m and n, the greatest common divisor of m and n is the greatest integer that divides both of them. For example, the greatest common divisor of 8 and 12 is 4; the greatest common divisor of 12 and 15 is 3. The ancient Greek mathematicians knew an algorithm, today called the Euclidean algorithm, to find the greatest common divisor of two numbers. Look up the Euclidean algorithm and write a description of how it works.

7 Rationals vs. Irrationals

When they were first discovered, irrational numbers were an oddity. As you saw in this chapter, there are many irrational numbers. One natural question to ask is which kind of number is more common: rational or irrational? Together with a friend, prepare a debate about this question. One side should present the position that rational numbers are more common and the other side should present that irrational numbers are more common. You may use any material you can find on the subject. Be sure to include a discussion about what you mean by "more common." Present your debate before the class.

Chapter 13 Summary and Vocabulary

- Generalizations in mathematics include **assumptions** (assumed properties), **definitions** (meanings of terms or phrases), and **theorems** (statements deduced from assumptions, definitions, or other theorems). These generalizations are often presented as **if-then statements.** For example, one assumed property of real numbers is the Distributive Property of Multiplication over Addition. It can be written in if-then form as: If a, b, and c are real numbers, then $a(b + c) = ab + ac$.

- The **converse** of the statement, "If a, then b" is the statement, "If b, then a." The converse of a true statement is not necessarily true. When the converse is true, then the statement "a if and only if b" is true. Definitions are **if-and-only-if statements.** For example, x is an even number if and only if x can be written as $2n$, where n is an integer.

- By putting together if-then statements of assumptions and definitions, a **mathematical proof** can be created. From the definition of even number, you can prove that if the square of an integer is even, then the integer is even. You can also prove that $\sqrt{2}$ and square roots of other nonzero integers that are not perfect squares are **irrational numbers.** Using the definition of divisibility by any number and what it means for a number to be in base 10, you can prove divisibility tests and other interesting properties of numbers.

- Every equation or inequality that you solve showing steps and justifications can be thought of as a **proof.** Suppose you solve $8x + 50 = 2$ and obtain $x = -6$. If you can justify the steps that you used in your solution, you have proved: "If $8x + 50 = 2$, then $x = -6$." The check is the converse: "If $x = -6$, then $8x + 50 = 2$."

- Mathematical knowledge grows by deducing statements from those that are assumed to be true or have been proved earlier to be true. Among the oldest and most important theorems in all of mathematics are the **Quadratic Formula** and the **Pythagorean Theorem.** Proofs of the Quadratic Formula use the properties that are most associated with solving equations. The proofs of the Pythagorean Theorem that we show in this chapter use area formulas for triangles, squares, and trapezoids.

Vocabulary

13-1
if-then statement
antecedent
consequent
generalization

13-2
converse
equivalent statements
if and only if

13-3
justifications
proof argument
deduction

13-5
closed under an operation
even integer, even number
odd integer, odd number
semiperimeter

13-7
irrational number

Theorems and Properties

Irrationality of $\sqrt{n}$ Theorem (p. 819)

Chapter 13 Self-Test

Take this test as you would take a test in class. You will need a calculator. Then use the Selected Answers section in the back of the book to check your work.

1. State conclusions and justifications to prove that if $8(2y - 1) = y + 37$, then $y = 3$.

2. Determine the antecedent and consequent of the statement proved in Question 1.

3. Amalia says that xy equals 0 if and only if both x and y equal 0.
   a. Write the two if-then statements that are equivalent to Amalia's if-and-only-if statement.
   b. Is Amalia correct? Explain your answer.

4. Marcus is measuring the diagonal across a piece of paper. The paper is 7 in. by 8 in.
   a. What is the exact length of a diagonal of the paper?
   b. Explain why Marcus' ruler will not give him an exact measurement of the diagonal.

5. Consider the following statement: All algebra students can solve quadratic equations.
   a. Write the statement in if-then form.
   b. Identify the antecedent and consequent for Part a.
   c. Write the converse of the statement you wrote in Part a.
   d. Decide whether the statement you wrote in Part c is true. Explain your answer.

6. Prove or find a counterexample to the statement: If the tens digit of a 4-digit number is 4 and the units digit is 8, then the number is divisible by 4.

7. What algebraic relationship is pictured by the rectangles, given that $b < a$?

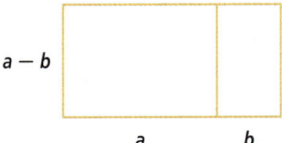

8. The product of two numbers is 717, and their sum is −242. What are the numbers?

9. a. Find the value of x in the diagram at the right.
   b. Is x rational or irrational?

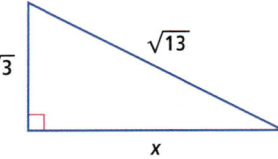

10. **True or False** Determine whether each of the following is true or false. Explain your answers.
    a. If a triangle is formed by cutting a square in half along one of its diagonals, then the triangle is isosceles.
    b. If a triangle is formed by cutting a square in half along one of its diagonals, then the triangle is equilateral.

11. **True or False** Determine whether the following statement is true or false and explain your answer: A person can be President of the United States if and only if he or she was born in the United States.

12. If a number is divisible by 3 and another number is divisible by 4, then their product is divisible by 12. Illustrate this statement with a picture and explain why your picture shows that the statement is true.

Chapter 13 Chapter Review

SKILLS
PROPERTIES
USES
REPRESENTATIONS

SKILLS Procedures used to get answers

OBJECTIVE A Show and justify the steps in solving an equation. (Lesson 13-3)

In 1 and 2, fill in the table for the proof.

1.

Conclusions	What Was Done	Justifications
$4x + 5 = 17$	?	?
$4x + 5 + -5 = 17 + -5$	?	?
$4x + 0 = 12$	?	?
$4x = 12$	?	?
$\frac{1}{4} \cdot 4x = 12 \cdot \frac{1}{4}$	?	?
$1 \cdot x = 3$	?	?
$x = 3$	?	?

2.

Conclusions	What Was Done	Justifications
$2n + 5 = 4n + 3$	?	?
$2n + 2 = 4n$	?	?
$2 = 2n$	?	?
$1 = n$	?	?

3. Prove: If $3t - 15 = 4t + 2$, then $t = -17$.
4. Prove: $\sqrt{16y - 16} = 2y$ if and only if $y = 2$.

OBJECTIVE B Find two numbers given their sum and product. (Lesson 13-4)

5. There are 26 students in a dancing class. If you know there are 165 possible boy-girl couples from this group, how many boys and how many girls are in the class?

In 6–9 find the two numbers that satisfy the given conditions.

6. $n + m = 10$, $nm = 24$
7. $xy = 2.3$, $x + y = 5.6$
8. $uv = 35$, $u + v = 12$
9. $p + q = -46$, $pq = 529$

10. Mrs. Violet doesn't know the dimensions of her rectangular garden, but she knows it has an area of 23.52 square meters. She also remembers that she needs 19.6 meters of fencing for her garden. Find the dimensions of Mrs. Violet's garden.

PROPERTIES The principles behind the mathematics

OBJECTIVE C Identify the antecedent and consequent of an if-then statement not necessarily given in if-then form. (Lesson 13-1)

In 11–14 identify the antecedent and the consequent.

11. If an animal has feathers then it is a bird.
12. It is spring if the trees are blooming.
13. No irrational number can be represented as the ratio of two integers.
14. James doesn't listen to music when he studies.

OBJECTIVE D Determine whether if-then and if-and-only-if statements in algebra or geometry are true or false. (Lessons 13-1, 13-2)

In 15–18, is the statement true or false?

15. A number is divisible by 3 if it is divisible by 9.

16. If $x = 7$ or $x = 3$, then $x^2 + 10x + 21 = 0$.

17. A triangle is equilateral if and only if two of its sides are equal and it has one 60° angle.

18. If only two outcomes are possible and they are equally likely, then the probability of each is 50%.

OBJECTIVE E Prove divisibility properties of integers. (Lessons 13-5, 13-6)

19. Prove that a 3-digit number abc is divisible by 7 only if the number $2a + 3b + c$ is divisible by 7.

20. Show that if n is even then n^3 is divisible by 8.

21. Show that all 6-digit integers of the form $xyzxyz$ are divisible by 13.

22. Show that if the 4-digit number $abcd$ written in base 10 is divisible by 11, then $b + d - (a + c)$ is divisible by 11.

OBJECTIVE F Apply the definitions and properties of rational and irrational numbers. (Lesson 13-7)

In 23–26, tell whether the number is rational or irrational.

23. $\sqrt{6}$ **24.** $0.\overline{142857}$

25. $\sqrt{169}$ **26.** $2\pi - 3$

27. Is it possible for two irrational numbers to have a product that is rational? Explain why or why not.

28. Is it possible for two rational numbers to have a product that is irrational? Explain why or why not.

USES Applications of mathematics in real-world situations

OBJECTIVE G Determine whether if-then and if-and-only-if statements in real-world contexts are true or false. (Lessons 13-1, 13-2)

In 29–32, a statement is given.
a. Is the statement true?
b. Is the converse true?
c. If either the statement or the converse is not true, change the statement so that both are true. Rewrite the new statement in if-and-only-if form.

29. A year with 366 days is a leap year.

30. If you live in France, you live within 10 kilometers of the Eiffel Tower.

31. If you are an eleventh grader, you are in high school.

32. All horses are four-legged animals.

REPRESENTATIONS Pictures, graphs, or objects that illustrate concepts

OBJECTIVE H Display or prove properties involving multiplication using areas of polygons or squares. (Lesson 13-8)

33. Picture the property that for all positive numbers a and b, $(a + b)^2 = a^2 + b^2 + 2ab$.

34. Square $ABCD$ is pictured below. Show that the area of $ABCD$ is equal to the sum of the areas of the four small triangles.

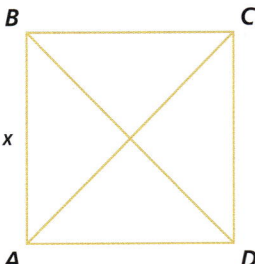

35. Draw a rectangle with dimensions a and b, and draw a diagonal from one corner to the other, making two triangles. Prove that the diagonal cuts the area of the rectangle in half.

36. Use the isosceles trapezoid below to show that $\frac{1}{2}(2b + 2x)h = xh + bh$.

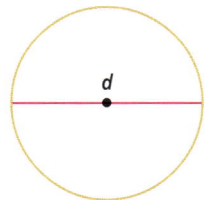

OBJECTIVE I Determine whether lengths of geometric figures are rational or irrational. (Lesson 13-7)

37. Consider the circle below. Its circumference is 32 inches.

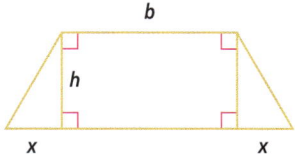

a. What is the exact radius of the circle?
b. Is this number rational or irrational?

In 38–40,
a. determine the missing length, and
b. determine whether your answer to Part a is rational or irrational.

38.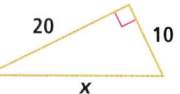

39.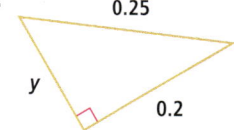

40.

41. a. Draw a segment whose length is $1 + \sqrt{8}$ centimeters.
b. Is that length rational or irrational?

Selected Answers

Chapter 1

Lesson 1-1 (pp. 6-12)
Guided Example 2: $\left(\frac{6.8 - w}{n + w}\right)^3$; $\left(\frac{6.8 - 0.5}{21 + 0.5}\right)^3$; $((6.8 - 0.5)/(21 + 0.5))\wedge 3$; 0.025

Questions: 1. 24   3. 81   5. −105   7. a. 32   b. −16   9. a. 500
b. 2,500   11. 6   13. false, −13 ≠ −9   15. a. $\frac{63}{225} = \frac{28}{100}$
b. yes, by the Transitive Property of Equality   17. $\frac{1}{-9} \cdot 4$
19. $\frac{6}{55}$ 21. −$5   23. 70   25. $617.50   27. 22   29. 15.08

Lesson 1-2 (pp. 13-19)
Guided Example 1:
a.

Months Since Beginning of School Year	Calculation	Pattern	Magazines in Library
0	3,600	3,600 + 22(0)	3,600
1	3,600 + 22	3,600 + 22(1)	3,622
2	3,600 + 22 + 22	3,600 + 22(2)	3,644
3	3,600 + 22 + 22 + 22	3,600 + 22(3)	3,666

b. Let m = the number of months since the beginning of the school year; $3,600 + 22m$

Questions: 1. Answers vary. Sample answer: $7 - 7 = 0$ and $2 - 2 = 0$ 3. Answers vary. Sample answer: $5 \cdot 5 = 5^2$ and $6 \cdot 6 = 6^2$ 5. $(3 + x) - 2 = 1 + x$ 7. a. $1,155; $1,110; $1,065   b. $1,200 - 45w$ dollars   9. C
11. a.

Istu's Age	Christine's Age
9	4
16	11
25	20
89	84

b. $i - 5$ c. $(i - 5) + 3$ or $i - 2$ 13. a. For 1 cut there are 2 pieces; for 2 there are 4; for 3 there are 6; for 4 there are 8; and for 5 there are 10.
b. Let c be the number of cuts and p be the number of pieces. Then, $p = 2c$.   15. a. 0   b. No; $xy = yx$ by the Commutative Property of Multiplication.   17. −309
19. a. −15   b. 420   21. $7

Lesson 1-3 (pp. 20-26)
Guided Example 1:

Alf

n	$(n + 1) + 3 + (n + 1)$
1	7
2	9
3	11
10	25
20	45
35	75

Beth

n	$5 + 2n$
1	7
2	9
3	11
10	25
20	45
35	75

Questions: 1. a. For $n = 1$, $3n - 2 = 1$; for $n = 2$, $3n - 2 = 4$; and for $n = 3$, $3n - 2 = 7$ b. 298 tiles

3.

x	$3x - 17$
10	13
9	10
8	7
7	4
6	1
5	−2

x	$x - 6 - (11 - 2x)$
10	13
9	10
8	7
7	4
6	1
5	−2

Because the tables have the same values, the two expressions seem to be equivalent.   5. a. Answers vary. Sample answer:

x	$25 + (x - 5)(x + 5)$	x^2
0	0	0
2	4	4
3	9	9
−1	1	1

b. Yes, they appear equivalent.   7. a. Answers vary. Sample answer:

x	$x^2 - 4x - 3$	$(x - 3)(x + 1)$
0	−3	−3
2	−7	−3
3	−6	0
−1	2	0

b. No, they are different.   9. Answers vary. Let $a = 0$ and $b = 1$. Then $(a - 5) + b = -4$ and $a - (5 + b) = -6$.   11. $x^2 - x = x(x - 1)$ 13. a. Answers vary. Sample answer: $2^2 \cdot 2 = 2^3$ and $(-1)^2 \cdot (-1) = (-1)^3$ b. Answers vary. Sample answer: $3(-1) - (-1) - (-1) = -1$ and $3(0) - 0 - 0 = 0$ c. Answers vary. Sample answer: $3(3 + 8) = 3(3) + 8(3)$ and $45(3 + 8) = 45(3) + 8(45)$

15. >   **17.** =   **19.** $37.89

Lesson 1-4 (pp. 27-32)
Questions:

1. a.

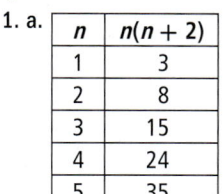

n	n(n + 2)
1	3
2	8
3	15
4	24
5	35

b.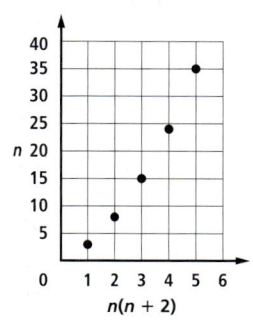

3.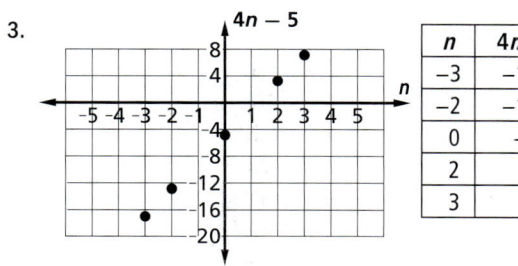

n	4n − 5
−3	−17
−2	−13
0	−5
2	3
3	7

5.

x	Value of Expression
−3	4
0	−2
2	−6
−1	0
−4	6
1	−4

7. a. 5 **b.** 3 and 5
9. Answers vary. Sample answer: $\{x : x$ is an integer and $x < -8\}$ **11.** Scatterplot corresponds to b; Connected graph corresponds to a.

13. d **15.**

n	2(n − 3)
−5	−16
−3	−12
1	−4
2	−2
5	4

n	2n − 3
−5	−13
−3	−9
1	−1
2	1
5	7

The expressions are not equivalent, since for given values of n, the expressions give different values.
17. a. For $t = 0$, $t − 2t + 3t =$ and $2t = 0$; for $t = 1$, $t − 2t + 3t = 2$ and $2t = 2$; and for $t = -2$, $t − 2t + 3t = -4$ and $2t = -4$. **b.** The pattern holds for all real numbers because the expressions are equivalent.
19. 150.28 cm³   **21.** 18 students

Lesson 1-5 (pp. 33-41)
Questions: **1.** $-7 \leq x \leq 6$ and $-10 \leq y \leq 20$

3.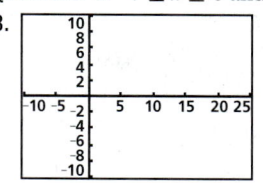

5.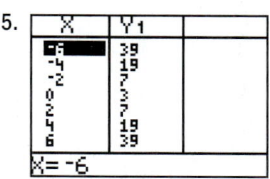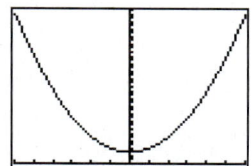

7. a. 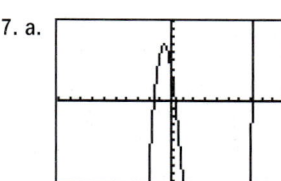 **b.** Xmin: −4, Xmax: 10, Ymin: −80, Ymax: 10

9. a. 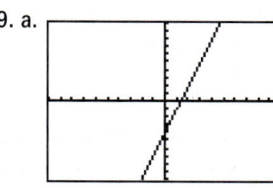 **b.** Answers vary. Sample answer: (2, 2), (0, −4), (−1, −7) **11.** Answers vary. Sample answer: $-10 \leq x \leq 10$ and $-40 \leq y \leq 10$ **13.** The expressions are not equivalent because the graphs do not overlap entirely.

15. a. $54 + 26m$ dollars **b.** Answers vary. Sample answer:

m	54 + 26m
0	$54
1	$80
2	$106
3	$132
4	$158
5	$184

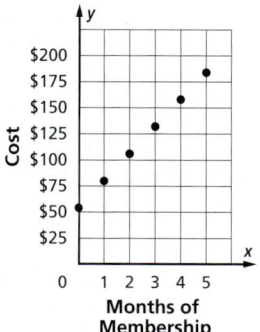

17. d   **19. a.** 9.2 **b.** 8.5 **c.** The air conditioner in Part a is more efficient because it has a higher EER.   **21.** 0   **23.** −4   **25.** −19

Student Handbook

Lesson 1-6 (pp. 42-46)

Questions: **1. a.** 2,833 ft **b.** −2,907 ft **c.** 2,907 ft
3. a. 75 **b.** 75 **c.** 5 **d.** 75 **5.** 1 **7.** 1
9. Answers vary. Sample answer: $t = -3$
11. a. 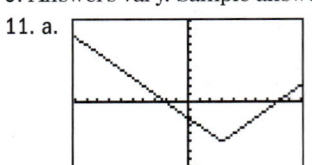 **b.**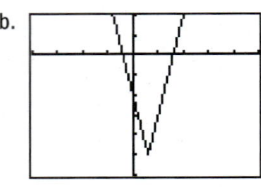

c. Xmin: −15, Xmax: 0, Ymin: −6, Ymax: 10 **13.** 5 **15.** $\frac{1}{2}$
17. a. 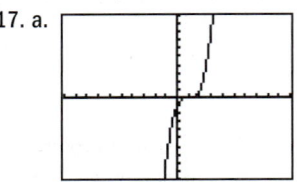 **b.** (1, 0) **19.** {5, 6, 7, 8, 9, 10} **21. a.** 12 **b.** 1

21. a.

Term Number	Number of Toothpicks
1	3
2	5
3	7
4	9
5	11

b. $2n + 1$ **c.** 201

Self-Test (pp. 58-59)

1. a. $9 \cdot 4 + 9 \cdot 6 = 9 \cdot 10; 9 \cdot 4 + 9 \cdot 7 = 9 \cdot 11; 9 \cdot 4 + 9 \cdot 8 = 9 \cdot 12$ **b.** Generally, for any x, $9 \cdot 4 + 9 \cdot x = 9 \cdot (x + 4)$. **2.** Answers vary. Sample answer: $3 \cdot 17 = 17 \cdot 3$
3. $3x \div 7y = 3x \cdot \frac{1}{7y}$ **4.** The total cost for the jerseys is the cost of a single jersey times the number of jerseys plus the cost of a single T-shirt times the number of T-shirts, so Total Cost = $179j + 24t$ dollars.

5. a.

n	$\frac{6n - 12}{3}$	$-4 + 2n$
−5	−14	−14
−3	−10	−10
0	−4	−4
2	0	0

b. Yes. For all values on this table, the two expressions are equal. **6. a.** $10 + 7 + 7 + 7 = 31$ tiles

b.

n	1	2	3	4	5	6
Tiles	10	17	24	31	38	45

c. The original design has 10 tiles, and each nth design has an additional $(n - 1) \cdot 7$ tiles, so the nth design has $10 + (n - 1) \cdot 7$, which simplifies to $3 + 7n$.

6. d.

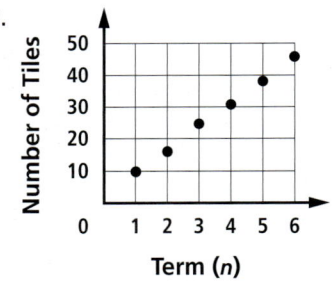

Lesson 1-7 (pp. 47-54)

Questions: **1.** about 2.46 runs

3.

Score	Mean	Deviation	|Absolute Deviation|
21	22	−1	1
15	22	−7	7
25	22	3	3
22	22	0	0
27	22	5	5

m.a.d. = 3.2 **5.** mean = 3, m.a.d. = 1 **7. a** **9. c**
11. a. For Presidents, the mean is about 5.0 years with a m.a.d. of about 2.1 years and range of 12, while for the English rulers, the mean is about 21.9 years with a m.a.d. of about 12.8 years and range of 63. **b.** Answers vary. Sample answer: Presidents have set term-lengths and a limit to the total amount of terms, whereas rulers, who may be very old or very young when crowned, typically reign until death. **13.** Answers vary. Sample answer: 3, 4, 7, 9, 10, 11, 12 **15.** San Diego **17.** x-min = −120; x-max = 60; y-min = −1.5; y-max = 0.5

19.

n	$n^2 - n$
−3	12
−2	6
0	0
2	2
3	6

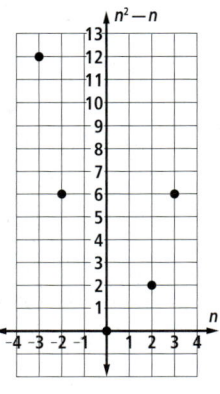

7. a. Answers vary. Sample answer: $m = 0$, since $\frac{m}{2} + \frac{3}{2} = \frac{0}{2} + \frac{3}{2} = 0 + \frac{3}{2} = \frac{3}{2}$ but $\frac{3 + m}{4} = \frac{3 + 0}{4} = \frac{3}{4}$
b. Answers vary. Sample answer: By the Distributive Property of Multiplication, we can factor out a $\frac{1}{2}$, and the definition of division gives the expression $\frac{m + 3}{2}$.
8. a. They appear to be equivalent. **b.** No. For instance, the value $x = 1$ gives 3.01 for the first expression and 3 for the second.

9. a. 3 b.

x	0	1	2	3	4	5
y	3	2	1	2	3	4

c. B. 10. Answers vary. Sample answer: x-min = –2; x-max = 20; y-min = –20; y-max = 2 11. skewed left; The tail is to the left.

12. $\mu = \dfrac{6+7+9+8+10+2+7+10+10+9+8+8+9+9+10}{15}$,

$\mu = \dfrac{1(2)+1(6)+2(7)+3(8)+4(9)+4(10)}{15} \approx 8.1$

m.a.d. = $\dfrac{|2-\mu|+|6-\mu|+|27-\mu|+|38-\mu|+|49-\mu|+|410-\mu|}{15}$,

m.a.d. = $\dfrac{6.1+2.1+2(1.1)+3(0.1)+4(0.9)+4(1.9)}{15} \approx 1.5$

13. The m.a.d. is what determines the spread, and since we have a lower m.a.d. in the original data set, there is less spread in the original set.

14. a. 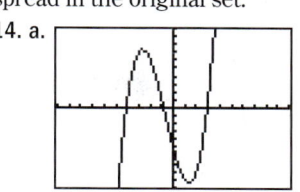 b. Answers vary. Sample answer: Xmin = –4, Xmax = –1, Ymin = 3, Ymax = 8

The chart below keys the **Self-Test** questions to the objectives in the **Chapter Review** on pages 60–63 or to the **Vocabulary (Voc)** on page 57. This will enable you to locate those **Chapter Review** questions that correspond to questions missed on the **Self-Test**. The lesson where the material is covered is also indicated on the chart.

Question	1	2	3	4	5	6	7	8	9	10
Objective(s)	B	G	F	H	A, C	B	C	L	D, J, K	M
Lesson(s)	1-2, 1-3	1-1, 1-2	1-1	1-2	1-1, 1-3	1-4	1-3	1-4, 1-5, 1-6	1-4, 1-5, 1-6	1-5

Question	11	12	13	14
Objective(s)	I	E, I	I	M
Lesson(s)	1-7	1-7	1-7	1-5

Chapter Review (pp. 60-63)

1. 0.8 3. 2.15 5. 676 7. $-\dfrac{251}{14}$, or about –17.93 9. 43.6
11. $4x + 3x = 7x$ 13. a. 25 dots b. Answers vary. Sample answer: There is one central dot and four "spokes," each with one with one less dot than the term number.
c. Answers vary. Sample answer: $4n - 3$ 15. They appear not to be equivalent. 17. 5 19. –10 21. 2
23. $r = 2.25$, m.a.d. = 0.89 25. $r = 28$ points, m.a.d. = about 9.42 points 27. $-8 + -y + -32$
29. true 31. $6.21 \cdot \dfrac{1}{3.14}$ 33. Commutative Property of Multiplication 35. Transitive Property of Equality
37. $1{,}225a + 1{,}405b$ dollars
39. a.

Week (w)	Total (t)
0	50
1	70
2	90
3	110
4	130

b.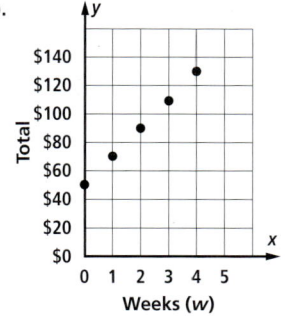

41. a.

n	$2n + (n+2)$
1	5
2	8
3	11
4	14
5	17

b.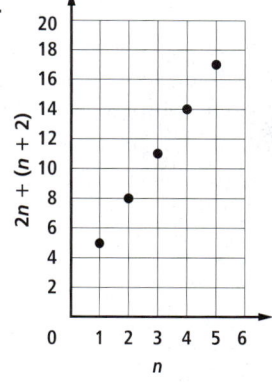

c. 35
43. They are not equivalent.

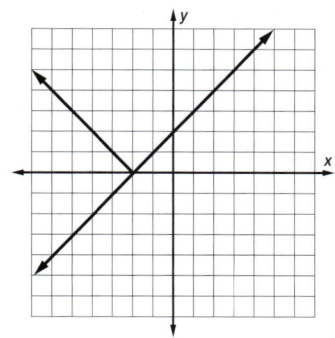

45. They are not equivalent.

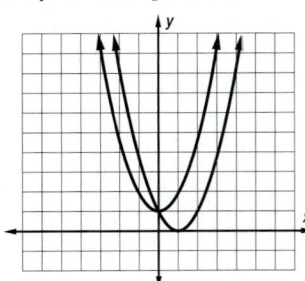

47.

x	y
0	–5
2	3
3	4
6	–5

Chapter 2

Lesson 2-1 (pp. 66–71)
Guided Example 3: $2x(5x - 3) = 2x \cdot 5x - 2x \cdot 3 = 10x^2 - 6x$
Questions: 1. a. $nk + nw$ **b.** $gd - ge$ **c.** $\frac{n}{r} + \frac{p}{r}$
3. Answers vary. Sample answer: $45(1.00 + 0.03) = 45 + 1.35 = \$46.35$ **5.** $5m + 20$ **7.** $12k - 2$ **9.** $18v - 48w + 54z^3$ **11.** $0.12 \cdot (17 + 6); 0.12 \cdot 17 + 0.12 \cdot 6 = \$2.76$ **13.** $k; m$ **15.** Answers vary. Sample answer: $9 \cdot \left(13\frac{1}{2}\right) = 9 \cdot \left(13 + \frac{1}{2}\right) = 9 \cdot 13 + 9 \cdot \frac{1}{2} = 117 + 4.5 = 121.5$ minutes. **17. a.** 40.3 films **b.** 9.3 films; On average, they have been within 9.3 films of their mean of 40.3.
19. 99 **21.** \$1,442 **23.** –2, 5, 8

Lesson 2-2 (pp. 72–78)
Guided Example 2: $4f^2 + f + 9 + 120f - 10f^3 - 30$ $(4f^2 + -10f^2) + (f + 120f) + (9 + -30) = -6f^2 + 121f + -29$
Guided Example 4: $m; 2p$
Questions: 1. c and d are like terms. **3.** $12x$
5. $8x + 3y$ **7.** $\frac{x+5}{3m}$ **9.** $26f + 3h$ **11.** $a^2; 4b$
13. $8p^2 + p$ **15.** $5(3ab + 8c - 2)$ **17.** $9f$ **19.** Answers vary. Sample answer: Let $x = 0; 2 \neq 1$ **21.** Answers vary. Sample answer: Timothy, you did not distribute –4 over all the terms. **23.** Answers vary. Sample answer: $8(40 - 0.05) = 320 - 0.40 = \$319.60$ **25. a.** $JM = x$ cm; $JI = x + 5$ cm; $IM = 2x - 7$ cm **b.** $4x - 2$ cm **27.** b
29. a **31. a.** \$6,200 **b.** \$27,900 **c.** \$41,850 **d.** \$9,300y
e. \$775m **33.** about \$60.77

Lesson 2-3 (pp. 79–84)
Guided Example 1: Answers vary. Sample answer: 17; 18; 36; 108; 104; 109; 7 **Guided Example 3:** $n - 4; 8n - 32;$ $8n - 32 + 8n; 16n - 32; \frac{16n - 32}{16}; n - 2; n - 2 + 8;$ $n + 6; n + 6 - n; 6$
Questions: 1. Answers vary. Sample answer: Variables are used to see how the puzzle works for *any* number.
3. The result is 97. **5.** The result is 192.
7. The result is 7. **9. a.** 7; –1; 6; 36; 41; 45; 15; 1
b. –2.9; –10.9; –3.9; –23.4; –18.4; –14.4; –4.8; 1 **c.** $n; n - 8;$ $n - 1; 6n - 6; 6n - 1; 6n + 3; 2n + 1; 1$ **11.** Answers

vary. Sample answer: 1. Pick a number. 2. Subtract 11. Multiply by 3. 4. Add 5 times your original number. 5. Add 1. 6. Divide by 8. 7. Add 4. 8. Your answer should be the number you picked. **13. a.** yes **b.** 39 **c.** Answers vary. Sample answer: Yes, suppose the numbers in a certain row, column, or diagonal are x, y, and z, which add to 15. Their new sum is $(x + k) + (y + k) + (z + k) = x + y + z + 3k$. This will be the sum in every row, column, and diagonal. **15.** Answers vary. Sample answer:

Spectator's Choice	Number of Cards in Small Pile
10	10 – 1 = 9
11	11 – 2 = 9
12	12 – 3 = 9
13	13 – 4 = 9
14	14 – 5 = 9

17. unlike **19.** unlike **21. a.** $\frac{29}{15}$ **b.** $\frac{11x}{14}$ **c.** $\frac{6}{5y}$
23. $33n^2 - 132n; 33(5)^2 - 132(5) = 165;$ $5(11(3(5 - 4))) = 165$ **25. a.** Eddie has $100 - 4w$ dollars.
b. Liseta owes $-350 + 5w$ dollars. **27.** 45 **29.** 78

Lesson 2-4 (pp. 85–90)
Guided Example 2: $-7; -14x + 21; -13x + 27$
Questions: 1. a. $48 - f - n$ ounces and $48 - (f + n)$ ounces
b. $48 - 12 - 5 = 31$ ounces and $48 - (12 + 5) = 31$ ounces
3. $2n$ **5.** $2a^2 - 28a + 15$ **7.** C **9.** $-x - 15$ **11.** -2
13. $-4k^4 + 13$ **15.** $2b - 2c$
17. a.

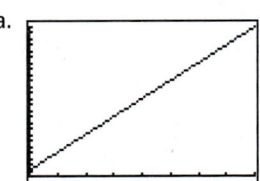

Yes, the expressions appear to be equivalent.
b. $4n - (n - 1) = 4n - n + 1 = 3n + 1 = 2n + n + 1$
19. a. True, they are both equal to –125. **b.** False. $(-5)^4$ is positive, -5^4 is negative. **c.** True, they are both equal to –625. **21.** Answers vary. Sample answer: We can think of "number of about-faces" as n and "facing direction" as $(-1)^n$, where a positive result is forward and a negative result is reverse. **23. a.** $\frac{1}{3}$ **b.** $-\frac{1}{4}$ **c.** $\frac{1}{5}$ **d.** $-\frac{1}{10}$ **25.** 16
27. $\frac{17}{3y}$ **29.** $6(15g - 12) = 90g - 72$ dollars
31. a. Answers vary. Sample answer: 13 and 56 **b.** Answers vary. Sample answer: –1.3 and –56 **c.** Answers vary. Sample answer: 1.3 and 5.6

Lesson 2-5 (pp. 91-97)

Guided Example 2: Answers vary. Sample answer: $x = 1$ gives $4x - x = 4(1) - 1 = 4 - 1 = 3 \neq 4$

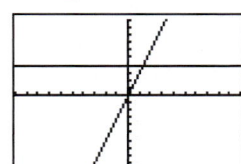

The graphs are not identical;

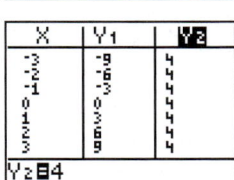

Guided Example 3: $-(6)^2 = -36$; $(-6)^2 = (-6)(-6) = 36$; -36 and 36 **Questions: 1. a.** 21 **b.**

Design Number	$1 + 3n + 2n$	$6n - (n - 1)$
4	21	21
5	26	26
6	31	31

c.

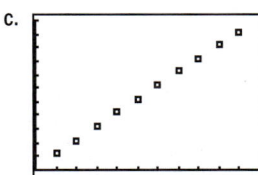

d. Dion's expression is $1 + 3n + 2n = 5n + 1$. Ellis's expression is $6n - (n - 1) = 6n - n + 1 = 5n + 1$. They are equivalent. **3.** They are not equivalent: $3x^2 + 6x(x + 2) = 3x^2 + 6x^2 + 12x \neq 3x^2 + 6x^2 + 2$
5. Answers vary. Sample answer: $x = 2$
7. $2x + 3 + 2x + 3 + 2x + 3 + 2x + 3 = 8x + 12 = 4(2x + 3)$ **9. a.** Answers vary. Sample answer:

x	$20(3x) - 2(8x)$	$20x + 4x + 20x$
0	0	0
1	44	44
2	88	88
3	132	132
4	176	176

b. $20(3x) - 2(8x) = 60x - 16x = 44x = 20x + 4x + 20x$, which shows the expressions are equivalent.   **c.** 220 units squared   **11.** $0.8v + 5.6$ **13. a.** yes **b.** no **c.** no **d.** yes
15. a. p^3 **b.** $3p$ **c.** $8p^3$ **d.** $6p$ **17.** Answers vary. Sample answer: $m = 1, 6 + 1 = 7, 2(1) - 3(1 - 2) = 5$

Lesson 2-6 (pp. 98-104)

Questions: 1. $35x^2 - 45xy$ **3.** Answers vary. Sample answer: The CAS recognized that $5t$ and $-5t$ are additive inverses and applied the Additive Inverse Property.
5. Answers vary. Sample answer: $3k(3k - 1)$; $10k^2 - 3k - k^2$; $9k\left(k - \frac{1}{3}\right)$ **7.** Answers vary. Sample answer: $-2 \cdot 3 \cdot 4y$; $0 - 24y$; $6 + 6y - 3(10y + 2)$; $\frac{-120y^2}{5y}$

9. Answers vary. Sample answer: First, add $-3y$ and $3y$ to get $5x - 7y - 3y + 3y$. Then regroup terms to get $(5x - 10y) + 3y$. Finally, factor 5 out of the first part of the expression.
11. Answers vary. Sample answer:

n	$3n - 15$	$3(n - 4) - 3$
-2	-21	-21
0	-15	-15
1	-12	-12
5	0	0
10	15	15

$3(n - 4) - 3 = 3n - 12 - 3 = 3n - 15$ They are equivalent.
13. a. Multiplying any real number by -1 gives the opposite of that number.
b. Answers vary. Sample answer: $-5 \cdot -1 = 5$ **15. a.** 3 folds has a paper thickness of 8, and 4 folds has a paper thickness of 16. **b.** 64 times the original thickness **c.** 2^n **17.** $5\frac{1}{2}L$ **19.** 33%

Lesson 2-7 (pp. 105-111)

Questions: 1. $0.23 + 0.44 = 0.67$, $0.44 + 0.23 = 0.67$, $0.67 - 0.44 = 0.23$, $0.67 - 0.23 = 0.44$ **3.** $-\frac{1}{8} = -4 + 3\frac{7}{8}$, $-\frac{1}{8} = 3\frac{7}{8} + -4$, and $-\frac{1}{8} - (-4) = 3\frac{7}{8}$

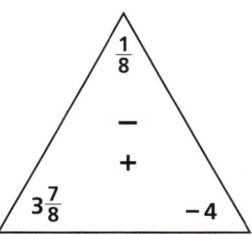

5. a. $y = 14 - (-6) = 20$ **b.** $y - 6 = 14$, so $y = y - 6 + 6 = 14 + 6 = 20$ **7.** -10
9. Answers vary. Sample answer: 629 and -629 **11.** x
13. Answers vary. Sample answers:

15. a.

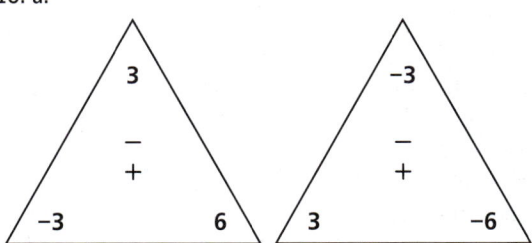

15. a. (continued)

15. b.

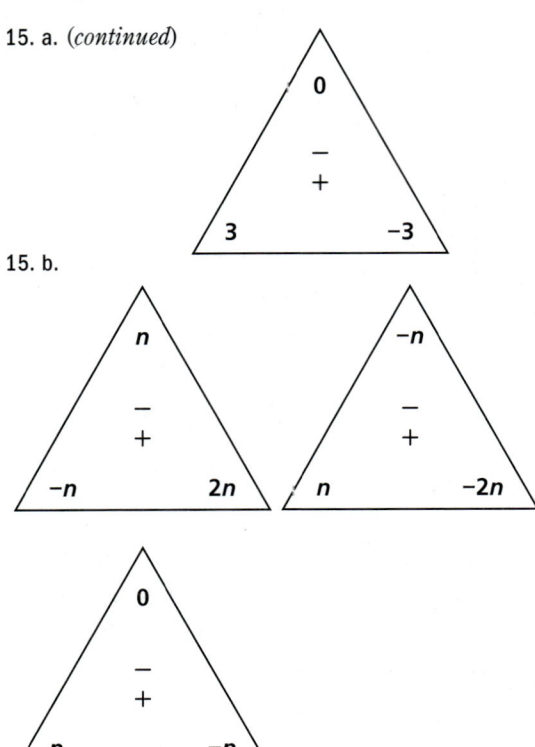

17. 19°F **19.** Answers vary. Sample answer: $350 \cdot 200 - W - C$ yd³ **21.** $p - 8$ **23.** $6m + 15$ **25. a.** $-\frac{376}{3}$ **b.** 5

Lesson 2-8 (pp. 112–120)
Guided Example 3: 0.64; $\frac{1}{\frac{1}{4}} = 1 \cdot \frac{15}{4}$; $\frac{1}{\frac{-34}{15}}$
Questions: **1.** $-14 \cdot -2 = 28, -2 \cdot -14 = 28, \frac{28}{-14} = -2$, and $\frac{28}{-2} = -14$ **3.** $cb = a, \frac{a}{b} = c$, and $\frac{a}{c} = b$ **5.** $6b$ **7.** $-\frac{1}{6}$ **9.** 1
11. Answers vary. Sample answer: Division by 0 is not defined. **13.** $x = 0$ **15.** Any number is a solution.
17. Multiplicative Property of Zero

19.

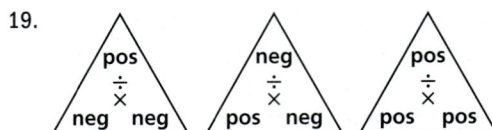

The product of two negative numbers or two positive numbers is positive. The product of a negative number and a positive number is negative. **21.** 6 **23. a.** Yes, it is possible. **b.** The width of the rectangle is $\frac{1}{50}$ cm.
25. a. $d = tr, r = \frac{d}{t}$, and $t = \frac{d}{r}$ **b.** about 3.81 hr
27.

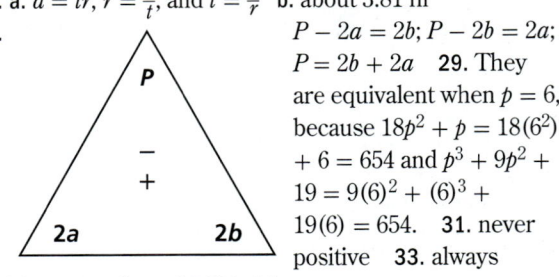

$P - 2a = 2b; P - 2b = 2a;$
$P = 2b + 2a$ **29.** They are equivalent when $p = 6$, because $18p^2 + p = 18(6^2) + 6 = 654$ and $p^3 + 9p^2 + 19 = 9(6)^2 + (6)^3 + 19(6) = 654$. **31.** never positive **33.** always positive **35.** about 1,588 bricks

Self-Test (p. 124)
1. $3w$ **2.** $\frac{6}{7}(4v + 78) = \frac{24v + 468}{7}$ **3.** $-5h - 8$
4. $3k + 30 - 11 - (-4k) = 7k + 19$ **5.** $-r$ **6.** $\frac{7x + 7}{6}$
7. $\frac{2 - 3}{3x} = -\frac{1}{3x}$ **8.** $7(20 - 0.02) = 140 - 0.14 = \$139.86$
9. $L \cdot W$ of the entire rectangle $= x(x + 3 + 2)$, the sum of the areas of both rectangles $= 2x + x(x + 3)$ **10.** 24
11. $d = \frac{C}{\pi}, \pi = \frac{C}{d}$ **12.** $d = \frac{C}{\pi}$
13.

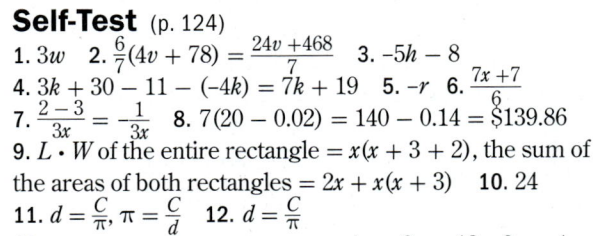

$-4 + -8 = -12; -8 + -4 = -12, -12 - (-4) = -8, -12 - (-8) = -4$
14. $3p$ **15.** $-b - 2$
16. $\frac{1}{-2.50} = -\frac{2}{5}$ **17.** $\frac{1}{\frac{4}{11d}} = \frac{11d}{4}$ **18.** False; A negative number to an odd power is negative. $(-2)^5 = -32$ and $2^5 = 32$
19. (1) n; (2) $10n$; (3) $10n + 30$; (4) $2n + 6$;

The chart below keys the **Self-Test** questions to the objectives in the **Chapter Review** on pages 125–127 or to the **Vocabulary (Voc)** on page 123. This will enable you to locate those **Chapter Review** questions that correspond to questions missed on the **Self-Test**. The lesson where the material is covered is also indicated on the chart.

Question	1	2	3	4	5	6	7	8	9	10
Objective(s)	B	A	A	B	A	F	A	D	C	C
Lesson(s)	2-4	2-1, 2-2	2-1, 2-2	2-4	2-1, 2-2	2-1	2-1, 2-2	2-4, 2-8	2-7, 2-8	2-7, 2-8
Question	11	12	13	14	15	16	17	18	19	20
Objective(s)	C	E	E	D	D	B	H	G	G	J
Lesson(s)	2-7, 2-8	2-7	2-7	2-4, 2-8	2-4, 2-8	2-4	2-1, 2-2	2-3	2-3	2-5, 2-6
Question	21	22	23							
Objective(s)	I	H	I							
Lesson(s)	2-5	2-1, 2-2	2-5							

(5) $n + 6$; (6) $n + 7$; (7) n **20.** No, they do not seem to be equivalent. **21.** They are not equivalent. $w + (2w - 1) + 3(w + 6) = 3w - 1 + 3w + 18 = 6w + 17 \neq 6w - 5$ **22.** Let t be the amount of time Darryl worked. Then, Carol worked $2t$, Beryl worked $3t$, Errol worked $4t$. We solve $t + 2t + 3t + 4t = \$150$. So, $t = 150$ dollars. Darryl got \$150, Carol got \$300, Beryl got \$450, and Errol got \$600. **23.** Answers vary. Sample answer:

x	2x + 1	|−2x − 1|
−2	−3	3
−1	−1	1
0	1	1
1	3	3
2	5	5

Chapter Review (pp. 125-127)

1. $3x + 12$ **3.** $4x - 31$ **5.** $4y - 52$ **7.** $18x$ **9.** $\frac{22x - 3}{8}$ **11.** $\frac{n}{3} + 2$ **13.** $-7a - 4$ **15.** z **17.** $\frac{3}{4} + y$ **19.** -8 **21. a.** 81 **b.** −81 **c.** −243 **d.** −243 **23.** negative; Of the two factors one is negative, yielding a negative product. **25.** There are no solutions. **27.** all real numbers **29.** $\frac{1}{2} = \frac{1}{3} + \frac{1}{6}, \frac{1}{6} = \frac{1}{2} - \frac{1}{3}$, and $\frac{1}{3} = \frac{1}{2} - \frac{1}{6}$ **31.** $317.23 = 317.23 \cdot 1, \frac{317.23}{1} = 317.23$ and $1 = \frac{317.23}{317.23}$ **33.** $x = 70$ **35.** $-\frac{1}{5}$ **37.** $8x$ **39.** Subtraction Property of Equality **41.** 0 **43.** 6 **45.** 0 **47.** 7.536 **49.** $-x$ **51.** Distributive Property of Multiplication over Addition **53.** First, find 36 times 100. Then, add 36 times 3 to get 3,708. **55.** $\$3.50(10 + 1) = \$3.50(10) + \$3.50(1) = \$38.50$ **57.** After Step 2, you have $4n$. After Step 3, you have $4n + 10$. After Step 4, you have $4n + 12$. After Step 5, you have $n + 3$. After Step 6, you have n. **59.** \$29,000 **61.** $\frac{2F}{5}$ **63.** Answers vary. Sample answer: For $x = 1, x^2 = 1$ but $2x = 2$. **65.** They are not equivalent. **67.** They are not equivalent.

Chapter 3

Lesson 3-1 (pp. 130-134)
Questions:

1. a.
Time x	Height y
0	18
1	15
2	12
3	9
4	6

b. 2 hr **c.** 9 in. **d.** 6 hr **3. a.** 102.5 ft **b.** 145 ft **c.** C

d.
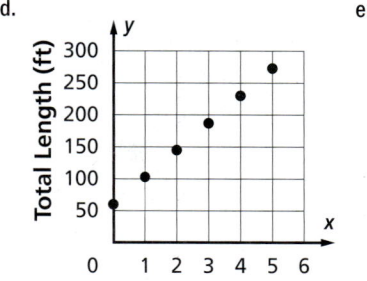

e. 570 ft

5. a–b.
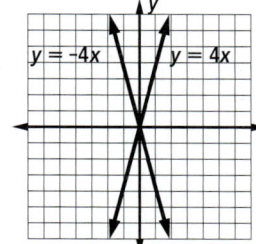

c. (0, 0) **d.** Answers vary. Sample answer: They are reflections of each other over the y-axis. **7.** linear **9.** linear **11.** $\frac{9}{11}h$ **13.** 0 **15.** $\frac{1}{6}$ mi or about 0.17 mi **17.** giraffe **19. a.** 36 **b.** −36 **c.** 36 **d.** −216 **e.** −216 **f.** −1,296

Lesson 3-2 (pp. 135-138)
Questions: **1. a.** no **b.** yes **c.** yes

3.
a	−5a + 7
0	7
2	−3
4	−13
(6)	(−23)

5.
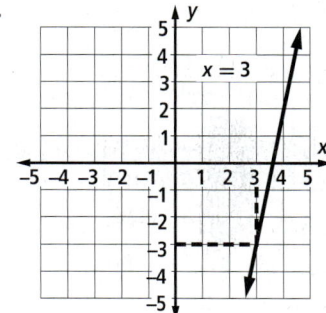

7. Let p be the price of a piece of pizza. Then, $20 - 5p = 7.65$. **9.** Let y be the number of years they save. Then, $5{,}275 + 950y = 20{,}000$.

11.
h	Savings
0	250
10	360
20	470
(21)	(481)

13. a.
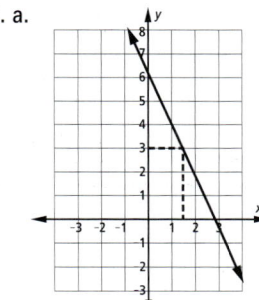

b. about 1.5 **15.** $2(3x - 6) + 2x + 1 = 97; x = 13.5$ **17.** It is not a line.

19. a.
Years	Height
0	28
1	31.5
2	35
3	38.5

b.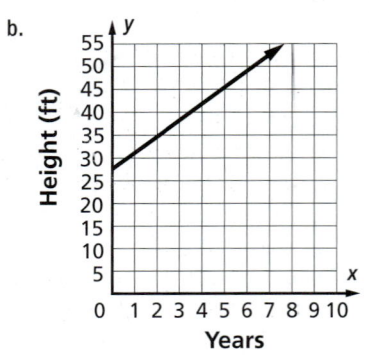

c. 101.5 ft d. 66 yr 21. a. −8.3 b. $\frac{1}{8.3} \approx 0.12$ 23. $\frac{142}{3}$ or about 47.33

Lesson 3-3 (pp. 139–143)

Questions: 1. All three equations have the same solution, 3. 3. a. add 27; divide by 5 b. $x = 8$ 5. added 11 7. multiplied by $\frac{5}{3}$ 9. Answers vary. Sample answer: Add 100. 11. a. $y + \frac{-14}{3} = \frac{-93}{3}$ b. Answers vary. Sample answer: No, subtracting 14 would have avoided fractions. 13. a. $x = \frac{c - 5y}{3}$ b. Answers vary. Sample answer: Let $y = 2$, $c = 3$. $3x + 5(2) = 3$; $x = -\frac{7}{3}$ and $x = 3 - \frac{5(2)}{3} = -\frac{7}{3}$ 15. a. Let c be the number of cards in a pack. $200 + c + c + c + c + c + 2c = 284$ or $200 + 7c = 284$ b. 12 cards

17. a.

Time of day	Temperature (°F)
9:00 A.M.	85
10:00 A.M.	92
11:00 A.M.	99
12:00 P.M.	106
1:00 P.M.	113
2:00 P.M.	120
3:00 P.M.	127
4:00 P.M.	134

b.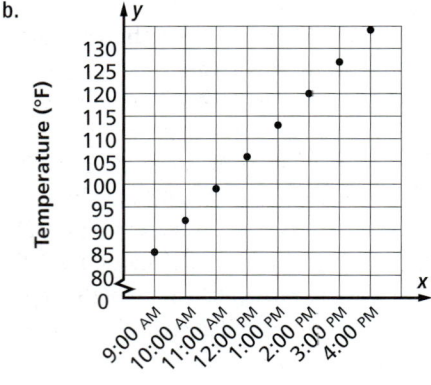

19. a. symmetric b. The mean is about 19.73 mosquitoes, and the m.a.d. is 1.52 mosquitoes.

Lesson 3-4 (pp. 144–148)

Guided Example 1: 53; 53; 53; 60; $-\frac{2}{3}$; $-\frac{2}{3}$; −40
Guided Example 3: 4; 3; 4; $(n + 10)$; $4n + 40$; 4; 52; 4; 52; −52; −52; $\frac{1}{4}$; $4n$; $\frac{1}{4}$; 42; 10.5
Questions: 1. a. 57; multiply; $\frac{1}{7}$ b. $t = 22$; $7(22) - 57 = 154 - 57 = 97$ 3. $a = 73$; $b = -432$; $c = 1{,}101$ 5. A 7. $A = 5$; $11(5) - 24 = 55 - 24 = 31$ 9. $B = \frac{2}{5}$; $16 + 5(\frac{2}{5}) = 16 + 2 = 18$ 11. $n = 2$; $2.4(2) - 2.4 = 4.8 - 24 = 2.4$ 13. $m = -56$; $4 - \frac{7}{2}(-56) = 4 + 196 = 200$ 15. a. Let g be the number of gallons the driver bought. Then, $1.50 + 2.39g = 31.15$. b. g is about 12.4 gal. 17. a. Let r be the number of seats in a row. Then, $3 + 7r = 80$. b. $r = 11$ seats 19. $x = \frac{1{,}889}{121} \approx 15.61$; $\frac{11}{17}(\frac{1{,}889}{121}) + \frac{17}{11} = \frac{1{,}889}{187} + \frac{17}{11} = 11\frac{11}{17}$ 21. $a = 3$; $\frac{4}{3}(3 - \frac{1}{2}) = \frac{4}{3}(\frac{5}{2}) = 3\frac{1}{3}$ 23. a. $7x + 1 = 15$ b. Remove 1 ounce from each side. Leave one seventh of the weight on each side. c. 2 oz

25. a.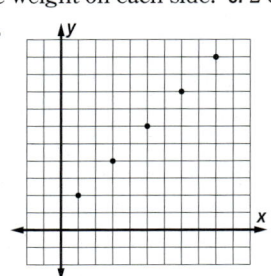

b. $a = \frac{10}{7} \approx 1.43$
c. $m + 1$ 27. $a + 3b$
29. $y = 128$

Lesson 3-5 (pp. 149–154)

Guided Example 5: $6k + 8 - 9k + 7$; −3; −3; Subtract 15 from each side; −3; −9; $\frac{-3k}{-3} = \frac{-9}{-3}$; 3; 3; 3

Questions: 1. 2 3. $\frac{8}{3}$ 5. $\frac{3}{2}$ 7. −7 9. 31 11. a. Answers vary. Sample answer: $15{,}000 + 4E + \frac{1}{2}E = 150{,}000$ b. $E = \$30{,}000$ and the head mechanic received $\$15{,}000$. 13. a. 36 in² b. 8 in. c. 16 in. 15. 23 cm 17. $\$40{,}000$ in the CD, $\$60{,}000$ in the savings account 19. −2.81; $3.4 + -(-2.81) = 3.4 + 2.81 = 6.21$ 21. a. 8 in. b. 6:40 A.M. 23. 6:15 P.M. 25. $\frac{2}{7}$

Lesson 3-6 (pp. 155–161)

Guided Example 2: a. $\geq$; $\geq$; $\leq$
b.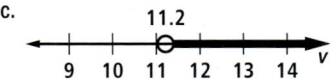
c. Step 1: 20; 5; Yes Step 2: 20; 0; 20; 0; $x \leq 5$
Questions: 1. a. $v \geq 11.2$ b. 11.2
c.
```
        11.2
←——+——+——○——+——+——+——→ v
   9  10  11  12  13  14
```
3.
```
←——●——+——+——+——+——●——→ w
  120 122 124 126 128 130
```
5.
```
←——+——●——+——+——+——+——→ y
  −5 −4 −3 −2 −1  0  1
```

7. $x > \frac{1}{3}$; $15\left(\frac{1}{3}\right) = 5$; Pick $x = 1$. $15(1) = 15 > 5$ **9.** $y \leq 25$; $\frac{4}{5}(25) = 20$; Pick $y = 0$. $\frac{4}{5}(0) = 0 \leq 20$ **11.** $a \geq 2$; $-3(2) = -6$; Pick $a = 3$. $-3(4) = -12 \leq -6$ **13. a.** $\frac{1}{2} > \frac{1}{3} > \frac{1}{4} > \frac{1}{5} > \frac{1}{6}$ **b.** $-4 < -\frac{8}{3} < -2 < -\frac{8}{5} < -\frac{4}{3}$ **c.** Yes, because we multiplied by a negative number.   **15.** 30; 31   **17. a.** Let r be the number of rows. Then, $80r \geq 2{,}200$.   **b.** $r \geq 27.5$ rows   **19.** $x = -\frac{13}{6}$ **21. a.** Let p be the number of bottles of soda Grafton buys. Then, $1.99p + 2.99(2p) + 0.15 = 40$.   **b.** 5 bottles of soda, 10 bags of chips   **23.** $5(x + 1) + (2 - x) = 31$, $(2 - x) + 5(x + 1) = 31$, $31 - 5(x + 1) = 2 - x$, and $31 - (2 - x) = 5(x + 1)$; $x = 6$ **25.** 198.25 mi

Lesson 3-7 (pp. 162–166)

Questions: 1. For real numbers a, b, and c, if $a > b$, then $a + c > b + c$.   **3.** $x < 5$

5. $n \leq 20$ **7.** 39 shirts

9. $y \leq 0.1$ **11.** $d > -108$ **13.** $q > -22$ **15. a.** $x = 7$ **b.** Answers vary. Sample answer: $x = 3$ **c.** Answers vary. Sample answer: $x = 11$ **d.** $x < 7$ **17.** 11 mo or fewer   **19.** Except for 2, which is a solution to both, every real number is a solution to exactly one of these inequalities.   **21. a.** $800 > -200$ **b.** $-40 < 10$ **c.** $4 > -1$ **d.** No inequality results.   **23.** $j > \frac{17}{5}$ **25.** $x \leq \frac{72}{5}$ **27.** 1,480 mi

Lesson 3-8 (pp. 167–173)

Guided Example 2: 12; 12; 12; 12; 12; 3; 96; 3; 98; $\frac{98}{3}$
Questions: 1. a. $3w + 10 = 130$ **b.** $w = 40$ **c.** $\frac{3}{5}(40) + 2 = 26$; $24 + 2 = 26$ **3. a.** 1,000; 10; and 100   **b.** $15.2 = 30m - 43$; $m = 1.94$ **c.** $\frac{152}{1{,}000} = \frac{3}{10}m - \frac{43}{100}$; $m = 1.94$ **d.** Converting decimals to fractions does not change the solution.
5. a. $10a + 3a \geq 315$ **b.** $a \geq \frac{315}{13}$ **c.** [number line from 23 to 27 with point at $\frac{315}{13}$]   **d.** $\frac{2}{3}\left(\frac{315}{13}\right) + \frac{\left(\frac{315}{13}\right)}{5} = \frac{210}{13} + \frac{63}{13} = 21$; Pick $a = 30$. $\frac{2}{3}(30) + \frac{30}{5} = 20 + 6 = 26 \geq 21$
7. $y = \frac{64}{9}$; $\frac{3}{4}\left(\frac{64}{9}\right) - \frac{1}{3} = \frac{16}{3} - \frac{1}{3} = 5$ **9.** $c = 49$; $138{,}000 - 2{,}000(49) = 40{,}000$ **11.** $n \leq 18$; $1 - \frac{18}{10} = \frac{-4}{5}$; Pick $n = 0$. $1 - \frac{0}{10} = 1 \geq -\frac{4}{5}$ **13.** $m \geq \frac{305}{286}$, $\frac{\left(\frac{305}{286}\right)}{5} - \frac{1}{3} = \frac{61}{286} - \frac{1}{3} = \frac{3}{22}$; Pick $m = 5$. $\frac{5}{5} - \frac{1}{13} = \frac{12}{13} \geq \frac{3}{22}$
15. Yes, because it simplifies the equation.   **17.** $x = \frac{269}{3}$
19. $t > -6$ **21.** $y > 13$ **23.** $w = 9$ **25.** $\frac{31}{7t}$ **27.** $\frac{x^2}{18}$

Self-Test (p. 177)

1. $t = 4.5$. First, add 5 to both sides. Then divide by 4.
2. $t = -4$. First, distribute to get $10 + 5t = -10$. Then subtract 10 from both side and divide by 5.   **3.** $f = \frac{109}{7}$. First, distribute and combine like terms to get $101 = 13f - 8 - 6f = 7f - 8$. Then add 8 to both sides and divide by 7.   **4.** $x \geq 8$ **5.** $x < -8$ **6.** $x < 8$
7. a. Question 6   **b.** Question 4   **c.** Question 5
8. $2n - n = 24$, so $n = 24$ **9.** Let $2x$ be the value of the grand prize. Then, $x + x + x + 2x = 3{,}500$. Combining like terms gives $5x = 3{,}500$, and dividing by 5 gives $x = 700$, so the grand prize is $1,400.   **10.** Subtract 7 from both sides and then divide both sides by -2. $x = \frac{-15 - 7}{-2} = 11$ Answers vary. Sample answer: $-15 = -2(11) + 7$; $-15 = -15$

11.

Months	Money in Account
0	$350
1	$330
2	$310
3	$290
8	$190

12.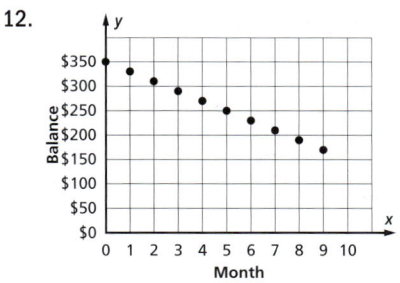

13. after 8 months

14.

x	$2x - 6$
12	18
14	22
16	26
18	30
20	34

So, $30 < 2x - 6$ when $x > 18$.   **15.** Subtract 5, then divide by -3. The solution is $x < -4$.   **16.** Let w be the number of weekends Toni collects. Toni has enough leaves when $9 + 7w \geq 37$. Subtracting 9 and dividing by 7 gives $w \geq 4$. So, she must collect for at least 4 weeks.   **17. a.** $P = 0.39 + 0.24(w - 1)$ **b.** Let $P = 3.27$. Then solve $3.27 = 0.39 + 0.24(w - 1)$ by subtracting 0.39 from both sides, which implies $2.88 = 0.24(w - 1)$. Then divide by 0.24, which implies $w - 1 = 12$. So $w = 13$ oz.

The chart below keys the **Self-Test** questions to the objectives in the **Chapter Review** on pages 178–179 or to the **Vocabulary (Voc)** on page 176. This will enable you to locate those **Chapter Review** questions that correspond to questions missed on the **Self-Test**. The lesson where the material is covered is also indicated on the chart.

Question	1	2	3	4	5	6	7	8	9	10
Objective(s)	A	A	A	B	B	B	F	C	D	C
Lesson(s)	3-4, 3-5	3-4, 3-5	3-4, 3-5	3-7, 3-8	3-7, 3-8	3-7, 3-8	3-6, 3-7	3-3, 3-6, 3-8	3-2, 3-4, 3-5, 3-7, 3-8	3-3, 3-6, 3-8

Question	11	12	13	14	15	16	17
Objective(s)	E	E	E	B	C	D	D
Lesson(s)	3-1, 3-2	3-1, 3-2	3-1, 3-2	3-7, 3-8	3-3, 3-6, 3-8	3-2, 3-4, 3-5, 3-7, 3-8	3-2, 3-4, 3-5, 3-7, 3-8

Chapter Review (pp. 178–179)

Guided Example 4: P; $0.15P$; $1P$; $0.15P$; $0.85P$; 0.85; 320; 272
1. $t = 3$ **3.** $n = 7$ **5.** $y = 46.15$ **7.** $W = 5$ **9.** $z = -15$
11. $w = 2.6$ **13.** $x < 94$ **15.** $y \leq 13$ **17.** 17 was added to each side. **19.** Sample answer: You could enter $(2x - 3 = 7) + 3$. The result would be $2x = 10$. **21.** Yes. The student multiplied both sides by -1, which changes the sense of the inequality. **23.** more than 26°C
25. 32 hr **27.** approximately 1,015 billion barrels
29. a.

Years	Radius (cm)
0	12
1	12.5
2	13
3	13.5
x	$12 + 0.5x$

b.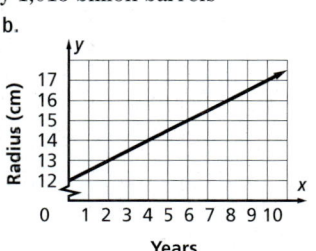

c. 6 yr **d.** 16 cm

31. a. $b = 55 + 20w$
b.

Weeks	Balance
0	$55
1	$75
2	$95
3	$115

c.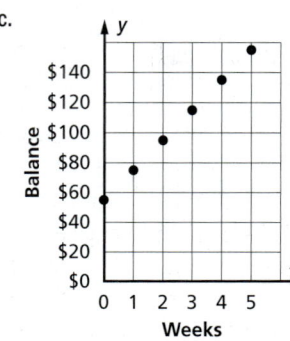

d. Answers vary. Sample answer: After 6 weeks find $175 on the y-axis and draw a horizontal line to the graph, then a vertical line down to the x-axis to find when Darnell has $175.

33.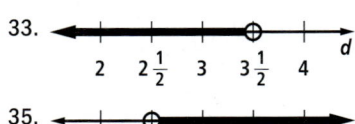

35.

Chapter 4

Lesson 4-1 (pp. 182-187)

Questions: **1.** 959.4 **3.** 20% **5. a.** 90% **b.** $21.15 **7.** $0.6T$
9. $16.84 **11.** 7.25% **13.** 59,670,000 U.S. citizens
15. about 51.5% **17. a.** $148.13 **b.** The regular price is not 120% of the sale price. The sale price is 80% of the regular price. **19.** $k = -59.5$ **21.** $a = -13$ **23.** $322.23
25. a. 13 cubes **b.** Answers vary. Sample answer: $4n - 3$
c. 9 **27.** 0

Lesson 4-2 (pp. 188-195)

Questions: **1. a.** Answers vary. Sample answer: $(0, 4)$, $(-3, 4)$, and $(-0.6, 4)$
b.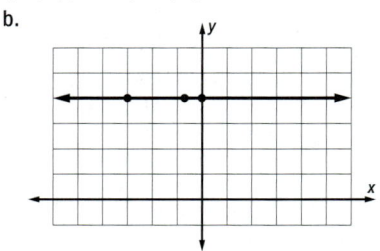

c. $y = 4$ **3.** $x = -4$ **5.** x
7. a.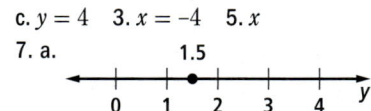

7.b.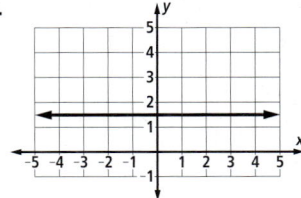

9. a. $S = 98.5$
b. 1997 11. a. 2000 and 2001 b. Answers vary. Sample answer: Bad news; snow depths below normal could affect snow tourism and also result in lower water supply from runoff.

13. $x = -6$ 15. $x; y$ 17. $(7, 5)$ 19. 10 A.M.
21. a. $y = 35 + 70x$

b. 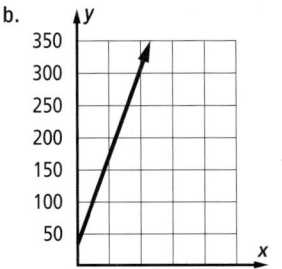 c.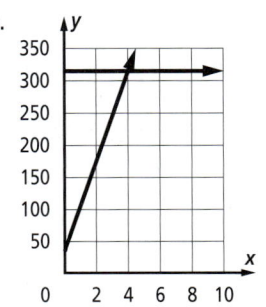

d. 4 hours e. Solving $35 + 70x \leq 315$ yields $x \leq 4$ hours.
23. $16,037.74 25. a. $18,550 > \frac{1}{4}I$ b. $I < 74,200$
27.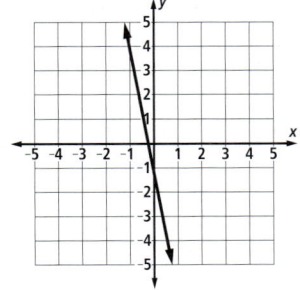

Lesson 4-3 (pp. 196–201)
Guided Example: Solution 1:

1.
Number of Copies x	Acme's Price $250 + 0.01x$	Best's Price $70 + 0.03x$
0	250	70
2,000	270	130
4,000	290	190
6,000	310	250
8,000	330	310
10,000	350	370
12,000	370	430
14,000	390	490
16,000	410	550
18,000	430	610
20,000	450	670

2. 10,000; 8,000

3. The break-even point is between 8,000 and 10,000 copies.
4.
Number of Copies x	Acme's Price $250 + 0.01x$	Best's Price $70 + 0.03x$
8,000	330	310
8,500	335	325
9,000	340	340
9,500	345	355
10,000	350	370

5. 9,000; $340 6. a. 9,000 b. $340 Solution 2:
2. 9,000; 340 3. 9,000; both companies charge the same amount for 9,000 copies. 340; the companies both charge $340 at the break-even point.
Questions: 1. a. 4 or more copies b. 1 or 2 copies
c.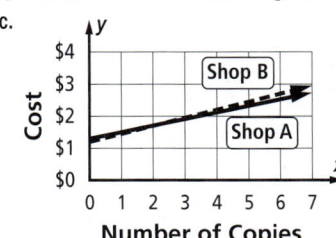

The graph of Shop A is higher for 2 copies. The graph of Shop B is higher for 7 copies, assuming the patterns hold. d. (3, 1.80)

3. a. Answers vary. Sample answer: around (70, 40)
b. number of miles driven for which the two costs are equal; cost of the rental c. Rhodes is less expensive when 80 miles are driven because at $x = 80$, the Rhodes' line is below the Extra Value line.
5. a. $30 + 6w$ b. $150 - 5w$ c. after about 11 wk

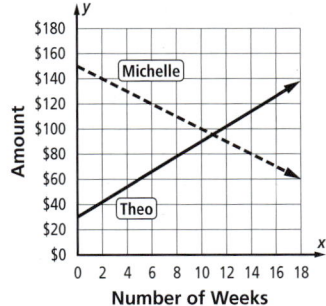

7. $x = 1$ 9.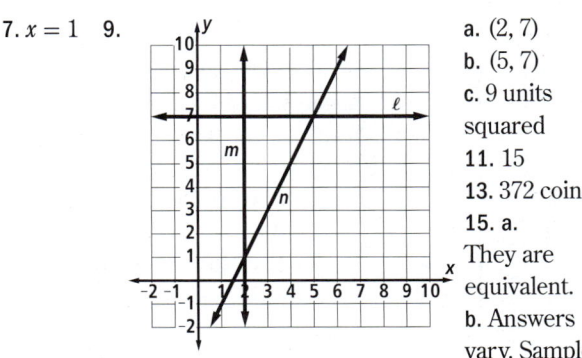

a. (2, 7)
b. (5, 7)
c. 9 units squared
11. 15
13. 372 coins
15. a. They are equivalent.
b. Answers vary. Sample answer: Distributing the left-hand side of both equations

yields $-18x - 18 + 6x$, which simplifies to $-12x - 18$, in the first equation and $-12x - 18$ in the second equation. These expressions are identical.

Lesson 4-4 (pp. 202–209)

Guided Example 3: 1. $-x - 5; 4x + 44$ 2. $2; 5x + 44; -42;$ $5x; -8.4$ 3. $-8.4; -8.4; -3.4; 2.6$

Questions: 1. $a = 9.3; b = -4; c = 2; d = 11$ 3. a. $-4t$ or $-10t$ b. $t = \frac{1}{3}$ 5. $n = 9,000$ 7. $k = 11; 4(11) + 39 = 83, 7(11) + 6 = 83$ 9. $y = -1; 14(-1) + 5 = -9, 8(-1) - 1 = -9$ 11. $m = \frac{1}{2}; 7 - \frac{1}{2} = \frac{13}{2}, 8 - 3(\frac{1}{2}) = \frac{13}{2}$ 13. $d = -\frac{1}{2}; 3(-\frac{1}{2}) + 4(-\frac{1}{2}) + 5 = \frac{3}{2}, 6(-\frac{1}{2}) + 7(-\frac{1}{2}) + 8 = \frac{3}{2}$ 15. about 8.15 years after 2000 (late February 2008) 17. a. $4W + 6 = 2W + 10$ b. First, remove 6 ounces from each side. Then, remove $2W$ from both sides. Then, remove half the weight on each side. c. $W = 2$ oz d. because you cannot have a negative number of boxes on the balance scale 19. a. $53.84 - 0.32x$ b. $48.17 - 0.18x$ c. about 40 yrs

21. a.

Hours	Chris's Distance (mi)	Lance's Distance (mi)
0	24	0
1	33	13
2	42	26
3	51	39
4	60	52
5	69	65
6	78	78

b. 6 hr c. $13x = 9x + 24; x = 6$; 6 hr 23. $y = d$ 25. $c > 6$ 27. $-5\frac{5}{6}$ 29. 3

Lesson 4-5 (pp. 210–215)

Guided Example 2: Solution 1. $11x; 11x; 11x; 12; 12; 12; \frac{-5}{15}; \frac{15x}{15}$; divide; $\frac{-1}{3}$ **Solution 2.** $4x; 4x; 4x; 7; 7; 7; \leq$; reverse; $\leq$ **Check.** $\frac{-1}{3}; \frac{-1}{3}; \frac{-1}{3}; \frac{-1}{3}$; Sample answer: $-1; -1$

Questions: 1. a. Answers vary. Sample answer: 2, 4, and 9. b. $t < 10$ c. $t > 10$ 3. a. $m < -\frac{10}{3}$ b. $m < -\frac{10}{3}$ c. Yes, the solutions are the same regardless of how the sentence is solved. d. In Part a, the second step is subtract 14 from both sides, and the third step is to divide both sides by 3. In Part b, the second step is to subtract 4 from both sides, and the third step is to divide both sides by -3. 5. a. $x \geq -9$ b. Step 1: Is $-53 + 15 \cdot -9 = -8 + 20 \cdot -9$? Yes, $-188 = -188$. Step 2: Answers vary. Sample answer: for $x = 0$, is $-53 + 15 \cdot 0 < -8 + 20 \cdot 0$? Yes, $-53 < -8$. 7. a. $n < -\frac{3}{2}$ b. Step 1: Is $14 - 4 \cdot -\frac{3}{2} = 29 + 6 \cdot -\frac{3}{2}$? Yes, $20 = 20$. Step 2: Answers vary. Sample answer: For $n = -2$, is $14 - 4 \cdot -2 > 29 + 6 \cdot -2$? Yes, $22 > 17$. 9. $x > \frac{1}{14}$ 11. It determines what side of the inequality x is on and could result in a negative coefficient of x, which affects the sense of the inequality. 13. Let h be the number of hits.

Then, Elevator Tunes' ads are more profitable when $5.00 + 0.06h > 3.50 + 0.10h$. This occurs when there are 37 hits or fewer. 15. $z = -\frac{280}{39}$ 17. $r = \frac{619}{380}$

19. a.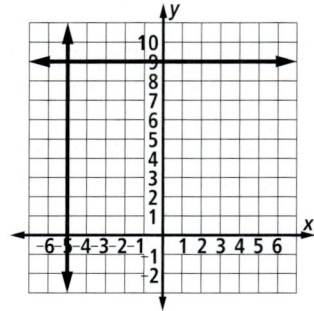

b. $(-5, 9)$ 21. B

Lesson 4-6 (pp. 216–220)

Guided Example 2: $42k - 42k < 42k + 6 - 42k$; Simplify; $42k < 80k + 6 - 38k$

Questions: 1. a. Answers vary. Sample answer:

Year	Hamburger Heaven	Video King
0	6.90	7.50
1	7.90	8.50
2	8.90	9.50
3	9.90	10.50
4	10.90	11.50
5	11.90	12.50
6	12.90	13.50

b. Let y be the number of years worked at either place. $6.90 + 1y < 7.50 + 1y$ c. never

3. a. $40 = 40$ b. Any real number is a solution to the equation.

c.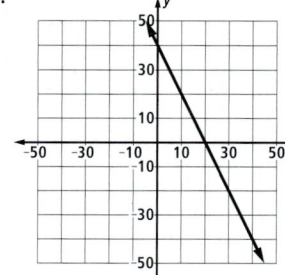

5. no solutions 7. Any real number is a solution. 9. a. $40,000 + 800y$ b. $200,000 + 800y$ c. never 11. Answers vary. Sample answer: $x + 7 = x + 9$ 13. $p < -2$ 15. Brokefast is cheaper, because $4.75 + 16(0.10) < 3.50 + 16(0.25)$. 17. about 16%

Lesson 4-7 (pp. 221-226)

Guided Example 1: $F - 32 = \frac{9}{5}C + 32 - 32; \frac{5}{9}(F - 32) = \frac{5}{9}(\frac{9}{5}C); \frac{5}{9}F - \frac{160}{9} = C; 212 = 1.8 \cdot 100 + 32; 212 = 180 + 32; 212 = 212; 100 = \frac{5}{9}(212 - 32); 100 = \frac{5}{9}(180); 100 = 100$

Guided Example 2: Multiply each side by 3; Divide each side by π; Divide each side by r^2.

Questions: 1. $1.8(-40) + 32 = -72 + 32 = -40$ and $\frac{5}{9}(-40 - 32) = \frac{5}{9}(-72) = -40$ 3. a. $\ell = \frac{p}{2} - w$

4. b. $\ell; p; w$ c. Answers vary. Sample answer: let $\ell = 1$, $w = 2$, and $p = 6$. $6 = 2(1) + 2(2)$, which checks with $\ell = \frac{p}{2} - w = \frac{6}{2} - 2 = 3 - 2 = 1$ 5. $y = \frac{1}{2}x + 5$
7. $n = \frac{S}{180} + 2$ 9. $h = \frac{2A}{b_1 + b_2}$ 11. Both students are correct, and their expressions are equivalent.
13. a. $\pi = \frac{C}{d}$ b. Measure a circle's circumference and divide it by the diameter. c. Answers vary. Sample answer: A circle with circumference of 22 cm has a diameter of about 7 cm, meaning $\pi \approx \frac{22}{7} \approx 3.143$.   15. They are parallel.   17. no solutions   19. a. $y = -3$ b. $x = 5$
21. a. $x = 2$ b. It would have sides of negative length.
23. $4x - 6 + 2x + 10 = 40; 40 - 4x + 6 = 2x + 10;$ $40 - 2x - 10 = 4x - 6; x = 6$

Lesson 4-8 (pp. 227-233)
Guided Example 4: left column: 2; −84; −7. right column: 0; −9; >; 3
Questions: 1. and   3. a   5. b   7. Let i be the size of the incision. $1 \leq i \leq 2$

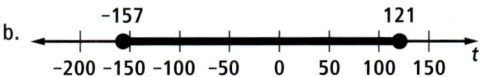

9. a. Let t be the temperatures that astronauts can endure. $-157°C \leq t \leq 121°C$
b.

11. $0.9 \leq x < 20$

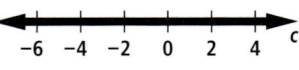

13. all real numbers

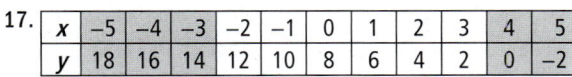

15. 6.5 ± 20.5

17.
x	−5	−4	−3	−2	−1	0	1	2	3	4	5
y	18	16	14	12	10	8	6	4	2	0	−2

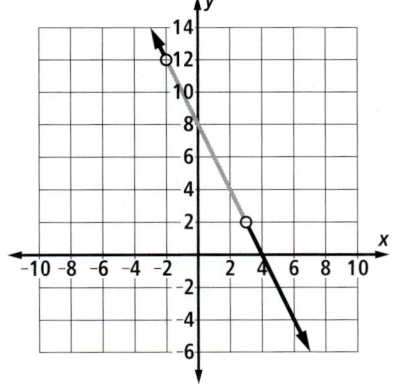

19. a. $r = \frac{d}{t}$ b. 61.25 mph   21. all real numbers
23. the second store

Lesson 4-9 (pp. 234-239)
Guided Example 1: Solution 1. (−6, 18); (12, 18); −6; 12; 12
Solution 2. 18; −18; 18; −18; −12; 24; −6; 12
Guided Example 2: −64; 16; −8; 2

x	$\|8x + 24\|$
−10	56
−8	40
−6	24
−4	8
−2	8
0	24
2	40
4	56
6	72

Questions: 1. C   3. $A = 6$ or $A = -6$ 5. $C = 0$ 7. a. $x = -2$ or $x = 5$ b. $-2 < x < 5$
c. $x < -2$ or $x > 5$ 9. C   11. $a = 2$ or $a = -22$
13. $-1 < c < 17$ 15. $g \geq -\frac{2}{5}$ or $g \leq -2$ 17. a. $x = -1$ or $x = 8$
b. $x = 3.5$ c. no solutions d. $x < 1$ or $x > 6$
19. a.

b. Let x be the amount in the box. $|x - 24.4375| \leq 0.5625$
21. $|x - 15.6| > 3.8$ 23. a. Let t be the time her commute takes. $12 \leq t \leq 20$ b. 16 ± 4 25. $m \leq -2$ or $m > \frac{37}{4}$
27. horizontal: $y = \frac{9}{11}$; vertical: $x = -4.25$

Self-Test (pp. 243-244)
1. The dress is cheaper online, because it costs $262 + $262(0.09) = $285.58 and $285.58 > $285, which is the price of the dress online.   2. Answers vary. Sample answer: $8t$, since it would isolate the variable on one side   3. Answers vary. Sample answer: 40
4. $-5m + 21 + 5m = 6m - 56 + 5m; 21 + 56 = 11m - 56 + 56; \frac{77}{11} = \frac{11}{11}m; m = 7$ 5. $0.73v + 37.9 + v = 16 - v + v; 1.73v + 37.9 - 37.9 = 16 - 37.9; \frac{1.73}{1.73}v = \frac{-21.9}{1.73}; v \approx -12.659$ 6. $\frac{1}{4}x + \frac{3}{5} - \frac{1}{4}x = \frac{1}{2}x - \frac{1}{4}x; 4\left(\frac{3}{5}\right) = 4\left(\frac{1}{4}x\right); x = \frac{12}{5}$ 7. $-6 + 2d > 6d - 9; -6 + 2d - 2d > 6d - 9 - 2d; -6 + 9 > 4d - 9 + 9; \frac{3}{4} > \frac{4}{4}d; d < \frac{3}{4}$ 8. $-29 + 44 < 7.5p - 44 + 44 \leq 28 + 44; \frac{15}{7.5} < \frac{7.5}{7.5}p \leq \frac{72}{7.5}; 2 < p \leq 9.6$ 9. $(2x - 3) = 21$ or $(2x - 3) = -21; 2x - 3 + 3 = 21 + 3$ or $2x - 3 + 3 = -21 + 3; \frac{2}{2}x = \frac{24}{2}$ or $\frac{2}{2}x = -\frac{18}{2}; x = 12$ or $x = -9$
10. $32 - 16n = -16n + 16; 32 - 16n + 16n = -16n + 16 + 16n; 32 = 16$; no solutions   11. a. $x \leq 5$ b. $x \geq 5$
12. a. Using the formula $30{,}000 + 0.05s$ for TurboTV and $25{,}000 + 0.08s$ for Sparkle Cable, where s is her sales, we construct the following table:

Sales ($)	Total Salary from TurboTV	Total Salary from Sparkle Cable
$25,000	$31,250	$27,000
$50,000	$32,500	$29,000
$75,000	$33,750	$31,000
$100,000	$35,000	$33,000
$125,000	$36,250	$35,000
$150,000	$37,500	$37,000
$175,000	$38,750	$39,000

b. $s \geq \$175,000$. c. TurboTV, because she would earn more there

13.

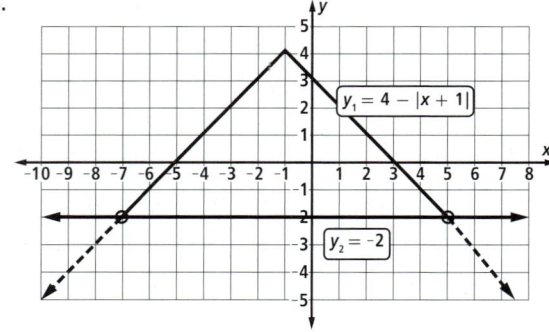

14. a. $y = 3$ b. $x = -2$

15. $-6x + 7y + 6x = -84 + 6x$; $7y + 84 = -84 + 6x + 84$; $x = \frac{7}{6}y + 14$

16. There is no solution, since the lines are parallel, but not the same. 17a. 1990 b. 1996, which tells us that the deviation was smallest that year. c. Answers vary. Sample answers: bad, since they would rather produce more ice cream

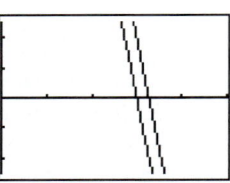

18. a. Let e be the number of strokes Elena took. $\frac{9}{10}e - 9 = e - 18$; $\frac{9}{10}e - 9 - \frac{9}{10}e = e - 18 - \frac{9}{10}e$; $-9 + 18 = \frac{1}{10}e - 18 + 18$; $10(9) = 10\left(\frac{1}{10}e\right)$; $e = 90$ b. $\frac{9}{10}(90) = 81$ strokes c. $90 - 18 = 72$ strokes 19. a. Let n be the number of CDs produced. $452.54 + 5.25n$ b. $17.99n$ c. $17.99n > 452.54 + 5.25n$; $12.74n > 452.54$; $n > 35.52$; 36 CDs or more 20. $0.3 \cdot x = 7$, so $x = \frac{7}{0.3} = 23\frac{1}{3}$

Chapter Review (pp. 245-249)
1. $A = -6$ 3. $n = -1$ 5. $f = \frac{7}{8}$ 7. $a = -\frac{3}{8}$ 9. $w \leq 129$
11. $n > -\frac{7}{3}$

13. $1 \leq p < 9$

The chart below keys the **Self-Test** questions to the objectives in the **Chapter Review** on pages 245–249 or to the **Vocabulary (Voc)** on page 242. This will enable you to locate those **Chapter Review** questions that correspond to questions missed on the **Self-Test**. The lesson where the material is covered is also indicated on the chart.

Question	1	2	3	4	5	6	7	8	9	10
Objective(s)	J	F	F	A	A	A	B	B	E	G
Lesson(s)	4-1	4-4, 4-5	4-4, 4-5	4-4	4-4	4-4	4-5, 4-8	4-5, 4-8	4-9	4-6
Question	11	12	13	14	15	16	17	18	19	20
Objective(s)	L	J	N	K	C	M	I	H	H	D
Lesson(s)	4-3	4-1	4-9	4-2	4-7	4-6	4-1	4-4, 4-5	4-4, 4-5	4-1

15. $-2 < x < -4$

17. $p = \frac{2A}{a}$ 19. $h = \frac{S}{2\pi r} - r$ 21. $y = 7 - 2x$
23. $\$2.40$ 25. 43% 27. 200 29. $x = 12$ or $x = 16$
31. no solutions
33. Let t be the temperatures at which Lisel's dog will go outside. $|t - 60| < 15$

35. a. $x = -5$ b. $x = -5$ c. They are the same.

37. a. Multiplication Property of Equality b. Distributive Property c, d. Addition Prperty of Equality e. Multiplication Property of Equality or Division Property of Equality 39. a. She subtracted $90n$ from the left side of the equation, but added $90n$ to the right side. b. $190n + 10 = 4$; $n = \frac{-3}{95}$ 41. no solutions 43. If you subtract x from both sides, you get $5 > 6$, which is never true. 45. a. Calling Cabs b. It does not matter; they cost the same. c. 15 mi 47. 24 signs

49. a.

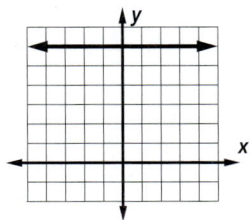

b. mean = 31.9 **c.** bags 4 and 9 **d.** bag 5
51. about 1,447,000 people **53.** $1,280
55.

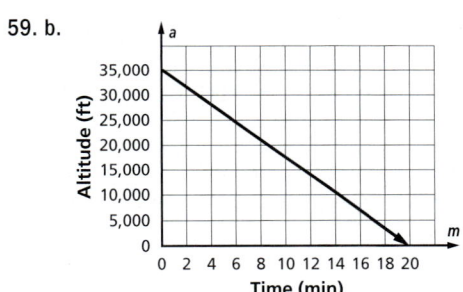

57. $x = 5$ **59. a.** $a = 35,000 - 1,750m$

59. b.

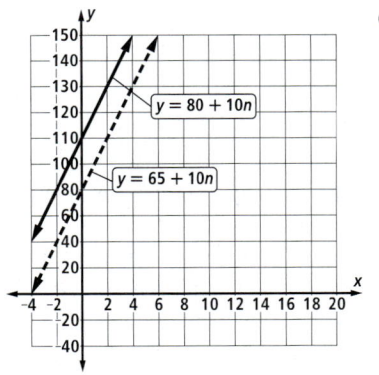

c. 20 min **61.** $x \leq 0$ **63. a.** $80 + 10n$ dollars
b. $65 + 10n$ dollars
c.

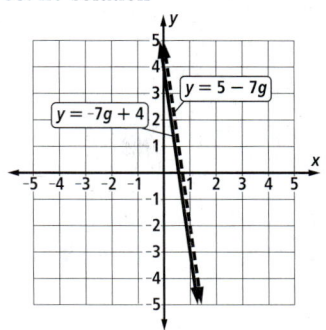

d. never

65. no solution

67. a.

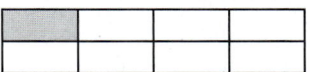

b. $t \leq 65$ or $t \geq 75$

Chapter 5

Lesson 5-1 (pp. 252–257)
Questions: **1.** For all real numbers a, b, c, and d, with b and d not zero, $\frac{a}{b} \cdot \frac{c}{d} = \frac{ac}{bd}$. **3.** $\frac{ab}{14}$ **5.** always true **7.** D **9.** $\frac{-1}{6n}$
11. 1.44 **13.** $\frac{2}{9ab^2}$ **15. a.** One rectangle has one eighth as much area as the other. **b.** Answers vary. Sample answer:

▨		

17. a. The Brocks' garden is half the area of the Peases' garden. **b.** Answers vary. Sample answers: The Peases have a 12-by-12 foot garden that has 144 ft² area; the Brocks' 8-by-9 foot garden has a 72 ft² area, which is half of 144 ft². **19.** 4 **21.** $\frac{1}{r}$ **23.** Answers vary. Sample answers: $\frac{6a^3}{5}$ and $\frac{6}{ax}$ **25. a.** $\frac{L^3}{6}$ units³
b. $(12)(6)(4) = 288 = \frac{(12)^3}{6}$ **c.** 6, stack them 2 high and 3 deep **27.** $x = 40$ **29.** 3,025 ft²

Lesson 5-2 (pp. 258–262)
Guided Example 2: $\frac{6x}{5}, \frac{9x}{10}, \frac{9x}{10}, \frac{10}{9x}$, 4; 3; 24; 18; yes
Questions: **1.** $\frac{5}{3} \approx 1.67, \frac{7}{11} \approx 0.64, 1.67 \div 0.64 \approx 2.61 \approx \frac{55}{21}$
3. a. n **b.** $\frac{1}{n}$ **5.** $\frac{24}{25}$ **7.** 3 **9.** $\frac{14m}{5n}$ **11.** $\frac{1}{10}$ **13.** $m = 92$
15. a. $200 \div \frac{1}{4}$ **b.** 800 **17.** b^2 **19.** $\frac{48}{n}$ **21.** $\frac{5d^3}{169}$
23. a. $V = 13{,}574.7$ **b.** 6%; $14,289.16; the car before taxes or discounts **25. a.** $3B + 2 = 10$ **b.** $\frac{8}{3}$ kg

Lesson 5-3 (pp. 263–268)
Guided Example 3: **a.** denominator; $x + 4$; −4; −4 **b.** numerator; x; 0
Questions: **1.** hundreds of thousands of people per year; $40 per ticket; $150 per ticket; $20 per hour; 50 dollars per hour to park; 230 miles per hour; 354 kilometers per hour; 2.5 miles per lap **3.** $\frac{6}{t}$ blocks per minute

5. about 7.77 books per person   7. a. $\frac{-5 \text{ degrees}}{8 \text{ hr}} = -\frac{5}{8}$ degrees per hour   b. $-\frac{5}{8}$ degrees per hour · 8 hours = −5 degrees   9. a. $w = 12$ b. $w = -5$
11. a. $y = -2$ b. none
13.

x	y
−3	$-\frac{2}{5}$
−2	$-\frac{1}{4}$
−1	0
0	$\frac{1}{2}$
1	2
2	undefined
3	−4

15. a. 1,000 people/mile^2 b. about 6.1 people/mile^2 17. a. $0.10/min   b. $6/hr   c. $70 - 0.10m$ dollars   19. $\frac{4b^2}{21}$ 21. $8n$ 23. $\frac{5}{7}$ 25. $\frac{2}{15}$ 27. a. $y < -10$ b. $x > -2$ c. $A \geq -3.5$ d. $B \leq -0.25$

Lesson 5-4 (pp. 269-273)
Guided Example 2: feet; seconds; 60 min; 1 min; 5,280 ft; 5,280; sec; about 100 ft/sec
Questions: 1. a. 22 mi   b. $\frac{11m}{15}$ mi   3. $\frac{1,370 \text{ pages}}{1} \cdot \frac{1 \text{ day}}{12 \text{ pages}} \cdot \frac{1 \text{ mo}}{30 \text{ days}} \approx 3.8$ months   5. a. Answers vary. Sample answer: Those delicious tomatoes cost $4/kg.   b. 0.25 kg/dollar   c. Answers vary. Sample answer: With one dollar you can buy 0.25 kg of those delicious tomatoes.   7. $\frac{36 \text{ in.}}{1 \text{ yd}}, \frac{1 \text{ yd}}{36 \text{ in.}}$ 9. 48 km/hr   11. Answers vary. Sample answer: The ant crawls one foot per minute. How far does the ant go, in inches, in 16 minutes and 40 seconds?   13. a. $2k$ dishes per min   b. $\frac{1}{2k}$ min per dish   15. a. 25 mi per day   b. about 11.5 marathons   17. 3.1 mi per day   19. 27 boxes   21. 73 in.   23. B   25. B

Lesson 5-5 (pp. 274-279)
Questions: 1. a. $\frac{9}{4}$ b. 225%   3. $6\frac{2}{3}$ qt of fruit punch and $13\frac{1}{3}$ qt of ginger ale   5. a. 0.684   b. $\frac{11}{17+n}$ c. 1   7. about 143 children   9. a. 20 min   b. 384 words   11. approximately 441 mi/hr   13. a. no   b. no   c. no   d. no   15. $0 < h < 432$ cm

Lesson 5-6 (pp. 280-288)
Questions: 1. the probability that event E occurs   3. a. 20%   b. 80%   5. a. $\frac{1}{36}$ b. 0   c. $\frac{1}{6}$ 7. a. 11.1%   b. about 84.7%   c. about 15.3%   d. conditional   9. a. 5 to 7   b. 7 to 5   11. A diamond, spade, or club is chosen from a standard deck of playing cards.   13. a. $\frac{1}{52}$ b. $\frac{1}{13}$ c. $\frac{2}{13}$ 15. $\frac{1}{3}$ 17. a. $\frac{1}{200}$ b. $\frac{101}{200}$ c. 0   19. Answers vary. Sample answer: Complementary events do not necessarily have equal probabilities.   21. a. $513   b. rate   23. $D = \frac{64}{3}$ 25. $149.50

Lesson 5-7 (pp. 289-295)
Guided Example 1: a. 198   b. 21   c. 95; 198; 0.48   d. 70; about 35%
Questions: 1. The event did not occur in any trial.   3. The event will never happen.   5. a. $\frac{5}{26}$ b. 12 vowels; 24 consonants   c. $\frac{1}{3}$ d. Answers vary. Sample answer: because most words have a vowel in them   7. 46% of tests were given this or a lower score.   9. 82 in. About 80.3% of the NBA players in the sample answer are 82 in. tall or shorter.   11. $a = 28$ and $b = 37$ 13. A   15. about 37%   17. a. $\frac{3}{19}$ b. $\frac{1}{578}$ c. twice   19. 8.13 in.
21. $x = -3$ or $x = 7$

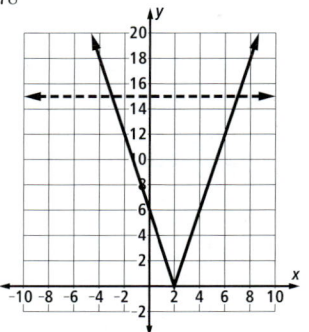

Lesson 5-8 (pp. 296-300)
Questions: 1. a. 0.49π ft^2 b. 5.29 ft^2 c. 29%   d. 71%   3. a. $\frac{1}{6}$ b. $\frac{1}{12}$ c. $\frac{1}{3}$ 5. approximately 60.5%   7. $\frac{5}{9}$ 9. a. approximately 29.2%   b. approximately 70.8%   11. Answers vary. Sample answer:

13. $\frac{9\pi}{160}$ 15. 9.8 ft and 4.2 ft   17. a. 0   b. The Zero Product Property   19. $\frac{7}{9}$

Lesson 5-9 (pp. 301-307)
Questions: 1. 54   3. a statement that two ratios are equal   5. a. $p = -48$ b. $\frac{64}{-48} = \frac{-4}{3}, \frac{-28}{21} = \frac{-4}{3}$ 7. a. about 1.85 million people   b. about 150,000   9. 280 wolves   11. $m = 16$ 13. $c = \frac{3b}{2}$ 15. $a = \frac{bx}{y}$ 17. 570 min   19. a. 114 bpm   b. no   21. 56 earned runs   23. a. $\frac{4}{29}$ b. $\frac{17}{29}$ c. $\frac{21}{29}$ 25. a. $n = \frac{S}{180} + 2$ b. 9 sides   27. a. your seven-digit phone number   b. If we let x be the first three digits of your phone number and y be the last four, then your phone number can be written as $10,000x + y$. We can also write the result of the puzzle as an algebraic expression involving x and y: $\frac{(80x+1) \cdot 250 + y + y - 250}{2} = 10,000x + y$, which is the same as the expression for your phone number.

Lesson 5-10 (pp. 308–314)

Guided Example 2: $\overline{LM}$; $\overline{BC}$; $\overline{KL}$; $\overline{LM}$; $\overline{BC}$; $\overline{KL}$; 12.6; 3; 4.8; 4.8; 3; 12.6; 7.875; 4.8

Questions: 1. Corresponding angles have the same measure. Ratios of lengths of corresponding sides are equal.
3. a. $\overline{XZ}$ **b.** $\overline{YZ}$ **c.** $\overline{XY}$ **5. a.** 7.5 **b.** 19.5 **7. a.** 1.5 in.
b. $\frac{1}{24}$ **c.** 1 in. **d.** 24 in. **9.** $\frac{40}{3}$ **11. a.** $x = 21$ **b.** 96
13. a.

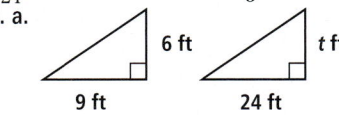

b. $\frac{6}{9} = \frac{t}{24}$ **c.** 16 ft
15. $x = 48$ **17.** 75% **19.** about 69.3 g **21. a.** 3 **b.** $\frac{1}{12}$
c. There are 6 ways to achieve a sum of 7, and each other sum has fewer than 6 ways in which it can be obtained.
23. 3,705,386 mi²

Self-Test (pp. 318–319)

1. $\frac{75c}{8p} \cdot \frac{2}{15c} = \frac{5}{4p} \cdot \frac{1}{1} = \frac{5}{4p}$ **2.** $\frac{5}{a} \div \frac{9}{3a^2} = \frac{5}{a} \cdot \frac{3a^2}{9} = \frac{5}{1} \cdot \frac{a}{3} = \frac{5a}{3}$
3. $\frac{2x}{5} \div \frac{x}{5} = \frac{2x}{5} \cdot \frac{5}{x} = \frac{2}{1} \cdot \frac{1}{1} = 2$ **4.** B **5.** $h = \frac{6}{101} \cdot 17 = \frac{102}{101}$
6. $4u = \frac{9}{5} \cdot 3$; $u = \frac{27}{5} \cdot \frac{1}{4} = \frac{27}{20}$ **7.** $24\left(\frac{x}{8}\right) = 24\left(\frac{2x-3}{24}\right)$;
$3x = 2x - 3$; $x = -3$ **8. a.** $8x = 105$ **b.** $x = \frac{105}{8} = 13.125$
9. $\left(6 \text{ boxes} \cdot \frac{10 \text{ pencils}}{\text{box}}\right) \div 4 \text{ children} = 60 \text{ pencils} \div$
$4 \text{ children} = 15 \frac{\text{pencils}}{\text{child}}$ **10.** $\frac{221}{1,563} \approx 0.141 = 14.1\%$
11. $\frac{7.5 \text{ min}}{1 \text{ mi}} \cdot \frac{1 \text{ mi}}{1.6 \text{ km}} \cdot \frac{60 \text{ sec}}{1 \text{ min}} = 281.25 \frac{\text{sec}}{\text{km}}$; 281.25 sec
12. $\frac{x}{x-5} = \frac{12}{4}$; $4x = 12(x-5)$; $4x = 12x - 60$; $60 = x$;
$x = \frac{60}{8} = 7.5$ **13.** The sector between 3 and 4 represents $\frac{1}{12}$ of the clock, so the probability is $\frac{1}{12}$.
14. The discount is $4.50 off $22.50, so $4.50/$22.50 = 0.2, or 20%. **15. a.** $50 \text{ km}/\left(1\frac{47}{60}\text{hr}\right) \approx 28$ km/hr **b.** about 0.036 hr/km **16.** 18%, because $7 + 2 = 9$ people out of 50.
17. 50%, because $50 - 7 - 16 - 2 = 25$ people out of 50. **18.** 46%, because $7 + 16 = 23$ people out of 50.
19. k cannot be 2, because then the denominator would be 0, and division by zero is undefined. **20.** $\frac{c}{f+s+c+d}$ **21.** Let m be the number of miles. Then $\frac{350}{20} = \frac{m}{35}$; $20m = 12,250$; $m = 612.5$ mi. **22.** Area of rectangle = 4 ft · 7 ft = 28 ft². Area of circle = $\pi \cdot \left(21 \text{ in.}/ 12 \frac{\text{in.}}{\text{ft}}\right)^2 \approx 9.62$ feet². So the probability is about $\frac{9.62}{28} \approx 34\%$. **23.** $\frac{1,210}{x} = \frac{4}{6}$; $x = 1,815$ ft **24.** 24; The 75th percentile implies 25% performed better. $\frac{6}{x} = \frac{1}{4}$; $x = 24$

The chart below keys the **Self-Test** questions to the objectives in the **Chapter Review** on pages 320–323 or to the **Vocabulary (Voc)** on page 317. This will enable you to locate those **Chapter Review** questions that correspond to questions missed on the **Self-Test**. The lesson where the material is covered is also indicated on the chart.

Question	1	2	3	4	5	6	7	8	9	10
Objective(s)	A	B	B	A	C	C	C	D	E	H
Lesson(s)	5-1	5-2, 5-3	5-2, 5-3	5-1	5-9	5-9	5-9	5-9	5-3	5-6, 5-7
Question	11	12	13	14	15	16	17	18	19	20
Objective(s)	F	L	I	E	F	H	H	H	B	G
Lesson(s)	5-4	5-10	5-8	5-3	5-4	5-6, 5-7	5-6, 5-7	5-6, 5-7	5-2, 5-3	5-5
Question	21	22	23	24						
Objective(s)	J	I	J	K						
Lesson(s)	5-9, 5-10	5-8	5-9, 5-10	5-7						

Chapter Review (pp. 320–323)

1. $\frac{14a}{5}$ **3.** $\frac{12}{p}$ **5.** $11b$ **7.** $\frac{c}{27}$ **9.** $\frac{-3}{25}$ **11.** -8
13. $k = -\frac{5}{6}$ **15.** $t = -1$ **17.** 486 **19.** $\frac{15}{6}$ **21.** $\frac{u}{m}$
23. a. $0.075 per oz **b.** $0.05625 per oz **c.** the 32 oz box
25. $4w$ words in $6m$ min, since $\frac{4w}{6m} = \frac{2w}{3m} > \frac{w}{2m}$
27. Answers vary. Sample answer: 1.5 ft per year **29. a.** 161,280 breaths **b.** The cat breathes at a faster rate, since humans take $16 \cdot 60 = 960$ breaths per hour **31. a.** $2n$ envelopes **b.** $\frac{1}{2n}$ min **33.** D **35.** 8 **37.** 12.5 gal of blue paint, 7.5 gal of yellow paint **39.** 5 hits **41.** $\frac{1}{3}$

Student Handbook

43. 69% **45.** $\frac{11}{36}$ **47. a.** $\frac{\pi}{64}$ **b.** $\frac{\pi}{8}$ **49.** $1,176 **51.** 40 moose **53.** 25th **55. a.** 3 **b.** 6 **c.** 16 **57. a.** Answers vary. Sample answer: $\frac{x}{6}, \frac{y}{4}$ **b.** $y = \frac{128}{9}$ **c.** $x = \frac{27}{16}$ **59.** 93.75 units^2

Chapter 6

Lesson 6-1 (pp. 326–332)
Questions: 1. dollars; hours **3. a.** 1920s **b.** 1910–1930 **c.** 1970s **d.** (B10 − B9)/(A10 − A9) **5.** increase **7.** decrease **9.** $-\frac{2}{3}$ **11. a.** $\frac{7}{60} \frac{m}{hr}$ **b.** negative **13.** −$131 per day **15.** True; $\frac{9}{15} = \frac{\frac{3}{5}}{1}$ **17.** 850 + 48x coins

Lesson 6-2 (pp. 333–340)
Guided Example 3: 3; −2; $-\frac{1}{2}$; 3; 2; $-\frac{1}{2}$; $\frac{9-5}{-10--2}$; $-\frac{1}{2}$; $-\frac{1}{2}, -\frac{1}{2}$

Questions: 1. line **3. a.** Answers vary. Sample answer: (6, 3.9) and (8, 3.3) **b.** Answers vary. Sample answer: $\frac{5.4-5.7}{1-0} = -0.3$ **c.** 0.3 cm of the candle burns every second. **5.** 5 **7. a.** $\frac{10}{3}$ **b.** $\frac{29}{9}$ **c.** No, because the line connecting the first two points has a different slope than the line connecting the next two. **9.** 2 **11. a.** Answers vary. Sample answer: (0, −6) and (5, 0) **b.** $\frac{6}{5}$ **c.**

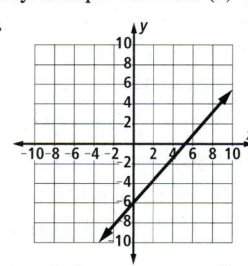

13. a. $-\frac{1}{5}$ **b.** Water is removed from the bathtub at a rate of one gallon every five seconds. **15. a.** degrees per hour **b.** The rate of change describes the variations in temperature over time. **17.** y = 11 **19. a.** 3 toothpicks per term **b.** Answers vary. Sample answer: Note that (0, 1) and (1, 4) lie on the line and $\frac{4-1}{1-0} = 3$ **c.** It is the rate at which the number of toothpicks increases from one term to the next. **21.** no solution **23.** 2b

Lesson 6-3 (pp. 341–347)
Questions: 1. a.

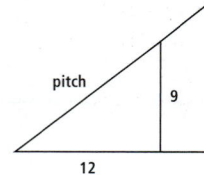

b. No, because $\frac{9}{12} > \frac{8}{12}$, and $\frac{8}{12}$ is considered the limit for the pitch of a safe roof. **c.** 0.75 **3.** Not possible because $\frac{12}{0}$ is undefined. **5.** 1

7. a.

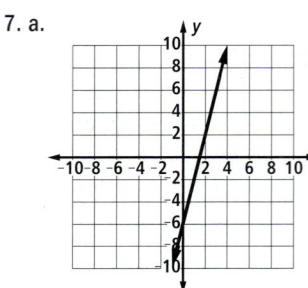

such as (1, 4) and (1, 2). If we calculate the slope of the line using the formula $\frac{y_2 - y_1}{x_2 - x_1}$, the denominator is 1 − 1 = 0. Division by zero is undefined.

13.

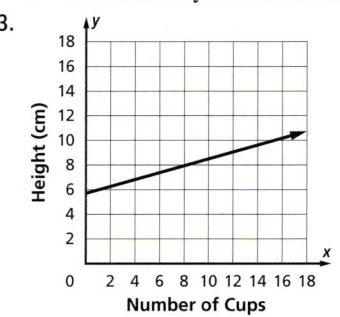

15.

x	y
2	0
3	−4
4	−8
5	−12

17. 37 ft

19.

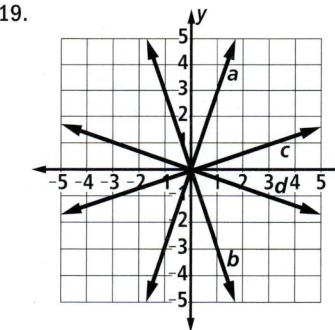

21. 1.5 **23. a.** 2001 and 2002 **b.** 2001 and 2002 **25.** b = −13.75

b. Answers vary. Sample answer: (3, 6) **c.** $\frac{6-2}{3-2} = \frac{4}{1} = 4$ **9.** Answers vary. Sample answer: (0, −5) **11.** Answers vary. Sample answer: A vertical line is one that passes through points with the same x-coordinate,

a. lines a and c **b.** lines b and d **c.** The lines with positive slope are slanted up to the right; the lines with negative slope are slanted down to the right. **d.** lines a and b **e.** Answers vary. Sample answer: 4

Lesson 6-4 (pp. 348-355)

Questions: 1. m; b 3. a. 2 b. 3
c.
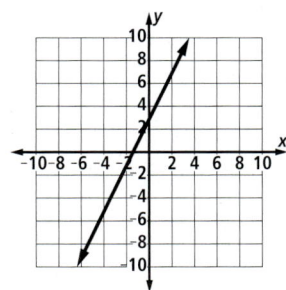

5. a. $y = -1.3x + 5.6$
b. -1.3 c. 5.6 7. a. $h = 2t + 3$

b.
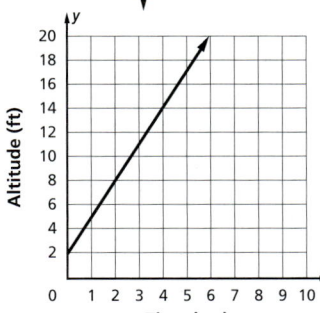

c. $m = 2$; $b = 3$ d. 123 ft
9. $y = 4x + 3$
11. $y = 2x - 5$
13. c; i 15. b; ii

17. a.
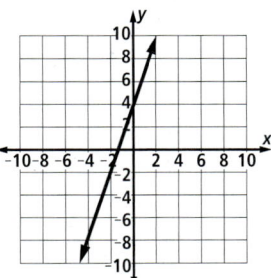

b.
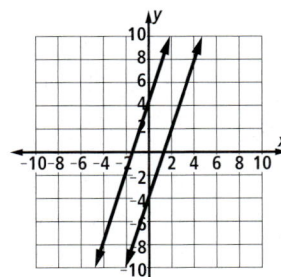

c. $y = 3x - 4$ d. They have the same slope.
19. $\frac{10}{27} \approx 0.37$
21. $x = \frac{3}{2}$ 23. -3 and -2

Lesson 6-5 (pp. 356-360)

Guided Example 2: k; h; 0; -5; 22
Questions: 1. a. $y = 9{,}700x + 143{,}560$ b. 289,060 people
c. 2027 3. $y - 6 = -2(x + 8)$ 5. $y = -\frac{1}{2}$ 7. $y - 15 = -2(x - 20)$ 9. $y + 3 = -4.3(x - 6.8)$ 11. a. $y = 1.5x + 2$
b. 74 cm 13. a. $m = 0.3$; $x = 5$; $y = 12$ b. $y - 12 = 0.3(x - 5)$ c. 14.1 lb 15. $y = 4x + 3$ 17. a. $m = 1.75$; (15, 29.25)
b. $y - 29.25 = 1.75(x - 15)$

19.
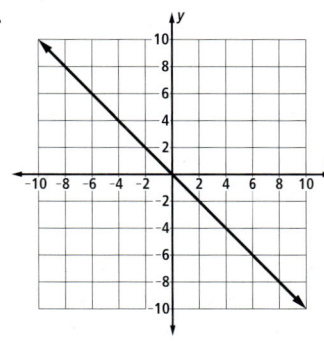

21. $2.31 per bushel; the price of each bushel 23. a. $\frac{6}{5}$
b. $\frac{6}{5}$ c. $\frac{z}{y}$

Lesson 6-6 (pp. 361-367)

Guided Example 3:
(10, 2.23) and (20, 4.12); slope; $\frac{4.12 - 2.23}{20 - 10}$; 2.23; 10 (or 4.12; 20); $b = 0.34$; 0.189; 0.34; 0; cost per minute; y; x; about 130
Questions: 1. a. $y - 70 = \frac{25}{54}(x - 70)$ b. about 79°F
3. $y - 1 = -1.5(x - 4)$ 5. $y - 11 = -(x - 4)$
7. $y - 27 = 0.5 \cdot (x - 30)$ 9. a. $y - 20 = -\frac{389}{60}(x - 1943)$
where x is the year and y is the price b. about $-$70.77
c. Answers vary. Sample answer: No, since the price of penicillin should never be negative. 11. a. (2.1, 59.24) and (3.7, 70.08) b. 6.775
c. The slope represents how much longer you will have to wait for the next eruption for every minute of an eruption.
d. $y = 6.775x + 45.0125$ e. 7.23 min 13. $y = 1.60x + 2.5$

15.
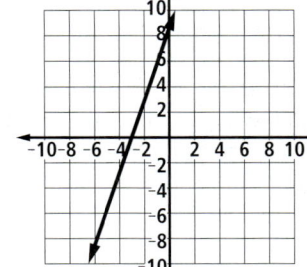

17. a. between the 47 mi and 72 mi signs
b. about 52.5 ft per mi
19. a. $l = 2w + 3$
b. 6 cm and 15 cm
c. 90 cm^2
21. $12g + 7.25h$

Student Handbook

Lesson 6-7 (pp. 368–373)
Guided Example: 11.02; 10.56
Questions: 1. a. no b. yes
3. a.

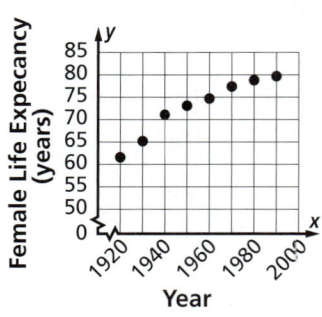

b. Answers vary. Sample answer:

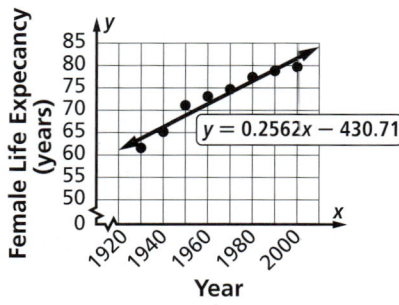

c. approximately 63.8, 66.3, 68.9, 71.4, 74.0, 76.6, 79.1, and 81.7 yr d. about 18.82 e. 86.67 yr

5. a.

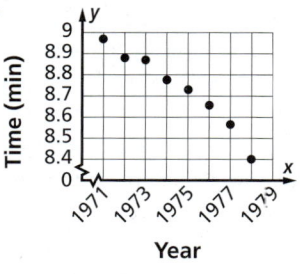

b. about 8.33 min
c. 0.6 min d. 0.51 min
e. Answers vary. Sample answer: Yes, the regression equation predicts values that are within 4 sec (for 1989) and within about 7 sec (for 1919), and that is not a very large difference in an 8-min race.

7. a.

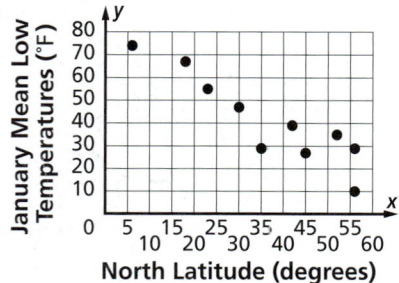

b. $y \approx -1.086x + 81.139$ c. decrease by about one degree Fahrenheit d. Tokyo, Japan e. about $-16°F$ f. about $71°F$
9. no 11. no 13. a. $y - 4 = \frac{5}{3}(x - 6)$ b. Answers vary. Sample answer: $(0, -6)$ 15. $-\frac{1}{6}; 0$

17. a. Answers vary. Sample answer:

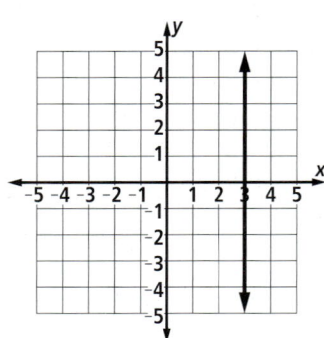

b. not defined
19. $\frac{1}{8}$ 21. a. 0
b. all numbers c. no solutions

Lesson 6-8 (pp. 374–380)
Questions: 1. standard form for an equation of a line
3. $A = 5; B = -3; C = 9$ 5. a. x-intercept = 10; y-intercept = 4 b.

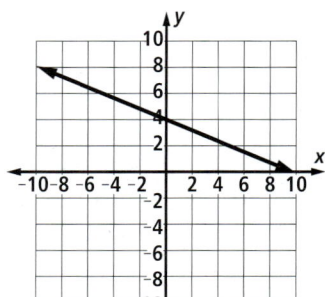

7. a. Sample answer: $(0, 9); (3, 7); (6, 5)$
b. $2x + 3y = 27$, where x are the 2-point shots and y are the 3-point shots c. 9 d. 12

e.

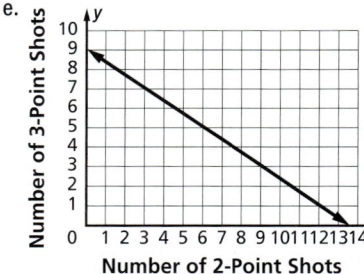

9. a. $9x - 8y = 0$
b. $A = 9; B = -8; C = 0$ 11. $y = \frac{4}{7}x - 44$ 13. a. Answers vary. Sample answer: $C = 62, W = 0; C = 65, W = 12; C = 68, W = 24$ b. $C - \frac{1}{4}W = 62$

c.

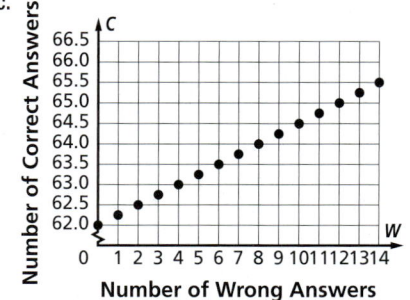

S22 Student Handbook

15. a. Answers vary. Sample answer: **b.** Answers vary. Sample answer: (1900, 11.2), (2000, 9.6); $y - 9.6 = -0.016(x - 2000)$; $y = -0.016x + 41.6$ **c.** Answers vary. Sample answer: 9.408 seconds; The prediction may not be correct because the trend in the data may not continue in the future.

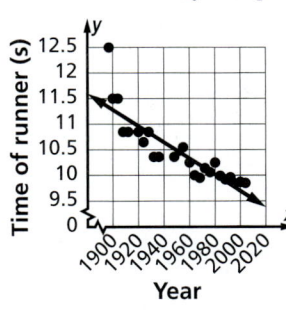

17. a. $y = 4x + 8$; with $y \leq 72$ **b.** 4 **c.** 16 hr **19.** $\frac{8}{11}$
21. no

Lesson 6-9 (pp. 381–386)

Questions: **1. a.**

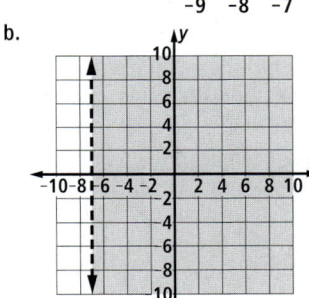

b.

3. $x \leq 1$ **5. a.** half-planes **b.** boundary **7.** The first is the graph of the half-plane under $y = -2x + 6$ *excluding* the line $y = -2x + 6$; the second is the same half-plane *including* the line.

9.

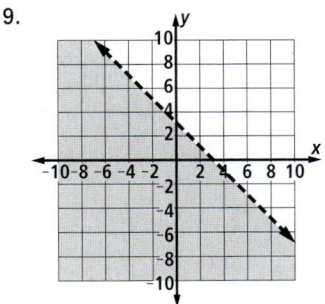

11. The points in the shaded area with integer coordinates are solutions.
13. a. It would imply that someone made more field goals than he attempted! **b.** $y \leq x$ **c.** Answers vary. Sample answer: The points would lie farther away from the x-axis. **15.** Answers vary. Sample answer: (6, 0) **17. a.** $8x + 12y = 288$ **b.** 24 **c.** 14 **19. a.** $y = \frac{1}{6}x$ **b.** 30 lb **c.** 37.2 lb **21. a.** 0.01 **b.** 0.006 **c.** 0.000012

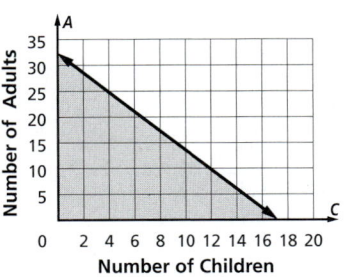

Self-Test (pp. 390–391)

1. a. −2 **b.** 3 **c.** $\frac{-2-0}{0-3} = \frac{-2}{-3} = \frac{2}{3}$ **2.** No, because the slope between (8, 2) and (18, −6) is $\frac{2-(-6)}{8-18} = \frac{8}{-10} = -\frac{4}{5}$, while the slope between (18, −6) and (12, −1) is $\frac{-1-(-6)}{12-18} = \frac{5}{-6} = -\frac{5}{6}$. All points on a line have the same slope between them.
3. $y = \frac{8}{9}x - 12$ **4.** $T = 3h + 58$, where h is hours after 8 A.M. **5.** The slope of a vertical line is undefined.
6. $3x + y = 8$; $A = 3$, $B = 1$, and $C = 8$ **7.** It will go down $\frac{1}{2}$ of a unit. **8.** $y = \frac{1}{2}x + -\frac{1}{4}$; slope = $\frac{1}{2}$, y-intercept = $-\frac{1}{4}$ **9. a.** $6,400 \geq 300p + 400b$
b. The values are points with integer coordinates in the shaded area.

10. 2003 and 2004
11. $\frac{103,111 - 96,500}{2004 - 1999} = 1,322.2$ people per year

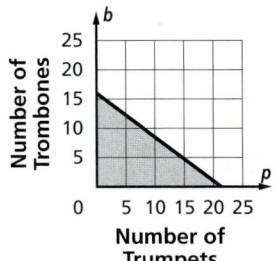

12. **13.**

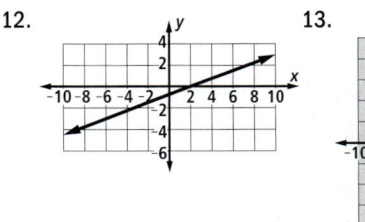

14. The points (22, 14) and (132, 9) imply $m = \frac{14-9}{22-132} = \frac{-5}{110} = -\frac{1}{22}$. So, $y = -\frac{1}{22}x + b$. Substituting (22, 14), $14 = -\frac{1}{22}(22) + b$, so $b = 15$, and $y = -\frac{1}{22}x + 15$. **15.** Answers vary. Sample answer: $y = \frac{5}{4}x + b$, $0 = \frac{5}{4}(-4)$, $0 = -5 + b$, $b = 5$; an equation of the line is $y = \frac{5}{4}x + 5$, so another point on the line is (0, 5). **16. a.** Answers vary. Sample answer: (2, 35), (15, 78) **b.** $m = \frac{78-35}{15-2} = \frac{43}{13} \approx 3.31$; Substitute (2, 35), so $35 = 3.31(2) + b$, $35 = 6.62 + b$, so $b = 28.38$; $y = 3.31x + 28.38$ **c.** $y = 3.31(25) + 28.38 = 111.13$; $111,130 **17.** $y = 2.98x + 33.28$ **18. a.** c **b.** b
19. a. a **b.** d **20.** $10x + 5y < 100$; the points with integer coordinates in the shaded area are solutions.

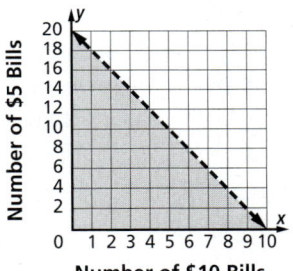

21. a. $y - 10 = -3(x - 4)$
b. Let $x = 0$. Then $y - 10 = -3(0 - 4)$ so $y - 10 = 12$, so $y = 22$. Patrick lives on the 22nd floor.

Student Handbook

The chart below keys the **Self-Test** questions to the objectives in the **Chapter Review** on pages 392–395 or to the **Vocabulary (Voc)** on page 389. This will enable you to locate those **Chapter Review** questions that correspond to questions missed on the **Self-Test**. The lesson where the material is covered is also indicated on the chart.

Question	1	2	3	4	5	6	7	8	9	10
Objective(s)	A	A	C	F	D	C	D	C	I	E
Lesson(s)	6-2	6-2	6-4, 6-8	6-4, 6-5, 6-6, 6-8	6-2, 6-3	6-4, 6-8	6-2, 6-3	6-4, 6-8	6-9	6-1, 6-3

Question	11	12	13	14	15	16	17	18	19	20
Objective(s)	E	H	I	F	B	G	G	D	D	I
Lesson(s)	6-1, 6-3	6-3, 6-4, 6-8	6-9	6-4, 6-5, 6-6, 6-8	6-4, 6-5, 6-6	6-7	6-7	6-2, 6-3	6-2, 6-3	6-9

Question	21
Objective(s)	F
Lesson(s)	6-4, 6-5, 6-6

Chapter Review (pp. 392-395)

1. $\frac{6}{5}$ 3. $-\frac{5}{2}$ 5. $y = 3$ 7. $y = 2.6x + 7$
9. $y - 20 = -0.5(x + 5)$ 11. $y - 0.25 = 3(x - 30)$
13. $y - 4 = \frac{3}{2}(x - 3)$ 15. $x = -3$ 17. Answers vary.
Sample answer: $x + -2y = -14, A = 1, B = -2, C = -14$
19. $y = -2x + \frac{5}{2}$ 21. slope = 12, y-intercept = -6.2
23. slope = -1, y-intercept = 0 25. y, change 27. u
29. 0 31. Any two points on a line can be used to determine the slope of the line. Choosing any two of the points must yield the same slope as any other two points. Therefore, if the slope between any two of the three points is different from the slope of the third point and either of the two points used, the points are not collinear.
33. $-\frac{1}{5}$ 35. 27.4 subscribers per year 37. –13.6°F/month
39. October and November 41. slope = –$100/day, y-intercept = $1,000 43. $y = 0.25x + 1.75$
45. $p = 50m + 25,350$ 47. $820 = 8S + 7B$
49. a. Answers vary. Sample answer:

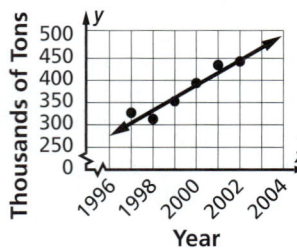

b. Answers vary. Sample answer: 29.2 c. It indicates how many additional tons of fish, crustaceans, and mollusks are being caught in the Philippines each year. d. Answers vary. Sample answer: $y = 29.2x - 58,012.4$ where y is the thousands of tons and x is the year e. $T = 28.2y - 56,008$ f. Answers vary. Sample answer: The equation in Part d predicts a value 6 thousand tons less (680) than the regression line (674).

51.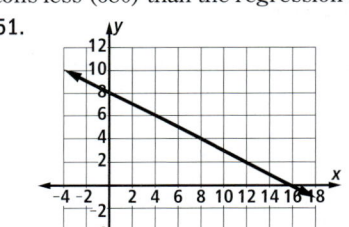

53.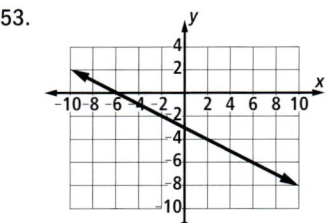

55.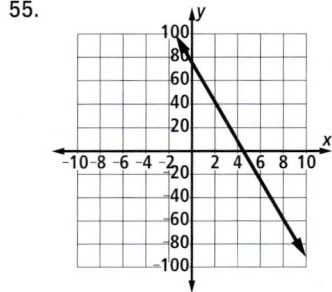

57.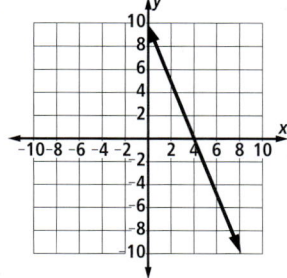

59. The possible values are points (n, q) with integer coordinates in the shaded area.

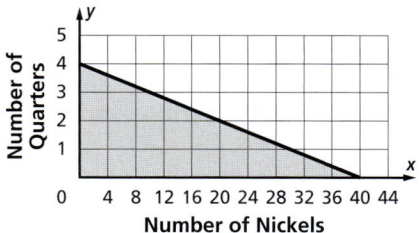

61.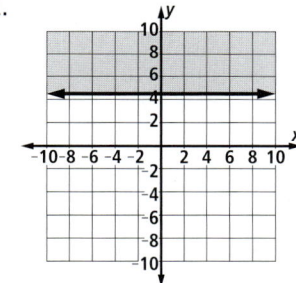

63.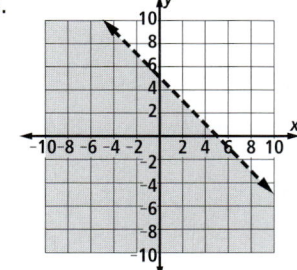

65.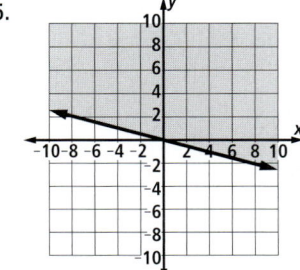

Selected Answers

Chapter 7

Lesson 7-1 (pp. 398–403)
Guided Example 3: 2,000; 0.054; 18; 2000; 1.054; 18; 5154.196734; $5,154.19; $5,154.19
Questions: 1. "4 to the 10th power" or "4 to the 10th"
3. a. 343 **b.** 343; 7 ^ 3 ENTER **5.** $18 \cdot (-3)^4$
7. a. 100,000 people **b.** 250,000 people **c.** about 269,159 people **9.** $1.02P$ **11. a.** $A = P(1 + r)^t$ **b.** total amount including interest **c.** starting principal **d.** annual yield **e.** number of years **13.** $2,529.55 **15. a.** Susana earns $20.40; Jake earns $41.60 **b.** No. After the first year, he has more money on which to earn interest than Susana does.
17. a. $A = 100 \cdot 1.1^t$
b.

Time since investment (in years)	Danica's account ($)
0	100.00
2	121.00
4	146.41
6	177.16
8	214.36
10	259.37
12	313.84
14	379.75
16	459.50
18	555.99
20	672.75

19. around year 15 **21. a.** $3W + T > 3$
b.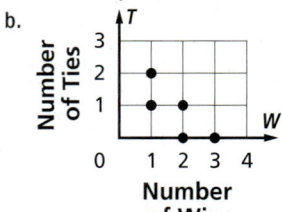
23. $\frac{5}{6}$ **25.** all real numbers **27.** $t = 5$

Lesson 7-2 (pp. 404–410)
1. a. $y = 11 \cdot 3^x$, where y is the population and x is the number of years after 1995. **b.** 99 round gobies **c.** 2,673 trillion round gobies **d.** about 2,265 trillion round gobies **3. a.** 1.045 **b.** $3,276.07 **5. a.** 1 **b.** 17 **c.** 1
7. $(-5)^0 = 1$, but $-5^0 = -1$, by the order of operations

9. a.

x	y
0	0.5
1	1
2	2
3	4
4	8

b.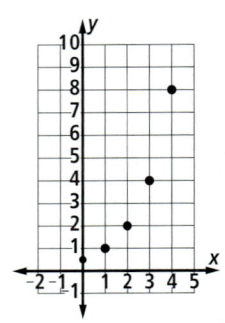

11. a. by about 3.585 million people **b.** about 56.190 million people **13. a.** 1.04 is the growth rate; 34,277 is the 1860 population **b.** 1,731,158 people; 22,789 people **c.** about 26,956,933 people
15. a.

b. 1.5 **c.** 58.59375 **d.** This graph is not a line. Its rate of change increases as x increases. **17. a.** $\frac{1}{31}$ **b.** $\frac{9}{31}$
19. 25.74

Lesson 7-3 (pp. 411–418)
Guided Example 3: a. 10; 0.97; 10; 0.97 **b.** 23; 23
c. 240; 240; 10; 0.97; 240
Questions: 1. 0.83 **3.** 1 **5. a.** $27,200 **b.** $25,600
c. $32,000 \cdot (1 - 0.01d)$ dollars **7.** about 6 units; $15 \cdot 0.97^x$ units **9.** constant **11.** decay
13. a. $y = 2,500 \cdot 0.98^x$ **b.** about 2,043 students

15. a.

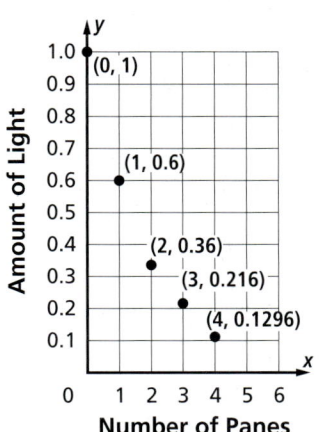

Light-Panes Relation

b. No, it will get infinitely close to zero.

17. a.

x	Decimals	Fractions
0	1	1
1	0.5	$\frac{1}{2}$
2	0.25	$\frac{1}{4}$
3	0.125	$\frac{1}{8}$
10	0.000976563	$\frac{1}{1{,}024}$
20	0.0000009536	$\frac{1}{1{,}048{,}576}$

b. no solutions

19. a.

[Scatter plot: Gallons per Person vs Year, 1999–2006, values approximately 17, 18, 19, 20, 21, 22, 24]

b. Answers vary. Sample answer: $y = 1.7x - 3{,}359.8$
c. Answers vary. Sample answer: $y = 1.7x - 3{,}359.8$
d. about 29.1 gal of bottled water per person

Lesson 7-4 (pp. 419–424)

Guided Example: a. $2.241 \cdot 1.218^x$ b. $2.241 \cdot 1.218^{24} \approx 255$; 255; 165; 165 MHz c. 2.241; 1.218; 44; 13,150
Questions: 1. a. 6 ft **b.** 55% **3.** 8.4 ft; about 0.83 ft **5.** about 80,817 MB **7.** c **9.** a

11. a.

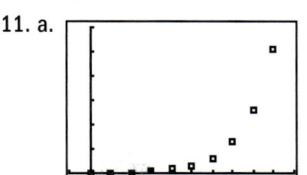

b. Let p be the number of pennies and t be the trial number. $p = 1.57 \cdot 1.49^t$ **13.** Answers vary. Sample answer: If 35% of a 14-kg block of ice melts every day, how much ice, y, remains after x weeks?

15.

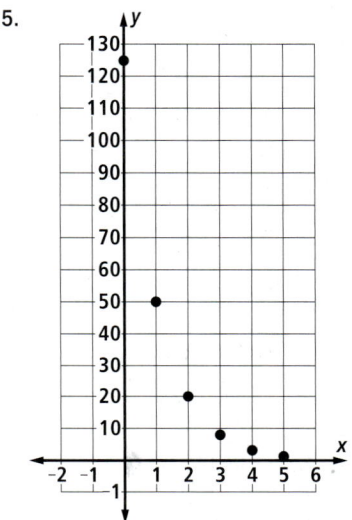

17. $23 **19. a.** $270 **b.** $225 **c.** $200

Lesson 7-5 (pp. 425–431)

Guided Example 2: all real numbers; all real numbers
Questions: 1. a. 9 **b.** 4 is the input and 16 is the output. **c.** 4 is the independent variable and 16 is the dependent variable. **d.** {1, 4, 9, 16, 25} **3. a.** −1 **b.** $\frac{2}{7}$ **5.** No real number has 0 as its reciprocal because if $0 = \frac{1}{x}$, then $x = \frac{1}{0}$, which is undefined. **7. a.** {x: −5 ≤ x ≤ 5} **b.** {y: 0 ≤ y ≤ 5} **9.** C **11.** No; The input $x = 3$ corresponds to both outputs $y = 1$ and $y = 4$. **13. a.** false **b.** 5 **c.** the set of all real numbers **d.** {y: $y > 0$} **e.** 4

15. Answers vary. Sample answer: time is input and height is output

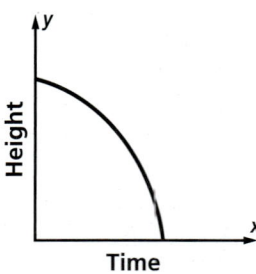

17. $\{y: 0 \leq y \leq 2\}$ 19. domain: the set of all real numbers; range: the set of all real numbers   21. a. Answers vary. Sample answer: As the value of the input increases, that of the output decreases.   b. Khalid Khannouchi is associated with two record-setting times; that is, one input produces two outputs.   23. $\frac{21}{160}$ 25. $pw \leq 1{,}500; p \leq \frac{1{,}500}{w}$

Lesson 7-6 (pp. 432–438)

Guided Example 2: 10; 121,899; 100,000 + 3,000(10); 130,000; 10; constant; 100,000

Questions: 1. f of x 3. a. $E(25) \approx 164{,}061$; $L(25) = 175{,}000$; $C(25) = 100{,}000$ b. They are the population estimates for 25 yr from the present.   5. 12,800   7. a. $45   b. $135   c. 20   9. a. 506.25; It is the estimated value of the computer in four years.
b.

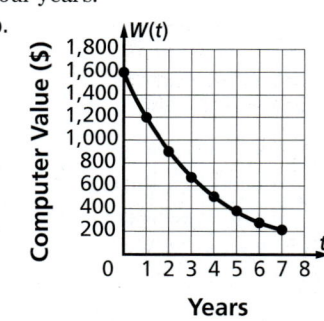

c. $t > 1.63$. This means after about 1.63 years the computer's value is less than $1,000.
11. a. about 325°
b. The temperature of the oven after it was on for 25 minutes was 325°.
c. $t \approx 6$ d. The oven reached 200° after being on for about 6 minutes.
13. No. Answers vary. Sample answer: $x = 1$ results in both $y = -2$ and $y = 0$.   15. a. $5x - 4y = -12$
b. $y = \frac{5}{4}x + 3$

Lesson 7-7 (pp. 439–446)

Guided Example: Option 1: $60; $110; Option 2: $15; $22.50
Questions: 1. Answers vary. Sample answer: The rate of change varies in an exponential growth situation, but remains constant in a constant increase situation.

3.

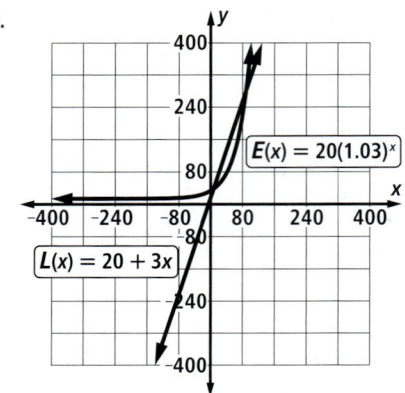

5. Answers vary. Sample answer: $E(x)$ could represent the value of a $20 investment in a bank account with an annual yield of 3% after x years. $L(x)$ could represent the amount in a bank account after x years if $20 is initially invested and $3 is added each year.   7. a. A2 + 13; A3 + 13
b. 54; 67   c. They will decrease to 21 and 34.   d. constant increase; The outputs are represented by the linear function $f(x) = 13x + 28$.   9. a. Next = Now + 270
b. Next = Now · 1.15   c. $L(x) = 85{,}000 + 270x$
d. $E(x) = 85{,}000(1.15)^x$ e. Answers vary. Sample answer: Yes, $L(8) = 87{,}160$.   11. a. Constant increase. Answers vary. Sample answer: Reading at 25 pages per hour is a constant rate increase.   b. $f(x) = 67 + 25x$ c. 33.32 hr
13. a   15. d
17. a.

	Constant	Exponential
0	2,410	2,410
1	2,270	2,270
2	2,130	2,138
3	1,990	2,014
4	1,850	1,897
5	1,710	1,787
6	1,570	1,683
7	1,430	1,585
8	1,290	1,493
9	1,150	1,406
10	1,010	1,325
11	870	1,248
12	730	1,175
13	590	1,107
14	450	1,043
15	310	982

b. 1,570, approximately 1,683; by 113 **c.** 170, approximately 925; by 755   **19.** true   **21.** $24x + 5y = -3$ **23.** $\frac{24}{73}$

Self-Test (pp. 450–451)

1. $\left(\frac{1}{5}\right)^2 + \left(\frac{1}{5}\right)^0 = \frac{1}{25} + 1 = \frac{26}{25}$ **2.** $8^4 d^6$ **3.** $f(1,729) = 3(1,729)^0 = 3 \cdot 1 = 3$ **4.** $g(-2) = 3(-2) - (-2)^2 = -6 - 4 = -10$ **5.** $400(1.044)^7 \approx 540.70$ **6.** Tyrone will have $400(1.044)^{10} \approx 615.26$ dollars and Oleta will have $400 + 22(10) = 620$ dollars, so Oleta will have more.
7. After 25 years, Tyrone will have $400(1.044)^{25} \approx 1,173.74$ dollars, and Oleta will have $400 + 22(25) = 950$ dollars, so Tyrone will have more money.   **8.** The value of the car is depreciating 16%, so the growth factor is $1 - 0.16 = 0.84$.
9. $m(x) = 34,975(0.84)^x, x \geq 0$. Because x represents years, it cannot be negative.   **10.** $m(5) = 34,975(0.84)^5 \approx 14,626.96$ dollars   **11.** $f(1) = 5 \cdot 0.74^1 = 3.7$
12. $f(5) = 5 \cdot 0.74^5 \approx 1.11$ **13.** $f(7) = 5 \cdot 0.74^7 \approx 0.61$
14. $f(12) = |-12 - 3| = |-15| = 15$ **15.** From the graph and knowledge of the absolute value function, you can see only positive values and 0 are in the range. The range is all nonnegative numbers.   **16.** From the graph you can see the only values in the range are those greater than or equal to $f(5)$. $f(5) = |5 - 3| = 2$ so the range is all real numbers ≥ 2.

17.

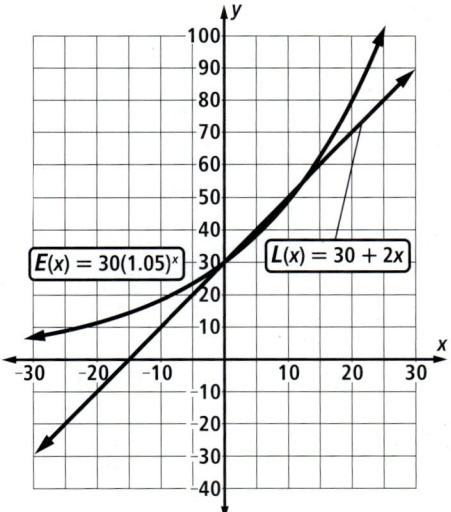

18. $L(9) = 30 + 2 \cdot 9 = 48$, $E(9) = 30(1.05)^9 \approx 46.5$; $L(9)$ is greater.   **19.** Answers vary. Sample answer: $x = 20$; $E(20) \approx 79.60$, $L(20) = 70$, $70 < 79.60$ **20.** An increase of 2.5% tells you that the growth factor is 1.025. The beginning population is 76 million, so the population p, in millions, k years after 1980, is $p(k) = 76(1.025)^k$.

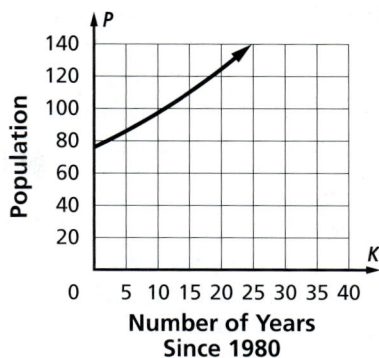

21. A decrease of 1% tells you that the growth factor is 0.99. The beginning circulation is 880,000, so the circulation is $c(x) = 880,000(0.99)^x$.

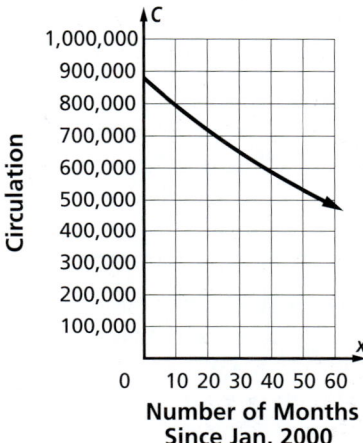

22.

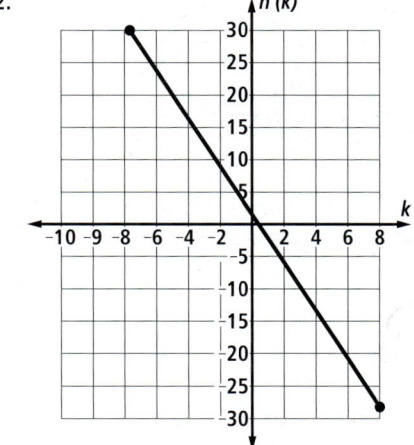

23.

24. a. f is a linear function with negative slope, so it describes a constant decrease situation, ii. b. g is an exponential function with growth factor 5. Since $5 > 1$, g describes an exponential growth situation, iii. c. h is an exponential function with growth factor 0.4. Since $0.4 < 1$, h describes an exponential decay situation, iv. d. m is a linear function with positive slope, so it describes a constant increase situation, i. 25. a. The year increases by 1, so add 1 to the value in A2. Input "=A2 + 1". b. Answers vary. Sample answer: In cell B3, input "=1.04*B2", and then replicate the formula from B3 to B4 through B22.

The chart below keys the **Self-Test** questions to the objectives in the **Chapter Review** on pages 452–455 or to the **Vocabulary (Voc)** on page 449. This will enable you to locate those **Chapter Review** questions that correspond to questions missed on the **Self-Test**. The lesson where the material is covered is also indicated on the chart.

Question	1	2	3	4	5	6	7	8	9	10
Objective	A	A	A	A	D	G	G	E	C	E
Lesson(s)	7-6	7-6	7-6	7-6	7-1	7-7	7-7	7-2, 7-3, 7-4	7-5, 7-6	7-2, 7-3, 7-4
Question	11	12	13	14	15	16	17	18	19	20
Objective	A	A	A	A	C	C	I	G	G	H
Lesson(s)	7-6	7-6	7-6	7-6	7-5, 7-6	7-5, 7-6	7-5, 7-6	7-7	7-7	7-2, 7-3
Question	21	22	23	24	25					
Objective	H	I	H	F	B					
Lesson(s)	7-2, 7-3	7-5, 7-6	7-2, 7-3	7-4	7-7					

Chapter Review (pp. 452–455)

1. 4 3. 17 5. $\frac{121}{36}$ 7. –4 9. a. Answers vary. Sample answer: "= B2*1.05" b. 1,050 c. Replicate the formula in B3 down to B12. 11. a. x b. f 13. a. $\{2, 4, 5, 10\}$ b. $\{200, 400, 500, 1,000\}$ c. $f(x) = 100x; x = 2, 4, 5, 10$ 15. $918.66 17. (b), $1.04^8 > 1.08^4$ 19. a. $20^5 = 3{,}200{,}000$ b. C 21. constant increase 23. exponential growth 25. neither 27. a. $A(n) = 10{,}000{,}000\,(1.02)^n$ b. $B(n) = 20{,}000{,}000 + 1{,}000{,}000n$ c. country B; $B(30) = 50{,}000 > A(30) = 18{,}113{,}616$ d. country B; $B(100) > A(100)$

29.

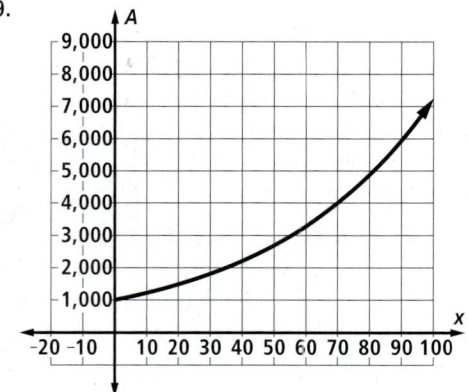

Answers vary. Sample answer: $1,000 is invested at 2% interest per year.

31.

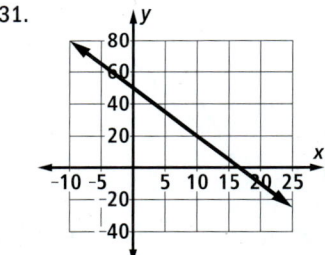

Answers vary. Sample answer: You have $50 in a bank account, and each day you take out $3.

33.

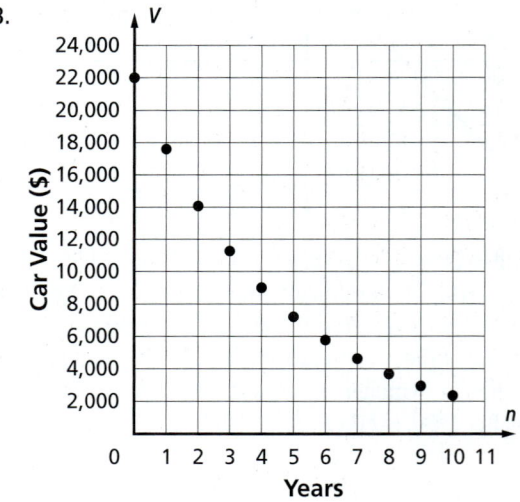

35.

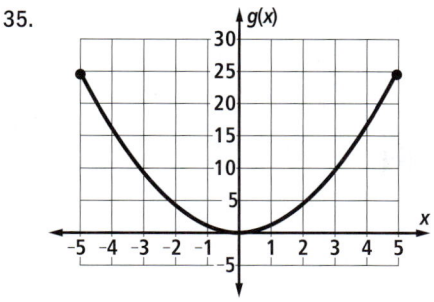

37.

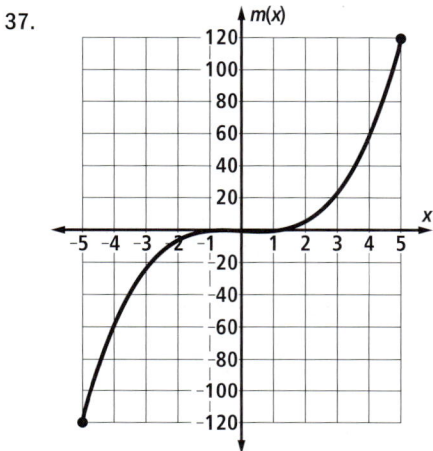

39. a. 1 b. 2 c. {$x: 0 \leq x \leq 4$} d. {$y: 0 \leq y \leq 2$}

Chapter 8

Lesson 8-1 (pp. 458–463)
Guided Example 3: 4^{20}; 5^{10}; $(4^{20})(5^{10})$; $\frac{1}{10,737,418,240,000,000,000}$
Questions: 1. a. AA, AE, AI, AO, AU, EA, EE, EI, EO, EU, IA, IE, II, IO, IU, OA, OE, OI, OO, OU, UA, UE, UI, UO, UU b. 2; 5   3. 15,625; 1.5625 · 10^4 5. a. 90 b. To label each cell, we choose each cell among 15 columns and 6 rows, so there are 6 · 15 choices, thus 90 cells.   7. a. 2^{10}; 1,024; 1.024 · 10^3 b. $\frac{1}{1,024}$ c. 2^Q; $\frac{1}{2^Q}$ 9. a. $10^3 = 1,000$ b. $10^{-12} = 0.000000000001$ 11. $\frac{1}{32}$ 13. 17,069 people
15. No, because for each value of x, there are several values of y.   17. 108 units^2 19. $x = 6$ 21. 2.8 · 10^9

Lesson 8-2 (pp. 464–468)
1. 18^9 3. w^7; $2^4 \cdot 2^3 = 16 \cdot 8 = 128$; $2^7 = 128$ 5. 2^7 7. m^8
9. m^{10} 11. $15a^6$ 13. d^{14} 15. $a^8 b^3$ 17. 65,536 sequences
19. $x = 1$ 21. a. $P \cdot 3^4$ b. 16 days   23. a. 30 ways
b. 6 more choices   25. a. $y = -51.6x + 3,058$ b. 2,026 tickets   27. 3.24 × 10^{-3}

Lesson 8-3 (pp. 469–473)
1. 2^3 3. 3^{m-n} 5. 1   7. $6ab^5$ 9. $4a^6$ 11. The bases differ.   13. 1   15. 2^{-3} 17. $7 + 3m$ 19. $4a$
21. a. about 12,410.8 barrels of oil per person per day
b. about 158,547,936 gallons   23. $3x^3$ 25. $24h^7$
27. a. $\frac{1}{4^5}$ b. $\frac{243}{1,024}$ 29. a. $23.92 b. $52.50

Lesson 8-4 (pp. 474–480)
Guided Example 2: a. $\left(\frac{4}{5}\right)^2$; $\frac{16}{25}$; b. $\left(\frac{m^2}{1}\right)^3$; $(m^2)^3$; m^6
Guided Example 4: –7; 2; –1; 2; 7; –1; $\frac{b^2}{3a^7c}$
Questions: 1. $\frac{1}{3}, \frac{1}{9}, \frac{1}{27}, \frac{1}{81}, \frac{1}{243}$ 3. $\frac{1}{125}$ 5. $\frac{1}{y^{24}}$ 7. 3^{-4} 9. 10^{-4}
11. a. $\frac{1}{w}$ b. $\frac{1}{wx^2}$ c. $\frac{y^3}{w}$ d. $\frac{5y^3}{wx^2}$ 13. 1   15. $\frac{4a^2c^2}{b^3}$
17. The values of y seem to get extremely close to 0.
19. $x^{-2}(x^{-3} + x^4) = x^{-2}(x^{-3}) + x^{-2}(x^4) = x^{(-2+-3)} + x^{(-2+4)} = x^{-5} + x^2 = x^2 + \frac{1}{x^5}$ 21. $m = -1$; $5^{-1} \cdot \frac{1}{25} = 5^{-1} \cdot 5^{-2} = 5^{-3}$
23. a^6; 64   25. $2a^b 3a$; 62,500   27. a. $m(x) = 8x + 25$
b. whole numbers less than 13 c. 121 and 25

Lesson 8-5 (pp. 481–487)
Guided Example 2: 3; 3; 3; 3; 3; 6; 9; 3; $-125x^6 y^9 z^3$
Questions: 1. a. $216x^3$ b. $(6 \cdot 2)^3 = 1,728$; $216(2)^3 = 1,728$
3. $-2,744x^3y^3$ 5. $-t^{93}$ 7. $\frac{16}{81}$ 9. $\frac{6,859}{8y^3}$ 11. The area is multiplied by 36.   13. $\frac{x^5}{y}$ 15. $144w^{10}$ 17. 0   19. 225
21. $\frac{17}{3,125r^2s^2}$; $\frac{17}{7,031.25}$ 23. $-\frac{1}{5} < p$

Lesson 8-6 (pp. 488–496)
1. a. 256 units^2 b. 4 units   3. 6   5. 7.07   7. 31.623   9. 5
11. a. 3.31662 b. 11 c. Square of the Square Root Property   13. 52   15. 1; 8; 27; 64; 125; 216; 343; 512; 729; 1,000   17. a. $\sqrt[3]{1,700}$ b. 11.935   19. 45   21. 100,995; Answers vary. Sample answer: The population 6 months from the original date   23. a. Answers vary. Sample answer: The scarecrow's statement is an incorrect statement about isosceles triangles, while the Pythagorean Theorem is a true statement about right triangles. The statement is about the square roots of the sides, while the Pythagorean Theorem is about the squares of the lengths of the sides. b. No, Answers vary. Sample answer: Let an isosceles triangle with $a = 9$, $b = 9$, and $c = 4$ cm. Then $\sqrt{a} + \sqrt{b} = \sqrt{9} + \sqrt{9} = 3 + 3 = 6$; while $\sqrt{c} = \sqrt{4} = 2$, so $\sqrt{a} + \sqrt{b} \neq \sqrt{c}$.   25. $(x^2)(y^2) = (xy)^2$ 27. 8
29. 865,177 yr

Lesson 8-7 (pp. 497–504)
Guided Example 2: a. 7; 7; $\sqrt{98}$ b. 49; 49; 2; 7; $7\sqrt{2}$
Guided Example 3: 16; 4; 4 Check 6,912; 48
Questions: 1. 4   3. 2   5. $x = 18, y = 3, z = 2$ 7. B   9. $3\sqrt{2}$
11. $5\sqrt{2}$ 13. a. $5m\sqrt{6n}$ b. $4m^2$ 15. Quotient of Square Roots Property   17. $3\sqrt{3}$, 5.20   19. $x = \sqrt{81 - y^2}$, $y = \sqrt{81 - x^2}$ 21. Answers vary. Sample answer: Because $\sqrt{49} = 7$ 23. $-8\sqrt{3}$ 25. $29\sqrt{2}$ 27. (250, 100)
29. x^5 31. $(6^4)^2$ 33. Slope = –1.5, y-intercept = 46; The slope describes how many floors it descends per second, and the y-intercept is where the elevator is at 0 sec.

Lesson 8-8 (pp. 505–510)

Guided Example 2: 23; 16; 31; –11; 31; 23; –11; 16; 8; –27; 64; 729; 793; 28.160

Questions: 1. a. 2 b. 6 c. $\sqrt{40} \approx 6.325$ 3. 7 5. 22
7. $\sqrt{29} \approx 5.385$ 9. 13 11. $\sqrt{65} \approx 8.062$
13. $\sqrt{0.0833} \approx 0.288617$ 15. $\sqrt{11.25} \approx 3.35$ miles
17. $\sqrt{b^2 + d^2}$ 19. $[(x_2 - x_1)^2 + (y_2 - y_1)^2]^{\frac{1}{2}}$ 21. 12
23. 40 25. $f(x) = \frac{8}{5}x - 10$

Lesson 8-9 (pp. 511–516)

Guided Example 4: Solution 1 –7; 35; –7; 21; –7; 14; –7; 7; 14; 7; 7; 14; 7; 14; 7; 14 Solution 2 1. $(9q^{-5}) = \left(\frac{9}{6}q^{-2}\right)^{-7}$
2. $\left(\frac{3}{2}q^{-2}\right)^{-7}$ 3. $\left(\frac{3}{2}\right)^{-7} q^{-2 \cdot -7} = \left(\frac{2}{3}\right)^{-7} \cdot q^{14}$

Questions: 1. B 3. C 5. A special case for which the answer is false 7. true 9. no 11. yes 13. Product of Powers 15. Negative Exponent 17. $\frac{y^6}{9x^4}$ 19. 6
21. Answers vary. Sample answer: $\left(\frac{x^6}{x^3}\right)^{-2} = \frac{x^{-12}}{x^{-6}} = x^{-6} = \frac{1}{x^6}$; $\left(\frac{x^6}{x^3}\right)^{-2} = (x^3)^{-2} = x^{-6} = \frac{1}{x^6}$ 23. $2^{75} 3^{50} 5^{25}$ 25. 66.7
27. Yes; Answers vary. Sample answer: In an isosceles right triangle with legs of length s and a hypotenuse of length h, $h^2 = s^2 + s^2 = 2s^2$. So, $h = \sqrt{2s^2} = \sqrt{2} \sqrt{s^2} = s \cdot \sqrt{2}$.
29. $-\frac{5}{4}x + 15$

Self-Test (p. 520)

1. A, $x^{4+7} = x^{11}$ 2. $5^{-3} = \frac{1}{5^3} = \frac{1}{125}$ 3. $(-4)(-3) = 12$; $(-4)^{-3} = -\frac{1}{64}$; $(-3)^4 = 81$; So, from least to greatest: $(-4)^{-3}$, $(-4)(-3)$, $(-3)^4$ 4. $\sqrt{600} = \sqrt{100 \cdot 6} = \sqrt{100} \cdot \sqrt{6} = 10\sqrt{6}$
5. $\sqrt{25x} = \sqrt{25 \cdot x} = \sqrt{25} \cdot \sqrt{x} = 5\sqrt{x}$
6. $2^{\frac{1}{2}} \cdot 50^{\frac{1}{2}} = (2 \cdot 50)^{\frac{1}{2}} = 100^{\frac{1}{2}} = 10$ 7. $y^4 \cdot y^2 = y^{4+2} = y^6$
8. $(10m^2)^3 = 10^3 (m^2)^3 = 10^3 m^6 = 1{,}000 m^6$
9. $\frac{a^{15}}{a^3} = a^{15-3} = a^{12}$ 10. $\left(\frac{m}{6}\right)^3 = \frac{m^3}{6^3} = \frac{m^3}{216}$
11. $g^4 \cdot g \cdot g^0 = g^{4+1+0} = g^5$
12. $\frac{6n^2}{4n^3 \cdot 2n} = \frac{6n^2}{8n^4} = \frac{3}{4} n^{2-4} = \frac{3}{4} n^{-2} = \frac{3}{4n^2}$ 13. $\frac{4w^2}{y^3}$, by the Negative Exponent Property 14. $\frac{2}{x^2} \cdot \frac{5}{x^5} = \frac{2 \cdot 5}{x^2 x^5}$ Multiplication of fractions $= \frac{10}{x^{(2+5)}}$ Arithmetic and Product of Powers Property $= \frac{10}{x^7}$ Arithmetic 15. The prime factorization of $10(288)^2$ is $2 \cdot 5 \cdot (2^5 \cdot 3^2)^2 = 2 \cdot 5 \cdot (2^5)^2 \cdot (3^2)^2 = 2 \cdot 5 \cdot 2^{10} \cdot 3^4 = 5 \cdot 2^{11} \cdot 3^4$ 16. $\left(\frac{3}{y^2}\right)^{-3} \left(\left(\frac{3}{y^2}\right)^{-1}\right)^3 = \left(\frac{y^2}{3}\right)^3 = \frac{(y^2)^3}{3^3} = \frac{y^6}{27}$ 17. 3.107 18. $1{,}000(1.06)^{-3} = 1{,}000 \cdot 0.84 \approx 840$ 19. a. Power of a Quotient Property b. Power of a Power Property c. Quotient of Powers Property d. Negative Exponent Property 20. Answers vary. Sample answer: $\frac{1}{4}$ 21. $\sqrt{(1-9)^2 + (-10-5)^2} = \sqrt{289} = 17$ 22. The upper right corner of the paper has coordinates (297, 210), the lower left corner has coordinates (0, 0) so using the distance formula gives $\sqrt{(297-0)^2 + (210-0)^2} = \sqrt{132{,}309} \approx 363.7$ mm.
23. $V = s^3$, so $s = \sqrt[3]{v}$. Therefore $s = \sqrt[3]{30} \approx 3.107$ in.
24. The diagonal and two consecutive sides of the square form an isoscels triangle. The Pythagorean Theorem gives $s^2 + s^2 = 12^2$, where s is the length of a side of the square. So, $2s^2 = 144$ or $s^2 = 72$. Since the area of the square is s^2, the area is 72 m². 25. $26^2 \cdot 10^4 = (676) \cdot (10{,}000) = 6{,}760{,}000$

The chart below keys the **Self-Test** questions to the objectives in the **Chapter Review** on pages 521–523 or to the **Vocabulary (Voc)** on page 519. This will enable you to locate those **Chapter Review** questions that correspond to questions missed on the **Self-Test**. The lesson where the material is covered is also indicated on the chart.

Question	1	2	3	4	5	6	7	8	9	10
Objective	A	B	B	D	D	D	A	C	A	C
Lesson(s)	8-2, 8-3, 8-4, 8-5	8-4, 8-5	8-4, 8-5	8-6, 8-7	8-6, 8-7	8-6, 8-7	8-2, 8-3, 8-4, 8-5	8-5, 8-9	8-2, 8-3, 8-4, 8-5	8-5, 8-9

Question	11	12	13	14	15	16	17	18	19	20
Objective	A	A	G	G	A	C	E	B	G	F
Lesson(s)	8-2, 8-3, 8-4, 8-5	8-2, 8-3, 8-4, 8-5	8-2, 8-3, 8-4, 8-5	8-2, 8-3, 8-4, 8-5	8-2, 8-3, 8-4, 8-5	8-5, 8-9	8-6	8-4, 8-5	8-2, 8-3, 8-4, 8-5	8-9

Question	21	22	23	24	25
Objective	J	U	I	I	H
Lesson(s)	8-8	8-8	8-8	8-8	8-1

Chapter Review (pp. 521-523)

1. $24m^9$ 3. $\frac{7y^4}{3x}$ 5. $\frac{1}{v^{24}}$ 7. $5a^{20} - 7a^{13}$ 9. a. approximately $1.33x^4$ b. $\frac{4}{3}x^4$ 11. $\frac{42}{x^7y^3}$ 13. $\frac{1}{36}$ 15. $\frac{243}{32}$ 17. $129,000
19. $n = -3$ 21. a. negative b. negative 23. $3,600m^4n^6$
25. $\frac{2,187}{16,384}$ 27. $\frac{t^8}{4,096s^4}$ 29. $\frac{9z^{14}}{y^{10}}$ 31. $8\sqrt{2}$ 33. $10\sqrt{21}$
35. $\frac{2x}{y}$ 37. -2 39. 5.848 41. 0.368; $0.368^3 \approx 0.0498 \approx 0.05$ 43. a. yes b. yes c. no d. It is not always true. Part c is a counterexample. 45. Answers vary. Sample answer: $a = 1$ 47. Power of a Quotient Property
49. Power of a Power Property or Zero Exponent Property 51. Product of Powers Property 53. Answers vary. Sample answer: By first applying the Negative Exponent Property for Fractions and then the Power of a Quotient Property, $\left(\frac{12}{13}\right)^{-4} = \left(\frac{13}{12}\right)^4 = \frac{13^4}{12^4} = \frac{28,561}{20,736}$. By applying the Power of Quotient Property and then the Negative Exponent Property, $\left(\frac{12}{13}\right)^{-4} = \frac{12^{-4}}{13^{-4}} = \frac{13^4}{12^4} = \frac{28,561}{20,736}$.
55. a. 243 answer sheets b. $\frac{1}{243}$ c. $\frac{32}{243}$
57. 54 different pizzas 59. $2\sqrt{2}$ units 61. x^2y^2
63. 1 m 65. a. 5 b. 3 c. $\sqrt{34}$ 67. $9\sqrt{2}$
69. $\sqrt{(a+1)^2 + (b-4)^2}$
71. a. yes b.

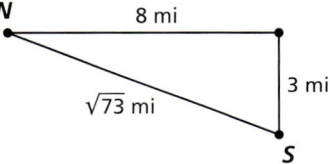

Chapter 9

Lesson 9-1 (pp. 526-531)

Guided Example 1: Step 1

A and r can only assume positive values.

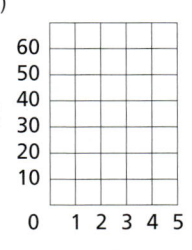

Step 2

r	A
1	3.14
2	12.57
3	28.27
4	50.27

Steps 3 and 4

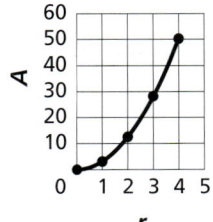

Step 5: about 2 units

Questions: 1. a.

x	$g(x) = \frac{1}{2}x^2$
−4	8
−3	4.5
−2	2
−1	0.5
0	0
1	0.5
2	2
3	4.5
4	8

b. 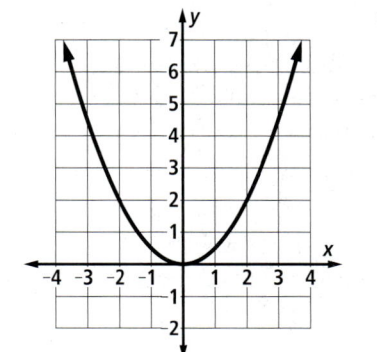 c. up

3. a. down b. maximum value c. $(0, 0)$ d. $x = 0$ 5. a. iii.
b. i. c. ii. 7. a. up b. down
9. a.

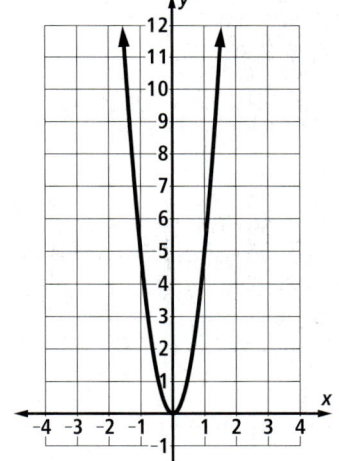

b.

x	$y = 5x^2$
−2	20
−1.5	11.25
−1	5
−0.5	1.25
0	0
0.5	1.25
1	5
1.5	11.25
2	20

c.

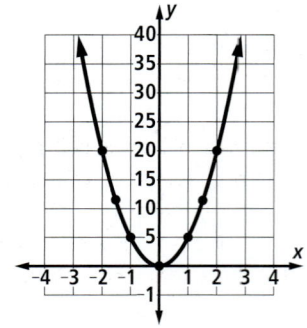

d. ≈ 1.7 and −1.7   11. a. $y = 0$ b. $x = 1$ and $x = −1$
c. $x = 2$ and $x = −2$ 13. a. negative   b. zero   c. negative
d. The graph is has vertex (0, 0) and opens down.   15. 0
17. $a = −1.5$ 19. a. $t = 5$ b. about 3.5 sec
21. $\sqrt{244} \approx 15.6$ in.   23. $2,250

Lesson 9-2 (pp. 532-536)
1. $x = \pm 5$ 3. $x = \sqrt{40} \approx \pm 6.32$ 5. $v = \sqrt{5} \approx \pm 2.24$
7. $a = −3$ or $a = −7$ 9. a. about 92 ft   b. about 2 sec
11. 5.64 units   13. $v \pm 2.5$ 15. about 8.50 in.
17. $m^5 n^6$ 19. $\frac{9}{25a^2}$ 21. a. the set of all real numbers
b. the set of nonnegative real numbers
23. a. $6x + 4y \geq 975$
b.

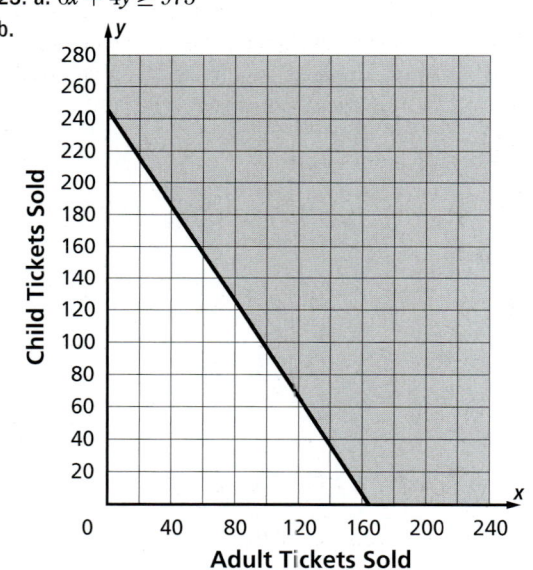

c. 121 tickets   25. B

Lesson 9-3 (pp. 537-543)
1. the total distance traveled in the time it takes for a car to stop   3. 206.25 ft   5. $a = 0.05; b = 1; c = 0$

7. a.

x	y
−3	3
−2	−2
−1	−5
0	−6
1	−5
2	−2
3	3

b.
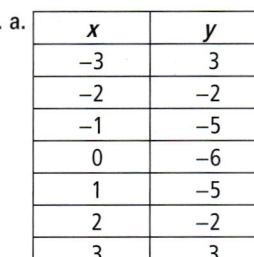

c. y-intercept: −6; x-intercepts: $\pm\sqrt{6}$; vertex: (0, −6)
d. $y \geq −6$

9. a.

x	y
−3	31
−2	16
−1	7
0	4
1	7
2	16
3	31

b.
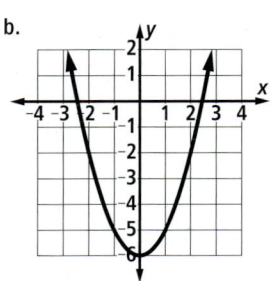

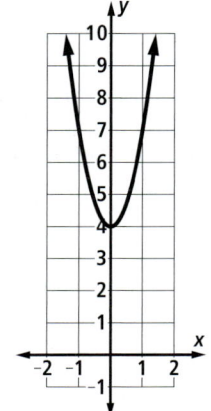

c. y-intercept: 4; x-intercepts: none; vertex: (0, 4)   d. $y \geq 4$
11. a. (−4, 36)   b. (−2, 20), (−1, 0), (0, −28)   13. B; Answers vary. Sample answer: When $x = 2, y = 0$. Only graph B intersects the x-axis at $x = 2$.   15. d   17. b
19. a. $t = \sqrt{\frac{350}{16}} \approx 4.68$ b. Answers vary. Sample answer: If you drop a stone off of a 350-ft cliff, how long will it take for the stone to hit the ground?   21. a. 288   b. 352   c. $4\sqrt{22}$ or about 18.76

Lesson 9-4 (pp. 544-551)
Guided Example 2: a. 0; 90; 90   b. 90; 90; 18.37; 4.3
Guided Example 3: a. $−4.9t^2 + 22t + 2$ b. 1.1; 3.4; 1.1; 3.4
Questions: 1. Answers vary. Sample answer: A projectile is an object that is dropped or launched and travels through the air to get to a target. Cannonballs, baseballs, and tennis balls can all be considered projectiles.   3. a. $h = −16t^2 + 30t + 5$ b. 1 ft   c. 19.1 ft   5. a. 0 ft/sec; 40 ft   b. $h = −16t^2 + 40$ c. about 1.6 sec   7. No; at 40 yards from the kicker, the ball is only 8.6 ft high.

9. a.
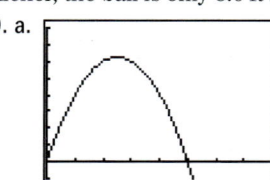

b. 31.25 m   c. 20 m
d. between 1 and 4 sec after launch   e. −30 m   f. 5 sec

11. a.

x	–3	–2	–1	0	1	2	3
y	2.25	1	0.25	0	0.25	1	2.25

b.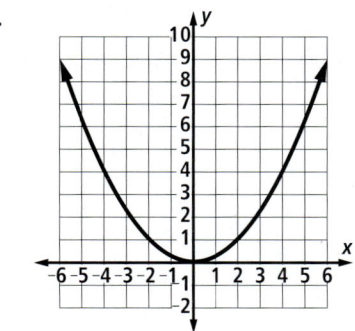

13. $A = (2, 2)$, $B = (3, 5)$ 15. D

Lesson 9-5 (pp. 552-557)

Guided Example 2: 4.3; –; –4.9; 10; –4.9; 214.49; 214.49; 214.49; 14.65; 14.65

Questions: 1. If $ax^2 + bx + c = 0$ and $a \neq 0$, then $x = \frac{-b \pm \sqrt{b^2 - 4ac}}{2a}$
3. 3 and –6   **5.** $t = -2$; Check: Does $(-2)^2 + 4(-2) + 4 = 0$? $4 + -8 + 4 = 0$ Yes, it checks.   **7.** $y = -4$ or $y = \frac{25}{3}$; Check: Does $3(-4)^2 = 13(-4) + 100$? $3(16) = -52 + 100$ Yes, it checks; Does $3(\frac{25}{3})^2 = 13(\frac{25}{3}) + 100$? $3(\frac{625}{9}) = \frac{325}{3} + 100$; Yes, it checks.   **9.** $p = -5.48$ or $p = -0.85$
11. a. $x = -3$ or $x = 5$

b.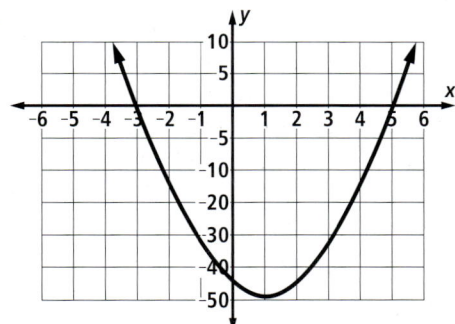

c.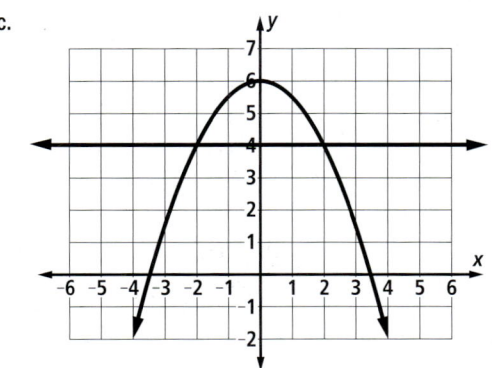

13. a. $t = 1.91$ sec and $t = 13.09$ sec   **b.** 15 sec   **15.** 2
17. 5 units   **19. a.** 2   **b.** 3   **c.** $\frac{5x - 1}{x + 1}$

Lesson 9-6 (pp. 558-564)

Guided Example 1: b. 1, 1, 1   **c.** 0, 0
Guided Example 3: 2; 16; 32; $(16)^2 - 4(2)(32)$; 256 –256; zero; 1

Questions: 1. a. $-x^2 + 2x = 0$ **b.** yes, twice   **3. a.** 0   **b.** 1   **c.** 2
5. a. 1   **b.** 0   **c.** 2   **d.** 2   **7. a.** 0   **b.** There is one real solution.
c. $n = 7$ **9. a.** 120   **b.** 2   **c.** $x = \frac{6 - \sqrt{30}}{2}$ or $x = \frac{6 + \sqrt{30}}{2}$
11. negative   **13. a.** 25   **b.** 2   **15.** 1   **17.** No; All parabolas of the form $y = ax^2 + bx + c$ contain the point $(0, c)$.
19. $x = \frac{-2\sqrt{5}}{3}$ or $x = \frac{2\sqrt{5}}{3}$ **21. a.** 48 ft   **b.** about 3.5 ft
23. down

Lesson 9-7 (pp. 565-570)

1. The cable can be placed either 87.87 ft away or 512.13 ft away from the left side of the bridge.   **3.** No, a polygon cannot have exactly 21 diagonals.   **5.** 120   **7. a.** $x + 7$ **b.** 9
9. 5.7 units   **11.** It is equal to 0.   **13.** 1   **15.** 0
17. a. $a = 36$ **b.** $b = 28$ **c.** $c = 10.25$ **19.** 26 ft
21. a. Answers vary. Sample answer:
$m = \frac{1 + 24}{2 - 7} = -5$
$y - 1 = -5(x - 2)$
$(31) - 1 = -5((-4) - 2)$
$30 = 30$
Since we can find the slope of the line between (2, 1) and (7, –24), we can find an equation for the line containing those two points. We then check to make sure that (–4, 31) is also on the line, which it is.   **b.** $5x + y = 11$

Self-Test (pp. 574-575)

1. $x^2 = 81$; $\sqrt{x^2} = \sqrt{81}$; $x = 9, -9$
2. $n^2 - 8n - 10 = 0$
$n = \frac{-b \pm \sqrt{b^2 - 4ac}}{2a}$
$n = \frac{-(-8) \pm \sqrt{(-8)^2 - 4(1)(-10)}}{2(1)}$
$n = \frac{8 \pm \sqrt{64 + 40}}{2}$
$n = \frac{8 \pm \sqrt{104}}{2}$
$n = \frac{8 \pm 2\sqrt{26}}{2}$
$n = 4 + \sqrt{26}, 4 - \sqrt{26}$
$n \approx 9.10, n \approx -1.10$

3. $5y^2 - 11y - 1 = 0$

$$y = \frac{-b \pm \sqrt{b^2 - 4ac}}{2a}$$

$$y = -(-11) \pm \frac{\sqrt{(-11)^2 - 4(5)(-1)}}{2(5)}$$

$$y = \frac{11 \pm \sqrt{121 + 20}}{10}$$

$$y = \frac{11 + \sqrt{141}}{10}, \frac{11 - \sqrt{141}}{10}$$

$$y \approx 2.29, y \approx -0.09$$

4. $\quad 24 = \frac{1}{6}z^2$

$\quad 6 \cdot 24 = \frac{6 \cdot 1}{6z^2}$

$\quad 144 = z^2$

$\quad z = 12, -12$

5. $v^2 - 16v + 64 = 0$
$(v - 8)(v - 8) = 0$
$v = 8$

6. $3p^2 - 9p + 7 = 0$ If there are any real solutions, then $b^2 - 4ac$ must be greater than or equal to 0.

$b^2 - 4ac = (-9)^2 - 4(3)(7)$
$= 81 - 84$
$= -3$ There are no real solutions. **7.** 2 because the discriminant is positive **8.** A

9. a.

x	$2x^2$
-3	18
-2	8
-1	2
0	0
1	2
2	8
3	18

b.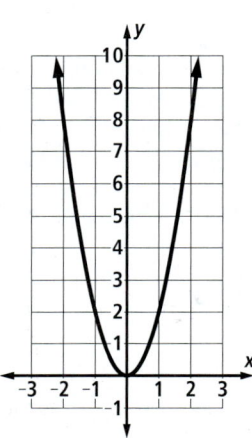

10. a.

x	$-x^2 + 4x - 3$
-3	-24
-2	-15
-1	-8
0	-3
1	0
2	1
3	0

b.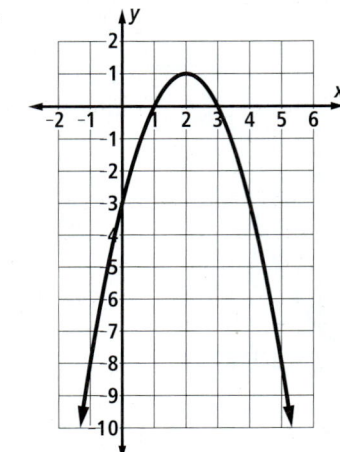

11. $h = 0.049(20)^2$; 19.6 m **12.** $44 = 0.049v^2$; $897.96 = v^2$; 29.97 m/sec **13.** $h = 0.049(35)^2$; 60 m
14. $n(n + 1) = 1{,}722$; $n^2 + n = 1{,}722$; $n^2 + n - 1{,}722 = 0$; $n = 41$ or $n = -42$; $n = -42$ or $n + 1 = -41$
15. $2x + 2(x + 6) = 24$; $4x = 12$; $x = 3$; $A = 3 \cdot 9 = 27$; πr^2; $8.59 \approx r^2$; $2.93 \approx r$ **16.** -1.6 **17.** -3, 1 **18.** $x = -1$
19. $0 = -16t^2 + 40t + 50$;

$$t = \frac{-b \pm \sqrt{b^2 - 4ac}}{2a};$$

$$t = \frac{-40 \pm \sqrt{(40)^2 - 4(-16)(50)}}{2(-16)};$$

$$t = \frac{-40 \pm \sqrt{1{,}600 + 3{,}200}}{-32};$$

$$t = \frac{-40 \pm \sqrt{4{,}800}}{-32};$$

$$t = \frac{-40 \pm 40\sqrt{3}}{-32}$$

$t = -0.92$, $t = 3.42$

$t = 3.42$ sec

20. $0 = -16t^2 + 40t - 20$

$$t = \frac{-b \pm \sqrt{b^2 - 4ac}}{2a}$$

$$t = \frac{-40 \pm \sqrt{(40)^2 - 4(-16)(-20)}}{2(-16)}$$

$$t = \frac{-40 \pm \sqrt{1{,}600 - 1{,}280}}{-32}; t = \frac{-40 \pm \sqrt{320}}{-32}$$

$$t = \frac{-40 \pm 8\sqrt{5}}{-32}$$

$t = 0.69$, $t = 1.81$
21. false **22.** $b^2 - 4ac = 1$; $(-5)^2 - 4(a)(3) = 1$
$\qquad\qquad\qquad\qquad 25 - 12a = 1$
23. $b^2 - 4ac$; $12a = 24$; $a = 2$
$= (12)^2 - 4(-3)(-7)$
$= 144 - 84$
$= 60$; There are two real solutions since the discriminant has a value greater than 0.

24. $b^2 - 4ac$;
$= (-4)^2 - 4(1)(4)$
$= 16 - 16$
$= 0$; There is one real solution since the discriminant has a value of 0.

The chart below keys the **Self-Test** questions to the objectives in the **Chapter Review** on pages 576–579 or to the **Vocabulary (Voc)** on page 573. This will enable you to locate those **Chapter Review** questions that correspond to questions missed on the **Self-Test**. The lesson where the material is covered is also indicated on the chart.

Question	1	2	3	4	5	6	7	8	9	10
Objective	A	B	B	A	B	B	C	G	G	H
Lesson(s)	9-2	9-5, 9-6	9-5, 9-6	9-2	9-5, 9-6	9-5, 9-6	9-6	9-1	9-1	9-3
Question	11	12	13	14	15	16	17	18	19	20
Objective	F	F	F	F	E	H	H	H	D	D
Lesson(s)	9-7	9-7	9-7	9-7	9-2, 9-7	9-3	9-3	9-3	9-2, 9-4	9-2, 9-4
Question	21	22	23	24						
Objective	C	C	C	C						
Lesson(s)	9-6	9-6	9-6	9-6						

Chapter Review (pp. 576-579)

1. $x = 13, x = -13$ 3. $k = \pm\sqrt{85}$ 5. $m = 3, m = -9$
7. $v = 4.75, v = 3.25$ 9. $m = -3, m = -4$ 11. $y = -0.46$, $y = 6.46$ 13. $p = -5$ 15. $n = -2, n = 0.2$ 17. $b = -7.9$, $b = 2.5$ 19. $x = \frac{-b \pm \sqrt{b^2 - 4ac}}{2a}$ 21. 48 23. $b = 4$ or $b = -4$
25. 0 27. 0 29. a. 3,600 ft b. about 19 sec
31. a. 15.1 m b. about 1.0 sec and 3.1 sec c. 4.1 sec
33. a. $A = 60x - x^2$ b. $x = 30$
35. about 98°C 37. a. Yes, the company's profits will exceed $100 million. b. 4.8 yr after 2005, or in late 2009
39. a.

x	$\frac{3}{5}x^2$
−2	2.4
−1	0.6
0	0
1	0.6
2	2.4

b.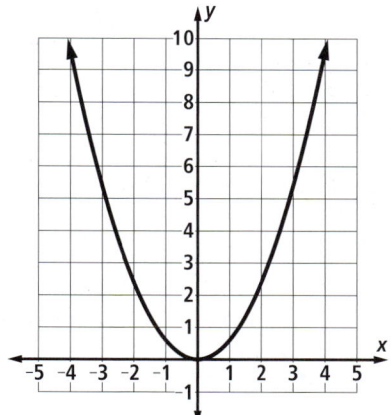

49. a.

x	$-x^2 - 4x + 3$
−5	−2
−4	3
−3	6
−2	7
−1	6
0	3
1	−2

41. B 43. a. 5
b. $x = 8$
c. $A = (6, 1)$, $B = (7, 4)$
45. true
47. $(24, -50)$

b.

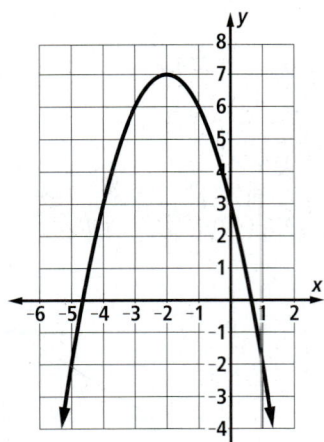

c. maximum

50. A

13. a. Answers vary. Sample answer: men: $y = -0.1235x + 522.17$; women: $y = -0.311x + 680.23$ Yes; both times will be equal in 2074. In 2076, the women's time will pass the men's time.

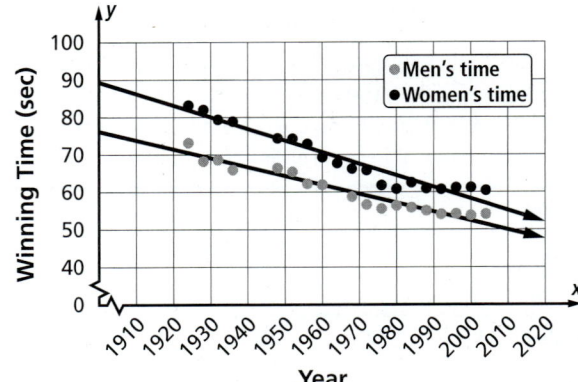

15. a. $-\frac{3}{5}$ b. -1 c. $-\frac{1}{2}$, 3 d. -1.137, 2.637 17. $625x^{28}y^{36}$
19. 21. 3 days

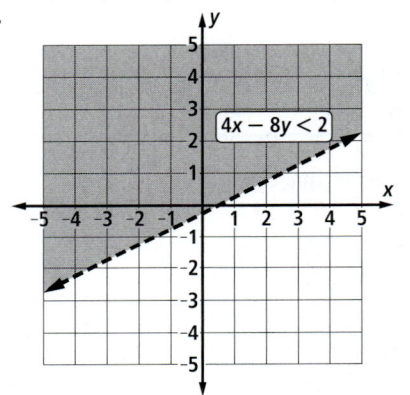

Chapter 10

Lesson 10-1 (pp. 582–588)
Guided Example 3: 1. zero 2. -1 3. parallel 4. no
Questions: 1. true 3. a. $-8 = 4(-2)$ and $2(8) + 3(-2) = 16 + -6 = 10$ b. Answers vary. Sample answer: $x = 8$ and $y = -2$; $(x, y) = (8, -2)$
5. a. $\begin{cases} 3x - y = 3 \\ y = -2x + 7 \end{cases}$ b. $(2, 3)$ c. $3(2) - 3 = 3$ and $2(2) + 3 = 7$

7. a. $(6, 6)$

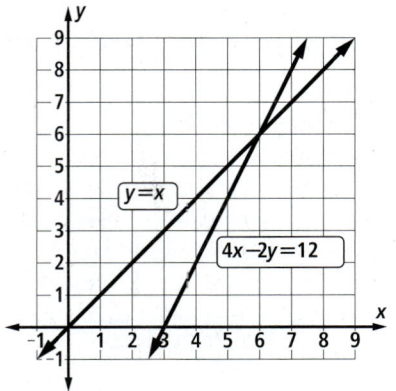

b. $6 = 6$; $24 - 12 = 12$ 9. a. $\begin{cases} y = 3{,}800 + 4.25x \\ y = 12.5x \end{cases}$
b. Answers vary. Sample answer: Xmin $= 400$, Xmax $= 800$; Ymin $= 3{,}800$, Ymax $= 8{,}000$ c. about $(461, 5{,}758)$ d. yes
11. (5, 15)

Lesson 10-2 (pp. 589–593)
Guided Example 3: $4t$; $10 + 3t$; $4t$; 10; $3t$; 10
Questions: 1. a. $(3, 5)$ b. $3(3) - 4 = 5$; $5(3) - 10 = 5$
3. a. $(-18, 8)$ b. $-\frac{1}{9}(-18) + 6 = 8$; $\frac{5}{3}(-18) + 38 = 8$
5. a. $(-144.6, -22.2)$ b. $8(-22.2) + 33 = -144.6$; $3(-22.2) - 78 = -144.6$ 7. $d = 15 + 3t$, $d = 4t$; The solution is $(15, 60)$, so Bart's sister will win.
9. a. $\begin{cases} d = 60t \\ d = 65\left(t - \frac{1}{10}\right) \end{cases}$ b. $(1.3, 78)$; after 1.3 hours
c. 78 miles 11. a. about 8.42 yr before 2000 (1991)
b. about 4,189,474 people 13. $2\frac{1}{2}$ hr 15. a. $(1, 4)$ and $(-8, 85)$ b. $2(1^2) + 5(1) - 3 = 4$ and $1^2 - 2(1) + 5 = 4$; $2(-8)^2 + 5(-8) - 3 = 85$ and $(-8)^2 - 2(-8) + 5 = 85$

17. (−18, 8)

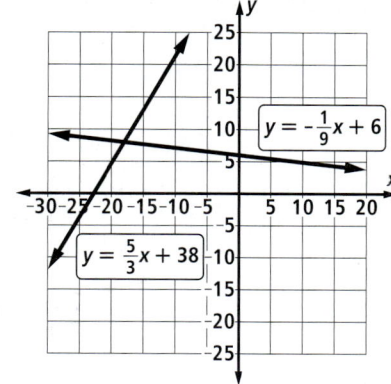

19. a. $y^2 − 5y + 1$ b. $y = −0.076$ and $y = 13.076$
21. $18x + 25y$ dollars

Lesson 10-3 (pp. 594-600)

Guided Example 3: $100 − x$; $100 − x$; $100 − x$; 3,000; $30x$; $10x$; $−10x$

Questions: 1. 1,190 adults, 2,380 children   **3.** $56.00 by the drama club, $224 by the service club   **5. a.** $n = 16, w = −2$ **b.** $16 + 5(−2) = 6$ and $−8(−2) = 16$ **7. a.** $x = 6, y = 5$ **b.** $6 − 1 = 5$ and $4(6) − 5 = 19$ **9. a.** (−4, 12)
b.

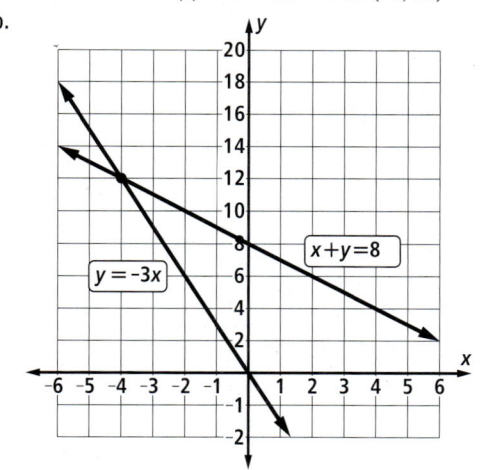

11. (40, −50)   **13.** $T = \$870{,}000, L = \$750{,}000$
15. a. $\begin{cases} m = v + 40 \\ m + v = 1{,}230 \end{cases}$ **b.** $m = 635, v = 595$
17. no solution   **19.** $x = −2$
21. a.

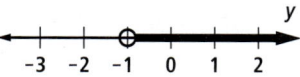

Lesson 10-4 (pp. 601–607)

Guided Example 3: $3N + 2M = 488$; $−3N − 2M = −488$; $4M = 76$; 19; 150; 19; $150; $19

Questions: 1. a. when the coefficients of the same variable are opposites   **b.** to eliminate one variable from a system   **3. a.** $x = \frac{-35}{4}, y = \frac{45}{4}$ **b.** $3\left(-\frac{35}{4}\right) + 9\left(\frac{45}{4}\right) = 75$; $-3\left(-\frac{35}{4}\right) - \frac{45}{4} = 15$ **5.** 1,634 and 142   **7.** when one of the coefficients for a variable in one equation is the same as the variable's coefficient in another equation
9. $(x, y) = (6, 15)$ **11.** $3(150) + 6(19) = 564$, $3(150) + 2(19) = 488$ **13.** $(x, y) = \left(12, -\frac{5}{3}\right)$ **15. a.** Yes; by the Generalized Addition Property of Equality   **b.** Answers vary. Sample answer: Yes; because $\frac{3}{5} = 60\%$, they are simply different ways of writing the same value.
17. $(x, y) = (0.9, −1.2)$ **19. a.** $x = −7, x = 4$ **b.** −7, 4
21. a. $\{x : x \geq \frac{9}{2}\}$ **b.** All nonnegative numbers   **23.** $-\frac{b}{a}$
25. $x = 212$

Lesson 10-5 (pp. 608-615)

1. a. Answers vary. Sample answer: second; −3, d
b. $(x, d) = (69, −112)$ **3. a.** Answers vary. Sample answer: The first equation can be multiplied by 2 and the second equation can be multiplied by −7.   **b.** The first equation can be multiplied by 5 and the second equation can be multiplied by 3.   **c.** $(r, s) = (3, 4)$ **5. a.** 31 musicians and 26 flag bearers   **b.** No, it is not possible.   **7.** $(x, y) = \left(-\frac{5}{4}, \frac{10}{3}\right)$
9. $(x, y) = \left(-\frac{13}{5}, -\frac{3}{5}\right)$ **11.** $(x, y) = (-5.\overline{1}, -8.47\overline{2})$ **13.** 5 ER, 30 MC   **15. a.** 11 cows, 16 chickens   **b.** Answers vary. Sample answer: By assuming there will be 14 chickens and 13 cows, a person would count 80 legs. For every cow replaced by a chicken, we lose 2 legs. Therefore, someone could conclude that there were 16 chickens and 11 cows.
17. $(x, y) = (12, 4)$ **19. a.** 15 weeks   **b.** $1,050   **21.** $-\frac{1}{4}$
23. a. true   **b.** true   **c.** true

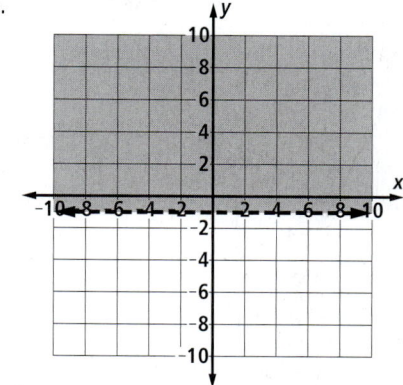

Lesson 10-6 (pp. 616-621)

Guided Example 2: 1. $36x - 30y = 6$ 2. $-36x + 30y = -6$ 3. $0 = 0$

Questions: 1. They are equal.

3. a.
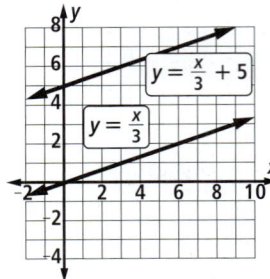

b.
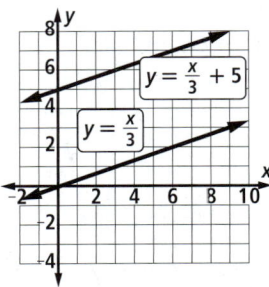

c. $y = \frac{1}{3}x$ 5. Answers vary. Sample answer:
$\begin{cases} y = 2x + 3 \\ 2y = 4x + 6 \end{cases}$ 7. a. coincident

b.
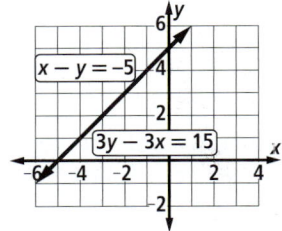

9. a. none b. Answers vary. Sample answer: The system $\begin{cases} x + y = -2 \\ \frac{x+y}{2} = 1 \end{cases}$ has no solution.   11. coincident lines
13. two intersecting lines   15. Yes, she got her 15% discount.   17. a. $t + u = 16, t + 3u = 28$
b. $t = 10, u = 6$ c. $6 + 10 = 16, 10 + 18 = 28$ 19. a. $2d^4$
b. $\sqrt{2^8 + 3(2)^8} = 32 = 2(2)^4$ 21. a. $x = -6$ b. $y = 18$

Lesson 10-7 (pp. 622-628)

1. a. 2×3 b. a, b, c c. f 3. a. $\begin{bmatrix} 5 & -2 \\ 3 & 4 \end{bmatrix}$ b. $\begin{bmatrix} -4 \\ 34 \end{bmatrix}$

5. -4 7. $\begin{bmatrix} 18.5 & -32 \\ -20 & 44 \end{bmatrix}$ 9. Answers vary.

Sample answer: $\begin{bmatrix} 2 & -1 \\ 1 & 2 \end{bmatrix} \cdot \begin{bmatrix} 3 & -1 \\ -3 & 2 \end{bmatrix} = \begin{bmatrix} 9 & -4 \\ -3 & 3 \end{bmatrix}$,

$\begin{bmatrix} 3 & -1 \\ -3 & 2 \end{bmatrix} \cdot \begin{bmatrix} 2 & -1 \\ 1 & 2 \end{bmatrix} = \begin{bmatrix} 5 & -5 \\ -4 & 7 \end{bmatrix}$ 11. -10 13. Answers

vary. Sample answer: $M = \begin{bmatrix} 1 & 1 \\ 1 & 1 \end{bmatrix}, N = \begin{bmatrix} 2 & 1 \\ 1 & 2 \end{bmatrix}$,

$P = \begin{bmatrix} 0 & -1 \\ 3 & 1 \end{bmatrix}$ a. $MN = \begin{bmatrix} 3 & 3 \\ 3 & 3 \end{bmatrix}$ b. $MN(P) = \begin{bmatrix} 9 & 0 \\ 9 & 0 \end{bmatrix}$

c. $NP = \begin{bmatrix} 3 & -1 \\ 6 & 1 \end{bmatrix}$ d. $M(NP) = \begin{bmatrix} 9 & 0 \\ 9 & 0 \end{bmatrix}$

e. It might be associative.   15. two intersecting lines, by comparing the slopes and y-intercepts   17. $(2.5, 0)$
19. a. $(0.5, 3), (5, 3), (5, -6)$ b. $4.5, 9,$ and approximately 10.06 c. 20.25 units^2 21. approximately 27.9%

Lesson 10-8 (pp. 629-634)

1. $(7, -3)$ 3. $\begin{bmatrix} 1 & 0 \\ 0 & 1 \end{bmatrix} \begin{bmatrix} x \\ y \end{bmatrix} = \begin{bmatrix} 7 \\ -3 \end{bmatrix}$

5. $\begin{bmatrix} 1 & -2 \\ 5 & 4 \end{bmatrix} \cdot \begin{bmatrix} \frac{2}{7} & \frac{1}{7} \\ -\frac{5}{14} & \frac{1}{14} \end{bmatrix} = \begin{bmatrix} \frac{2}{7} & \frac{1}{7} \\ -\frac{5}{14} & \frac{1}{14} \end{bmatrix} \cdot \begin{bmatrix} 1 & -2 \\ 5 & 4 \end{bmatrix} = \begin{bmatrix} 1 & 0 \\ 0 & 1 \end{bmatrix}$

7. a. $\begin{bmatrix} 3 & 5 \\ 2 & 3 \end{bmatrix} \begin{bmatrix} x \\ y \end{bmatrix} = \begin{bmatrix} 27 \\ 17 \end{bmatrix}$ b. $\begin{bmatrix} -3 & 5 \\ 2 & -3 \end{bmatrix}$

c. $(x, y) = (4, 3)$ 9. a. $\begin{bmatrix} 2 & -6 \\ 7.5 & -15 \end{bmatrix} \begin{bmatrix} m \\ t \end{bmatrix} = \begin{bmatrix} -6 \\ -37.5 \end{bmatrix}$

b. $\begin{bmatrix} -1 & 0.4 \\ -0.5 & 0.13 \end{bmatrix}$ c. $(m, t) = (-9, -2)$ 11. a. $\begin{bmatrix} 0.992 & -.413 \\ -1.281 & 0.950 \end{bmatrix}$

$\begin{bmatrix} -5.5 \\ -1.1 \end{bmatrix}$ b. $(x, y) = (-5, 6)$ c. $2.3(-5) + 6 = -5.5; 3.1(-5) +$

$2.4(6) = -1.1$ 13. a. III; $\begin{bmatrix} 4 & -5 \\ 12 & -15 \end{bmatrix}$ b. Answers vary.

Sample answer: Error: Singular matrix.   15. $\begin{bmatrix} -20 & 17 \\ -8 & -4 \end{bmatrix}$

17. $a = 8$ 19. intersecting at only one point
21. a. $x = 12, x = -12$ b. $x = 10, x = -10$ c. $x = 5, x = -5$
d. $x = -5.5$ 23. $\$2{,}154.76$

Lesson 10-9 (pp. 635-639)

Guided Example 3: 1. parallel

2.

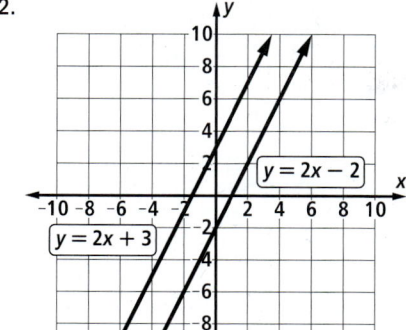

3. below; above. No, because the half-planes can intersect even if the lines do not.

4.

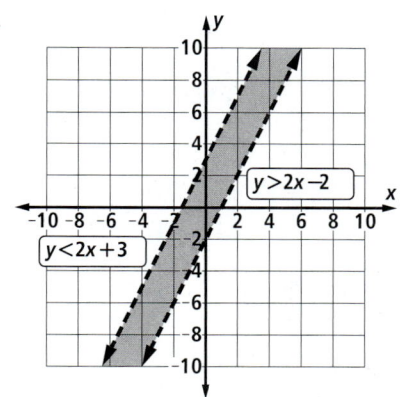

Questions: 1. IV **3. a.** It is the intersection of the half-planes below or on the line $y = 4x + 1$ and above $y = 2x + 1$. **b.** half-plane **c.** Because ≤ means less than or equal to, not just less than **d.** No, it is not because it is on the boundary line $y = 2x + 1$.

5.

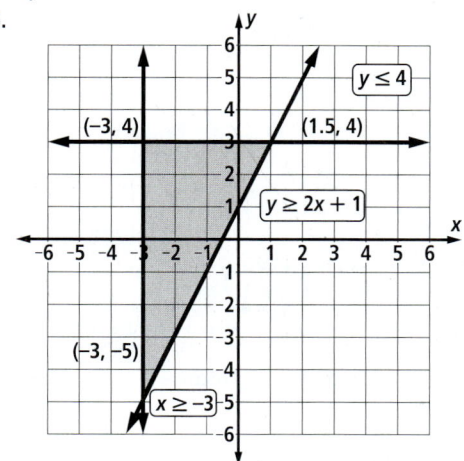

7. Yes it is possible. If there are no solutions satisfying both conditions, then the solution is ∅.

9. $\begin{cases} y \leq 0 \\ 3x + 7y < 10 \end{cases}$ **11. a.** $\begin{cases} 10L + 8P \leq 60 \\ P \geq 0 \\ L \geq 0 \end{cases}$

b.

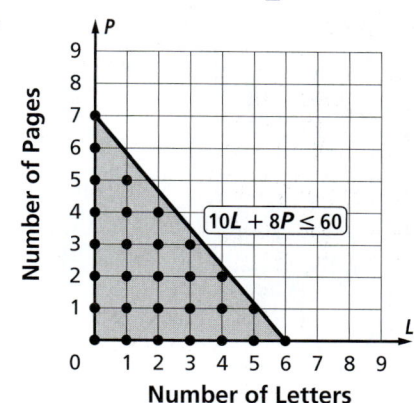

13. a.

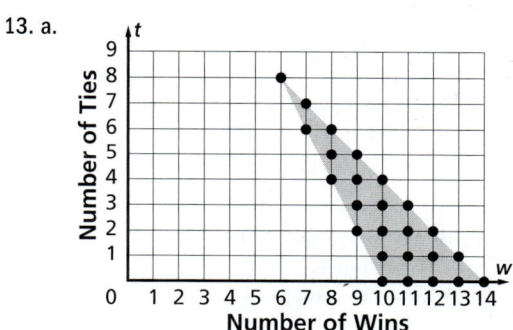

b. 25 **15. a.** $\begin{bmatrix} 6 & 4 \\ -2 & -3 \end{bmatrix} \cdot \begin{bmatrix} x \\ y \end{bmatrix} = \begin{bmatrix} 14 \\ -18 \end{bmatrix}$

b. $\begin{bmatrix} \frac{3}{10} & \frac{2}{5} \\ -\frac{1}{5} & -\frac{3}{5} \end{bmatrix}$ **c.** (−3, 8) **17.** (6.5, −10.25) **19. a.** −10 ft

b. The rocket has already landed.

Lesson 10-10 (pp. 640-644)

Guided Example 1: 7; 12; –2, 7; 3, 12
Questions: 1. (–3, –1) **3.** (0, 0) **5.** (4, 61), (–3, –2)
7.

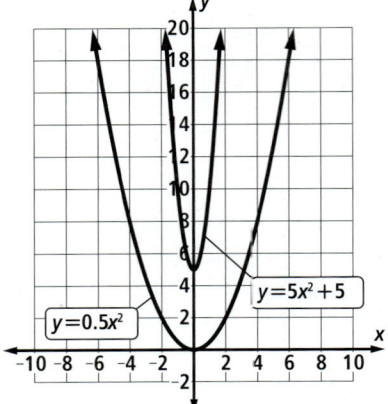

no solution
9. (3, 0)

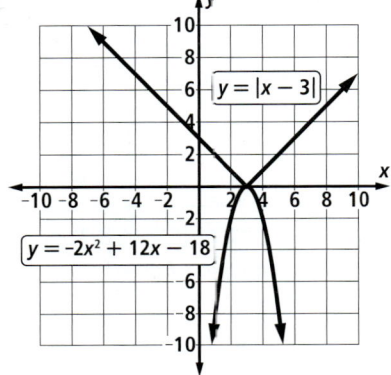

11. $\begin{cases} y = x^2 + 6x + 11 \\ y = -x^2 + 6x - 9 \end{cases}$; multiply the second equation

by –1: $\begin{cases} y = x^2 + 6x + 11 \\ -y = x^2 - 6x + 9 \end{cases}$; then add: $0 = 2x^2 + 20$.

So, $2x^2 = -20$, $x^2 = -10$ which there is no solution.

13. a. The system must have exactly one solution.
b. 0.41 sec **c.** about 2.76 ft **d.** Answers vary. Sample answer: No, the graph represents the height with respect to time; if the graph were of height with respect to distance traveled, then the graph would be of the ball's flight. **15.** $n = 5$ **17.** $45,604

Self-Test (pp. 648-649)

1. $(x, y) = (-18, -25)$; $x - 7 = 1.5x + 2 - 0.5x = 9$; $x = -18$; $y = -25$ **2.** $(d, f) = (6, 3)$; $4f = 12$; $f = 3$; $d = 6$

3. $(g, h) = \left(-\frac{6}{13}, \frac{10}{13}\right)$; $\begin{cases} 7h + 3g = 4 \\ 6h - 3g = 6 \end{cases}$ $13h = 10$; $h = \frac{10}{13}$; $g = -\frac{6}{13}$

4. (15, 7)

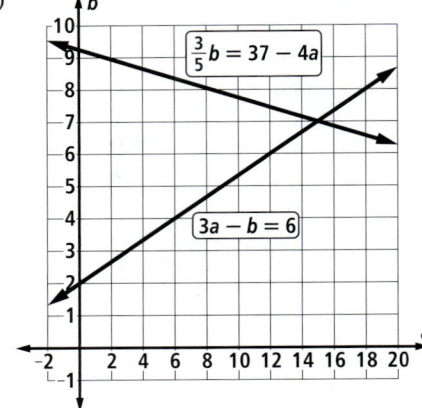

5. (1, –1) and (2, 5). Answers vary. Sample answer: $6x - 7 = x^2 + 3x - 5$, $x^2 - 3x + 2 = 0$, $x = 1$ or $x = 2$. When $x = 1$, $y = -1$. When $x = 2$, $y = 5$. **6.** This is a linear system of 2 lines with different slopes. Thus there is one solution.

7. $\begin{bmatrix} 2 \cdot 3 + 7 \cdot 4 \\ 1 \cdot 3 + 0 \cdot 4 \end{bmatrix} = \begin{bmatrix} 34 \\ 3 \end{bmatrix}$

8. $\begin{bmatrix} 3 \cdot 2 + 5 \cdot 1 & 3 \cdot 8 + 5 \cdot 7 \\ 4 \cdot 2 + 6 \cdot 1 & 4 \cdot 8 + 6 \cdot 7 \end{bmatrix} = \begin{bmatrix} 11 & 59 \\ 14 & 74 \end{bmatrix}$

9. $\begin{bmatrix} 3 & 5 \\ 1 & -1 \end{bmatrix} \begin{bmatrix} p \\ q \end{bmatrix} = \begin{bmatrix} 5 \\ 7 \end{bmatrix}$

$\begin{bmatrix} \frac{1}{8} & \frac{5}{8} \\ \frac{1}{8} & -\frac{3}{8} \end{bmatrix} \begin{bmatrix} 3 & 5 \\ 1 & -1 \end{bmatrix} \begin{bmatrix} p \\ q \end{bmatrix} = \begin{bmatrix} \frac{1}{8} & \frac{5}{8} \\ \frac{1}{8} & -\frac{3}{8} \end{bmatrix} \begin{bmatrix} 5 \\ 7 \end{bmatrix}$

$\begin{bmatrix} p \\ q \end{bmatrix} = \begin{bmatrix} 5 \\ -2 \end{bmatrix}$

10. $\begin{cases} 3d + 6t = 6{,}795 \\ 4d + 4t = 4{,}860 \end{cases}$, $\begin{cases} -6d - 12t = -13{,}590 \\ 12d - 12t = 14{,}580 \end{cases}$, $6d = 990$, $d = 165$, $t = 1{,}050$; DVD players cost $165 each; high-definition televisions cost $1,050 each. **11.** Answers vary. Sample answer: $m = n = 5$, $c = d = 4$ **12.** All points with integer coordinates in the shaded region below.

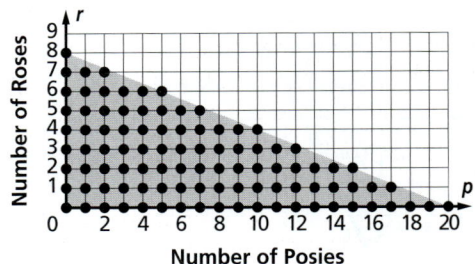

13. (0.4, 6.2)

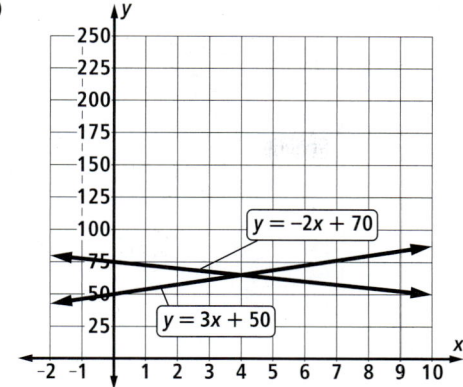

16.

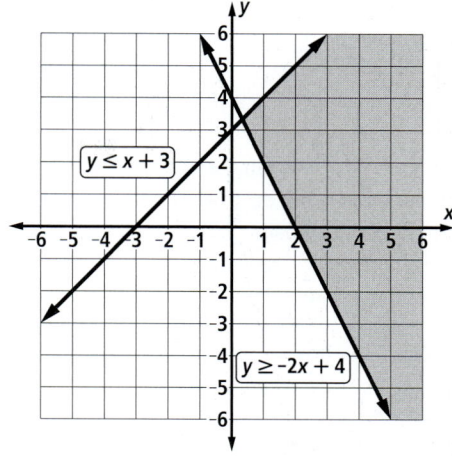

14. $\begin{cases} 2j + 2w = 1{,}000 \\ 2.5j - 2.5w = 1{,}000 \end{cases}$, $\begin{cases} 2j + 2w = 1{,}000 \\ 2.5j - 2.5w = 1{,}000 \end{cases}$;
$j = 500 - w$, $2.5(500 - w) - 2.5w = 1{,}000$, $1{,}250 - 2.5w - 2.5w = 1{,}000$, $5w = 250$, $w = 50$, $j = 450$ airplane's speed: 450 mph; speed of jet stream: 50 mph

15. a. 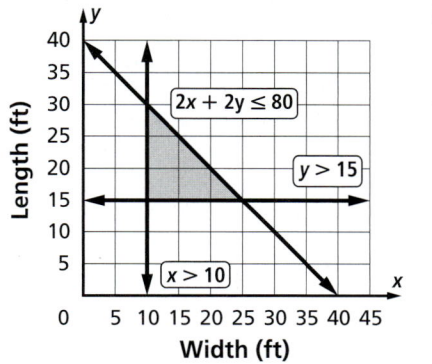 b. 25 ft

17. $\begin{cases} q + d = 15 \\ 0.25q + 0.10d = 2.40 \end{cases}$;
$q = 15 - d$, $0.25(15 - d) + 0.10d = 2.40$, $3.75 - 0.25d + 0.10d = 2.40$, $-0.15d = -1.35$, $d = 9$, $q = 6$; 9 dimes and 6 quarters

18. $\begin{cases} y < 7 \\ y \geq 2x - 7 \\ y \geq -\frac{3}{4}x + 4 \end{cases}$

The chart below keys the **Self-Test** questions to the objectives in the **Chapter Review** on pages 650–653 or to the **Vocabulary (Voc)** on page 647. This will enable you to locate those **Chapter Review** questions that correspond to questions missed on the **Self-Test**. The lesson where the material is covered is also indicated on the chart.

Question	1	2	3	4	5	6	7	8	9	10
Objective(s)	A	B	B	I	E	F	C	C	D	G
Lesson(s)	10-2, 10-3	10-4, 10-5	10-4, 10-5	10-1, 10-6, 10-10	10-10	10-6	10-7	10-7	10-8	10-2, 10-3, 10-4, 10-5, 10-6

Question	11	12	13	14	15	16	17	18
Objective(s)	F	J	I	H	J	J	G	K
Lesson(s)	10-6	10-9	10-1, 10-6, 10-10	10-9	10-9	10-9	10-2, 10-3, 10-4, 10-5, 10-6	10-9

Chapter Review (pp. 650–653)

1. $m = 2, n = 2$ 3. $(-0.2, 4.6)$ 5. $(a, b) = \left(-\frac{1}{3}, -\frac{13}{9}\right)$

7. $(f, g) = \left(\frac{102}{7}, -\frac{19}{7}\right)$ 9. $(v, w) = \left(-\frac{29}{27}, \frac{44}{27}\right)$ 11. $\begin{bmatrix} -2 \\ 41 \end{bmatrix}$

13. $\begin{bmatrix} 26 & -2 \\ 19 & -13 \end{bmatrix}$ 15. a. $\begin{bmatrix} 3 & 2 \\ 5 & 7 \end{bmatrix} \begin{bmatrix} x \\ y \end{bmatrix} = \begin{bmatrix} 7 \\ 9 \end{bmatrix}$

b. $\begin{bmatrix} \frac{7}{11} & -\frac{2}{11} \\ -\frac{5}{11} & \frac{3}{11} \end{bmatrix}$ c. $\begin{bmatrix} x \\ y \end{bmatrix} = \begin{bmatrix} \frac{31}{11} \\ -\frac{8}{11} \end{bmatrix}$ 17. a. $\begin{bmatrix} 5 & -7 \\ 4 & -8 \end{bmatrix} \begin{bmatrix} p \\ q \end{bmatrix} = \begin{bmatrix} 20 \\ 14 \end{bmatrix}$

b. $\begin{bmatrix} \frac{2}{3} & -\frac{7}{12} \\ \frac{1}{3} & -\frac{5}{12} \end{bmatrix}$ c. $\begin{bmatrix} p \\ q \end{bmatrix} = \begin{bmatrix} \frac{31}{6} \\ \frac{5}{6} \end{bmatrix}$ 19. (−1, −2), (1, −2)

21. 0 solutions 23. 1 solution 25. When $a \ne b$, since equating y and subtracting mx from both sides yields $a = b$. 27. slope 29. Austin: 280 mi, Antonio: 70 mi 31. after the fourth year 33. $\frac{8}{3}$ pints of the 15% solution and $\frac{16}{3}$ pints of the 30% solution

35.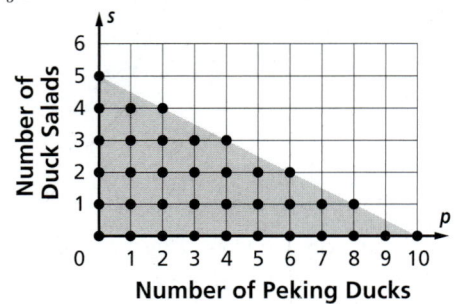

37. a.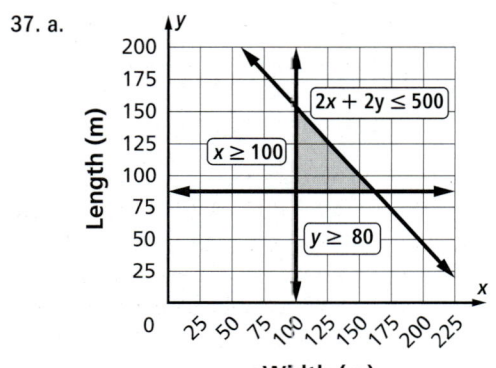

b. 650 ft c. 670 ft 39. (5, −4.5) 41. infinitely many solutions 43. (−4.2, −2.8), (2.7, 7.6)

45.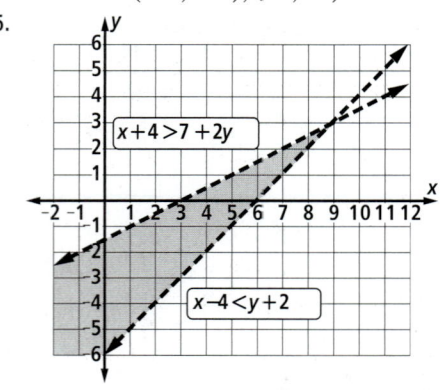

47.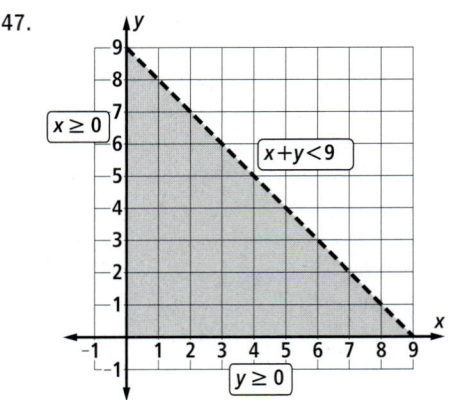

49.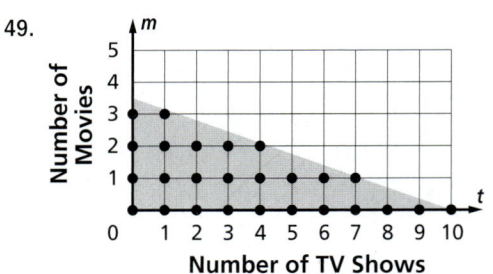

51. $\begin{cases} y \le 0 \\ y \ge -5 \\ y \le -2x + 7 \\ y \ge -2x - 5 \end{cases}$

Chapter 11

Lesson 11-1 (pp. 656–662)
Guided Example 2: $100; 100; 3; $120; $120x^2$; $140; $140x^1$; $160
Questions: 1. a. $4,071 b. $125.90 3. a. $75x^3 + 85x^2 + 95x + 105$ b. $225x^3 + 255x^2 + 285x + 315$ 5. Kelsey is correct. 7. $84,267 = 8 \cdot 10^4 + 4 \cdot 10^3 + 2 \cdot 10^2 + 6 \cdot 10^1 + 7 \cdot 10^0$
9. $515 11. a. $200x^2 + 300x + 250$ b. $200x^3 + 300x^2 + 250x + 300$ c. $200x^4 + 300x^3 + 250x^2 + 300x$ d. $200x^5 + 300x^4 + 250x^3 + 300x^2$ 13. D 15. $54n^2 + 65n - 21$
17. $6x^3 + 2x^2 - 2x + 2$ 19. $x = \frac{2}{9}$ 21. $7y^2 + 4y - 22$
23. $0.56 25. a. $h = \frac{1}{2}$ b. h can be any real number but $\frac{1}{2}$.
27. a. cross-fertilized: skewed left; self-fertilized: symmetric b. Cross-fertilize the plants because the mean is greater than that of self-fertilized plants.

Lesson 11-2 (pp. 663–668)
Guided Example: 1. $22w + 8$ 2. 2 3. $x^3 - x^2 - 12$; 3
4. $2x - 2$; 1 Questions: 1. $3x^2 + 4$ is a sum of monomials, while $\frac{3}{x^2} + 4$ includes a quotient of monomials.
3. a. a monomial b. 11 5. a. a monomial b. 2 7. xyz is not a trinomial, it is a monomial because there is only one term. 9. $-3x^5 + 8x^2 - 4x + 12$ 11. $3x^2 + 2x + 5$

13. never; Polynomials with the same degree are like terms; to find their sum you add the coefficients. Thus, the degrees of the sum may be less than or equal to, but will never be greater than the degree of either addend.
15. a. $-140x^2$ b. 2 17. a. $2xy$ b. 2 19. a. 3 b. 3 c. 3
d. 3 21. 1 23. 2 25. a. Answers vary. Sample answer: x^{70} b. Answers vary. Sample answer: $x^{35}y^{35}$
27. a. Answers vary. Sample answer: $x^5 + x + 1$, $x^5 + x + 6$
b. Answers vary. Sample answer: $x^5 + 6x + 8$, $-x^5 + 4x + 2$
29. a. $1{,}000x^{18}$ b. $2,406.61 31. 2 m 33. $n^2 + 52n$

Lesson 11-3 (pp. 669–674)

1. $55x^2$ 3. a. $3h^2 + 15h$
b.

h	1	1	1	1	1
h	h^2	h	h	h	h	h
h	h^2	h	h	h	h	h
h	h^2	h	h	h	h	h

5. a. $4x^2 + 2x$ b. $2x(2x + 1)$
c. $4x^2 + 2x = 2x(2x + 1)$
7. $ab - ac + ad$
9. $-25x^3 - 5x^2 + 31x$
11. $-2ab^2 + a^2b - 5ab$

13. a. $(2 + h)C + (2 + h)B + (2 + h)E$ b. $h(C + B + E) + 2(C + B + E)$ c. $(2 + h)(C + B + E)$ 15. $4x^2$
17. $2m^4 + 2m^3 + 6m^2$ 19. $4xy$ 21. $20n^2 + 9n - 10$
23. a. not a polynomial b. The term $2a^{-2}$ is not a monomial and polynomials are all monomials or sums of monomials. 25. Answers vary. Sample answer: $x^4 + x + 1$
27. $c > \frac{4}{3}$ 29. $\frac{10}{9m}$ 31. $y = \frac{1}{2}(x - 1)$

Lesson 11-4 (pp. 675–679)

Guided Example 3: 4; a; $4a$; $5a^2b$; 2; $-3a^4b^2$; $5a^2b$; 2; $3a^4b^2$
Questions: 1. 1, 3, 11, 33, x, x^2, x^3, x^4, $3x$, $3x^2$, $3x^3$, $3x^4$, $11x$, $11x^2$, $11x^3$, $11x^4$, $33x$, $33x^2$, $33x^3$, $33x^4$ 3. a^2b
5. a. The greatest common factor of $15c^2 + 5c$ is $5c$. So $15c^2 + 5c = 5c(3c + 1)$.
b.

c	c	c	1
c	c^2	c^2	c^2	c
c	c^2	c^2	c^2	c
c	c^2	c^2	c^2	c
c	c^2	c^2	c^2	c
c	c^2	c^2	c^2	c

7. The factor of both x^2 and xy is x and thus, it is factorable. 9. $6n^4 + 5n^2$
11. $33(a - b + ab)$
13. $4v^9(3 + 4v)$ 15. $7rh$

17. a. 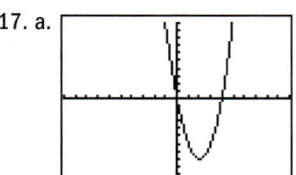 b.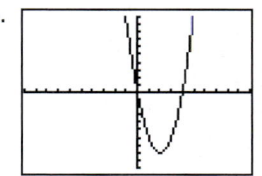

c. The two graphs are the same because $2x(x - 4)$ is a factored form of $2x^2 - 8x$, so the two equations describe the same graph. 19. $2\pi r(r + h)$

21. $-50n^{98} - 40n^{78} + 30n^{58}$ for $n \neq 0$
23. $2k^3 - kn + 4n^2$ 25. B 27. 121

Lesson 11-5 (pp. 680–684)

Guided Example 1: $5x$; (-3); $5x$; (-3); $5x$; (-3); $5x^3$; $(-3x^2)$; $(-20x^2)$; $12x$; $40x$; (-24); $5x^3$; $-23x^2$; $52x$; -24
Questions: 1. a. $(y + 12) \cdot (y^2 + 5y + 7)$ b. $y^3 + 17y^2 + 67y + 84$ 3. a. $12x^2 - 22x - 20$ b. $(2(10) - 5) \cdot (6(10) + 4) = 960$; $12(10)^2 - 22(10) - 20 = 960$
c.

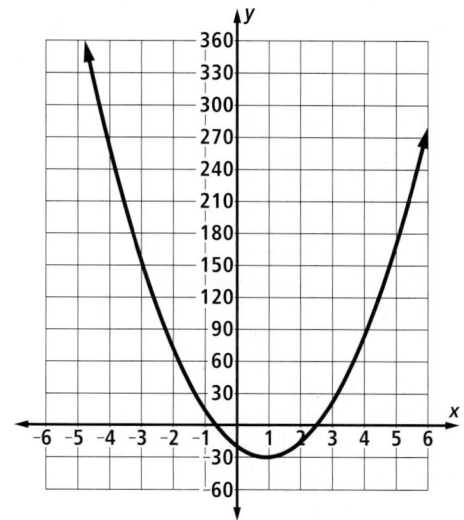

5. $2n^3 + 5n^2 + 2n - 1$ 7. $4x^4 + 16$ 9. $5c^2 + 28d^2 - 39cd + c - 7d$ 11. a. $12x^2 + 26x + 10$ b. $12x^2 - 26x + 10$
c. The expansion of $(a - b)(c - d)$ is the same as the expansion of $(a + b)(c + d)$, except the second term has the opposite sign. 13. $\frac{1}{25}x^2 - 1.08x + 7.29$ 15. $n = 5$ or $n = -\frac{31}{2}$ 17. $-5x^2 - 8$ 19. a. $\frac{33}{16}$ b. 6 c. 69

Lesson 11-6 (pp. 685–690)

Guided Example 2: Solution 1: $2 \cdot 7c \cdot 5$; 5^2; 49; 70; 25
Solution 2: $7c$; 5; $7c$; $7c$; 5; 5; 49; 70; 25
Solution 3:

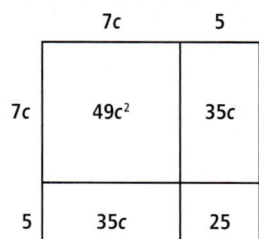

$49c^2 + 70c + 25$
Check: 26; 676; 49; 70; 25; 49; 70; 25; 676

Questions: 1. $g^2 + 2gh + h^2$ 3. $g^2 - h^2$ 5. Answers vary. Sample answer: $x^2 + 2x + 1$

7. a.

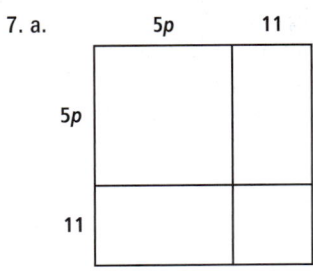

b. $(5p + 11)^2$ c. $25p^2 + 110p + 121$ 9. $x^2 - 10x + 25$ 11. $n^4 - 16$ 13. $81 - 36x + 4x^2$ 15. $9x^2 - y^2z^2$ 17. a. 900; Answers vary. Sample answer: $30^2 = (3 \cdot 10)^2 = 3^2 \cdot 10^2 = 900$ b. 899; Answers vary. Sample answer: $29 \cdot 31 = (30-1)(30+1) = 30^2 - 1^2 = 899$ c. 896; Answers vary. Sample answer: $28 \cdot 32 = (30-2)(30+2) = 30^2 - 2^2 = 896$ d. 891; Answers vary. Sample answer: $27 \cdot 33 = (30-3)(30+3) = 30^2 - 3^2 = 891$ 19. 40,401; Answers vary. Sample answer: $201^2 = (200+1)(200+1) = 200^2 + 2 \cdot 200 \cdot 1 + 1^2 = 40{,}401$ 21. perfect square trinomial 23. perfect square trinomial 25. difference of squares 27. If the sum is 100, the two numbers can be written as $(50 + x)$ and $(50 - x)$, and the product is $2{,}500 - x^2$. That value is greatest when $x^2 = 0$, or $x = 0$, so the numbers are 50 and 50. 29. $18x^2 + 2y^2$ 31. a. $2{,}000 b. $800 c. Answers vary. Sample answer: $x = 1.03$; $10{,}453.31 33. Answers vary. Sample answer: the volume of a cube with side e 35. Answers vary. Sample answer: the area of the region between a circle with radius s that is inside a circle with radius r

Lesson 11-7 (pp. 691-696)

Guided Example 2: without; 11; 10; 9; 11; 10; 9; 990
Questions: 1. a. Answers vary. Sample answer: ABCD, ABCE b. 360 permutations 3. $n(n-1)$ 5. a. Answers vary. Sample answer: VWXYZ, ZYXWV b. 120 permutations 7. $n!$ 9. 56 11. $\frac{1}{272}$ 13. Yes, $n! = n \cdot (n-1)!$, for $n \geq 1$ 15. a. 24 b. 3157; 3159; 3175; 3179; 3195; 3197; 3517; 3519; 3571; 3579; 3591; 3597; 3715; 3719; 3751; 3759; 3791; 3795; 3915; 3917; 3951; 3957; 3971; 3975
17. ABCD, ABDC, ACBD, ACDB, ADBC, ADCB 19. The smaller circle has $8\pi(r-2)$ square units less area.
21. $6x^4 + 2x + 4, x \neq 0$ 23. Answers vary. Sample answer: $-3x^3 + 4x^2 + 7$

Lesson 11-8 (pp. 697-702)

1. how different a set of actually observed scores is from a set of expected scores 3. for expected frequencies less than 5 5. a. 23.7 b. Yes, the chi-square value of 23.7 for 7 events occurs with probability less than 0.001. 7. The spinner seems to be fair. 9. The chi-square value of 44.9 for 4 events occurs with probability less than 0.001, so there is evidence for the view that more tornadoes occur at certain times of the year (spring) than at others.
11. a. 120 b. 720 13. a. $4x^2 + 4xy + y^2$ b. $4x^2 - 4xy + y^2$ c. $4x^2 - y^2$ 15. $9k^4 - 36k^3m + 54k^2m^2 - 36m^3k + 9m^4$
17. a. 3.6 cm/day b. 25.2 cm/wk

Self-Test (pp. 706-707)

1. $3x(10 - 4x + x^3) = 30x - 12x^2 + 3x^4 = 3x^4 - 12x^2 + 30x$ 2. $(2b - 5)^2 = (2b - 5)(2b - 5) = (2b)^2 - 2(2b)(5) + 5^2 = 4b^2 - 20b + 25$ 3. $(8z + 3)(8z - 3) = (8z)^2 - 3^2 = 64z^2 - 9$ 4. $6a(2a^2 + 9a - 1) = 12a^3 + 54a^2 - 6a$ 5. $(5a^2 - a)(5a^2 - a) = (5a^2)^2 - 2(5a^2)(a) + a^2 = 25a^4 - 10a^3 + a^2$ 6. $(2 - 6c)(4 + 3c) = 2(4 + 3c) - 6c(4 + 3c) = 8 + 6c - 24c - 18c^2 = -18c^2 - 18c + 8$ 7. 3 8. In standard form the polynomial is $19x^3 - 9x^2 + 2x - 5$ and has four terms, so it is not a monomial, binomial, nor a trinomial.
9. $6x^2y^2(2x - 4y + 5y^2)$ 10. $(20n^2 - 8n - 12) + (16n^3 - 7n^2 + 5) = 16n^3 + (20n^2 - 7n^2) - 8n + (5 - 12) = 16n^3 + 13n^2 - 8n^2 - 7$ 11. $9p^4 + p^2 - 5 - p(3p^3 + p - 2) = 9p^4 + p^2 - 5 - 3p^4 - p^2 + 2p = 6p^4 + 2p - 5$ 12. False; the expression has three terms so it is a trinomial, but the highest power of a variable term is 2 so it has degree 2.
13. $\frac{28w^3 - 18w}{2w} = \frac{28w^3}{2w} - \frac{18w}{2w} = 14w^2 - 9$ 14. area of rectangle: $(2n + 2)(2n - 2)$; area of triangle: $\frac{1}{2}(n + 3)(n - 1)$; area of shaded region $= 4n^2 - 4 - \frac{1}{2}n^2 - n + \frac{3}{2} = \frac{7}{2}n^2 - n - \frac{5}{2}$ 15. 840; There are 7 swimmers to choose from to swim first, then 6 left to choose the second swimmer, 5 left to choose the third swimmer, and 4 swimmers left to pick the final racer. Thus, there are $7 \cdot 6 \cdot 5 \cdot 4 = 840$ possible orders. 16. 13,800; There are 25 students who could get 1st place, then 24 remaining who could get 2nd place, and 23 remaining to get 3rd place. $25 \cdot 24 \cdot 23 = 13{,}800$ different permutations. 17. For each year that she has had a certain amount of money she receives interest on that amount at the rate of x, thus M dollars received n years ago will be worth Mx^n dollars. So on her 21st birthday she will have $200x^5 + 150x^4 + 300x^3 + 300x^2$ dollars. 18. $200(1.03)^5 + 150(1.03)^4 + 300(1.03)^3 + 300(1.03)^2 = 1{,}046.77$; she would have $1,046.77 on her 21st birthday. 19. a. $9x^2 + 18x$ b. $3x(3x + 6)$ c. No, 3 can be factored out of the parentheses and it can be factored completely to get $9x(x + 2)$.

20.

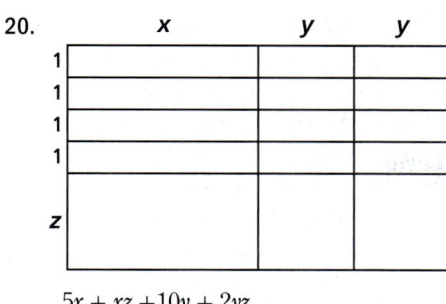

$5x + xz + 10y + 2yz$

21. The probability of the chi-square value of 11.03 for 6 events is between 0.10 and 0.05, so the results aren't very far from a random distribution; the die is probably not weighted.   22. $x^2 + 8x + 16$; See students' work.

The chart below keys the **Self-Test** questions to the objectives in the **Chapter Review** on pages 708–711 or to the **Vocabulary (Voc)** on page 705. This will enable you to locate those **Chapter Review** questions that correspond to questions missed on the **Self-Test**. The lesson where the material is covered is also indicated on the chart.

Question	1	2	3	4	5	6	7	8	9	10
Objective	B	D	B	B	D	B	E	E	C	A
Lesson(s)	11-3, 11-5, 11-6	11-6	11-3, 11-5, 11-6	11-3, 11-5, 11-6	11-6	11-3, 11-5, 11-6	11-2	11-2	11-4	11-1, 11-2

Question	11	12	13	14	15	16	17	18	19	20
Objective	A, B	E	C	I	G	G	F	F	C, I	I
Lesson(s)	11-1, 11-2, 11-3	11-2	11-4	11-3, 11-5, 11-6	11-7	11-7	11-1	11-1	11-3, 11-4, 11-5	11-3, 11-5, 11-6

Question	21	22
Objective	H	I
Lesson(s)	11-8	11-3, 11-5, 11-6

Chapter Review (pp. 708–711)

1. $4k^3 - 17k$ 3. $5.4s^4 + 3.7x^3 + 9.8s^2 - 5.2s - 4$ 5. -21
7. $9p^4 + 9p^3 - 54p^2 + 45p$ 9. $\frac{1}{7}q^{11} - 2q^{10} + 16q^9 - \frac{5}{7}q^8$
11. $b^2 - 9$ 13. $-72w^2 + 2$ 15. $n^3 - 9n^2 + 26n - 24$
17. $\frac{51}{2}x^2 - \frac{11}{2}x - 1$ 19. $uv(u - v)$ 21. $15a^3b^2(3a^6b^3 + 4a^3b^2 - a^2b + 28)$ 23. $p^2 + 12p + 36$ 25. $2,025z^2 - 990z + 121$ 27. B   29. Answers vary. Sample answer: $xy^3z + 3y + 9$ 31. b and c   33. c   35. a. $1,200x^3 + 700x^2 + 500x$ b. $373.05   37. 2,520 different ways
39. 10,000 different ways   41. 4,536 different ways
43. 504 different ways   45. a. The company expected 36 pieces of mail each day Tuesday through Friday and 109 pieces on Monday. b. 5.523 c. There is no evidence for the belief that there was too much mail coming on Monday. The probability of a chi-square value of 5.523 for 5 events is greater than 0.1, so it is likely the deviation from the expected amount of mail was due to random chance.
47. a. $ab + 2a + 6b + 12$ b. $(a + 6)(b + 2)$ c. Yes, the answers are equal.   49. a. $3x(2x + 1)$ b. $6x^2 + 3x$

51.

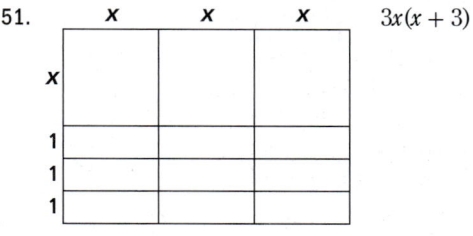

$3x(x + 3)$

Chapter 12

Lesson 12-1 (pp. 714–722)

Guided Example 4: a. $24 - 2L$; $12 - L$; $12 - L$; $(12 - L)$
c. 9 units; $3 \cdot 9$; 27 units^2; 2 units; $10 \cdot 2$; 20 units^2; 36 units^2; 6 units; 6 units
Questions: 1. a. $(-8, -16)$ b. $x = -8$ 3. because $(x - 5)^2$ is never negative, $-4(x - 5)^2$ is never positive, and the greatest nonpositive number is 0, and so the greatest value of $-4(x - 5)^2 + 7$ is 7.   5. a. $(15, 21)$ b. $x = 15$ c. up
7. a. $(-6, -0.5)$ b. $x = -6$ c. down   9. All 3 graphs have $(4, -8)$ as their vertex, c opens up wider than a, and a opens up wider than b.   11. $y - 9 = -(x + 12)^2$

13. $y + 6 = -0.5x^2$ 15. Answers may vary. Sample: $y + 18 = -(x-5)^2$ and $y + 18 = -2(x-5)^2$ 17. (23, 5)
19. Answers may vary. Sample:

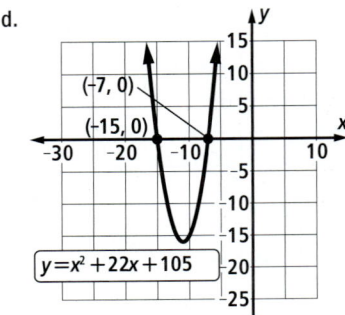

21. $m = \pm\sqrt{3}$ or $m = \pm 2\sqrt{3}$ 23. 5:20 P.M.   25. a. $x = 0$ b. $x < 0$

Lesson 12-2 (pp. 723–728)
1. $x^2 + 14x + 49$ 3. $100; x + 10$ 5. a. 1 b. $x + 1$
7. a. 12.25 b. $r - 3.5$ 9. a. $\frac{1}{4}b^2$ b. $w - \frac{1}{2}b$
11. a. $y + 1.25 = (x - 1.5)^2$ b. -1.25 13. a. (1, 22)
b. maximum   15. No, because the minimum value of y is 0.75 since the vertex is (0.5, 0.75).   17. $x = \sqrt{137} - 12$ or $x = -\sqrt{137} - 12$ 19. The data do support the hypothesis. The chi-square statistic yielded a value of approximately 4. For $n = 3$, this occurs with a probability greater than 0.10. So there is no reason to question the expected values of the hypothesis.   21. 7.5   23. $120.83

Lesson 12-3 (pp. 729–735)
1. 8; −4   3. $x = 32$ or $x = -89.326$ 5. $p = 0$ or $p = -19$
7. a. −15 and −7 b. (−11, −16)
c.

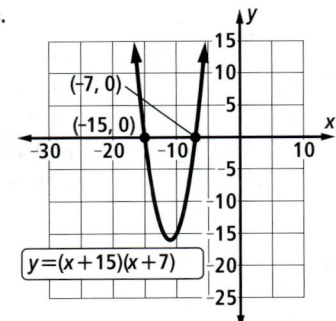

d.

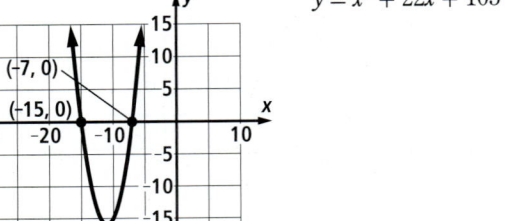

9. a. 0 and $-\frac{11}{4}$ b. $\left(-\frac{11}{8}, \frac{121}{16}\right)$
c.

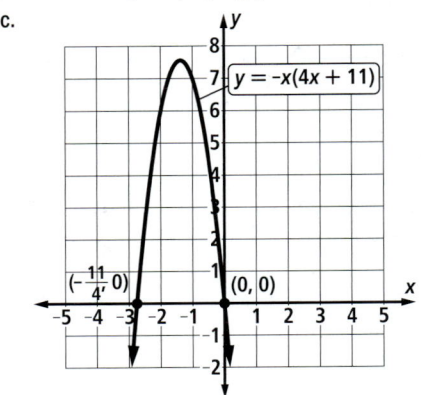

d.

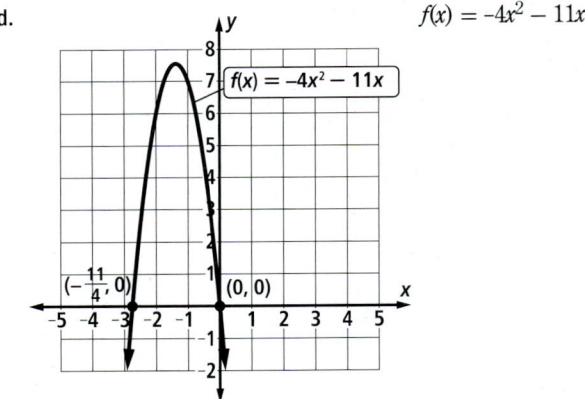

11. a. $x = -6.5$ b. Answers may vary. Sample: $y = 3(x + 11) \cdot (x + 2), y = (x + 11)(x + 2), y = 50(x + 11) \cdot (x + 2)$
13. a. −5 b. $y = 2(x - 1)(x + 5); y + 18 = 2(x + 2)^2;$ $y = 2x^2 + 8x - 10$ 15. a. $y - 5 = -(x - 5)^2$ b. (5, 5)

c.

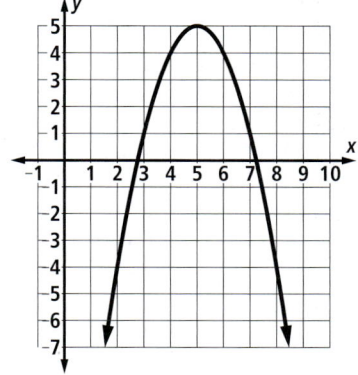

17. $25n^2 - 64$ 19. a. $(2, -4)$ b. $(a, 0)$ c. (r, s)

Lesson 12-4 (pp. 736–741)

Guided Example 2:

Product is –30	Sum of Factors
–1, 30	29
–2, 15	13
–3, 10	7
–5, 6	1
–6, 5	–1
–10, 3	–7
–15, 2	–13
–30, 1	–29

–6; 5; 6; 5 Check 1: 6; 5; 5; 6 Check 2: 6; 5

Guided Example 4: 1. –24; 5   2. 1   3. $m + 8$; $m - 3$
Check: $x + 8$; $x - 3$

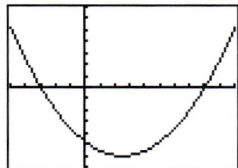

Questions: 1. a. factors: 1, 24; 2, 12; 3, 8; 4, 6; sums: 25, 14, 11, 10   b. $(x + 4)(x + 6)$ c. $(x + 4)(x + 6) = x^2 + 6x + 4x + 24 = x^2 + 10x + 24$ 3. Simona's factorization is correct. The b term in Sandra's factorization is $6 - 8 = -2$, not 2 as desired. The c term in Steve's factorization is $(-6)(-8) = 48$, not –48 as desired.   5. $(q + 1)(q + 19)$ 7. $(v - 1)(v - 101)$
9. $(m - 2)(m + 19)$ 11. a. $(x - 3)(x - 5)$ b. 3, 5
13. a. $y = (x - 2)(x - 16)$

b.

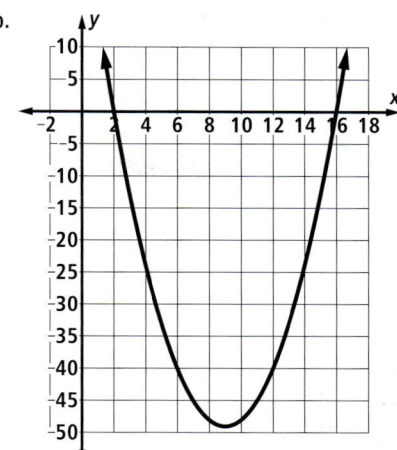

15. $-(x - 8)(x - 5)$ 17. $n = 10$ or $n = -12$. Check: If $n = 10$, $(n - 10)(\frac{1}{2}n + 6) = 0 \cdot 11 = 0$. If $n = -12$, $(n - 10)(\frac{1}{2}n + 6) = -22 \cdot 0 = 0$.   19. 20.25
21. $25a^2 - 9$ 23. $9z^2 + 10, z \neq 0$ 25. $4(7b^4 + 2b^2 + 10)$
27. slope $= \frac{1}{4}$, y-intercept $= 0$

Lesson 12-5 (pp. 742–747)

Guided Example 3: 15; 15; 1; –7; 1; 7; –1; $15y^2 + 32y - 7$; $15y^2 - 32y - 7$; $15y^2 + 16y - 7$; $15y^2 - 16y - 7$; 1; 15; $15y^2 + 104y - 7$; 1; 15; $15y^2 - 104y - 7$; 1; 15; $15y^2 - 8y - 7$; 1; 15; $15y^2 + 8y - 7$; $(3y + 1)(5y - 7)$
Guided Example 4: $3y + 1$; $5y - 7$; $3y + 1$; $5y - 7$; $-\frac{1}{3}$; $\frac{7}{5}$
Questions: 1. a. $8x^2 + 22x + 15$ b. $8x^2 + 26x + 15$
c. $8x^2 + 34x + 15$ d. $8x^2 + 62x + 15$ e. All of these would be found in the process of trying to factor the trinomial.
3. a. $2(x^2 + 7x + 1)$ b. $5(n^2 + 7n - 10)$ 5. $-(x - 3)(3x - 2)$
7. prime   9. $(x - 1)(17x - 19)$ 11. $x = -\frac{3}{4}$ or $x = \frac{1}{5}$
13. $(\frac{3}{8}, -\frac{1}{8})$ 15. a. $n = -2$ or $n = \frac{1}{3}$ b. The solution works.
17. $p^2(5p + 3)^2$ 19. a. $(x^4 - 4)(x^4 + 4)$ b. $(x^2 - 2)(x^2 + 2)$
$(x^4 + 4)$ 21. $4 - \sqrt{13}, 4 + \sqrt{13}$ 23. A

Lesson 12-6 (pp. 748–753)

Guided Example 2: 1. $2x^2 + 5x - 10$ 2. 2; 5; –10   3. 105
4. no   5. The expression is not factorable with integer coefficients.
Questions: 1. 169; factorable; $(x - 11)(x + 2)$ 3. 0; factorable; $(2n - 3)(2n - 3)$ 5. 337; prime

7. The x-intercepts will be rational numbers when the expression is factorable. This occurs when $b^2 - 4ac$ is a perfect square. 9. $n = 999$ 11. Answers vary. Sample answer: $k = 8$ 13. $(r-4)(r-1)$ 15. a. $c = 20.25$ b. $(w + 4.5)(w + 4.5)$ 17. $\frac{4m^7}{3n^3}$ 19. $\frac{x^2}{25y^2}$

Lesson 12-7 (pp. 754–760)
Guided Example 1: $(x - 7.8)$; $(x + 46)$; $(x + 200)$; $x - 7.8$; $x + 46$; $x + 200$
Questions: 1. a. -5; 1.5
b.

c. $f(x) = 2x^2 + 7x - 15$ 3. a. 1; -0.5
b.
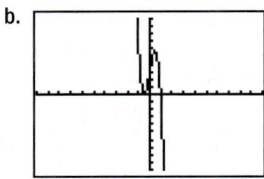
c. $y = -12x^3 + 9x + 3$ 5. Answers vary. Sample answer: $y = (x + 9)(x - 4)$ 7. a. Answers vary. Sample answer: $y = (x - 2)(x + 2) \cdot (x - 5)(x + 5)$ b. $y = x^4 - 29x^2 + 100$
9. Answers vary. Sample answer: $y = (x + 3)(x - 2)(x - 7)$
11. Answers vary. Sample answer: $y = (x + 5)^2(x - 4)^2$
13. a. 55; $1 + 4 + 9 + 16 + 25 = 55$ b. $(n + 1)^2$
c. $S(n) = \frac{1}{3}n^3 + \frac{1}{2}n^2 + \frac{1}{6}n$
15. a.
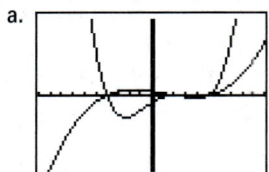

b. Answers vary. Sample answer: The graph crosses the x-axis at the x-intercept 5. c. Answers vary. Sample answer: The graph crosses the x-axis at the x-intercept 5. d. Answers vary. Sample answer: If a polynomial in factored form has a term of the form $(x - a)^3$, then the graph will cross the x-axis at $(a, 0)$. 17. $4(a + 2b)(a - 2b)$
19. $a = -11$ or $a = 5$ 21. B

Lesson 12-8 (pp. 761–767)
Guided Example 1: 1. $(2n + 1)^2$ 2. $(2n + 1)(2n - 1)$
3. 2; 1; 2; 1; 2; 1; 2; 1 4. 2; 1; 2; 1
Guided Example 3: d; b; $ad + bc$
Questions: 1. A rational expression is the written quotient of two polynomials. 3. $\frac{5b}{2c^2}$, $a \neq 0$, $b \neq 0$, $c \neq 0$ 5. a. $\frac{2x-3}{x+2}$
b. 2 and -2 c. $\frac{2(10)-3}{(10)+2} = \frac{17}{12} = \frac{2(10)^2 - 7(10) + 6}{(10)^2 - 4}$
7. $\frac{1 \cdot 15 + 9 \cdot 7}{9 \cdot 15} = \frac{78}{135} = \frac{26}{45}$ 9. Answers vary. Sample answer: Let $k = 2$; then $\frac{6}{2+1} - \frac{3(2)+7}{4-1} = -\frac{7}{3} = \frac{3(2)-13}{(2-1)(2+1)}$. 11. $\frac{15}{8n}$;
let $n = 1$, $\frac{3}{2} + \frac{3}{8} = \frac{12}{8} + \frac{3}{8} = \frac{15}{8}$. 13. $\frac{24x^2 - 29}{15(x-1)}$; $x = 0$, then we have $\frac{8}{5} - \frac{1}{-3} = \frac{29}{15} = \frac{-29}{-15}$. 15. $\frac{(x-6)(x-4)(x-2)}{(x+2)(x+4)(x+6)}$ 17. a. $S = 819$, $F = 91$, $\frac{S}{F} = 9$ b. $\frac{S}{F} = \frac{2n+1}{3}$ c. $\frac{S}{F}$ will only be an integer if $2n + 1$ is a multiple of 3. However, we know for some n, such as $n = 2$, $2n + 1$ will not be a multiple of 3 so $\frac{S}{F}$ is not an integer for that n. 19. a. $t = \frac{v}{4.9}$ b. v is a multiple of 4.9. Answers vary. Sample answer: t is an integer if $v = 4.9$ m/sec and t is not an integer if $v = 10$ m/sec.
21. $0.09x^2$ 23. 17 questions right

Self-Test (pp. 771–772)
1. The product is -40, so possible factors include 10 and -4, -10 and 4, 8 and -5, and because $8 + -5 = 3$, the two factors are 8 and -5. Thus, $(x + 8)(x - 5)$. 2. The product is 72, and the sum is negative. Therefore, possible factors include -18 and -4, -24 and -3, and -8 and -9, and since $-8 + -9 = -17$, the two factors are -8 and -9. Thus, $(m - 8)(m - 9)$. 3. $a = -9$, so possible factors are 3 and -3, 9 and -1, or -9 and 1. $c = -2$, so the factors are either 2 and -1 or -2 and 1. Because $dg + ef = 9$, $d = -3$, $e = 2$, $f = 3$, and $g = -1$ as this is the only combination that works. Thus, $(-3h + 2)(3h - 1)$. 4. B 5. To complete the square, we add $\left(\frac{1}{2}b\right)^2$, so because $(-6)^2 = 36$, add 36.
6. $\frac{3x^2 - 75}{2x^2 - 7x - 15} = \frac{3(x^2 - 25)}{(2x+3)(x-5)} = \frac{3(x+5)(x-5)}{(2x+3)(x-5)} = \frac{3(x+5)}{2x+3}$, $x \neq 5$; $x \neq -\frac{3}{2}$ 7. D 8. a. By the Factor Theorem, the polynomial must have the factors $(x - 0) = x$, $(x + 3)$, and $(x - 9)$. One such polynomial is $y = x(x + 3)(x - 9)$.
b. $y = x^3 - 6x^2 - 27x$ 9. Let x be the length of a side of the frame. Since the area of the painting is 12 square feet, we can write $(x - 1)(x - 2) = 12$. We then put the equation in standard form, so $x^2 - 3x + 2 = 12$, which gives $x^2 - 3x - 10 = 0$. This factors as $(x - 5)(x + 2)$, so we have $(x - 5)(x + 2) = 0$. We cannot have a frame that has length -2, so the frame must have side length 5 feet. 10. To put the equation in $y - k = a(x - h)^2$ form, we must complete the square. Thus, since $\left(\frac{1}{2}b\right)^2 = 1$, we must add 4 to both sides to get $y + 4 = x^2 - 2x + 1$. This then factors as $y + 4 = (x - 1)^2$, so the vertex is at $(1, -4)$.
11. To put the equation in $y - k = a(x - h)^2$ form, we must complete the square. First, we move the 7 to the other side and factor a -2 out of the right hand side, which gives $y - 7 = -2(x^2 + 2x)$. Thus, because $\left(\frac{1}{2}b\right)^2 = 1$, we must add 1 into the $(x^2 + 2x)$ quantity. This gives $y - 9 = -2(x^2 + 2x + 1)$, which factors as $y - 9 = -2(x + 1)^2$.

Thus, the vertex is at $(-1, 9)$.   **12.** By the equation, the vertex is at $(3, 4)$ and a is positive, so the graph opens up.

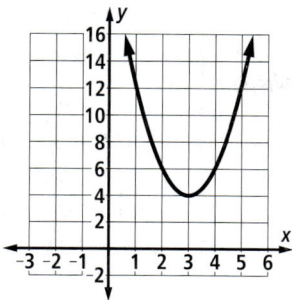

13. By the equation, the zeros of the function are at $x = 5$ and $x = -4$. Also, the axis of symmetry is $x = 0.5$, so the vertex is at $(0.5, -20.25)$.

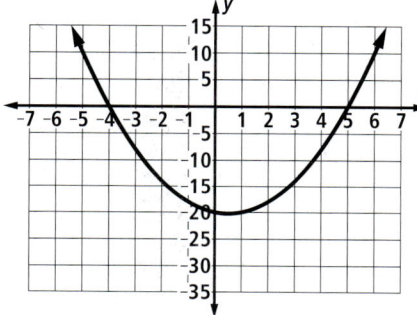

14. First, factor out $3n^2$ to get $3n^2(n^2 - 5n + 6)$. Now deal with what is in the parentheses. Because $c = 6$, possible factors include 3 and 2, -3 and -2, 6 and 1, and -6 and -1. Because $-3 + -2 = -5$, we have $3n^2(n - 3)(n - 2)$.
15. Let x be the length of the shorter side. Then $x(x + 9) = 486$, which gives $x^2 + 9x = 486$. Solving this equation gives x to be -27 or 18, but because length is positive, $x = 18$, so the dimensions are 18 cm by 27 cm.   **16.** False. A quadratic expression is only factorable over the integers when the discriminant is a perfect square, and not all positive numbers are perfect squares.   **17.** Answers vary. Sample answer: $y = 0.04(x + 3)^2(x - 1)(x - 4)$
18. a. Yes, they are equivalent because expanding each equation gives $x^2 + 4x + 3$.   **b.** the vertex of the parabola

The chart below keys the **Self-Test** questions to the objectives in the **Chapter Review** on pages 773–775 or to the **Vocabulary (Voc)** on page 770. This will enable you to locate those **Chapter Review** questions that correspond to questions missed on the **Self-Test**. The lesson where the material is covered is also indicated on the chart.

Question	1	2	3	4	5	6	7	8	9	10
Objective	B	B	B	E	A	D	J	F	G	I
Lesson(s)	12-4, 12-5	12-4, 12-5	12-4, 12-5	12-4, 12-5, 12-6	12-2	12-8	12-3	12-7	12-1	12-2

Question	11	12	13	14	15	16	17	18
Objective	I	H	J	C	G	E	F	H
Lesson(s)	12-2	12-1	12-3	12-7	12-1	12-4, 12-5, 12-6	12-7	12-1

Chapter Review (pp. 773–775)
1. a. 4 b. $(x+2)^2 - 4$ 3. a. $\left(\frac{1}{2}b\right)^2$ b. $\left(\frac{z+1}{2}b\right)^2 - \left(\frac{1}{2}b\right)^2$
5. $(x-3)(x+2)$ 7. $(m-6)(m+4)$ 9. $(3x+4)(x-2)$
11. $2(3d+2)(d-2)$ 13. a. $x=3, x=-1, x=\frac{11}{2}$
b. $y = 12x^3 - 90x^2 + 96x + 198$ 15. a. $x=4, x=\frac{7}{2}, x=-2$
b. $f(x) = 2x^3 - 11x^2 - 2x + 56$ 17. $4(n-2m); n \neq 0,$
$m \neq 0$ 19. $\frac{3-4n}{n-3}$ 21. $\frac{-(m-4)}{(2m-7)}$ 23. perfect squares
25. $3(x+3)(5x+8)$ 27. $2(4x^2 + 17x + 91)$
29. a. Answers vary. Sample answer: $f(x) = (x-2)(x+2)$
b. $f(x) = (x-2)(x+2) = x^2 - 4$ 31. Answers vary.
Sample answer: $y = 4(x+4)(x-1)(x-2)$
33. a. $A = 10x - x^2; P = 20$ b. $5 - \sqrt{5}$ units by $5 + \sqrt{5}$
units 35. a. $A = x(195 - x)$ b. $9{,}000$ yd^2
37. 90 m by 140 m
39.

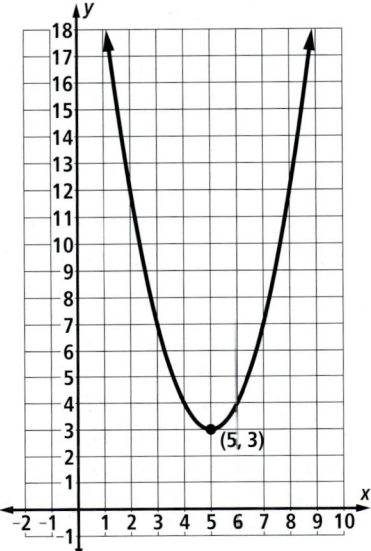

41.

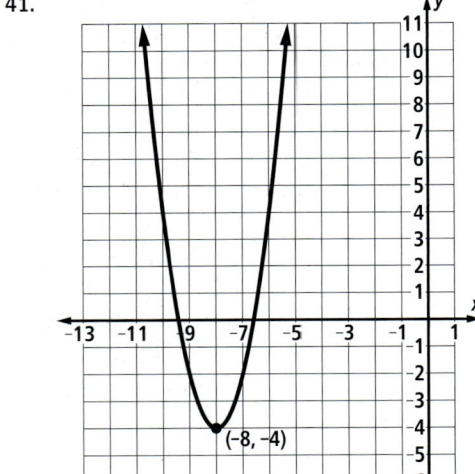

43. ii 45. iii 47. $(0, 0)$ 49. $\left(\frac{1}{3}, -\frac{14}{3}\right)$

51.

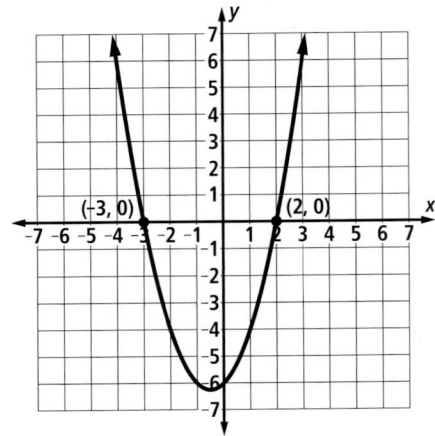

53.

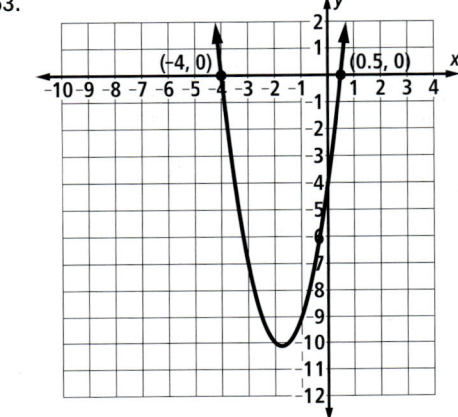

Chapter 13

Lesson 13-1 (pp. 778-783)
Guided Example 4: a. 60; 60 b. true, a rectangle c. false
d. A rectangle has 3 right angles, but it does not have to
be a square.
Guided Example 5: a. x is a real number b. P is a person
born in the United States c. p^g is not negative
Questions: 1. the sun shines this afternoon; I will be
happy 3. If x is an integer greater than 1, then x is a prime
number or the product of prime numbers. 5. Solving the
inequality for x gives $x > -2$, so if the original inequality
is true, then $x > -2$ must also be true. 7. Answers vary.
Sample answer: LW is the area of the rectangle; $2L + 2W$
is the perimeter of the rectangle. 9. Answers vary.
Sample answer: n is a multiple of 10; n is an integer.

11. a. false b.

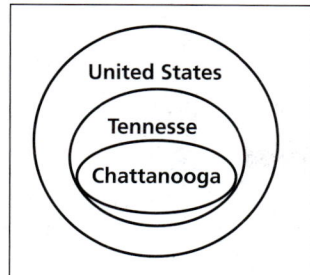

13. a. false b.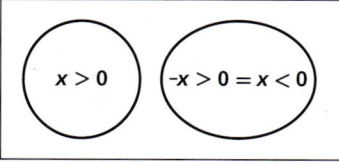
15. a.
Answers vary. Sample answer: Let $t = 0$. Then $t^4 = 0$, and 0 is not positive. b. Answers vary. Sample answer: t can be any real number except 0. 17. a. Answers vary. Sample answer: Let $x = -\frac{1}{2}$. Then $x^2 + x = \left(-\frac{1}{2}\right)^2 + \left(-\frac{1}{2}\right) = \frac{1}{4} - \frac{1}{2} = -\frac{1}{4}$, so $y < 0$. b. $x \geq 0$ or $x \leq -1$
19. Answers vary. Sample answer:

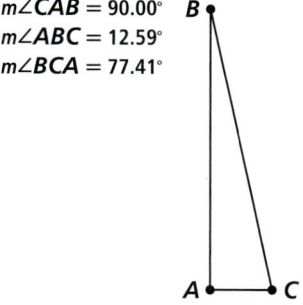

$m\angle CAB = 90.00°$
$m\angle ABC = 12.59°$
$m\angle BCA = 77.41°$

21. If something is a sentence, then it has a subject, verb, and object; Answers vary. Sample answer: The sentence "He went." has no object. 23. a. $(a - 6)(a + 6)$ b. $(n - 9)(n + 4)$ c. $(x - 9y)(x + 4y)$ 25. 2,643 adult tickets and 507 child tickets 27. a. 53 b. -67 c. Answers vary. Sample answer: The graph is the line with slope $\frac{3}{4}$, y-intercept -7.

Lesson 13-2 (pp. 784-787)

Guided Example 1: a. a rectangle; it has 4 sides of the same length and its diagonals have the same length; a rectangle can have different width and length. **b.** you live in the United States; you live in the state of North Dakota; you could live in any of the other 49 states **Guided Example 2: a.** is a square; it has 4 sides of the same length and its diagonals have the same length; it is a square **b.** You live in Brazil; you live in the largest country in South America
Questions: 1. If there is fire, then there is smoke. **3. a.** no **b.** If an integer is divisible by 24, then it is divisible by 3 and 4. **c.** yes **d.** Answers vary. Sample answer: If an integer is divisible by 3 and 4, then it is divisible by 12. If an integer is divisible by 24, then it is divisible by 3 and 8. **5. a.** no **b.** If $u > 8$, then $7u < 56$. **c.** no **d.** If $7u > 56$, then $u > 8$; If $u > 8$, then $7u > 56$. **7.** If a quadrilateral has four right angles, then it is a rectangle; If a quadrilateral is a rectangle, then it has four right angles. **9.** A function is a linear function if and only if it has an equation of the form $f(x) = ax + b$. **11.** yes **13.** No, the statement "If $z^2 + 48 = 14z$, then $z = 8$" is not true, because $z^2 + 48 = 14z$ is also true for $z = 6$. **15. a.** yes **b.** If $2x^2 - 7x + 30 = 90$, then $(x - 5)(2x + 3) = 45$. **c.** yes **d.** n/a **17. a.** no **b.** If the x-intercepts of a parabola are -5 and 3, then it has an equation of the form $y = (x - 5)(x + 3)$. **c.** no **d.** If a parabola has an equation of the form $y = a(x - 5)(x + 3)$, then its x-intercepts are 5 and -3; If the x-intercepts of a parabola are -5 and 3, then it has an equation of the form $y = a(x + 5)(x - 3)$. **19. a.** true **b.** Answers vary. Sample answer: $x = 10$. **21. a.** false **b.** A person can be a U.S. citizen if and only if the person was born in the United States or the person was naturalized. **23.** False; the graph of $y = ax^2$ crosses the x-axis only once at $(0, 0)$ for all $a \neq 0$. **25.** $0.56 **27.** $\frac{27}{8}$

Lesson 13-3 (pp. 788-794)

Guided Example 1: 1. ii. Addition Property of Equality; iii. $-6x = 118 + 14$; v. both sides multiplied by $-\frac{1}{6}$; vi. Multiplicative Inverse Property; vi. $-\frac{1}{6}(132) = -22$; Identity **Guided Example 2:** i. Given; ii. Addition Property of Equality; iii. Additive Inverse Property and arithmetic; iv. Multiplication Property of Equality; v. Multiplicative Inverse Property and arithmetic
Questions: 1. a. $3(6) = 18$, $3(13x) = 39x$; Distributive Property **b.** Add -12 to both sides; Addition Property of Equality **c.** $12 + -12 = 0$; Additive Inverse Property **d.** $40x + 0 = 40x$; Additive Identity Property **e.** $18 + -12 = 6$; arithmetic **f.** If $40x + 12 = 3(6 + 13x)$, then $40x = 6 + 39x$. **3.** Inductive reasoning is used to make a general conclusion out of a specific instances, while deduction is used to prove a specific instance of a general case, using known facts. **5. a.** Sula **b.** Lana **7. a.** (i.) $12m + -3m = 3m + -3m + 5$; Addition Property of Equality (ii.) $9m = 0 + 5$; Additive Inverse Property and arithmetic (iii.) $m = \frac{5}{9}$; Additive Identity Property and Multiplication Property of Equality **b.** Show that if $m = \frac{5}{9}$, then $12m = 3m + 5$. **9.** Multiply both sides by $\frac{1}{t}$ (Multiplication Property of Equality), and $t \cdot \frac{1}{t} = 1$ by the Multiplicative Inverse Property. **11.** Because the equation is quadratic, the Quadratic Formula can be applied to solve for n. **13.** $3x - 4y = -18$ can be added to $3x + 4y = 6$ to get $6x = -12$; Addition Property of Equality **15.** (i.) $ax + b = c$ Given (ii.) $ax + b - b = c - b$ Addition

Property of Equality (iii.) $ax + 0 = c - b$ Additive Inverse Property (iv.) $ax = c - b$ Additive Identity Property (v.) $\frac{1}{a} \cdot ax = \frac{1}{a} \cdot (c - b)$ Multiplication Property of Equality (vi.) $x = \frac{(c-b)}{a}$ Multiplicative Inverse Property, Multiplicative Identity Property   17. $x = \frac{c-b}{a}$ if and only if $ax + b = c$ and $a \neq 0$.   19. a. (i.) $a(x - h)^2 = k$ Given (ii.) $\frac{1}{a} \cdot a(x - h)^2 = \frac{1}{a} \cdot k$ Multiplication Property of Equality (iii.) $\sqrt{(x-h)^2} = \sqrt{\frac{k}{a}}$ definition of square root (iv.) $x - h = \pm\sqrt{\frac{k}{a}}$ definition of square root (v.) $x - h + h = \pm\sqrt{\frac{k}{a}} + h$ Addition Property of Equality (vi.) $x = h \pm\sqrt{\frac{k}{a}}$ Additive Inverse Property b. (i.) $x = h \pm\sqrt{\frac{k}{a}}$ Given (ii.) $x - h = \pm\sqrt{\frac{k}{a}} + h - h$ Addition Property of Equality (iii.) $x - h = \pm\sqrt{\frac{k}{a}}$ Additive Inverse Property (iv.) $(x - h)^2 = \left(\sqrt{\frac{k}{a}}\right)^2$ Multiplication Property of Equality (v.) $(x - h)^2 = \frac{k}{a}$ definition of square root (vi.) $a \cdot (x - h)^2 = \frac{k}{a} \cdot a$ Multiplication Property of Equality (vii.) $a(x - h)^2 = k$ Multiplicative Inverse and Identity Properties
c. $a(x - h)^2 = k$ if and only if a and k are both positive and $x = h \pm\sqrt{\frac{k}{a}}$.   21. a. If you are under 8 years of age, then you receive a reduced fare on the metro city bus.
b. If you receive a reduced fare on the metro city bus, then you are under 8 years of age. c. No. Answers vary. Sample answer: Other groups of people might receive a reduced fare as well.   23. a. $3(x - 1)(x + 4)$ b. Answers vary. Sample answer: $-\frac{3}{2} + \sqrt{\frac{119}{12}}$ 25. $\sqrt{ab} = (\sqrt{a})(\sqrt{b})$ 27. $(a^x)^y = a^{xy}$ 29. $-1 \cdot a = -a$ 31. $AB = 9\sqrt{29}$ cm ≈ 48.47 cm, $XZ = 10$ cm, $XY = 5\sqrt{29}$ cm ≈ 26.93 cm   33. $\frac{4}{663} \approx 0.006033$

Lesson 13-4 (pp. 795-801)
Guided Example: $2L$; $2W$; L; W; $2W$; $2L$; L; $150 - L$; $150 - L$; 40; 110; 110; 40; 40; 110
1. D   3. Al-Khwarizmi created a general method similar to today's Quadratic Formula.   5. $x + y = 53$ so their average is 26.5. Let $M = 26.5 + x$ and $N = 26.5 - x$. $MN = 612$, so $(26.5 + x)(26.5 - x) = 612$; $702.25 - x = 612$; $x^2 = 90.25$, $x = 9.5$ or $x = -9.5$. If $x = 9.5$, $M = 26.5 + 9.5 = 36$, $N = 26.5 - 9.5 = 17$. If $x = -9.5$, $M = 26.5 - 9.5 = 17$, $N = 26.5 + 9.5 = 36$. The two numbers are 17 and 36.
7. $L = 25 - W$; $W(25 - W) = 144$, $25W - W^2 = 144$, $W^2 - 25W + 144 = 0$; $(W - 16)(W - 9) = 0$; 16 yards by 9 yards   9. $\frac{15 + \sqrt{185}}{2}$ and $\frac{15 - \sqrt{185}}{2}$ 11. $x = -\frac{1}{7}$ or $x = 1$
13. The door is 9.6 units high and 2.8 units wide.   15. a. If a number is divisible by 8, then it is also divisible by 4.
b. true   c. If a number is divisible by 4, then it is also divisible by 8.   d. false; Answers vary. Sample answer: 12 is divisible by 4, but not by 8.   17. $s = \pi r(r + 2h)$
19. about 603mL   21. a. $2\sqrt{2} + \sqrt{5}$ b. $2\sqrt{10}$ c. $2\sqrt{5}$

Lesson 13-5 (pp. 802-808)
Guided Example 2: $2p + 1$; $2q + 1$; $p - q$; $p - q$; 2
Questions: 1. an integer that can be written as $2n$, where n is an integer   3. Answers vary. Sample answer: 6 and 4 are both divisible by 2, but 10 is not divisible by 4.
5. a. Answers vary. Sample answer: Let p and q be odd integers such that $p = 2m + 1$ and $q = 2n + 1$, where m and n are integers. Then $p + q = 2m + 1 + 2n + 1 = 2m + 2n + 2 = 2(m + n + 1)$. Since m, n, and 1 are integers, $m + n + 1$ is an integer. Thus $p + q$ is an even integer. b. No. Answers vary. Sample answer: From Part a, the sum of the two odd integers is an even integer, not an odd one.
7.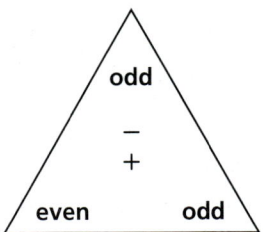
9. even + odd = odd; odd + even = odd; odd − even = odd; odd − odd = even

```
        odd
         −
         +
   even     odd
```

11. Answers vary. Sample answer: Let m be an even number. Then there is an integer p such that $m = 2p$, so $m^2 = (2p)^2$. Thus $m^2 = 4p^2$, and since p is an integer, p^2 is an integer. Thus m^2 is a multiple of 4.   13. Answers vary. Sample answer: Let m and n be numbers such that $m + n = 35a$ and $n = 70b$, where a and b are integers. Then $m + n = m + 70b = 35a$, so $m = 35a - 70b$, and by the Distributive Property, $m = 35(a - 2b)$. Because a and $2b$ are integers, their difference is an integer, so m is divisible by 35.   15. Answers vary. Sample answer: Let m and n be numbers such that $m = 4p$ and $n = 6q$, where p and q are integers. Then $mn = (4p)(6q) = 24pq$ by the Commutative and Associative Properties of Multiplication. Because p and q are integers, pq is an integer.   17. Answers vary. Sample answer: $b = 13$ 19. $x = 7, x = 4$ 21. $a = 3$
23. $x = -2.3$ 25. a. 18 units b. 54 units^2

Lesson 13-6 (pp. 809-815)
Guided Example 3: 1. $100h$; $10t$; u 2. $999T + 99h + 9t$
4. $999T$; $99h$; $9t$; $k + 333T + 33h + 3t$; $k + 333T + 33h + 3t$
Questions: 1. 57 3. $1{,}000A + 100B + 10C + D$
5. a. $1{,}000T + 100h + 10t + u$ b. $1{,}000u + 100t + 10h + T$
7. A five digit number in base 10 can be written as $N = 10{,}000D + 1{,}000T + 100h + 10t + u$, where D, T, h, t, u are all digits. Separate the sum of the digits from the value of

the number. $N = (D + T + h + t + u) + (9{,}999D + 999T + 99h + 9t)$. The sum of the digits is divisible by 9, so there is an integer k with $D + T + h + t + u = 9k$. Substituting, $N = 9k + (9999D + 999T + 99h + 9t) = 9(k + 1{,}111D + 111T + 11h + t)$. Since $k + 1{,}111D + 111T + 11h + t$ is an integer, N is divisible by 9. **9. a.** No, the units digit is odd. **b.** No, the units digit is neither 5 nor 0. **c.** No, the digits do not sum to a number divisible by 9. **11. a.** Yes, the units digit is even. **b.** No, the units digit is neither 5 nor 0. **c.** Yes, the digits sum to 9. **13. a.** Answers vary. Sample answer: $2{,}346 - 6{,}432 = -4{,}086$, which, when divided by 99 gives about -41.28, which is not an integer. **b.** A four digit number in base 10 can be written as $N = 1{,}000T + 100h + 10t + u$, where T, h, t, u are all digits. Moreover, the number with reversed digits is $1{,}000u + 100t + 10h + T$. The difference between these two numbers is $1{,}000T + 100h + 10t + u - (1{,}000u + 100t + 10h + T) = 999T + 90h - 90t - 999u = 9(111T + 10h - 10t - 111u)$. Since $111T + 10h - 10t - 111u$ is an integer we know that this difference is divisible by 9. **15.** A six digit number in base 10 can be written as $N = 100{,}000H + 10{,}000D + 1{,}000T + 100h + 10t + u$, where H, D, T, h, t, u are all digits. Our given conditions mean $H = h$, $D = t$ and $T = u$, and so our number can be rewritten as $100{,}000h + 10{,}000t + 1{,}000u + 100h + 10t + u = 100{,}100h + 10{,}010t + 1{,}001u$. We can factor 13 from this expression to get $N = 13(7{,}700h + 770t + 77u)$, and since $7{,}700h + 770t + 77u$ is an integer, N is 13 times an integer, and thus divisible by 13. **17.** 285 **19.** Answers vary. Sample answer: As a counterexample, consider 6 and 8, where $6 \cdot 8 = 48$ which is not divisible by 7. **21.** $h = 5$ **23. a.** 0 **b.** 1 **25.** $y > 0.5$ **27.** $x = 45$

Lesson 13-7 (pp. 816-822)
1. Answers vary. Sample answer: $\frac{1}{3}$ **3.** Answers vary. Sample answer: 2 **5.** $\frac{28}{33}$ **7.** D **9.** irrational **11.** rational

13.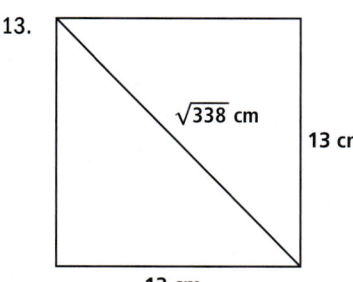

15. Yes, Answers vary. Sample answer: $\sqrt{2} + (-\sqrt{2}) = 0$, and both of them are irrational, while 0 is rational.

17. a.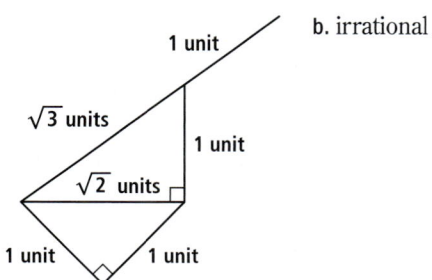

b. irrational

19. $84\sqrt{2}$ cm; irrational **21. a.** $a = \sqrt{1{,}606}$ **b.** irrational

23.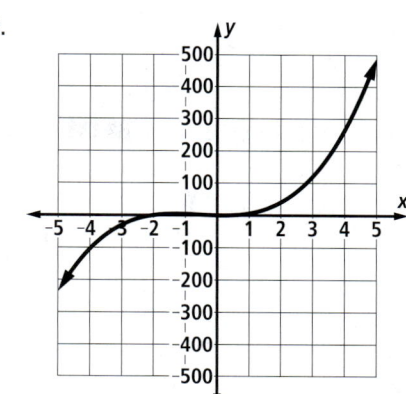

25. $f(x) = x(x + 2)(3x - 1)$ **27. a.** $9y$ **b.** $b = -4y$, $c = 4, d = 5y$ **29.** $25 - x^2$ **31.** 70.56 ft

Lesson 13-8 (pp. 823-828)
1. Answers vary. Sample answer: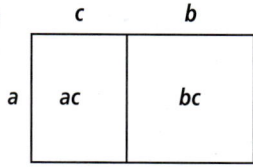

3. a.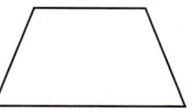

b. 64 units² **c.** Answers vary. Sample answer: Subtract four times the area of a triangle with base 6 and height 2 from the total area of 64. **d.** $2\sqrt{10}$

5. a. Answers vary. Sample answer: **b.** 1.5 in.² **7. a.** 2,664.5 units² **b.** Subtract twice the area of a rectangle with base a and height b from the total area, i.e., the trapezoid with bases a and b and height $a + b$. **c.** 53 **d.** yes **9.** The Extended Distributive Property **11. a.** $\sqrt{2a^2 + b^2 - 2ab}$ **b.** $\frac{1}{2}(ab - a^2)$ **c.** $\frac{1}{2}c^2$ **d.** $\frac{1}{2}(ab - a^2 + c^2)$ **e.** $\frac{1}{2}(ba + b^2)$ **f.** $\frac{1}{2}(ab - a^2 + c^2) = \frac{1}{2}(ba + b^2)$; $ab - a^2 + c^2 = ba + b^2$; $c^2 = a^2 + b^2$ **13.** A **15. a.** 362,880 orders **b.** 60,480 ways **17.** about $9,095.32

Self-Test (p. 832)

1.

Conclusions	Justifications
$8(2y - 1) = y + 37$	Given
$16y - 8 = y + 37$	Distributive Property
$16y - 8 - y = y + 37 - y$	Addition Property of Equality
$15y - 8 = 0 + 37$	Arithmetic, Additive Inverse Property
$15y - 8 = 37$	Additive Identity Property
$15y - 8 + 8 = 37 + 8$	Addition Property of Equality
$15y + 0 = 45$	Arithmetic, Additive Inverse Property
$15y = 45$	Additive Identity Property
$\frac{1}{15} \cdot 15y = \frac{1}{15} \cdot 45$	Multiplication Property of Equality
$1y = 3$	Arithmetic, Multiplicative Inverse Property
$y = 3$	Multiplicative Identity Property

2. The antecendent is $8(2y - 1) = y + 37$ and the consequent is $y = 3$. 3. a. If $xy = 0$, then both x and y equal 0; If both x and $y = 0$, then $xy = 0$. b. No, Amalia is not correct. Answers vary. Sample answer: The statement "If $xy = 0$, then both x and y equal 0" is not true because, for example, when $x = 1$ and $y = 0$, $xy = 0$ but x does not equal 0. 4. a. $\sqrt{7^2 + 8^2} = \sqrt{49 + 64} = \sqrt{113}$ in. b. $\sqrt{113}$ is an irrational number, so its decimal is infinite and does not repeat. Marcus's ruler is not accurate for the smallest length that it measures. 5. a. If a student is taking algebra, then the student can solve quadratic equations. b. The antecedent is a student is taking algebra, the consequent is the student can solve quadratic equations. c. If a student can solve quadratic equations, then the student is taking algebra. d. The statement is not true. For example, a student who knows how to solve quadratic equations could be a student in geometry. 6. True. Answers vary. Sample answer: If the tens digit of a four-digit number is 4 and the units digit is 8, then the number can be written as $1,000n + 48$, where n is a whole number. Then $1,000n + 48 = 4(250n + 12)$, and because $250n$ and 12 are integers, $250n + 12$ is an integer, so 4 divides $100n + 48$. 7. The rectangles picture the equation $(a - b)(a + b) = (a - b)a + (a - b)b = a^2 - b^2$.
8. $ab = 717$ and $a + b = -242$. Then $a = -242 - b$, so substitution gives $(-242 - b)b = 717$, so $b^2 + 242b + 717 = 0$. This is a quadratic equation, so solving for b gives $b = -239$ or $b = -3$. Thus, the numbers are -239 and -3.
9. a. $x^2 + (\sqrt{3})^2 = (\sqrt{13})^2$, so $x^2 + 3 = 13$. $x^2 = 10$ so $x = \sqrt{10}$ b. x is irrational 10. a. True. Because the sides of a square are all equal, two of the sides of the triangle will have equal length, so the triangle will be isosceles. b. False. If the length of one side of the square is a, then the length of the diagonal will be $a\sqrt{2}$, so the lengths of the sides are not all equal. 11. False. Any person born in the United States cannot necessarily become president, since the person also needs to be at least 35 years old and have lived in the United States for at least 14 years.
12. Answers may vary. Sample: If a number is divisible by 3 and 4, then it can be represented as such and broken up into rectangles of 12 dots each. Thus, it is divisible by 12.

The chart below keys the **Self-Test** questions to the objectives in the **Chapter Review** on pages 833–835 or to the **Vocabulary (Voc)** on page 831. This will enable you to locate those **Chapter Review** questions that correspond to questions missed on the **Self-Test**. The lesson where the material is covered is also indicated on the chart.

Question	1	2	3	4	5	6	7	8	9	10
Objective	B	B	B	E	A	E	H	B	A	D
Lesson(s)	13-3	13-1	13-1, 13-2	13-4	13-1, 13-2	13-5, 13-6	13-8	13-4	13-3	13-1, 13-2

Question	11	12	13	14
Objective	G	E	F	I
Lesson(s)	13-1, 13-2	13-5, 13-6	13-7	13-7

Chapter Review (pp. 833-835)

1.

	Conclusions	What Was Done	Justifications
i.	$4x + 5 = 17$		Given
ii.	$4x + 5 + -5 = 17 + -5$	-5 added to both sides.	Addition Property of Equality
iii.	$4x + 0 = 12$	$5 + -5 = 0;$ $17 + -5 = 12$	Additive Inverse Property; Arithmetic
iv.	$4x = 12$	$4x + 0 = 4x$	Additive Identity Property
v.	$\frac{1}{4} \cdot 4x = 12 \cdot \frac{1}{4}$	Both sides were multiplied by $\frac{1}{4}$.	Multiplication Property of Equality
vi.	$1 \cdot x = 3$	$\frac{1}{4} \cdot 4 = 1;$ $12 \cdot \frac{1}{4} = 3$	Multiplicative Identity Property; Arithmetic
vii.	$x = 3$	$1 \cdot x = x$	Multiplicative Identity Property

3.

	Conclusions	What Was Done	Justifications
i.	$3t - 15 = 4t + 2$	Given	Given
ii.	$3t - 17 = 4t$	-2 added to both sides.	Addition Property of Equality
iii.	$-17 = t$	$-3t$ added to both sides.	Addition Property of Equality

5. 15 boys, 11 girls or 11 boys, 15 girls **7.** $\frac{5.6 + \sqrt{22.16}}{2}$, $\frac{5.6 - \sqrt{22.16}}{2}$ **9.** -23 and -23 **11.** antecedent: an animal has feathers: consequent: it is a bird **13.** antecedent a number is irrational; consequent: it cannot be represented as the ratio of two integers **15.** true **17.** true **19.** Suppose that $2a + 3b + c$ is not divisible by 7. Then $2a + 3b + c + (14)(7a) + 7b$ is not divisible by 7. Hence, $100a + 10b + c$ is also not divisible by 7. So the three-digit number abc is not divisible by 7. **21.** Consider a six-digit integer of the form $xyzxyz$. This is equal to $100{,}000x + 10{,}000y + 1{,}000z + 100x + 10y + z$. And equivalently, this is $(7{,}692)(13x) + 4x + (769)(13y) + 3y + (76)(13z) + 12z + (7)(13x) + 9x + 10y + z$. Combining and factoring, we obtain $xyzxyz = (13)(7{,}700x + 770y + 77z)$. So all six-digit integers of the form $xyzxyz$ are divisible by 13.

23. irrational **25.** rational **27.** Yes, for example $\sqrt{2} \cdot \sqrt{2} = 2$. **29. a.** yes **b.** yes **c.** n/a **31. a.** yes **b.** no **c.** You are in high school if and only if you are in grades 9–12. **33.**

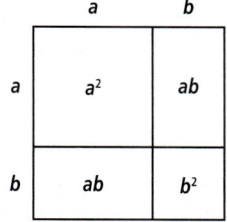

35. The area of each triangle is $\frac{1}{2}ab$. Because the area of the entire rectangle is ab, and $\frac{1}{2}ab + \frac{1}{2}ab = ab$, each of the triangles must occupy exactly half the area of the rectangle. So the diagonal cuts the area of the rectangle in half. **37. a.** $\frac{16}{\pi}$ in. **b.** irrational **39. a.** 0.15 **b.** rational **41 a.**

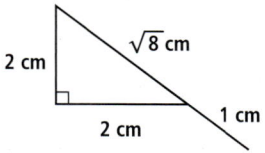

b. irrational

Selected Answers

Glossary

A

absolute value (A) The distance a number x is from 0 written $|x|$. (B) $|x| = x$ if $x \geq c$; $|x| = -x$ if $x < c$. (**43**)

addition method for solving a system Given two equations in a system, applying the Addition Property if $a = b$ and $c = d$, then $a + c = b + d$, to obtain another equation that is satisfied by the solution to the system. (**602**)

Addition Property of Equality For all real numbers a, b, and c, if $a = b$, then $a + c = b + c$. (**106**)

Addition Property of Inequality For all real numbers a, b, and c, if $a < b$, then $a + c < b + c$. (**162**)

Additive Identity Property For any real number a, $a + 0 = 0 + a = a$. (**108**)

Additive Inverse Property For any real number a, $a + -a = -a + a = 0$. (**109**)

additive inverses Two numbers whose sum is zero; also called opposites. (**85**)

Algebraic Definition of Division For all real numbers a and b with $b \neq 0$, $a \div b = \frac{a}{b} = a \cdot \frac{1}{b}$. (**7**)

Algebraic Definition of Subtraction For all real numbers a and b, $a - b = a + -b$. (**7**)

algebraic expression An expression that includes one or more variables. (**6**)

algebraic fraction A fraction with a variable in the numerator, in the denominator, or in both. (**252**)

annual yield The percent of interest that money on deposit earns per year. (**398**)

antecedent The clause following *if* in an if-then statement. (**778**)

Arrangements Theorem If there are n ways to select each object in a sequence of length L, then n^L different sequences are possible. (**459**)

Associative Property of Addition For any real numbers a, b, and c, $(a + b) + c = a + (b + c)$. (**10**)

Associative Property of Multiplication For any real numbers a, b, and c, $(ab)c = a(bc)$. (**9**)

axis of symmetry Given a figure, a line over which the reflection image of the figure is the figure itself. (**526**)

B

base The number x in the power x^n. (**398**)

binomial A polynomial that has two terms. (**663**)

boundary line A line that separates two sets, e.g., solutions from nonsolutions in the graph of a linear inequality. (**381**)

boundary A point or number that separates solutions from nonsolutions on a number line. (**155**)

C

capture-recapture method The use of proportions to estimate the total number of objects in a set (e.g., the number of deer in a forest or the number of fish in a pond). (**303**)

Celsius scale The temperature scale of the metric system, developed by Anders Celsius, in which the freezing point of water is 0° and the boiling point of water is 100°; also known as the *centigrade scale*. (**221**)

centigrade scale See *Celsius scale*. (**221**)

chi-square statistic A number calculated from data to determine whether the difference in two frequency distributions is greater than that expected by chance. (**698**)

circular permutation An ordering of objects around a circle. (**695**)

clearing fractions Multiplying each side of an equation or inequality by a constant to get an equivalent equation without fractions as coefficients. (**170**)

closed under an operation A set is closed under an operation if the result of the operation always lies within the particular set. (**802**)

coefficient A number that is a factor in a term containing a variable. (**73**)

coefficient matrix In the matrix equation $\begin{bmatrix} a & b \\ c & d \end{bmatrix} \cdot \begin{bmatrix} x \\ y \end{bmatrix} = \begin{bmatrix} e \\ f \end{bmatrix}$, the coefficient matrix is $\begin{bmatrix} a & b \\ c & d \end{bmatrix}$. (**622**)

coincident Two or more lines or other sets of points that are identical. (**618**)

collinear points Points that all lie on the same line. (**130**)

Commutative Property of Addition For all real numbers a and b, $a + b = b + a$. (**16**)

Commutative Property of Multiplication For all real numbers a and b, $ab = ba$. (16)

complementary events Two events that have no elements in common, but together they contain all the possible outcomes. (283)

complete factorization The representation of a polynomial as a product of prime polynomials. (676)

completing the square Converting a quadratic equation in standard form to one in vertex form. (724)

complex fraction A fraction that has a fraction in the numerator or a fraction in the denominator, or both. (259)

compound interest The money a bank pays on the principal and earned interest in an account. (399, 400)

compound sentence A single sentence consisting of two or more sentences linked by the words *and* or *or*. (227)

conditional probability The probability that an event will occur given that another event has occurred. (283)

consequent The clause following *then* in an if-then statement. (778)

constant-decrease situation A quantity that decreases at a constant amount in a period of time. (130)

constant-increase situation A quantity that increases at a constant amount in a period of time. (130)

constant matrix In the matrix equation $\begin{bmatrix} a & b \\ c & d \end{bmatrix} \cdot \begin{bmatrix} x \\ y \end{bmatrix} = \begin{bmatrix} e \\ f \end{bmatrix}$, the constant matrix is $\begin{bmatrix} e \\ f \end{bmatrix}$. (622)

constant term A term in a polynomial without a variable. (736)

converse An if-then statement in which the antecedent and the consequent of the statement have been switched. (784)

conversion rate A rate determined from an equality between two quantities with different units. (269)

coordinates The numbers x and y that locate a point (x, y) in the coordinate plane. (27)

counterexample An instance for which a general statement is not true. (23)

cube The third power of a number x, or x^3. (493)

Cube of the Cube Root Property For any nonnegative number x, $\sqrt[3]{x} \cdot \sqrt[3]{x} \cdot \sqrt[3]{x} = \sqrt[3]{x^3} = x$. (493)

cube root If $V = s^3$, then s is a cube root of V. (493)

cubic polynomial A polynomial of degree 3. (756)

D

deduction Using a proof to show that one if-then statement follows from another. (789)

define a variable The process of describing the quantity a variable represents. (14)

degree of a monomial The sum of the exponents of the variables in the monomial. (664)

degree of a polynomial The highest degree of any of the monomial terms of a polynomial. (664)

dependent variable A variable (y) whose value is determined by the value of at least one other variable in a given function. (426)

deviation (A) The difference between a member of a data set and the mean of that data set. (42) (B) The difference between an expected number and an actual observed number. (697)

difference of squares An expression of the form $x^2 - y^2$. For all real numbers x and y, $x^2 - y^2 = (x + y)(x - y)$. (687–688)

dimensions (of a matrix) The number of rows and columns in a matrix. (622)

direct variation A number y varies directly with a number x if y is a constant multiple of x. (353)

discount The amount by which the original price of an item is lowered. (184)

discount rate The ratio of the discount to the original price. (275)

discriminant The value of $b^2 - 4ac$ in the quadratic equation $ax^2 + bx + c = 0$. (561)

Discriminant Property Suppose $ax^2 + bx + c$ and a, b, and c are real numbers with $a \neq 0$. Let $D = b^2 - 4ac$. Then when $D > 0$, the equation has two real solutions. When $D = 0$, the equation has exactly one real solution. When $D < 0$, the equation has no real solutions. (561)

Discriminant Theorem When a, b, and c are integers, with $a \neq 0$, either all three of the following conditions are true or none are true. 1. $b^2 - 4ac$ is a perfect square. 2. $ax^2 + bx + c$ is factorable over the set of polynomials with integer coefficients. 3. The solutions to $ax^2 + bx + c = 0$ are rational numbers. **(749)**

Distance between Two Points in a Coordinate Plane The distance AB between the points $A = (x_1, y_1)$ and $B = (x_2, y_2)$ in a coordinate plane is $AB = \sqrt{(x_2 - x_1)^2 + (y_2 - y_1)^2}$. **(507)**

Distributive Property of Multiplication over Addition For all real numbers a, b, and c, $c(a + b) = ca + cb$. **(66)**

Distributive Property of Multiplication over Subtraction For all real numbers a, b, and c, $c(a - b) = ca - cb$. **(67)**

Dividing Fractions Property For all real numbers a, b, c, and d, with $b \neq 0$, $c \neq 0$, and $d \neq 0$, $\frac{a}{b} \div \frac{c}{d} = \frac{a}{b} \cdot \frac{d}{c}$. **(258)**

Division Property of Equality For all real numbers a, b, and all real nonzero numbers c, if $a = b$, then $\frac{a}{c} = \frac{b}{c}$. **(113)**

domain of a function The set of possible values of the first (independent) variable. **(426)**

domain of a variable All the values that may be meaningfully substituted for a variable. **(28)**

double inequality An inequality of the form $a < x < b$. (The $<$ may be replaced with $>$, $\leq$, or $\geq$.) **(156, 227)**

E

elements (of a matrix) The objects in a rectangular array. **(622)**

empty set A set that has no elements in it, written as { } or ∅. **(584)**

endpoints (A) The points A and B in the segment $\overline{AB}$. (B) The coordinates of those points on a number line. **(156)**

Equal Fractions Property For all real numbers a, b, and k, if $b \neq 0$ and $k \neq 0$, then $\frac{a}{b} = \frac{ak}{bk}$. **(253)**

equivalent equations Equations with exactly the same solutions. **(139)**

equivalent expressions Expressions that have the same value for *every* number that can be substituted for the variable(s). **(22)**

equivalent formulas Two or more formulas in which every set of values that satisfies one of the formulas also satisfies the others. **(222)**

equivalent statements When an if-then statement and its converse are both true, then the antecedent and consequent are equivalent. **(784)**

equivalent systems Systems with exactly the same solutions. **(608)**

evaluating an expression The process of finding the numerical value of an expression. **(6)**

even integer (even number) An integer that can be written as $2n$, where n is an integer. **(802)**

event A set of possible outcomes. **(280)**

expected number The mean frequency of a given event that is predicted by a probability. **(697)**

exponent The number n in the power x^n. **(398)**

exponential decay A situation in which $y = bg^x$ and $0 < g < 1$. **(411)**

exponential growth A situation in which $y = bg^x$ and $g > 1$. **(404)**

exponential growth equation If the amount at the beginning of the growth period is b, the growth factor is g, and y is the amount after x time periods, then $y = b \cdot g^x$. **(405)**

exponential regression A method to determine an equation of the form $y = b \cdot g^x$ for modeling a set of ordered pairs. **(419)**

Extended Distributive Property To multiply two sums, multiply each term in the first sum by each term in the second sum, and then add the products. **(680)**

extremes The numbers a and d in the proportion $\frac{a}{b} = \frac{c}{d}$. **(301)**

F

factors (A) A number or expression that is multiplied. (B) If $ab = c$, then a and b are factors of c. **(15)**

factored form of a quadratic function A quadratic function $y = ax^2 + bx + c$ is in factored form when it is written as $y = a(x - r_1)(x - r_2)$. **(730)**

factoring The process of expressing a given number or expression as a product. **(75, 675)**

factorization The result of factoring a number or polynomial. **(676)**

Factor Theorem Let r be a real number and $P(x)$ be a polynomial in x. If $x - r$ is a factor of $P(x)$, then $P(r) = 0$; that is, r is an x-intercept of the graph of P. If $P(r) = 0$, then $x - r$ is a factor of $P(x)$. (**755**)

Factor Theorem for Quadratic Functions The x-intercepts of the graph of $y = a(x - r_1)(x - r_2)$ are r_1 and r_2. (**731**)

fact triangle A triangle in which any pair of numbers in the triangle can be added, subtracted, multiplied, or divided to produce the third number. (**105, 112**)

Fahrenheit scale A temperature scale, developed by Gabriel Fahrenheit, in which the freezing point of water is 32° and the boiling point of water is 212°. (**221**)

fair A situation in which each outcome has the same probability; also called *unbiased*. (**281**)

Fundamental Property of Similar Figures If two polygons are similar, then the ratios of corresponding lengths are equal. (**309**)

function A set of ordered pairs in which each first coordinate corresponds to *exactly one* second coordinate. (**426**)

function notation Notation to indicate a function, such as $f(x)$, and read "f of x." (**435**)

f(x) notation Notation indicating the value of a function f at x. When a function f contains the ordered pair (x, y), then y is the value of the function at x, and we may write $y = f(x)$. (**435**)

G

general formula for the height of a projectile over time Let h be the height (in feet) of a projectile launched from Earth's surface with an initial upward velocity v feet per second and an initial height of s feet. Then, after t seconds, $h = -16t^2 + vt + s$. (**546**)

generalization An if-then statement in which there is a variable in the antecedent and in the consequent. (**778**)

Generalized Addition Property of Equality For all numbers or expressions $a, b, c,$ and d: If $a = b$ and $c = d$, then $a + c = b + d$. (**601**)

general linear equation An equation of the form $ax + b = cx + d$, where $a \neq 0$. (**202**)

greatest common factor (GCF) For two or more integers, the greatest integer that is a common factor. For two or more monomials, the GCF is the product of the greatest common factor of the coefficients and the greatest common factor of the variables. (**675**)

growth factor In exponential growth or decay, the positive number which is repeatedly multiplied by the original amount. (**404**)

growth model for powering If a quantity is multiplied by a positive number g (the growth factor) in each of x time periods, then, after the x periods, the quantity will be multiplied by g^x. (**405**)

H

half-life The time it takes for one half the amount of an element to decay. (**414**)

half-plane In a plane, the region on either side of a line. (**381**)

horizontal line A line with the equation $y = k$, where k is a real number. (**189**)

I

if and only if A phrase used to connect equivalent if-then statements. (**785**)

if-then statement A statement that contains an antecedent and a consequent. (**778**)

independent variable A variable whose value does not rely on the values of other variables. (**426**)

inductive reasoning The process of arriving at a general conclusion (not necessarily true) from specific instances. (**776**)

inequality A mathematical sentence with one of the verbs < (is less than), > (is greater than), ≤ (is less than or equal to), ≥ (is greater than or equal to), or ≠ (is not equal to). (**155**)

initial height The starting height of a projectile. (**545**)

initial upward velocity The velocity of a projectile when it is first launched, assuming no gravity effects. (**545**)

input A number substituted for the independent variable in a function. (**221, 426**)

instance A special case of a general pattern. (**13**)

interest The amount that a bank or other financial institution pays on money in an account, based on a percentage of the principal. (**398**)

intersection of sets The set of elements in both set A and set B and written as $A \cap B$. (**227**)

interval The set of numbers between two numbers a and b, possibly containing a or b. (**156**)

inverse (of a matrix) For a matrix A, the matrix B such that AB and BA are the identity matrix. (**629**)

Irrationality of $\sqrt{n}$ Theorem If n is an integer that is not a perfect square, then $\sqrt{n}$ is irrational. (**819**)

irrational number A real number that is not a rational number. For example, the square roots of integers that are not perfect squares are irrational. (**817**)

J

justification A statement explaining why each step in a proof follows from preceding statements. (**787**)

L

least squares line The line whose squares of deviations from data set points are least. Also called the *line of best fit*. (**370**)

like terms Two or more terms in which the variables and corresponding exponents are the same. (**72**)

linear combination An expression of the form $Ax + By$, where A and B are fixed numbers. (**375**)

linear inequalities Inequalities of the form $Ax + By < C$ or $Ax + By \leq C$, where A, B, and C are constants. (> and ≥ can be substituted for < and ≤.) (**383**)

linear polynomial A polynomial of degree one. (**664**)

linear regression The fitting of a straight line through a given set of points according to specific criteria, such as least squares. (**369**)

linear term A term containing one variable with a power equal to 1. (**736**)

line of best fit A line whose equation is determined by the method of least squares and represents a linear relationship between data values. (**369**)

lowest terms (A) A fraction whose numerator and denominator have no common factors other than 1. (**761**) (B) A rational expression with no polynomial being a factor of both its numerator and denominator. (**762**)

M

markup A percent by which the original price of an item is raised. (**184**)

matrix (matrices) A rectangular array, such as $\begin{bmatrix} 3 & -4 \\ 15 & 0 \end{bmatrix}$. (**622**)

matrix form A way of expressing a system of equations using matrices. The matrix form for $\begin{cases} ax + by = e \\ cx + dy = f \end{cases}$ is $\begin{bmatrix} a & b \\ c & d \end{bmatrix} \cdot \begin{bmatrix} x \\ y \end{bmatrix} = \begin{bmatrix} e \\ f \end{bmatrix}$. The coefficient matrix is $\begin{bmatrix} a & b \\ c & d \end{bmatrix}$, the variable matrix is $\begin{bmatrix} x \\ y \end{bmatrix}$, and the constant matrix is $\begin{bmatrix} e \\ f \end{bmatrix}$. (**623**)

mean absolute deviation (m.a.d.) The average difference between individual measurements and the mean. (**48**)

means The numbers b and c in the proportion $\frac{a}{b} = \frac{c}{d}$. (**301**)

Means-Extremes Property For all real numbers a, b, c, and d (with $b \neq 0$ and $d \neq 0$), if $\frac{a}{b} = \frac{c}{d}$, then $ad = bc$. (**302**)

monomial A polynomial with 1 term. (**663**)

Multiplication Counting Principle If one choice can be made in m ways and a second choice can be made in n ways, then there are mn ways of making the first choice followed by the second choice. (**459**)

multiplication method for solving a system Given two equations in a system, applying the Multiplication Property of Equality to obtain another equation that is satisfied by the solution to the system. (**609**)

Multiplication Property of Zero For any real number a, $a \cdot 0 = 0 \cdot a = 0$. (**115**)

Multiplication Property of -1 For any real number a, $a \cdot -1 = -1 \cdot a = -a$. (**86**)

Multiplication Property of Equality For all real numbers a, b, and c, if $a = b$, then $ca = cb$. (**113**)

Multiplication Property of Inequality If $x < y$ and a is positive, then $ax < ay$. If $x < y$ and a is negative, then $ax > ay$. (**157**)

Multiplicative Identity Property of 1 For any real number a, $a \cdot 1 = 1 \cdot a = a$. (**116**)

Multiplicative Inverse Property For any real number a, where $a \neq 0$, $a \cdot \frac{1}{a} = \frac{1}{a} \cdot a = 1$. (**116**)

Multiplying Fractions Property For all real numbers a, b, c, and d, with $b \neq 0$ and $d \neq 0$, $\frac{a}{b} \cdot \frac{c}{d} = \frac{ac}{bd}$. (**252**)

N

n factorial (n!) The product of the integers from 1 to n. (693)

Negative Exponent Property For any nonzero b and all n, $b^{-n} = \frac{1}{b^n}$, the reciprocal of b^n. (474)

Negative Exponent Property for Fractions For any nonzero x and y and all n, $\left(\frac{x}{y}\right)^{-n} = \left(\frac{y}{x}\right)^n$. (475)

nonlinear system A system of equations or inequalities in which at least one of the equations or inequalities is nonlinear. (640)

nth power The number x^n is the nth power of x. (398)

null set See *empty set*. (584)

O

oblique A line that is neither horizontal nor vertical. (377)

odd integer (odd number) An integer that can be written as $2n + 1$, where n is an integer. (803)

odds of an event The ratio of the probability that an event will not occur to the probability that an event will occur. (284)

Opposite of a Difference Property For all real numbers a and b, $-(a - b) = -a + b$. (87)

Opposite of a Sum Property For all real numbers a and b, $-(a + b) = -a + -b = -a - b$. (86)

Opposite of Opposites Property For any real number a, $-(-a) = a$. (85)

opposites Two numbers that add to zero; also called additive inverses. (85)

order of operations The correct order of evaluating numerical expressions: perform operations within parentheses or other grouping symbols. Then evaluate powers from left to right. Next multiply or divide from left to right. Then add or subtract from left to right. (6)

origin The point (0, 0) on a coordinate graph. (44)

outcomes A result of an experiment. (280)

output A number that is returned by a function after it is evaluated. (221, 426)

P

parabola The curve that is the graph of an equation of the form $y = ax^2 + bx + c$, where $a \neq 0$. (526)

Parabola Vertex Theorem The graph of all ordered pairs (x, y) satisfying the equation $y - k = a(x - h)^2$ is a parabola with vertex (h, k). (716)

pattern A general idea for which there are many instances. (13)

P(E) The probability of event E or "P of E." (280)

percent (%) A number times $\frac{1}{100}$ or "per 100." (182)

percentile The pth percentile of a data set is the smallest data value that is greater than or equal to p percent of the data values. (292)

perfect square trinomial A trinomial that is the square of a binomial. $a^2 + 2ab + b^2 = (a + b)^2$ and $a^2 - 2ab + b^2 = (a - b)^2$. (687)

period of a pendulum The time it takes a pendulum to complete one swing back and forth. On Earth, the formula $p = 2\pi\sqrt{\frac{L}{32}}$ gives the time p in seconds for one period in terms of the length L (in feet) of the pendulum. (501)

permutation An ordered arrangement of letters, names, or objects. (691)

± notation (A) $\pm x$ means (x or $-x$). (B) $a \pm b$ means ($a + b$ or $a - b$). (553)

point-slope form An equation of a line in the form $y - k = m(x - h)$, where m is the slope and (h, k) is a point on the line. (357)

polynomial An expression that is either a monomial or a sum of monomials. (663)

polynomial in x A sum of multiples of powers of x. (657)

population The set of individuals or objects to be studied. (302)

power An expression written in the form x^n. (398)

Power of a Power Property For all m and n, and all nonzero b, $(b^m)^n = b^{mn}$. (466)

Power of a Product Property For all nonzero a and b, and for all n, $(ab)^n = a^n b^n$. (481)

Power of a Quotient Property For all nonzero a and b, and for all n, $\left(\frac{a}{b}\right)^n = \frac{a^n}{b^n}$. (482)

prime polynomial A polynomial that cannot be factored into polynomials of lower degree. (676)

polynomial over the integers A polynomial with integer coefficients. (738)

principal Money deposited in an account. (398)

probability of an event A number from 0 to 1 that measures the likelihood that an event will occur. (280)

probability distribution The set of ordered pairs of outcomes and their probabilities. (281)

Probability Formula for Geometric Regions Suppose points are selected at random in a region and some of that region's points represent an event E of interest. The probability $P(E)$ of the event is given by $\frac{\text{measurement of region in event}}{\text{measure of entire region}}$. (296)

Product of Powers Property For all m and n, and all nonzero b, $b^m \cdot b^n = b^{m+n}$. (465)

Product of Square Roots Property For all nonnegative real numbers a and b, $\sqrt{a} \cdot \sqrt{b} = \sqrt{ab}$. (498)

proof argument A sequence of justified conclusions, starting with the antecedent and ending with the consequent. (789)

proportion A statement that two fractions are equal. (301)

Pythagorean Theorem In any right triangle with legs of lengths a and b and hypotenuse of length c, $a^2 + b^2 = c^2$. (492)

Q

quadratic equation An equation that can be written in the form $ax^2 + bx + c = 0$ with $a \neq 0$. (552)

Quadratic Formula If $ax^2 + bx + c = 0$ and $a \neq 0$, then $x = \frac{-b \pm \sqrt{b^2 - 4ac}}{2a}$. (553)

quadratic polynomial A polynomial of degree 2. (664)

Quotient of Powers Property For all m and n, and all nonzero b, $\frac{b^m}{b^n} = b^{m-n}$. (469)

Quotient of Square Roots Property For all positive real numbers a and c, $\frac{\sqrt{c}}{\sqrt{a}} = \sqrt{\frac{c}{a}}$. (499)

R

radical sign (A) ($\sqrt{}$) The symbol for square root. (489) (B) ($\sqrt[3]{}$) The symbol for cube root. (493)

radicand The quantity under the radical sign. (499)

randomly (chosen) Every member of a population has an equal chance of being chosen. (302)

range (A) The difference between the maximum value M and minimum value m of a data set. (48) (B) The set of possible values of the second (dependent) variable. (426)

rate The quotient of two quantities with different units. (263)

rate of change The difference of values of a quantity divided by the amount of time between the values. The rate of change between points (x_1, y_1) and (x_2, y_2) is $\frac{y_2 - y_1}{x_2 - x_1}$. (328)

rate unit The unit of a rate. (328)

ratio A quotient of two quantities with the same units. (274, 275)

rational expression A quotient of two polynomials. (761)

rational number A number that can be expressed as a simple fraction. (816)

ratio of similitude The ratio of the lengths of corresponding sides of two similar figures. (309)

reciprocal rates Two rates in which the quantities are compared in both orders. (264)

reflection-symmetric The property held by a figure that coincides with its image under a reflection over a line. (526)

Related Facts Property of Addition and Subtraction For all real numbers a, b, and c, if $a + b = c$, then $b + a = c$, $c - b = a$, and $c - a = b$. (106–107)

Related Facts Property of Multiplication and Division For all nonzero real numbers a, b, and c, if $ab = c$, then $ba = c$, $\frac{c}{b} = a$, and $\frac{c}{a} = b$. (113–114)

relation Any set of ordered pairs. (428)

relative frequency The ratio of the number of times an event occurs to the total number of possible occurrences. (280, 289)

Repeated Multiplication Property of Powers When n is a positive integer, $x^n = x \cdot x \cdot \ldots \cdot x$ for n factors. (398)

S

sample A subset taken from a set of people or things. (302)

scatterplot A two-dimensional coordinate graph of individual points. (**27**)

scientific notation A number represented as $x \cdot 10^n$, where n is an integer and $1 \leq x < 10$. (**460**)

semiperimeter Half the perimeter of a figure. (**807**)

sequence A collection of numbers or objects in a specific order. (**20**)

simple fraction A fraction with integers in its numerator and denominator. (**816**)

skewed left A distribution in which the lower half of the values extends much farther to the left than the upper half, leaving a tail on the left. (**51**)

skewed right A distribution in which the upper half of the values extends much farther to the right than the lower half, leaving a tail on the right. (**51**)

slope The rate of change between points on a line. The slope of the line through (x_1, y_1) and (x_2, y_2) is $\frac{y_2 - y_1}{x_2 - x_1}$. (**334**)

slope-intercept form An equation of a line in the form $y = mx + b$, where m is the slope and b is the y-intercept. (**350**)

Slopes and Parallel Lines Property If two lines have the same slope, then they are parallel. (**616**)

solution to an equation Any value of a variable that makes an equation true. (**135**)

solution to a system In a system of equations with two variables, the solution is all ordered pairs (x, y) that satisfy all equations in the system. (**582**)

square The second power of a number x, or x^2. (**488**)

Square of the Square Root Property For any nonnegative number x, $\sqrt{x} \cdot \sqrt{x} = \sqrt{x^2} = x$. (**490**)

square root If $A = s^2$, then s is a square root of A. (**489**)

square term The terms containing a variable with a power equal to 2. (**736**)

squaring function A function defined by $y = x^2$. (**426**)

standard form of an equation of a line An equation in the form $Ax + By = C$, where A, B, and C are constants. (**375**)

standard form for a polynomial A polynomial written with the terms in descending order of the exponents of its terms. (**658**)

standard form of a quadratic equation An equation of the form $ax^2 + bx + c = 0$, where $a \neq 0$. (**555**)

standard window The common view on a graphing calculator. (**34**)

Subtraction Property of Equality For all real numbers a, b, and c, if $a = b$, then $a - c = b - c$. (**106**)

symmetric distribution Data that are centered around one point and in which the values on the left and right sides are roughly mirror images. (**51**)

system A set of equations or inequalities separated by the word *and* that together describe a single situation. (**582**)

T

tax rate The ratio of the tax to the amount being taxed. (**275**)

term A number, variable, or product of numbers and variables. (**15, 20, 663**)

Transitive Property of Equality For any real numbers a, b, and c, if $a = b$ and $b = c$, then $a = c$. (**10**)

trinomial A polynomial that has three terms. (**663**)

trivial factors In every expression, the factors 1 and the expression itself. (**675**)

2 × 2 identity matrix The matrix $\begin{bmatrix} 1 & 0 \\ 0 & 1 \end{bmatrix}$. (**626**)

U

unbiased A situation in which each outcome has the same probability; also called *fair*. (**281**)

uniform distribution A distribution that has roughly the same quantity for all events. (**51**)

union of sets The set of elements in either set A or set B (or in both) and written as $A \cup B$. (**228**)

Unique Factorization Theorem for Polynomials Every polynomial can be represented as a product of prime polynomials in exactly one way, disregarding order and integer multiples. (**677**)

V

value of a function The output of a function obtained for a given first variable. (**426**)

variable A letter or other symbol that can be replaced by any number (or other object) from a set. (6)

variable matrix In the matrix equation $\begin{bmatrix} a & b \\ c & d \end{bmatrix} \cdot \begin{bmatrix} x \\ y \end{bmatrix} = \begin{bmatrix} e \\ f \end{bmatrix}$, the variable matrix is $\begin{bmatrix} x \\ y \end{bmatrix}$. (622)

vertex The point of intersection of a parabola with its axis of symmetry. (526)

vertex form of an equation for a parabola An equation of the form $y - k = a(x - h)^2$, where (h, k) is the vertex. (716)

vertical line A line with the equation $x = h$, where h is a real number. (189)

W

window The part of a coordinate grid that is visible on a graphing calculator. (33)

X

x-intercept The *x*-coordinate of a point where a graph intersects the *x*-axis. (357)

Xmax The greatest *x*-value (right edge) displayed on the window screen of a graphing calculator. (34)

Xmin The least *x*-value (left edge) displayed on the window screen of a graphing calculator. (34)

Xscl The *x*-scale of a graphing calculator. (34)

Y

y-intercept The *y*-coordinate of a point where a graph intersects the *y*-axis. (350)

Ymax The greatest *y*-value (top edge) displayed on the window screen of a graphing calculator. (34)

Ymin The least *y*-value (bottom edge) displayed on the window screen of a graphing calculator. (34)

Yscl The *y*-scale of a graphing calculator. (34)

Z

Zero Exponent Propery If x is any nonzero real number, then $x^0 = 1$. (405)

Zero Product Property For any real numbers a and b, if $ab = 0$, then either $a = 0$, $b = 0$, or both a and b equal 0. (115)

Index

A

"*a* plus (or minus) *b*, quantity squared," 685
abbreviated proof, 789–790
absolute difference in elevations, 43
absolute value, 43
    of deviation in altitude, 43
    of deviation of guess, 234
    expressions with, 44
    solving equations, 234–235, 241
    solving inequalities, 236–237
acronyms, 458–459, 691
Activities, 20–21, 33–37, 44, 100–102, 131, 140, 156, 168, 191–192, 222–223, 254, 274, 297, 302–303, 309, 348–349, 377, 419–420, 435, 440–442, 464, 469, 483, 488, 490–491, 497, 498, 527, 539, 540, 544–545, 671, 676, 687, 688, 714, 716–718, 724, 730, 732, 738–739, 751–752, 754–755, 757, 761, 803, 805, 809–810
actual frequencies, 697–700
addition
    fact triangle, 105
    of fractions, 74
    of inequalities, 162
    of like terms, 72–74
    perfect squares: square of a sum, 685–686
    properties
        additive identity, 108
        additive inverse, 109
        associative, 10
        commutative, 16
        distributive, of multiplication over, 66
        of equality, 106
        of equality, generalized, 601
        of inequality, 162, 211, 230
        opposite of a sum, 86
        related facts, of addition and subtraction, 106–107
    of rational expressions, 764–765
    representing integers as sum of powers of 2 or 3, 703–704
    solving a system of equations, 601–604
    special numbers for, 108–109
    of square root(s), 503
    of two even numbers, 804
addition method for solving a system, 602
Addition Property of Equality, 106
    generalized, 601
Addition Property of Inequality, 162, 211, 230
Additive Identity Property, 108
additive inverse, 85, 109
Additive Inverse Property, 109
add-ons, in percents, 183–184
algebra, word origin, 797
algebraic definition
    of division, 7
    of subtraction, 7
algebraic descriptions, 5
algebraic expression, 6
algebraic fractions, See also *fractions*.
    complex, simplifying, 259–260
    defined, 252
    division of, 258–259
    multiplication of, 252–253
    properties
        dividing fractions, 259–260
        equal fractions, 253–254
        multiplying fractions, 252–253
    simplifying, 254
algebraic order of operations, 6
algebraic processes, solving
    $ax + b = cx + d$, 203–204
algebraic properties, See also *properties*.
    comparing with tables and graphs, 198
algebraic riddle, for Diophantus, 173
algorithm, word origin, 797
al-Karkhi (al-Karaji), 55
al-Khwarizmi, Muhammad bin Musa, 797
always true, 216–218
and
    in compound sentence, 227
    solving inequalities, 230
    in system of equations, 582
angles
    in polygons, sum of measure of interior angles, 307
    supplementary, 278
annual yield, 398, 400
antecedent, 778
applications *(Applications are found throughout the text. The following are selected applications.)* See also *mathematics in other disciplines*.
    acronyms in different languages, 458–459, 461, 463
    adult height, 272, 307, 340
    age, guessing, 188
    air conditioning, 40
    animals
        beetles, 779
        cricket chirping, 362–363, 364
        elephants, similar figures of, 308, 310, 311
        giant tortoises, 543
        growth of, 358
        land speed, 134
        nesting count for bald eagles, 445
        non-native species, 404, 407
        pigeon messenger system, 600
        snake lengths, 277
        wildlife management, 305, 442, 444
    appliances and energy, 174
    astrology, 704
    bamboo, 702
    bicycling, 208, 275
    birthdays, 660
    birthrates, 280, 284, 326, 697
    blood types, 283, 285, 428
    body temperature, 229
    books, 270, 271, 346, 445
    bouncing ball, 419–423
    bricklaying, 120
    buying and selling CDs, 174
    carbon-14 dating, 418
    card trick, 64–65, 794
    chairs, stacking, 343
    charitable contributions, 598
    coffee consumption, 25, 84, 472
    cog railway, 355
    coins
        collecting, 128, 332
        counterfeit, 646
        guessing, 594, 633
    combination locks, 535
    computers
        Moore's Law, 690
        passwords, 480
    consumer spending, 154
    debt per capita, 268
    depth of wells, 821
    drought and inequalities, 164
    Earth areas, 299
    earthquakes, 447
    elections, 220, 304, 614
    elevator capacity, 383–385, 431
    employment, 760
    exercise, 166, 272, 306
        calories burned, 551
        rollerblading, 516
        folding paper, 465

football players in high school, 110
frostbite time, 438
game of *Dish*, 462
gardening, 256, 299
Gateway Arch, 572
GDP per capita, 470, 472
gratuities (tips), 183
gross domestic product (GDP), 470
hardware, 627
hopscotch, 673
hot air balloons, 213, 353, 599
house painting, 78, 276, 278
initial upward velocity, 11
insulin and diabetes, 414, 416
knitting, 458
Koch snowflake, 447
lake depth, 262
lawn services, 67
light bulbs, 180–181
magazines in school library, 14
manufacturing T-shirts, 585
map distances, 314
marriage age, 387
mountain carvings, 312
mountains, 42, 156, 722
movies, 495
music
    jazz bands, 692
    marching bands, 610–611, 613, 620
Neptune volume, 18
Old Faithful geyser, 365
paper towels: price vs. absorbency, 387
parachutes, 430, 577
passports, 787
penicillin, 365, 408
pet ownership, 294, 404
planets, volumes of, 513
population
    foreign cities, 331
    foreign countries, 431, 470, 472, 785
    U.S. cities, 324–325, 327, 358, 592, 651
    U.S. states, 215, 356, 406–407, 409
    world, 472–473
population density, 267
population questions, using tables and graphs, 199, 219
postal rates, 122
puzzles, See *puzzles*.
racing, 55, 152, 266, 271
raffles, 294, 695
rainfall in Hawaii, 306
redwood trees, 138

Richter scale, 447
robot competitions, 505, 510
sale prices, 184, 186
saving vs. spending, 200
savings accounts, 398, 399, 400, 402
school supplies, 570
science fair projects, 641
shaded squares, 479
shoe sizes, 365
slope of road (grade), 346
solutions of chemicals, 651
strength of spaghetti, 175
Strontium-90 half-life, 417
Sudoku puzzle, 830
targets and probabilities, 296, 297, 298
taxicab fares, 76, 590, 592
temperature
    conversions, 221–222
    outdoors, 143, 194
    of oven, function of time, 437
    rate of change, 264, 394
tests
    standardized, 460
    types of questions on, 614
theater arts, 536, 620, 638
tornadoes, 701
toys, 586
tree growth, 210, 213, 240
triangle height, 814
triangle side lengths, 17
trip planning, 240, 295, 350, 353
unemployment rate, 347
water emergency, 164
water flood, 130, 132
water intake, 270
water parks, 15
weight percentiles, 293
wind chill, 438
Arab mathematicians, 797
Archimedes, 316
area
    of geometric shapes, See *formula(s)*.
    probabilities from, 296
    providing proofs for Pythagorean Theorem, 823
    of squares, 488
Area Model for Multiplication, 669, 680
arithmetic, 790
Arrangements Theorem, 459
Associative Property of Addition, 10
Associative Property of Multiplication, 9
assumed property, 790
axis of symmetry, 526

B

Babylonian mathematicians, 492, 795–796
Babylonian method, 517
bar graph, 47
base, 398
    dividing powers with different bases, 471
    dividing powers with same base, 469
    multiplying powers with different bases, 466
    multiplying powers with same base, 464–465
Bhaskara's proof, 824
binomials, 663, See also *monomial; polynomials*.
    difference of two squares, 687–688
    perfect squares: square of a difference, 687
    perfect squares: square of a sum, 685–686
bisection method, 769
Bobbitt, Shannon, 606
boundary line, 381
boundary point, 155
brackets, on matrix, 622
Braille letters, 518
break-even point, 196, 584–585
Buffon, George, 315
Buffon's needle, 315

C

Cagnotto, Tania, 554
calculators, See also *computer algebra system (CAS); graphing calculators*.
    entering complex fractions, 260
    evaluating with technology, 8–9
    finding inverse of coefficient matrix, 630
    INTERSECT command, 585
calendars, 19
capture-recapture method, 303, 305
card tricks, 64–65
Cardano, Girolamo, 797
Carlsbad Caverns, 42, 45
CAS, See *computer algebra system (CAS)*.
catching up, 590
catenaries, 565, 572
Celsius, Anders, 221
Celsius scale, 221–222, 224
    linear relationship with Fahrenheit, 366
centigrade scale, 221

Chamberlain, Wilt, 394
change, See *rate of change*.
Chapter Review, 60–63, 125–127, 178–179, 245–249, 320–323, 392–395, 452–455, 521–523, 576–579, 650–653, 708–711, 773–775, 833–835, See also *Review questions*.
check, of a proof, 790–791
Chinese mathematicians, 492
chi-square statistic, 697–700
    calculating, 698
    reading table, 699
choosing from *n* objects repeatedly, 459–460
chunking, 533
circle
    circumference, 835
    equation, 644
circular permutation, 695
clearing decimals, 170–171
clearing fractions to solve equations, 167–170
closed under an operation, 802
closure properties, 802
codes, encoding and decoding using formulas, 56
coefficient, 73
    distinguished from exponents, 74
    negative, solving inequalities with, 164
    positive, solving inequalities with, 163
coefficient matrix, 622
    inverse of, 629
coincide, 618
coincident lines, 617–618
collinear points, 130
common denominator, 74
Commutative Property of Addition, 16
Commutative Property of Multiplication, 16
complementary events, 283
complements, 283
complete factorization, 677
completing the square, 723–725
complex fractions, 259–260
    infinite repeating continued, 768
complex numbers, 797
compound inequalities, 227–231
    intersection and union of sets, 227–228
    intervals, 229
    solving with *and* and *or*, 230
compound interest, 399
    calculation of, 399–400

formula, 405
formula, with negative exponents, 476
and polynomials, 659
Compound Interest Formula, 400
compound sentences, 227–231
computer algebra system (CAS), 98
    EXPAND command, 168, 203
    EXPAND feature, 681
    multiplying both sides of an equation, 167
    SIMPLIFY command, 168
    SOLVE command, 597, 612
    solving general linear equations, 202–203
    solving linear equations, 140
    solving systems of equations, 597–598, 612
    string variable, 435
    substitution, 435
    testing for equivalence, 100–102
computers
    memory, 422
    processing speed, 420–421
    value of, 436
conditional probability, 283
cone, volume of, 222, 667
conjectures, 830
consequent, 778
consistency, measuring with mean absolute deviation, 53
constant function, 432
constant growth, graph of, 396–397, 415
constant matrix, 622
constant of variation, 353
constant slope of line, 334–335
constant term, 736
constant-decrease situations, 333
    compared to exponential decay, 439–442
    patterns in, 130
constant-increase situations, 356–357
    compared to exponential growth, 439–443
    patterns in, 130
continuous graph, 130
converse, as check of proof, 790–791
converses of true statements, 784–785
conversion of temperatures, 221–222
conversion rates, 269–270
coordinate plane
    distance between two points in, 506–507
    moving in, 518

coordinates, 27
cord of wood, 661
cosine, 279
cost equation, 584–585
counterexample, 23, 512, 779
counting
    diagonals in convex polygon, 567
    multiplication principle, 458–459
crossword puzzles, and mathematics, 518
cryptography, public-key, 769
Cseh, Laszlo 582
Cube of the Cube Root Property, 493
cube of *x*, 493
cube roots, 493
cubic polynomial, 756
Cyclopedia of Puzzles (Loyd), 209
cylinders
    surface area of, 667, 753, 800
    volume of, 486, 800

D

Darwin, Charles, 662
data, fitting a line to, 368–370
data set
    mean absolute deviation of, 48–50
    mean of, 47–48
    range of, 48
Death Valley, California, 143
decay, See *exponential decay*.
decimal expansion, 819
decimals
    clearing, 170–171
    finite, 816
    multiplying by power of 10, 671–672
    ratios as, 275
    repeating finding simple fractions for, 607
    rational numbers vs. irrational, 816, 818
decreasing function, 431
deduction, 789
define a variable, 14
defined property, 790
degree of a monomial, 664
degree of a polynomial, 664
    checking operations with, 665
density of object, calculation of, 316
dependent variable, 426
depreciation, 412
Descartes, René, 144, 797
deviation
    between actual and predicted, 370
    in altitude, 43
    of *a* from *e*, 697
    from the mean, 191–192

mean absolute, 48
diagonal game, 388
diagonals, in polygons, 567, 735
dice, tossing, 281–282, 289–290, 300, 747
difference, See *subtraction*.
Difference of Two Squares, 687–688
digits, See *numbers*.
dimensions of matrix, 622
Diophantus, 173
direct variation, 352–353
discount rate, 275
discounts, in percents, 183–184, 186
discrete points, 130
discriminant of quadratic equation, 560–561, 749
Discriminant Property, 561
Discriminant Theorem, 749, 751–752
dissection proof, 828
distance, See also *absolute value*.
   along vertical and horizontal lines, 505–506
   stopping, of a car, 537, 540, 606
   between two points in a plane, 506–507
   between two points in space, 510
distance formula, 270
   for falling objects, 531, 533, 545
   between two points in a plane, 507
   between two points in space, 510
distribution, shape of, 50–51
Distributive Property
   extended, 680
   multiplying monomial by polynomial, 669, 670
   removing parentheses with, 150–152
   reversing sides of, 72
   solving equations with, 149–152
Distributive Property of Multiplication over Addition (DPMA), 66
Distributive Property of Multiplication over Subtraction (DPMS), 67
Dividing Fractions Property, 258–259
divisibility properties
   depending on rightmost digits of numbers, 810–811
   project challenge, 829
   proofs of, 802–806
   tests based on sum of digits, 811–812
divisible by 2, 802–803
divisible by 3, 805
divisible by 4, 805

division
   algebraic definition of, 7
   of inequalities, 211
   of polynomials, 703
   power of a quotient, 482–483, 485
   of powers
      with different bases, 471
      with same base, 469
      with same base and negative exponents, 476
   properties
      dividing fractions, 258–259
      of equality, 113
      powers of a quotient, 482–483
      quotient of powers, 469–470
      quotient of square roots, 499
      related facts, of multiplication and division, 113–114
   rate calculation, 263–264
   of rates, 269–270
   situations for using, 250–251
   of square roots, 497–502
   using distributive property, 68
   by zero, 113, 114–115, 265
division expressions, 7–8
Division Property of Equality, 113
Dobbs, W.J., 827
domain of a function, 426
   not set of numbers, 428
domain of the variable, 28, 29
dot plot, 47, 50
double inequalities, 156, 227
Dydek, Margo, 606
dynamic graphing software, 348

E

Eames, Charles and Ray, 448
earned run average (ERA), 306
Ederle, Gertrude, 371
electromagnetic spectrum, 456–457
elements of matrix, 622
elevations
   measuring, 42–43
   rate of change, 366
empty set, 584, 638
endpoints of an interval, 156
energy efficient ration (EER), 40
Enigma cypher machine, 56
equal fractions, 253–254
Equal Fractions Property, 253–254
equality properties
   addition, 106
   addition, generalized, 601
   division, 113
   multiplication, 113, 167
   subtraction, 106
equally likely outcomes, 282, 296

equations, See also *solving equations; system of equations*.
   of absolute value, 234–235, 241
   adding and subtracting, 646
   for circle, 644
   clearing fractions in, 169
   collecting like terms, 149–150
   equivalent, 139
   exponential growth, 405, 411
   for falling objects, 531, 533, 545
   of form $ax + b = c$, 140, 144–148
   of form $ax + b = cx + d$, 202–206
   of form $p \cdot q = r$, 182–183
   of form $y = ax^2$, 527
   of form $y = b \cdot g^x$, 411, 414
   of form $y = mx + b$, 350
   of form $y = x^2$, 526
   general linear, 202
   for horizontal lines, 188–189
   linear, 129
   for lines
      finding from graphs, 351–352
      finding slope with, 336–337
      intercept form of, 380
      point-slope form, 356–357
      rewriting in slope-intercept and standard form, 376–377
      slope-intercept form, 350–351
      standard form, 375–376
      summary of three common forms, 608
      through two points, 361–363
   with no solution, 616–617
   for parabolas
      converting from standard form to vertex form, 724–725
      of form $ax^2 = b$, 532–534
      of form $y = ax^2 + bx + c$, 537–538, 552–553
      related to factoring quadratic expressions, 749–750
      vertex form, 714–719
   for projectile height over time, 545–548
   for projectile paths, 544–545
   quadratic, 553, 555
   setting up with ratios, 276
   situations leading to, 136
   solving as proofs, 788–791
   solving percent problems with, 182–184
   standard form of
      for a line, 375–376
      for a polynomial, 658
      quadratic, 555
   variations of form $ax + b = c$, 146

for vertical lines, 189
equivalence, 91
   testing for, 91–94
   using CAS to test for, 100–102
equivalent equations, 139
   quadratic equation and Quadratic Formula, 553
   rewriting equations of lines in slope-intercept and standard form, 376
equivalent expressions, 22
   finding, 20–23
   solving equations with, 139–140
   summary of, 712
   with technology, 98–102
equivalent formulas
   in geometry, 98–99
   solving, 221–224
equivalent statements, 785
equivalent systems of equations, 608
Eratosthenes, 768
estimating square roots, 517
Euclidean algorithm (Euclid's algorithm), 830
Euler, Leonhard, 435
evaluating expressions in formulas, 9
evaluating the expression, 6
even integer (even number), 802–803
event, 280, See also *probability of an event, P(E)*.
expanding a fraction, 68–69
expanding the expression, 66
expected frequencies, 697–700
expected number, 697
Exploration questions, 12, 19, 26, 32, 41, 46, 54, 71, 78, 84, 90, 97, 104, 111, 120, 134, 138, 143, 148, 154, 161, 166, 173, 187, 195, 201, 209, 215, 220, 226, 233, 239, 257, 262, 268, 273, 279, 288, 295, 300, 307, 314, 332, 340, 347, 355, 360, 367, 373, 380, 386, 403, 410, 418, 424, 431, 438, 446, 463, 468, 473, 480, 487, 496, 504, 510, 516, 531, 536, 543, 551, 557, 564, 570, 588, 593, 600, 607, 615, 621, 628, 634, 639, 644, 662, 668, 674, 679, 684, 690, 696, 702, 722, 728, 735, 741, 747, 753, 760, 767, 783, 787, 794, 801, 808, 815, 822, 828
exponential decay, 411
   compared to constant-decrease, 439–442

   examples of, 411–414
   graphs of, 415
   modeling, 419–420
   points on graph, 413
exponential function, 432
exponential growth, 397, 404
   compared to constant-increase, 439–443
   equation form, 411
   graphs of, 405–406, 407, 415
   modeling, 420–421
exponential growth curves, 405
exponential growth equation, 405
exponential population growth, 406–407
exponential regression, 419
exponents, 398
   distinguished from coefficients, 74
   on fractions, 517
   negative
      for fractions, 475–477
      value of power with, 474–475
expressions, See also *equivalent expressions; quadratic expressions; rational expressions.*
   with absolute value, 44
   comparing with technology, 36
   evaluating, 6
   evaluating in formulas, 9
   expanding with distributive property, 66
   picturing, 27–29
   undefined, 265
Extended Distributive Property, 680
extremes, 301–302
eyeballing a line of fit, 368–369

F

fact triangle, 105, 112, 116
   for product of powers property, 465
   for quotient of powers property, 470
   square roots, products and quotients, 499
Factor Theorem, 755
Factor Theorem for Quadratic Functions, 731
factored form of polynomials, 754–757
   converting to standard form, 755–756
factored form of quadratic function, 729–733
   display of *x*-intercepts, 730–731
   theorem for, 731

factorial notation, 693–694
factoring, 75, 675
   monomials, 675–678
   quadratic trinomials of form $ax^2 + bx + c$, 742–745
   related to solving equation, 749–750
   quadratic trinomials of form $x^2 + bx + c$, 736–739
   and rational expressions, 761–765
factorization, 676
   complete, 677
factors, 15
   related to *x*-intercepts, 754–755
   trivial, 675
Fahrenheit, Gabriel, 221
Fahrenheit scale, 221–222, 224
   linear relationship with Celsius, 366
fair situation, 281
false sentences, 216–218
   if-then statements, 779–780
Fermat's Last Theorem, 830
50th percentile, 291
figurate numbers, 56
fitting a line to data, 368–370
fixed costs, 584–585
Fleming, Alexander, 408
f(x) notation, 432–435
focus of parabola, 571
force of gravity, 545
formula(s)
   adult height, 272
   area
      of cylinder surface, 667, 753, 800
      of isosceles right triangle, 486
      of octagon, 97
      of rectangle, 823
      of right triangle, 823
      of square, 823
      of trapezoid, 667, 823
      of triangle, 823
   compound interest, 400, 405
   diagonals in polygon with *n* sides, 567, 735
   distance, 270
      for stopping a car, 537, 606
      between two points in a coordinate plane, 507
   equivalent, 221–224
   for height of projectile over time, 11, 545–546
   for inverse of 2 × 2 matrix, 634
   perimeter of triangle, 667
   period of pendulum, 501

probability for geometric regions, 296
Quadratic, 552–553
square root of positive integer, 55
sum
of cubes of integers, 766
of integers, 572, 766
of interior angles of polygon, 307
of squares of integers, 766
volume
of cone, 222, 667
of cylinders, 486, 800
of rectangular solid, 32, 484
of sphere, 18, 513
Formula 1 game, 55
fractions, See also *algebraic fractions.*
addition of, 74
algebraic, division of, 258–259
algebraic, multiplication of, 252–253
choosing a multiplier for clearing, 167–168
clearing, in equations, 167–169
clearing, in inequalities, 170, 277
complex, simplifying, 259–260
equal, 253–254
expanding, 68–69
with exponents, 517
infinite repeating continued, 768
negative, 265
with negative exponents, 476
power of a quotient property, 483, 485
product of powers property, 468, 469
ratios as, 275
for repeating decimals, 607
rule for multiplying, 4
simple, 816
frequency
in birthrates, 280, 284, 697
deviation of actual from expected, 697–700
frequency analysis, 56
frequency table, 47
function notation, 432–435
representing polynomials in variables, 664
functions, See also *quadratic function.*
constant, 432
decreasing, 431
domain, 426
domain or range is not set of numbers, 428
exponential, 432
linear, 432

logistic, 448
polynomials of higher degree, 754–757
range, 426
reciprocal, 427, 447
squaring, 425–426
types of, 432
Fundamental Property of Similar Figures, 309

G

Galilei, Galileo, 501, 524
Garfield, James, 824
Garfield's proof, 824
Gauss, Carl Friedrich, 373
general linear equation, 202
generalization, 778
Generalized Addition Property of Equality, 601
geometric regions, probability formula, 296
geometry, See also *areas.*
equivalent formulas in, 98–99
similar figures, 308–311
GFC (greatest common factor), 675–676
given information, 790
Goddard, Robert H., 11
Gödel, Kurt, 829
Gougu Theorem, 492
grade of a road (gradient), 346
graphing calculators, 8, 33
equivalent formulas, 223–224
INTERSECT command, 206
mean absolute deviation procedure, 49
Now/Next method, 406
standard window, 34
table feature, 92
tables from, 36
TRACE command, 35, 136, 224
VALUE command, 35
window, 33
Xmax (x-max), 34
Xmin (x-min), 34
Xscl (x-scale), 34
Ymax (y-max), 34
Ymin (y-min), 34
Yscl (y-scale), 34
ZOOM menu, 34
graphing linear inequalities, 383
involving horizontal or vertical lines, 381–382
involving oblique lines, 382–383
graphing lines, See *line(s), equations for.*

graphing with technology, 33–35
graphs
compared to tables and algebraic properties, 198
comparing constant increase/decrease and exponential growth/decay, 440
connected, 28–29
of exponential growth, 405–407
and growth factor, 414–415
of linear inequalities
and horizontal or vertical lines, 381–382
and oblique lines, 382–383
of lines, See *line(s), equations for.*
on a number line, 155–156
of parabolas, 526–527, 532, 538
vertex form of equation, 714–719
parallel lines, 616–617
of polynomial functions of higher degree, 754–757
slope calculation, 335
solving a system of equations, 583–584
solving by comparing values, 197–198
solving equations with, 135
solving inequalities with, 211
solving quadratic equations, 558–559
of squaring function, 425
systems of inequalities, 635–637
of $y = ax^2 + bx + c$, 537–538
of $y = x^2$, 426, 526
gravity force, 545
greater than, 155
greatest common factor (GFC), 675–676
Greek mathematicians, 173, 316, 488–490, 492, 768, 817, 830
growth factor, 404
and exponential decay, 411
and graphs, 414–415
Growth Model for Powering, 405
Guadalupe Mountains, 42
Guerrero, Vladimir, 549

H

half-life (decay), 414, 417, 418
half-planes, 381
Hancock, Winfield Scott, 824
heights, finding with similar figures, 311

Hero of Alexandria, 807
Heron's formula (Hero's formula), 807
horizontal distance, 506
horizontal lines
   distances along, 505–506
   equations for, 188–189
   graphing linear inequalities, 381–382
   and linear patterns, 190
   slope of, 335, 343
hours per mile, 264
Huygens, Christiaan, 501
hypothesis, 830

I

I Can Guess Your Age puzzle, 188
identity
   2×2 identity matrix, 626, 629–630
   3×3 identity matrix, 628
   properties
      additive, 108
      multiplicative, 116
if and only if, 785
if-then statements, 778–781
   converse of, 784–785
   defined, 778
   determination of true or false, 778–780
   examples, 776–777, 778
   in games, 830
   in mathematics, 780
   putting into if-then form, 781
   Venn diagrams, 779–780
Incompleteness Theorems, 829
independent variable, 426
Indian mathematicians, 492
inequalities, See also *compound inequalities; inequality.*
   of absolute value, 236–237
   clearing fractions in, 170, 277
   compound, 227–231
   double, 156, 227
   of form $ax + b < c$, 162–164
   of form $ax + b < cx + d$, 210–212
   graphs with horizontal or vertical lines, 381–382
   graphs with oblique lines, 382–383
   linear, 383
   on a number line, 155–156
   solving, 157–159
   systems of, 635–637
inequality, 155
   changing the sense of, 157
   properties
      addition, 162, 211, 230
      multiplication, 156–157, 159, 211, 230

inequality sign, 156, 211
infinite repeating continued fractions, 768
initial height, 546
initial upward velocity, 545
input
   in formulas, 221
   in a function, 426, 434
instances, 13
integers, 29
   properties of, 809–813
   representing as sum of powers of 2, 703
   representing as sum of powers of 3, 704
   sum of cubes of, from 1 to n, 766
   sum of, from 1 to n, 572, 766
   sum of squares of, from 1 to n, 766
   systems with integer solutions, 645
intercept form of equation for a line, 380
   slope-intercept form, 350
interest, 398, See also *compound interest.*
   calculation of, 398
   investments, 656–659
   receiving, on savings from bank, 400–401
International Space Station, 232
Internet World Stats, 138
intersection of sets, 227–228
intervals, 156, 229
inverse 2×2 matrices, 631
   formula for, 634
inverse of coefficient matrix, 629
inverse of matrix symbol, 631
inverse properties
   additive, 109
   multiplicative, 116
invert and multiply rule, 258
investments, and polynomials, 656–659
irrational numbers, 817–818, 830
   proof of, 818–819
   as solutions to quadratic equation, 752
Irrationality of Square Root of n Theorem, 819
isosceles trapezoid, 835
isosceles triangle, See *right triangles.*

J

Johnson, Randy, 643
justifications, 788
   compared to what was done, 789
   and properties, 790

K

Kaiser Family Foundation, 71
karat, 313
Kasparov, Garry, 830
Katz, Victor, 795
key sequences on calculators, 8
Kilby, Jack, 420
Koch snowflake, 447

L

Lambert, Johann, 819
Landsteiner, Karl, 428
least common denominator, 764
least squares line, 370
Ledecky, Katie, 371
left-hand brace symbol, 582
Legendre, Adrien-Marie, 373
lengths
   probabilities from, 297
   in similar figures, 309–310
   in similar figures, finding without measuring, 310–311
less than, 155
like terms, 72
   adding coefficients of, 73
   collecting, to solve equations, 149–150
limiting factors, 442, 444
Lincoln, Abraham, 237, 301
line(s), See also *horizontal lines; parallel lines; slope of line; vertical lines.*
   of best fit, 369–370
   coincident, 617–618
   equations for
      finding slope, 336–337
      from graph, 351–352
      intercept form, 380
      point-slope form, 356–357
      rewriting in slope-intercept and standard form, 376–377
      slope-intercept form, 350
      standard form, 375–376
      from two points, 361–363
   of fit, 368
   fitting to the data, 368–370
   least squares, 370
   oblique, 377
line graph, 326
line of best fit, 369–370
line of fit, 368
linear combination of x and y, 375
linear combination situations, 374–375
linear equations, See also *system of equations.*
   combining, 175

defined, 129
integer solutions to, 175
solving form $ax + b = c$, 140, 144–148
solving with CAS, 140
linear expression, 129
linear function, 432
linear increase, See *constant-decrease situation*; *constant-increase situation*.
linear inequalities, 383, See also *inequalities*.
graphs with horizontal or vertical lines, 381–382
graphs with oblique lines, 382–383
linear patterns, graphing constant-increase/decrease, 130–132
linear polynomial, 664
linear regression, 369–370, See also *exponential regression*.
linear relationship, between Fahrenheit and Celsius temperatures, 366
linear sentences, always or never true, 216–218
linear systems, 580
linear term, 736
list of elements, 29
Lochte, Ryan, 582
logistic function, 448
lowest terms, 762
Loyd, Sam, Jr., 209

M

m.a.d., See also *mean absolute deviation*.
magic square, 83
magic squares, 122
Marion, Shawn, 208
markups, in percents, 184
mathematics in other disciplines.
(Examples of math in other disciplines are found throughout the text. The following are selected examples.) See also *applications*.
agriculture, 110, 287, 596
archeology, 796, 800
aviation, 25
banking
and ATMs, 190, 709
and interest, 398, 399, 400, 402
and investing, 674, 690, 696, 709
botany, 662
chess, 373, 830
communications, 569
construction, 214, 257, 340
parabolic curves in, 565–566, 569
pitch of roof, 341, 344, 570
cryptography, public-key, 769
dog breeding, 205
dog-walking business, 70
energy industry, 147, 226
farming, 110
fishing industry, 356
food production, 163
forestry, 661
household movers, 151
jewelry and precious materials, 313
landscaping, 31, 828
military, 89
Olympic games, 379, 394
swimming, 207, 580–582, 587
radio stations, 462
rental of property, 219
rocket technology, 11
space, 577
distances in, 496
golf ball on the moon, 556
planets, volumes of, 513
space industry, 159, 232
sports
baseball, 195, 306, 549, 643
basketball, 208, 290–291, 378, 385, 551, 606, 722, 727
football, 169, 549, 643, 701
golf, 90, 192, 722
hockey, 638, 783
hurling, 774
soccer, 32, 305, 403, 550
softball, 50
swimming and diving, 371, 554, 556, 558, 562
tennis, 531
track and field, 364, 379, 431, 576, 590
volleyball, 385, 694
stocks and dividends, 172
transportation, 28, 240
airports and airplanes, 279, 602, 605, 621
car and computer numbers, 645
car purchases, 468
car trips, 536
driver age and accidents, 542
school buses, 446
speed limits, 781
traffic lights, 537
trains, 133
zookeeping, 145
matrices (matrix), 622
2×2 identity, 626, 629–630
3×3 identity, 628
inverse 2×2, 631
inverse 2×2, formula for, 634
multiplication of 2×2, 623–624
solving a system, 629–632
types of, 622
matrix form of a system, 623
matrix method
solving a system, 629–632
summary of, 631–632
matrix multiplication, 623–625
mean
of a data set, 47–48
deviation from, 191–192
mean absolute deviation (m.a.d.), 48
algorithm for finding, 50
of a data set, 48–50
measuring consistency, 53
means, in proportions, 301–302
Means-Extremes Property, 302
median, and percentiles, 291, 292
method of testing a point, 382
miles per hour, 264
mixed number, 816
models
exponential decay, 411, 419–420
exponential growth, 420–421
growth for powering, 405
of life expectancy, 370
population, See *population models*.
for weather forecasting, 284
monomial, 663, See also *binomials*.
degree of, 664
factoring, 675–678
multiplication by polynomial, 669–671
Moore, Gordon, 690
Morton, J. Sterling, 210
most likely outcome, 281–282
mu **(mean),** 47
multiplication
of algebraic fractions, 252–253
counting principle, 458–459
of decimal by power of 10, 671–672
fact triangle, 112
by growth factor, 404
of inequalities, 156–157, 211
matrices, 2×2, 624–625
matrix, 623–624
of polynomial by monomial, 669–671
of polynomials, 680–682
power of a power, 466
power of a product, 481–482
of powers with different bases, 466

of powers with same base, 464–465
properties
of –1, 86
associative, 9
commutative, 16
distributive, over addition, 66
distributive, over subtraction, 67
of equality, 113, 167
of inequality, 156–157, 159, 211, 230
multiplying fractions, 252–253
power of a power, 466
power of a product, 481–482
product of powers, 465
product of square roots, 498
related facts, of multiplication and division, 113–114
repeated, of powers, 398
of rates, 269–270
solving a system of equations, 608–612
of square roots, 497–502
of sum and difference of two numbers, 687–688
of two odd numbers, 804–805
by zero, 114–115
Multiplication Counting Principle, 458–459
in permutations, 691
in polynomials, 681
multiplication method for solving a system, 609
Multiplication Property of –1, 86
Multiplication Property of Equality, 113
clearing fractions with, 167
Multiplication Property of Inequality, 156–157, 159, 211, 230
Multiplication Property of Zero, 115
Multiplicative Identity Property, 116
Multiplicative Inverse Property, 116
multiplier, for clearing fractions, 167–168
Multiplying Fractions Property, 252–253

N

n **factorial,** 693–694
$n!$ **symbol,** 693
Nash, Steve, 208
National Center for Health Statistics, 280
negative coefficients, solving inequalities with, 164

Negative Exponent Property, 474
Negative Exponent Property for Fractions, 475
negative exponents
applying power of a power property, 477–478
for fractions, 475–476
value of power with, 474–475
negative first power, 475
negative fractions, 265
negative rate of change, 327, 330
negative root, 489
negative slopes, 335
never true, 216–218
nonintersecting parallel lines, 616–617
nonlinear systems, 640–642
nonnegative real numbers, 29
± notation, 229
notation, scientific, See *scientific notation*.
Now/Next method, 406, 439
Noyce, Robert, 420
nth power, 398
null set, 584, 638
number line, graphs and inequalities on, 155–156
number puzzles, 79–84, 809–813, See also *puzzles*.
creating with algebra, 81–82
divisibility properties, 808, 810–812
horizontal line equations as solutions, 188
magic square, 83
phone numbers, 307
repeating, 121
reversing digits of number, 812–813
Seven is Heaven, 80–81, 188
square array of numbers, 84
Sudoku, 830
writing expressions for steps, 828
numbers, See also *decimals; irrational numbers; one; square numbers; zero.*
complex, 797
even, 802–803
expected, 697
figurate, 56
negative, and rates, 264–265
nonnegative real, 29
odd, 803–805
pentagonal, 56
prime, 768
real, 29
reversing digits of, 812–813

rightmost digits, and divisibility properties, 810–811
square, 56
sum of digits, and divisibility tests, 811–812
testing, 93
triangular, 56
whole, 29

O

oblique lines, 377
graphing linear inequalities, 382–383
octagon, area of, 97
odd integer (odd number), 803–805
odds, 284
odds of an event, 284
one, See also *identity*.
properties of, 116
operations, order of, in evaluating expressions, 6
Opposite of a Difference Property, 87
Opposite of a Sum Property, 86
Opposite of Opposites Property, 85
opposites, 85, 109
as additive inverses, 85
multiplying by –1, 86
or
in compound sentence, 227
solving inequalities, 230
order of operations in evaluating expressions, 6
precedence of powers, 482
ordered pairs, set of, 426
ordered triplets, 510
origin, 44
Otis, Elisha Graves, 383, 504
outcomes
defined, 280
equally likely, 296
most likely, 281–282
output
in formulas, 221
in a function, 426, 434

P

$P(E)$, See *probability of an event, $P(E)$*.
par (in golf), 192
Parabola Vertex Theorem, 716
parabolas, 526
converting from standard form to vertex form, 724–725
creating by folding paper, 564
diving curves as, 552, 554, 556, 558, 562
equations for heights of

projectiles over time, 545–548
equations for projectile paths, 544–545
focus of, 571
graph of $y = ax^2 + bx + c$, 538
graph of $y = x^2$, 526
graphing nonlinear systems, 641
graphs of $y = ax^2$, 527
importance in construction, 565–566
points on, 527–528, 572
vertex form of equation for, 714–719
parallel lines
nonintersecting, 616–617
properties and slopes, 616
solving systems of equations, 616–619
systems of inequalities, 636–637
parentheses, removing, 66
Pascal's triangle, 122
patterns, 13
of change, using algebra to describe, 396
graphing constant-increase/decrease, 130–132
if-then statement with variable, 779
looking at with tables, 13–14
showing to be not always true, 512
two or more variables, 15
Pearson, Karl, 698
pendulums, formula for period, 501
pentagonal numbers, 56
people per year, 328
per, 263
per capita, 268
percent, 182
describing growth factor, 415
ratio as, 275
percent problems
add-ons and discounts, 183–184
putting into equation form, 182–183
solving with equations, 182–184
percentiles, 291–292
perfect square trinomials, 687
perfect squares
in quadratic equations with rational solutions, 748–749
square of a difference, 687
square of a sum, 685–686
Perfect Squares of Binomials, 687
period of pendulum formula, 501
permutations
circular, 695
order of actions, importance of, 703
with replacement, 458–459, 691

switches and sorting, 704
using all the items, 693–694
without replacement, 691–694
Phelps, Michael, 582
pi (π)
approximating with Buffon's needle, 315
proof of irrationality, 819
picture, worth of, 4
picturing expressions, 27–29
pitch of roof, 341, 344
plane, distance in, 505–507
plus or minus symbol, 553
points
distance between two, in a plane, 506–507
distance between two, in space, 510
equation for lines through two points, 361–363
on graph of parabola, 527–528, 572
point-slope form of equation of line, 357
polygons
diagonals in, 567, 735
sum of measure of interior angles, 307
polynomial in x, 657
polynomials, See also *binomials*.
approximating with, 769
and base-10 arithmetic, 654
checking operations with degree of, 665
classifying by degree, 664–665
classifying by numbers of terms, 663
converting from factored form to standard form, 755–756
cubic, 756
defined, 663
degree of, 664
in geometry, 655
of higher degrees, graphs of, 754–757
and investments, 656–659
linear, 664
long division of, 703
multi-digit base-10 numbers, 684
multiplication by monomial, 669–671
multiplying, and picturing with area, 680–682
and permutations without replacement, 692–693
prime, 677, 768
prime factorization of, 677
properties

distributive, 669, 670
extended distributive, 680
quadratic, 664
as rational expressions, 761
standard form for, 658
tile representation, 665
unique factorization theorem, 677
population density, 267, 315
population growth, 396–397, 404
exponential, 406–407
population, in statistics, 302
population models
in function notation, 432, 433
growth functions, 396–397
logistic function, 448
positive coefficients, solving inequalities with, 163
positive rate of change, 327, 330
positive root, 489
positive slopes, 335
Power of a Power Property, 466
applying with negative exponents, 477–478
Power of a Product Property, 481
Power of a Quotient Property, 482
powering, and population growth, 404
powers, 398
contrasted with like terms, 74
dividing with different bases, 471
dividing with same base, 469
multiplying with different bases, 466
multiplying with same base, 464–465
negative exponents, 474–477
and positive square roots, 490–491
of a product, 481–482
properties
negative exponent, 474
negative exponent for fractions, 475
powers of a power, 466
product of powers, 465
quotient of powers, 469–470
summary of, 511
of a quotient, 482–483
repeated multiplication property of, 398
as squares, 488
third, 493
of three, 487
value of, with negative exponents, 474–475
of zero, 470
powers of ten
multiplying decimals by, 671–672

polynomials as multi-digit numbers in base-10, 684
prefixes in metric system for, 468
Powers of Ten (movie), 448
prefixes for powers of 10 in metric system, 468
President Garfield's proof, 824
previously-proved property, 790
prime factorization, 515
  of a polynomial, 677
prime numbers, sieve of Eratosthenes, 768
prime polynomial over the integers, 738, 750
prime polynomials, 677
  sieve of Eratosthenes, 768
principal, 398
Principles, See also *Properties*; *Theorems*.
  multiplication counting, 458–459
probability, 280–284
  from areas, 296
  of blood types, 283, 285
  conditional, 283
  defined, 280
  determination of, 280–281
  differences with relative frequency, 289–290
  from lengths, 297
probability distribution, 281–282
  from relative frequencies, 282–283
Probability Formula for Geometric Regions, 296
probability of an event, P(E), 280
  complementary event and, 283
  with equally likely outcomes, 282, 296
Product of Powers Property, 465, 468, 469
Product of Square Roots Property, 498
products, See *multiplication*.
projectiles, 544
  equations for heights of, over time, 11, 545–548
  equations for paths of, 544–545
  verifying motion of, 571
Projects
  Absolute Value Equations, 241
  Adding and Subtracting Equations, 646
  Appliances and Energy, 174
  Bisection Method, 769
  Buffon's Needle, 315
  Calculating Density, 316
  Can Everything Be Proved?, 829
  Cars and Computers, 645
  Catenaries, Parabolas, and a Famous Landmark, 572
  Checking Whether Points Lie on a Parabola, 572
  Combining Linear Equations, 175
  Combining Solutions, 768
  Conjectures, 830
  Converting Ratios, 316
  Copy Machine Puzzle, 316
  Counting Braille Letters, 518
  Diagonal Game, 388
  Differences of Higher Powers, 704
  Dividing Polynomials, 703
  Divisibility Tests, 829
  Encoding and Decoding Using Formulas, 56
  Estimating Square Roots, 55, 517
  Euclid's Algorithm, 830
  Examining Pi, 55
  Famous Snowflake, 447
  Figurate Numbers, 56
  Finding the Counterfeit, 646
  Focus of a Parabola, 571
  Formula 1, 55
  Formulas for Sums of Consecutive Integers, 572
  Fraction Exponents, 517
  Getting Closer and Closer to Zero, 241
  Growing Trees, 240
  Hybrid Cars, 240
  If-Then Statements in Games, 830
  Infinite Repeating Continued Fractions, 768
  Integer Solutions to Linear Equations, 175
  Interview with Pythagoras, 517
  Magic Squares, 122
  Marriage Age, 387
  Mathematics and Crossword Puzzles, 518
  Modeling Buying and Selling, 174
  Moving in the Coordinate Plane, 518
  New Operation, 121
  Paper Towels: Price vs. Absorbency, 387
  Pascal's Triangle, 122
  Planning Your Trip, 240
  Population Densities, 315
  Postal Rates, 122
  Powers of Ten, 448
  Prime Numbers, Prime Polynomials, 768
  Programming the Quadratic Formula, 571
  Public-Key Cryptography, 769
  Rationals vs. Irrationals, 830
  Reciprocal Functions, 447
  Repeating Number Puzzle, 121
  Representing Positive Integers Using Powers, 703–704
  Richter Scale, 447
  Right Order, 703
  Slopes of Perpendicular Lines, 388
  Squares Surrounding Triangles, 829
  Strength of Spaghetti, 175
  Switches and Sorting, 704
  Systems with More Variables, 645
  Testing Astrology, 704
  Using Polynomials to Approximate, 769
  Verifying Projectile Motion, 571
  When Do Systems Have Integer Solutions?, 645
proof, 788
  abbreviated, 789–790
  dissection, 828
  of divisibility properties, 802–806
  of everything?, 829
  of irrational numbers, 818–819
  of Pythagorean Theorem, 823–825, 827, 828
  of Quadratic Formula, 797, 800
proof argument, 788–789
Properties, See also *Principles*.
  addition, of equality, 106
  addition, of equality, generalized, 601
  addition, of inequality, 162, 211, 230
  additive identity, 108
  additive inverse, 109, 113
  associative, of addition, 10
  associative, of multiplication, 9, 10
  closure, 802
  commutative, of addition, 16
  commutative, of multiplication, 16
  cube of the cube root, 493
  discriminant, 561
  distributive, extended, 680
  distributive, of multiplication over addition, 66
  distributive, of multiplication over subtraction, 67
  distributive, reversing sides of, 72
  dividing fractions, 258–259
  division, of equality, 113
  equal fractions, 253–254
  of graph of $y = ax^2$, 527
  of integers, 809–813
  means-extremes, 302

multiplication of –1, 85, 86
multiplication, of equality, 113, 167
multiplication, of inequality,
    156–157, 159, 211, 230
multiplication, of zero, 115
multiplicative identity, 116
multiplicative inverse, 116
multiplying fractions, 252–253
negative exponent, 474
negative exponent, for fractions,
    475
opposite of a difference, 87
opposite of a sum, 85, 86
power of a power, 466
power of a product, 481
power of a quotient, 482
of powers, summary of, 511
product of powers, 465
product of square roots, 498, 499
quotient of powers, 469–470
quotient of square roots, 499
related facts, of addition and
    subtraction, 106–107
related facts, of multiplication
    and division, 113–114
repeated multiplication of
    powers, 398
similar figures, fundamental, 309
of slope, 341–344
slopes and parallel lines, 616
square of the square root, 490
of square roots, summary of, 511
subtraction, of equality, 106
testing for equivalence, 93–94
transitive, of equality, 10
types of, 790
zero exponent, 405
proportions
    solving, 301
    and statistics, 302–304
    using division, 251
***p*th percentile**, 292
public-key cryptography, 769
puzzles, See also *number puzzles*.
    algebraic riddle, 173
    copy machine enlargements, 316
    crossword, 518
    Cyclopedia of Puzzles (Loyd), 209
    hen and egg, 273
    logic table, 201
    St. Ives nursery rhyme riddle, 600
Pythagoras
    interview with, 517
    and square roots, 490
Pythagorean Theorem, 491–492
    finding distance in a plane, 505
    proofs of, 823–825, 827, 828

Q

quadratic equations, 525, 552
    discriminant of, 560–561
    distance formula for falling
        objects, 531, 533
    of form $y = x^2$, 526
    heights of projectiles, 545
    with rational solutions, 748–749
    solutions to, 558–560
    standard form of, 555
    sum and product of solutions, 801
quadratic expressions, 525
    factorable over the integers,
        748–750
    table of, 739
Quadratic Formula, 552–553
    applying, 553–555
    finding number of real solutions,
        559–560
    history of, 795–797
    programming, 571
    proofs of, 797–798, 800
quadratic function, 525, 526
    applications of, 565–567
    factor theorem for, 731
    factored form of, 729–733
    factoring $ax^2 + bx + c$, 742–745
    factoring $x^2 + bx + c$, 736–739
quadratic polynomial, 664
quadrilaterals
    perimeter, 73
    similar lengths in, 310
quartiles, 292
Quotient of Powers Property,
    469–470
Quotient of Square Roots Property,
    499
quotients, See *division*.

R

radical sign, 489
    of cube root, 493
radicals, simplifying, 499–501
radicand, 499
randomly-taken samples, 302
range of a data set, 48
range of a function, 426
    not set of numbers, 428
rate(s), 263–266
    calculating, 263–264
    conversion, 269–270, 316
    cricket chirping, 362–363
    defined, 263
    discount, 275
    distance formula, 270
    dividing, 269–270
    division by zero, 265
    multiplying, 269–270
    negative numbers and, 264–265
    tax, 275
    using division, 251
rate of change, 326, 328
    calculating with spreadsheet,
        328–329
    negative, 327, 330
    positive, 327, 330
    zero, 329, 330
rate unit, 263, 328
ratio(s), 274–277
    in adult body vs. infant body, 308
    of angle measures and
        probabilities, 297
    birthrates, 280, 284
    changing, 277
    defined, 274–275
    mixing paint, 276, 278
    and percents, 275
    in right triangles, 279
    setting up equations, 276
    of similitude, 308
    using division, 251
rational expressions, 761
    adding and subtracting, 764–765
    writing in lowest terms, 761–764
rational numbers, 816–817, 830
**rational solutions of quadratic
    equations**, 748–749
real numbers, 29
reciprocal functions, 427, 447
reciprocal rates, 264, 270
reciprocals
    negative exponents, 475
    of numbers, 258
    of slope of line, 793
rectangle, area of, 823
rectangular arrays, See *matrices*.
rectangular solid, volume of, 32, 484
reflection image, 526
reflection-symmetric to *y*-axis, 526
reflective property of parabolas, 565
regression lines, 581
**related facts for addition and
    subtraction**, 105–108, 111
**related facts for multiplication and
    division**, 112–114
**Related Facts Property of Addition
    and Subtraction**, 106–107
**Related Facts Property of
    Multiplication and Division**,
    113–114
relation, 428

Student Handbook

relative frequency, 289–292
  in birthrates, 280, 284
  comparing to probabilities, 297
  defined, 289
  differences with probability, 289–290
  distributions, 290–291
  probability distribution from, 282–283
repeated multiplication in power of a product, 481
Repeated Multiplication Property of Powers, 398
repeating decimals
  finding simple fractions for, 607
  rational vs. irrational, 816, 818
replacement
  permutations with, 458–459, 691
  permutations without, 691–694
revenue equation, 584–585
Review questions, 12, 18, 25–26, 32, 40–41, 46, 53–54, 70–71, 76–78, 83–84, 89–90, 96–97, 103–104, 110–111, 119–120, 134, 137–138, 142–143, 147–148, 153–154, 160–161, 166, 172–173, 186–187, 195, 200–201, 208–209, 215, 219–220, 225–226, 232–233, 239, 257, 261–262, 267–268, 272–273, 279, 287–288, 294–295, 300, 306–307, 313–314, 332, 340, 346–347, 355, 359–360, 366–367, 373, 379–380, 386, 403, 410, 418, 424, 431, 437–438, 446, 462–463, 468, 472–473, 480, 487, 495–496, 503–504, 509–510, 516, 531, 535–536, 542–543, 551, 557, 563–564, 569–570, 588, 593, 599–600, 606–607, 614–615, 620–621, 627, 633–634, 639, 644, 661–662, 667–668, 674, 679, 684, 690, 695–696, 702, 721–722, 727–728, 735, 741, 746–747, 753, 760, 767, 782–783, 787, 793–794, 800–801, 807, 814–815, 821, 827–828, See also *Chapter Review*.
Richter scale, 447
right triangles
  area, 823
  and irrational numbers, 817–818
  isosceles area, 486
  isosceles definition, 516
  Pythagorean Theorem, 491–492
  ratios in, 279
roots, See *cube roots*; *square root(s)*.

roster, 29
row-by-column multiplication, 624, 628
rule of 72, 418
Rule of Three, 301
Rusanov, Nikolai, 213
Rutan, Dick, 621

S

sales tax, 136
sample, 302
scale factor, 657
scatterplots, 27–28
  creating with technology, 37
  exponential equations and, 419
scientific calculators, 8
scientific notation, 460
  dividing numbers with quotient of powers property, 470–471
Self-Test, 58–59, 124, 177, 243–244, 318–319, 390–391, 450–451, 520, 574–575, 648–649, 706–707, 771–772, 832
semiperimeter of triangle, 807
sequence, 20
  choosing from n objects repeatedly, 459–460
set of ordered pairs, 426
set-builder notation, 29
  coordinate plane vs. number line, 382
  using with inequalities, 155
sets
  closed under an operation, 802
  intersection of, 227–228
  union of, 227–228
Shepard, Alan, 556
sieve of Eratosthenes, 768
similar figures, 308–311
  finding lengths in, 309–310
  finding lengths without measuring, 310–311
  fundamental property of, 309
  ratios in adult body vs. infant body, 308
similitude, ratio of, 309
simplifying algebraic fractions, 254
simplifying complex fractions, 259–260
simplifying equations
  to solve $ax + b = c$, 144–145
  to solve $ax + b = cx + d$, 204–206
simplifying problems, counting combinations of acronyms, 458–459
simplifying radicals, 499–501
sine, 279

situations, comparing, 216–218
skewed left distribution, 51
skewed right distribution, 51
sliders, 348–349
slope of line, 334
  calculating from graph, 335
  constant, 334–335
  finding from equation of the line, 336–337
  grade of road, 346
  graphing a line using, 342–343
  of horizontal lines, 335, 343
  negative, 335
  of perpendicular lines, 388
  pitch of roof, 341, 344
  point-slope form of equation, 356–357
  positive, 335
  properties of, 341–344
  parallel lines and, 616
  and reciprocal of slope, 793
  undefined, 344
  of vertical lines, 344
  zero, 329, 330, 335, 343
slope-intercept form of equation of line, 350–351
  rewriting in standard form, 376–377
  solving systems with substitution, 589, 595
Slopes and Parallel Lines Property, 616
solution to a system, 582–584
  with different parallel lines, 638
solution to an equation, 135
solving a system, 581
  by addition, 601–604
  graphing solution, 583–584
  having parallel lines, 616–619
  matrix method, 629–632
  by multiplication, 608–612
  using substitution, 589–591
solving equations, 135–136
  of absolute value, 234–235, 241
  algebraically, 139
  algebraically, compared to graphs and tables, 198
  with a balance, 139
  by clearing fractions, 167–170
  by collecting like terms, 149–150
  by creating equivalent equations, 139–140
  general linear equation, 202–206
  linear, 140, 144–146
  for parabolas, 532–534
  as proofs, 788–791
  quadratic, 558–560

by removing parentheses, 150–152
with tables and graphs, 135–136
solving inequalities, 157–159
of absolute value, 236–237
of form $ax + b < c$, 162–164
of form $ax + b < cx + d$, 210–212
with negative coefficients, 164
with positive coefficients, 163
solving nonlinear systems, 640–642
special case, testing for, 511–512
sphere, volume of, 18, 513
splitting into parts, 251
spread of a data set, 47
spreadsheet
calculating rate of change, 328–329
comparing constant increase/decrease and exponential growth/decay, 440–442
mean absolute deviation procedure, 49
replication, 441
square numbers, 56
completing, 723–725
difference of two, 687–688
of odd integer, 805–806
perfect squares: square of a difference, 687
perfect squares: square of a sum, 685–686
Square of the Square Root Property, 490
square of x, 488
square root(s), 488–492
addition of, 503
Babylonian method, 517
definition of, 489, 533
estimating, 517
irrational numbers, 817
multiplying and dividing, 497–501
origin of, 489
of positive integer, formula, 55
and power of x, 490–491
products of, 498
quotients of, 499
simplifying radicals, 499
summary of properties, 511
that are not whole numbers, 490
square term, 736
squares
area of, 823
basis of squared numbers and square roots, 488–489
of small positive integers, 489
surrounding triangles, 823, 829
squaring function, 425–426
standard form for a polynomial, 658

standard form of equation of a line, 375–376
rewriting in slope-intercept form, 376–377
standard form of quadratic equation, 555
standard window, 34
statistics
chi-square, 697–700
and proportions, 302–304
Stevens, John, 133
stopping distance of a car, 537, 540, 606
subscripts, on coordinates, 327
substitution
solving a system of equations, 589–591
using in other situations, 594–598
subtraction
algebraic definition of, 7
difference of two squares, 687–688
differences of higher powers, 704
fact triangle, 105
perfect squares: square of a difference, 687
properties
distributive, of multiplication over, 67
of equality, 106
opposite of a difference, 87
related facts, of addition and subtraction, 106–107
of rational expressions, 764–765
of two odd numbers, 804
subtraction expressions, 7
Subtraction Property of Equality, 106
sum, See *addition*.
summaries
constant increase and exponential growth, 443
for equations of lines, three common forms, 608
equivalent expressions, 712
of matrix method, 631–632
polynomial expressions for permutations, 696
polynomial names, 663
properties of powers, 511
solutions to a system, four ways, 584
system of equations, types of solutions for, 618
table of quadratic expressions, 739
Summary/Vocabulary, 57, 123, 176, 242, 317, 389, 449, 519, 573, 647, 705, 770, 831
supplementary angles, 278

surveys, 303
symbols
brackets, on matrix, 622
for inverse of matrix, 631
left-hand brace, 582
for n factorial, 693
for null set, 584
plus or minus, 553
radical sign, 489, 493
symmetric distribution, 51
system, nonlinear, 640–642
system of equations, 582
equivalent systems, 608
integer solutions, 645
with more than two variables, 645
no solutions, 584
nonlinear, 640–642
parallel lines and, 616–619
solution to, 582–584
solving by addition, 601–604
solving by graphing, 583–584
solving by multiplication, 608–612
solving with matrix method, 629–632
systems of inequalities, 635–637

T

tables
chi-square, 698–699
compared to graphs and algebraic properties, 198
finding vertex of parabola, 539–540
frequency, 47
from graphing calculator, 36, See also *graphing calculators*.
looking at patterns, 13–14
solving by comparing values, 196–198
solving equations with, 135
solving quadratic equations, 558
of values, 22
of values for parabola equation, 532
tangent, 279, 568
tax, on purchases, 184
tax rate, 275
technology
comparing expressions, 36
and equivalent expressions, 98–102
evaluating with, 8–9
scatterplot creation, 37
temperature
rate of change, 264
scales and conversions, 221–222, 224
terms, 15, See also *like terms*.
lowest, 762

of the sequence, 20
in trinomial, 736
testing a special case, 511–512
testing equivalence, 91–94
testing numbers, 93
tetrahedron, and related facts, 111
Theorem of the Three Squares, 492
Theorems, 789, See also *Principles*.
  arrangements, 459
  discriminant, 749, 751–752
  factor, 755
  factor, for quadratic functions, 731
  Fermat's last, 830
  incompleteness, 829
  irrationality of square root of *n*, 819
  parabola vertex, 716
  Pythagorean, 491–492
  unique factorization, for polynomials, 677
third power of *x*, 493
three, powers of, 487
time-distance graphs, 131
toothpick sequence, 20–21, 27, 340
Transitive Property of Equality, 10
trapezoid
  area of, 667, 823
  isosceles, 835
triangles
  area of, 823
  right
    area, 823
    area of isosceles, 486
    and irrational numbers, 817–818
    isosceles definition, 516
    Pythagorean Theorem, 491–492
    ratios in, 279
  semiperimeter of, 807
  squares surrounding, 823, 829
triangular numbers, 56
trinomials, 663
  factoring $ax^2 + bx + c$, 742–745
  factoring $x^2 + bx + c$, 736–739
  perfect square, 687
trivial factors, 675
true sentences, 216–218
  converses of, 784–785
  if-then statements, 778–779
Truman, Harry S., 700

U

unbiased (fair) situation, 281
undefined expression, 265
undefined number, zero power of zero, 405, 410
undefined slope of line, 344
uniform distribution, 51

union, math definition, 228
union of sets, 227–228
Unique Factorization Theorem for Polynomials, 677
unit cost, 267
units, in ratios, 274–275

V

value of the function, 426, 433
variable costs, 584–585
variable matrix, 622
variables, 5, 6
  defining, 14
  domain of, 28, 29
  in if-then statements, 778
  independent and dependent, 426
  input and output, 426
  in number puzzles, 80–82
  in patterns with two or more, 15
  polynomials in function notation, 664
  systems with multiple, 645
Venn diagrams, 779–780
vertex form of an equation for a parabola, 716
  converting from standard form, 724–725
vertex of parabola, 526
  finding with tables, 539–540
vertical distance, 506
vertical lines
  distances along, 505–506
  equations for, 189, 351
  graphing linear inequalities, 381–382
  slope of, 344
Viète, François, 801
vinculum, 489
volume of geometric shapes, See *formula(s)*.

W

weather forecasting probabilities, 284
whole numbers, 29
Wiles, Andrew, 830
window, in graphing calculators, 33
Woods, Tiger, 722
World Almanac and Book of Facts, 220, 221, 226, 248, 701

X

x cubed, 493
x squared, 488
x-coordinates, 34
  and domain, 426

x-intercept(s), 357
  bisection method, 769
  displayed by factored form of quadratic function, 730–731
  related to factors, 754–755
Xmax (*x*-max), 34
Xmin (*x*-min), 34
Xscl (*x*-scale), 34

Y

y-coordinates, 34
Yeager, Jeana, 621
y-intercept, 350
Ymax (*y*-max), 34
Ymin (*y*-min), 34
Yscl (*y*-scale), 34

Z

zero
  division by, 113, 114–115, 265
  getting closer and closer, 241
  multiplication properties, 115
  as power, 470
  as special number for addition, 108–109
Zero Exponent Property, 405
zero power, 470
  of any number, 405
  of zero, 405, 410
Zero Product Property, 115
  used in factored form, 731
zero rate of change, 329, 330
zero slope, 335, 343
zero the dive, 346
zooming in and out, Powers of Ten (movie), 448

Photo Credits

Chapters 1–6

Cover: ©Scott McDermott/Corbis; **vi** (l) ©ivansmuk/iStock, (r) ©JOE CICAK/iStock; **vii** (l) ©JoeDphoto/iStock, (r) ©DK.samco/Shutterstock; **viii** (l) ©Alan Lagadu/iStock, (r) ©dibrova/iStock; **ix** (l) ©Christopher Futcher/iStock, (r) ©anankkml/iStock; **x** (l) ©Fly_Fast/iStock, (r) ©IPGGutenbergUKLtd/iStock; **xi** (l) ©mddphoto/iStock, (r) ©athuric/iStock; **xii** ©ErikdeGraaf/iStock; **3** ©megasquib/iStock; **4–5** ©ivansmuk/iStock; **6** ©KatarzynaBialasiewicz/iStock; **8** Courtesy Texas Instruments; **11** ©Kameleon007/iStock; **12** ©alvarez/iStock; **15** ©DragonImages/iStock; **18** (l) ©Vasko/iStock; (c) ©Vasko/iStock; (r) ©Vasko/iStock; **20** ©Steve Debenport/iStock; **25** ©wesmyles/iStock; **26** Charles Mann/iStock; **28** ©barsik/iStock; **31** ©Artfoliophoto/iStock; **32** ©IPGGutenbergUKLtd/iStock; **40** ©Paha_L/iStock; **42** ©Fred LaBounty/Shutterstock; **45** ©megasquib/iStock; **50** ©John Barry de Nicola/Shutterstock; **52** Public Domain; **54** ©LeventKonuk/iStock; **55** (l) ©sidsnapper/iStock; (r) ©braverabbit/iStock; **56** Public Domain; **64–65** ©JOE CICAK/iStock; **67** ©t-lorien/iStock; **70** ©Ray Hems/iStock; **71** ©Pamela Moore/iStock; **76** ©andresr/iStock; **78** ©photoL/iStock; **84** ©PeopleImages/iStock; **86** ©ByeByeTokyo/iStock; **88** ©adlifemarketing/iStock; **90** ©warrengoldswain/iStock; **110** ©MaxPreps; **120** ©RosetteJordaan/iStock; **121** ©damircudic/iStock; **122** ©carterdayne/iStock; **128–129** ©JoeDphoto/iStock; **130** ©satori13/iStock; **133** ©anyaivanova/iStock; **134** ©GP232/iStock; **136** ©Juanmonino/iStock; **138** ©yenwen/iStock; **143** ©davidchoophotography/iStock; **144** ©traveler1116/iStock; **145** ©IS_ImageSource/iStock; **147** ©lisafx/iStock; **151** ©Susan Chiang/iStock; **152** ©Sjoerd van der Wal/iStock; **153** ©middelveld/iStock; **154** ©MACIEJ NOSKOWSKI/iStock; **156** ©mbarrettimages/iStock; **159** ©NASA; Goddard Space Flight Center; **163** ©silviacrisman/iStock; **164** ©MIMOHE/iStock; **165** ©JoeinQueens from Queens, USA/Wikimedia Commons; **166** ©urbancow/iStock; **169** ©Olga Bogatyrenko/Shutterstock; **172** ©EdStock/iStock; **173** ©Todor Tsvetkov/iStock; **174** (l) ©stephanie phillips/iStock; (r) ©Juanmonino/iStock; **175** Courtesy Texas Instruments; **180–181** ©DK.samco/Shutterstock; **183** ©mediaphotos/iStock; **185** ©Mlenny/iStock; **186** ©IS_ImageSource/iStock; **190** ©ElsvanderGun/iStock; **195** ©Marcio Silva/iStock; **201** ©byryo/iStock; **205** ©101cats/iStock; **207** McSmit/Wikimedia Commons/GFDL; **210** ©matka_Wariatka/iStock; **213** ©Frau Siebenschläfer/Wikimedia Commons/GFDL; **214** ©shaunl/iStock; **218** ©toddmedia/iStock; **219** ©Jitalia17/iStock; **221** ©eyedias/iStock; **226** ©TomasSereda/iStock; **227** ©www.Alibaba.com; **229** ©Mirko_Rosenau/iStock; **232** ©NASA; **237** ©Roel Smart/iStock; **240** ©EdStock/iStock; **250–251** ©Alan Lagadu/iStock; **256** ©cjp/iStock; **261** ©eAlisa/iStock; **262** ©avdeev007/iStock; **266** ©carroteater/Shutterstock; **269** ©James Anderson/iStock; **270** ©BONNINSTUDIO/iStock; **271** ©Eldad Carin/iStock; **272** ©Cathy Yeulet/iStock; **274** (l) ©Csanad Kiss/Shutterstock; (c) ©master1305/iStock; (r) ©Eric Isselee/Shutterstock; **275** ©ClarkandCompany/iStock; **276** ©RTimages/iStock; **277** ©Sjoerd van der Wal/iStock; **279** ©stevecoleimages/iStock; **280** ©P_Ntagios/iStock; **282** ©monkeybusinessimages/iStock; **283** ©Clint Spencer/iStock; **287** ©simazoran/iStock; **299** ©NASA/NOAA/GOES Project; **301** Courtesy Library of Congress, Prints and Photographs Online Catalog [LC-USZC4-2472]; **303** ©Natalie Fobes/CORBIS; **305** ©fstockfoto/Shutterstock; **306** ©Lee Prince/Shutterstock; **308** ©Britta Kasholm-Tengve/iStock; **309** ©Ana Vasileva/Shutterstock; **312** (t) ©SallyFelton/iStock; (b) ©North Wind Picture Archive/AP Images; **313** ©spxChrome/iStock; **314** ©CrackerClips/iStock; **315** ©Robert Simmon/NASA Earth Observatory/Chris Elvidge/NOAA National Geophysical Data Center; **316** ©visionchina/iStock; **324–325** ©dibrova/iStock; **326** ©veronicagomepola/iStock; **329** ©franckreporter/iStock; **332** ©vitapix/iStock; **340** ©PeopleImages/iStock; **346** (t) ©BjarteSorensen/Public Domain/Wikimedia Commons; (b) ©Snaprender/iStock; **353** ©GeorgeBurba/iStock; **355** ©Photawa/iStock; **356** ©filo/iStock; **358** ©Craig Dingle/iStock; **362** ©Lytninbug/iStock; **365** ©mjunsworth/iStock; **366** ©sestevens/iStock; **371** ©Michael Ledecky/Wikimedia Commons/CC-BY-SA-4.0; **372** ©Daniel Mayer/Wikimedia Commons/GFDL; **373** ©PeopleImages/iStock; **378** ©monkeybusinessimages/iStock; **383** ©FlairImages/iStock; **385** ©Christopher Futcher/iStock; **387** ©digitalskillet/iStock

Photo Credits

Chapters 7–13

396–397 ©Christopher Futcher/iStock; 400 ©kickstand/iStock; 403 ©AGIF/Shutterstock; 404 ©DamianKuzdak/iStock; 407 Public Domain; 408 ©Olha Rohulya/iStock; 409 ©PeopleImages/iStock; 412 ©Charlesimage/Shutterstock; 417 ©SurkovDimitri/iStock; 421 ©arcoss/iStock; 424 ©SolisImages/iStock; 428 ©fotokostic/iStock; 430 ©mashabuba/iStock; 436 ©fotostorm/iStock; 442 ©MikeLane45/iStock; 444 ©gsagi/iStock; 446 ©PenelopeB/iStock; 447 ©Matt Katzenberger, www.flickr.com/photos; 448 ©Jeff Schmaltz/LANCE/EOSDIS MODIS Rapid Response Team/NASA GSFC; 456–457 ©anankkml/iStock; 458 ©omgimages/iStock; 460 ©VIPDesignUSA/iStock; 462 ©RAUSINPHOTO/iStock; 463 ©Phototreat/iStock; 468 ©monkeybusinessimages/iStock; 470 ©Dougall_Photography/iStock; 472 ©svetikd/iStock; 480 ©Cathy Yeulet/iStock; 487 ©microgen/iStock; 496 ©Andrea_Hill/iStock; 501 ©PhilipCacka/iStock; 504 ©Holger Mette/iStock; 505 ©Ralf Roletschek/Wikimedia Commons/CC-BY-3.0/GFDL; 510 ©RicAguiar/iStock; 516 ©gbh007/iStock; 517 ©tupungato/iStock; 518 ©Marti157900/iStock; 524–525 ©Fly_Fast/iStock; 531 ©orangelinemedia/iStock; 536 ©ferrantraite/iStock; 537 ©Vilches/iStock; 543 ©Elena Kalistratova/iStock; 549 ©Mike Liu/Shutterstock.com; 550 ©strickke/iStock; 551 ©monkeybusinessimages/iStock; 554 ©Paolo Bona/Shutterstock; 558 ©rickeyre/iStock; 564 ©jemsgems/iStock; 565 ©compassandcamera/iStock; 569 ©Allkindza/iStock; 571 (t) ©ermingut/iStock; (b) ©Dario Lo Presti/iStock, ©Paha_L/iStock; 572 ©astrolin/iStock; 580–581 ©IPGGutenbergUKLtd/iStock; 582 ©Eric Draper/Public Domain; 583 ©skynesher/iStock; 585 ©Joe_Potato/iStock; 586 ©Floortje/iStock; 588 ©joaofilipe/iStock; 590 ©microgen/iStock; 592 ©aamorim/iStock; 594 ©monkeybusinessimages/iStock; 598 ©gchutka/iStock; 599 ©joyride/iStock; 602 ©georgejurasek/iStock; 604 ©BrettCharlton/iStock; 606 (t) ©Donald Barnat/Wikimedia Commons/CC-BY-SA-3.0; 611 ©GarysFRP/iStock; 614 ©IS_ImageSource/iStock; 620 ©lore/iStock; 621 ©Mark Greenberg; 627 (l) ©mofles/iStock; (r) ©DNY59/iStock, ©kolosm/iStock; 638 ©oei1/iStock; 643 ©Debby Wong/Shutterstock; 645 ©pagadesign/iStock; 646 ©FOTOGRAFIA INC./iStock; 654–655 ©mddphoto/iStock; 658 ©Roberto A Sanchez/iStock; 660 ©kirin_photo/iStock; 661 ©stockstudioX/iStock; 692 ©leminuit/iStock; 694 ©Christopher Futcher/iStock; 695 ©Juanmonino/iStock; 700 Public Domain; 702 ©sandsun/iStock; 704 ©sololos/iStock; 712–713 ©athuric/iStock; 722 ©DavorLovincic/iStock; 727 ©Keith Allison from Owings Mills, USA/Wikimedia Commons/CC-BY-SA-2.0; 741 ©zennie/iStock; 768 Public Domain; 769 ©Oxford/iStock; 776–777 ©ErikdeGraaf/iStock; 779 ©EarnestTse/iStock; 781 ©narvikk/iStock; 783 ©Unclweed/Flickr/CC-BY-SA-2.0; 785 ©crazycroat/iStock; 787 ©davincidig/iStock; 794 ©vadimrysev/iStock; 796 Public Domain; 797 Public Domain; 807 Public Domain; 819 Public Domain; 821 ©JordiRamisa/iStock; 824 ©traveler1116/iStock; 828 ©LifesizeImages/iStock; 830 (l) ©blackred/iStock; (r) ©ASSOCIATED PRESS

Illustrations: Ron Carboni

Acknowledgements: It is impossible for UCSMP to thank all the people who have helped create and test these books. We wish particularly to thank Carol Siegel, who coordinated the use of the test materials in the schools; Kathleen Andersen, Aisha Bradshaw, Paul Campbell, Jena Dropela, Meri Fohran, Lisa Hodges, Rachel Huddleston, Evan Jenkins, Nurit Kirshenbaum, Lindsay Knight, Nathaniel Loman, Matthew McCrea, Jadele McPherson, Erin Moore, Dylan Murphy, Gretchen Neidhardt, Jennifer Perton, Daniel Rosenthal, Luke I. Sandberg, Sean Schulte, Andrew L. Shu, Emily Small, John Stevenson, James Thatcher, Alex Tomasik, Erica Traut, Alex Yablon, Melissa Yeung, Elizabeth Olin, Don Reneau, and Loren Santow.

We wish to acknowledge the generous support of the Amoco Foundation and the Carnegie Corporation of New York in helping to make it possible for the first edition of these materials to be developed, tested, and distributed, and the additional support of the Amoco Foundation for the second edition.

We wish to acknowledge the contribution of the text *Algebra Through Applications with Probability and Statistics*, by Zalman Usiskin (NCTM, 1979), developed with funds from the National Science Foundation, to some of the conceptualizations and problems used in this book.

Symbols

>	is greater than	$0.\overline{a}$	repetend bar
<	is less than	{ }, Ø	empty or null set
≥	is greater than or equal to	:, |	such that
≤	is less than or equal to	$P(E)$	probability of an event E
=	is equal to	∩	intersection of sets
≠	is not equal to	∪	union of sets
≈	is approximately equal to	$\overleftrightarrow{AB}$	line through A and B
±	plus or minus	$\overrightarrow{AB}$	ray with endpoint at A and containing B
+	plus sign	$\overline{AB}$	segment with endpoints A and B
−	minus sign	AB	length of segment from A to B; distance between A and B
×, ·, *	multiplication signs	∠ABC	angle ABC
÷, /	division signs	m∠ABC	measure of angle ABC
%	percent	△ABC	triangle with vertices A, B, C
π	pi	⌐	right angle symbol
|n|	absolute value of n	$n°$	n degrees
√	radical sign	(x, y)	ordered pair
$\sqrt[3]{}$	cube root	$\frac{a}{b}$	a divided by b
A'	image of point A	a^b, $a\wedge b$	a to the bth power
( )	parentheses	b_1	superscript variable ("b sub 1")
[ ]	brackets	$f(x)$	function notation "f of x"
{ }	braces	$-x$	opposite of x
. . .	continuing pattern	$n!$	n factorial
		Q_1, Q_2, Q_3	first, second, and third quartiles
		μ	mean